CIMNE
BARCELONA

The finite element method in the 1990's

A BOOK DEDICATED TO O.C. ZIENKIEWICZ

EDITED BY E. OÑATE, J.PERIAUX, A.SAMUELSSON

CIMNE
BARCELONA

Springer-Verlag Berlin Heidelberg GmbH

First edition, October 1991

Originally published by Springer-Verlag Berlin, Hidelberg, New York, London, Paris, Tokyo, Hong Kong, Barcelona, Budapest in 1991.
Softcover reprint of the hardcover 1st edition 1991

Cover designed by Jordi Palló.

Printed by Artes Gráficas Torres S.A., Morales 17, 08029 Barcelona
Depósito legal: B-35590-91

ISBN 978-3-662-10328-9 ISBN 978-3-662-10326-5 (eBook)
DOI 10.1007/978-3-662-10326-5

Cover: Unstructured mesh of triangular finite elements for analysis of hypersonic air flow past a double ellipse. From O. C. Zienkiewicz, K. Morgan, J. Peraire, J. Peiro and L. Formaggia, «Finite elements in fluid mechanics. Compressible flow, shallow water equations and transport», in *ASME Conference on Recent Development in Fluid Dynamics,* AMD 95, American Society of Mechanical Engineers, December 1988.

O. C. Zienkiewicz. at class.

THIS BOOK IS DEDICATED TO

OLGIERD C. ZIENKIEWICZ

WITH OCCASION OF HIS

70TH BIRTHDAY

PREFACE

This book is dedicated to Olgierd C. Zienkiewicz with occasion of his 70th birthday.

Olgierd C. Zienkiewicz , (Olek for his many friends) was born on May 18, 1921 in England of a British mother and Polish father.

Already 1922 the family moved to Poland where they settled after a few years in Katowice. Here and in Poznan Olek got his school education. When this finished in 1939, he went to Warsaw for entrance examination to the Polytechnic.

Just one of these days the Germans invaded Poland and Olek got involved in the defense of Warsaw. After a few days, howewer, he joined his family and together they moved to France. There, in Angiers, they stayed for about a year until the Germans also invaded France and the Zienkiewicz family had to move again now to England. In the autumn Olek started the course in Civil engineering at Imperial College. He graduated in 1943 with first class honours. Two years later he got his PhD on a dam analysis project under supervision of Prof. Richard Southwell, famous for the method of relaxation based on finite differences.

Now Olek started professional work. For four years he took part in surveying, design and construction of dam projects in Scotland. In 1949 he accepted an offer to become a lecturer in Civil Engineering at Edinburgh. Here he stayed for seven years during which time he started research in hydraulics resulting in series of papers.

In Edinburgh he met a Canadian chemist named Helen who in 1952 became his wife. While in Edinburgh they got two sons, Andrew and David. In 1956 Olek was offered a position as Associate Professor at Northwestern University. He accepted it and settled in Evanston with his family that soon was enlarged with a daugter, Krystina. After a year he was promoted to full professor. However he did not stay long here. In 1961 he applied for and got the chair of Professor of Civil Engineering in Swansea.

In Swansea Olek has with great skill, energy and enthusiasm created an international centre for Computational Mechanics - a combination of engineering, mechanics, numerical analysis and computer science. The centre has had and still has a tremendous importance for the development of this new science. Under Olek's supervision a great number of research students from all over the world have got their education there and senior researchers from all countries have spent long or short periods at the centre.

Olek has given important contribution to almost all disciplines within Computational Mechanics. His special interest is the development of The

Finite Element Method to a universal tool for solution of all problems that can be modelled with partial differential equations. He has published more than 400 papers of which a large number has opened new avenues for research in structural and solid mechanics, in compressible and incompressible flow, in geomechanics, biomechanics, electromagnetics and so on.

Olek has spread his ideas not only by publishing papers but also by giving lectures for engineers and scientists all around the world, by organizing conferences - often several every year - and by writing books. His famous book The Finite Element Method, first edition published in 1967, has probably meant more than anything else to make the method and the new science known at industries and at universities.

In 1969 Olek started the International Journal of Numerical Methods in Engineering which still is the leading journal in its field, now on its 31st volume. Olek has also been very active in bringing the computational mechanics society together by initiating IACM, the International Asociation for Computational Mechanics of which he was the first president. Olek's great accomplishments have been recognized in many ways. He has received honorary degrees and fellowships from a large number of universities and has been rewarded with many medals. He is a fellow of the Royal Society and Commander of the British Empire.

Today, Olek has left management of the department in Swansea is as active as ever. He shares his time between Swansea, Barcelona, Austin and Göteborg. His presence at international conferences is strongly apreciated. His lectures and comments to contributions from others are always the highlights.

This book contains papers written by some of Olek's numerous friends. Most of them were his students at one time.To all of them the editors want to expess their gratitude for their contribution in this volume, which undoubtly will be a point of reference for the development of the finite element method in the next decade.

Finally, the editors want to thank the valuable help of Mercè Alberich, Edith Wolf and Adriana Hanganu in the edition of this book.

E. OÑATE J.PERIAUX A. SAMUELSSON

Barcelona, Paris, Göteborg

October 1991

SECTION 1
PERSONAL ADDRESSES

SECTION 2
STRUCTURAL AND SOLID MECHANICS

SECTION 3
FLUID MECHANICS

SECTION 4
GENERAL APPLICATIONS OF THE FINITE ELEMENT METHOD

SECTION 5
MATHEMATICAL ASPECTS OF THE FINITE ELEMEND METHOD

SECTION 1

PERSONAL ADDRESSES

FINITE ELEMENTS IN THE 90's, E. Oñate, J. Periaux, A. Samuelsson (Eds.)
Springer-Verlag/CIMNE, Barcelona, 1991

ON OCZ

T. Oden
TICOM
The University of Texas at Austin
USA

In May of this year, a few friends, students, and colleagues from the Texas Institute for Computational Mechanics were present on the very day that Olek Zienkiewicz reached the age of 70 years. We celebrated that occasion on my ranch, west of Austin, near Marble Falls, Texas, and we all reveled and enjoyed the hiking, swimming and snorkeling in the lake, barbecue, and the general merriment with Olek and Helen, while marveling at their stamina, energy and zest for life. We were meeting there with a Chaired Professor at The University Texas, O. C. Zienkiewicz himself, the Joe C. Walter, Jr. Chair of Engineering and a member of TICOM. Olek had made his second extended visit to Austin, had taught once again his successful and highly popular course on Advanced Topics in Finite Elements, had given a rash of penetrating seminars on his research, had participated in proposal writing and lectures, and held down a variety of national and international committee assignments, while finding time in the afternoons to swim in Barton Creek, to climb the hills outside of Austin, to sail in Lake Travis, and to generally be on the go every day with something new and challenging. So it is with Olek, who has apparently found a way to avoid frailties we normally associate with threescore and ten years in the heetic business of finite element research. But indeed it is his energy and his love for challenge and competition that have not only made him the leader in this subject, but have also added to his energy and enthusiasm.

Today Olek is just as energetic, competitive and creative as he was almost thirty years ago when I met him in the sequence of early conferences on finite element methods. The first may have been the first meeting in Wright-Patterson Air Force Base, regarded by many as an important milestone in the early development of finite element methods. His name was known to most of us then for his book entitled, *Stress Analysis*, in which important works on finite element methods were brought to the attention of the research community. It was not many years later when his book with Kai Cheung appeared, and this early and timely publication established Olek once and for all as a leader and an important figure in the world of finite element research. In those days, we were actually working on many of the same ideas, including finite elements in time, discrete Kirchhoff elements, curvilinear elements, special types of plates and shell elements, and various kinds of nonlinear problems. A few years later, we were also concerned with convergence and what it meant for an element to pass the patch test with regard to its convergence properties. I recall discussing these issues at length with him during the various Wright-Patterson conferences and finally, in the late 1960s, we all began to be interested in finite elements in computational fluid mechanics.

I was very fortunate to have the opportunity to work with Olek and Dick Gallagher in planning the first of the popular series of conferences on finite elements and flow problems. The first was held in 1974 in Swansea, and was an extremely successful meeting. The proceedings of that meeting represent the first one or two volumes on finite elements in CFD and were the beginnings of the development of a new field which has since had a dramatic impact on the computer simulation of flow phenomena. But that was by no means the only interactions with Olek over the years. We worked together on proceedings of these various flow meetings, editorial assignments for *IJNME*, and finally in the late '70s and early '80s, on the creation of the International Association for Computational Mechanics, of which he served as its first President and me, the second.

We have a lot of adventures together, including swimming off the coast of Portofino in 1976, hiking down the streets of Lisbon, Paris, Swansea and Banff, Canada, Tokyo, and many other placer; a two-weeks in China in the early '80s with our wives, which included some difficulties at the border which we all joke about, but which were not funny at the time; days together in Hong Kong, long trips back through the Middle East and Wales, but most happily the pleasant times in Austin that are now a periodic event.

I take great personal satisfaction and special pride in having Olek and Helen as members of our University family at TICOM each spring. As a Professor at The University of Texas, he has not only brought visibility and strength to our work, but he has invigorated all of us by his style and energy, particularly the younger faculty who marvel at his intuition and knowledge of mechanics and finite element methods.

A lot of ingredients make Olek a great man; his charm and humanity, his intellectual ability, his stubbornness and competitiveness, and his wife Helen, who has not only provided great support for him during his professional career, but who with him makes a unique and charming duo that is a credit to the profession and a joy to anyone who knows them.

May this day mark just another beginning of a period of contributions and activities of Olek Zienkiewicz. We continue to look for and expect more from him in research and we especially hope to enjoy the good company and friendship of him and Helen for many years to come.

T. Oden

FINITE ELEMENTS IN THE 90's, E. Oñate, J. Periaux, A. Samuelsson (Eds.)
Springer-Verlag/CIMNE, Barcelona 1991

OUR FIRST ENCOUNTER

M. Kwiecinski
Institute of Structural Mechanics,
Warsaw University of Technology,
al.Armii Ludowej 16, 00-637 Warsaw,
POLAND

Being no outstanding specialist in the finite element method per se, I decided to contribute this short note of a rather personal than technical character to commemorate Olek's 70th birthday - unbelievable, judging from his never ceasing energy and passion for research, great enjoyment from the travel, arts, nature and hobbies. Realizing that all due tributes will have deservedly been paid to him in earnest by some most distinguished men of science and technology, let me be very informal in what I have an urge to say.

At the beginning of 1964 I was working under Professor Wacław Olszak in the Chair of Elasticity and Plasticity of the Faculty of Civil Engineering, Warsaw University of Technology. It was at that point in my academic life that I desired to go abroad for a year or so to acquaint myself with other institutions of higher education and do some useful post-doctoral research. My application to the British Council was received favourably and a scholarship was granted to me to spend the 1964-65 academic year in Britain. My suggestion was to stay at a place where research was going on connected with the development work in plasticity and plastic behaviour of structures.

Let me quote an excerpt from the British Council letter to me, dated 8th June 1964: "You can be safely assured that Swansea has the right facilities for you and you would do well to accept the place (...) They are doing research which would seem to be precisely what you want and I understand that Professor Zienkiewicz is keen on having you". That was the first time I came across Olek's name and, suspecting - or rather hoping - that he must have been in some way of Polish origin, I gladly accepted the offer.

It was at the end of September 1964 that, 35 years of age and a new doctor of Warsaw University I entered Professor Zienkiewicz's room on the Swansea campus, introduced myself, got immensely impressed by the handsomeness and extremely nice manners of that 43 years old, already of international repute, Professor and Head of the Civil Engineering Department. About his Polish ancestry and political convictions I did not have to ask. Just after shaking hands I glanced at the wall behind Olek's desk and saw a bronze relief of Marshal Józef Piłsudski, who had fought for Poland's independence during and after the First World War and was considered by all patriotic Poles to be the national hero of the 20th century. After the Second World War, due to the well known circumstances connected with the Jalta agreement, he was not recognized by the pro-Russian regime in Warsaw and it would have been impossible to hang his portrait in any public place in Poland at the time I went to Swansea.

Needless to say how nice I was received at the Department and how agreeable conditions I got to work and enjoy myself. I shared a small room in a very modest barrack near a narrow gate to Singleton Park with the then Mr Yau Kai Cheung, shortly Kai and a friend of mine and now the Pro-Vice Chancellor of Hong Kong University and Head of the Civil and Structural Department.

In 1964 Olek was in the course of looming as the major figure in the numerical analysis of problems in engineering science, especially as a co-founder of the finite element philosophy and technique. My research was mainly devoted to the ultimate load theory of engineering structures, plastic beams and plates in particular, with special reference to reinforced concrete slabs that were the subject of my doctoral thesis defended in Warsaw before I came to Swansea. However, my approach was rather analytical and, moreover, my plans were to extend my studies while in Britain to write the so-called habilitation doctor's thesis which would have enabled me, after return to my country, to be promoted to the position of "docent" - associate professor.

Olek was so kind and understanding as not to push me too hard in the direction of strictly numerical studies and was generous enough to give me free hand in the subject of my future publications. Thus I continued my research on the plastic behaviour of slabs which paved the way to my final degree in Warsaw after my return.

Although I did not go into a deeper and formal study of the FEM as such, we both Olek and me seemed to share the same kind of a "lumpy" approach to modelling structural situations. We agreed that it was reasonable and intuitively sound to simplify any mechanical entity as a set of lumps - or finite elements if you like - more or less deformable in themselves and connected in some specific ways at nodes and edges which were to determine the overall behaviour of the whole system under consideration, be it elastic, plastic, viscous or whatever. An example of our cooperation can be found in the paper "Ultimate load behaviour of Mangla Dam baffle blocks: analysis and model tests", coauthored by Olek, Brian Preece and myself, which appeared in The Structural

Engineer in No 7 of July 1971. The problem was to make a preliminary collapse load analysis of the huge baffle blocks of the Mangla Dam spillway. We predicted a number of kinematically admissible collapse, of failure, modes of the blocks treated as undeformable lumps capable of rigid rotations and translations associated with the development of certain systems of cracks in concrete and yield forces in the reinforcement. Then, from the virtual work principle, we estimated the most relevant collapse mechanism leading to the lowest upper bound on the collapse thrust. Model tests, on which I will not dwell, showed that the predicted failure mode did, in fact, occur. Thus, a rather rough model of brittle-plastic collapse of a reinforced concrete structure , backed by thorough knowledge of bounding theorems of the limit analysis resulted in an interesting and relatively simple solution of a very complicated three-dimensional situation in the area of fluid-structure interaction.

Let me make an additional remark at this point. The collapse load theory, to which I have modestly contributed over the years, does not seem, at first glance, to be a F.E. technique at all. However, on second thought, I venture to say that the yield line method is a peculiar, and a very natural, procedure based on the F.E. philosophy!

Namely, due to the fact that such material as, for instance, mild steel, behaves initially very rigidly and then becomes rather suddenly very ductile, what is created is a system of actual finite elements, i.e. rigid portions of a structure interacting with their neighbours not at fictitious nodes, but along real edges where narrow plastic zones develop known as yield hinges in which all deformations are thus assumed to concentrate. Virtual work equation is thus employed not for a mentally discretized system, but for the real system taking place in nature as soon as the applied load approaches the collapse load.

Towards the end of my short note on our deep and everlasting friendship I would like to stress how wonderful memories, both of research and personal character, brings to me every thought of Swansea. Olek and Helen's hospitality to me was so warm and marvellous - let me just recall Christmas Day in 1964. I was of course invited to join the family party. Before we were asked by the charming mistress of the house to take places at the table, Olek seemed to show some signs of uneasiness followed by a question asked in a rather subdued voice: "By the way, Marek, do you in any particular way like to eat skin of a roast goose?". With some embarrassment I ventured to tell him the naked truth that, alas, the roast skin of any flying creature was not my favourite specialty. Olek's faced beamed, he became completely relaxed and shouted with joy: "That is great because I do love skin so you can safely have all the rest of the bird!". This is just a substantial, and very tasty, proof that we always agreed on each important issue - and tissue! Moreover, he always knew what he wanted and played fair to achieve his purpose. Incidentally, I still fear to think what would have happened to me had I been fond of the skin and made a frank and naive confession of that!

Dear Olek, all the best wishes from the bottom of my heart and many happy returns. To dear Helen, your most faithful companion, go also my deepest appreciation and best regards.

FINITE ELEMENTS IN THE 90's, E. Oñate, J. Periaux, A. Samuelsson (Eds.)
Springer-Verlag/CIMNE, Barcelona 1991

PERSONAL ADDRESS FOR OLEK C. ZIENKIEWICZ AT THE OCCASION OF HIS 70TH BIRTHDAY

Professor Dr.-Ing. E. Stein
Institute of Structural and Computational Mechanics,
University of Hannover,
GERMANY

The whole Community in the wide field of Computational Mechanics is celebrating Olek Zienkiewicz's birthday. We are very proud of him due to his scientific achievements in the Finite-Element-Method and also for devoting a big lot of his energy to the nucleation, the ,growth and the strengthening of the new fascinating field of simulating complicated processes within engineering systems in computers in a reliable and efficient way.

I congratulate you also on behalf of the GAMM being a member of the board and on behalf of the German Committee of Mechanics (DEKOMECH), the official board of mechanicians in Germany.

Furthermore you will be glad to hear that GACM- the German Association for Computational Mechanics, affiliated to IACM- is now really alife, and as I hope will be going strong

The first conference will be a

GAMM-GACM-Seminar on
ADAPTIVE METHODS For FINITE DOMAIND and BOUNDARY ELEMENTS
on March 5 and 6, 1992

at my Institute in Hannover. The co-chairman will be Professor Rannacher from the University of Heidelberg.

Of course it is a great pleasure to remember the wonderful ceremony in Braunschweig in May 1987, when the Gauss-Medal was awarded to you. In your substantial lecture "The Challenge of computational mechanics integrated with the computer aided design will be a helpful tool for engineering intuition.

So in continuation of *von Kàrmàn's* words which you cited in Braunschweig, namely

"Engineering creates what never has been",

one could add

"Computing realizes what never was seen"

If one realizes the results in damage mechnics or in non-linear dynamical systems, the meaning of this sentence becomes clear.

Dear Olek, I hope that the Lord will give you many years of health, of new ideas and of love and happiness together with Helen

Yours cordially,

E. Stein

FINITE ELEMENTS IN THE 90's, E. Oñate, J. Periaux, A.Samuelsson (Eds.)
Springer-Verlag/CIMNE, Barcelona 1991

UNIVERSITY COLLEGE OF SWANSEA : 1962-1966

G. S. Holister
Universidad Politécnica de Cataluña,
08034 Barcelona,
SPAIN

I first joined O.C.Z.'s team at University college, Swansea, in 1962. I had recently returned from several yours in industry in the United States, and Swansea was my first academic appointment. If my memory serves me correctly, I was also the first new member of Olek's staff, since he had only recently been appointed to the Chair of Civil Engineering himself.

Olek's reputation in numerical methods was already well established and he was looking for someone to develop an effective research group in experimentally-based stress analysis in order to give balance to his own work in finite elements, which in those days was almost exclusively applied to stress analysis problems in the building industries. The fact that I was appointed as a Lecturer in civil engineering, whereas I had graduated as a physicist and had spent most of my subsencuent career in mechanical engineering research, did not seem to worry Olek in the slightest, although it naturally caused me great trepidation.

Nevertheless, I survived my first year, successfully teaching mechanics and materials science to first-year students. My growing confidence in my teaching abilities, however, was slattered the following year when Olek assigned me to a third - year class in soil mechanics. My protestations of complete ignorance of the subject were brushed aside by Olek, who advised me to read the text-books and be sure to "stay a week ahead of the students". Unconvinced by his assurances that I could cope, I decided on a policy of absolute honesty with my students, and proposed to them that we study the subject together, week by week, as a means of group therapy. Since I had always been a slow learner, this resulted in me being consistently a week behind my students rather than ahead of them. However, the difficulties faced by my (fortunately small) class in explaining the subjects to me ensured that by the end of the year they all understood the syllabus far better than if I had attempted to teach them. As a result they all passed the end-of-year exams with marks that were considerably better that average. Olek simply said, "I told you so"!

Our research activities were equally successful, but much less traumatic. We rapidly developed a small, competent team, working in the fields of Moiré methods, photoelasticity and photoelastic coatings and, with Olek's encouragenent I took my first hesitant steps in the field of consulting. The next few years were the most productive in my whole career, mainly due to Olek's unique "hands-off" method of looking after his team. But it was when the time came for me to move on that Olek's help became crucial. I had been offered the Chair of Engineering Mechanics at Pennsylvania State University, subjects to my obtaining a Ph. D.- something that I had avoided doing earl'ier. Once again, Olek's advice turned out to be invaluable. When I first joined his staff he had registered me to submit a Ph. D. thesis based on my research at some time in the future. Since, at that time, neither of us had any idea what direction such research would take, the thesis title submitted was a masterpiece of generality. Although highly esoteric, it could loosely be translated as "The Application of Experimental Stress Analysis Methods to Something-or other". It worked. Since, as I have said, this was a highly productive period in my career, I had already published a fairly extensive list of research papers. With a thesis title already registered these publications, with the liberal use of paste and scissors were rapidly transformed into a relatively acceptable (Olek's comment: "I've seen worse") thesis. One final problem: following submission of the thesis only a few weeks remained before I was to take up my new post-insufficient-time, under normal circumstances, to choose a second thesis supervisor (Olek's was the first) and arrange for the customary 'viva'. Once more Olek rode to the rescue, finding for me within a few days a most distinguished professor as my external supervisor. The fact he was Polish was, I am sure, quite coincidental. More important, he held similar views to Olek an university regulations and their relevance to academic merit. As a result, within ten days my thesis had been read (I assume) and pronounced of "sufficient merit" to permit the waiving of the viva formality. Thus was established what I belive to be an all-time record for a Ph. D. : from submission to award within ten weeks! I arrived at Penn State, properly qualified, with days to spare. Looking back, I now realise that this highly irregular (if quite legal) state of affairs could not have been accomplished without the application of heavy doses of Olek's considerable charm, authority and distinction.

Many years later, when I joined UNESCO as Director of their Technology Division, I was able to repay, in part, my great debt to Olek by helping create, in collaboration with Eugenio Oñate, the Internacional Centre for Numerical Methods in Engineering in Barcelona - an original idea of Olek's. Subsequently, I created the "UNESCO Chairs" programe and was, of course, delighted (but not surprised) when Professor O.C. Zienkiewicz, F.R.S., became the first Unesco Professor of Numerical Methods - a post he still, happily, holds.

Working with Olek is a unique experience. His friends and colleagues have somehow inevitably become nine. His influence has been largely responsible for whatever success I have had. But most important of all, working with Olek has always been and, I am sure, will continued to be, great fun.

G. S. Holister

SECTION 2

STRUCTURAL AND SOLID MECHANICS

FINITE ELEMENTS IN THE 90's, E. Oñate, J. Periaux, A. Samuelsson (Eds.)
Springer-Verlag/CIMNE, Barcelona 1991

A NEW BOUNDARY-TYPE FINITE ELEMENT FOR 2D- AND 3D- ELASTIC SOLIDS

E. Stein and K. Peters
Institut für Baumechanik und Numerische Mechanik,
Appelstr. 9A,
D-W-3000 Hannover 1,
GERMANY

Summary

Starting from the variational principle with the so-called hybrid stress method the trial functions for the stresses have now to fulfill the *Beltrami*-equations, that means also the compatibility equations for the strains. So the divergence theorem can be applied, and one arrives at a pure boundary formulation in the sense of *Trefftz*'s method. Beside the resulting variational formulation different regularizations of the interelement conditions are investigated by numerical tests. Two 3d-examples are presented, one to show the effects on distorted elements and the other for numerical comparison of the test of a composite box girder with inner steel flanges. The numerical efficiency in comparing with competing finite elements is promising and stimulates further research.

1 Survey on hybrid stress elements – finite domain and boundary elements

Pians [1] hybrid stress method needs trial functions for stresses which fulfill the static field equations and, additional, trial displacements at the interelement boundaries in order to fulfill the weak conditions for the continuity of the stress vectors. By this the divergence of the stresses is approximated in an L_2-space so that the reciprocity condition is fulfilled in a weak sense. Therefore these elements are numerically stable - similar to properly balanced mixed hybrid approaches.

By taking trial stresses which even fulfill the *Beltrami* differential equations, i.e. $\triangle \triangle \boldsymbol{\sigma} = \mathbf{0}$, the field integral for the stress energy can be transformed into a boundary integral in each element, and one arrives at a *Trefftz*-type formulation [2]. Coupling *Trefftz*-domains with finite displacement elements and using Galerkin techniques were given by *Stein* [3]. As outlined in [3], additional regularization terms can be necessary at the interfaces to achieve convergence, e.g. by using squared error integrals.

In this paper different regularization terms are numerically tested in order to get information about the computational efficiency and the behaviour in the case of disturbed 3d–meshes.
A similar strategy of so–called hybrid Trefftz elements was applied by *Jirousek* [4] for *Kirchhoff*-plates in bending.

2 Functional, trial and test functions

The starting point is the complementary energy functional for the body $\mathcal{B}$ which is discretized into n elements $\mathcal{B}_e \subset R^3$ or $\mathcal{B}_e \subset R^2$

$$\mathring{\pi}_h \;=\; \frac{1}{2}\int_{\cup\mathcal{B}_e} \boldsymbol{\sigma}_h^T \mathbf{C}^{-1}\boldsymbol{\sigma}_h \; d\mathbf{x} \;+\; \pi_a \;\rightarrow\; \text{min.} \tag{1}$$

In this functional the shape functions for $\boldsymbol{\sigma}_h$ have to fulfill the statical field equations and the statical continuity conditions between adjacent element domains. The continuity requirement for the interface stress vectors are essential conditions in the variational calculus whereas the continuity of the displacements are natural conditions.

The Trefftz elements follow the main idea of the so–called hybrid stress elements, i.e. to release the statical continuity equations. Then the following extended functional is written [1]

$$\mathring{\pi}_{Eh} \;=\; \mathring{\pi}_h \;-\; \int_{\cup\Gamma_e} \mathbf{u}_h^T(\mathbf{t}_h - \bar{\mathbf{t}}_h)\; d\mathbf{x} \qquad \rightarrow \text{stat.} \qquad [\text{Term } 0] \tag{2}$$

In the sequel the following regularization terms, here written as a sum, are investigated at the interelement boundaries by numerical tests

$$\begin{aligned} F_h \;=\; \mathring{\pi}_h \;&+\; \frac{1}{\alpha_1}\;\sum_{(t)}^{n} |\mathbf{u}_k - \bar{\mathbf{u}}_k| && [\text{Term } 1]\\ &+\; \frac{1}{\alpha_2}\int_{\cup\Gamma_e} (\mathbf{u}_h - \bar{\mathbf{u}}_h)^2 \;\; d\mathbf{x} && [\text{Term } 2]\\ &+\; \frac{1}{\alpha_3}\int_{\cup\Gamma_e} (\mathbf{t}_h - \bar{\mathbf{t}}_h)^2 \;\; d\mathbf{x} && [\text{Term } 3]\\ &-\; \alpha_4 \int_{\cup\Gamma_e} \mathbf{t}_h^T(\mathbf{u}_h - \bar{\mathbf{u}}_h) \;\; d\mathbf{x} \quad \rightarrow \text{stat.} && [\text{Term } 4] \end{aligned} \tag{3}$$

where k are nodal points, $\boldsymbol{\sigma}_h$ are the trial stresses in the element domain, satisfying $\triangle\triangle\,\boldsymbol{\sigma}_h = \mathbf{0}$, and $\mathbf{t}_h(\boldsymbol{\sigma}_h)$ and $\mathbf{u}_h(\boldsymbol{\sigma}_h)$ are calculated from the stress trial function according to

$$\boldsymbol{\sigma}_h = \mathbf{N}_\sigma\;\boldsymbol{\beta} \qquad \text{in} \quad \mathcal{B}_e \tag{4}$$

using the constitutive law and the compatible integration of the strains

$$\begin{aligned} \boldsymbol{\varepsilon}_h &= \mathbf{C}^{-1}\mathbf{N}_\sigma\boldsymbol{\beta}\\ \mathbf{u}_h &= \int_{[S]} \boldsymbol{\varepsilon}_h ds = \mathbf{Z}\boldsymbol{\beta}\\ \mathbf{t}_h &= \mathcal{R}\boldsymbol{\sigma}_h = \mathcal{R}\mathbf{N}_\sigma\boldsymbol{\beta} = \mathbf{R}\boldsymbol{\beta}. \end{aligned} \tag{5}$$

The trial functions marked with a bar in equ. (3) are defined at the interfaces of the Trefftz–element domains. They depend on the unknown nodal displacement parameters $\mathbf{v}$.

The trial interface displacements are given by

$$\bar{\mathbf{u}}_h = \mathbf{N}_u \mathbf{v} \qquad \text{on } \Gamma_e \tag{6}$$

and the derived interface stresses $\mathbf{t}_h(\mathbf{u}_h)$ are derived from

$$\bar{\mathbf{t}}(\mathbf{u}_h) = \bar{\mathbf{C}}\,\mathbf{v} \qquad \text{on } \Gamma_e, \tag{7}$$

where the matrix $\bar{\mathbf{C}}$ is a traction trial function.

For the formulation of Trefftz–elements different regularizations in equ. (3) can be chosen. All additional terms are implemented in our computer program within INA–SP and can be activated separately.

If we choose [Term 1, 2 or 3] with α_1, α_2 or $\alpha_3 \neq \infty$, we get a penalty regularisation on the surfaces of the Trefftz elements with the penalty factors α_1, α_2 and α_3.

[Term 4] in equ. (3) with $\alpha_4 = 1$ is similar to a Lagrange–extension. The regularization is obtained by using the known derived trial function $\mathbf{t}_h$ of the domain as a test function $\boldsymbol{\lambda}_t$.

If two or more of the additional terms are used, for example $\alpha_2 \neq \infty$, $\alpha_1, \alpha_3 = \infty$, $\alpha_4 = 1$ one gets augmented methods which are not investigated in this paper.

3 Nodal and weak coupling of the boundary elements

The additional terms in equ. (3) are nodal and weak coupling conditions of the Trefftz element domains for displacements or stresses.
[Term 1] postulates that the continuity of the natural boundary condition is fulfilled exactly in selected node points k at the interfaces of adjacent Trefftz elements via the interelement displacement functions $\bar{\mathbf{u}}_h$. This is a strong coupling condition for selected points.

[Term 2] in equ. (3) postulates—in contrast to the first one—the minimization of the displacements $\mathbf{u}$ and $\bar{\mathbf{u}}$ in a quadratic average sense so that the defect on the surface is minimized. This is a regularization for the natural interface condition $\mathbf{u}^+ = \mathbf{u}^-$. Another weak coupling condition for the boundary condition $\mathbf{t}_h^+ = \mathbf{t}_h^-$ leads to [Term 3] in equ. (3).

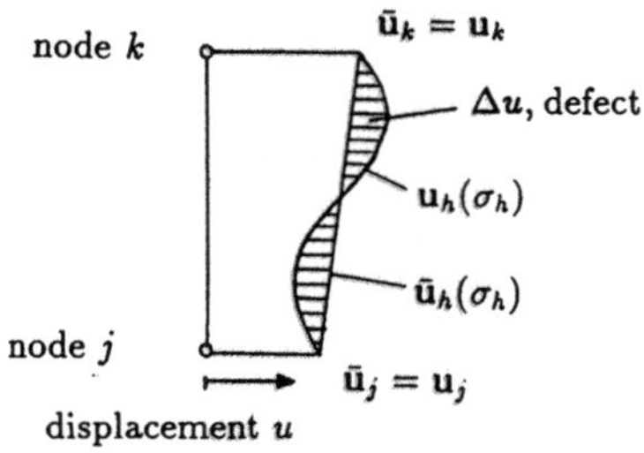

Figure 1: Nodal coupling of the displacements

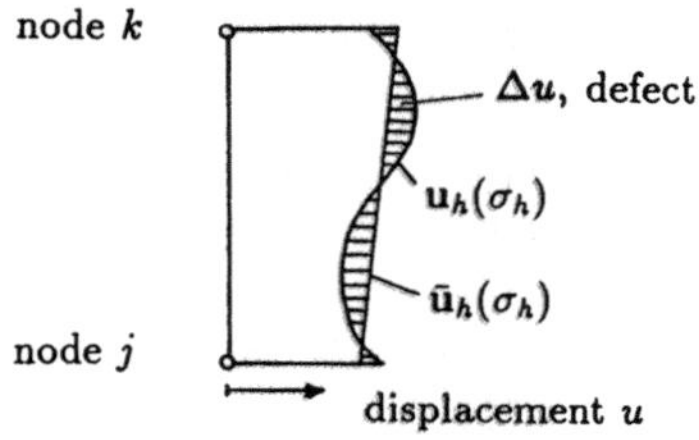

Figure 2: Weak coupling of the displacements

4 Element matrices, elimination of the stress parameters

The complementary stress potential

$$\overset{*}{\pi}_i = \frac{1}{2}\int_{\cup B_e} \boldsymbol{\sigma}_h^T \mathbf{C}^{-1} \boldsymbol{\sigma}_h d\mathbf{x} \tag{8}$$

can be written as the surface integral

$$\overset{*}{\pi}_i = \frac{1}{2}\int_{\Gamma_e} \mathbf{u}_h^T(\boldsymbol{\sigma}_h)\ \mathbf{t}_h(\boldsymbol{\sigma}_h)\, d\mathbf{x} \tag{9}$$

because the static field equations are fulfilled by the trial function $\boldsymbol{\sigma}_h$.

From the symmetric quadratic form for the stress energy of an element

$$\overset{*}{\pi}_i \;=\; \frac{1}{2}\boldsymbol{\beta}^T \mathbf{H}\, \boldsymbol{\beta} \tag{10}$$

we can derive the matrix $\mathbf{H}$ from eqns. (5) and (9)

$$\mathbf{H} = \frac{1}{2}\int_{\Gamma_e} \left(\mathbf{Z}^T\mathbf{R} + \mathbf{R}^T\mathbf{Z}\right) d\mathbf{x}. \tag{11}$$

To get a nodal displacement formulation with the element stiffess matrix $\hat{\mathbf{k}}$,

$$\pi_i = \frac{1}{2}\mathbf{v}^T\hat{\mathbf{k}}\mathbf{v}, \tag{12}$$

one of the extensions in equ. (3) is used.

The transformation matrix $\mathbf{G}$ that connects the stress parameters $\boldsymbol{\beta}$ with the nodal displacements $\mathbf{v}$

$$\boldsymbol{\beta} = \mathbf{G}\ \mathbf{v} \tag{13}$$

is calculated as shown below. The order of the row regularity must correspondent to the number of the independent stress parameters.
The Trefftz–stiffness matrix is given by

$$\hat{\mathbf{k}} = \mathbf{G}^T\mathbf{H}\,\mathbf{G}. \tag{14}$$

In the sequel the matrix $\mathbf{G}$ is derived for the different terms of equ. (3):

From [Term 1]

$$J_1 = \sum_{(k)} \mid \mathbf{u}_h^k - \bar{\mathbf{u}}_h^k \mid = 0 \tag{15}$$

with the trial functions

$$\mathbf{u}_h^k = \mathbf{Z}^k\boldsymbol{\beta} \quad \text{in } \mathcal{B}_e; \qquad \bar{\mathbf{u}}_h^k = \mathbf{N}_u^k\mathbf{v} \quad \text{on } \Gamma_e, \tag{16}$$

written for every node k

$$\begin{bmatrix} \mathbf{Z}^1 \\ \mathbf{Z}^2 \\ \vdots \\ \mathbf{Z}^k \end{bmatrix} \begin{bmatrix} \beta_1 \\ \beta_2 \\ \vdots \\ \beta_m \end{bmatrix} = \begin{bmatrix} \mathbf{H}_u^1 \\ \mathbf{H}_u^2 \\ \vdots \\ \mathbf{H}_u^k \end{bmatrix} \begin{bmatrix} \mathbf{v}^1 \\ \mathbf{v}^2 \\ \vdots \\ \mathbf{v}^k \end{bmatrix}$$

$$\mathbf{Z}_G \quad \boldsymbol{\beta} \quad = \quad \mathbf{1} \quad \mathbf{v}. \tag{17}$$

one obtains the transformation–matrix $\mathbf{G}^{(1)}$

$$\boldsymbol{\beta} = \mathbf{Z}_G^{-1}\mathbf{v} = \mathbf{G}^{(1)}\mathbf{v}. \tag{18}$$

For [Term 2]

$$J_2 = \int_{\Gamma_e} (\mathbf{u}_h - \bar{\mathbf{u}}_h)^2\, d\mathbf{x} \longrightarrow \min. \tag{19}$$

with the trial functions

$$\mathbf{u}_h(\boldsymbol{\sigma}_h) = \mathbf{Z}\boldsymbol{\beta} \quad \text{in } \mathcal{B}_e; \qquad \bar{\mathbf{u}}_h = \mathbf{N}_u\mathbf{v} \quad \text{on } \Gamma_e \tag{20}$$

the stationary condition

$$\frac{\partial J_2}{\partial \boldsymbol{\beta}} = 0 \tag{21}$$

yields the transformation matrix

$$\boldsymbol{\beta} = \mathbf{Q}_F^{-1}\mathbf{L}_F\mathbf{v} = \mathbf{G}^{(2)}\,\mathbf{v} \tag{22}$$

with

$$\mathbf{Q}_F = \int_{\Gamma_e} \mathbf{Z}^T\mathbf{Z}dx \qquad \text{and} \qquad \mathbf{L}_F = \int_{\Gamma_e} \mathbf{Z}^T\mathbf{N}_u dx. \tag{23}$$

[Term 4] in equ. (3)

$$F_4 = \frac{1}{2}\int_{\Gamma_e} \mathbf{u}_h^T\mathbf{t}_h dx - \int_{\Gamma_e} \mathbf{t}_h^T\left(\mathbf{u}_h - \bar{\mathbf{u}}_h\right)dx \longrightarrow \text{stat.} \tag{24}$$

with

$$\mathbf{u}_h(\boldsymbol{\sigma}_h) = \mathbf{Z}\boldsymbol{\beta}, \quad \mathbf{t}_h(\boldsymbol{\sigma}_h) = \mathbf{R}\boldsymbol{\beta}, \quad \bar{\mathbf{u}}_h = \mathbf{N}_u\,\mathbf{v} \quad \text{on } \Gamma_e \tag{25}$$

and

$$\delta F_4 = 0 \tag{26}$$

yields the transformation matrix $\mathbf{G}^{(4)}$

$$\mathbf{H}\boldsymbol{\beta} - \mathbf{L}_E\,\mathbf{v} = \mathbf{0}, \quad \boldsymbol{\beta} = \mathbf{H}^{-1}\mathbf{L}_E\mathbf{v} = \mathbf{G}^{(4)}\,\mathbf{v} \tag{27}$$

with

$$\mathbf{H} = \int_{\Gamma_e} \mathbf{Z}^T\mathbf{R} + \mathbf{R}^T\mathbf{Z}dx \quad \text{and} \quad \mathbf{L}_E = \int_{\Gamma_e} \mathbf{R}^T\mathbf{N}_u dx. \tag{28}$$

By elimination of the rigid body modes in $\mathbf{H}$ the reduced matrix $\mathbf{H}$ is positive definite. The rank of $\mathbf{H}$ is equal to the rank of $\mathbf{Z}_G$ and $\mathbf{Q}_F$ reduced by the number of rigid body modes.

The [Term 3] in equ. (3) of tractions

$$J_3 = \int_{\Gamma_e} \left(\mathbf{t}_h - \bar{\mathbf{t}}_h\right)^2 d\Gamma \longrightarrow \text{min.} \tag{29}$$

with

$$\mathbf{t}_h(\boldsymbol{\sigma}_h) = \mathbf{R}\,\boldsymbol{\beta}, \qquad \bar{\mathbf{t}}_h = \bar{\mathbf{C}}\,\mathbf{v} \quad \text{on } \Gamma_e \tag{30}$$

and

$$\frac{\partial J_3}{\partial \boldsymbol{\beta}} = 0 \tag{31}$$

leads to

$$\mathbf{Q}_S\boldsymbol{\beta} = \mathbf{L}_S\mathbf{v}, \qquad \boldsymbol{\beta} = \mathbf{Q}_S^{-1}\mathbf{L}_S\mathbf{v} = \mathbf{G}^{(3)}\,\mathbf{v}, \tag{32}$$

$$\mathbf{Q}_S = \int_{\Gamma_e} \mathbf{R}^T\mathbf{R}dx \quad \text{and} \quad \mathbf{L}_S = \int_{\Gamma} \mathbf{R}^T\bar{\mathbf{C}}dx. \tag{33}$$

The additional term in equ. (2) the Lagrangian extension of the complementary energy, yields the functional

$$F_{Eh} = \frac{1}{2}\int_{\Gamma_e} \mathbf{u}_h^T\mathbf{t}_h dx - \int_{\Gamma_e} \mathbf{u}_h^T\left(\mathbf{t}_h - \bar{\mathbf{t}}_h\right)dx \longrightarrow \text{stat.} \tag{34}$$

with the trial functions

$$\mathbf{t}_h(\boldsymbol{\sigma}_h) = \mathbf{R}\boldsymbol{\beta}, \qquad \mathbf{u}_h(\boldsymbol{\sigma}_h) = \mathbf{Z}\boldsymbol{\beta}, \qquad \bar{\mathbf{t}}_h = \bar{\mathbf{C}}\,\mathbf{v} \quad \text{on } \Gamma_e \tag{35}$$

and the condition

$$\delta F_{Eh} = 0 \tag{36}$$

results into the transformation matrix

$$\mathbf{H}\boldsymbol{\beta} - \mathbf{L}_L \mathbf{v} = 0, \quad \boldsymbol{\beta} = \mathbf{H}^{-1}\mathbf{L}_L\mathbf{v} = \mathbf{G}^{(5)}\,\mathbf{v}, \tag{37}$$

$$\mathbf{H} = \int_{\Gamma_e} \mathbf{Z}^T\mathbf{R} + \mathbf{R}^T\mathbf{Z}dx \quad \text{and} \quad \mathbf{L}_L = \int_{\Gamma_e} \mathbf{Z}^T\bar{\mathbf{C}}dx. \tag{38}$$

5 Numerical integration of surface integrals, trial functions for 3d–problems

The trial functions for the stresses can be gained from the stress tensor $\boldsymbol{\chi}$

$$\boldsymbol{\sigma} = \text{Ink}\ \boldsymbol{\chi} = -e_{ijk}e_{rst}\chi_{kt,js} \tag{39}$$

and has to fulfill the *Beltrami*–equation

$$\text{Ink}\left(\mathbf{C}^{-1}\left(\text{Ink}\boldsymbol{\chi}\right)\right) = 0. \tag{40}$$

This ensures that the equilibrium conditions, the constitutive law and the integrability conditions for the strains are fulfilled by the stress trial functions in the element domain.

Several solutions for isotropic and anisotropic elastic materials are found for 2d–problems where a characteristic equation has to be solved which is dependent on the constitutive law.

For the use of higher order stress shape function for anisotropic 3d materials a nonlinear system of equations has to be solved where the system matrix depends on the constitutive law.
In this paper linear 3d–stress functions are chosen which fulfill the homogenous differential equation independent of the constitutive law:

$$\begin{array}{rclll}
\sigma_x &=& 1, & \sigma_x = y, & \sigma_{xy} = z \\
\sigma_y &=& 1, & \sigma_x = z, & \sigma_{yz} = x \\
\sigma_z &=& 1, & \sigma_y = x, & \sigma_{xz} = y \\
\sigma_{xy} &=& 1, & \sigma_y = z, & \sigma_x = 2x \ \wedge \ \sigma_{xy} = -y \ \wedge \ \sigma_{xz} = -z \\
\sigma_{yz} &=& 1, & \sigma_z = x, & \sigma_y = 2y \ \wedge \ \sigma_{xy} = -x \ \wedge \ \sigma_{yz} = -z \\
\sigma_{xz} &=& 1, & \sigma_z = y, & \sigma_z = 2z \ \wedge \ \sigma_{yz} = -y \ \wedge \ \sigma_{xz} = -x
\end{array} \tag{41}$$

In the case of quadratic shape functions the difference of *Pians* method and the presented strategy becomes more efficient.

To avoid spurious energy modes, the number of stress parameters has to be greater or equal than the number of nodal displacements surrounding the domain minus the rigid body modes

$$m \geq n - r. \tag{42}$$

Furthermore, the trial stress functions have to be selected such that the linear dependence of the displacement field—which is integrated from the trial stress-field in the element domain—is avoided on the surface of each element.

The computation of an element stiffness matrix is performed numerically with a surface integration technique, using the isoparametric projection.
In contrast to the conventional isoparametric concept, where the trial functions are defined firstly in the ξ, η, ζ-coordinate system of the unit surface or unit cube and then are transformed into the real shape, the trial functions of the Trefftz elements are defined in the real element domain and then have to be integrated over the real element surfaces.

The integration is done by a *Gauss*-point integration. The position of the Gauss-points on the surfaces is found by the isoparametric projection of the surfaces into unit lines or surfaces. The integration is performed similarly to the conventional isoparametric concept using the stress and displacement values of the real surface, the Jacobian determinant and the weights of the isoparametric projection. The assembling of the Trefftz element stiffness matrices is organized as

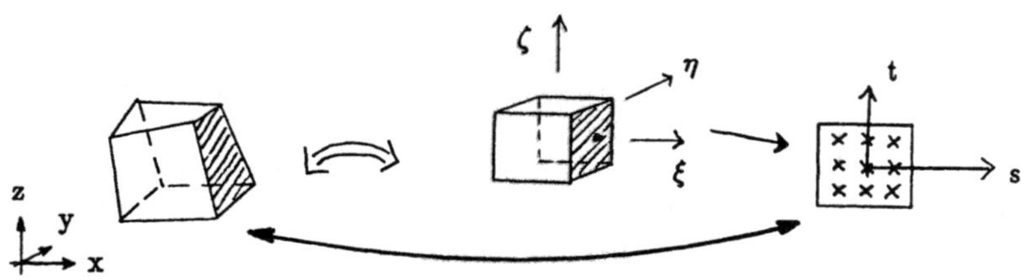

Figure 3: Integration concept

for the displacement method. One gets a positive definite global stiffness matrix. The Trefftz elements can be incorporated in any available finite element program for displacement elements using the pre- and post processing utilities.

6 Examples for 3D–problems

6.1 Disturbed meshes with convergence studies for different coupling techniques

The problem of a cantilever beam under pure bending will be used as example 1 (Figure 4). Different element meshes (Figure 5) and the results of different Trefftz

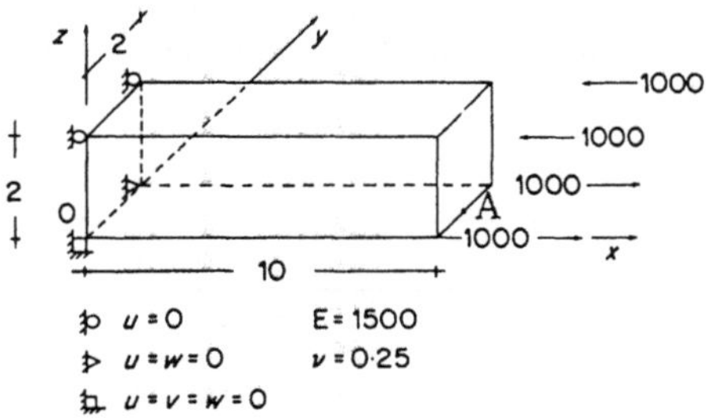

Figure 4: System: cantilever beam

element formulations are shown. The stress distribution in the beam, especially the effect of geometric element distortions on the accuracy of the stresses, will be investigated in detail.

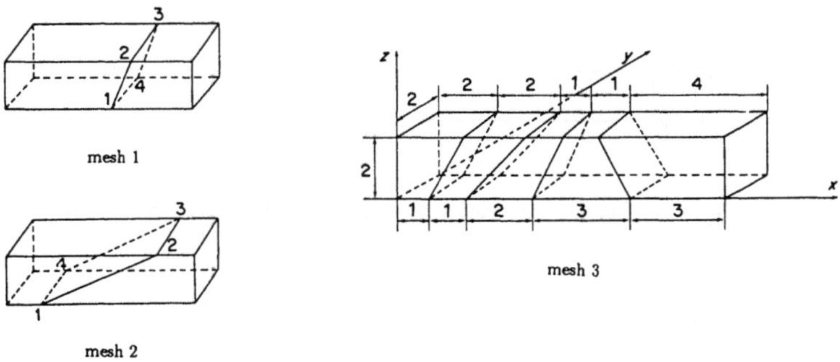

Figure 5: Discretisation mesh 1, 2 and 3

The bending stresses within the beam are computed for the Trefftz elements, as shown in section 4:

[Term 0]	Lagrangian extension.
[Term 1]	Nodal errors of the relative displacements.
[Term 2]	Squared error integral of relative displacements on Γ_e.
[Term 3]	Squared error integral of relative tractions on Γ_e.
[Term 4]	Weak regularisation of the relative displacements.

The results are compared with hybrid elements in [5], marked **Q**.
The Figures 6 to 8 show the stresses at the top and at the bottom of the beam. From the results it can be concluded that:

(a) The Trefftz elements are more or equal accurate than other hybrid or displacement elements of the same order for the shape functions.
(b) The nodal interelement condition [Term 1] appears to be the most accurate formulation.

(c) The elements are not sensitive to geometric distortions. This can be seen from the results of the second mesh in which the elements are severely distorted.

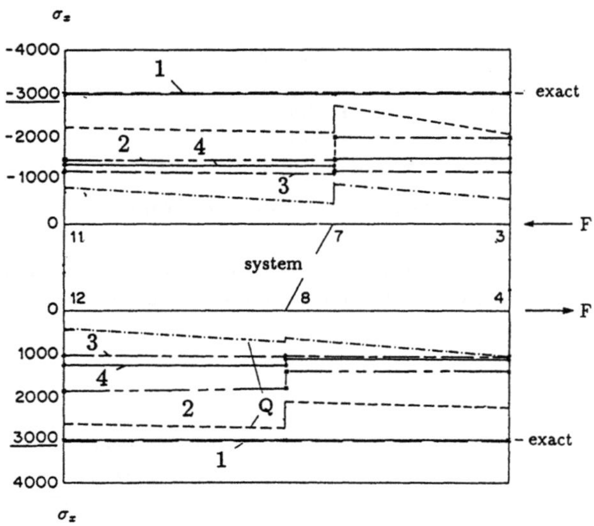

Figure 6: Stress variations at the top and bottom fibres, mesh 1

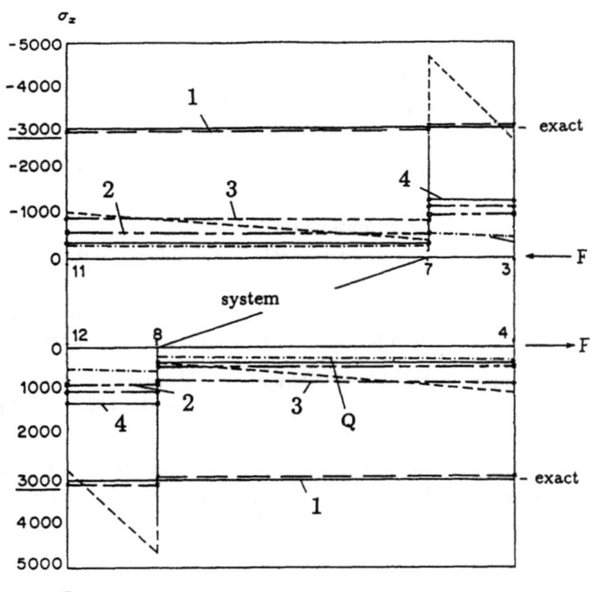

Figure 7: Stress variations at the top and bottom fibres, mesh 2

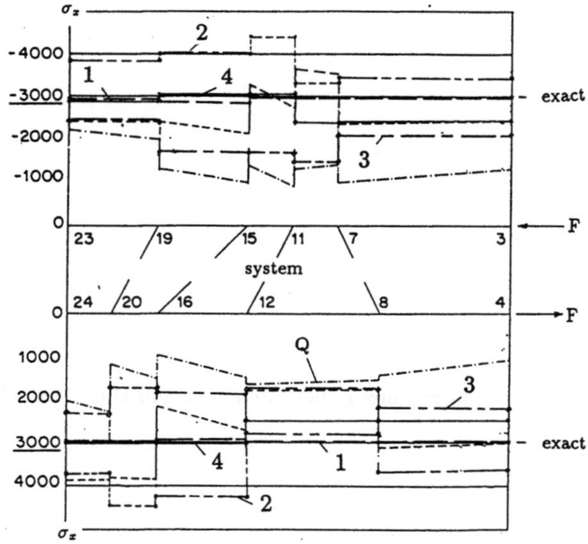

Figure 8: Stress variations at the top and bottom fibres, mesh 3

For formulations [Term 1, 2 and 0] the somehow problematic *Irons* patch test is not fulfilled if the elements are arbitrary shaped brick elements. An improvement of these elements could be done by enriching the trial and test functions or choosing alternative functions.

6.2 Connection of a composite glass fiber reinforced box girder with inner steel flanges

The stress distribution in adjacent Trefftz element domains and the displacements of the interelement trial functions are shown for an example of greater complexity. The connection of a composite glass fiber box girder with steel flanges was tested in an experiment, see Figure 9.

Figure 10 shows the shaded connection detail and the orientation of the single glass fiber layers. The laminate is connected with the steel flange by a fusion adhesive.

Figure 11 shows the interelement displacements of the subsector where the external angle of the steel flange generates large stress concentrations on the upper 0°–layer for which the displacements and stresses are shown. The discretization was made with the formulation of [Term 4]. The displacements are continuous through the whole section. Figure 12 shows the results for the stresses of a rough element mesh. There are large differences between the stresses of the adjacent Trefftz domains. This is typical for significant stress gradients and rough meshes.

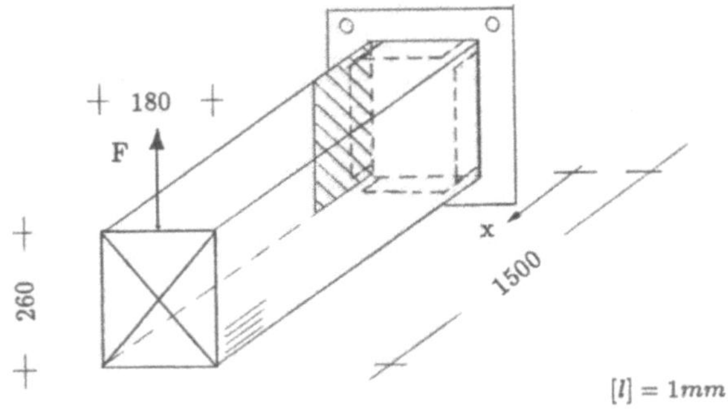

Figure 9: Reinforced box girder, mounting and static load F

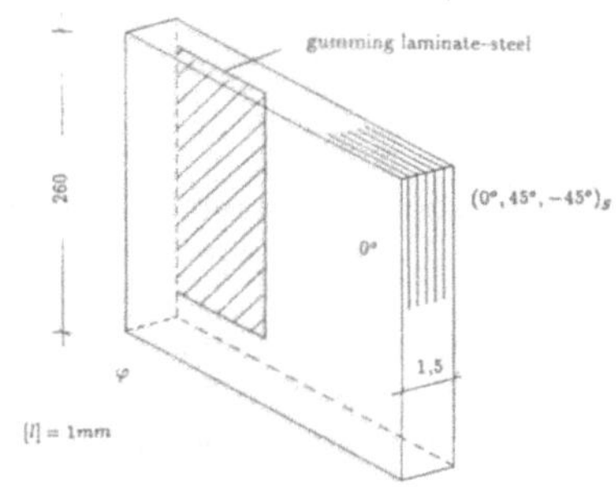

Figure 10: Connection detail

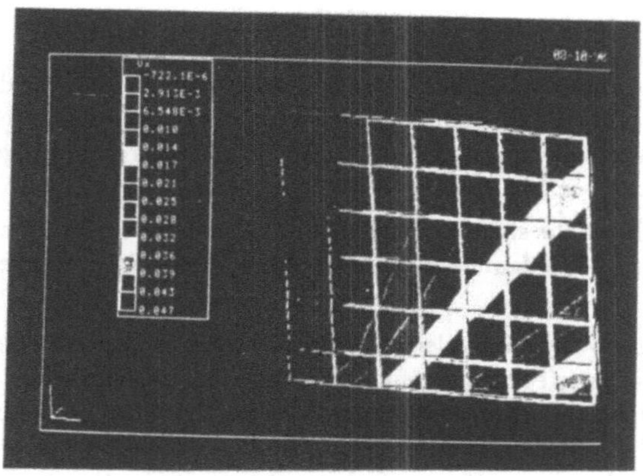

Figure 11: Interelement displacements

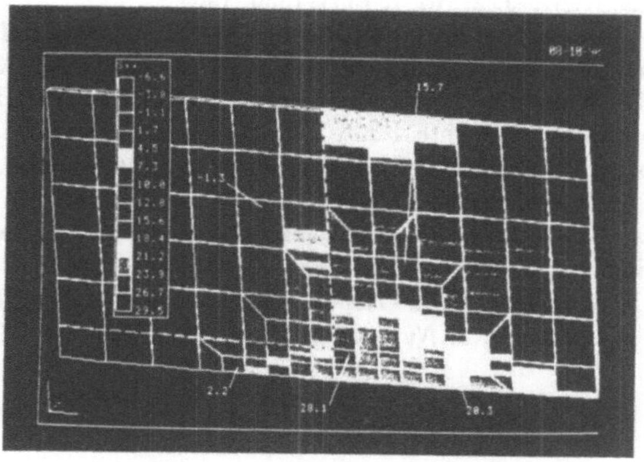

Figure 12: Trefftz stresses σ_x in Trefftz domains

The maximum difference between the measured datas of the experiment and the mean values of the adjacent Trefftz stresses was about 5%.

7 Concluding remarks

Five different Trefftz–element formulations for 3d isotropic and anisotropic elastic problems are presented which yield stable elements of hybrid stress type with comparatively good approximation properties.

The advantages with respect to *Pians* or different hybrid stress elements are:
(a) The integration is only necessary over the element surfaces and not within the element domain,
(b) Special trial stress functions with singularities inside the elements or special boundary conditions can be applied,
(c) For higher order trial stresses (at least quadratic) the accuracy of the Trefftz elements is higher than of comparable hybride elements.

Further research, mainly the use of quadratic trial functions and analytical investigations of the properties of the functional equ. (3) [Terms 1 to 4] are necessary. Coupling strategies of Trefftz elements with boundary elements are investigated at the time in another project at our institute.

References

[1] Pian, T. H. H. - Derivation of element stiffness matrices by assumed stress distributions. AIAA-J. 2, No. 7, 1333–1336, 1964

[2] Trefftz, E. - Ein Gegenstück zum Ritzschen Verfahren. *Verhandlungen des 2. internationalen Kongresses für technische Mechanik Zürich,* 131–137, 1926.

[5] Stein, E. - Die Kombination des modifizierten Trefftzschen Verfahrens mit der Methode der finiten Elemente. Finite Elemente in der Statik, Ed. Bruck, K.E. and Scharpf, D.W. and Stein, E., Verlag Wilhelm Ernst & Sohn, Berlin 1973, p. 172–185 and (with Ruoff) 242–259.

[4] Jirousek, J. - The hybrid-trefftz finite element model and its application to plate bending. *Int. J. Num. Meth. Eng.,* 23, 651–693, 1986.

[5] Cheung, Y.K. and Wanji, J.C. - Isoparametric hybrid hexahedral elements for 3–d stress analysis. *Int. J. Num. Meth. Eng.,* 26, p. 677–693, 1988.

FINITE ELEMENTS IN THE 90's, E. Oñate, J. Periaux, A. Samuelsson (Eds.)
Springer-Verlag/CIMNE, Barcelona 1991

NEW TRENDS IN HYBRID-TREFFZ p-ELEMENT APPROACH

J. Jirousek
LSC, Dept. of Civil Engineering,
Swiss Federal Institute of Technology,
CH-1015 Lausanne,
SWITZERLAND

The Hybrid-Trefftz (HT) finite element model, initiated in 1977-78 [1,2] and based on enforcing the conformity on an assumed element displacement field chosen so as to a priori satisfy the governing differential equations of the problem has now become a powerful and well established tool [3]. The paper shortly overviews and assesses the past an recent developments and outlines some new promising directions of the research under way.

1. INTRODUCTION

An obvious alternative to the Rayleigh-Ritz method as a basis for a finite element (FE) formulation is the method of Trefftz [4]. Here the assumed element displacement field,

$$\mathbf{u} = \mathring{\mathbf{u}} + \sum_j \mathbf{N}_j c_j = \mathring{\mathbf{u}} + \mathbf{N}\mathbf{c} \tag{1}$$

is such that a priori the governing differential equations (and, as a consequence the equations of equilibrium and constitutive relations) are satisfied in the element subdomain Ω^e. If

$$\mathcal{L}\mathbf{u} = \bar{\mathbf{b}} \tag{2}$$

are the governing differential equations (in solid mechanics $\mathcal{L} = -\mathbf{L}^T(\mathbf{D}\mathbf{L}\mathbf{u})$ and $\varepsilon = \mathbf{L}\mathbf{u}$, $\sigma = \mathbf{D}\varepsilon$, $\mathbf{L}^T\sigma + \bar{\mathbf{b}} = \mathbf{0}$ on Ω), the satisfaction of (2) for any value of the undetermined parameters c_j implies

$$\mathcal{L}\mathring{\mathbf{u}} = \bar{\mathbf{b}} \quad \text{and} \quad \mathcal{L}\mathbf{N}_j = \mathbf{0} \ \text{on } \Omega^e \tag{2a,b}$$

All what is left is to determine the parameters $\mathbf{c}$ so as to enforce on $\mathbf{u}$ the conformity ($\mathbf{u}^e = \mathbf{u}^f$ on $\Gamma^e \cap \Gamma^f$) and the boundary conditions ($\mathbf{u}^e = \bar{\mathbf{u}}$ on Γ_u^e, $\mathbf{t}^e = \bar{\mathbf{t}}$ on Γ_t^e).

One of the possible ways [1-3] is to link the Trefftz type solutions (1) through an interface displacement frame surrounding the element which is approximated in terms of the same degrees of freedom (DOF), $\mathbf{d}$, as used in the conventional elements:

$$\tilde{\mathbf{u}} = \tilde{\mathbf{N}}\mathbf{d} \quad \text{on} \quad \Gamma^e \equiv \partial\Omega^e . \tag{3}$$

This makes it possible to eliminate the internal parameters $\mathbf{c}$ and derive for the element the customary force-displacement relationship,

$$\underline{\mathbf{r} = \overset{\circ}{\mathbf{r}} + \mathbf{k}\mathbf{d}}\,, \tag{4}$$

where **r** are the equivalent nodal forces associated with the nodal DOF, **d**, contributed by element e. Clearly, such elements can be handled exactly as the conventional elements for which the standard assembly procedure of the matrix displacement method is valid.

The relation (4) can most easily be obtained by first enforcing in a weak sense the conformity on **u**,

$$\int_{\Gamma^e} \delta\mathbf{t}^T(\mathbf{u}-\tilde{\mathbf{u}})d\Gamma = 0\,, \tag{5a}$$

and then using the equivalency of virtual work

$$\int_{\Gamma^e} \delta\tilde{\mathbf{u}}^T\mathbf{t}d\Gamma = \int_{\Gamma^e_t} \delta\tilde{\mathbf{u}}^T\bar{\mathbf{t}}d\Gamma + \delta\mathbf{d}^T\mathbf{r}\,, \tag{5b}$$

to express the equivalent nodal forces **r**.

These statements where, **t** stands for boundary tractions derived from **u**,

$$\mathbf{u} = \overset{\circ}{\mathbf{u}} + \mathbf{N}\mathbf{c} \quad \rightarrow \quad \mathbf{t} = \overset{\circ}{\mathbf{t}} + \mathbf{T}\mathbf{c}\,,$$

and where the virtual quantities are defined as

$$\delta\mathbf{u} = \mathbf{N}\delta\mathbf{c}\,, \quad \delta\mathbf{t} = \mathbf{T}\delta\mathbf{c}\,, \quad \delta\tilde{\mathbf{u}} = \tilde{\mathbf{N}}\delta\mathbf{d}\,,$$

lead to a symmetric system of matrix equations

$$\begin{bmatrix} \mathbf{H} & -\mathbf{G}^T \\ -\mathbf{G} & \mathbf{0} \end{bmatrix} \begin{Bmatrix} \mathbf{c} \\ \mathbf{d} \end{Bmatrix} = \begin{Bmatrix} -\mathbf{h} \\ \mathbf{g}-\mathbf{r} \end{Bmatrix} \tag{6}$$

involving a symmetric matrix

$$\mathbf{H} = \int_{\Gamma^e} \mathbf{T}^T\mathbf{N}_e d\Gamma = \int_{\Gamma^e} \mathbf{N}^T\mathbf{T}d\Gamma \tag{6a}$$

and three other matrix quantities

$$\mathbf{G} = \int_{\Gamma^e} \tilde{\mathbf{N}}^T\mathbf{T}d\Gamma\,, \quad \mathbf{g} = \int_{\Gamma^e} \tilde{\mathbf{N}}^T\overset{\circ}{\mathbf{t}}d\Gamma - \int_{\Gamma^e_t} \tilde{\mathbf{N}}^T\bar{\mathbf{t}}d\Gamma, \quad \mathbf{h} = \int_{\Gamma^e} \mathbf{T}^T\,\overset{\circ}{\mathbf{u}}d\Gamma\,. \tag{6b-d}$$

The symmetry of **H** is obvious if one observes that in absence of body forces $\bar{\mathbf{b}}$, the strain energy

$$U = \frac{1}{2}\int_{\Omega^e} \varepsilon^T\mathbf{D}\varepsilon d\Omega = \frac{1}{2}\int_{\Gamma^e} \mathbf{u}^T\mathbf{t}d\Gamma = \frac{1}{2}\int_{\Gamma^e} \mathbf{t}^T\mathbf{u}d\Gamma\,.$$

After elimination of the internal parameters in (6),

$$\mathbf{c} = -\mathbf{H}^{-1}\mathbf{h} + \mathbf{H}^{-1}\mathbf{G}^T\mathbf{d}\,, \tag{7}$$

the load dependent part $\overset{\circ}{\mathbf{r}}$ of **r** and the stiffness matrix **k** are readily evaluated as

$$\overset{\circ}{\mathbf{r}} = \mathbf{g} - \mathbf{G}\mathbf{H}^{-1}\mathbf{h} \quad \text{and} \quad \mathbf{k} = \mathbf{G}\mathbf{H}^{-1}\mathbf{G}^{T}. \tag{8a,b}$$

Several other equivalent formulations are available for the HT elements. In [2], the equations (6) to (8) have been obtained from a stationary variational principle. Zienkiewicz [3] has presented the HT formulation as a particular case in the more general context of linking amongst mixed form subdomains. Still another equivalent approach [4] has shown the condition (5a) to actually be equivalent to

$$\|\mathbf{u}-\tilde{\mathbf{u}}\|^2 = \frac{1}{2} \int_{\Omega^e} (\varepsilon^T - \tilde{\varepsilon}) \mathbf{D} (\varepsilon - \tilde{\varepsilon}) d\Omega = \text{Min.} \tag{9}$$

thus demonstrating that the conformity is optimally enforced in the energy norm sense.

Finally, it should be stressed that the rigid body modes must be discarded from **N** to arrive at a non-singular matrix **H** (the missing rigid body motion modes can however be evaluated later, after the nodal degrees of freedom **d** have been calculated) and that the necessary (but not sufficient) stability condition for **k** to have full rank is

$$m \geq NDOF - NRIG \tag{10}$$

(m = number of internal parameters, **c** ; NDOF = number of frame function DOF, **d** ; NRIG = number rigid body motion modes).

Though presented for solid mechanics, the HT element formulation clearly also holds for other problems. In particular for the Laplace's equation in 2-D, $\nabla^2 u = \bar{b}$ (for which an independent formulation has been presented by Zielinski and Zienkiewicz [5]), equivalent relations can be obtained through simple replacement in (5a,b) of **t** by $\partial u/\partial n$ (n = external normal to element boundary).

Very soon it has been recognized [6] that various singularities and/or stress concentrations induced by both the geometry and the loading, can efficiently be handled without troublesome mesh refinement, by providing the FE subroutine with a library of optional special purpose functions **N** and $\overset{\circ}{\mathbf{u}}$.

The idea of forming p-version HT elements has been suggested already in [6], but the p-method capabilities have been practically implemented much later [7] and their properties have thoroughly been studied only quite recently [8]. For HT p-elements a simple stress error estimator has also been presented [9] and a simple adaptive approach devised [10,11].

Thus far, the HT approach has been applied to plane elasticity [6], to isotropic [1,12] and orthotropic [13] Kirchhoff plates, moderately thick plates [14], flat slabs with drops [15], thin shells with constant principal curvatures [16] and to Poisson's equation [5,7,8].

In the following sections the HT approach is first shortly assessed and then the developments under way are briefly outlined.

2. SHORT ASSESSMENT OF HT APPROACH

Already the earliest applications [1,2,6] have clearly shown that the HT elements are more accurate than the conventional elements with which they can have an external resemblance. A thorough assessment of plane elasticity and plate bending HT elements is available in [6] and [12]. Later, when the HT p-elements have been implemented, it has become obvious that not only such elements (Fig. 1) are very easy to generate for both C^0 and C^1 conformity problems, but also that their performance is largely superior to that of the earlier h-elements.

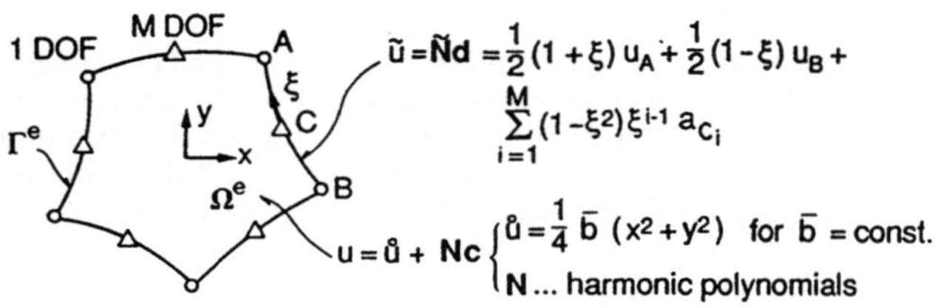

Fig. 1. Typical HT p-element for Poisson equation : $\nabla^2 u = \bar{b}$.

Furthermore, compared to the conventional p-elements (Table 1) with which they again may have an external resemblance, these elements exhibit an appreciable improvement in accuracy and a marked enhancement in insensitivity to mesh distortion.

Table 1. Comparison of conventional and HT p-elements on Dirichlet problem of Poisson equation : $\nabla^2 u = 2$.

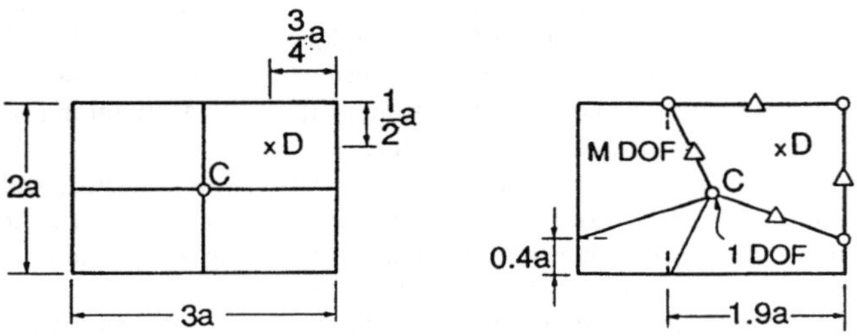

M*)	$u_{FE} : u_{EX}$				$\frac{u_{irregular} - u_{regular}}{u_{regular}} \times 100\%$			
	C		D		C		D	
	Conv.	HT	Conv.	HT	Conv.	HT	Conv.	HT
0	1.319	1.319	0.514	0.910	-7.9	-2.0	-41.3	8.5
1	0.908	0.965	0.952	1.001	-4.4	3.4	-26.8	0.1
2	0.930	1.004	0.955	1.000	5.1	-0.4	-24.8	-0.1
3	1.019	0.999	1.004	1.000	1.3	0.0	-4.3	0.0
4	0.992	1.000	1.004	1.000	-0.1	0.0	0.4	0.0
5	1.004	1.000	0.999	1.000	0.1	0.0	0.4	0.0
6	0.998	1.000	0.999	1.000	0.0	0.0	0.0	0.0
7	1.001	1.000	1.000	1.000	0.0	0.0	-0.1	0.0

*) M = number of hierarchic side-mode DOF (see Fig. 1).

Recently, a thorough study [8] including both the theoretical and the practical comparison of various aspects of the two alternative p-formulations has been completed and the main conclusions are summarized hereafter:

• While exhibiting at various p-refinement levels the same h-convergence rate, the p-convergence is faster than in the conventional case (see Fig. 2, where the ratio of errors in strain energy of the conventional to HT element passes from 1 for M=0 to 42.1 for M=7).

• The HT elements exhibit a completely different type of error distribution (Fig. 3) in comparison with the conventional elements, a distribution which is typical for the boundary solution (BS) approaches.

• While the efficiency of optional special purpose functions (such as the Williams eigenfunctions for various corner singularity problems) is not a surprise (Table 2b), excellent results have also been obtained with only the standard T-functions (biharmonic polynomials in plate bending applications) without any local mesh refinement (Table 2a).

• One of the most important advantages in engineering practice is the fact that (thanks to the load term $\mathring{u}$, and its use as a global or 'semi-global' function) the accuracy depends little on the load condition. By integration of the Green function, a collection of load terms has been derived [6,12,17] for various loads (including patch load, line load,...) which now makes it possible to handle concentrated loads accurately without any mesh adjustment (Fig. 4).

• Unlike the conventional p-method (aiming at reaching the specified accuracy in energy norm) a direct stress error estimator is available [9] for the HT p-elements and a simple adaptive approach based on a uniform p-refinement enables the specified stress accuracy to be reached in a single trial [10].

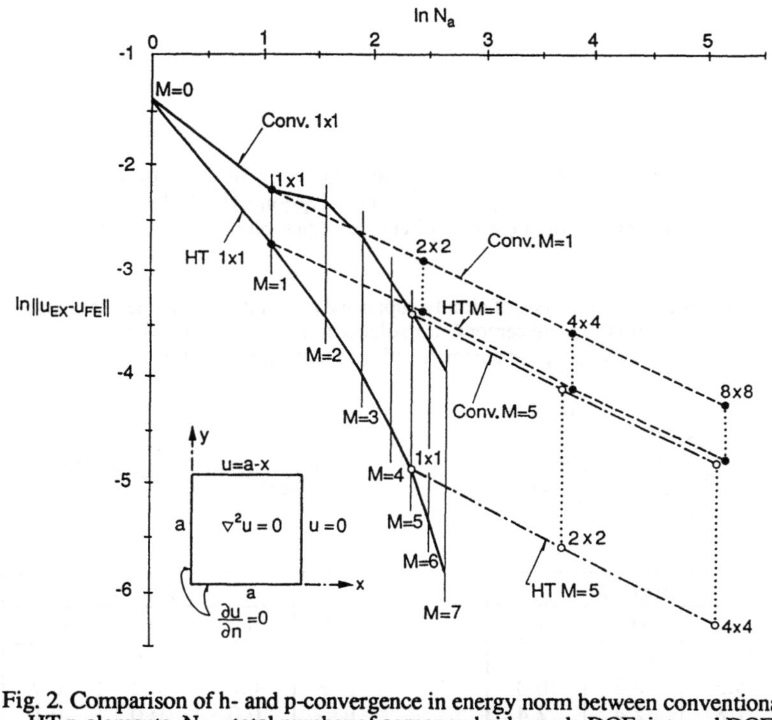

Fig. 2. Comparison of h- and p-convergence in energy norm between conventional and HT p-elements. N_a = total number of corner and side mode DOF; internal DOF of conventional element (assumed statically condensed) not included.

• **Finally, if several independent load cases are considered and the p-refinement level is adaptively adjusted for the one presenting the largest initial error, then as a rule a satisfactory stress accuracy is warranted for all load cases.**

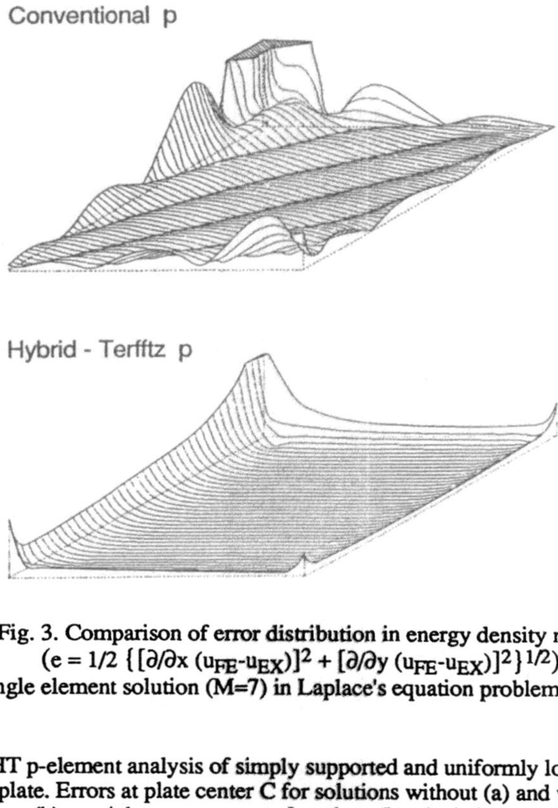

Fig. 3. Comparison of error distribution in energy density norm
$(e = 1/2\ \{[\partial/\partial x\ (u_{FE}-u_{EX})]^2 + [\partial/\partial y\ (u_{FE}-u_{EX})]^2\}^{1/2})$
for single element solution (M=7) in Laplace's equation problem of Fig. 2.

Table 2. HT p-element analysis of simply supported and uniformly loaded rhombic plate. Errors at plate center C for solutions without (a) and with (b) special purpose corner functions (hatched elements).

a)

M	Error %		
	w_c	M_{xc}	M_{yc}
1	24.0	13.0	24.7
3	5.3	6.3	10.0
5	1.8	2.4	2.4
7	-0.3	0.5	-1.8

b)

M	Error %		
	w_c	M_{xc}	M_{yc}
1	2.1	-0.9	-4.5
3	-0.2	-0.2	-0.9
5	0.0	0.0	-0.1
7	0.0	0.0	0.0

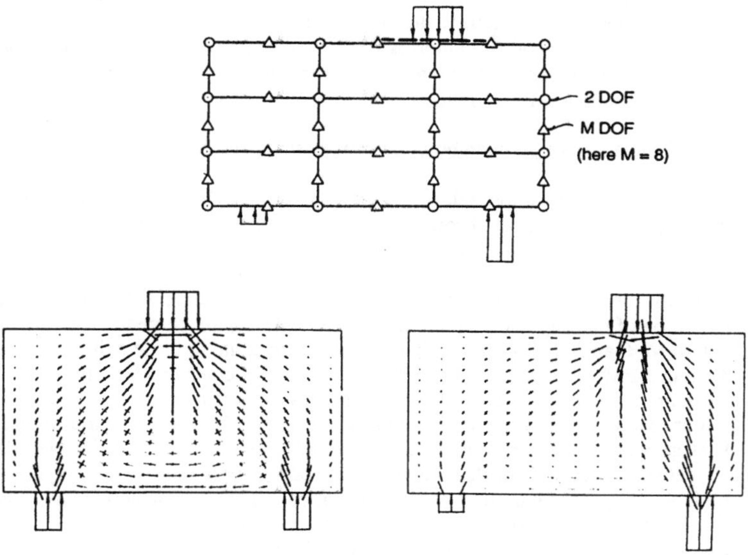

Fig. 4. Deep beam subjected to a moving load.
Analysis with HT p-version plane elasticity elements.

3. NEW TRENDS

3.1. Extension to partial differential equations with variable coefficients

Although in some cases the Trefftz functions may also be generated for problems with variable coefficients (or a step-wise variation can be used in place of continuous one [15]), the difficulties arising in more general cases confine the standard HT approach practically to problems with constant coefficients. Some of the advantages of this approach (great liberty in element geometry, facility for generation of p-version elements for both the C^0 and C^1 conformity problems and, possibly in part also the improved accuracy) can however be preserved, if the internal field is generated for a related problem with suitably chosen constant coefficients. For a shell element for example, the variable principal curvatures may be replaced for that purpose by their average values. The conformity is then again enforced with the help of the weighted statement (5) which yields (after the elimination of the internal parameters **c**) a quasi-conforming internal field given in terms of the conventional FE parameters **d** :

$$\mathbf{u} = (\mathring{\mathbf{u}}-\mathbf{N}\mathbf{H}^{-1}\mathbf{h}) + (\mathbf{N}\mathbf{H}^{-1}\mathbf{G})\mathbf{d} = \underline{\mathring{\mathbf{u}}^{*} + \mathbf{N}^{*}\mathbf{d}} \qquad (11)$$

where **H**, **G** and **h** are the matrix quantities (6a,b) and (6d).

The quasi-conforming field (11) may further be used to generate the element matrices $\mathring{\mathbf{r}}$ and **k** following the standard FE technology of the conventional assumed displacement model.

One of the advantages of the outlined approach is for example, the fact that very general thin shell p-elements may now easily be generated whereby the governing differential equations are already a priori satisfied in an average sense. Note that additional internal modes analogous to the "bubble modes" of a conventional p-element can easily be generated as additional terms $\mathring{\mathbf{u}}^*$ for a hierarchic set of loads (uniform, linearly variable with respect to x or y axes, varying quadrically,...). Since a standard FE technology is used, an extension to non linear problems is also possible.

3.2. Optimal substructure oriented boundary solution approach

By its nature, the HT approach is related to both the hybrid FE methods and the boundary solution (BS) methods. From the stand-point of the latter, this approach may be considered as a particular form of a symmetric substructure oriented BEM formulation where, the sides of the element frame may be assimilated to a special form of boundary elements. In this context it is of interest to observe that for the single element solution of the Laplace's equation problem (Fig. 2), the percentage errors in u, obtained with M=0 and M=7 at the left lower corner, are equal to those exhibited by the standard direct BEM solution with 4 (-23%) and 32 (0.3%) constant elements.

The main advantages of the HT approach are symmetric equations, low cost of generation of results inside the domain and absence of the algorithmic complexities of the BEM (sophisticated quadrature rules for evaluation of strongly singular boundary integrals, special procedures for dealing with non homogeneous differential equations, double node technique for handling of discontinuities,...). The drawback is the cost of inversion of the **H** matrix. But, if only a small number of large subdomains is to be assembled, as usual in the BS procedure, then this drawback can easily be removed by confining the frame to interelement portion of the boundary (Fig. 5). The aim is to develop again an efficient formulation which can be implemented into the usual FE packages.

One possibility is the earlier presented variational formulation (the functional $J(\mathbf{u},\tilde{\mathbf{u}})$ in [2]) which yields as a special case (if $\Gamma_I^e \equiv \Gamma^e$) the standard HT model. The augmented form of this functional, as discussed in [6], is particularly convenient for practical use since here all rigid body motion modes are involved in the internal displacement field.

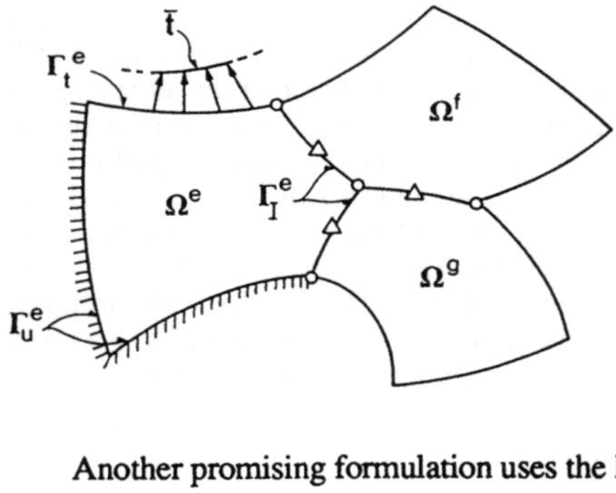

Fig. 5.
Substructure oriented BS approach with displacement frame confined to substructure interface.

Another promising formulation uses the least square statement

$$E\int_{\Gamma_u^e}(\mathbf{u}-\bar{\mathbf{u}})^T(\mathbf{u}-\bar{\mathbf{u}})d\Gamma + \frac{\beta}{E}\int_{\Gamma_t^e}(\mathbf{t}-\bar{\mathbf{t}})^T(\mathbf{t}-\bar{\mathbf{t}})d\Gamma + E\int_{\Gamma_I^e}(\mathbf{u}-\tilde{\mathbf{u}})^T(\mathbf{u}-\tilde{\mathbf{u}})d\Gamma = \text{Min.} \qquad (12a)$$

to enforce the boundary conditions on $\Gamma_u^e \cup \Gamma_t^e$ in addition to the displacement continuity on Γ_I^e and combines it with the virtual work statement

$$\int_{\Gamma^e} \delta \mathbf{u}^T \mathbf{t} = \int_{\Gamma_t^e} \delta \mathbf{u}^T \bar{\mathbf{t}} d\Gamma + \delta \mathbf{d}^T \mathbf{r} . \tag{12b}$$

The quantities **u**, **t** and $\tilde{\mathbf{u}}$ have the same definitions as in Section 1, E is Young modulus and β an optional non dimensional factor which serves the purpose of tuning the weights set upon the displacement and traction conditions. With

$$\mathbf{A} = E \int_{\Gamma_u^e \cup \Gamma_I^e} \mathbf{N}^T \mathbf{N} d\Gamma + \frac{\beta}{E} \int_{\Gamma_t^e} \mathbf{T}^T \mathbf{T} d\Gamma , \quad \mathbf{Q} = \int_{\Gamma_I^e} \mathbf{N}^T \tilde{\mathbf{N}} d\Gamma , \tag{13a,b}$$

$$\mathbf{a} = E \int_{\Gamma_u^e} \mathbf{N}^T (\bar{\mathbf{u}} - \mathring{\mathbf{u}}) d\Gamma + \frac{\beta}{E} \int_{\Gamma_t^e} \mathbf{T}^T (\bar{\mathbf{t}} - \mathring{\mathbf{t}}) d\Gamma - E \int_{\Gamma_I^e} \mathbf{N}^T \mathring{\mathbf{u}} d\Gamma , \tag{13c}$$

the least square statement (12a) yields for the internal parameters **c** the following system of linear equations :

$$\mathbf{A}\mathbf{c} = \mathbf{a} + \mathbf{Q}\mathbf{d} . \tag{14a}$$

Rather than performing the expensive inversion of the large m × m matrix **A** (in order to express the m internal parameters in terms of a much smaller number N of frame function DOF involved in **d**) the standard elimination process with 1+N right hand side terms is used and yields

$$\mathbf{c} = \mathring{\mathbf{c}} + \mathbf{C}\mathbf{d} . \tag{14b}$$

Finally, from (12b) the standard force-displacement relationship (4) is obtained with

$$\mathring{\mathbf{r}} = \mathbf{C}^T (\mathbf{H}\mathring{\mathbf{c}} - \mathbf{p}) \quad , \quad \mathbf{k} = \mathbf{C}^T \mathbf{H} \mathbf{C} , \tag{15a,b}$$

where **H** is the symmetric and positive definite matrix (6a) and

$$\mathbf{p} = \int_{\Gamma_t^e} \mathbf{N}^T \bar{\mathbf{t}} d\Gamma - \int_{\Gamma^e} \mathbf{N}^T \mathring{\mathbf{t}} d\Gamma . \tag{16}$$

It is interesting to observe that this formulation involves as particular cases, the alternative Trefftz element formulation derived for $\Gamma_I^e = \Gamma^e$ from Ref. [18] and, for $\Gamma_u^e \cup \Gamma_t^e = \Gamma^e$ the earlier presented least square formulation of Ref. [19].

The outlined approaches represent only the most obvious symmetric formulations of which many possibilities exist. Some of such possibilities may be derived from references [19,20] which are seminal papers in BEM/FEM coupling. Such approaches can be applied either with the regular Trefftz functions (as in the standard HT elements) or with the singular fundamental solutions of the type used in the boundary element method (with singular points taken outside the domain Ω^e).

3.3. Other developments

In the standard HT approach the displacement conformity is much better enforced (see (9)) than the continuity of the associated stress fields. An alternative formulation involving independent 'frames' for boundary displacement and boundary tractions is presently being studied with the aim of optimally tuning the satisfaction of the displacement and traction continuities. Another alternative formulation attempts to attenuate the concentrations of errors at the element corners (Fig. 3). The development of infinite or semi-infinite HT elements based on T-complete solutions of the type presented in [21] is also being examined.

REFERENCES

1. Jirousek, J. and Leon, N. — A Powerful Finite Element for Plate Bending, Comp. Meth. Appl. Mech. Eng., **12**, 77-96 (1977).
2. Jirousek, J. — Basis for Development of Large Finite Elements Locally Satisfying All Field Equations, Comp. Meth. Appl. Mech. Eng., **14**, 65-92 (1978).
3. Zienkiewicz, O.C. and Taylor, R.L. — **The Finite Element Method**, Vol. 1, McGraw-Hill, London, 1989.
4. Trefftz, E. — 'Ein Gegenstück zum Ritzschen Verfahren', Proc. 2nd Int. Congr. Appl. Mech., Zürich, 1926, p. 131.
5. Zielinski, A.P. and Zienkiewicz, O.C. — Generalized Finite Element Analysis with T-complete Solution Functions, Int. Journal for Num. Meth. in Eng., **21**, 509-528 (1985).
6. Jirousek, J. and Teodorescu, P. — Large Finite Element Method for the Solution of Problems in the Theory of Elasticity, Computers and Structures, **15**, 575-587 (1982).
7. Jirousek, J. — Hybrid-Trefftz Plate Bending Elements with p-method Capabilities, Int. Journal for Num. Meth. in Eng., **24**, 1367-1393 (1987).
8. Jirousek, J., Venkatesh, A., Zielinski, A.P. and Rabemanantsoa, H., — Comparative Study of p-extensions Based on Conventional Assumed Displacement and Hybrid-Trefftz FE Models, Internal Report LSC 91/10, Swiss Federal Institute of Technology, Lausanne, 1991.
9. Jirousek, J. and Venkatesh, A. — A Simple Stress Estimator for Hybrid-Trefftz p-version Elements, Int. Journal for Num. Meth. Eng., **28**, 211-236 (1989).
10. Jirousek, J. and Venkatesh, A. — Adaptivity in Hybrid-Trefftz Finite Element Formulation, Int. Journal for Num. Meth. in Eng., **29**, 391-405 (1990).
11. Jirousek, J. and Venkatesh, A. — A New FE Approach for Adaptive Reliability Assurance, Computers and Structures, **37**, 217-230 (1990).
12. Jirousek, J. and Guex, L. — The Hybrid-Trefftz Finite Element Model and its Application to Plate Bending, Int. Journal for Num. Meth. Eng., **23**, 651-693 (1986).
13. Jirousek, J. and N'Diaye, M. — Solution of Orthotropic Plates Based on p extension of the Hybrid-Trefftz Finite Element Model, Computers and Structures, **34**, 51-62 (1990).
14. Petrolito, J. — Hybrid-Trefftz Quadrilateral Elements for Thick Plate Analysis, TICOM Report 89-08, Texas Inst. for Comput. Mech., The University of Texas in Austin, 1989.
15. Jirousek, J. and N'Diaye, M. — Hybrid-Trefftz p-method Elements for Analysis of Flat Slabs with Drops, Computers and Structures (to appear).
16. Vörös, G.M. and Jirousek, J. — 'Application of the Hybrid-Trefftz Finite Element Model to Thin Shell Analysis', European Conf. on Advances in Comp. Struct. Mech., Giens, France, 1991, p. 547.
17. Venkatesh, A. and Jirousek, J. — 'Accurate FE Analysis of Thin Plates Under Concentrated Loading', IREM Internal Report 89/6, Swiss Federal Institute of Technology, Lausanne, 1989.
18. Piltner, R. — Special Finite Elements with Holes and Internal Cracks, Int. Journal for Num. Meth. in Eng., **21**, 1471-1485 (1985).
19. Zienkiewicz, O.C., Kelly, D.W. and Bettess, P. — The Coupling of the Finite Element Method and Boundary Solution Procedures, Int. Journal for Num. Meth. in Eng., **11**, 355-375 (1977).
20. Zienkiewicz, O.C., Kelly, D.W. and Bettess, P. — 'Marriage à la Mode - the Best of Both Worlds (finite elements and boundary integrals)', Engineering Methods in Finite Element Analysis, Eds. R. Glowinksi, E.Y. Rodin and O.C. Zienkiewicz, Wiley, 1979, p. 81.
21. Jin, W.G., Cheung, Y.K. and Zienkiewicz, O.C. — Application of the Trefftz Method in Plane Elasticity Problems, Int. Journal for Num. Meth. in Eng., **30**, 1147-1161 (1990).

FINITE ELEMENTS IN THE 90's, E. Oñate, J. Periaux, A. Samuelsson (Eds.)
Springer-Verlag/CIMNE, Barcelona 1991

THERMO MECHANICAL COUPLING IN ELASTO-PLASTIC ANALYSIS

S. Glaser and B. Kröplin
Department for Statics and Dynamics of Aerospace Structures,
University of Stuttgart,
7000 Stuttgart 80,
GERMANY

Summary

The numerical solution of coupled nonlinear field problems is increasingly gaining importance. One of these problems is the thermo mechanical problem, where the temperature field and the displacement field of a structure underlying thermal and mechanical boundary conditions is determined. Within the context of mechanics and thermodynamics the field equations of the coupled thermo mechanical problem are developed. The constitutive equations of an elastic plastic material are extended to include thermal effects. The nonlinear system of equations is solved employing the Newton-Raphson solution technique for each increment. Therefore the complete linearization of the coupled problem is performed.

1 Introduction

In metal forming processes materials are used which show inelastic material behaviour within areas of large deformation. The behaviour of these materials is characterized by small elastic and thermal strains and large inelastic, i.e., plastic strains. The material shows elastic behaviour with the stresses within a certain yield surface and plastic behaviour when the stresses reach the yield surface boundary. In the latter case large deformations could be obtained without increasing forces.

The mechanical energy is converted into heat by means of plastic dissipation. Due to plastic dissipation varying with the location the temperature is non-uniformly distributed. This causes heat conduction in the structure, which is in addition influenced by the thermal boundary conditions.

The material propertys of metals are in general temperature dependent. The yield stress of most metals decreases with the rise of temperature whereas the density decreases due to thermal strains.

Therefore the temperature field influences the plastification and hence the forming process. On the other hand the deformation determines energy dissipation and hence the temperature distribution. Note, that within large deformation analysis the heat conduction equation is a function of the deformation, because the shape of the structure changes.

The physical description of the thermo mechanical problem by the first law of thermodynamics and of the balance of momentum with the constitutive laws leads to a system of two field equations, the thermal equation where the temperature field is unknown, and the mechanical equation with the displacement field to be determined. This system of equations is strongly coupled, i.e., the mechanical equation is a function of the temperatur field and the thermal equation is a function of the displacement field.

For the solution of this coupled field problem a method named 'staggered solution' is often used. One increment of this method for example in a first step solves the mechanical equation with the temperatur distribution obtained from the previous increment. Then the thermal equation is solved using the results of the first step. The method is started with an initial temperature distribution.

Another strategy to treat the coupled problem is to solve both the thermal and the mechanical equation within one iteration loop, but in general convergence of that method can not be expected.

In this work a coupled strategy is developed which treats both equations simultaneously. The full derivation of the equations can be found in [9]. The nonlinear system of equations is solved employing the Newton-Raphson solution technique for each increment. Therefore the complete linearization of the coupled problem is necessary, i.e., the entire tangent of the coupled problem consisting of a mechanical, a thermal, a mechanical-thermal and a thermal-mechanical part needs to be determined. The advantage of this method is that only one nonlinear system of equations has to be solved, therefore it is possible to use a standard nonlinear Finite Element code. A disadvantage is that the tangent is non-symmetric.

2 Kinematics

For the kinematics of the thermo elastic plastic deformation the body has to be thougt of as cut into pieces with each piece made stress free and brought to the initial temperature. The remaining configuration is the plastic intermediate configuration which is not continuous [3]. This leads to the multiplicative split of the deformation gradient $\boldsymbol{F}$ into an elastic part $\boldsymbol{F}^e$ and a plastic part $\boldsymbol{F}^p$

$$\boldsymbol{F} = \boldsymbol{F}^e \boldsymbol{F}^p \tag{1}$$

The thermal strains are treated as elastic strains that originate from stresses due to temperature changes. This avoids the introduction of a thermal intermediate configuration.

3 Gouverning Equations

3.1 Balance of Momentum

The balance of momentum for continua is

$$\rho \dot{\boldsymbol{v}} = \rho \boldsymbol{b} + \operatorname{div} \boldsymbol{\sigma} \tag{2}$$

with the mass density ρ, the velocity $\boldsymbol{v}$, the volume forces $\boldsymbol{b}$ and the cauchy stress tensor $\boldsymbol{\sigma}$. This equation is also known as Cauchy's first law of motion [4].

From the balance of moment of momentum follows the symmetry of the Cauchy stress tensor.

3.2 Balance of Energy

The first law of thermodynamics states that the change of kinetic and internal energy of any portion of a body equals the rate of work done on that portion plus the rate of heat supply and heat flow across the boundary. Thus local form of the first law of thermodynamics reads

$$\rho \dot{e} + \operatorname{div} \boldsymbol{q} = \boldsymbol{\sigma} : \boldsymbol{d} + \rho r \tag{3}$$

where e is the internal energy per unit mass, $\boldsymbol{q}$ is the heat flux vector, r is the heat supply per unit mass, and $\boldsymbol{d}$ is the rate of deformation tensor which can be written using Lie derivatives

$$\boldsymbol{d} = L_v \boldsymbol{e} = \frac{1}{2} L_v \boldsymbol{g} \tag{4}$$

of the Eulerian strain tensor $\boldsymbol{e}$ or the spatial metric tensor $\boldsymbol{g}$. The Eulerian strain tensor can be written in terms of the deformation gradient F:

$$\boldsymbol{e} = \frac{1}{2} \left(\boldsymbol{g} - (\boldsymbol{F}^{-T} \boldsymbol{F}^{-1})^{\flat} \right) \tag{5}$$

The symbol $\flat$ denotes lowering of all indices.

4 Constitutive Law

4.1 Elastic and Thermal Material Propertys

The isotropic elastic behaviour of the material is determined by the free energy

$$\psi = \frac{\lambda}{2\rho_{\text{ref}}} (\ln J^e)^2 + \frac{\mu}{2\rho_{\text{ref}}} (\boldsymbol{g} : \bar{\boldsymbol{b}}^e - 3) \quad , \tag{6}$$

which is composed of two uncoupled terms for the volumetric and isochoric responses, where λ is the bulk modulus and μ the shear modulus. J^e is the elastic part of the Jacobian

$$J = \det \boldsymbol{F} \frac{\sqrt{\det \boldsymbol{g}}}{\sqrt{\det \boldsymbol{G}}} = J^\theta J^e J^p \tag{7}$$

of the deformation, which can be divided into elastic, thermal and plastic contributions J^e, J^θ and J^p, respectivly. The modified elastic left Cauchy Green tensor is

$$\bar{\boldsymbol{b}}^e = J^{-\frac{2}{3}}\, \boldsymbol{b}^e = J^{-\frac{2}{3}}\, (\boldsymbol{F}^e \boldsymbol{F}^{eT})^\sharp \tag{8}$$

with the the index $\sharp$ denoting rising of all indices, and $\rho_{\text{ref}} = J\rho$ is the density of the reference configuration.

The behaviour of the material with respect to temperature is isotropic. Assuming linear elongation in one direction with the thermal expansion coefficient α with respect to temperature θ leads to the thermal volume change

$$J^\theta = e^{3\alpha(\theta-\theta_0)} \quad , \tag{9}$$

which is an approximation obtained from a truncated series expansion of the natural logarithm.

The heat capacity can be shown to be a derivative of the free energy,

$$c_v = -\theta \frac{\partial^2 \psi}{\partial \theta^2} \quad . \tag{10}$$

Assuming constant heat capacity c_v the free energy of the thermo elastic plastic problem is

$$\begin{aligned} \psi \; = \; & \frac{\lambda}{2\rho_{\text{ref}}} \left(\ln \left(J e^{-3\alpha(\theta-\theta_0)}\right)\right)^2 + \frac{\mu}{2\rho_{\text{ref}}} \left(J^{-\frac{2}{3}}\, \boldsymbol{g} : \boldsymbol{b}^e - 3\right) - \frac{\lambda}{2\rho_{\text{ref}}} (3\alpha\theta)^2 \\ & + c_v \left(-\theta \ln \frac{\theta}{\theta_0} + \theta - \theta_0\right) - \eta_0 (\theta - \theta_0) + \psi_0 \end{aligned} \tag{11}$$

For this free energy the entropy $\eta = -\frac{\partial \psi}{\partial \theta}$ evaluates to

$$\eta = \frac{3\alpha\lambda}{\rho_{\text{ref}}} \ln \left(J e^{-3\alpha(\theta-\theta_0)}\right) + c_v \ln \frac{\theta}{\theta_0} + \eta_0 \tag{12}$$

and the Kirchoff stress tensor $\boldsymbol{\tau} = 2\rho_{\text{ref}} \frac{\partial \psi}{\partial \boldsymbol{g}}$ reads

$$\boldsymbol{\tau} = \lambda \ln \left(J e^{-3\alpha(\theta-\theta_0)}\right) \boldsymbol{g}^{-1} + \mu \operatorname{dev} \bar{\boldsymbol{b}}^e \tag{13}$$

4.2 Plastic Material Property

For the plastic part of the constitutive equations a model proposed by Simo [6, 7] is adopted. This model uses the v. Mises yield condition

$$\|\operatorname{dev} \boldsymbol{\tau}\| - \sqrt{\frac{2}{3}} \kappa(e^p, \theta) \le 0 \tag{14}$$

in connection with an associative flow rule, which is derived for the isothermal case from the principle of maximum plastic dissipation. The flow stress is a function of the equivalent plastic strain e^p and the temperature θ. The term

$$\|\operatorname{dev} \boldsymbol{\tau}\| = \sqrt{(\operatorname{dev} \boldsymbol{\tau})^{ij} (\operatorname{dev} \boldsymbol{\tau})^{kl} g_{ik} g_{jl}} \tag{15}$$

is the norm of the deviator of the Kirchhoff stress tensor $\boldsymbol{\tau}$. Under the assumption of small elastic deformations, the associative flow rule takes the form

$$-\,L_v\boldsymbol{\tau}^p = \mu J^{-\frac{2}{3}}\operatorname{dev}[L_v\boldsymbol{b}^e] = -2\bar{\mu}\dot{\gamma}\boldsymbol{n} \tag{16}$$

with

$$\bar{\mu} = \frac{1}{3}\mu J^{-\frac{2}{3}}\operatorname{tr}\boldsymbol{b}^e \quad \text{and} \quad \boldsymbol{n} = \frac{\operatorname{dev}\boldsymbol{\tau}}{\|\operatorname{dev}\boldsymbol{\tau}\|} \quad . \tag{17}$$

Here, $\dot{\gamma}$ is the plastic consistency parameter and $L_v\boldsymbol{\tau}^p$ is called the plastic relaxation stress. This equation can be solved for the deviatoric part of $L_v\boldsymbol{b}^e$. To determine $L_v\boldsymbol{b}^e$ completely the isochoric plastic deformation must be taken into account. For small elastic deformations the relation

$$\operatorname{tr}(L_v\boldsymbol{b}^e) = \boldsymbol{g} : L_v\boldsymbol{b}^e = 0 \tag{18}$$

holds.

The plastic dissipation is given by

$$\mathcal{D} = \sqrt{\frac{2}{3}}\dot{\gamma}\kappa \quad . \tag{19}$$

From this equation the evolution equation of the equivalent plastic strain

$$\dot{e}^p = \sqrt{\frac{2}{3}}\dot{\gamma} \tag{20}$$

follows.

The integration of the plastic flow rule is done by the projection method [2]. The stresses read

$$\operatorname{dev}\boldsymbol{\tau}_{n+1} = \boldsymbol{s}_{n+1} = \boldsymbol{s}_{n+1}^{trial} - 2\bar{\mu}_{n+1}\Delta\gamma\boldsymbol{n}_{n+1} \quad , \tag{21}$$

where $\boldsymbol{s}_{n+1}^{trial} = \mu\operatorname{dev}\bar{\boldsymbol{b}}_{n+1}^{e\,trial}$ is the elastic predictor. The plastic consistency parameter $\Delta\gamma$ is determined by satisfying the yield condition (14) at the end of each increment. This leads to finding the root of a nonlinear equation which is done numerically employing the Newton method. The flow stress κ can be choosen arbitrarily as a function of the equivalent plastic strain e^p and the temperature $\boldsymbol{\theta}$.

The modified elastic left Cauchy Green tensor $\bar{\boldsymbol{b}}^e$ is determined by

$$\bar{\boldsymbol{b}}_{n+1}^e = \bar{\boldsymbol{b}}_{n+1}^{e\,trial} - \frac{2\bar{\mu}\Delta\gamma}{\mu}\boldsymbol{n}_{n+1} \quad . \tag{22}$$

4.3 Heat Conduction

The relation between the heat flux $\boldsymbol{q}$ and the temperature θ is discribed by Fourier's law. For isotropic materials it is given by

$$\boldsymbol{q} = -\frac{k}{J}\operatorname{grad}\theta \quad . \tag{23}$$

The conductivity k is material dependent.

5 Functionals of the Thermomechanical Problem

5.1 Thermal equation

The first law of thermodynamics in combination with Fourier's law leads to the thermal equation in differential form

$$\rho c_v\dot{\theta} - \operatorname{div}(\frac{k}{J}\operatorname{grad}\theta) - \frac{1}{J}\mathcal{D} = 0 \quad . \tag{24}$$

The time domain integration uses the Euler backward scheme. Therefore, the first time derivative of the temperature is substituted by the Euler backward difference step. The weak form of the thermal equation discrete in time can then be written as

$$\int_{\Omega_0} \left(\frac{\rho_{\text{ref}} c_v}{\Delta t} (\theta - \theta_n)\, \delta\theta + k \operatorname{grad} \theta \cdot \operatorname{grad} \delta\theta - \mathcal{D}\, \delta\theta \right) dV = 0 \tag{25}$$

5.2 Mechanical Equation

For the mechanical equation a mixed variational principle is used. A new field variable $\hat{J}$ is introduced which describes the volumetric deformation. In the special case of an uncoupled free energy $\Psi = \Psi_D + \Psi_H$ where Ψ_D and Ψ_H are the volumetric and isochoric terms, respectively, the mechanical problem can be reduced to the task of making the following Lagrange functional stationary

$$\Pi = \int_{\Omega_0} \left(\rho_{\text{ref}} \Psi_D(\boldsymbol{C}, \boldsymbol{C}^{p-1}, \Theta) + \rho_{\text{ref}} \Psi_H(\hat{J}, \Theta) + p \left(\hat{J} - J \right) \right) dV$$

$$- \int_{\Omega_0} \rho_{\text{ref}}\, \boldsymbol{B} \cdot \Phi \, dV - \int_{\partial\Omega_0} \boldsymbol{T} \cdot \Phi \, dA, \tag{26}$$

with the thermal variables temperature and entropy held constant. The Lagrange parameter p, can be interpreted physically as the hydrostatic pressure. $\boldsymbol{C}$ is the right Cauchy Green Tensor and $\boldsymbol{C}^{p-1}$ is the inverse plastic right Cauchy Green Tensor which is the pull back of the elastic left Cauchy Green tensor $\boldsymbol{b}^e$. The last two integrals in this equation describe the work done by external volume forces $\boldsymbol{B}$ and surface tractions $\boldsymbol{T}$, respectively. Inertia forces are not considered in this equation.

5.3 Functionals of the Coupled Problem

The system F of field equations describing the thermomechanical problem consists of the four equations

$$0 = F_\Phi = \int_{\Omega_0} \left(\rho_{\text{ref}} \frac{\partial \psi_D}{\partial \boldsymbol{F}} \boldsymbol{F} : \operatorname{grad} \delta\Phi - p\, J \operatorname{div} \delta\Phi \right) dV - \ldots \tag{27}$$

$$0 = F_p = \int_{\Omega_0} \left(\left(\hat{J} - J \right) \delta p \right) dV \tag{28}$$

$$0 = F_J = \int_{\Omega_0} \left(\left(\rho_{\text{ref}} \frac{\partial \Psi_H}{\partial \hat{J}} + p \right) \delta\hat{J} \right) dV \tag{29}$$

$$0 = F_\theta = \int_{\Omega_0} \left(\frac{\rho_{\text{ref}} c_v}{\Delta t} (\theta - \theta_n)\, \delta\theta + k \operatorname{grad} \theta \cdot \operatorname{grad} \delta\theta - \mathcal{D}\, \delta\theta \right) dV. \tag{30}$$

The first three equations describe the mechanics, the last one describes the thermal energy balance of the problem.

6 Solution Method

The nonlinear system of field equations is discretized utilizing the Finite Element method [8]. Isoparametric four noded mixed elements for plane strain and axisymmetry are implemented. The shape functions of the displacements are bilinear, and $\hat{J}$ is constant over the element.

The resulting nonlinear system of equations of the coupled problem is solved simultaneously with Newton's method. Therefore the tangent of the coupled problem must be derived. The tangent matrix can be split into four sections, see figure 1.

The upper left section is the mechanical tangent which consists of the derivation of the mechanical equation with respect to the mechanical variables.

The lower right section is the derivation of the thermal equation with respect to temperature.

Due to coupling effects two coupling sections are obtained. The upper right one is the derivation of the mechanical equation with respect to temperature, it is named mechanical–thermal part. This part is due to thermal strains and the temperature dependance of the flow stress.

The lower left section, named the thermal–mechanical part, is the derivation of the thermal equation with respect to deformation. The dissipated energy is a function of the deformation and the spatial gradient grad is also deformation dependent.

$$
\left[\begin{array}{c|c|c||c}
\begin{array}{l}\int\limits_{\Omega_0}\Big(\delta\Phi^a{}_{|c}\Big(J\mathbf{c}_D^{cedf}\,g_{ae}g_{bf}\\ +\tau^{cd}g_{ab}\\ -pJ\,\delta^c_{\;a}\,\delta^d_{\;b}\\ +pJ\,\delta^c_{\;b}\,\delta^d_{\;a}\Big)\,\Delta\Phi^b{}_{|d}\Big)\;dV\end{array} & -\int\limits_{\Omega_0} J\,\delta\Phi^a{}_{|a}\,\Delta p\,dV & 0 & \int\limits_{\Omega_0}\frac{\partial\tau^{ab}}{\partial\theta}\,\delta\Phi_{a|b}\,\Delta\theta\,dV \\
\hline
-\int\limits_{\Omega_0} J\,\delta p\,\Delta\Phi^b{}_{|b}\,dV & 0 & \int\limits_{\Omega_0}\delta p\,\Delta\hat{J}\,dV & 0 \\
\hline
0 & \int\limits_{\Omega_0}\delta\hat{J}\,\Delta p\,dV & \int\limits_{\Omega_0}\frac{\partial^2\Psi_H}{\partial\hat{J}^2}\,\delta\hat{J}\,\Delta\hat{J}\,dV & \int\limits_{\Omega_0}\frac{\partial^2\psi_H}{\partial J\partial\theta}\,\delta\hat{J}\,\Delta\theta\,dV \\
\hline\hline
\begin{array}{l}-\int\limits_{\Omega_0}\Big(k\,\Big(\delta\theta_{|c}\,g^{ca}\,\theta_{|b}\,\Delta\Phi^b{}_{|a}\\ +\delta\theta_{|b}\,\theta_{|c}\,g^{ca}\,\Delta\Phi^b{}_{|a}\Big)\\ +\delta\theta\,2\frac{\partial\mathcal{D}}{\partial g_{ac}}\,g_{ab}\,\Delta\Phi^b{}_{|c}\Big)\,dV\end{array} & 0 & 0 & \begin{array}{l}\int\limits_{\Omega_0}\Big(\frac{\rho_{ref}\,c_v}{\Delta t}\,\delta\theta\,\Delta\theta\\ +k\,\delta\theta_{|b}\,g^{ba}\,\Delta\theta_{|a}\\ -\frac{\partial\mathcal{D}}{\partial\theta}\,\delta\theta\,\Delta\theta\Big)\,dV\end{array}
\end{array}\right]
\begin{array}{c}\Phi\\ \\ p\\ \\ \hat{J}\\ \\ \theta\end{array}
$$

Figure1: Coefficient matrix of the tangent

The consistent elastoplastic material tensor $J\mathbf{c}_D$ which appears in the mechanical part of the tangent is

$$
\begin{aligned}
J\mathbf{c}_D = & \left[1-\frac{2\bar{\mu}\Delta\gamma}{\|\mathbf{s}^{\text{trial}}\|}\right]J\mathbf{c}_D^{el} \\
& -\left[2\bar{\mu}\left(\frac{1}{1+\frac{1}{3\bar{\mu}}\frac{\partial\kappa}{\partial e^p}}-\frac{2\bar{\mu}\Delta\gamma}{\|\mathbf{s}^{\text{trial}}\|}\right)+\frac{4}{3}\Delta\gamma\|\mathbf{s}^{\text{trial}}\|\left(1-\frac{1}{1+\frac{1}{3\bar{\mu}}\frac{\partial\kappa}{\partial e^p}}\right)\right]\mathbf{n}\otimes\mathbf{n} \\
& -\left[2\|\mathbf{s}^{\text{trial}}\|\left(\frac{1}{1+\frac{1}{3\bar{\mu}}\frac{\partial\kappa}{\partial e^p}}-\frac{2\bar{\mu}\Delta\gamma}{\|\mathbf{s}^{\text{trial}}\|}\right)\right]\mathbf{n}\otimes\operatorname{dev}\left[\mathbf{n}^2\right] \quad . \qquad (31)
\end{aligned}
$$

The derivation of the dissipated energy with respect to the metric of the spatial configuration which is used in the mechanical–thermal part of the tangent is found to be

$$
\frac{\partial\mathcal{D}}{\partial g}=\frac{\sqrt{\frac{2}{3}}\left(\kappa_{n+1}+\sqrt{\frac{2}{3}}\Delta\gamma\frac{\partial\kappa}{\partial e^p}|_{n+1}\right)}{\Delta t\left(2\bar{\mu}+\frac{2}{3}\frac{\partial\kappa}{\partial e^p}|_{n+1}\right)}\left[\left(\bar{\mu}-\frac{2}{3}\|\mathbf{s}^{\text{trial}}\|\,\Delta\gamma\right)\mathbf{n}+\|\mathbf{s}^{\text{trial}}\|\operatorname{dev}\left[\mathbf{n}^2\right]\right] \; . \qquad (32)
$$

In the thermal part the derivation $\frac{\partial \mathcal{D}}{\partial \theta}$ evaluates to

$$\frac{\partial \mathcal{D}}{\partial \theta} = \frac{\sqrt{\frac{2}{3}}}{\Delta t}\left(\frac{\partial \Delta\gamma}{\partial \theta}\kappa_{n+1} + \Delta\gamma\frac{\partial \kappa}{\partial \theta}|_{n+1} + \Delta\gamma\frac{\partial \kappa}{\partial e^p}|_{n+1}\frac{\partial e^p}{\partial \Delta\gamma}\frac{\partial \Delta\gamma}{\partial \theta}\right) \tag{33}$$

with

$$\frac{\partial \Delta\gamma}{\partial \theta} = \frac{-\sqrt{\frac{2}{3}}\frac{\partial \kappa}{\partial \theta}|_{n+1}}{2\bar{\mu} + \frac{2}{3}\frac{\partial \kappa}{\partial e^p}|_{n+1}} \quad . \tag{34}$$

The partial derivatives of the flow stress with respect to temperature and equivalent plastic strain must be calculated from the hardening rule

$$\kappa = \kappa(e^p, \theta) \tag{35}$$

which is a general function of temperature and equivalent plastic strain.

7 Example

As an example the thermomechanical simulation is applied to the necking of a bar with circular cross section. Making use of the symmetry a quarter of the longitudinal cross section of the bar is discretized with 260 axisymmetric elements. The bar has a lenght of 40 mm and a diameter of 10 mm.

Due to symmetry conditions the displacements in the mid cross section are supressed in axial direction and the displacements along the axis are supressed in radial direction. The heat flux across the mid surface has to be zero.

As initial condition a constant temperature distribution of 20°C is used. To simulate the clamping the temperature of the front side is held constant. Convective heat transfer across the surface is neglected.

The structure is loaded by means of a prescribed displacement velocity of the front side.

For the hardening rule the following nonlinear function is used:

$$\kappa = \left(\kappa_0 + \kappa' e^p + (\kappa_\infty - \kappa_0)\left(1 - e^{-\omega_{e^p} e^p}\right)\right) e^{-\omega_\theta(\theta - \theta_0)} \quad . \tag{36}$$

The material parameters are given in table (1).

ρ_{ref}	=	$7.87 \cdot 10^3 \frac{kg}{m^3}$	κ_0	=	$200 \cdot 10^6 \frac{N}{m^2}$
c_v	=	$0.465 \cdot 10^3 \frac{J}{kg K}$	κ_∞	=	$350 \cdot 10^6 \frac{N}{m^2}$
α	=	$1.2 \cdot 10^{-5} \frac{1}{K}$	κ'	=	$100 \cdot 10^6 \frac{N}{m^2}$
k	=	$58 \frac{W}{m K}$	ω_{e^p}	=	16
λ	=	$175000 \cdot 10^6 \frac{N}{m^2}$	ω_θ	=	$2.5 \cdot 10^{-4} \frac{1}{K}$
μ	=	$80769 \cdot 10^6 \frac{N}{m^2}$			

Table1: Material parameters

In contrast to purely mechanical simulation [1], in thermomechanical analysis no imperfection is needed to achieve necking. Necking occures automatically due to nonuniform temperature distribiution which is caused by heat conduction in the bar towards the front side.

In figure (2) the temporal development of the temperature distribution along the axis is shown for a displacement velocity of 0,01 $\frac{\text{m}}{\text{s}}$. The contour lines of the temperature on the deformed geometry are shown in figure (3) and figure (4) for elongations of 25 and 35 percent, respectively.

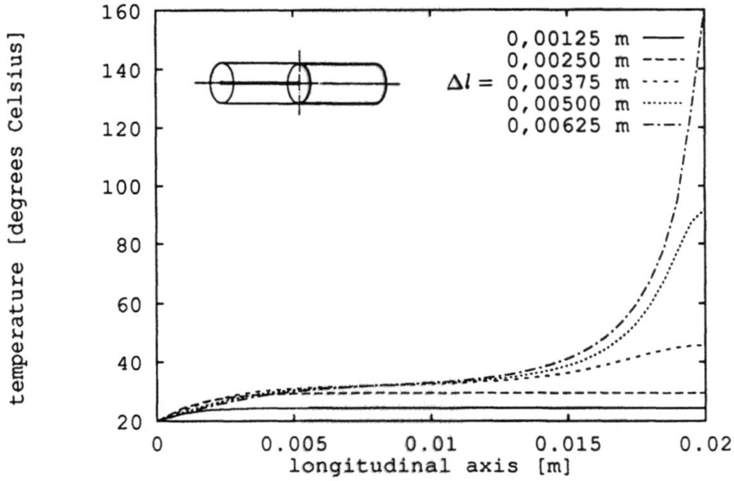

Figure2: Temporal development of the temperature distribution along the half axis for a displacement velocity of $0,01\,\frac{\mathrm{m}}{\mathrm{s}}$.

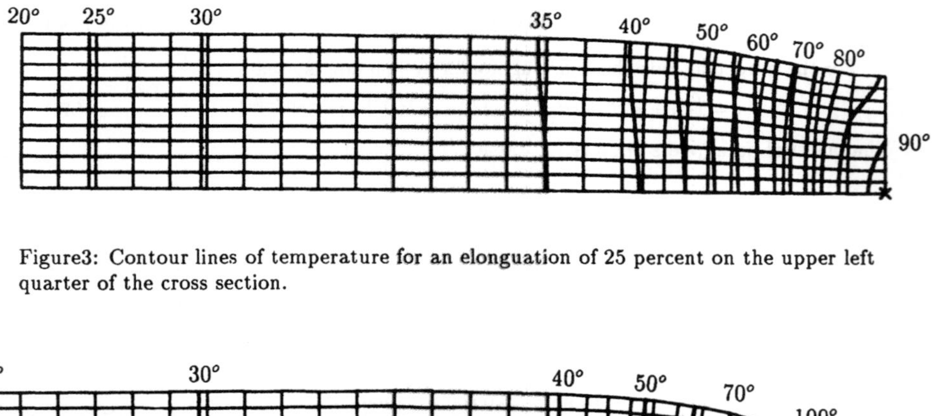

Figure3: Contour lines of temperature for an elonguation of 25 percent on the upper left quarter of the cross section.

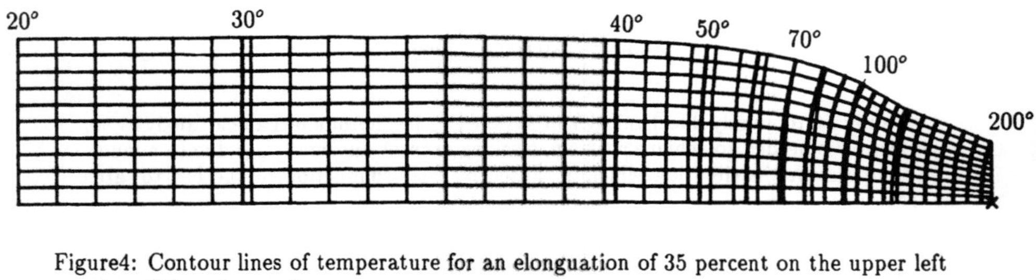

Figure4: Contour lines of temperature for an elonguation of 35 percent on the upper left quarter of the cross section.

References

[1] Herrmann, M.. Beitrag zur Berechnung von Vorgängen der Blechumformung mit der Methode der Finiten Elemente. Proze"ssimulation in der Umformtechnik, Springer-Verlag Berlin Heidelberg New York, 1991.

[2] Krieg, R.D. und Krieg, D.B..Accuracies of Numerical Solution Methods for the Elastic-Perfectly Plastic Model. ASME 99, 1977.

[3] Lee, E.H.. Elastic-plastic deformation at finite Strains. Journal for Applied Mechanics, 36:1-6, 1969.

[4] Marsden, J.E., Hughes, T.H.. Mathematical foundations of elasticity. Prentice Hall, Inc., Englewood Cliffs, 1983.

[5] Matties, H. and Strang, G.. The solution of nonlinear finite element equations, Internat. J. Numer. Meths. Engrg. 14, 1613-1626, 1979.

[6] Simo, J.C.. A framework for finite strain elastoplasticity based on maximum plastic dissipation and the multiplicative decomposition: Part I: continuum formulation, Comput. Meths. Appl. Mech. Engrg. 66 (1988) 199 - 219.

[7] Simo, J.C.. A framework for finite strain elastoplasticity based on maximum plastic dissipation and the multiplicative decomposition: Part II: computational aspects, Comput. Meths. Appl. Mech. Engrg. 68 (1988) 1 - 31.

[8] Zienkiewcz, O.C.. The finite element method. McGraw Hill, 1977.

[9] Glaser, S.. Phd-Thesis, to appear, Bericht aus dem ISD Nr. 91/3.

FINITE ELEMENTS IN THE 90'S, E. Oñate, J. Periaux, A. Samuelsson (Eds.)
Springer-Verlag/CIMNE, Barcelona 1991

A VARIATIONAL FORMULATION OF SHALLOW SHELLS[1]

A. Ibrahimbegovic and F. Frey
Department of Civil Engineering, Swiss Federal Institute of Technology, Lausanne,
SWITZERLAND

G. Fonder and C. Massonnet
Department of Civil Engineering, University of Liège, Liège,
BELGIUM

Abstract

A sound variational framework applicable to shallow shell analysis is proposed in this work. Addition of so-called drilling degrees of freedom, usually absent in the shell analysis, is facilitated within the proposed framework at the continuum level. Therefore, it provides a solid theoretical basis for development of thick and thin curved shell elements with six degrees of freedom per node.

1. INTRODUCTION

The seminal work of Ahmad, Irons and Zienkiewicz [1] established the importance of setting the curved shell elements formulation in the Cartesian coordinate system. In that case, capturing rigid body modes becomes implicit, which is in contrast to early work on curved shell elements formulated in curvilinear coordinates [5], which needed an explicit addition of rigid body modes to be successful. However, all developments patterned after [1], usually called 'degenerated shell element procedure', exclude the rotational degrees of freedom around a chosen fiber (usually the axes initially perpendicular to the shell surface), the so-called 'drilling degrees of freedom' [9].

In this work we propose a shallow shell variational formulation which incorporates the addition of drilling degrees of freedom at the continuum level. There are at least two important reasons for incorporating drilling degrees of freedom in a shell element. First, in nearly co-planar structural configurations, the lack of drilling degrees of freedom causes ill-conditioning of the structure stiffness matrix. More importantly, shell elements with six nodal degrees of freedom are needed in modeling of shell branches and connections to beams and stiffeners. It is fair to say that possessing the shell elements with six degrees of freedom per node represents an indispensable tool for any general purpose structural analysis computer program. This has been recognized much earlier, but in the absence of a better solution, the ad-hoc modification of adding a rotational spring with a small stiffness in the drilling degree of freedom is utilized instead. However, an adequate calibration of the spring stiffness is not easily reached and this modification can produce an unpredictable error, especially in nonlinear problems and distorted shell configurations [7].

[1]Dedicated to Prof. O.C. Zienkiewicz on the occasion of his 70th anniversary

The successful addition of drilling rotations to a shell element, in our opinion, should be based on a sound variational theory, rather than an ad-hoc device. This point of view has proved successful in creating membrane elements with rotational degrees of freedom (see [10] and [11]), which are free of any deficiencies such as spurious modes, adjusted parameters, etc. This is in sharp contrast with the vast majority of recent engineering works on membrane elements with rotational degrees of freedom, which are based on break-through ideas of 'free formulation' of Bergan and Felippa [4] on one side or non-conventional interpolation of Allman [2] on the other.

In this work we aim at providing a sound variational basis for the curved shell elements with six degrees of freedom per node. The formulation is based on the shallow shell theory and an unconventional variational principle of Fraeijs de Veubeke [6]. By the consistent linearization and introduction of Kirchhoff constraint, we are able to recover the shallow shell theory of Marguerre. However, we favor a generalization of the well-known Reissner - Mindlin plate theory to the shallow shell formulation, as a starting point of the finite element discretization. This is motivated by an easier task to construct C^0 interpolations required in the discrete formulation, as well as the possibility to include the contribution of shear deformation.

The proposed approach should be able to cure the deficiencies of the shallow shell element of Jetteur and Frey [13], which is based on Allman-type interpolation, in the same manner as the membrane elements of Allman [2] are cured in [10]. A 4-node geometry interpolation shell element with six degrees of freedom per node has already been developed [11] and has exhibited a very satisfying performance in solving the standard benchmark problems. The formulation proposed in this work should facilitate development of the shell elements with a higher order approximation of the curved geometry, for further enhancement of their accuracy. In addition, the proposed formulation should provide a sound basis for an extension to nonlinear analysis.

A short outline of the paper is given as follows. In the next section, we briefly recall the basic derivation of Marguerre's shallow shell theory. Our developments, however, proceed from the variational principle of Fraeijs de Veubeke given for large displacements. In Section 3 we demonstrate how, by using the consistent linearization, we can recover the variational formulations of Reissner [18] and Naghdi [15] used as a starting point in the previous work on continuum with independent rotation field. In Section 4, the variational principle of Fraeijs de Veubeke is used to derive a shallow shell variational formulation which considers independent rotation fields and, in Section 5, the same variational formulation is recast in terms of stress resultants. In the last section we give some closing remarks.

2. SHALLOW SHELL THEORY OF MARGUERRE

For the sake of completeness, a very brief discussion of Marguerre's shallow shell theory is given in this section (e.g., see [20]).

Let the Cartesian coordinate system (X_1, X_2, X_3) be placed so that the shell surface can be described by a unique functional relationship

$$X_3 = f(X_1, X_2) \tag{1}$$

It is considered that the shell reference surface is shallow, so that $(\partial f/\partial X_1)^2$ and $(\partial f/\partial X_2)^2$ are small compared to 1. Therefore, it holds

$$ds_1 = \sqrt{1+(\partial f/\partial X_1)^2}dX_1 \approx dX_1; \quad ds_2 = \sqrt{1+(\partial f/\partial X_2)^2}dX_2 \approx dX_2 \quad (2)$$

and the metric of the plane can be used instead of the metric of the shell. The unit normal vector in the shell reference configuration is determined by

$$\mathbf{n} = \begin{pmatrix} -(\partial f/\partial X_1) \\ -(\partial f/\partial X_2) \\ 1 \end{pmatrix} \quad (3)$$

which is indeed a unit vector according to the approximation in (2).

If the displacement vector $\mathbf{v}(X_1, X_2) = < v_1; v_2; v_3 >^T$ of the shell mid-surface is such that $v_3(X_1, X_2)$ is of the same order of magnitude as $f(X_1, X_2)$, i.e. if

$$(\partial v_3/\partial X_1)^2 \approx (\partial v_3/\partial X_2)^2 \ll 1 \quad (4)$$

then the unit normal vector in the deformed configuration is

$$\mathbf{m} = \begin{pmatrix} -(\partial f/\partial X_1 + \partial v_3/\partial X_1) \\ -(\partial f/\partial X_2 + \partial v_3/\partial X_2 \\ 1 \end{pmatrix} \quad (5)$$

Using Kirchhoff constraint, we construct position vectors $\mathbf{p}$ and $\mathbf{r}$ in the reference and deformed configurations, respectively

$$\mathbf{p} = \begin{pmatrix} X_1 - \zeta\ (\partial f/\partial X_1) \\ X_2 - \zeta\ (\partial f/\partial X_2) \\ X_3 + \zeta \end{pmatrix} \ ; \ \mathbf{r} = \begin{pmatrix} X_1 + v_1 - \zeta(\partial f/\partial X_1 + \partial v_3/\partial X_1) \\ X_2 + v_2 - \zeta(\partial f/\partial X_2 + \partial v_3/\partial X_2) \\ X_3 + v_3 + \zeta \end{pmatrix} \quad (6)$$

where $\zeta \in [-h/2, h/2]$ and h is the shell thickness. The displacement vector $\mathbf{u}$ is defined as

$$< u_1;\ u_2;\ u_3 >^T = \mathbf{u} = \mathbf{r} - \mathbf{p} \quad (7)$$

The Green-Lagrange finite strain measure (e.g., see [16]) is then obtained by

$$2\ E_{\alpha\beta} = \frac{\partial \mathbf{r}}{\partial X_\alpha}\frac{\partial \mathbf{r}}{\partial X_\beta} - \frac{\partial \mathbf{p}}{\partial X_\alpha}\frac{\partial \mathbf{p}}{\partial X_\beta}\ ; \quad \alpha, \beta \in \{1,2\} \quad (8)$$

By linearizing the expression above, we get a component form for the strain measure used in Marguerre's shell theory

$$\epsilon_{\alpha\beta} = \frac{1}{2}(\frac{\partial v_\beta}{\partial X_\alpha} + \frac{\partial v_\alpha}{\partial X_\beta}) + \frac{1}{2}(\frac{\partial f}{\partial X_\alpha}\frac{\partial v_3}{\partial X_\beta} + \frac{\partial f}{\partial X_\beta}\frac{\partial v_3}{\partial X_\alpha}) - \zeta\frac{\partial^2 v_3}{\partial X_\alpha \partial X_\beta} \quad (9)$$

In the foregoing we give a generalization of the Marguerre's shallow shell theory.

3. THREE-DIMENSIONAL VARIATIONAL PRINCIPLE

The fundamental starting point in the present development is provided by the variational principle of Fraeijs de Veubeke [6] which uses polar decomposition of the deformation gradient. The polar decomposition of the deformation gradient $\mathbf{F}$ is given as (e.g., see [8])

$$\mathbf{F} = \mathbf{RU} \; ; \; \frac{\partial x_i}{\partial X_j} = R_{ik} U_{kj} \; ; \; i,j \in \{1,2,3\} \tag{10}$$

where $\mathbf{R}$ is a proper orthogonal tensor which represents (locally) rigid body rotation, and $\mathbf{U}$ is the right stretch tensor which represents pure stretch. In equation (10), X_j are coordinates of the material point in the reference configuration, and $x_i(X_j)$ denote the mapping of the components of the same point to the current (deformed) configuration. For convenience which does not conflict with the later developments, both sets of coordinates are referred to the Cartesian coordinate system. Accordingly, no distinctions need to be made between covariant and contravariant components.

The polar decomposition can be rewritten in the form

$$\mathbf{F} = \mathbf{R}(\mathbf{I} + \mathbf{H}) \; ; \; \frac{\partial x_i}{\partial X_j} = R_{ik}(\delta_{kj} + H_{kj}) \tag{11}$$

where $\mathbf{H} = \mathbf{U} - \mathbf{I}$ is referred to the Biot strain tensor (Ogden [16], p. 119). The Biot strain tensor $\mathbf{H}$ can be related to the Green-Lagrange strain tensor $\mathbf{E}$ by using

$$\mathbf{E} = \frac{1}{2}(\mathbf{C} - \mathbf{I}) \; ; \; \mathbf{C} = \mathbf{F}^T \mathbf{F} \tag{12}$$

and the polar decomposition form (11), to get

$$\mathbf{E} = \frac{1}{2}[(\mathbf{I} + \mathbf{H})^T \mathbf{R}^T \mathbf{R} (\mathbf{I} + \mathbf{H}) - \mathbf{I}] = \mathbf{H} + \frac{1}{2}\mathbf{H}^2 \tag{13}$$

where, in (13) above, we have used that

$$\mathbf{R}^T \mathbf{R} = \mathbf{I} \; ; \; \mathbf{H}^T = \mathbf{H} \tag{14}$$

The standard form of the variational formulation for finite displacements (e.g., see [20]) is given by

$$\Pi_W = \int_V W(H_{ij}) \, dV - \int_V u_i f_i \, dV \tag{15}$$

where $W(H_{ij})$ is the internal strain energy, given in terms of a chosen finite strains measure $\mathbf{H}$. For brevity, in (15) we limit ourselves to the Dirichlet boundary value problem, with zero displacements prescribed over the boundary of the domain V, and disregard inertia effects.

The polar decomposition (11) can be removed as a subsidiary condition of the standard variational principle and enforced in the weak sense, via use of the Lagrange multiplier procedure (e.g., see [21]). We set

$$D = \int_V \mathbf{P} \cdot [\mathbf{F} - \mathbf{R}(\mathbf{I} + \mathbf{H})] \, dV \; ; \; D = \int_V P_{ij} [\frac{\partial x_j}{\partial X_i} - R_{jm}(\delta_{mi} + H_{mi})] \, dV \tag{16}$$

If we add (16) above to the standard variational formulation for large displacements, we get the variational formulation of Fraeijs de Veubeke [6]

$$\Pi_{FV} = \Pi_W + D = \int_V \{W(H_{ij}) + P_{ij}[\frac{\partial x_j}{\partial X_i} - R_{jm}(\delta_{mi} + H_{mi})]\} \, dV - \int_V u_i f_i \, dV \tag{17}$$

In equations (16) and (17), the Lagrange multiplier P_{ij} can be recognized as the non-symmetric Piola-Kirchhoff stress tensor which is energy conjugate (see [14] or [16]) to the deformation gradient.

We make a small digression in order to give an interesting justification of the polar decomposition variational principle for finite strains (17) being a proper framework for studying continuum with independent rotation fields. Namely, if we apply the consistent linearization procedure (see [14]) to the variational principle (17), we recover the variational principle of Naghdi [15] and Reissner [18], which was used as a starting point for successful developments in the linear theory (see [9], [10] and [11]).

Using consistent linearization in equation (13), we can obtain that both finite strain measures, the Green-Lagrange strain $\mathbf{E}$ and the Biot strain $\mathbf{H}$ reduce to the conventional engineering infinitesimal strain measure $\boldsymbol{\epsilon}$, given in terms of the symmetric part of the displacement gradient

$$H_{ij} \xrightarrow{lin.} \epsilon_{ij} = \frac{1}{2}(\frac{\partial u_i}{\partial X_j} + \frac{\partial u_j}{\partial X_i}) \tag{18}$$

The orthogonal tensor $\mathbf{R}$, by the consistent linearization procedure, transforms into

$$R_{ij} \xrightarrow{lin.} \delta_{ij} + \omega_{ij} \tag{19}$$

where ω_{ij} is a skew-symmetric tensor, i.e. $\omega_{ij} = -\omega_{ji}$, which corresponds to the infinitesimal rotations.

Introducing into Fraeijs de Veubeke variational principle (17) the definition of the displacement vector, $u_i = x_i - X_i$, and the linearized measures (18) and (19), we obtain

$$\Pi_{FV} = \int_V \{W(\epsilon_{ij}) + \sigma_{ij}[\delta_{ji} + \frac{\partial u_j}{\partial X_i} - (\delta_{jm} + \omega_{jm})(\delta_{mi} + \epsilon_{mi})]\} \, dV - \int_V u_i f_i \, dV \tag{20}$$

where σ_{ij} is a non-symmetric stress tensor, which corresponds to the non-symmetric Piola-Kirchhoff stress P_{ij}.

If we linearize further by neglecting the product of the symmetric (ϵ_{im}) and skew-symmetric tensor (ω_{jm}) as a higher order term, then (20) takes the form of the variational principle of Naghdi [15]

$$\Pi_N = \int_V \{W(\epsilon_{ij}) + \sigma_{ij}(\frac{\partial u_j}{\partial X_i} - \omega_{ji}) - \sigma_{ij}\epsilon_{ij}\} \, dV - \int_V u_i f_i \, dV \tag{21}$$

By using the Legendre transformation to define the complementary energy density

$$\sigma_{ij}\epsilon_{ij} - W(\epsilon_{ij}) = \Sigma(\sigma_{ij}) \tag{22}$$

we recover the variational formulation of Reissner [18]

$$\Pi_R = \int_V \{-\Sigma(\sigma_{ij}) + \sigma_{ij}(\frac{\partial u_j}{\partial X_i} - \omega_{ji})\} \, dV - \int_V u_i f_i \, dV \tag{23}$$

The variational formulations (21) and (23) were applied to membrane elements ([10] and [11]), using the isotropic linear elastic material with quadratic forms for $W(\epsilon_{ij})$ and $\Sigma(\sigma_{ij})$.

The Euler-Lagrange equation obtained from the linearized variational principle (21), by taking variation with respect to the infinitesimal rotation ω_{ij}, is

$$skew\{\sigma_{ij}\} \equiv \frac{1}{2}(\sigma_{ij} - \sigma_{ji}) = 0 \implies \sigma_{ij} = \sigma_{ji} \tag{24}$$

which is an angular momentum balance, or the symmetry condition of the stress tensor.

Coming back to the Fraeijs de Veubeke variational principle (17), we find that the Euler-Lagrange equation obtained by taking the variation with respect to rotations R_{ij}, also leads to angular momentum balance

$$P_{ip}\frac{\partial x_q}{\partial X_i} = P_{iq}\frac{\partial x_p}{\partial X_i} \tag{25}$$

which is a well-known rotational equilibrium equation in terms of non-symmetric Piola-Kirchhoff stress tensor (e.g., see [14], pp. 142).

The transition from the four field principle (21) of Naghdi [15] to the three field principle (23) of Reissner [18], performed in the linear theory, can be performed in the nonlinear theory as well, if we assume that the Legendre transformation

$$H_{mi}T_{mi} - W(H_{mi}) = \Sigma(T_{mi}) \tag{26}$$

is possible. In equation (26), T_{ij} is used to denote the symmetric part of the Biot stress tensor (yet called the Jaumann stress tensor), which is energy conjugate to the Biot strain tensor (see [16]). Namely, by taking the variation of (17) with respect to the Biot strain tensor H_{ij} and utilizing the symmetry of the Biot strain tensor (14), we obtain the definition of the symmetric part of Biot stress tensor in terms of the non-symmetric Piola-Kirchhoff stress

$$T_{ij} = \frac{\partial W(H_{ij})}{\partial H_{ij}} = \frac{1}{2}(P_{ik}R_{kj} + P_{jk}R_{ki}) \tag{27}$$

By using the results in (27) and (26), the result in (17) of Fraeijs de Veubeke can be recast in the form of a three-field variational principle

$$\Pi_{FV} = \int_V \{-\Sigma(T_{ij}(P_{ij}, R_{ij})) + P_{ij}(\frac{\partial u_j}{\partial X_i} + \delta_{ji}) - P_{ij}R_{ji}\}\, dV - \int_V u_i f_i \, dV \tag{28}$$

4. APPLICATION TO SHALLOW SHELLS

When applying the variational principle (28) to derivation of the shallow shell theory, we assume that it holds in the shallow shell configuration. If $(\hat{\bullet})$ denotes the quantities measured in the shallow shell configuration, then (28) can be rewritten as

$$\Pi_{FV} = \int_{\hat{V}} \{-\Sigma(\hat{T}_{ij}(\hat{P}_{ij}, R_{ij})) + \hat{P}_{ij}(\frac{\partial u_j}{\partial \hat{X}_i} + \delta_{ji}) - \hat{P}_{ij}R_{ji}\}\, d\hat{V} - \int_{\hat{V}} u_i \hat{f}_i \, d\hat{V} \tag{29}$$

We next assume that the shallow shell configuration is obtained from the plate configuration by an adequate (locally defined) isometric transformation, which preserves length, the area and the volume element, so that the metric of the plate can be used rather than the metric of the shell. Therefore, the deformation gradient of such a transformation consists of pure rotation Q_{ij}. In other words, if $\{\hat{X}_i\}$ is a set of Cartesian coordinates of a material point in the shallow shell configuration and $\{X_i\}$ is a set of Cartesian coordinates of the same point in the plate configuration, then the deformation gradient is defined as

$$\frac{\partial \hat{X}_i}{\partial X_j} = Q_{ij} \; ; \quad Q_{ki}Q_{kj} = \delta_{ij} \tag{30}$$

Under the transformation given in (30), the non-symmetric Piola-Kirchhoff stress tensor changes into (recall that $\hat{P}_{ij}$ is a two-point tensor, see [14])

$$\hat{P}_{ij} = Q_{ik}P_{kj} \tag{31}$$

where no-hat quantities are measures in the flat (plate) configuration. By using the chain rule and (30)

$$\frac{\partial u_j}{\partial X_l} = \frac{\partial \hat{X}_i}{\partial X_l}\frac{\partial u_j}{\partial \hat{X}_i} = Q_{il}\frac{\partial u_j}{\partial \hat{X}_i} \implies \frac{\partial u_j}{\partial \hat{X}_i} = Q_{li}\frac{\partial u_j}{\partial X_l} \tag{32}$$

By using (30) to (32), we can rewrite the variational principle (29) as

$$\Pi = \int_V \{-\Sigma(T_{ij}(P_{ij}, R_{ij})) + P_{ij}\frac{\partial u_j}{\partial X_i} - P_{ij}(R_{jk}Q_{ki} - Q_{ji})\}\, dV - \int_V u_i f_i \, dV \tag{33}$$

where all the quantities are now measured in the plate configuration.

The present consideration until this point allows for arbitrary size of rotations, and, for that reason, it is often referred to as geometrically exact theory (e.g., see [19]). We next consider an approximate theory, introduced by restricting the size of the rotations.

If we recall the result of Argyris [3], we can state that any orthogonal tensor can be represented as a power series of skew-symmetric tensors converging to what is called exponential map (see [19]). In other words, by exponentiating a skew-symmetric tensor the orthogonal tensor is recovered. If the orthogonal tensors Q_{ij} and R_{ij} are obtained by exponentiating, respectively, the skew-symmetric tensors θ_{ij} and ω_{ij}, then we can write

$$Q_{ji} \approx \delta_{ji} + \theta_{ji} + \frac{1}{2}\theta_{jk}\theta_{ki} + O(\theta^2) \tag{34}$$

and

$$R_{jk}Q_{ki} \approx \delta_{ji} + (\theta_{ji} + \omega_{ji}) + \frac{1}{2}(\theta_{jk} + \omega_{jk})(\theta_{ki} + \omega_{ki}) + O((\theta + \omega)^2) \tag{35}$$

as the second order approximation for those orthogonal tensors.

If we use (35) to linearize the expression for Biot stress, which appears in the complementary energy density in (33), then we get

$$T_{ij} \xrightarrow{lin.} \frac{1}{2}(\sigma_{ij} + \sigma_{ji}) \equiv symm\{\sigma_{ij}\} \tag{36}$$

where σ_{ij} is the non-symmetric stress tensor which corresponds to the linearized Piola-Kirchhoff stress. Considering (36) above, an approximate quadratic form of the complementary energy for a linear elastic body can be defined as

$$\Sigma(T_{ij}) \xrightarrow{lin.} \Sigma(\sigma_{ij}) = \frac{1}{2} symm\{\sigma_{ij}\} C^{-1}_{ijkl} symm\{\sigma_{kl}\} \tag{37}$$

With the introduced approximations, the variational principle (33) can be rewritten as

$$\Pi = \int_V \{-\frac{1}{2} symm\{\sigma_{ij}\} C^{-1}_{ijkl} symm\{\sigma_{kl}\} + skew\{\sigma_{ij}\}(skew\{\frac{\partial u_j}{\partial X_i}\} - \omega_{ji})$$

$$+ symm\{\sigma_{ij}\}[symm\{\frac{\partial u_j}{\partial X_i}\} + \frac{1}{2}(\theta_{jk}\omega_{ik} + \theta_{ki}\omega_{jk})]\} \, dV - \int_V u_i f_i \, dV \tag{38}$$

where, in (38) above, $symm\{\bullet\}$ and $skew\{\bullet\}$ are the operators which jointly perform Euclidean decomposition of the second order tensor into its symmetric and skew-symmetric part (see (36) and (24) for their application on the stress tensor σ_{ij}).

Taking variation of (38) with respect to the stress σ_{ij}, we get

$$\epsilon_{ij} = C^{-1}_{ijkl} symm\{\sigma_{kl}\} = symm\{\frac{\partial u_j}{\partial X_i}\} + \frac{1}{2}(\theta_{jk}\omega_{ik} + \theta_{ki}\omega_{jk}) \tag{39}$$

associated with the symmetric part, and

$$\omega_{ij} = skew\{\frac{\partial u_j}{\partial X_i}\} \tag{40}$$

associated with the skew-symmetric part, which provides us with the definitions of the rotation field.

The Euler-Lagrange equation (39) can be used to eliminate the symmetric part of the stress tensor, $symm\{\sigma_{ij}\}$, from the variational formulation (38), before making transition to the resultant form (see Section 5),

$$\Pi = \int_V \{\frac{1}{2} symm\{\epsilon_{ij}(u_j, \omega_{ji})\} C_{ijkl} symm\{\epsilon_{kl}(u_l, \omega_{lk})\}$$

$$+ skew\{\sigma_{ij}\}(skew\{\frac{\partial u_j}{\partial X_i}\} - \omega_{ji})\} \, dV - \int_V u_i f_i \, dV \tag{41}$$

where ϵ_{ij} is given in terms of (39).

5. VARIATIONAL PRINCIPLE IN TERMS OF STRESS RESULTANTS

In this section we make the final transition from the three-dimensional variational principle (41) to the appropriate two-dimensional form given in terms of stress resultants. Rather than using the Kirchhoff constraint, as in (6), the derivation is performed under the Reissner-Mindlin kinematic assumption, so that the displacement field is defined as

$$u_\alpha(X_\beta, \zeta) = v_\alpha(X_\beta) - \zeta \phi_\alpha(X_\beta)$$

$$u_3(X_\beta, \zeta) = v_3(X_\beta) \; ; \quad \alpha, \beta \in \{1,2\} \tag{42}$$

The rotation vector ϕ_α is related to the right-hand-rule rotation vector θ_α via the two-dimensional alternating tensor $e_{\alpha\beta}$

$$\phi_\alpha = -e_{\alpha\beta}\, \theta_\beta \tag{43}$$

The three-dimensional independent rotation field ω_{ij} is represented by its axial vector ω_k (see [6]) via the three-dimensional alternating tensor e_{ijk}

$$\omega_k = e_{ijk}\, \omega_{ij} \tag{44}$$

The components of the axial vector are assumed in the form similar to the one proposed by Naghdi [15]

$$\omega_\alpha(X_\beta, \zeta) = \phi_\alpha(X_\beta) \; ; \quad \omega_3(X_\beta, \zeta) = \psi(X_\beta) + \zeta\varphi(X_\beta) \tag{45}$$

The assumptions introduced on the stress tensor components are the same as those of Reissner [17], i.e.

$$\sigma_{33} = 0 \; ; \quad \sigma_{13} = \sigma_{31} \; ; \quad \sigma_{23} = \sigma_{32} \tag{46}$$

The first assumption in (46) introduces plane stress condition and requires that the constitutive modulus $C_{\alpha\beta\gamma\delta}$ be modified accordingly. The remaining assumptions in (46), which disregard inequality of selected shear stress components, are justified by the fact that out-of-plane shear stress is much smaller than the in-plane shear stress. Therefore, unknown stress components are: σ_{11}, σ_{22}, σ_{12}, σ_{21}, $\sigma_{13} \equiv \sigma_{31}$ and $\sigma_{23} \equiv \sigma_{32}$.

If we define symmetric stress resultants

$$n_{\alpha\beta} = \int_{-h/2}^{h/2} symm\{\sigma_{\alpha\beta}\}\, d\zeta \tag{47}$$

$$m_{\alpha\beta} = \int_{-h/2}^{h/2} symm\{\sigma_{\alpha\beta}\}\, \zeta\, d\zeta \tag{48}$$

and

$$q_\alpha = \int_{-h/2}^{h/2} \sigma_{\alpha 3}\, d\zeta \tag{49}$$

the associated strain measures are given by

$$\epsilon_{\alpha\beta} = symm\{\frac{\partial v_\alpha}{\partial X_\beta}\} + \frac{1}{2}(\frac{\partial f}{\partial X_\alpha}\phi_\beta + \frac{\partial f}{\partial X_\beta}\phi_\alpha) \tag{50}$$

$$\kappa_{\alpha\beta} = symm\{\frac{\partial \phi_\alpha}{\partial X_\beta}\} \tag{51}$$

and

$$\gamma_\alpha = \frac{\partial v_3}{\partial X_\alpha} - \phi_\alpha \tag{52}$$

In equation (50), we have approximated the fictitious rotation field θ_{ij} used to define shallow shell from plate configuration by the partial derivatives of the shell reference surface, in the same way as it was introduced in Marguerre's theory (see Section 2).

The constitutive equations in terms of stress resultants become

$$n_{\alpha\beta} = h \; C_{\alpha\beta\gamma\delta} \; \epsilon_{\gamma\delta} \tag{53}$$

$$m_{\alpha\beta} = -\frac{h^3}{12} \; C_{\alpha\beta\gamma\delta} \; \kappa_{\gamma\delta} \tag{54}$$

and

$$q_\alpha = h \; C_{\alpha\beta}\gamma_\beta \tag{55}$$

In addition, we define the resultants of the skew-symmetric part of the stress tensor as

$$p_{12} = \int_{-h/2}^{h/2} skew\{\sigma_{12}\} \; d\zeta \tag{56}$$

$$r_{12} = \int_{-h/2}^{h/2} skew\{\sigma_{12}\} \; \zeta \; d\zeta \tag{57}$$

and the loading vector

$$g_\alpha = \int_{-h/2}^{h/2} f_\alpha \; d\zeta \; ; \quad g_3 = \int_{-h/2}^{h/2} f_3 \; d\zeta \tag{58}$$

The variational equation (41), recast in terms of stress resultants, then becomes

$$\Pi = \int_A \{\frac{1}{2}(\epsilon_{\alpha\beta} \; n_{\alpha\beta} - \kappa_{\alpha\beta} m_{\alpha\beta} + \gamma_\alpha q_\alpha)$$

$$+p_{12}(skew\{\frac{\partial v_2}{\partial X_1}\} - \psi) + r_{12}(skew\{\frac{\partial \phi_2}{\partial X_1}\} - \varphi)\} \; dA - \int_A (v_\alpha g_\alpha + v_3 g_3) \; dA \tag{59}$$

where A is the projected (flat) two-dimensional domain. The variational formulation (59) serves as the starting point of the discrete formulation. Within the context of discrete approximation, however, the variational formulation (59) should be regularized (see [9] and [10]) in order to preserve the stability for arbitrary choice of interpolations.

The Marguerre's shallow shell theory can be recovered from (59), if, first, we neglect the drilling rotations and consequently the non-symmetry of stress tensor, and, second, enforce the Kirchhoff constraint that the shear deformation vanishes, i.e. for

$$\gamma_\alpha = 0 \implies \phi_\alpha = \frac{\partial v_3}{\partial X_\alpha} \tag{60}$$

Combining (60) above with (50) and (51), we recover the corresponding strain measure (9) used in Marguerre's theory.

6. CLOSING REMARKS

The extension of Marguerre's shallow shell theory to account for shear deformation and addition of drilling degrees of freedom, has been proposed in this work. The formulation is given in Cartesian coordinates, and hence, as argued in [12], the convergence to the deep shell solutions can be obtained.

The variational formulation is cast in the resultant form over the two-dimensional projected (flat) reference configuration. In the discrete approximation, the projected configuration is selected locally for each shell element. Possible extension of this work to account for large rotations of the shell element projected configuration, can be accomplished through the use of co-rotational formulation.

Acknowledgement

This work was supported by the Swiss National Science Foundation, Grant no. 21-28942.20.

7. REFERENCES

1. Ahmad S., B.M. Irons and O.C. Zienkiewicz, Analysis of Thick and Thin Shell Structures by Curved Elements, Int. J. Numer. Methods Eng., 2, 419-451, 1970

2. Allman D.J., A Compatible Triangular Element Including Vertex Rotations for Plane Elasticity Problems, Comput. Struct., 19, 1-8, 1984

3. Argyris J., An Excursion into Large Rotations, Comput. Methods Appl. Mech. Eng., 32, 85-155, 1982

4. Bergan P.G. and C.A. Felippa, A Triangular Membrane Element with Rotational Degrees of Freedom, Comput. Methods Appl. Mech., 50, 25-60, 1985

5. Fonder G. and R.W. Clough, Explicit Addition of Rigid Body Motion in Curved Finite Elements, AIAA J., 11, 305-315, 1973

6. Fraeijs de Veubeke B., A New Variational Principle for Finite Elastic Displacements, Int. J. Engng. Sci., 10, 745-763, 1972

7. Frey F., Shell Finite Elements with Six Degrees of Freedom per Node, in Analytical and Computational Models for Shells (eds. A.K. Noor, T. Belytschko, J.C. Simo), ASME, 291-317, 1989

8. Gurtin M., *An Introduction to Continuum Mechanics*, Academic Press, 1981

9. Hughes T.J.R. and F. Brezzi, On Drilling Degrees of Freedom, Comput. Methods Appl. Mech. Eng., 72, 105-121, 1989

10. Ibrahimbegović A., R.L. Taylor and E.L. Wilson, A Robust Membrane Quadrilateral Element With Drilling Degrees of Freedom, Int. J. Numer. Methods Eng., 30, 445-457, 1990

11. Ibrahimbegović A. and F. Frey, Quadrilateral Membrane Elements with Rotational Degrees of Freedom. Comput. Struct., submitted, 1991

12. Idelsohn S., On the Use of Deep, Shallow or Flat Shell Finite Elements for the Analysis of Thin Shell Structures, Comput. Methods Appl. Mech. Eng., 26, 321-330, 1981

13. Jetteur Ph. and F. Frey, A Four Node Marguerre Element for Nonlinear Shell Analysis, Eng. Comput., 3, 276-282, 1986

14. Marsden J.E. and T.J.R. Hughes, *Mathematical Foundations of Elasticity*, Prentice Hall, Englewood Cliffs, New Jersey, 1983

15. Naghdi P.M., On a Variational Theorem in Elasticity and Its Application to Shell Theory, J. Appl. Mech., 12, 647-653, 1964

16. Ogden R.W., *Nonlinear Elastic Deformations*, John Wiley, London, 1984

17. Reissner E., On the Form of Variationally Derived Shell Equations, J. Appl. Mech., 86, 233-238, 1964

18. Reissner E., A Note on Variational Principles in Elasticity, Int. J. Solids Struct., 1, 93-95, 1965

19. Simo J.C. and D.D. Fox, On a Stress Resultants Geometrically Exact Shell Model. Part I: Formulation and Optimal Parameterization, Comp. Methods Appl. Mech. Eng., 72, 267-304, 1989

20. Washizu K., *Variational Methods in Elasticity and Plasticity*, Pergamon Press, Oxford, 1982

21. Zienkiewicz O.C. and R.L. Taylor, *The Finite Element Method: Basic Formulation and Linear Problems*, vol I, McGraw-Hill, London, 1989

FINITE ELEMENTS IN THE 90's, E. Oñate, J. Periaux, A. Samuelsson (Eds.)
Springer-Verlag/CIMNE, Barcelona 1991

STIFFNESS IDENTIFICATION OF STRUCTURAL COMPONENTS BY MIXED EXPERIMENTAL AND NUMERICAL TECHNIQUES

W.P. De Wilde
Laboratory of Structural Analysis, Vrije Universiteit Brussel,
Pleinlaan, 2, B-1050 Brussel,
BELGIUM

Summary

The main objective of this paper is to give a brief introduction to the dynamical behaviour of (anisotropic) structural elements and the use of their natural vibration patterns in structural stiffness identification. After having set up the dynamic equilibrium equations, a finite element formulation of the problem can be derived and an analysis of the time dependent equations carried out. Numerical modal analysis and the modal superposition technique is discussed. Finally, the paper covers applications to laminated plates, in particular their structural stiffness characterisation, combining both numerical (Rayleigh-Ritz or finite element method) and experimental techniques (modal analysis).

1. INTRODUCTION TO STRUCTURAL VIBRATIONS

The *dynamic* (i.e. time dependent) behaviour of structural elements, either iso- or anisotropic, can quite easily be achieved with finite element techniques. For this purpose, one can start with the equations of *static* equilibrium of an elastic structure, subdivided into finite elements, which yields a set of linear algebraic equations (see e.g. Zienkiewicz, O.C. and Taylor, R.L.[1]) :

$$\mathbf{F} = \mathbf{K} \cdot \mathbf{q} \tag{1}$$

in which **F** stands for the nodal point generalised forces, **K** for the stiffness matrix and **q** for the nodal point generalised displacements[1].

Extending to the dynamic behaviour of the structure, one can apply the principle of *de d' Alembert* to these equations and replace the volume forces by equivalent forces, which are the applied (time dependent) volume forces (also called "excitations"), corrected by an *inertia reaction.* If one neglects damping, one thus makes following substitution:

$$\mathbf{f}(t) \quad \Rightarrow \quad \mathbf{f}(t) - \rho \frac{\partial^2 \mathbf{u}}{\partial t^2} \tag{2}$$

in which **u** stands for the displacement vector of an arbitrary point of the structure with mass density ρ, subject to the volume force **f**.

---

[1] **arrays and vectors are indicated with bold type characters.**

Applying standard finite element techniques, one ends up with a set of coupled differential equations of the second order :

$$\mathbf{K}\,\mathbf{q} + \mathbf{M}\,\frac{d^2\mathbf{q}}{dt^2} = \mathbf{F}(t) \tag{3}$$

in which **M** stands for the (*consistent*) mass matrix of the structure:

$$\mathbf{M} = \int_V \mathbf{N}^T \rho\, \mathbf{N}\, dV \tag{4}$$

ρ being the mass density and **N** the matrix of shape functions, linking the point displacements to the nodal point displacements; the ^{T} superscript indicates matrix transposition. The mass matrix **M** has properties which are completely similar to the stiffness matrix **K**: it is symmetric, positive definite and sparse, non-zero elements being found in the same entries [1].

In the case that *damping* occurs, the correction includes an additional term, depending on the velocity of the nodal points; this case has not been taken into account here, as the considered structures are showing very few damping, which thus can be neglected. Nevertheless, in a later section, we come back to this problem, which can take into account linear visco-elastic behaviour of the materials.

2. MODAL ANALYSIS WITH FINITE ELEMENTS

One of the ways to integrate the set of coupled differential equations consists in first extracting the eigenmodes of this structure: those eigenmodes are in fact vibration patterns which can occur in absence of external forces; they can thus be considered as patterns allowing for continuous interchange in forms of energy: from *potential* energy (i.e. deformation energy) into *kinetic* energy and vice versa. Those interchanges happen at discrete frequencies ω_i, (i=1,...n), found by solving the so-called associated (generalised) eigenvalue problem:

$$\mathbf{K} - \lambda\,\mathbf{M} = 0 \qquad \text{in which } \lambda = \omega^2 \tag{5}$$

As both matrices **K** and **M** are symmetric and positive definite, the eigenvalues λ are all real and positive. An interesting algorithm has been developed by Van Overmeire and Deconinck, based on the Sturm sequence properties of the principal minors of the matrix $\mathbf{K} - \lambda\,\mathbf{M}$. It allows the extraction of multiple eigenvalues, with their degree of multiplicity, as well as the extraction within a preset interval [2]. The eigenvectors are found with an inverse iteration technique. For very large problems, involving many thousands of degrees of freedom, it now seems clear that Lancsoz's algorithm cannot be ignored anymore.

The eigenvectors $\mathbf{q}^i$ show an interesting property, i.e. they are orthogonal with respect to both **K** and **M** :

$$\{\mathbf{q}^i\}^T\,[\,\mathbf{K}\,]\,\{\mathbf{q}^i\} = \lambda_{(i)}\,\delta_{ij} \tag{6}$$

$$\{\mathbf{q}^i\}^T\,[\,\mathbf{M}\,]\,\{\mathbf{q}^i\} = \delta_{ij} \tag{7}$$

in which δ_{ij} is the Kronecker delta. The modes are then considered to be *orthonormalised.*

As one can see, the *eigenpairs,* i.e. the set of eigenvalues and -vectors, can be considered as a "signature" of a structure, as they depend on mass and stiffness distribution, the latter being dependent on both material stiffnesses and the form of the structural part. One will use this characteristic in material or structural stiffness identification, explained in a further section.

Time response of a linear structure can be found, using the technique of modal superposition, based on the orthogonality of the modes; the response is written as a linear combination of eigenmodes:

$$\{ q \} = \sum_{j=m}^{p} x_i(t) \{ q^i \} \tag{8}$$

the limits of the sum depending on the modes which effectively participate to the response (this depends on the frequency content of the excitation; generally modes are taken in the interval of 1/5th of the lowest frequency of the excitation to 5 times the highest frequency). Substitution of equation (8) into equation (3) and premultiplication by the transposed of $\{q^i\}$ yields a set of (p-m) uncoupled equations of the type :

$$\lambda_i \, x_i(t) + \frac{d^2}{dt^2} x_i(t) = f_i(t) \tag{9}$$

which can be evaluated either numerically or by Duhamel's integral :

$$x_i(t) = \int_0^t f_i(t) \sin \omega_i (t-\tau) \, d\tau \tag{10}$$

3. APPLICATION TO LAMINATED STRUCTURES

The previously established equations can be applied to a laminated plate. The latter consists of m stacked thin anisotropic layers, often called *laminae*. Each of these layers shows in-plane stiffness Q_{ij} (i,j = 1,2,6), linking the stresses to the deformations:

$$\begin{Bmatrix} \sigma_x \\ \sigma_y \\ \tau_{xy} \end{Bmatrix} = \begin{bmatrix} Q_{11} & Q_{12} & Q_{16} \\ Q_{12} & Q_{22} & Q_{26} \\ Q_{16} & Q_{26} & Q_{66} \end{bmatrix} \begin{Bmatrix} \varepsilon_x \\ \varepsilon_y \\ \gamma_{xy} \end{Bmatrix} \tag{11}$$

For a symmetrically stacked plate, one can then find a relation between the bending and torsion moments and the curvatures and twist of the plate :

$$\begin{Bmatrix} M_x \\ M_y \\ M_{xy} \end{Bmatrix} = \begin{bmatrix} D_{11} & D_{12} & D_{16} \\ D_{12} & D_{22} & D_{26} \\ D_{16} & D_{26} & D_{66} \end{bmatrix} \begin{Bmatrix} \kappa_x \\ \kappa_y \\ \kappa_{xy} \end{Bmatrix} \tag{12}$$

In the latter relation:

$$D_{ij} = \frac{1}{3} \sum_{k=1}^{m} Q_{ij} \left(z_k^3 - z_{k-1}^3 \right) \tag{13}$$

which indicates that the bending and torsional stiffness are dependent on the stacking sequence and the individual layer properties. z_k is the distance from the middle plane of the laminated structure to the lower face of layer k.

In the case of *orthotropic* laminated plates, one also finds $D_{16} = D_{26} = 0$. This is the case which we will consider, most of the laminated structures being orthotropic. In this case, the free vibrations of the plate can be found by solving the following eigenvalue problem, formulated locally [3]:

$$D_{11} \frac{\partial^4 w}{\partial x^4} + 2 \left(D_{12} + 2D_{66} \right) \frac{\partial^4 w}{\partial x^2 \partial y^2} + D_{22} \frac{\partial^4 w}{\partial y^4} = \omega^2 \rho h w \tag{14}$$

in which ρ stands for the mass density of the plate, h its height and ω for the eigenfrequency of the plate; w is the deflection pattern (eigenmode, associated with the eigenfrequency ω) of the plate, perpendicular to its middle plane. Some of these variables are quite easily found or measured experimentally: the mass density, the thickness; the idea arose that perhaps a (limited) set of eigenpairs could yield some information on the bending stiffnesses D_{ij} of the plate, perhaps even allowing to establish them with more precision than traditional experimental techniques.

The idea of coupling numerical (i.e. finite element or alike) and experimental modal analysis techniques for the purpose of stiffness identification was first proposed by De Wilde et al. in 1984 [4] and later completely developed by Sol in his doctoral thesis [5]. Sol proposed some essential modifications to the original idea, thus allowing to develop a completely automated procedure. Next section is dedicated to this topic.

4. STRUCTURAL IDENTIFICATION BY MIXED EXPERIMENTAL-NUMERICAL TECHNIQUES.

These techniques make use of more complicated test structures than classical ones (e.g. the so-called "dog bone" specimen), in the sense that an analytical solution of the modal response is in general not available. The stress field is also not uniform. The elastic stiffness properties appear thus as "unknowns" in a numerical model of the test structure. The parameters are then estimated (or corrected) in order to have a perfect match between some experimentally measured responses and the corresponding values, calculated through the mathematical model.

The basic idea was first proposed by Kavanagh in 1972 [7], combining strain gauge measurements and finite elements. This work inspired Hermans et al. [6], who combined boundary elements and extensometry. De Wilde et al. [4], inspired by Wolf and Carne [9], explored dynamic behaviour. Finally, Sol [5], elaborating further on the latter idea, developed the method combining a Rayleigh-

Ritz model of a vibrating plate with 4 free boundaries and modal analysis of the same plate, taking *only* into account the resonant frequencies.

The latter may seem - at first sight - curious: in the procedure, developed by Sol, one does not consider potential information given by the modes. As he shows in his thesis, if one considers anisotropic plates with a given *aspect ratio* (i.e. ratio between length a and width b of the plate) given by :

$$\frac{a}{b} = \left(\frac{D_{11}}{D_{22}}\right)^{\frac{1}{4}} \qquad (15)$$

one ensures the *maximum sensitivity* in the measurement of the eigenfrequencies, which are then associated with eigenmodes which appear in an *a priori* known order. One thus gets the applied procedure, shown in fig. 1:

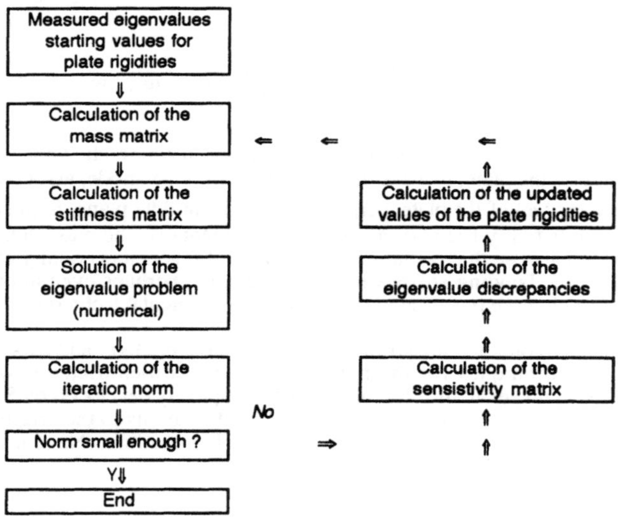

fig. 1. : Flow chart of the procedure

One of the key ideas of Sol was to include two techniques in the optimisation algorithm:

a) a *Bayesian estimation technique*, which allows to give a more important weight to stiffness values which are known to be easier to forecast: the algorithm is thus a kind of weighted least squares technique (one knows that in e.g. a unidirectional laminate E_1 is easier to forecast than, say, E_2);

b) *sensitivity analysis*, which allows to predict how the measured values will evolve with a small change of one of the unknown parameters; in this technique, one thus evaluates - through the numerical model - the sensitivity of the eigenfrequencies λ_j with respect to a stiffness parameter D_{kl} and thus gets a *sensitivity matrix* :

$$[\mathbf{S}] = [\frac{\partial \lambda_j}{\partial D_{kl}}] \tag{16}$$

It should be noted that this matrix is - generally speaking - rectangular (the number of measured frequencies is not necessarily equal to the number of sought parameters), so that a method of ***pseudo-inversion*** will be used [8].

An experimental procedure can now easily be imagined: a plate with the right a/b ratio (established previously by two simple vibration tests on beams taken in the two main directions of anisotropy) is hung and excited by a slight impulse excitation (tick with a pencil); an accelerometer registers the vibration in a known point; the time signal is convoluted through a digital FFT algorithm and yields the eigenfrequencies of the plate. These eigenfrequencies are matched with the ones obtained through a numerical model, based on guessed - or corrected - values of the plate rigidities. The latter is done using Bayesian estimation and sensitivity analysis, included in an optimisation algorithm, based on a method of pseudo-inversion (see fig. 1).

The whole software procedure has been implemented on a PC, which also registers the time signal, convolutes it and compares these eigenfrequencies with calculated ones.

The latter also is done in the PC; indeed, Sol showed that one can use a simple Rayleigh-Ritz model, which is in fact a ***superelement*** for thin plates, developed for this purpose: the plate is considered as one element with 49 (i.e. 7 by 7) nodes; to each node a "shape function" is associated :

$$N_{ij} = L_i(\xi) \cdot L_j(\eta) \tag{17}$$

in which i and j indicate the "coordinates" of the considered nodal point (i,j = 1,...,7) and L_i is a Lagrange polynomial function of the 6-th degree. As the latter is C_1 - continuous, one only takes w, the plate deflection, as unknown and the resulting stiffness matrix and mass matrix are (49x49). The calculation of the associated eigenvalue problem can thus easily be carried out with a PC: on a 386 processor based PC it takes roughly 1 à 2 minutes to achieve all the necessary calculations.

The whole procedure is thus quite simple and does not require expensive equipment; on the other hand it only provides the elastic stiffness of the plate (no information is obtained about strength).

During the past years, attention has been paid to two main directions of further developments:

a) extension of these basic ideas to use the eigenvalues of a complex elastic structural element as a "fingerprint" of this structure and to monitor the ***evolution*** of these frequencies, thus trying to link them to "damage" or "quality". This promising (and exciting) orientation is now fully investigated, in collaboration with industry: it presents spin-off in quality control (either during production or during service). It seems that the measurement of the modes (e.g. through laser interferometry) also provides important information on the ***quality*** of the structural elements;

b) extension of the developed techniques to establish visco-elastic properties of the materials, through the measurement of damping (the latter can be related to the complex moduli of the structure); extension of the software, combined with acoustic excitation of the plate, is now fully implemented. Here again, information on the vibration pattern, i.e. the eigenmodes (and their anomalies) is of paramount importance.

The information obtained with the latter techniques can be used in a finite element analysis, based on analysis in the frequency domain [9]. When the excitation on the structure is narrow-banded, only a few modes are sufficient.

5. COMPARISON ON CROSS-PLY LAMINATES

The following tables give a comparison of values of E_1, E_2 and G_{12} (resp. the elasticity moduli in the 0° and 90°-directions and the in-plane shear modulus) for cross-ply laminated orthotropic structures in graphite epoxy systems: the values are a) calculated with micro-mechanical models, obtained with Mic Mac, a program written and documented by Tsai [10] and Colap, written by Dascotte and Sol (based on slightly different micromechanical assumptions), b) measured with classical traction experiments and c) measured with the Resonalyser ®, equipment based on the principles which were discussed in the previous paragraph.

System	E1[GPa] MM flex.	E1[GPa] MM inpl.	E1[GPa] Colap flex	E1[GPa] Colap inpl	E1[GPa] Instron	E1[GPa] Reson
$[0_8]_s$	130	130	130	130	131	130
$[0_6,90_2]_s$	128,2	100,4	129	105	108	122
$[0_4,90_2,0_2]_s$	117,2	100,4	122	106	105	118
$[0_3,90_2,0_3]_s$	107,4	100,4	112	105		112
$[0_3,90_4,0_1]_s$	101,3	70,2	106	74		96
$[0_2,90_2,0_4]_s$	94,7	100,4	99	105	107	94,5
$[0_2,90_2,0,90,0,90]_s$			95	74	60	86
$[0,90]_{4s}$	81,5	70,2	86	74		81
$[0_2,90_6]_s$	79,6	100,4	84	105	105	79
$[0,90_2,0_5]_s$	79,2	100,4	83	105		82

Table 1: Comparison of E_1 for various stacking sequences

System	E2[GPa] MM flex	E2[GPa] Colap flex	E2[GPa] Reson	G12[GPa] Colap	G12[GPa] Reson
$[0_8]_s$	9,8	9,8	9,8	3,2	3,2
$[0_6,90_2]_s$	11,7	11,8	8,3	3,2	4,5
$[0_4,90_2,0_2]_s$	23,0	23,9	22,1	3,2	4,7
$[0_3,90_2,0_3]_s$	32,9	34,4	33,8	3,2	4,4
$[0_3,90_4,0_1]_s$	39,1	41,0	36,1	3,2	5,0
$[0_2,90_2,0_4]_s$	45,7	48,0	43,1	3,2	4,7
$[0_2,90_2,0,90,0,90]_s$		53,0	49,6	3,2	4,3
$[0,90]_{4s}$	58,9	62,0	56,0	3,2	4,8
$[0_2,90_6]_s$	60,8	64,0	60,0	3,2	4,8
$[0,90_2,0_5]_s$	61,2	64,5	62,0	3,2	

Table 2: Comparison of E_2 and G_{12} for various stacking sequences

The systems are indicated with the stacking sequence: a numerical subscript indicates the number of plies in the corresponding direction, subscript "s" indicates that the system is symmetrically stacked. Only 0° and 90° plies were combined, but in various sequences; all the cross-ply laminates have 16 plies. For the Resonalyser, the most logical comparison is of course between values evaluated in bending (with MicMac (MM flex) and Colap flex); on the other hand, the Instron tests are to be compared with "in plane" values, obtained through similar mathematical models. Micro-mechanical models consider that the in-plane shear modulus is independent from the stacking sequence: this is clearly contradicted by experiments, and shown once more here.

6. CONCLUSIONS

As a conclusion, one can say that it is possible to identify structural properties by combining dynamic measurement techniques with numerical models of these vibrations, be it natural (*stiffness identification*), or forced (*damping measurements*, which in turn can yield information about the visco-elastic properties of the material). The whole procedure is quite easy to implement on a PC computer; as such it can thus be considered as economical, compared with traditional tests; it is also much less time-consuming.

7. ACKNOWLEDGMENTS

The author would like to thank his collaborators for the experimental data; in particular, the discussions with A. Cardon, H. Sol, L. Radisewski [12], J. Van Tomme and J. De Visscher are gratefully acknowledged.

The continuous financial support of the Belgian Foundation for Scientific Research, the IWONL (Institute for industrial research), the Belgian Scientific Policy Ministery and the Vrije Universiteit Brussel should also be mentionned. The author is very grateful for this support.

8. REFERENCES

1. ZIENKIEWICZ, O.C., TAYLOR, R.L., **The finite element method, fourth edition, vol.1 and 2**, Mac Graw Hill, London, 1990 & 1991.

2. VAN OVERMEIRE, M., DECONINCK, J., **Trillingen van constructies**, Engineering thesis, Structural Analysis, VUB, 1978.

3. LEKHNITSKII, S. G., **Anisotropic Plates**, Gordon and Breach Science, 1968.

4. DE WILDE, W.P., SOL, H., NARMON, B., ROOVERS, M., Determination of the material constants of an anisotropic lamina by free vibration analysis, Proc. 2nd Int. Conf. on Modal analysis, publ. by Univ. of Shenectady, N.Y., 1984.

5. SOL, H., **Identification of anisotropic plate rigidities using free vibration data**, Doctoral dissertation, Structural Analysis, VUB, 1986.

6. HERMANS,F., DE WILDE, W.P., HIEL,C., 'Boundary Integral Method applied in the characterisation of elastic materials', Proc. Int. Symp. on Computational Methods and Experimental Measurements, Washington DC, Springer Verlag, 1982.

7. KAVANAGH, K.T., 'Finite element applications in the characterisation of elastic solids', Journal of Experimental Mechanics, 1972.

8. DE WILDE, W.P, SOL, H., 'The combined use of Finite Elements and Sensitivity Analysis for the Optimization of the Dynamical Behaviour of Structures (Invited lecture)', Proc. 2nd Int. Conf. on Variational Methods in Engineering, Southampton, GB, Springer Verlag, 1985.

9. WOLF, J.A., CARNE, T.G., 'Identification of the elastic constants for composites using modal analysis', Proc. Annual Meeting SESA, San Francisco, May 1979.

10. DE WILDE, W.P., CARDON, A.H., 'Analysis of viscoelastic structures in the frequency domain', IJNME, (ed. R. Ohayon, O.C. Zienkiewicz), Dec. 1989.

11. TSAI, S., Composites design, 4th edition, Think Composites, Paris, 1988.

12. RADISEWSKI, L., Private communication, 1991.

FINITE ELEMENTS IN THE 90's, E. Oñate, J. Periaux, A. Samuelsson (Eds.)
Springer-Verlag/CIMNE, Barcelona 1991

RIGID FINITE ELEMENT METHOD IN STRUCTURAL ANALYSIS*

Q. LingXi and **Z. Xiong**
Research Instutute of Engineering Mechanics,
Dalian University of Technology,
Dalian 116023,
P. R. CHINA

ABSTRACT: The mathematical formulation of *Rigid Finite Element Method* (**RFEM**) is given in this paper. Its applications to elastic analysis and elastoplastic analysis as well as to limit and shakedown analysis will be presented here for illustrating its advantages: namely saving of the computational work and improving of the accuracy of stress as compared with traditional FEM. It is expected that RFEM will be useful in diverse fields of structural analysis, especially in the *Soil* and *Rock Mechanics* in which the constitutive law belong to the Coulomb type.

1. INTRODUCTION

Rigid Bodies-Spring Model[1] (**RBSM**) proposed by professor **KAWAI** in 1976 considerably reduces the computational work of finite element method. It will be shown here that **RBSM** has its unique advantages in different branches of structural analysis, especially when the constitutive law of material belong to the Coulomb type

$$f(\sigma)=|\tau_s|+\sigma_n tg\phi-C\le 0$$

In **RBSM** a given structure is divided into a proper number of rigid elements mutually connected by spring system distributed along the seams between elements(**Fig.1**). In such discrete models, displacement of each element can be completely described by rigid body movement of its centroid. The strain within each element is zero and the strain energy is only stored in spring system along seams between elements.

The displacement Φ of any point within each element can be completely described by rigid body movement $\Phi_g = [u_g \quad v_g \quad \theta_g]^T$ of corresponding centroid, such as

* The project is supported by *National Natural Science Foundation* **of China**

$$\Phi=[u(x,y)\ \ v(x,y)]^T=\begin{bmatrix} u_g+(y-y_g)\theta_g \\ v_g+(x_g-x)\theta_g \end{bmatrix}=N\Phi_g \tag{1}$$

in which $N=\begin{bmatrix} 1 & 0 & y-y_g \\ 0 & 1 & x_g-x \end{bmatrix}$ (2)

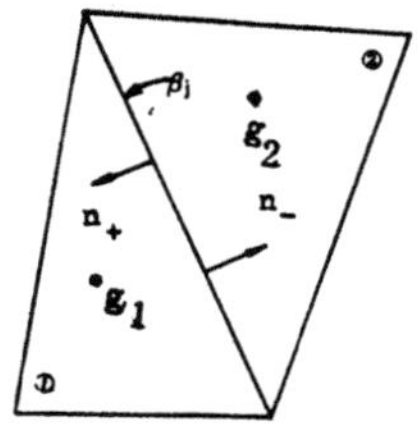

Fig.1 Two typical rigid finite elements

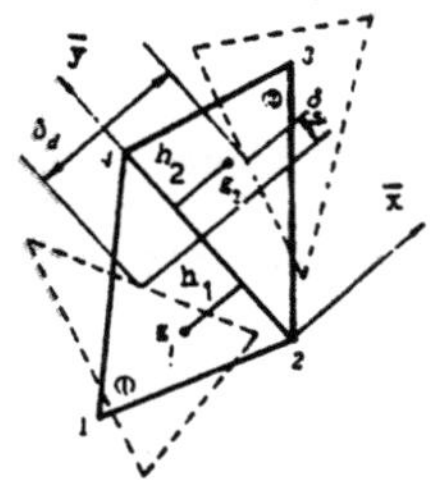

Fig.2 Relative displacement of rigid finite element

The size of the element stiffness matrices is only 3×3 for plane problem and the semi-band widths of overall stiffness matrices are greatly smaller than those of traditional FEM. The saving of computational work will be very significant for the solution of nonlinear problems which are in general reduced to a series of linear analysis steps.

2. RIGID BODY ELEMENT

Consider a rigid body element of the plane problem. Our purpose is to seek such displacement vector function $\Phi(x,y)=[u(x,y),v(x,y)]^T$ that it satisfies the partial differential equations of equilibrium

$$B^T(\frac{\partial}{\partial x},\frac{\partial}{\partial y})\sigma+F=0 \tag{3}$$

with the displacement boundary conditions

$$\begin{cases} u=0 \\ v=0 \end{cases} \quad \text{on } \Gamma_1 \tag{4}$$

and the traction boundary conditions

$$B^T(l,m)\sigma=P \quad \text{on } \Gamma_2 \tag{5}$$

where $B(\frac{\partial}{\partial x},\frac{\partial}{\partial y})=\begin{bmatrix} \frac{\partial}{\partial x} & 0 & \frac{\partial}{\partial y} \\ 0 & \frac{\partial}{\partial y} & \frac{\partial}{\partial x} \end{bmatrix}^T \quad B(l,m)=\begin{bmatrix} l & 0 & m \\ 0 & m & l \end{bmatrix}^T$

To solve the problem, the test function $\Phi(x,y)=[u(x,y),v(x,y)]^T$ which satisfies the boundary condition (4) is taken. By integrating Eq.(3) weighted by $\Phi^T(x,y)$ in Ω, we have

$$\iint_{\Omega}\Phi^T(x,y)[B^T(\frac{\partial}{\partial x},\frac{\partial}{\partial y})\sigma+F]dxdy=0 \tag{6}$$

Let $L=\begin{bmatrix} l & m \\ -m & l \end{bmatrix}$, $R=[\sigma_n \ \tau_s]^T$, $\sigma=[\sigma_x \ \sigma_y \ \tau_{xy}]^T$ (7)

then the relation between R and σ can be stated as

$$B^T(l,m)\sigma=L^T R \tag{8}$$

where σ_n and τ_s are the normal stress and tangential stress on the element boundary lines. Using Green's formulation, **Eq.(6)** and **Eq.(8)** give that

$$\oint_{\partial\Omega}\Phi^T L^T R ds-\iint_{\Omega}[B(\frac{\partial}{\partial x},\frac{\partial}{\partial y})\Phi]^T\sigma dxdy+\iint_{\Omega}\Phi^T F dxdy=0 \tag{9}$$

It can be observed that $B(\partial/\partial x,\partial/\partial y)\Phi=0$ in rigid body element. Hence **Eq.(9)** reduces to

$$\oint_{\partial\Omega}\Phi^T L^T R ds+\iint_{\Omega}\Phi^T F dxdy=0 \tag{10}$$

which is the equation of equilibrium of **Rigid Body Element**.

3. DISCRETIZATION WITH THE RIGID BODY ELEMENTS

After dividing region Ω into NE polygon elements we get NUMSP contact lines or seams between elements. We denote the boundary of element " e " by $\partial\Omega^e$, which includes traction boundary $\gamma_k=\partial\Omega^e\cap\Gamma_2$, displacement boundary $\alpha_i=\partial\Omega^e\cap\Gamma_1$ and contact lines $\beta_j=\partial\Omega^e-\alpha_i-\gamma_k$ between adjacent elements.

Assembling the NE polygon rigid elements according to **Eq.(10)**, the overall equilibrium equation will take following forms

$$\sum_{e=1}^{NE}[\oint_{\partial\Omega^e}\Phi^T L^T R ds+\iint_{\Omega^e}\Phi^T F dxdy]=0 \tag{11}$$

In line integration $\partial\Omega^e$, we should consider the fact that Φ is equal to zero on the displacement boundary α_i and the line integration along each seams β_j must be performed twice (backward and forthward). Substitutinging **Eq.(1)** into **Eq.(11)** and considering the arbitrarity of Φ_g lead to the overall equilibrium equation of the discrete system

$$-\sum_j\int_{\beta_j}N_G^T L_G^T R ds+\sum_k\int_{\gamma_k}N_G^T[P,0]^T ds+\sum_e\iint_{\Omega^e}N_G^T[F,0]^T dxdy=0 \tag{12}$$

where $N_G=\begin{bmatrix} N & 0 \\ 0 & N \end{bmatrix}$ $\quad L_G=[-L \quad L]$ (13)

4. ELASTIC ANALYSIS

The strain of rigid body $B(\partial/\partial x,\partial/\partial y)\Phi$ is equal to zero in region Ω, but it has no sense on the seams between elements where the discontinuity occurs. The relative movement δ between elements

(**Fig.2**) can be expressed as

$$\delta=[\delta_n \ \delta_s]^T=L_G N_G u_G \tag{14}$$

where $u_G=[u_{g1} \ v_{g1} \ \theta_{g1} \ u_{g2} \ v_{g2} \ \theta_{g2}]^T$ (15)

Eq.(14) is the strain-displacement relation in **RFEM**.

In general the relation between the boundary stress vector $R=[\sigma_n \ \tau_s]^T$ and the relative displacement vector $\delta=[\delta_n \ \delta_s]^T$ can be given as

$$R=D\delta \tag{16}$$

where D is the elastic matrix, which is

$$D=\begin{bmatrix} \dfrac{E}{(h_1+h_2)(1-\mu^2)} & 0 \\ 0 & \dfrac{E}{(h_1+h_2)(1+\mu)} \end{bmatrix} \tag{17}$$

in plane stress problem. In which h_1 and h_2 are perpendicular distances from centroids of two adjacent rigid elements to their contact line. **Eq(16)** is the stress-strain relation in **RFEM**.

Substituting **Eqs. (14)** and **(16)** into **Eq(12)** we finally get the overall equilibrium equation of structure for the elastic analysis

$$KU=P \tag{18}$$

where $K = \sum_j \int_{\beta_j} B^T DB ds$ (19)

$$P = \sum_k \int_{\gamma_k} N_G^T [P,0]^T ds + \sum_e \iint_{\Omega^e} N_C^T [F,0]^T dxdy \tag{20}$$

$$U = [u_1, v_1, \theta_1, \cdots, u_{NE}, v_{NE}, \theta_{NE}]^T \tag{21}$$

The Σ in **Eq.(19)** indicates assembling of all elements. In **Eq.(19)** assembling is performed on seam lines. Since a seam line is only concerned by two elements, so the symmetrical global stiffness matrix is highly sparse and its semi-band widths are much smaller than those of traditional **FEM**. After incorporating the boundary conditions (4), the assembled stiffness matrix K becomes positive-definite, so **Eq.(18)** has a unique solution.

Now the accuracy of displacement and stress of **RFEM** could be observed. Firstly, the infinitesimal factor (h_1+h_2) in the denominator of D may lead to that the accuracy of stress R seems to be inferior to that of displacement U. But when solving displacement $U=K^{-1}P$ from **Eq.(18)**, the infinitesimal factor (h_1+h_2) will appear in the numerator of displacement. Thus the factor h_1+h_2

appears simultaneously in numerator and denominator in the expression of stress R=Dδ so that its accuracy will not be inferior than that of displacement. Secondly, the stress is in proportion to the relative displacement δ between elements. In general the accuracy of relative displacement is better than that of displacement its self, so the accuracy of stress may be better than that of displacement. This will be illustrated in examples.

Elasticity analysis of the cantilevered beam shown in **Fig.(3.a)** is worked out. The displacement and stress results obtained for three different load conditions are compared with those given by simple beam theory in **Tab.1**. It can be found that the results of **RFEM** compare well with those of simple beam theory. The mesh used here is shown in **Fig.(3.b)**.

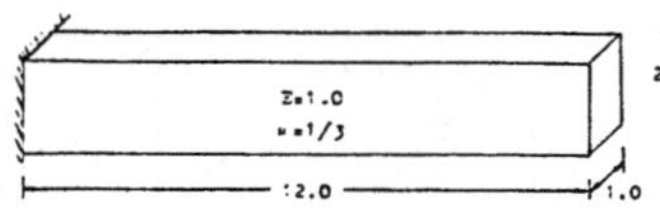

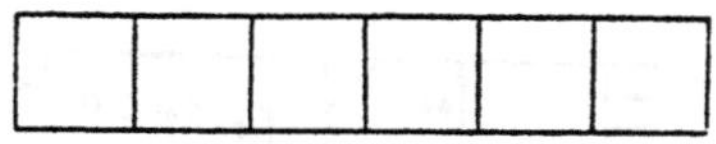

Fig.3.a Cantilevered square beam **Fig.3.b** Finite element mesh

Tab. 1 The cantilevered beam

	Simple beam theory			This paper		
	σ_{max}	δ_{max}	θ_{max}	σ_{max}	δ_{max}	θ_{max}
Concentrate load	108.0	5184.0	648.0	108.005	4848.3	576.04
Concentrate moment	108.0	7776.0	1296.0	108.007	6912.5	1152.10
uniform load	108.0	3888.0	432.0	108.004	3768.3	396.03

5. ELASTO-PLASTIC ANALYSIS

An efficient method "Parametric Variational Principle(PVP)" to solve non-linear problem has been published in paper [2] and applied successfully in diverse fields. The traditional elasto-plastic analysis consists of a series of incremental steps within each of them iteration are needed. But the **PVP** avoids iteration in each incremental step, so that the computer work is significantly reduced in comparing with the traditional method. Moreover when **RFEM** is used in the stress analysis in each step the efficiency of solving elasto-plastic problem would be once again improved considerably. **PVP** can deal with the strain hardening and strain softening as well as with the associated or non-associated flow rule. According to [2] using PVP reduces to a quadratic programming of **RFEM** for solving elasto-plastic problem

$$\text{Min} \quad \Pi=\frac{1}{2}dU^T K dU - dU(\phi\lambda+q)$$
$$\begin{cases} CdU-M\lambda-d+\nu=0 \\ \nu^T\lambda=0, \quad \nu\ge 0, \quad \lambda\ge 0 \end{cases} \tag{22}$$

in which K is assembled stiffness matrix of structure, dU is the incremental displacement vector, q is the incremental load vector, λ is the plastic flow factor vector. The matrices ϕ and C concern with the plastic potential and yield function respectively.

A 90° notched tension specimen shown in **Fig.4.a** is analyzed for yield condition $\sigma_n^2+3\tau_s^2=\sigma_s^2$. When the external load P reached 60.0 **Eq.(22)** will give a ray solution for any infinitesimal incremental load which means that the specimen collapsed. This coincides with the analytical solution. The load was subdivided from 39.68 to 60.0 to 12 steps and the finite element mesh is shown in **Fig.4.b**.

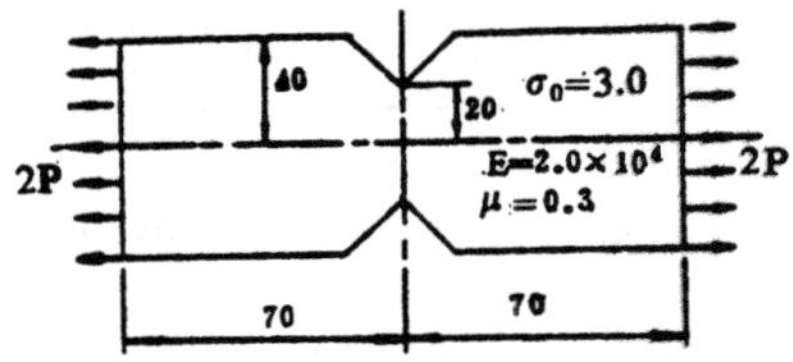

Fig.4.a 90° notched tension specimen

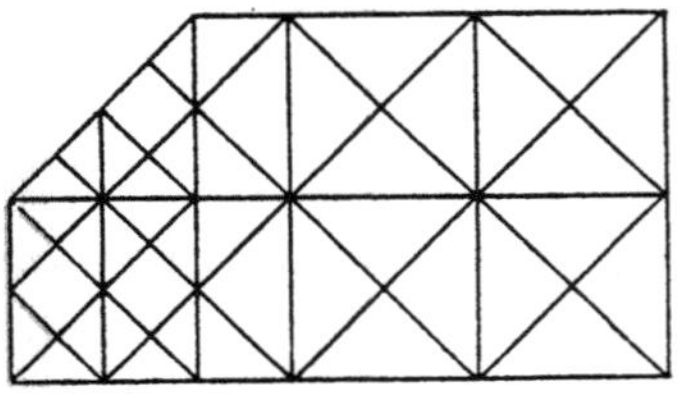

Fig.4.b Finite element mesh

6. LIMIT AND SHAKEDOWN ANALYSIS

The well-known Melan's theorem for plastic shakedown analysis states that if a self-equilibrating residual stress field R_0 can be found so that superimposing on the envelop of elastic stress R_E due to complex loading don't violate anywhere the yield condition $f(R_E+R_0)\le 0$, then the developing of plasticity would be shakedown. It can be mathematically expressed as

$$\begin{aligned} &\text{find } \lambda \text{ and } R_0 \\ &\text{to max } \lambda \\ &\text{s.t.} \begin{cases} \lambda R_E+NR_0-C\le 0 \\ \lambda \ge 0 \end{cases} \end{aligned} \tag{23}$$

in which λ is load multiplier factor which multiplying the basic load vector gets shakedown load vector, R_E is maximum elastic stress envelop value due to varying basic load and C is yield stress. The basic load vector P_0 is defined as a reference whose components could vary independently between limits

$$\mu_-^k P_0^k \le P_0 \le \mu_+^k P_0^k \tag{24}$$

When $\mu_{-}^{k}=\mu_{+}^{k}$ in **Eq.(24)** the formulation **(23)** reduces to the mathematical model of limit analysis.

The difficulty in solving **Eq.(23)** resides in finding a variable residual stress field R_0 which should be self-equilibrating. Reference [3] proposed to take the thermo-stress field as the residual stress field in assuming the temperature of each node as varying thermo-parameters. The **Thermo-Parameter Method** provides an efficient approach to the limit and shakedown analysis. If **RFEM** is used in the stress analysis instead of traditional **FEM**, the efficiency could be much raised.

Four square slabs with square cutout, oblique square opening, slit and circular hole shown in **Fig.5** are analyzed respectively and lower bounds solution are found(**Tab.2** and **Fig.6**). The yield condition used here is a linearization of $\sigma^2+3\tau^2-\sigma_0^2 \le 0$.

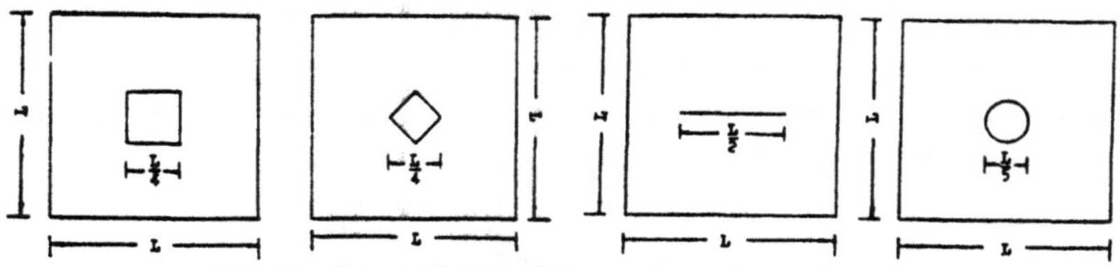

Fig.5 Four slabs with weakened opening

Tab.2 The results comparison of limit analysis

	$\rho^{-[6]}$	$\rho^{+[6]}$	this paper
slab with square hole	0.693	0.764	0.7246
oblique square hole	0.740	0.799	0.7354
slab with a slit	0.498	0.522	0.5000

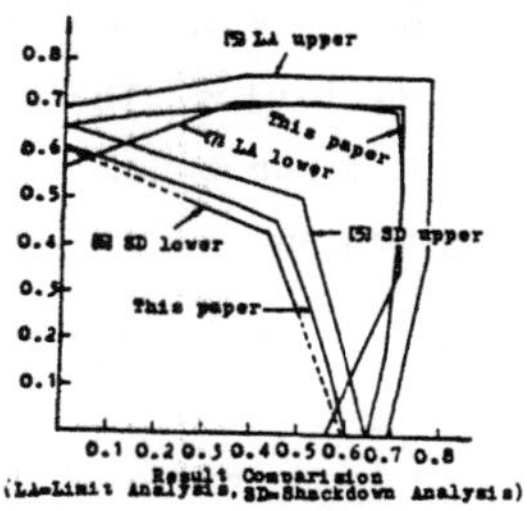

Fig.6 The shakedown of slab with circular hole

7. SOIL AND ROCK MECHANICS

The yielding condition used in *Soil* and *Rock Mechanics* is usually the Coulomb condition

$$f(\sigma)=\begin{bmatrix} -tg\phi & -1 \\ tg\phi & 1 \\ 1 & 0 \end{bmatrix}\begin{Bmatrix} \sigma_n \\ \tau_s \end{Bmatrix} + \begin{Bmatrix} C \\ C \\ R \end{Bmatrix} \tag{25}$$

This is piecewise linear yield condition in function of σ_n and τ_s (**Fig.6**). By **RFEM** the σ_n and τ_s are readily determined in the seams of rigid finite elements. The meshes of finite element should be arranged to follow the rock's joints or the soil's probable failure mode.

The problem of slope's stability, retaining wall's earth pressure and foundation's bearing capacity could be analysed readily by **RFEM** in using the limit analysis or elasto-plastic analysis based on the yield condition **(25)** with associated or non-associated flow rules.

The slope shown in **Fig.7** is stable in undisturbed state and its limit load factor λ can be determined as 7.6113 using **RFEM**. When the rock mass is subject to a seismic load $P=-2.0\times10^5$kg it becomes unstable and its limit load factor λ reduces to 0.65553 which means the slope can bear a seismic load $P=0.65553\times(-2.0\times10^5)=-1.31106\times10^5$kg. The coefficient of friction $tg\phi$ and cohesion C of rock joints are 0.43 and $4.0\times10^3 kg/m^2$ respectively and the unit weight of the rock mass $\gamma=2.7\times10^3 kg/m^3$.

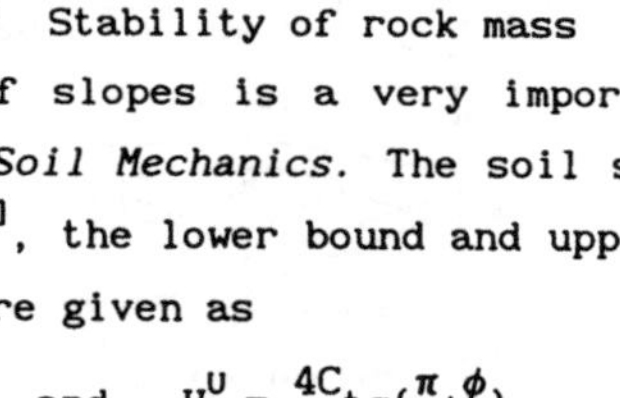

Fig.7 Stability of rock mass

The stability of slopes is a very important and well known problem in *Rock* and *Soil Mechanics*. The soil slope shown in **Fig.8** is analyzed by CHEN[4], the lower bound and upper bound of critical height of slope H_{cr} are given as

$$H^L_{cr} = \frac{2C}{r}tg(\frac{\pi}{4}+\frac{\phi}{2}) \quad \text{and} \quad H^U_{cr} = \frac{4C}{r}tg(\frac{\pi}{4}+\frac{\phi}{2})$$

respectively.

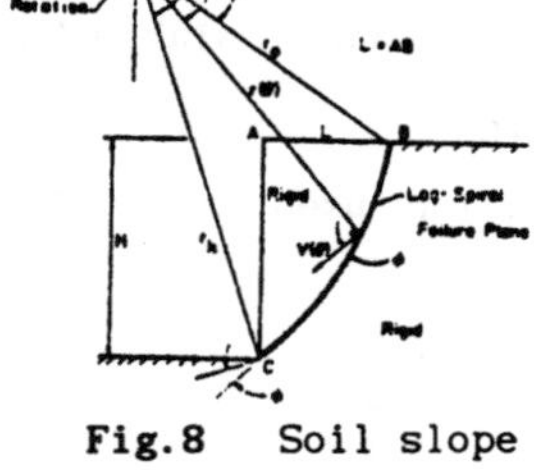

Fig.8 Soil slope

In this paper, a rotational mechanism (logarithmic spiral) is used to find the lower bound of the slope. ABC is considered as the unique rigid element in **RFEM** (**Fig.8**). In the case of $\Phi=20^{\circ}$ the lower bound is obtained approximately as

$$H_{cr}=\frac{2.98C}{r}tg(\frac{\pi}{4}+\frac{\phi}{2})$$

The frictionless strip footing shown in Fig.9.a was analyzed using Mohr-Coulomb yield condition for associated and non-associated case respectively and the load-settlement curve are shown in Fig.9.b.

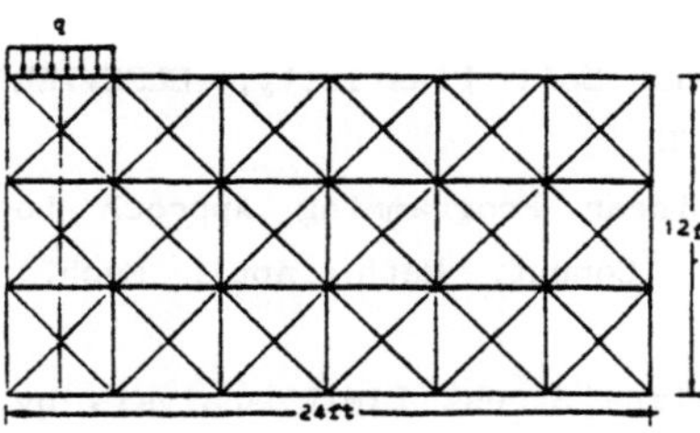

Fig.9.a Strip footing

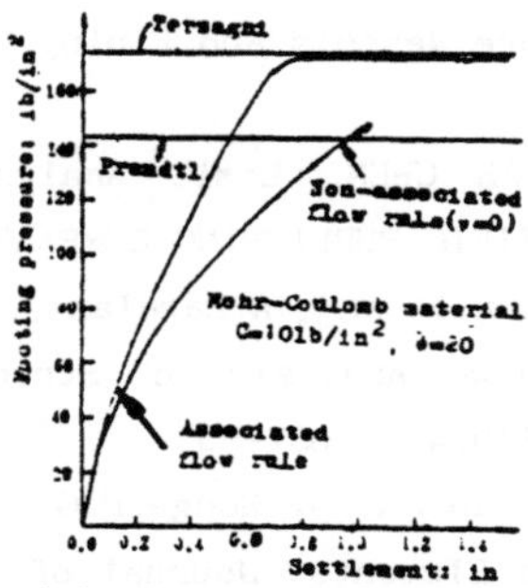

Fig.9.b Load-settlement curve

8. CONCLUSION

The semi-band width and scale of the global stiffness matrices of **RFEM** are smaller than those of traditional **FEM**, so **RFEM** reduces considerably the computational work especially in nonlinear analysis. All results in this paper are worked out on IBM-PC/AT computer.

In our experience the accuracy of stress by **RFEM** could be better than that by traditional **FEM**. The **RFEM** is incompatible in the displacement but it is continuous in the stress between rigid elements along the seam.

RFEM is tempting in *Soil* and *Rock Mechanics* because the constitutive law of material is Coulomb type which is expressed only by stresses on seams instead of by stress components within elements. **RFEM** is expected to be useful for the stability problem in geotechnical engineering.

REFERENCES

1. T. KAWAI, A New Element in Discrete Analysis of Plane Strain Problems, Seisan Kenkyu, Vol.29 (1977), No.4
2. Zhong,W.X. & Zhang, R.L., The Parametric Variational Principle for Elasto-Plasticity, Acta Mechanics Sinica, Vol.4,2(1988)
3. L.X.Qian & Z.B.Wang, Structure Limit Analysis and Shakedown Analysis — A Thermo-Parameters Method, Proceedings of 1990 Pressure Vessels and Piping Conference (PVP'90), Nashnille, TN., U.S.A.
4. WAI-FAH CHEN, Limit Analysis and Soil Plasticity, ELSEVIER SCIENTIFIC PUBLISHING COMPANY, 1975
5. L.Corradi and A.Zavelani, A Linear Programming Approch to Shakedown Analysis of Structure, Comput. Meth. Appl. Mech., vol.3(1974), p37-53
6. Belytschko,T. & Hodge,P.G.Jr., Plane Stress Limit Analysis by Finite Elements, Journal of the Engineering Mechanics Division, ASCE, EM6, 931-944
7. Nguyen Dang Hung, Sur La Plasticite et Le Calul Des Etats Limites Par Elements Finis, *These de Doctor Special en Scinces Appliques de le Universite de Liege ,1985*

FINITE ELEMENTS IN THE 90's, E. Oñate, J. Periaux, A. Samuelsson (Eds.)
Springer-Verlag/CIMNE, Barcelona 1991

SEISMIC FINITE ELEMENT ANALYSIS OF MASONRY BUILDINGS REINFORCED WITH FERROCEMENT

P. Ballesteros and **P. R. Reyes**
Ballesteros, S.A.,
Consulting Engineers
México, D.F.,
MEXICO

SUMMARY

The structural analysis by Finite Element Method, assembling the beam and plate elements, of six degrees of freedom per each nodal point is discussed. The rotational stiffness $K_{\theta z}$ is obtained and the aplication, to the analysis and design of a masonry building, which has been subjected to earthquake damage in its load bearing walls, is also presented.

NOTATIONS

$k_{\theta z}$	rotational stiffness
θ_x, θ_y, θ_z	rotations
u, v, w	displacement components
f_u, f_v, f_w	nodal element forces
$f_{\theta x}$, $f_{\theta y}$, $f_{\theta z}$	nodal element moments
a, b	dimensions of the plate element
t	thickness of the plate element
E	Young's modulus
σ_x, σ_y, σ_z	components of direct stresses
σ_o	yielding stress
σ_1, σ_2, σ_3	principal stresses
ϕ	internal friction angle
[D]	stress strain transformation matrix
A_i	displacement function coefficient
ν	poisson's ratio

INTRODUCTION

There are several types of masonry structural elements within a building, the one, most subjected to earthquake damages is the load bearing wall, to reinforced and readapted it to a new codes, the ferrocement has been used as interconected coating material.

Ferrocement is an orthotropic composite material, having a high strength cement mortar matrix, reinforced whith layers of fine steel wires forming a mesh, which results in a high composite compressive strength (345 to 550 kg/cm^2), which is dependent on the volume ratio of the reinforcement (0.5% to 5%), the mesh type and its orientation. The most significant properties of the composite are, its ductility and the high tensile strength, which ranges from 35 kg/cm to 140 kg/cm [1].

In the figures 1, 2, 3, and 4, the mechanical properties of the ferrocement used in the example, and the coating of a load bearing wall are presented.

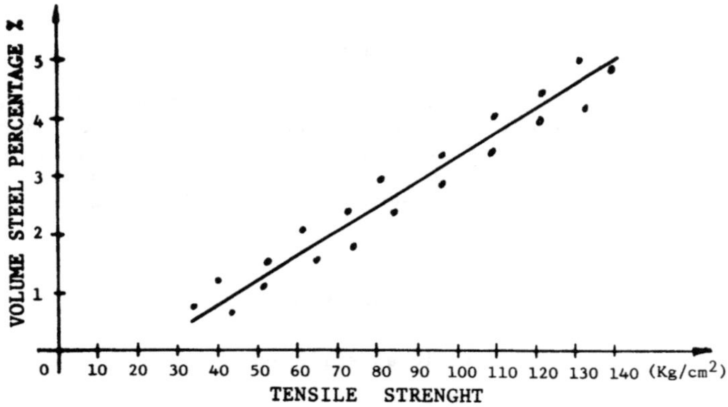

Figure 1.- Ferrocement, tensile strenght against volume steel percentage.

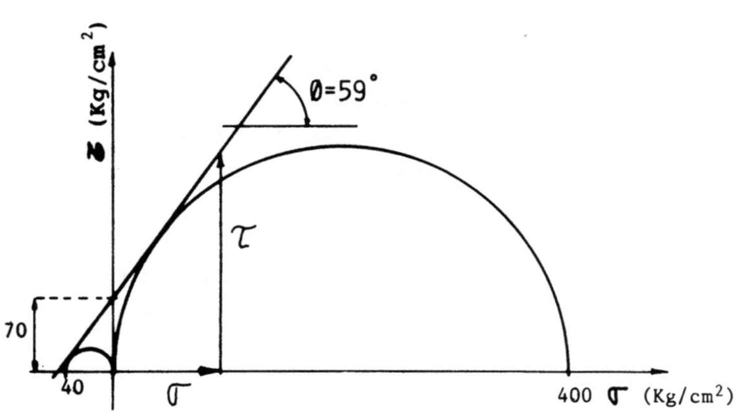

Figure 2.- Maximum shearing strees theory for the ferrocement used in the example, ($\tau = 70 + \sigma \tan 59°$).

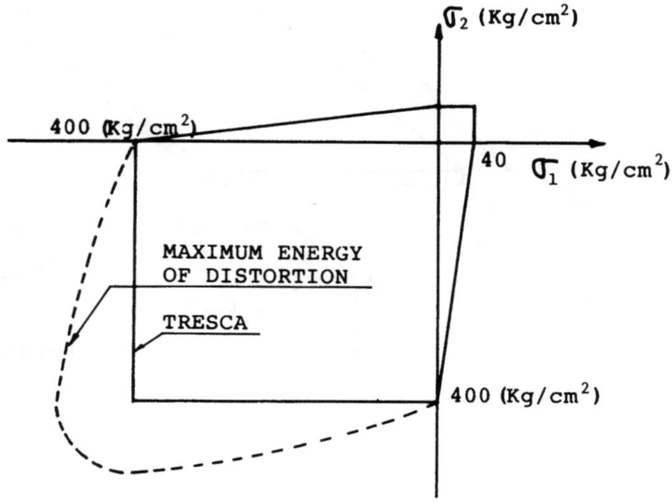

Figure 3.- Maximum energy of distortion theory and Tresca aproximation for the ferrocement used in the example. [2]

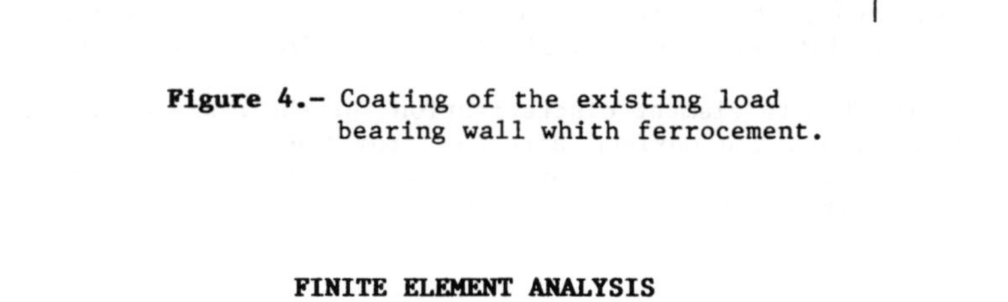

Figure 4.- Coating of the existing load bearing wall whith ferrocement.

FINITE ELEMENT ANALYSIS

In this particular case the Finite Element formulation is the usual well known [3], but to be able to obtain the connection between the plate and the beam elements with six degrees of freedom per nodal point (Fig. 5), it has been necessary to introduce the rotational stiffness $k_{\theta z}$, in the stiffness matrix of the plate element, such that

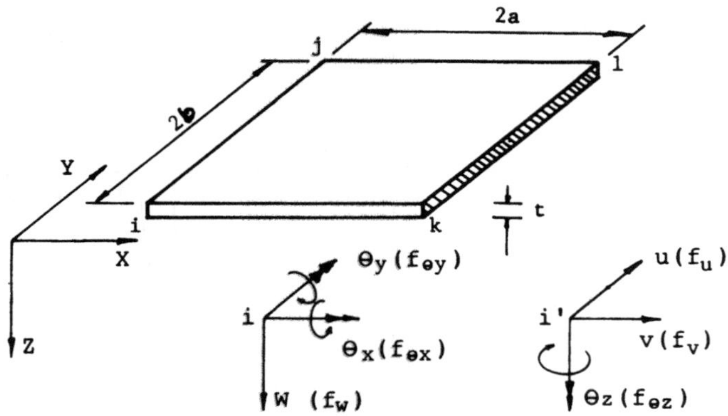

Figure 5.- A rectangular plate element with six degrees of freedom per nodal point.

$$f_{\theta z} = K_{\theta z}\, \theta_z \tag{1}$$

and the superposition of case i and i',will give the solution of this problem. Where we have found that, the rotational stiffness is

$$k_{\theta z} = E\ t\ ((a^4+b^4/ab))/6 \tag{2}$$

in this way, it has been possible to obtain the connection of the beam and plate element with six degrees of freedom per nodal point.

The finite element discretization for the example, is shown in figure 6, and the maximum stresses from the analysis are shown in table 1. For example for the element 351, the safety factors are; from maximum shearing stress theory

$$(70 + \sqrt{\tan 59^\circ})/6.82 = 12.5 \tag{3}$$

and from the maximum energy of distortion theory :

$$\sigma_o / (\sigma_1^2 - \sigma_1 \sigma_2 + \sigma_2^2) = 7.16 \quad (4)$$

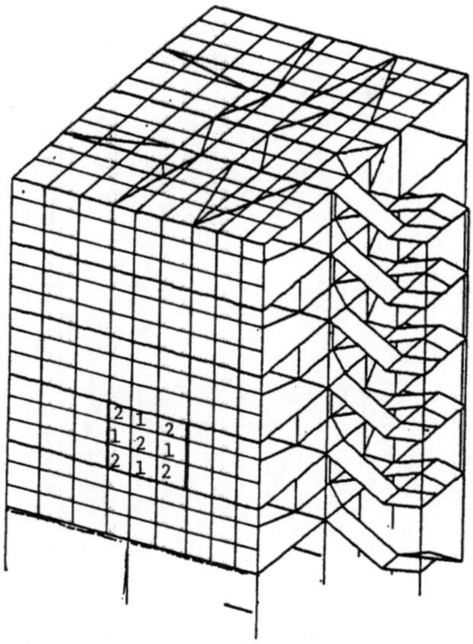

Figure 6.- Finite element discretization of slabs, stairs, beams and bearing walls.

Elem	σ_x	σ_y	τ_{xy}	σ_1	σ_2	τ_{max}
363	-2.62	-14.34	2.67	-2.61	-14.35	5.87
345	-2.49	-11.68	2.30	-1.95	-12.24	5.14
369	-2.53	-14.28	1.97	-2.53	-14.28	5.88
351	-3.53	-14.79	3.85	-2.34	-15.98	6.82

Table 1.- Maximum stresses from analysis in kg/cm^2

Pictures of the load bearing walls showing the reinforcement, before and during the placing of the cement mortar are shown in figures 7, 8, and 9.

CONCLUSIONS

The complete stiffness matrix of the beam element is 12 x 12, and in the square plate element is 24 x 24. This are the most used elements in building structures, its proper assembly is very important to obtain equilibrium and continuity, in each nodal point. This has been the reason to introduce the rotational stiffness $K_{\theta z}$. Most of the commercial programs do not explain this kind of six-degrees-of-freedom assembly per nodal point. The load bearing walls building showed, in the analysis were considered elastically linear, and in its design, its orthotropic material properties were taken in consideration.

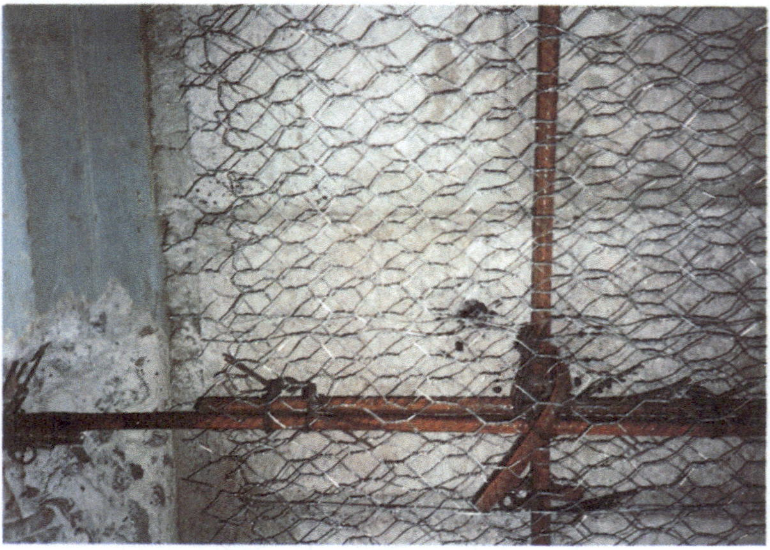

Figure 7.- Ferrocement reinforcement in bearing walls before placing the cement mortar.

Figure 8.- Ferrocement in Foundation Bearing Walls before placing the cement mortar.

Figure 9.- Ferrocement in a Bearing Wall during the placing of the cement mortar

REFERENCES

1.- Reinhorn, A. M., Prawel,S.P. and Zi-He Jia "Experimental Study of Ferrocement as a Seismic Retrofit Material for Mansonry Walls", Journal of Ferrocement , vol. 15, No. 3, July 1985.

2.- Ballesteros, P. and Alamo, R., "Ecuaciones Constitutivas del Concreto, División de Estudios de Posgrado, Facultad de Ingeniería U.N.A.M., 1986, México, D.F.

3.- Zienkiewicz, O.C. and Taylor, R.L.,"The Finite Element Method", Fourth Ed., McGraw Hill, 1989.

4.- MacLeod, I.A., " Panel Elements for Shear Wall Analysis", Research Report, Department of Civil Engineering, Glasgow University, May 1966.

APPENDIX

DERIVATION OF THE SUPERPOSITION STIFFNESS MATRIX. [4]

To achieve this, two types of elements are required as shown on figure 10(a) and 10(b). It can be seen, if the elements type numbers are allocated, so that no pair of elements with a common boundary has the same type number, then there will be a unique rotation at each node, as in figure 10(c).

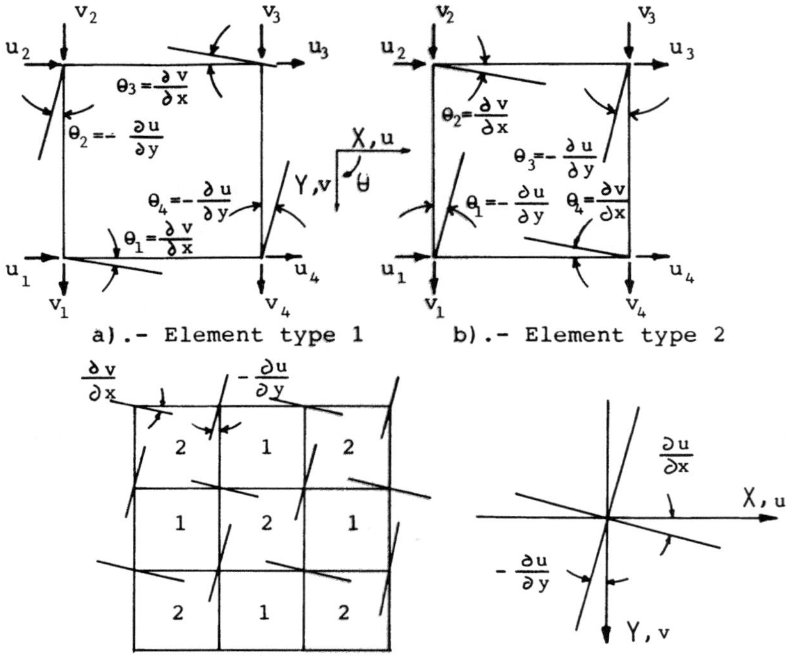

Figure 10.- Second order element.

The displacement functions are

$$u = A_4 + A_7x + A_6y + A_{12}xy + A_1y^2 + A_{10}xy^2 \tag{5}$$

$$v = A_5 + A_8x + A_2y + A_3xy + A_9x^2 + A_{11}x^2y \tag{6}$$

The coefficients A_i are numbered in the order given so as to easy the process of the inversion of the [A] matrices.

The strain $\{\varepsilon\}$ are related to the coefficients $[\alpha]$ by the relationship

$$\begin{bmatrix} \frac{\partial u}{\partial x} \\ \frac{\partial v}{\partial y} \\ \frac{\partial u}{\partial y}+\frac{\partial v}{\partial x} \end{bmatrix} = \begin{bmatrix} \varepsilon_x \\ \varepsilon_y \\ \gamma_{xy} \end{bmatrix} = \begin{bmatrix} & & & & 1 & & & Y^2 & & Y \\ & 1 & X & & & & & & X^2 & \\ 2Y & & Y & 1 & & 1 & 2X & 2XY & 2XY & X \end{bmatrix} [\alpha] \quad (7)$$

i.e., $[\varepsilon] = [B][\alpha]$ in which $[\alpha]^T = [A_1, A_2, A_3, \ldots\ldots A_{12}]$.

The stresses are related to the strains by

$$\begin{bmatrix} \sigma_x \\ \sigma_y \\ \tau_{xy} \end{bmatrix} = \frac{E}{1-\nu^2} \begin{bmatrix} 1 & \nu & 0 \\ \nu & 1 & 0 \\ 0 & 0 & (1+\nu)/2 \end{bmatrix} \begin{bmatrix} \varepsilon_x \\ \varepsilon_y \\ \gamma_{xy} \end{bmatrix} \quad (8)$$

i.e. $\{\sigma\} = [D]\{\varepsilon\}$

The nodal displacements are related to the coefficients by $[\delta] = [A][\alpha]$ in which [A] is a square matrix (12x12) in terms of the dimensions of the element (a,b). Inverting [A] gives $[\alpha] = [A^{-1}]\{\delta\}$

By the principle of virtual work or the Rayleigh-Ritz procedure, the stiffness matrix [K] defined by $\{P\} = [K]\{\delta\}$ is given by

$$[K] = [A^{-1}]^T [K][A^{-1}] \quad (9)$$

where

$$[K] = \int_{VOL} [B]^T [D][B] \, dxdydz \quad (10)$$

The vector $\{P\}$ is the vector of nodal actions (forces and moments) which correspond to $\{\delta\}$. The $[\ A^{-1}\ ]$ matrices are

$$
\begin{bmatrix}
1/b^2 & & & -1/b^2 & & 1/b & & & & & & \\
& 1/b & & & -1/b & & & & & & & \\
& & 1/b & & 2/ab & & & -2/ab & 1/b & & & \\
& & & 1 & & & & & & & & \\
& & & & 1 & & & & & & & \\
& & & & & -1 & & & & & & \\
& & & -1/a & & & 1/a & & & & & \\
& & & & -2/a & & & 2/a & -1 & & & \\
& & & & 1/a^2 & & & -1/a^2 & 1/a & & & \\
-1/ab^2 & & & 1/ab^2 & & & & & & -1/ab^2 & & -1/ab \\
& -1/ba^2 & -1/ab & & -1/ba^2 & & & & & & 1/ab^2 & \\
& & & & & 1/a & -2/ab & & & 2/ab & & 1/a
\end{bmatrix}
$$

Matrix $[A^{-1}]$ for Element type 1 (10)

$$
\begin{bmatrix}
-1/b^2 & & -1/b & 1/b^2 & & & & & & & & \\
& 1/b & & & -1/b & & & & & & & \\
& -2/ab & & & & -1/b & & & & & 2/ab & -1/b \\
& & & 1 & & & & & & & & \\
& & & & 1 & & & & & & & \\
2/b & & 1 & -2/b & & & & & & & & \\
& & & -1/a & & & 1/a & & & & & \\
& & & & & 1 & & & & & & \\
& & & & -1/a^2 & -1/a^2 & & 1/a^2 & & & & \\
1/ab^2 & & 1/ab & -1/ab^2 & & & -1/ab^2 & & 1/ab & 1/ab^2 & & \\
& 1/ba^2 & & & 1/ba^2 & 1/ab & & 1/ba^2 & & & -1/ba^2 & 1/ab \\
-2/ab & & -1/a & 2/ab & & & & & -1/a & & &
\end{bmatrix}
$$

Matrix $[A^{-1}]$ for Element type 2 (11)

FINITE ELEMENTS IN THE 90's, E. Oñate, J. Periaux, A. Samuelsson (Eds.)
Springer-Verlag/CIMNE, Barcelona 1991

ARE THERE ANY NEW ELEMENTS?

I. M. Smith
Engineering Department,
University of Manchester,
Manchester M13 9PL,
UNITED KINGDOM

Summary Shape functions for commonly used plane and three-dimensional finite elements were established about 20 years ago. Over this time, substantial advances have been made in the design of computer algebra packages so that new elements can in principle be formulated quite quickly. In this paper some new elements are discussed. Attention is also given to the question of how good a particular element is.

1. INTRODUCTION

If one compares the third and fourth editions of the text by Zienkiewicz [1], [2], the most "static" area between the two is possibly that of element shape functions. (The most "dynamic" relates to computer techniques!). One could conclude that all is known with respect to useful element shape function families and that Lagrangian and what has become known as "Serendipity" interpolation will suffice for all time.

For quadrilateral and brick-based elements, however, element shape function derivation has never been particularly logical as the "serendipity" label implies. Referring to the "Pascal pyramid" in Figure 1, Lagrangian elements such as the 27-node brick neglect many lower order terms at the expense of higher order ones while the "logical" serendipity brick is taken to have 20 nodes (eight corner and twelve mid-edge nodes). One can legitimately ask, why?

Irons [3], pp 384-385, speculated on the usefulness of a 14-node brick (eight corner and six mid-face nodes) without ever deriving appropriate shape functions. This element will be used to explore the power of computer algebra in determining shape functions for arbitrary element families.

2. COMPUTER ALGEBRA

As far as the writer is aware, Gupta [4] first applied computer algebra to finite element analysis. He used a package called MACSYMA which has since been supplemented by many more - REDUCE, MATHEMATICA, MAPLE, SMP etc. Gupta illustrated one powerful feature of these packages in that the error-prone process of shape function differentiation, including output in FORTRAN, could be completely automated. Kidger [5] demonstrated an even more powerful facet of the computer algebra system REDUCE when he used it to generate shape functions for 14-node brick elements. Figure 2 shows a typical program. All that has to be supplied to the program is the (symbolic) form of the generic shape function polynomial and the selected nodal coordinate positions. The output consists of a directly includable FORTRAN subroutine which can be readily incorporated into a NAG-Type library [6].

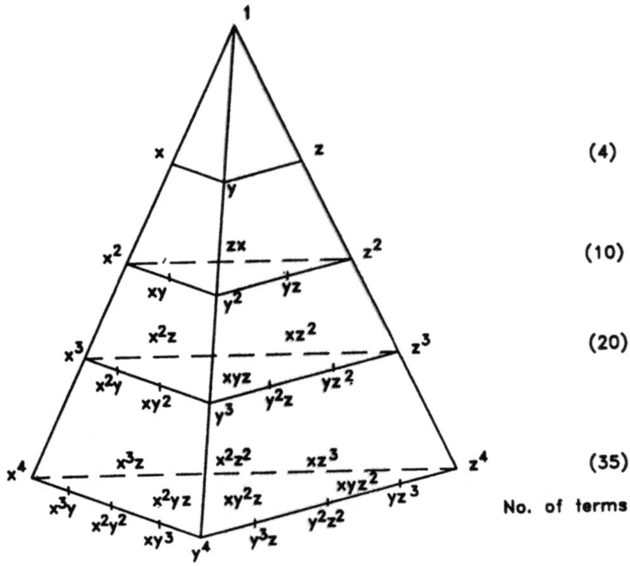

Figure 1 The Pascal Pyramid

Even though the resulting shape functions may not be factorisable, many alternative elements can be investigated cheaply and easily. For example one tactic in generating 14-node bricks is to treat as "obvious" candidates for inclusion in shape functions the terms $1,x,y,z,x^2,y^2,z^2,xy,yz,zx,xyz$ in Figure 1 (ie terms down to the second "plane" plus the central term in the third "plane"). This leaves three extra terms to be chosen.

Kidger [5] originally tried many choices, of which those we may call TYPE 1 to TYPE 4 are listed below:

TYPE	Extra Terms		
1	x^2y	y^2z	z^2x
2	xy^2	yz^2	zx^2
3	x^3	y^3	z^3
4	x^2yz	xy^2z	xyz^2

Completely inadmissible choices (like TYPE 3 : x^3, y^3, z^3) are detected at an early stage since the coefficient matrix is rank deficient. TYPE 4 is viable, as long as the local coordinates are not centred at the origin, and has the advantage that it does not exhibit directionality. TYPEs 1 and 2 do have such a directionality but were rapidly tested in real analyses and found to give better results than TYPE 4 [5].

```
      %%%%%%%%%%%%%%%%%%%%%%%%%%%%%%%%%%%%%%%%%%%%%%%%%%%%%%%%%%%%%%%%
      %          SHAPE FUNCTIONS FOR 14 NODE BRICK                    %
      %%%%%%%%%%%%%%%%%%%%%%%%%%%%%%%%%%%%%%%%%%%%%%%%%%%%%%%%%%%%%%%%
for all X,Y,Z let GSF(X,Y,Z)=                          %
A0+A1*X       +A2  *Y       +A3    *Z                  %    shape
+A4  *X*Y     +A5  *Y*Z     +A6    *Z*X                %    function
+A7  *X*Y*Z                                            %    polynomial
+A8  *X*X     +A9    *Y*Y   +A10   *Z*Z                %    terms
+A11 *X*X*Y   +A12   *Y*Y*Z +A13 *X*Z*Z;
clear SF,L;array SF(14),L(14);                         % set up arrays
for I:=1:14 do begin;                                  % loop nodes
for J:=1:14 do L(J):=0;L(I):=1;                        % set RHS
write AA:=(L( 1)=GSF(-1,-1,-1),L( 2)=GSF(-1, 1,-1),
     L( 3)=GSF( 1,-1,-1),L( 4)=GSF( 1, 1,-1),
     L( 5)=GSF(-1,-1, 1),L( 6)=GSF(-1, 1, 1),          % local
     L( 7)=GSF( 1,-1, 1),L( 8)=GSF( 1, 1, 1),          % nodal
     L( 9)=GSF( 0, 0,-1),L(10)=GSF( 0, 0, 1),          % co-ords
     L(11)=GSF( 0,-1, 0),L(12)=GSF( 0, 1, 0),
     L(13)=GSF(-1, 0, 0),L(14)=GSF( 1, 0, 0));
write LIS:=first solve (AA,                            % solve for
A0,A1,A2,A3,A4,A5,A6,A7,A8,A9,A10,A11,A12,A13);        % coeffs
write SF(I):=GSF(X,Y,Z);
for J:=1:14 do SF(I):=sub(part(LIS,J),SF(I));          % re-form
write "SF(",I,"):=",SF(I);                             % functions
;end;

%%%%   Now write out the  shape functions and derivatives  %%%%
%%%%                as a FORTRAN subroutine                 %%%%
CMS   "FILEDEF OUTF DISK FM14XXY FORTRAN A";ON FORT;
off    echo;off time;out outf;
write "       SUBROUTINE FM14XXY(DER,IDER,FUN,SAMP,ISAMP,I,J,K)";
write "C                                                   ";
write "C      SHAPE FUNCS AND DERIVS - 14 NODE BRICK       ";
write "C            (C) D.J.Kidger       MAY 1989          ";
write "C                                                   ";
write "       REAL DER(IDER,*),FUN(*),SAMP(ISAMP,*)        ";
write "       X=SAMP(I,1)                                  ";
write "       Y=SAMP(J,1)                                  ";
write "       Z=SAMP(K,1)                                  ";
for J:=1:14 do write "       FUN("  ,J,")=",   SF(J)    ;  %
for J:=1:14 do write "       DER(1,",J,")=",DF(SF(J),X) ;  % auto
for J:=1:14 do write "       DER(2,",J,")=",DF(SF(J),Y) ;  % diffn.
for J:=1:14 do write "       DER(3,",J,")=",DF(SF(J),Z) ;  %
write "       RETURN" ; write "      END"               ;
shut outf;end;
```

Figure 2 Brick subroutine produced by REDUCE

The next step [6] was to remove the directional bias in TYPEs 1 and 2 by averaging them to create an unbiassed element we may label TYPE 5. This element has many attractive features. Its eigenmodes, for full 27-point Gaussian quadrature, are illustrated in Figure 3. However it emerges that 8-point (2x2x2 Gauss) quadrature only very slightly under-integrates the element, leading to no spurious zero energy modes. This is important when one remembers that 27-point Gaussian quadrature is essential for 20-node bricks, unless special rules, which have rather unpredictable consequences, are used. The minimum number of integrating points for viable 20-node bricks is probably 14, making the 14-node brick almost twice as cheap to integrate (although in the limit there would be little saving in the total number of degrees of freedom). As pointed out by Irons [3] a 14-node mesh will also tend to have a frontwidth about 2/3 of its 20-node counterpart.

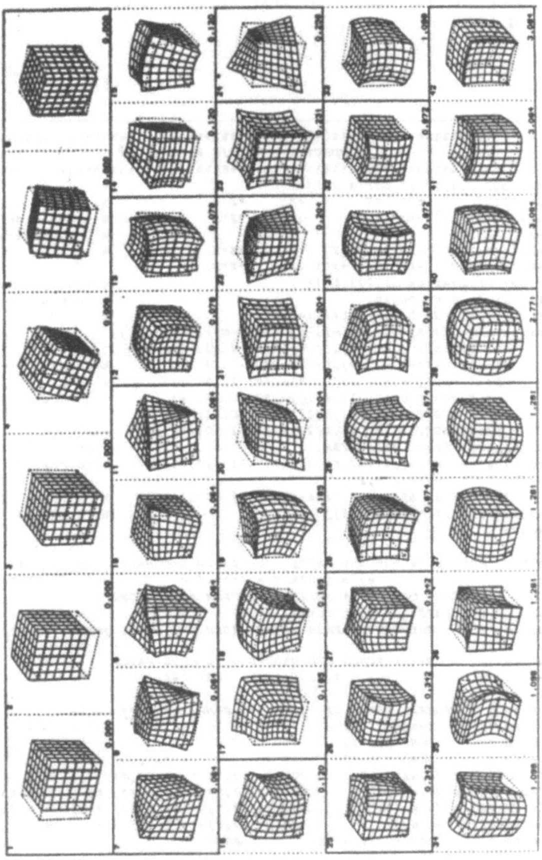

Figure 3 Eigenmodes of 14-node brick

3. PATCH TESTS

Irons [3] warned that too much should not be expected of 14-node elements, speculating that "it apparently cannot satisfy a quadratic patch test". It turns out that the TYPE 5 element described above does not pass the linear patch test because movement of a mid-face node causes small movements of adjacent faces. This can be rectified [7] and some quadratic tests passed, but it cannot be concluded that an entirely robust 14-node element has yet been devised. However there is little doubt that the element is "better" than 8-node or 20-node bricks for elastoplastic analysis of bulky solids (as distinct from structural members).

4. EXAMPLES OF ANALYSES USING 14-NODE BRICKS

A simple test of 14-node elements in elastoplastic problems is the "Prandtl" punch indentation. In plane strain, for a material with shear strength C_u, Prandtl gives an exact solution from limit plasticity of an ultimate punch pressure of 2+π (5.14) times C_u. Using a crude mesh of 8x4 14-node elements, and applying incremental displacements to the punch, a level plateau of 5.2 C_u in the load is achieved. This is very comparable to the results obtained for plane 8-node elements with reduced integration [6]. When 3-dimensional effects are included by making the punch square, for a 4x4x4 mesh the collapse load is found to be 5.88 C_u, that is 13% higher than the plane strain result. This accords with empirical evidence in soil mechanics for non-plane foundations on clays.

Extending this type of analysis to a more complicated geometry, Figure 4 shows a mesh modelling a "spud-can" which forms the foundation of each leg of a temporary "jack-up" structure used in offshore engineering. When a leg with a spud-can base penetrates into soft seabed soil near a production platform, it is of interest to study the effect of the spud-can penetration on the piled legs of the permanent production facility nearby. Figure 5 shows the results of such an analysis, leading to a computation of the bending moments and axial forces induced in the platform piles for various separations of the spud-can.

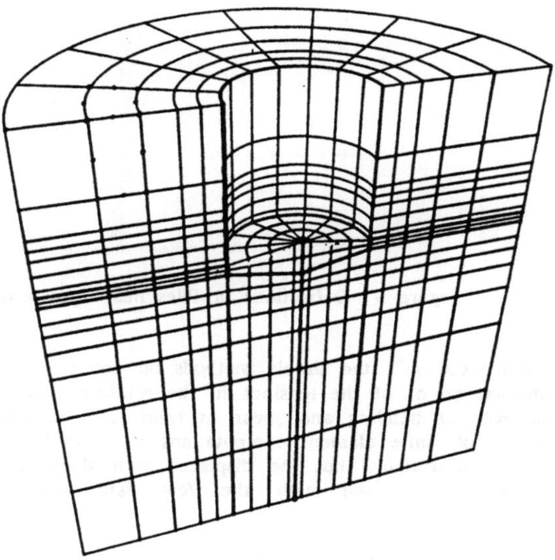

Figure 4 Spud Can Mesh

A final example is taken from the area of "reinforced soil". Figure 6 shows what is called a "nailed" soil wall. Soil is excavated and the exposed face stabilised with shotcrete and by drilling in or explosively inserting steel rods or "nails". A rather complicated interaction is set up between the nails and the soil which is experiencing stress relief. The nails have to be far enough apart (from economic considerations) for there to be plastification of the soil between the nails. Thus in reinforced concrete terminology, the soil tends to be "under-reinforced" compared with some built-up reinforced soil constructions which use relatively large and stiff plate inclusions and can be thought of as "over-reinforced". In these constructions, plastification is essentially restricted to the unreinforced soil, and methods of analysis based on sliding wedges at failure give reasonable results.

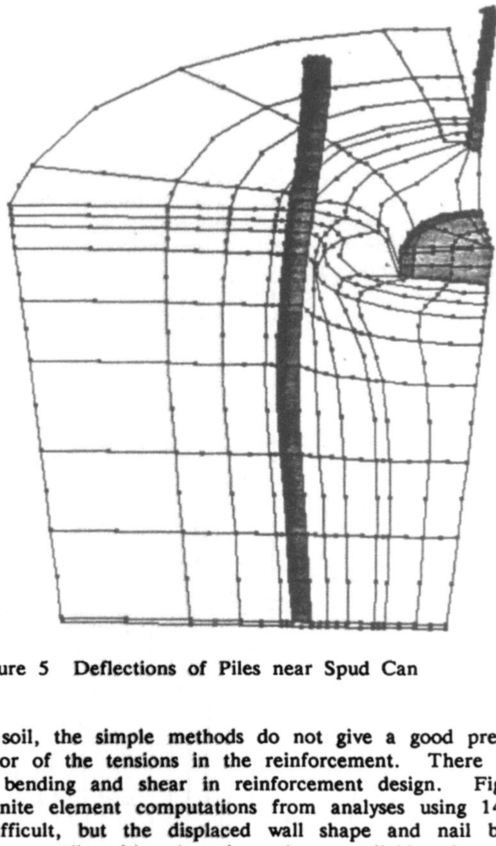

Figure 5 Deflections of Piles near Spud Can

For under-reinforced soil, the simple methods do not give a good prediction of the failure mechanism or of the tensions in the reinforcement. There is also a debate on the role of bending and shear in reinforcement design. Figure 7 shows typical results of finite element computations from analyses using 14-node bricks. Verification is difficult, but the displaced wall shape and nail bending moments at failure agree well with the few data available from field measurements.

5. OTHER CANDIDATE NEW ELEMENTS

Given the power of computer algebra, and the ability to compute and display element eigenmodes, new elements can be investigated (and discarded) very readily. Arising from the previously-described work on 14-node elements, a promising 3-dimensional element (much more efficient at populating space with nodes than the 32-node brick) contains 26 nodes as shown in Figure 8. The 26 terms in the shape function polynomials comprise every term in the Pascal pyramid of degree less than 3 in x,y and z. Thus 3 terms are excluded from the "third" plane in Figure 1, 9 are excluded from the "fourth" plane and the extra 3 terms from the "fifth" plane are x^2yz^2, x^2y^2z and xy^2z^2 (excluding 18 terms).

6. HOW "GOOD" IS A PARTICULAR ELEMENT?

In the foregoing, the eigenvalues of a finite element's stiffness matrix have

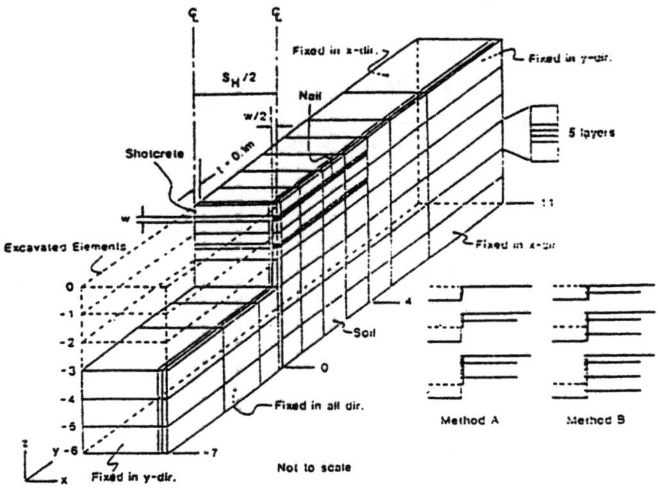

Figure 6 FE Idealisation of Nailed Soil Wall

been computed as a means of assessing element viability. Indeed, for simple elements, these eigenvalues have been determined explicitly [8],[9]. If the eigenvalues of, say, a quadrilateral piece of elastic solid deforming with linear edges could be evaluated analytically, a means would have been found of comparing all compatible 4-node finite elements.

As a first step in this process, the eigenvalues of 4-node elastic square elements whose shape functions have been augmented by bubble functions have been computed.

The only "problem" eigenmodes are the two hourglass modes which for full integration of the basic bilinear element have eigenvalues 7/9 G + 1/3 K. Thus "shear locking" occurs for this element as ν tends to 0.5 due to the presence of K. The effects of reduced and of selective-reduced integration are shown in Figure 9. As is well known, fully reduced integration results in the hourglass eigenvalues becoming zero, with consequent danger of the propagation of spurious deformation modes. Selective reduced integration, on the other hand, approximates the hourglass eigenvalues as the constant 2/3 G (see Figure 9). What are the true values?

These have been estimated [10] by adding bubble functions of progressively higher order to the basis shape functions. The added bubble functions of zero order (N = 0) take the form $(1-x^2)(1-y^2)(A_0)$ and for order 1 (N = 1) take the form $(1-x^2)(1-y^2)(A_0+ A_1x+A_2y)$ and so on. The scalars A_i are determined by minimising the element's strain energy with respect to them (they are effectively internal variables).

Again, computer algebra has been used to determine the eigenvalues explicitly for these assumptions. The only affected eigenvalues are associated with the hourglass modes and for N = 0 and N = 1, the relevant eigenvalues are given by

$$\lambda_4 = \lambda_5 = G \; \frac{\frac{2023}{54} + \frac{269}{6} R + \frac{27}{2} R^2 + \frac{1}{2} R^3}{49 + 42R + 9R^2} \qquad (1)$$

in which

$$R = \frac{K}{G} = \frac{2(1+\nu)}{3(1-2\nu)} \qquad (2)$$

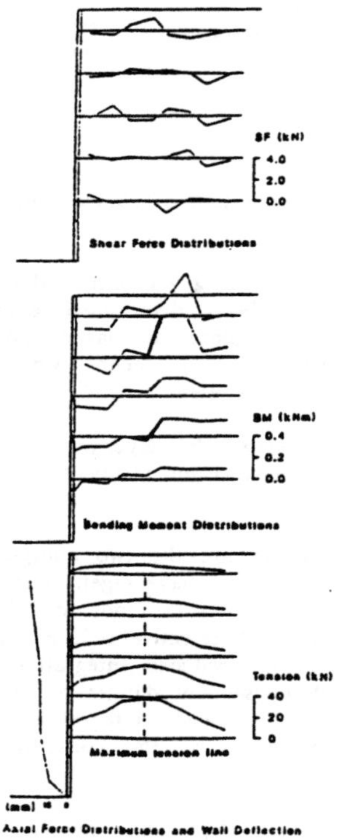

Figure 7 Wall Deflections; Shear, Tension and Bending in the Nails

The resulting eigenvalues for N = 0, 1 and 2 are plotted in Figure 9 as functions of Poisson's ratio ν. One may speculate that the true hourglass modes are not associated with unbounded values at ν = 0.5 but their precise magnitude awaits analytical evaluation.

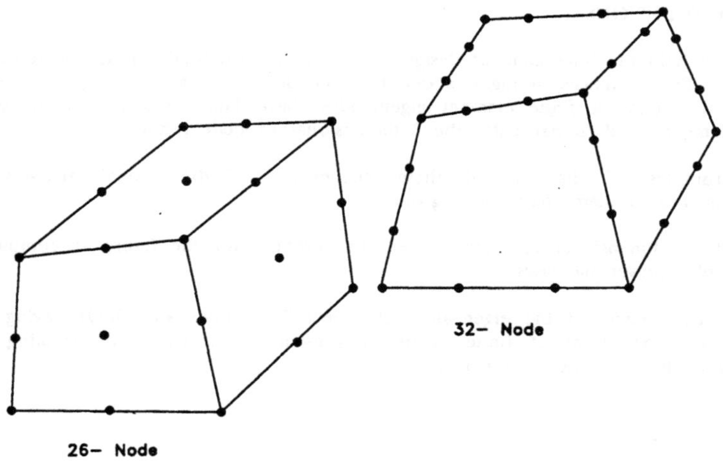

Figure 8 The 26-node and 32-node brick elements

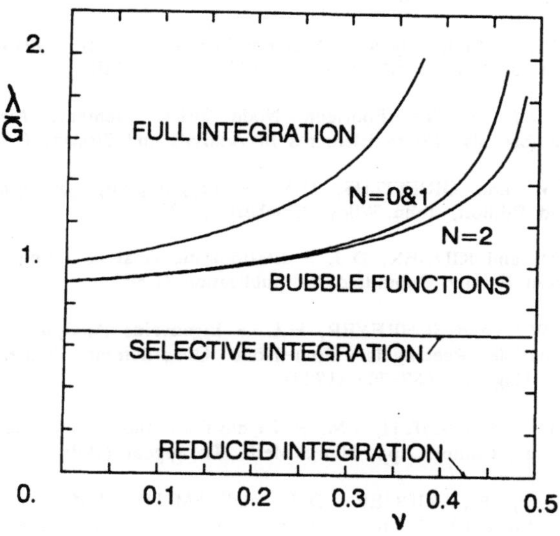

Figure 9 Normalised Eigenvalues for 'Hourglass' Modes of the Plane 4-node Element

7. CONCLUSIONS

The field of finite element design has remained essentially static for some 20 years. The advances brought about by computer algebra, and by the ready graphical display of finite element eigenmodes, have been used to explore some new elements, and in particular the 3-dimensional 14-node family.

Examples of the use of these elements in 3-dimensional analyses of geomechanics problems have been given.

The 3-dimensional 26-node element has promise for the future as computing and display power increases.

Analytic study of the eigenvalues of "pieces" of solids and fluids having the boundary deformations of finite elements, presents a means of determining the relative quality of various elements.

REFERENCES

1. ZIENKIEWICZ, O.C. - The Finite Element Method, 3rd Edition, McGraw-Hill, London, 1977.

2. ZIENKIEWICZ, O.C. and TAYLOR, R.L. - The Finite Element Method, 4th Edition, McGraw-Hill, London, 1989.

3. IRONS, B.M. and AHMAD, S. - Techniques of Finite Elements, Ellis Horwood, Chichester, 1980.

4. GUPTA, K.K. - Finite Dynamic Element Formulation for a Plane Triangular Element, Int. J. Num. Meth. Eng, 14, 1431-1448 (1979).

5. KIDGER, D.J. - The Fourteen Node Brick Element, Melosh Medal Competition (see also Finite Elements in Analysis and Design) (1990).

6. SMITH, I.M. and GRIFFITHS, D.V. - Programming the Finite Element Method, 2nd Edition, John Wiley, Chichester, 1988.

7. SMITH, I.M. and KIDGER, D.J. - Elastoplastic Analysis Using the 14-Node Brick Element Family, Submitted for publication (1991).

8. HACKER, W.L. and SCHREYER, H.L. - Eigenvalue Analysis of Compatible and Incompatible Rectangular Four-Node Quadrilateral Elements, Int. J. Num. Meth. Eng, 28, 687-704 (1989).

9. KIDGER, D.J. and SMITH, I.M. - Eigenvalues and Eigenmodes of 8-node Brick Elements, Comm. App. Num. Meth. To appear (1991).

10. MOLENKAMP, F., KIDGER, D.J. and SMITH, I.M. - Accuracy of Four-Node Standard Finite Element, Int. J. Num. Analytical Meth. Geomech, To appear, (1991).

FINITE ELEMENTS IN THE 90's, E. Oñate, J. Periaux, A. Samuelsson (Eds.)
Springer-Verlag/CIMNE, Barcelona 1991

GENERALIZED PLASTICITY MODELLING OF SATURATED SAND BEHAVIOUR UNDER EARTHQUAKE LOADING

M. Pastor
Sector de Ingeniería Computacional - CETA,
Centro de Estudios y Experimentación de Obras Públicas,
Alfonso XII 3, 28014 Madrid,
SPAIN

ABSTRACT

Modelling of soil behaviour has progressed very much during the past years. This paper addresses the problem of reproducing the response of saturated sands under dynamic loading such as caused by earthquakes,within the framework of Generalized Plasticity Theory, a summary of which is provided in the first part of the paper.

Next, a simple model for saturated sand is presented, where the unloading process is considered as a new loading with the advantage of a better modelling of cyclic mobilization phenomena.

Anisotropy of sand may result sometimes in an overestimated dynamic strength. Last part of the paper is devoted to show how isotropic models formulated in terms of invariants may be easily generalized to account for material structure.

1. INTRODUCTION

One of the mayor difficulties encountered in Computational Geotechnique is that of material nonlinearity. Irreversible plastic strain may develop at levels as low as 10^{-4}. If the loading process is continued the soil becomes softer ,reaching finally a residual state at which the stress does not varies anymore. These conditions take place on a surface which reduces to a come when represented in the space (σ'_1, σ'_2, σ'_3) of principal effective stresses. If the load is progressively increased and the material is of the softening type ,failure can be catastrophic after the deviatoric stress reaches a peak.

The situation becomes more complex under dynamic loading, as most of geomaterials densify. If the loading process is fast enough, this tendency results in a pore pressure build-up which may cause the stress path to approach the failure conditions, leading to catastrophic failure phenomena such as caused by very loose sand liquefaction.

To predict accurately the behaviour of soil structures and foundations, it is necessary to develop a mathematical model able to describe the interaction between soil skeleton and pore fluid in addition to a constitutive equation. To solve the resulting set of strongly coupled, highly non-linear equations, special discretization techniques should be used.

During the past decade, Prof. O.C.Zienkiewicz has contributed in a decisive manner to solve the three aspects of the problem [1-10]. This paper presents a review of a class of constitutive models for geomaterials which was the result of a long and fruitful collaboration.

GENERALIZED PLASTICITY

The increment of strain produced by a increment of effective stress $d\underset{\sim}{\sigma}$ may be written as

$$d\underset{\sim}{\varepsilon} = \underset{\sim}{C} : d\underset{\sim}{\sigma}' \tag{1}$$

where $\underset{\sim}{C}$ is a fourth-order constitutive tensor. Sometimes it can be assumed that deformation is caused by M independent mechanisms subjected to the same external action

$$d\underset{\sim}{\varepsilon} = \sum d\underset{\sim}{\varepsilon}^{(m)} \tag{2}$$

$$d\underset{\sim}{\varepsilon}^{(m)} = \underset{\sim}{C}^{(m)} : d\underset{\sim}{\sigma}'$$

Mechanisms may be identified as different physical causes of deformation such as crushing and rearrangement of sand grains. This is the case of classical plasticity models which introduce two or more field surfaces [11].

Multilaminate models [12,13] assume that soil deformation is caused by the sliding and dilation which takes place along all possible planes, and present the advantage of their ability to model anisotropy and rotation of principal axes phenomena. To reduce the computational effort, only a reduced number of representative planes is selected to perform a sort of "numerical integration".

Another possibility is to group together the responses corresponding to the three coordinate planes XY, YZ and XZ. [14] Mobilized plane models [15] can be also interpreted as multimechanisms.

The constitutive tensor $\underset{\sim}{C}$ depends on the state of stress and strain, and also on the past material history and the direction of the increment of stress $d\underset{\sim}{\sigma}$.

The last requirement is of particular importance for cyclic and dynamic problems in which material densifies and plastic shear strain accumulates.

If the predicted behaviour were not dependent on the direction of $d\underset{\sim}{\sigma}$, any infinitesimal cycle $\{d\underset{\sim}{\sigma}', -d\underset{\sim}{\sigma}'\}$ would result in zero accumulated strain.

To account for this dependence, . a unit norm tensor $\underset{\sim}{n}^{(m)}$ is postulated for all mechanisms to exist in the stress space. It can be taken as the normal to some defined surface or defined directly without reference to any given surface.

All possible stress increments $d\underset{\sim}{\sigma}'$ may be classified as "loading", "unloading" or "neutral loading" in the m-th mechanism according to:

$$\begin{aligned} d\underset{\sim}{\sigma}' : \underset{\sim}{n}^{(m)} &> 0 \quad \text{loading} \\ d\underset{\sim}{\sigma}' : \underset{\sim}{n}^{(m)} &= 0 \quad \text{neutral loading} \\ d\underset{\sim}{\sigma}' : \underset{\sim}{n}^{(m)} &< 0 \quad \text{unloading} \end{aligned} \tag{3}$$

and different constitutive tensors are assumed for loading and unloading situations

$$\begin{aligned} d\underset{\sim}{\varepsilon} &= \underset{\sim L}{C}^{(m)} : d\sigma' \\ d\underset{\sim}{\varepsilon} &= C_u^{(m)} : d\underset{\sim}{\sigma}' \end{aligned} \tag{4}$$

Now, if the condition of material response continuity is imposed, loading and unloading constitutive tensors have to be of the form

$$\begin{aligned} \underset{\sim L}{C}^{(m)} &= \underset{\sim e}{C}^{(m)} + \frac{1}{H_L} n_{gL}^{(m)} \otimes \underset{\sim}{n}^{(m)} \\ \underset{\sim u}{C}^{(m)} &= \underset{\sim e}{C}^{(m)} + \frac{1}{H_u} n_{\sim gu}^{(m)} \otimes \underset{\sim}{n}^{(m)} \end{aligned} \tag{5}$$

where $n_{\sim gL/U}^{(m)}$ are two unit tensors, $H_{L/U}^{(m)}$ two scalar functions which will be referred to as loading and unloading plastic moduli and $C_{\sim e}^{(m)}$ a constitutive tensor which characterizes the material behaviour under neutral loading.

As the material behaviour under neutral loading is reversible, $C_{\sim e}^{(m)}$ may be related to the elastic component of strain. It is therefore possible to decompose the strain increment into elastic and plastic components given by

$$d\underset{\sim}{\varepsilon}^{(m)} = d\varepsilon_{\sim e}^{(m)} + d\varepsilon_{\sim p}^{(m)} \tag{6}$$

with

$$\begin{aligned} d\varepsilon_{\sim e}^{(m)} &= C_{\sim e}^{(m)} : d\underset{\sim}{\sigma}' \\ d\varepsilon_{\sim p}^{(m)} &= \frac{1}{H_{L/U}^{(m)}} n_{gL/U}^{(m)} \otimes \underset{\sim}{n}^{(m)} \end{aligned} \tag{7}$$

It can be seen that the material response dependence on the direction of $d\underset{\sim}{\sigma}$ is multi-linear with 2^M tensorial zones. [16].

The material history of past events is introduced through a set of internal variables $\{\underset{\sim}{\phi}\}$, on which directions $n_{\sim gL/U}^{(m)}$ and $\underset{\sim}{n}^{(m)}$, plastic modulus $H_{L/U}^{(m)}$ and tensor $C_{\sim e}$ depend.

For instance, Classical Plasticity introduces two surfaces

$$f(\underset{\sim}{\sigma}', \underset{\sim}{\alpha}) = 0 \quad \text{(yield surface)}$$
$$g(\underset{\sim}{\sigma}', \underset{\sim}{\alpha}) = 0 \quad \text{(plastic potential)} \tag{8}$$

where $\underset{\sim}{\alpha}$ is a set of parameters defining the surface size, position, shape and orientation and which depend on the internal variables $\underset{\sim}{\phi}$. In above, a single mechanism has been considered for simplicity. Directions $\underset{\sim}{n}$ and $\underset{\sim}{n}_g$ can be obtained as the unit normals to f and g. The material behaviour is elastic for unloading and for all the stress points within the yield surface, and the plastic modulus H_L is obtained by applying de so-called consistency condition

$$df = 0$$

$$\frac{\partial f}{\partial \underset{\sim}{\sigma}'} \cdot d\underset{\sim}{\sigma}' + \frac{\partial f}{\partial \underset{\sim}{\alpha}} d\underset{\sim}{\alpha} = 0 \tag{9}$$

Kinematic hardening models allow these surfaces to travel in the stress space, and a family of them is introduced to account for a more precise description of material history [17-19].

Bounding Surface Models [20-21] are a particular case of kinematic hardening surfaces, in which the number of surfaces reduces to two corresponding to an elastic nucleus and a classical yield surface. Direction $\underset{\sim}{n}$ and plastic modulus H are obtained by a radial interpolation rule for stress points within the bounding surface domain.

Isotropic models cast in terms of stress invariants can provide many times an accurate response of soil behaviour. However, if the material structure is highly anisotropic, or if the loading process may cause a strong stress-induced anisotropy, other alternatives have to be considered. [22-27]

ISOTROPIC MODELS FOR GRANULAR SOILS

All tensorial relations presented above admit an equivalent vectorial form in which $\underset{\sim}{\sigma}'$ is expressed as $\{\sigma_{xx},\sigma_{yy},\sigma_{zz},\sigma_{xy},\sigma_{yz},\sigma_{xz}\}^T$ which reduces to $\{\sigma_1,\sigma_2,\sigma_3\}$ in the principal stress axes. If the material is isotropic, its behaviour may be described in terms of stress and strain invariants I_1', J_2 and J_3 given by

$$I'_1 = \mathrm{tr}\,(\underset{\sim}{\sigma}') \tag{10}$$

$$J_2 = \left\{ \frac{1}{2}\,\mathrm{tr}\left(\underset{\sim}{s}^2\right)\right\}^{1/2}$$

$$J_3 = \left\{ \frac{1}{3}\,\mathrm{tr}\left(\underset{\sim}{s}^3\right)\right\}^{1/3}$$

where the deviatoric stress tensor $\underset{\sim}{s}$ is

$$\underset{\sim}{s} = \underset{\sim}{\sigma}' - \frac{1}{3} I'_1\, \underset{\sim}{\delta} \tag{11}$$

and $\underset{\sim}{\delta}$ is the identity tensor.

Alternatively, other sets of invariants may be introduced for convenience. Here we will use {p', q, θ} given by

$$p' = \frac{1}{3} I'_1 \tag{12}$$

$$q = \sqrt{3\,J_2}$$

$$\theta = \frac{1}{3}\sin^{-1}\left\{ -\frac{\sqrt{3}}{2}\,\frac{J'_3}{J'^{\,3/2}_2}\right\}$$

$$|\theta| \le \pi/6$$

Associated strain invariants ε_p, ε_s and ε_θ are defined such that plastic work is conserved.

Working in a reduced space presents the advantage of simplicity as most of experimental results have been obtained using simple stress paths such as the cylindrical triaxial for which θ is kept constant.

To extend the laws obtained in reduced spaces to more general stress conditions, the following relations may be used [28]

$$\underset{\sim}{n} = \frac{\partial \hat{\underset{\sim}{\sigma}}'}{\partial \underset{\sim}{\sigma}'} \tag{13}$$

$$H_{L/U} = \hat{H}_{L/U}$$

where the superscript "^" refers to the reduced stress space.

The material behaviour under monotonic virgin loading may be specified by defining directions $\underset{\sim}{n}_g$ and $\underset{\sim}{n}$ and the plastic modulus H_L.

Direction of plastic flow $\underset{\sim}{n}_g$ can be obtained from experiments in which sand dilatancy d_g is measured along different stress paths. A simple law was proposed in Ref [9],

$$d_g = (1 + \alpha)(M_g - \eta) \tag{14}$$

where η is the stress ratio

$$\eta = q/p' \tag{15}$$

and α and M_g material parameters. The last characterizes the points at which plastic flow takes place at constant volume, and is referred to as Critical State or Characteristic State line when plotted in the (p'.q) space.

It depends on Lode's angle θ, and it may be assumed to vary according to the law

$$M_g(\theta) = \frac{6\ M_{gc}}{6 + f_g\ M_{gc}(1-\sin3\theta)} \tag{16}$$

where f_g controls the ratio between extension and compression values. The surface coincides with Mohr-Coulomb at $\sin\theta = \pm 1$ if f_g is taken as 1.

Direction $\underset{\sim}{n}_g$ is then obtained as

$$\underset{\sim}{n}_g = \underset{\sim}{V}_g / ||V_g||$$

where

$$\underset{\sim}{V}_g = \left(V_{gp},\ V_{gq}\right) \tag{17}$$

and

$$V_{gp} = (1 + \alpha)(M_g - \eta)$$

$$V_{gq} = 1$$

Loading-unloading direction can be assumed to be given by a similar law depending now on M_f. However, simpler relations can be used, and here $\underset{\sim}{n}$ will be taken as

$$\underset{\sim}{n} = \left(n_p\ n_q\right)^T$$

$$n_p = \frac{\eta}{(1+\eta^2)^{1/2}} \tag{18}$$

$$n_q = \frac{1}{(1+\eta^2)^{1/2}}$$

which are normals to the trace of a symmetric cone with the triaxial plane. [29]

To complete the description, H_{VL} is taken as

$$H_{VL} = H_o \, p' \; H_f \left\{ H_v + H_s \right\}$$

where

$$H_f = \left(1 - \eta/\eta_f\right)^4 \tag{19}$$

$$H_v = \left(1 - \eta/M_g\right)$$

and

$$\xi = \int d\xi = \int |d\varepsilon_s| \tag{20}$$

$$\eta_f = \left(1 + \frac{1}{\alpha}\right) M_f$$

$$M_f = \frac{6 \; M_{fc}}{6 + f_f M_{fc} (1 - \sin 3\theta} \tag{21}$$

To account for history of past events, two elements are kept

(i) A surface defining the maximum stress level reached

(ii) The point at which last reversal took place.

Fig. 1 shows the case of a stress path beginning at A, reaching a maximum stress level at B and continuing to a point C.

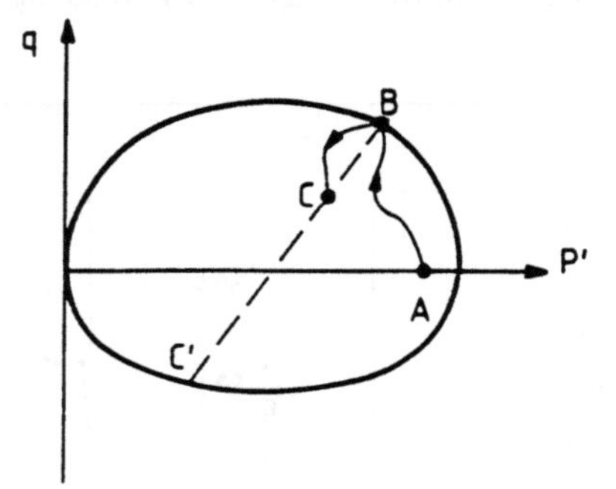

Fig. 1 Unloading and reloading : Proposed Interpolation

Directions $\tilde{n}$ and $\tilde{n}_g$, and the plastic moduli H depend on the relative position of C with respect to B and the image point C', and they are obtained by interpolation.

In particular, $\tilde{n}$ is interpolated from $-\tilde{n}_B$ to $\tilde{n}_{C'}$ using a linear law. Direction of plastic flow is obtained again by defining a suitable dilatancy at C, d_{gC} which is interpolated from

$$d_{gB} = \text{abs}\left\{ (1 + \alpha)M_g \right\} \tag{22}$$

to

$$d_{gC'} = (1 + \alpha)\left(M_g - \eta_{C'}\right)$$

and the plastic modulus is given by

$$H = H_{Bo} + f\left(H_{C'} - H_{Bo}\right)$$

where

$$H_{Bo} = H_o\, p' \left(\frac{\eta_B}{M_g}\right)^{-\gamma_1} \tag{23}$$

$$F = \left| \frac{(\eta - \eta_B)}{(\eta_{C'} - \eta_B)} \right|^{\gamma_3 \exp(-\gamma_2 \xi)}$$

This interpolation law provides a smooth transition between unloading to reloading as it can be seen in figs. 2 and 3. In fact, unloading may be considered as a new loading process between B and C'.

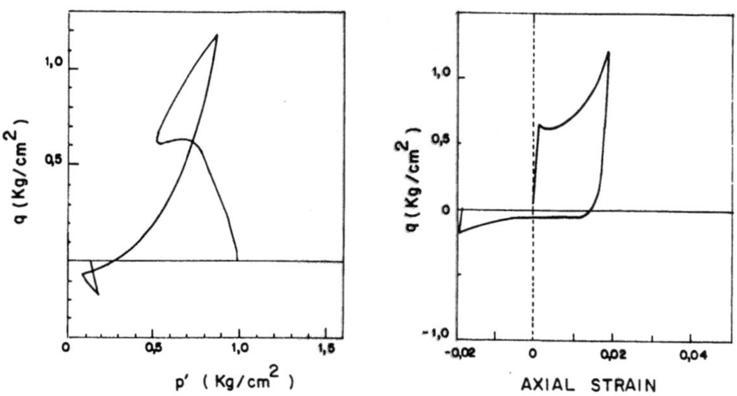

Fig. 2 Unloading Plasticity

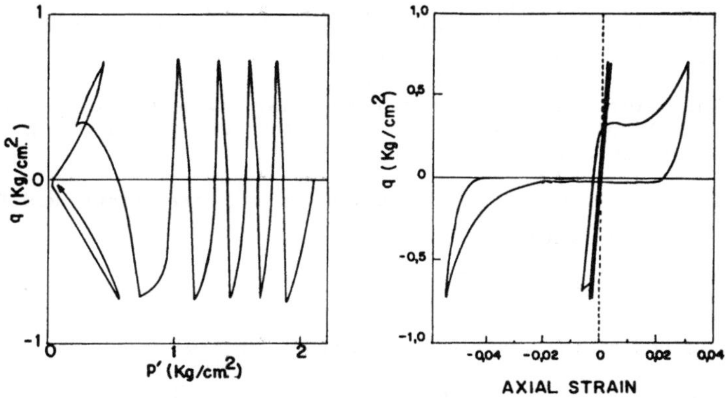

Fig.3 Cyclic mobility of a medium dense sand

ANISOTROPY OF SANDS

If the shape of the grains is not spherical, they can be arranged to form anisotropic structures, such as found in natural deposits of sand where the grains have their major axes on horizontal planes. This initial or fabric anisotropy may be altered when the material is sheared. Here, the material response will depend also on the relative orientation of the stress increment tensor and the structure, which may be approximately described by a second order tensor $\underset{\sim}{A}$.

If we consider for instance a saturated sand layer subjected to an earthquake, the stress path would consist on a horizontal cyclic shear stress superimposed to the initial stresses. Here we can see that the shear stress acts along the bedding planes. If this phenomenon is reproduced in a cylindrical triaxial testing machine, the shear stress will be applied along inclined planes, the material response will be much stiffer, and the dynamic resistence to liquefaction will be overestimated.

To account for the relative orientation of stress and structure, a set of modified invariants may be introduced [24,25,27]

$$\bar{I}'_1 = B^{I}_{ij} \cdot \sigma'_{ij}$$

$$\bar{J}'_2 = \sigma'_{ij} \, B^{II}_{ijkl} \, \sigma'_{kl} \qquad (24)$$

$$\bar{J}'_3 = \sigma'_{ij}\sigma'_{kl}\sigma'_{mn} B^{III}_{ijklmn}$$

where $\underset{\sim}{B}^{I}$, $\underset{\sim}{B}^{II}$ and $\underset{\sim}{B}^{III}$ are generated from tensorial products of the identity and the structure tensors. [27]

This formulation includes mixed invariants as suggested by Baker and Desai [26] if the material structure is defined by the plastic strain tensor.

For instance, if $\underset{\sim}{B}^{I}$ is taken as $\underset{\sim}{\varepsilon}^{P}$, the first modified invariant reduces now to

$$\bar{I}'_1 = \underset{\sim}{\varepsilon}^{P} : \underset{\sim}{\sigma}' \tag{25}$$

which is one of the invariants introduced in [26].

Another case is that of rotated surfaces [22,23], which are defined in terms of a modified deviatoric stress $\bar{\underset{\sim}{s}}$ given by

$$\bar{\underset{\sim}{s}}' = \underset{\sim}{\sigma}' - \bar{I}'_1 \underset{\sim}{A} \tag{26}$$

$$\bar{I}'_1 = \mathrm{tr}(\underset{\sim}{\sigma}' . \underset{\sim}{\alpha})$$

where $\underset{\sim}{A}$ is a unit norm tensor

Now, invariants $\bar{J}'_2$ and $\bar{J}'_3$ are introduced

$$\bar{J}'_2 = \frac{1}{2}\,\mathrm{tr}\left(\bar{\underset{\sim}{s}}'^{2}\right) \tag{27}$$

$$\bar{J}'_3 = \frac{1}{3}\,\mathrm{tr}\left(\bar{\underset{\sim}{s}}'^{3}\right)$$

It can be easily checked that this formulation is a particular case of (24) where

$$\underset{\sim}{B}^{I} = \underset{\sim}{\alpha}$$

$$\underset{\sim}{B}^{II} = (\underset{\sim}{\delta} \otimes \underset{\sim}{\delta} - \underset{\sim}{A} \otimes \underset{\sim}{A}) \tag{28}$$

$$\underset{\sim}{B}^{III} = \frac{1}{6}\, \underset{\sim}{\delta} \otimes \underset{\sim}{\delta} \otimes \underset{\sim}{\delta} - \frac{1}{9}(\underset{\sim}{A} \otimes \underset{\sim}{\delta} \otimes \underset{\sim}{\delta} + \underset{\sim}{\delta} \otimes \underset{\sim}{A} \otimes \underset{\sim}{\delta} + \underset{\sim}{\delta} \otimes \underset{\sim}{\delta} \otimes \underset{\sim}{A})$$

$$+ \frac{1}{9}\,(\underset{\sim}{\delta} \otimes \underset{\sim}{A} \otimes \underset{\sim}{A} + \underset{\sim}{A} \otimes \underset{\sim}{\delta} \otimes \underset{\sim}{A} + \underset{\sim}{A} \otimes \underset{\sim}{A} \otimes \underset{\sim}{\delta}) - \frac{1}{3}\,\mathrm{tr}(\underset{\sim}{A}^{3}).(\underset{\sim}{A} \otimes \underset{\sim}{A} \otimes \underset{\sim}{A})$$

If the initial structure of the material is characterized by $\underset{\sim}{A}^{o}$, it can be assumed that it will vary with the plastic strain as

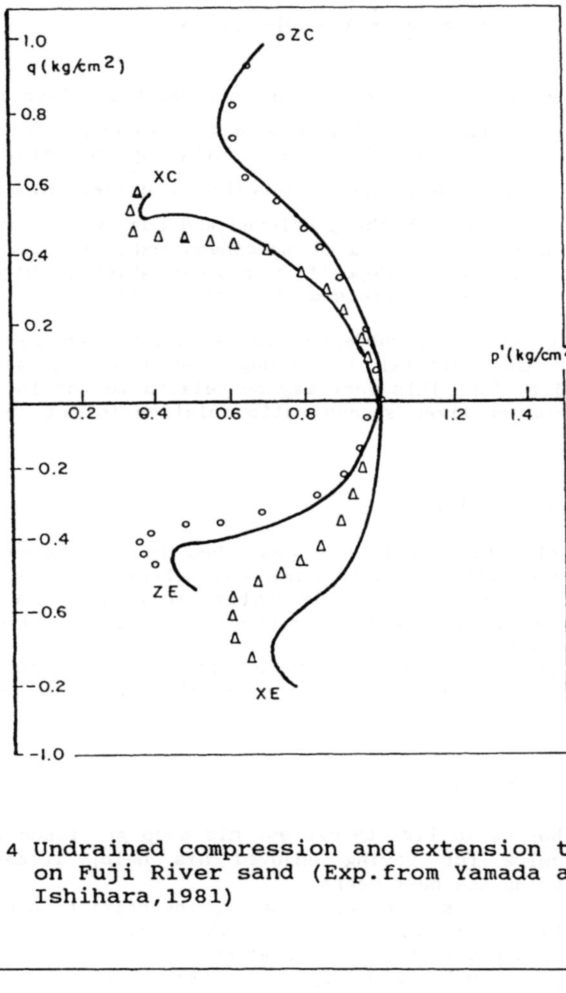

Fig. 4 Undrained compression and extension tests on Fuji River sand (Exp. from Yamada and Ishihara, 1981)

$$\underset{\sim}{A} = \underset{\sim}{A}^{o} + d\underset{\sim}{A}$$
$$d\underset{\sim}{A} = d\underset{\sim}{A}(d\underset{\sim}{\varepsilon}^{p}) \quad (29)$$

A simple evolution law was proposed in [27] as

$$d\underset{\sim}{A} = d\underset{\sim}{\varepsilon}^{p} \left\{ A_1 \exp\left(- A_o \xi \right) \right\} \quad (30)$$

where A_1 and A_o are two material constants and ξ is given in (20)

Now, any isotropic model can be easily extended to describe the anisotropic behaviour of granular soils, by substituting I'_1, J'_2 and J'_3 or p' q and θ by their equivalent modified invariants.

Fig. 4 presents both the predicted and observed behaviour of Fuji River sand [30]. The sand specimens exhibited an initial anisotropy caused by the deposition process which resulted in a stiffer behaviour when compressed along the vertical.

If the material is subjected to extension, the tendency is reversed and the material is stronger when tested along the horizontal direction. This fact may be related to the inclination of the maximum mobilized stress ratio relative to the anisotropy axes.

CONCLUDING REMARKS

The models which have been described in this paper may be applied together with suitable numerical schemes to predict most of fundamental aspects of saturated soil behaviour under earthquake loading, specially when phenomena like liquefaction or cyclic mobility may develop.

Comparison with results obtained in the centrifuge machine [6] confirms that simple isotropic models may provide enough accurate results.

ACKNOWLEDGEMENTS

The author would like to express his deep gratitude to Prof. O.C. Zienkiewicz for having enjoyed his help, guidance and friendship during the past years.

This gratitude is extended to Cedex, The British Council, the spanish Ministery of Education and the Eurogreco for having contributed to support this collaboration.

Finally, to M.D. Azcárraga for her technical help and assistance.

REFERENCES

1. Zienkiewicz, O.C, Chang, C.T. and Bettess, P. : "Drained, consolidating and dynamic behaviour assumptions in soils". Geotechnique, 30 no 4, 385-395 (1980).

2. Zienkiewicz, O.C., Chang, C.T. and Hinton, E. "Non-linear seismic response and liquefaction". Int.J.Num.Anal. Meth. Geomech., 2, 381-404, (1987)

3. Zienkiewicz, O.C., Leung, K.H., Hinton, E. and Chang, C.T. "Liquefaction and Permanent Deformation under Dynamic Conditions-Numerical Solution and Constitutive Relations". in G.N.Pande and O.C.Zienkiewicz, Soil Mechanics. - Transient and Cyclic Loads, Ch. 5, 71-102, Wiley, 1982.

4. Zienkiewicz, O.C. and Shiomi, T. "Dynamic behaviour of Saturated Porous Media : The generalised Biot formulation and its numerical solution", Int.J.Num.Anal.Meth.Geomech., 8 71-96, 1984.

5. Zienkiewicz, O.C. and Taylor, R.L., "Coupled problems a single time - stepping procedure", Comm.Appl.Numer.Meth., 1, 233-239., (1985).

6. Zienkiewicz, O.C., Chan, A.H.C:, Pastor, M., Paul, D.K. and Shiomi, T. "Static and Dynamic behaviour of geomaterials rational approach to quantitative solutions. I : Fully saturated problems". Proc. Royal Society Series A 429, 285-309 (1990)

7. Zienkiewicz, O.C. and Mroz, Z. "Generalized plasticity formulation and applications to geomechanics". Mechanics of Engineering Materials C.S.Desai and R.H.Gallagher (Eds) Wiley 665-679 (1984)

8. Zienkiewicz, O.C., Leung, K.H. and Pastor, M. "A simple model for transient soil loading in earthquake analysis. I : Basic model and its application". Int. J.Numer. Anal.Methods Geomech., 9, 953-976 (1985)

9. Pastor, M., Zienkiewicz, O.C. and Leung, K.H: "A simple model for transient soil loading in earthquake analysis. II : Non-associative models for sands". Int.J. Numer. Anal. Methods Geomech, 9, 977-998 (1986)

10. Pastor, M., Zienkiewicz, O.C. and Chan, A.H.C. "Generalized plasticity and the modelling of soil behaviour" Int.J. Number Anal.Methods Geomech, 14 (1990)

11. Poorooshasb, H.B. and Pietrusczak, S. "On yielding and flow of sand : a generalized two surface model", Computers and Geotechnics, 1, 35-38, 1985.

12. Pande, G.N. and Sharma, K.G. "Multi-laminate model of clays : a numerical evaluation of the influence of principal stress axes". Int.J. Numer. Anal.Methods Geomech, 7, 397-418 (1983)

13. Pande, G.N. and Pietrusczak, S., "Multilaminate framework of soil models plasticity formulation", Int.J.Number. Anal.Meth. Geomech., 11, 651-658 (1987).

14. Aubry, D., Hujeux, J.C., Lassoudiere, F. and Meimon, Y. "A double memory model with multiple mechanisms for cyclic soil behaviour" Int. Symp. Num. Models in Geomechanics, R. Dungar, G.N. Pande and J.A. Studer (Eds) Balkema, Rotterdam(1982)

15. Matsuoka, H. "Stress-strain relationships of sands based on the mobilized plane". Soils and Foundations Vol.14, N 2, pp 47-61 (1974)

16. Darve, F. and Labanieh, S. "Incremental constitutive law for sands and clays : simulation of monotonic and cyclic tests" Int.J. Numer. Anal. Methods Geomech, 6, 243-275 (1982)

17. Mroz, Z. "On the description of anisotropic work-hardening". J. Mech. Phys. Solids, 15, 163-175 (1967)

18. Iwan, W.D., "On a class of models for the yielding behaviour of continuous and composite systems", J. App. Mech., 34 : E3, 612-617

19. Hashiguchi, K. "Constitutive equations of elastoplastic materials with elastic plastic transition". J. Appl. Mech. ASME, 47, 266-272 (1980)

20. Dafalias, Y.F. and Herrmann, L.R., "Bounding Surface formulation of soil plasticity", Soil Mechanics-Transient and Cyclic loads, ch.10, 253-283, eds. G.N. Pande and O.C. Zienkiewicz, Wiley, 1982.

21. Anandarajah, A. and Dafalias, F. "Bounding surface plasticity III : Application to anisotropic cohesive soils" J. Eng. Mech. ASCE, 112, 12, 1292-1318 (1986)

22. Hirai, H. "An elastoplastic constitutive model for cyclic behaviour of sands" Int.J. Numer. Anal. Methods Geomech 11, 503-520 (1987)

23. Ghaboussi, J. and Momen, H. "Modelling and analysis of cyclic behaviour of sands" Soil Mechanics-Transient and Cyclic Loads, pp 313-342G.N. Pande and O.C. Zienkiewicz (Eds) Wiley (1982)

24. Hill, R. "The mathematical theory of plasticity". Oxford at the Clarendon Press (1950)

25. Nova, R. and Sacchi, G. "A model of the stress-strain relationship of orthotropic geological media". J. Mec. Theor. Appl, 1, 6, pp 927-949 (1982)

26. Baker, R. and Desai, C.S. "Induced anisotropy during plastic straining" Int.J. Numer. Anal. Methods Geomech, 8 167-185 (1983)

27. Pastor, M., Modelling of anisotropic sand behaviour, Computers and Geotechnics. (To appear 1991)

28. Chan, A.H.C., Zienkiewicz, O.C. and Pastor, M. "Transformation of incremental plasticity relation from defining space to general cartesian stress space". Comm. Appl. Num. Meth, 4, 577-580 (1988)

29. Pastor, M., Zienkiewicz, O.C. and Chan, A.H.C. "A Generalized Plasticity continuous loading model for Geomaterials", NUMETA, 1987.

30. Yamada, Y. and Ishihara, K. "Undrained deformation characteristics of loose sands under three-dimensional stress conditions", Soils & Foundations, Vol.21, n 1, 97-107 (1981)

FINITE ELEMENTS IN THE 90's, E. Oñate, J. Periaux, A. Samuelsson (Eds.)
Springer-Verlag/CIMNE, Barcelona 1991

ADAPTIVE MESH REFINEMENT TECHNIQUES FOR STRUCTURAL PROBLEMS

E. Oñate and **J. Castro**
International Center for Numerical Methods in Engineering,
Universidad Politécnica de Cataluña,
08034 Barcelona,
ESPAÑA

SUMMARY

In this paper some adaptive mesh refinement (AMR) strategies for finite element analysis of structural problems are discussed. Two mesh optimality criteria based on the equal distribution of: (a) the *global error*, and (b) the *specific error* over the elements are studied. It is shown that the correct evaluation of the rate of convergence of the different error norms involved in the AMR procedures is essential to avoid oscillations in the refinement process. The behaviour of the different AMR strategies proposed is compared in the analysis of some structural problems.

1. INTRODUCTION

The evaluation of discretization errors and the design of suitable meshes via adaptive mesh refinement (AMR) are nowadays two of the challenging issues in the analysis of structures using the finite element method (FEM).

The topic of error estimation and mesh adaptivity in the FEM is by no means new. For a comprehensive review of the topic see the reference list of Chapter 14 of Volume I of [20]. Zienkiewicz and Zhu [1-5] have introduced a successfull AMR strategy for elasticity problems using a simple error estimate based on the difference of the discontinuous finite element stress (or strain) field with an "improved"smooth solution. These authors and others have extended this AMR procedure for plate bending and shell analysis[6–10], [19]. This AMR strategy has also been successfully used for fluid flow problems [2], [19].

In this paper we present an overview of different AMR strategies based on the Zienkiewicz and Zhu error estimator [1-6] for structural analysis. We will show first how the standard AMR algorithm based on equal distribution of the error over all the elements requires a careful identification of the rate of convergence of the different error terms involved in the design of the new element size to avoid oscillations in the refinement process.

An alternative AMR strategy based on the equal-distribution of the "specific error", i.e. the *error per unit area* (or volume) in all the elements in the mesh will be presented next. We will see in the examples shown that this strategy allows to concentrate more and smaller elements in zones where stress concentrations occur, as it should be expected from the engineering point of view.

In next section some basic concepts of the error estimation process are given.

2. BASIC CONCEPTS OF ERROR ESTIMATION

In dealing with adaptive mesh refinement the following two concepts should be clearly defined:

(a) *Error estimator.* Since the "exact"solution is not known, a method to approximately evaluate the error of the finite element solution should be defined.

(b) *Acceptable solution.* A finite element solution is "acceptable" if the estimated error satisfies some prescribed *global* and *local* conditions.

Both concepts (a) and (b) are further explained in next sections.

2.1 Error estimator

One of the most popular error estimators for structural problems is based on the error energy norm expressed as [1–6]

$$\|e\| = [\int_\Omega [\boldsymbol{\sigma} - \hat{\boldsymbol{\sigma}}]^T \mathbf{D}^{-1} [\boldsymbol{\sigma} - \hat{\boldsymbol{\sigma}}] d\Omega]^{1/2} \qquad (1)$$

where $\boldsymbol{\sigma}$ are the exact stresses, $\hat{\boldsymbol{\sigma}}$ are the stress values obtained from the finite element solution, $\mathbf{D}$ is the constitutive matrix and Ω is the structure domain. For plate and shell problems $\boldsymbol{\sigma}$ should here be interpreted as the "resultant stresses" [10], [16].

Since the exact stresses are usually not known they are approximated by

$$\boldsymbol{\sigma} \simeq \boldsymbol{\sigma}^* = \mathbf{N}\bar{\boldsymbol{\sigma}}^* \qquad (2)$$

where $\bar{\boldsymbol{\sigma}}^*$ are nodal values obtained by simple nodal averaging of the finite element values, least squares local and global smoothing, or other adequate projection methods [1-5]. A simple approach is to use a global nodal smoothing with a lumped "mass" matrix giving the nodal smoothed values, $\bar{\boldsymbol{\sigma}}^*$, as

$$\bar{\boldsymbol{\sigma}}^* = \mathbf{M}_D^{-1} \int_\Omega \mathbf{N}_\sigma \hat{\boldsymbol{\sigma}} d\Omega \qquad (3)$$

where $\mathbf{N}_\sigma$ are the chosen stress interpolating functions giving a smooth stress field in terms of the nodal stress values obtained [1-5] and $\mathbf{M}_D$ is a diagonal matrix with $M_{D_{ii}} = \int_\Omega N_{\sigma_i} d\Omega$. Eq. (3) can be obviously applied to solve independently for each individual stress component. It can be verified that eq.(3) yields an accurate smoothed stress field for elements with linear interpolation and is adequate for quadratic elements providing correction factors are used [1]. Recently much improved methods have been devised by Zienkiewicz and Zhu [23] to obtain accurately recovered $\boldsymbol{\sigma}^*$ and these can be used as alternatives.

The strain energy of the exact solution is estimated as

$$\|U\| = [\int_\Omega \boldsymbol{\sigma}^{*T} \mathbf{D}^{-1} \boldsymbol{\sigma}^* d\Omega]^{1/2} \tag{4}$$

Both $\|e\|^2$ and $\|U\|^2$ can be evaluated as sum of their respective element contributions so that

$$\|e\|^2 = \sum_{i=1}^{n} \|e\|_i^2 \quad , \quad \|U\|^2 = \sum_{i=1}^{n} \|U\|_i^2 \tag{5}$$

where n is the total number of elements in the mesh.

2.2 Definition of acceptable solution

There is unfortunately no general consensus of what accuracy is acceptable. However, it is usually accepted that a solution is "correct" if the two following conditions are satisfied:

(a) The global error in energy norm is less than a specified percentage value of the total strain energy, i.e.

$$\|e\| \leq \eta \|U\| \tag{6}$$

where η is the user's specified value of the permissible relative global error.

Eq.(6) allows to define a global error parameter, ξ_g, as

$$\xi_g = \frac{\|e\|}{\eta \|U\|} \tag{7}$$

Clearly the values $\xi_g \leq 1$ denote satisfaction of the global error criterion, whereas $\xi_g > 1$ indicates that further refinement is necessary.

(b) The distribution of the elements in the refined mesh satisfies a "mesh optimality criterion". This local condition can be expressed as

$$\|e\|_i = \|e\|_{r_i} \tag{8}$$

where $\|e\|_i$ is the actual error norm in each element i and $\|e\|_{r_i}$ is the "required" error norm in the element.

From eq.(8) we can define a local error parameter $\bar{\xi}_i$ for each element i as

$$\bar{\xi}_i = \frac{\|e\|_i}{\|e\|_{r_i}} \tag{9}$$

Note that a value of $\bar{\xi}_i = 1$ defines an "optimal" element size, whereas $\bar{\xi}_i > 1$ and $\bar{\xi}_i < 1$ indicate that the size of element i needs refinement and de-refinement, respectively.

The definition of the required error in each element $\|e\|_{r_i}$ is a key issue and it strongly affects the distribution of element sizes in the mesh. This definition can be based on different mesh optimality criteria and some of these are presented in a later section.

2.3 Element refinement parameter

We can define now a single *element refinement parameter*, combining the satisfaction of the global and local conditions (a) and (b) of previous section as

$$\xi_i = \bar{\xi}_i \xi_g = \frac{\|e\| \; \|e\|_i}{\eta \|U\| \; \|e\|_{r_i}} \tag{10}$$

The element refinement parameter ξ_i was first introduced in ref[1] and since then it has been used by many authors as the basis for deciding the new element sizes in a general AMR strategy [1-9]. However, ξ_i can be also interpreted from (10) as the result of trying to satisfy both the global and local error conditions in a successive manner. Eq.(10) provides all the terms involved in this combined process and these could play individually a very different role as explained in next section.

3. MESH OPTIMALITY CRITERIA AND AMR PROCEDURES

3.1 Mesh optimality criterion based on the equal–distribution of the global error

A very popular mesh optimality criterion for structural analysis is based on the so called equal-distribution of the error , i.e. a mesh is defined as optimal if the global error is equally distributed over all the elements [1-10]. On the basis of this assumption we can define the required error for each element as the ratio between the global error and the total number of elements in the mesh. Thus noting that only the square of the error norm is additive (see eq.(5)) we have

$$\|e\|_{r_i} = \frac{\|e\|}{\sqrt{n}} \tag{11}$$

Combining (9) and (11) yields the expression of the local error parameter $\bar{\xi}_i$ as

$$\bar{\xi}_i = \frac{\|e\|_i}{\|e\| n^{-1/2}} \tag{12}$$

The element refinement parameter is now obtained viz. eq.(10) as

$$\xi_i = \bar{\xi}_i \xi_g = \frac{\|e\|_i}{\eta \|U\| n^{-1/2}} \tag{13}$$

The parameter ξ_i can now be readily interpreted as the ratio between the element error and the distributed value of the permissible error over the mesh.

Expression (13) is in fact identical to that used in ref[1]. However, the multiplicative form $\xi_i = \bar{\xi}_i \xi_g$ allows to derive the correct AMR strategy. Thus by noting that the convergence rates of the element and global error norms are

$$\|e\|_i \simeq O(h_i^m)\Omega_i^{1/2} \simeq O(h_i^{m+\frac{d}{2}}) \tag{14a}$$

$$\|e\| \simeq O(h^m) \tag{14b}$$

where h_i and h are the existing element size and the average size of all the elements in the mesh, respectively, m is the degree of the shape function polynomials ($m = 1$ for linear elements, $m = 2$ for quadratic elements, etc.), and d is the number of dimensions of the problem ($d = 1, 2, 3$ for 1D , 2D and 3D problems, respectively) we can deduce that the new element size $\bar{h}_i$ can be obtained in terms of the existing size using the expression

$$\boxed{\bar{h}_i = \frac{h_i}{\xi}} \tag{15}$$

with

$$\boxed{\xi = \bar{\xi}_i^{\frac{2}{2m+d}} \xi_g^{1/m}} \tag{16}$$

The expression of the element size parameter ξ as given by (16) takes into account the different convergence rates of the element and global error norms.

Zienkiewicz and Zhu, followed by others [1-10], use a simpler expression for ξ based directly on the element refinement parameter ξ_i as

$$\xi = (c\xi_i)^{1/m'} \tag{17}$$

where c is a relaxation factor and the new exponent m' is taken as m except for elements adjacent to singularities where $m = \lambda$ is used (λ being the singularity strength).

The authors have found that the computation of ξ as given by (17) with the very common choice of $c = 1$ and $m' = m$ [1-10], leads to a non consistent mesh refinement [16]. This is shown by an oscillatory re and de-refinement of the same mesh zones in the AMR process. This problem dissappears if the correct expression (16) for ξ used (see first example and [16]).

3.2 A mesh optimality criterion based on the equal-distribution of the specific error

An alternative criterion is to assume that a mesh is optimal if the square of the error per unit area (or volume) is the same over the whole mesh. It is clear then that in the optimal mesh

$$\frac{\|e\|_i}{\Omega_i^{1/2}} = \frac{\|e\|}{\Omega^{1/2}} \tag{18}$$

Obviously in (18) Ω_i and Ω denote the element and total area (or volume) respectively.

Comparing (8) and (18) yields the expression of the required error norm for each element as

$$\|e\|_{r_i} = \|e\|(\frac{\Omega_i}{\Omega})^{1/2} \tag{19}$$

The element error parameter $\bar{\xi}_i$ is obtained now using (9) and (19) as

$$\bar{\xi}_i = \frac{\|e\|_i}{\Omega_i^{1/2}}\left[\frac{\|e\|}{\Omega^{1/2}}\right]^{-1} = \frac{\|e\|_i}{\|e\|}\left(\frac{\Omega}{\Omega_i}\right)^{1/2} \tag{20}$$

The element refinement parameter is obtained from (9), (10) and (20) as

$$\xi_i = \bar{\xi}_i \xi_g = \frac{\|e\|_i}{\eta\|U\|}\left(\frac{\Omega}{\Omega_i}\right)^{1/2} \tag{21}$$

Note that eqs.(13) and (21) coincide if $\frac{\Omega}{\Omega_i} = n$ (i.e. all elements are equal in size). This is however not the case for unstructured meshes which results in quite different mesh distributions for each mesh optimality criterion as shown in the examples.

Moreover, the way we have defined now the element error eliminates its dependance with the element area. Therefore, the new convergence rate of the element error can be deduced from (14a) as

$$\frac{\|e\|_i}{\Omega_i^{1/2}} \simeq O(h_i^m) \tag{22}$$

Note the coincidence of the convergence rates of the element and global error norms (eqs.(14b) and (22)).

The new element size is obtained from (15) with ξ given now by

$$\xi = (\bar{\xi}_i \xi_g)^{1/m} = (\xi_i)^{1/m} \tag{23}$$

The new expression for the element refinement parameter coincides with that given by (17) with $c = 1$ and $m' = m$. Note, however, that (17) was introduced in the context the mesh optimality criterion based on the equal-distribution of the global error. The different forms obtained here clarify the correct expression to be used for each mesh optimality criterion chosen.

To our knowledge the criterion of equal-distribution of the specific error was introduced by Bugeda [17] and it has been used in the context of optimum structural design problems by Bugeda and Oliver [18] and for AMR analysis of plates and shells by Oñate, Castro and Kreiner [16]. This criterion can also be shown to be equivalent to that of equal error in stresses mentioned in [1].

4. EXAMPLES

4.1 Analysis of a cylindrical shell roof

The first example is the analysis of the classical Scordelis cylindrical roof under self weight shown in Figure 1 together with the initial mesh of 48 triangular facet shell elements used. The element formulation combines the 12 d.o.f. triangular Reissner-Mindlin plate element based on a linear interpolation for the lateral deflection, an incomplete quadratic interpolation for the rotations and an assumed linear shear strain field developed in [11–15] with a simple linear field for the in-plane displacements. The values of m and d in eq.(16) are $m = 1$ and $d = 2$. A value of the permissible global error $\eta = 10\%$ has been chosen.

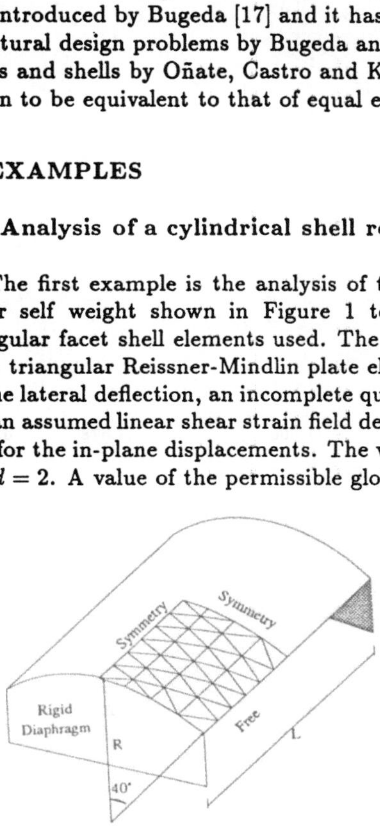

$L = 600.0$

$R = 300.0$

thickness $= 3.0$

$E = 3.0 \times 10^6$

$\nu = 0.0$

B.C. : in diaphragm $u = 0, \; w = 0$

Load : Self weight, $q = 0.625$

Initial global error parameter $\xi_g = 6.2295$

Figure 1. Cylindrical shell roof under self weight. Symmetric quadrant used in the analysis and initial mesh of 48 triangular assumed shear strain facet shell elements [11–15].

Figure 2 shows the sequence of refined meshes obtained with the three AMR strategies studied in the paper. First column (strategy A) shows the results obtained using the criterion of equal-distribution of the global error over all the elements and the inconsistent value of ξ defining the new element sizes as given by eq.(17) with $c = 1$ and $m' = m = 1$. Note the oscillations in the AMR process clearly shown by the alternative re and de-refinements of the same mesh zones.

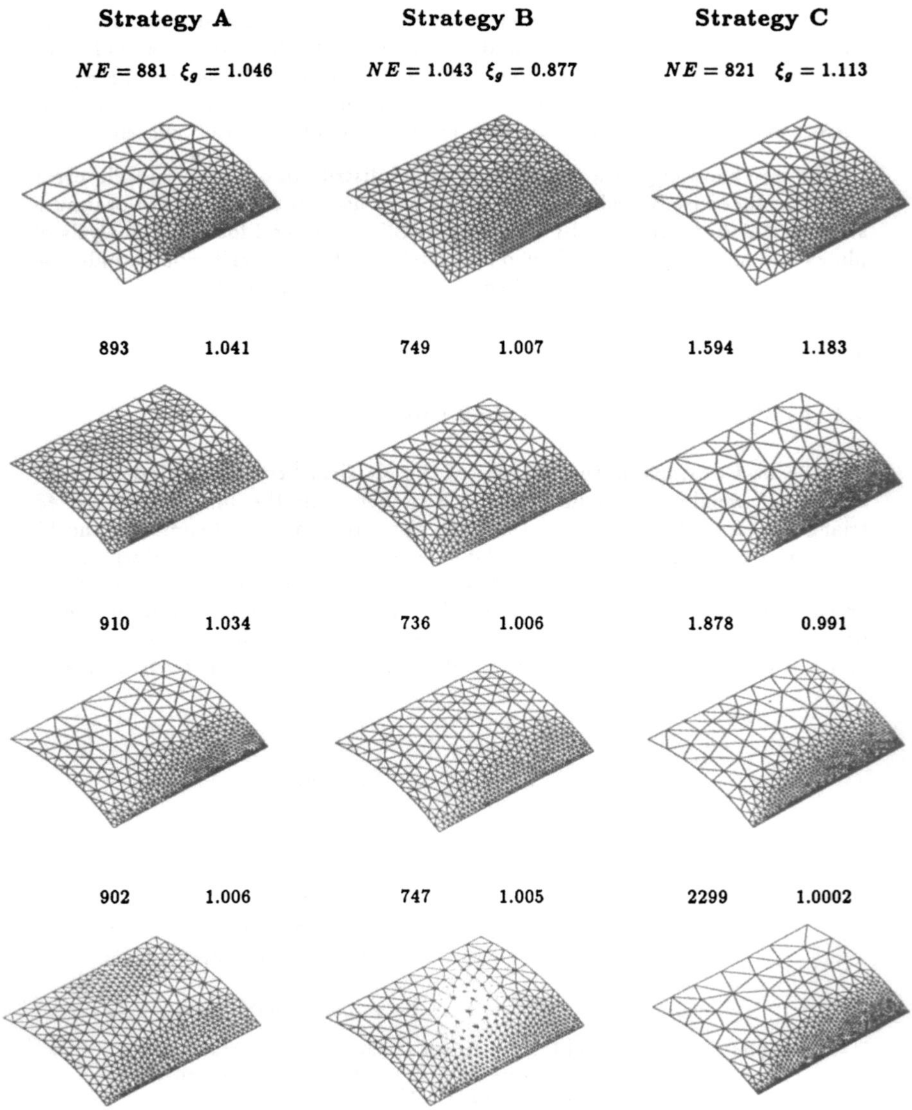

Figure 2. Symmetric quadrant of a cylindrical shell roof under self weight. Sequence of meshes obtained with AMR strategies based on: (A) Equal distribution of global error and inconsistent definition of parameter ξ (viz eq.(17)); (B) Idem with ξ consistently defined by eq.(16); and (C) Equal distribution of specific error. For each mesh is shown the number of elements (NE) and the value of the global error parameter ξ_g.

Results labelled as strategy B in Figure 2 have been obtained with the same mesh optimality criterion, but using now the correct expression for ξ as given by eq.(16). Note that the refinement oscillations dissapear and the AMR process converges in a consistent manner.

Results for strategy C have been obtained with the mesh optimality criterion based on the equal distribution of the specific error, with the element size parameter ξ as given by eq.(23). It can be clearly seen that: (a) The AMR process converges without oscillations, and (b) This AMR strategy concentrates more and smaller elements in the vicinity of the free edge (where the error is greater due to the higher membrane stress gradients), whereas in the rest of the mesh bigger elements than in the previous case are allowed. The prize to be paid is the increase in the total number of elements with respect to strategies A and B for the same global accuracy as shown in Figure 2.

Further details of this example and of the general AMR strategy for plates and shells can be found in [16].

4.2 Analysis of a cylindrical shell with a circular perforation under uniform traction

Figure 3 shows the geometry, material properties of the shell and the initial mesh used. The analysis has been performed using the same facet shell element as in the previous example. A value of the permissible global error $\eta = 10\%$ has been taken.

Figure 3 also shows the sequence of refined meshes obtained with the criterion of equal-distribution of the global error and the correct value for the element size parameter ξ as given by eq.(16) (Strategy B), and also with the criterion of equal distribution of the specific error with ξ given by eq.(23) (Strategy C).

Table I shows some characteristic results for each solution like the number of elements, the global error parameter ξ_g, the average value of the local error parameter $(\bar{\xi}_i^2)_a$ and its mean deviation $(\bar{\xi}_i^2)_\sigma$ over each mesh for the two AMR strategies used. From the numbers shown in this table we deduce

- Both AMR strategies converge to the global permissible error chosen
- Both AMR processes converge to an "optimal mesh", characterized by the appriate values $(\bar{\xi}_i^2)_a = 1.0$ and $(\bar{\xi}_i^2)_\sigma = 0.0$. However, the number of elements and its distribution in each mesh is very different for the two AMR strategies and again that based on the equal distribution of specific error (strategy C) tends to concentrate more and smaller elements in the vicinity of the central hole where stress gradients are higher (see Figure 3).

5. CONCLUDING REMARKS

We have shown that the correct evaluation of the rate of convergence of the different error norms involved in the AMR strategy is essential to avoid oscillations in the refinement process. Also, the mesh optimality criteria based on global and specific error distributions are conceptually very different. Thus, whereas the former leads to meshes with smaller number of elements, the second captures

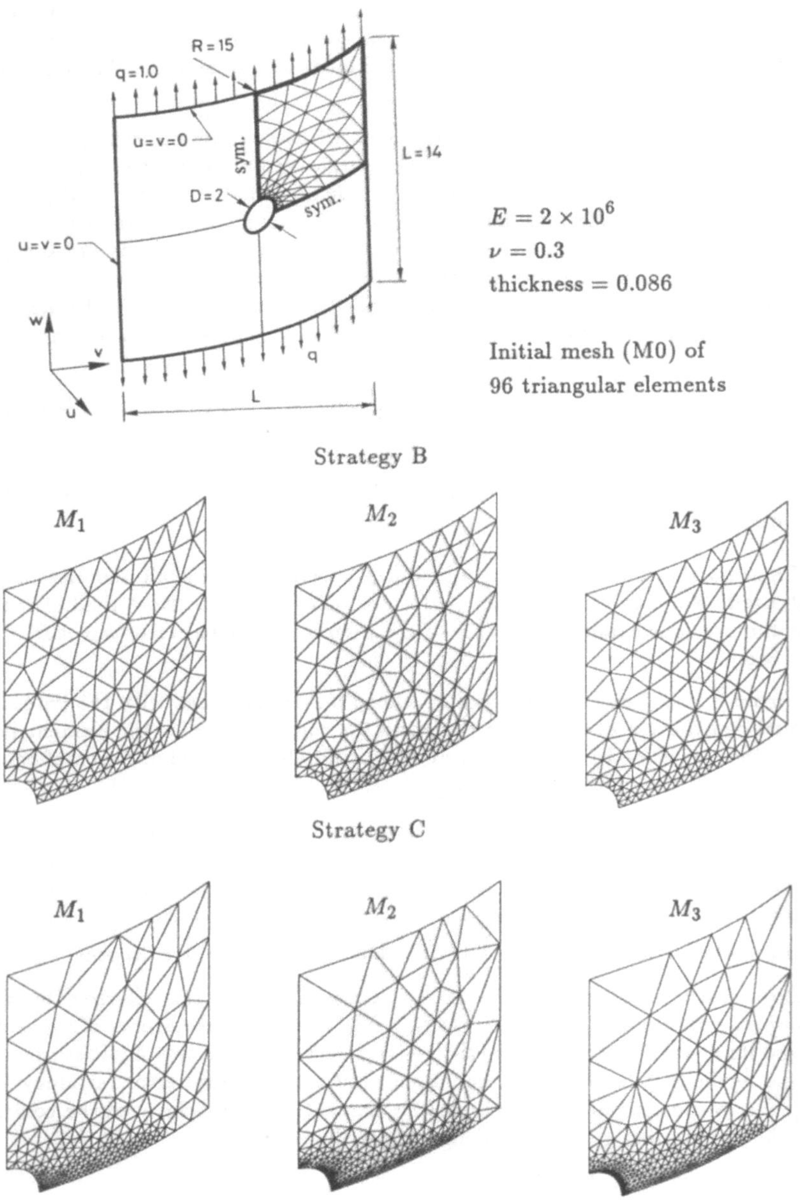

Figure 3. **Analysis of a cylindrical shell under uniform traction. Sequence of meshes obtained with AMR strategies based on: Equal distribution of global error (Strategy B) and Equal distribution of specific error (Strategy C).**

	Strategy B				Strategy C			
	NE	ξ_g	$(\bar{\xi}_i^2)_a$	$(\bar{\xi}_i^2)_\sigma$	NE	ξ_g	$(\bar{\xi}_i^2)_a$	$(\bar{\xi}_i^2)_\sigma$
M0	96	1.776	1.000	1.413	96	1.776	1.728	8.668
M1	234	1.078	1.000	0.613	393	0.993	1.545	3.432
M2	262	0.959	1.000	0.518	593	0.991	1.166	0.684
M3	266	0.968	1.000	0.384	729	0.979	1.050	0.381

TABLE I. Symmetric quadrant of a tractioned cylindrical shell. Some statistical results of the AMR processes based on the equal distribution of the global error (Strategy B) and the specific error (Strategy C).

better the effect of high stress gradients. Further research should allow to balance the possibilities of these two criteria for use in practical structural applications.

ACKNOWLEDGEMENTS

The authors thank Prof. O.C. Zienkiewicz for many useful discussions and Mr. R. Kreiner for his work in the generation of the meshes for the examples analized. The mesh generators used were based on ideas and software originally developed by Drs. J. Peraire [21], J. Peiró [22] and G. Bugeda [17], whose contribution is also gratefully acknowledged.

REFERENCES

[1] Zienkiewicz, O.C. and Zhu, J.Z., "A simple error estimator and adaptive procedure for practical engineering analysis", *Int. Num. Meth. Engrg.*, **24**, 337–357, 1987.

[2] Zienkiewicz, O.C., Zhu, J.Z., Liu, I.C., Morgan, K. and Peraire, J., "Error estimates and adaptive from elasticity to high speed compressible flow", in J.R. Whiteman, ed., MAFELAP 87, 483–512, Academic Press, New York, 1988.

[3] Zhu, J.Z. and Zienkiewicz, O.C., "Adaptive techniques in the finite element method", *Comm. Appl. Numer. Methods*, **4**, 197–204, 1988.

[4] Zienkiewicz, O.C., Zhu, J.Z. and Gong, N.G., "Effective and practical $h-p$ version adaptive analysis procedures for the finite element method", *Internat. J. Numer. Methods Engrg.*, **28**, 879–891, 1989.

[5] Zienkiewicz, O.C., Zhu, J.Z., "The three R's of engineering analysis and error estimation and adaptivity", *Comp. Meth. in Appl. Mech. and Engng*, **82**, 95-113, 1990.

[6] Zienkiewicz, O.C., Zhu, J.Z., "Error estimates and adaptive refinement for plate bending problems", *Internat. J. Numer. Methods Engrg.*, **28**, 2839–53, 1989.

[7] Atamaz-Sibai, W. and Hinton, E., "Adaptive mesh refinement with the Morley plate element", Proc. of NUMETA 90 conference held at Swansea 1990, **2**, 1044–57, Elserver App. Sc., London 1990.

[8] Atamaz-Sibai, W., Hinton, E. and Selman, A., "Adaptive mesh refinement with Mindlin-Reissner elements", Proc. *2nd Int. Conf. Computer Aided Analysis and Design of Concrete Structures*, Apr. 1990, Zell-am-see, Austria, **1**, 303–15, Pineridge Press, Swansea, U.K.

[9] Selman, A., Hinton, E. and Atamaz-Sibai, W., "Edge effects in Mindlin-Reissner plates using adaptive mesh refinement", *Eng. Comp.*, **7**. 3. 217–27, 1990.

[10] Hinton, E. Özakça, M. and Rao, N.V.R., "Adaptive analysis of thin shells using facet elements", Int. Report CR/950/90, Univ. College of Swansea, U.K. April 1990.

[11] Zienkiewicz, O.C. Taylor, R.L., Papadopoulus, P. and Oñate, E., "Plate bending with discrete constraints. New Triangular element", *Comp. and Struct.*, **35**, 4, 505–22, 1990.

[12] Oñate, E., Zienkiewicz, O.C. and Taylor R.L., "Consistent formulation of shear constrained Reissner-Mindlin plate Elements", in *Discretization Methods in Struct. Mech.*, G. Kuhn and H. Mang (Eds.), Springer-Verlag, 1990.

[13] Oñate, E., Zienkiewicz, O.C. Suárez, B. and Taylor, R.L., "A general methodology for deriving shear constrained Reissner-Mindlin plate elements", *Int. J. Num. Meth. Engng.*, To be published 1991.

[14] Papadopoulus, P. and Taylor, R.L., "A triangular element based on Reissner-Mindlin plate theory", *Int. J. Num. Meth. Engng.*, **5**, 1029–51. 1990.

[15] Oñate, E. and Castro, J., "Some new plate elements based on assumed shear strain fields", European Conf. on *New Advances in Computat. Struct. Mehc.*, Giens, France, April, 1991.

[16] Oñate, E., Castro, J. and Kreiner, R., "Error estimations and mesh adaptivity techniques for plate and shell problems", presented at the *3rd. International Conference on Quality Assurance and Standards in Finite Element Methods*, Stratford-upon-Avon, England, 10–12 September, 1991.

[17] Bugeda, G., "Utilización de técnicas de estimación de error y generación automática de mallas en procesos de optimización estructural", *Ph D. Thesis*, Universitat Polytècnica de Catalunya, 1990.

[18] Bugeda, G. and Oliver, J., "Automatic adaptives remeshing for structural shape optimization", European Conf. on New Advances in *Computat. Struct. Mech.*, Giens, France, April, 1991.

[19] Wu, J., Zhu, J.Z., Smelter, J. and Zienkiewicz, O.C., "Error estimation and adaptivity in Navier-Stokes incompressible flow", *Computational*

Mechanics, **6**, 4, 259–71, 1990.

[20] Zienkiewicz, O.C. and Taylor, R.L., "The finite element method", Mc Graw Hill, **I**, 1989; **II**, 1991.

[21] Peraire, J., "A finite element method for convection dominated flows", *Ph. D. Thesis*, Civil Eng. Dept. University College of Swansea, U.K. 1986.

[22] Peiro, J., "A finite elemente procedure for the solution of the Euler equations on unstructured meshes", *Ph.D. Thesis*, Civil Eng. Dept., University College of Swansea, U.K., 1989.

[23] Zienkiewicz, O.C. and Zhu, J.Z., "Superconvergence derivative recovery techniques and a posteriory error estimation in the finite element method. Part I and part II". *Inst. Num. Meth. Engng.*, Swansea CR/671/91, June 1991, (To appear in Int. J. Num. Meth. Engng.).

FINITE ELEMENTS IN THE 90's, E. Oñate, J. Periaux, A. Samuelsson (Eds.)
Springer-Verlag/CIMNE, Barcelona 1991

REMARKS ON SELECTION OF STRESSES TO SUPPRESS ZERO ENERGY DEFORMATION MODES IN HYBRID ELEMENT FORMULATIONS

T. H. H. Pian
Department of Aeronautics and Astronautics,
Massachusetts Institute of Technology,
Cambridge, Massachusetts 02139,
U.S.A.

1 INTRODUCTION

Mathematical basis for the existence and stability of numerical solutions of multi-field finite element methods has been studied by many authors [1 - 4]. For problems in elasticity it is well recognized that such stability problem is associated with the zero energy deformation modes. For example, for the formulation of hybrid elements a procedure has been proposed for the selection of assumed stresses to suppress such modes [5]. Examples in Reference [5] are plane stress, 3-D solid and axisymmetric solid elements. Reference [4] applies the same approach and includes an example of plate bending element with transverse shear effect. These examples are all concerning problems that involve homogeneous solids and also require only C^0 continuity conditions for the assumed displacements. The present paper is to extend this appoach to two problems of different nature. The first one is the Kirchhoff plate bending problem for which the displacements are required to be C^1 continuous. The second one is an element which contains two layers of different materials with assumed stresses that satisfy the interface equilibrium conditions.

2 KIRCHHOFF PLATE ELEMENT

Hybrid stress elements can be formulated by using the Hellinger-Reissner principle, which, for an element with prescribed boundary displacements $\tilde{u}$ is the stationary condition of the functional

$$\pi_R = \int_{V_n} [-(1/2)\sigma^T S\sigma + \sigma^T(Du)]dV - \int_{\partial V_n} T^T(u - \tilde{u})dS \tag{1}$$

where u = element displacements; σ =stresses; S =elastic matrix; T = boundary tractions in terms of σ; and

$$\epsilon = \mathrm{Du} \tag{2}$$

gives the strain-displacement relation. When the element displacements u satisfy boundary compatibility, the last integral term is absent.

In the finite element formulations, the assumed stresses σ are expressed in terms of stress parameters β, while the displacements u are interpolated in terms of nodal displacements q. The strains (Du) can also be expressed in terms of a set of independent strain modes with parameters α. The stability of individual element formulated by compatible assumed displacements is determined by the following integral I:

$$I = \int_{V_n} \sigma^T (D\mathbf{u}) dV \tag{3}$$

which is twice the strain energy of the element. In terms of the stress and strain parameters,

$$I = \beta^T \mathbf{G}_\alpha \alpha \tag{4}$$

The necessary and sufficient condition for the absence of zero energy deformation modes in the element level is [4].

$$G_\alpha \alpha = 0 \quad \rightarrow \quad \alpha = 0 \tag{5}$$

The procedure for selecting the assumed stresses is to make sure that each of the independent straining modes can be individually suppressed and the resulting matrix $\mathbf{G}_\alpha$ will have a rank of $(n - r)$ where n is the number of nodal displacements of the element and r, the number of rigid body degrees of freedom. With the stress terms so chosen the number of stress terms m is equal to $(n - r)$,which is the minimum required for maintaining the stability of the element [6].

For the Kirchhoff plate problem, the element displacement is the lateral displacement w which is required to be C^1 continuous along the interelement boundary. If such condition cannot be satisfied, the corresponding expression of I in Eq. (4) should include the contribution of the boundary integral of Eq. (1). In the finite element formulations both the element displacements u and the boundary displacements $\tilde{\mathbf{u}}$ are interpolated in terms of the nodal displacements q and, in turn, the integral I can be expressed in terms of either q or the corresponding strain parameters α. For illustrative purpose, we consider a square plate element of dimension 2 by 2 (Figure 1).

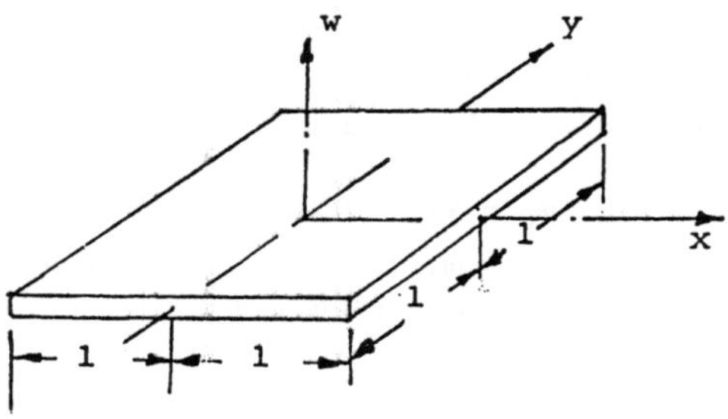

Figure 1. Four-node Plate Bending Element

Here, u is simply the lateral displacement w, and the nodal displacements q are w, $w_{,x}$ and $w_{,y}$ at the four corners. For compatibility of $w(x, y)$ with neighboring elements $w(s)$ is required to be cubic along all edges, while the normal derivative $w_{,n}(s)$ must be linear. The first conditon can be easily maintained by a suitable shape function, but it is not possible to satisfy the latter[7]. The expression of I for an element of area A_n, in this case, must be

$$I = \int_{A_n} M_{\alpha\beta} w_{,\alpha\beta} dA - \int_{\partial A_n} M_n (w_{,n} - \tilde{w}_{,n}) ds \tag{6}$$

where ($\alpha, \beta = 1, 2$) refers to Cartesian coordinates (x, y), $M_{\alpha\beta}$ are stress couples, and M_n is the bending moment along the element boundary ∂A_n. For the present rectangular element the obvious shape function for $w(x, y)$ is the one used by Adini and Clough [8] and Melosh [9] for their incompatible plate element. The corresponding polynomial expansion of this shape function is

$$w = \alpha_1 + \alpha_2 x + \alpha_3 y + \alpha_4 x^2 + \alpha_5 xy + \alpha_6 y^2 + \alpha_7 x^3 + \alpha_8 x^2 y + \alpha_9 xy^2 + \alpha_{10} y^3 + \alpha_{11} x^3 y + \alpha_{12} xy^3 \tag{7}$$

and the resulting strains, which are the curvatures, are,

$$\begin{aligned} w_{,xx} &= 2\alpha_4 + 6\alpha_7 x + 2\alpha_8 y + 6\alpha_{11} xy \\ w_{,yy} &= 2\alpha_6 + 2\alpha_9 x + 6\alpha_{10} y + 6\alpha_{12} xy \\ w_{,xy} &= \alpha_5 + 2\alpha_8 x + 2\alpha_9 y + 3\alpha_{11} x^2 + 3\alpha_{12} y^2 \end{aligned} \tag{8}$$

In applying the procedure of using one stress mode to suppress one independent straining mode, one of the possible sets of stress couples is,

$$\begin{aligned} M_{xx} &= \beta_1 + \beta_4 x + \beta_5 y + \beta_8 xy \\ M_{yy} &= \beta_2 + \beta_6 x + \beta_7 y + \beta_9 xy \\ M_{xy} &= \beta_3 \end{aligned} \tag{9}$$

To account for the boundary integral terms, we consider, for example, the normal derivative $w_{,x}$ along $x = +1$,

$$w_{,x} = \alpha_2 + 2\alpha_4 + 3\alpha_7 + (\alpha_5 + 2\alpha_8 + 3\alpha_{11})y + \alpha_9 y^2 + \alpha_{12} y^3 \tag{10}$$

It can be shown that, for $\tilde{w}_{,x}$ to be linear, and $w_{,x} = \tilde{w}_{,x}$ at $y = \pm 1$, we must have,

$$w_{,x} - \tilde{w}_{,x} = \alpha_9(y^2 - 1) + \alpha_{12} y(y^2 - 1) \tag{11}$$

along the edge $x = +1$, and this equation also holds for the edge $x = -1$. Similar expressions can be obtained for the other edges, and,

$$-\int_{\partial A_n} M_n(w_{,n} - \tilde{w}_{,n})ds = (8/3)(\beta_4\alpha_9 + \beta_5\alpha_8) + (8/15)(\beta_8\alpha_{12} + \beta_9\alpha_{11}). \tag{12}$$

The $\mathbf{G}_\alpha$ matrix corresponding to $\lfloor \beta_1 \ldots \beta_9 \rfloor$ and $\{\alpha_4\alpha_6\alpha_5\alpha_7\alpha_{10}\alpha_8\alpha_9\alpha_{11}\alpha_{12}\}$ is,

$$\begin{bmatrix} 8 & 0 & 0 & 0 & 0 & 0 & 0 & 0 & 0 \\ 0 & 8 & 0 & 0 & 0 & 0 & 0 & 0 & 0 \\ 0 & 0 & 8 & 0 & 0 & 0 & 0 & 0 & 0 \\ 0 & 0 & 0 & 8 & 0 & 0 & 8/3 & 0 & 0 \\ 0 & 0 & 0 & 0 & 8 & 8/3 & 0 & 0 & 0 \\ 0 & 0 & 0 & 0 & 0 & 8/3 & 0 & 0 & 0 \\ 0 & 0 & 0 & 0 & 0 & 0 & 8/3 & 0 & 0 \\ 0 & 0 & 0 & 0 & 0 & 0 & 0 & 8/3 & 8/15 \\ 0 & 0 & 0 & 0 & 0 & 0 & 0 & 8/15 & 8/3 \end{bmatrix}$$

The rank of this matrix is 9. Thus, zero energy deformation modes can all be suppressed, if the stress terms of Eq.(9) are used in the element formulation. The resulting element, however, is not optimal. As shown in Ref. 10, for improved performance, a 4-node Kirchhoff plate element should be formulated by keeping at least the linear stress terms complete. For rectangular shaped elements, it is desirable to add the stress terms $M_{xy} = \beta_{10}x + \beta_{11}y$ to those of Eq. (9).

3 TWO-LAYER ELEMENT THAT MAINTAIN INTERFACE EQUILIBRIUM

In using hybrid stress elements for the evaluation of interlaminar stresses for laminated structures typical in composites and adhesive joints, it is desirable to employ special elements that contain multi-layers and that the interlaminar stress equilibrium conditions can be maintained [11,12]. For the present illustration, a simple six-node two-layer rectangular plane element is considered. The two layers are all of 2 by 2 in dimension. They are defined by local Cartesian coordinates (x_1, y_1) and (x_2, y_2), respectively, as shown in Figure 2.

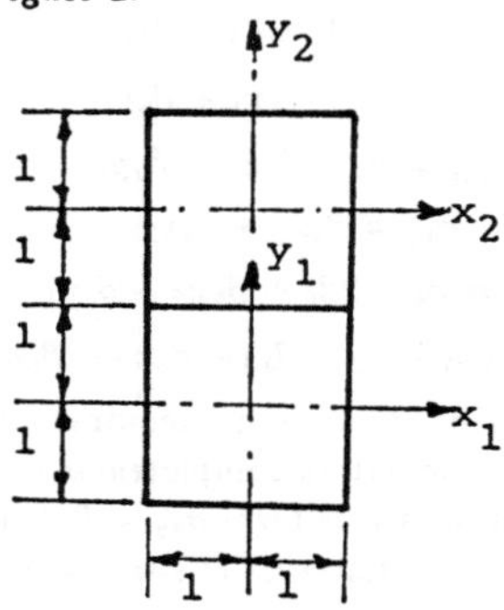

Figure 2. Six-node Two-layer Plane Element

In the finite element formulation, the displacements u and v in each layer are expanded by bilinear interpolations. It can be shown that, individually, the normal strains ϵ_x and ϵ_y and shearing strain γ_{xy} can be expressed in terms straining modes α as,

$$\begin{aligned}
\epsilon_{x_1} &= \alpha_1 + \alpha_2 y_1 \\
\epsilon_{y_1} &= \alpha_3 + \alpha_4 x_1 \\
\gamma_{xy_1} &= \alpha_5 + \alpha_2 x_1 + \alpha_4 y_1 \\
\epsilon_{x_2} &= \alpha_6 + \alpha_7 y_2 \\
\epsilon_{y_2} &= \alpha_8 + \alpha_9 x_2 \\
\gamma_{xy_2} &= \alpha_{10} + \alpha_7 x_2 + \alpha_9 y_2
\end{aligned} \tag{13}$$

However, the continuity of ϵ_x along the layer interface requires,

$$\alpha_6 = \alpha_1 + \alpha_2 + \alpha_7 \tag{14}$$

and the resulting element will have only 9 independent straining modes.

According to the procedure of assigning one stress mode for suppressing one independent straining mode, the needed assumed stress terms with 9 β-parameters are,

$$
\begin{aligned}
\sigma_{x_1} &= \bar{\beta}_1 + \bar{\beta}_2 y_1 \\
\sigma_{y_1} &= \bar{\beta}_3 + \bar{\beta}_4 x_1 \\
\tau_{xy_1} &= \bar{\beta}_5 \\
\sigma_{x_2} &= \bar{\beta}_6 y_2 \\
\sigma_{y_2} &= \bar{\beta}_7 + \bar{\beta}_8 x_2 \\
\tau_{xy_2} &= \bar{\beta}_9
\end{aligned} \tag{15}
$$

However, when the stress equilibrium conditions in each layer and the continuity of σ_y and τ_{xy} along the interface (at $y_1 = 1$ and $y_2 = -1$) are considered, additional stress terms must be introduced so that the same number of independent stress parameters can be maintained. The resulting required stress terms are as follows:

$$
\begin{aligned}
\sigma_{x_1} &= \beta_1 + \beta_2 x_1 + \beta_3 y_1 \\
\sigma_{y_1} &= \beta_4 + \beta_5 x_1 + \beta_6 y_1 \\
\tau_{xy_1} &= \beta_7 - \beta_6 x_1 - \beta_2 y_1 \\
\sigma_{x_2} &= \beta_8 x_2 + \beta_9 y_2 \\
\sigma_{y_2} &= \beta_4 + 2\beta_6 + \beta_5 x_2 + \beta_6 y_2 \\
\tau_{xy_2} &= (\beta_7 - \beta_2 - \beta_8) - \beta_6 x_2 - \beta_8 y_2
\end{aligned} \tag{16}
$$

These are the minimum number of stress terms required to suppress the zero energy modes. But, obviously, in order to maintain completeness in the polynomial expansion, it is mandatory to add the constant stress term $\sigma_{x_2} = \beta_{10}$. In this case,the $\mathbf{G}_\alpha$ matrix corresponding to $\lfloor \beta_1 \ldots \beta_{10} \rfloor$ and $\{\alpha_1 \alpha_{10} \alpha_2 \alpha_3 \alpha_4 \alpha_8 \alpha_5 \alpha_9 \alpha_7\}$ is,

$$
\begin{bmatrix}
4 & 0 & 0 & 0 & 0 & 0 & 0 & 0 & 0 \\
0 & -4 & 0 & 0 & -4/3 & 0 & 0 & 0 & 0 \\
0 & 0 & 4/3 & 0 & 0 & 0 & 0 & 0 & 0 \\
0 & 0 & 0 & 4 & 0 & 4 & 0 & 0 & 0 \\
0 & 0 & 0 & 0 & 4/3 & 0 & 0 & 4/3 & 0 \\
0 & 0 & -4/3 & 0 & 0 & 8 & 0 & 0 & -4/3 \\
0 & 4 & 0 & 0 & 0 & 0 & 4 & 0 & 0 \\
0 & -4 & 0 & 0 & 0 & 0 & 0 & -4/3 & 0 \\
0 & 0 & 0 & 0 & 0 & 0 & 0 & 0 & 4/3 \\
4 & 0 & 4 & 0 & 0 & 0 & 0 & 0 & 4
\end{bmatrix}
$$

It is clear that this matrix has a required rank of 9. Two-layer elements of different configurations have been constructed based on assumed stresses determined by the above mentioned procedure, for the analysis of interlaminar stresses of adhesive joints [13].

4 CONCLUDING COMMENTS

These two examples all indicate that, although in the formulation of hybrid stress elements, it is possible to use only minimum number of stress parameters to maintain the stability of the element, for optimal element performance it may be necessary to introduce additional stress terms.

References

[1] BABUSKA, I. - The Finite Element Methods with Lagrange Multipliers, Numer. Math., **20**, 179-192 (1973)

[2] BREZZI, F. - On the Existence, Uniqueness, and Approximation of Saddle-point Problems Arising from Lagrange Multipliers, R.A.I.R.C., Ser. Rouge **8**, 129-151 (1974)

[3] XUE, W-M., KARLOVITZ, L.A. and ATLURI, S.N. - On the Existence and Stability Conditions for Mixed-Hybrid Finite Element Solutions Based on Reissner's Variational Principle, Int.J. Solids and Structures, **21**, 97-116 (1985)

[4] WU, C-C. - Dual Zero Energy Modes in Mixed/Hybrid Elements —Definition, Analysis and Control, Comp. Meth. Appl. Mech. Eng., **81**, 39-56 (1990)

[5] PIAN, T. H. H. and CHEN, D. - On the Suppression of Zero Energy Deformation Modes, Int. J. Num. Meth. Eng., **19**, 1741-1752 (1983)

[6] TONG, P. and PIAN, T. H. H. - A Variational Principle and the Convergence of a Finite Element Method Based on Assumed Stress Distribution, Int. J. Solids and Structures, **5**, 463-472 (1969)

[7] IRONS, B.M. and DRAPER, K.J. - Inadequacy of Nodal Connections in a Stiffness Solution for Plate Bending, AIAA J., **3**, 961 (1965)

[8] ADINI, A. and CLOUGH, R. W. - **Analysis of Plate Bending by the Finite Element Method**. Report submitted to the Nat. Sci. Foundation, Grant G7337, Uni. of California, Berkeley, (1960)

[9] MELOSH, R. J. - Basis for Derivation of Matrices for the Direct Stiffness Method, AIAA J., **1**, 1631-1637 (1963)

[10] PIAN, T. H. H., KANG, D. and WANG, C. - 'Hybrid Plate Elements based on balanced Stresses and Displacements', in **Finite Element Methods for Plate and Shell Structures, Vol. 1: Element Technology**, Edited by T.J.R. Hughes and E. Hinton, Pineridge Press International, Swansea, U.K., 1986, pp.244-265

[11] MAU, S. T., TONG, P. and PIAN, T. H. H. - Finite Element Solutions for Laminated Thick Plates, J. Composite Materials, **6**, 304-311 (1972)

[12] PIAN, T. H. H. and LI, M. - 'Stress Analysis of Laminated Composites by Hybrid Finite Elements', in **Discretization Methods in Structural Mechanics**, IUTAM/IACM Symposium Vienna, Austria, 1989, Edited by G. Kuhn and H. Mang, Springer-Verlag, 1990, pp.363-372

[13] PIAN, T. H. H. and SAETHER, E. - **Analysis of Interlaminar Stress Distribution of Adhesive Joints by Special Hybrid Stress Finite Elements**, U.S. Army Materials Technology Laboratory Technical Report (to be published)

FINITE ELEMENTS IN THE 90's, E. Oñate, J. Periaux, A. Samuelsson (Eds.)
Springer-Verlag/CIMNE, Barcelona 1991

FINITE ELEMENT ADAPTIVITY IN DYNAMICS AND ELASTOPLASTICITY

A. Samuelsson and **N-E. Wiberg**
Chalmers University of Technology,
Dep. of Struct. Mechanics,
S-412 96 Göteborg,
SWEDEN

SUMMARY

The paper discusses adaptive procedures for FE-analysis of elastomechanic problems. Error indicators and estimators are based on interpolation theory or on post-processing measures. For the mesh generation and regeneration an isoline technique is utilized by which the error can be reasonably uniformly distributed by interpolation theory.

A new simple a posteriori local error estimator of time discretization for direct integration methods in structural dynamic analysis is described. It is derived from the difference of the solutions between an ordinary integration method (e.g. the Newmark scheme) and another higher-order one with linear derivatives of accelerations within each time step. It may be obtained directly without resolving new equations so the additional computational cost is small and the implementation is convenient.

For elastoplasticity error indicators of interpolation type and of postprocessing type are compared. In both cases the strain distribution obtained from the FE-equations is smoothed. The continuous stress distribution is then by iteration forced to be in equilibrium.

1. INTRODUCTION

During the early days of finite elements the errors were estimated from the jumps of the stresses between the elements. Where the jumps were too high the mesh was refined. Possibly also the stress field was smoothed by averaging. The mathematical analysis for elliptic problems was studied with the assumption that the mesh was refined in a regular fashion. Asymptotic error estimates were then obtained which gave information about the rate of convergence and the dependence upon the regularity of the true solution. Even if this information is of great value it could not directly be used for a practical error estimation.

In the early 80th two new concepts were developed which both were of big importance for the later development of adaptive methods. One was the p-method with finite element functions of hierarchical type allowing an easy improvement of the solution which also gave an a-posteriori estimation of the error, see [1], [2], [3]. The other idea was introduced by Babuska [4]. He calculated the residual error consisting of the error in the equilibrium at the interior of the elements and the interelement errors and used this for an adaptive h-method refinement aiming at getting a small and even error over the region. It was later found that a proper combination of h- and p-refinement gave the best convergence (exponential).

A very important step forward was taken in 1987 by Zienkiewicz and his coworkers [5] when they found a very simple way to obtain both local and global error estimates for linear elements. They found for problems in elasticity that the exact stress fields could be with surprising accuracy approximated by smoothening of the discontinuous stress fields obtained from the finite element equations. With the aid of an efficient generator for unstructured meshes they succeeded to develop an economic adaptive procedure of high accuracy. They have later extended this concept to higher order elements.

Another important development was started in 1988 by Johnsson and his coworker [6] who transformed the a priori asymptotic estimates to a-posteriori form. However, the estimates include a constant that is problem dependent. Its determination requires some effort.

We will here describe some developments in structural dynamics and elastoplasticity in which we use estimates similar to those proposed by Zienkiewicz and Johnson.

2. ADAPTIVE FE-PROCEDURES FOR DYNAMIC PROBLEMS

2.1. Basic equations

The elastomechanical problems are governed by the differential equation

$$\varrho\frac{\partial^2 u}{\partial t^2} + c\frac{\partial u}{\partial t} - \bar{\nabla}^t K^d \bar{\nabla} u = f(t), \quad u = u(x,t), \quad x = x,y,z \tag{1}$$

where u is the displacement, $\bar{\nabla}$ the strain operator, K^d the elasticity matrix, c the material damping, ϱ the mass density and f the time-dependent body force. By an approximation $u^h(x,t) \approx \phi(x)U(t)$ with basis functions ϕ and nodal parameters U and by application of the Galerkin procedure the weak form of (1) leads to a linear system of ordinary differential equations of semidiscrete form

$$M\ddot{U} + C\dot{U} + KU = F\,, \quad \dot{U}(0) = \dot{U}_0 \text{ and } U(0) = U_0 \tag{2}$$

The integration in time is made either by modal superposition based on solutions of eigenproblems or by direct integration.

2.2. Discretization error

The finite element solution is denoted by u^h. The error of the solution and its first time derivative are

$$e = u^h - u, \quad \dot{e} = \partial e/\partial t = \dot{u}^h - \dot{u} \tag{3}$$

The error comes both from the spatial discretization and the time discretization. This means that for an optimal solution both the spatial mesh and the time step ought to be refined in an adaptive way.

The interpolation theory [6] is used to smooth the spatial discretization error. By partition of the scalar function $|\nabla u|$ over the domain by use of isolines, the local element size is found to be the distance between the isolines [7]. For a FE-approximation the spatial error in kinetic energy ΔE_k is two orders smaller than the error in strain erergy ΔE_s which means that error in total energy $E_t \approx \Delta E_s$ and the same error estimation can be used both for static and dynamic problems [8]. The Zienkiewicz-Zhu method [5] is applied to measure the error in energy norm (4) for linear triangles where σ^* is a postprocessed stress field of one order higher than the FE-solution σ^h.

$$\| e \|_E = \left(\int_V e^t \bar{\nabla}^t K^d \bar{\nabla} e dV\right)^{1/2} = (2\Delta E_s)^{1/2} = \left[\int_V (\sigma^h - \sigma^*) K^{d-1} (\sigma^h - \sigma^*)\right]^{1/2} dV \tag{4}$$

Different methods for automatic time-step control have been described in the literature. Generally, the global error is difficult to obtain, and a local error method is often used in order to estimate the time discretization error and to control the step size. Traditionally, the local error estimates were obtained either from the difference of approximate solutions between a full step and two half steps or from the comparison of the results by two different methods. They are often unpractical because of the substantial additional computational cost. Considering computational efficiency and convenience of implementation, Zienkiewicz et al.[9] introduced a simple expression to indicate the local error. Unfortunately, it does not give an accurate approximation of the real local error. Recently, Zienkiewicz and Xie[10] proposed another local error estimator for the Newmark scheme by comparing the numerical solution with the exact solution expanded in Taylor series. Also recently, Zeng et al.[11] derived a similar expression in a more simple and intuitive way based on the so-called "post-processing" technique. This error estimator is simple and asymptotically convergent for the original Newmark scheme, however, it is invalid for the other integration schemes which assume that the accelerations vary linearly within each time interval.

A new simple and reliable local error estimator which can overcome this problem is presented by Li et al.[12]. The error of solution is here based on the difference of a more accurate solution and the solution itself, in the same way as for the error estimation of spatial discretization in (4). Starting from the basic assumption of a direct integration method as stated above, we expect a more accurate solution to be at hand, if a higher order variation of accelerations within each time interval is assumed. We here assume that the derivatives of the accelerations vary linearly. In practice, we don't really need to calculate this higher-order solution, instead we calculate the difference of solutions directly by the "post-processing" technique. By this approach, we derive an a posteriori error estimator, which is very simple and shows a quite high accuracy and reliability for the Newmark scheme. The Newmark scheme consists of the following equations

$$\mathbf{M}\ddot{\mathbf{U}}_{t+\Delta t} + \mathbf{C}\dot{\mathbf{U}}_{t+\Delta t} + \mathbf{K}\mathbf{U}_{t+\Delta t} = \mathbf{F}_{t+\Delta t} \ , \quad \dot{\mathbf{U}}_{t+\Delta t} = \dot{\mathbf{U}}_t + [(1-\delta)\ddot{\mathbf{U}}_t + \delta\ddot{\mathbf{U}}_{t+\Delta t}]\Delta t \ , \tag{5a}$$

$$\mathbf{U}_{t+\Delta t} = \mathbf{U}_t + \dot{\mathbf{U}}_t \Delta t + \ddot{U}_t^N \Delta t^2/2, \quad \ddot{U}_t^N = [(1-2\beta)\ddot{\mathbf{U}}_t + 2\beta \ddot{U}_{t+\Delta t}] \tag{5b,c}$$

where $\mathbf{U}_{t+\Delta t}$, $\dot{\mathbf{U}}_{t+\Delta t}$ and $\ddot{\mathbf{U}}_{t+\Delta t}$ are the approximations of $\mathbf{U}$, $\dot{\mathbf{U}}$ and $\ddot{\mathbf{U}}$ at time station t+Δt, respectively, and $\ddot{\mathbf{U}}_t^N$ is the approximation of the acceleration.

An error estimate of postprocessing type [11] can be obtained if we assume that the acceleration $\ddot{U}_t^N$ is constant in each time intervall.
A one order better approximation is to assume a linear variation of $\ddot{U}$

$$\ddot{U}_\tau^* = \ddot{U}_t + \frac{\tau}{\Delta t}(\ddot{U}_{t+\Delta t} - \ddot{U}_t)\,, \quad \tau \in 0, \Delta t \tag{6}$$

A post-processed error can be written as

$$\ddot{e}(\tau) = \ddot{U}_t^N - \ddot{U}_\tau^* \tag{7}$$

Integration two times and letting τ to be Δt gives the local error estimate, which is given in a certain norm

$$\| e_1^N \| = (-\frac{1}{6} + \beta)\Delta t^2 \, \| (\ddot{U}_{t+\Delta t} - \ddot{U}_t \| \tag{8}$$

This is exactly the same estimate as given in [10], where the derivation was made by Taylor expansion. For the constant-average-acceleration method ($\beta = 1/4$) we get the error $\| e^0 \| = (\Delta t^2/12) \, \| \ddot{U}_{t+\Delta t} - \ddot{U}_t \|$. This estimate cannot be used for the linear acceleration method ($\beta = 1/6$) because $\ddot{U}^N$ is linear and $\ddot{U}^*$ in (6) is then not a higher order approximation. For many parameters the Newmark scheme is of second order, so the local error should be of order Δt^3 and not Δt^2 as in (8).
The Newmark scheme contains an approximation of the derivative of the acceleration which is $\dddot{U}_t^N = 6\beta\dddot{U}_t^A$, where $\dddot{U}_t^A = (\ddot{U}_{t+\Delta t} - \ddot{U}_t)/\Delta t$ is the derivative of the acceleration which keeps constant within each time step. A higher order aproximation is then a linear variation of the derivative of the acceleration in the time step

$$\dddot{U}_\tau^* = \dddot{U}_t + \frac{\tau}{\Delta t}(\dddot{U}_{t+\Delta t} - \dddot{U}_t), \quad \tau \in 0, \Delta t \tag{9}$$

Fig. 1 shows the variation of the derivative of acceleration for the case of the linear acceleration method. Naturally, the derivatives of accelerations in the linear acceleration method ($\beta = 1/6$) keeps constant within each time interval, which means that $\dddot{U}_t^N = \dddot{U}_t^A$.

Now we construct a post-processed error of the derivative accelerations [12]

$$\dddot{e}(\tau) = \dddot{U}_t^N - \dddot{U}_\tau^* \quad , \quad \dddot{U}_t^N = 6\beta\dddot{U}_t^A$$

Since the values at time station t are assumed to be exact, integrating (10) three times and letting τ be Δt, we obtain the local error estimate which is given in a certain norm either as (11a) or (11b)

$$\| e^N \| = \frac{1}{12}\Delta t^3 \, \| (12\beta - 1)\dddot{U}_t^A - \dddot{U}_t \| \, , \quad \| e^N \| = \frac{1}{12}\Delta t^3 \, \| (6\beta - \frac{1}{2})\dddot{U}_{t+\Delta t} - (\frac{3}{2} - 6\beta)\dddot{U}_t \| \tag{11a,b}$$

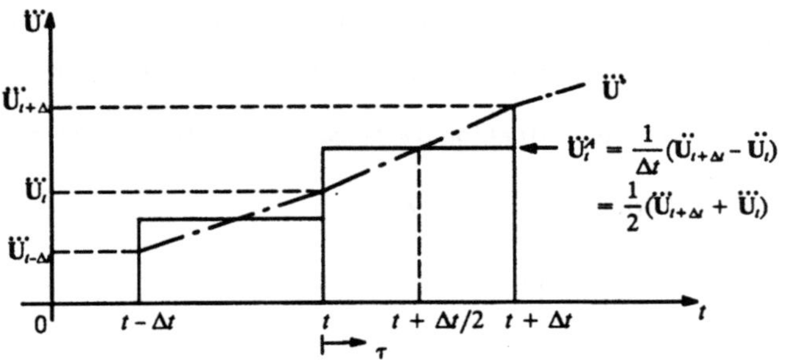

Fig.1 The derivative of acceleration: linear acceleration method (———) and the post-processed continuous derivative (—·—)

When $\beta = 1/6$ we obtain the error (12a) for the linear acceleration method as

$$\| e^L \| = \frac{1}{12} \Delta t^3 \| \dddot{U}_t^A - \dddot{U}_t \| , \tag{12}$$

When $\beta = 1/4$ we obtain the error (13a) for the constant-average-integration method, which for $\dddot{U}_t \equiv 0$ equals the error estimate $\| e^{00} \|$ in (13b) given in [9].

$$\| e^A \| = \frac{1}{12} \Delta t^3 \| 2\dddot{U}_t^A - \dddot{U}_t \| , \quad \| e^{00} \| = \frac{1}{6} \Delta t^2 \| \ddot{U}_{t+\Delta t} - \ddot{U}_t \| \tag{13a,b}$$

As shown in [12], this simple local error estimator (11) may also be obtained either by Taylor expansion as or by the difference between the solution of (5) and a solution based on the assumption (9).

The reliablity of the error estimation is measured by an effectivity index $\theta = \| \bar{e} \| / \| e \|$ where $\bar{e}$ and e represent the predicted and actual local error, respectively, or by an absolute error tolerance $\eta = \| \bar{e} \| / \| U \|_{max}$, when $(| U \|_{max}$ is the maximum value of the corresponding norm of the displacement solution recorded so far during the computation.

2.3. Numerical examples for dynamics

(a) Dynamic problem, direct integration, spatial discretization error

Fig. 2a shows a rectangular domain subjected to a half-sinusoidal pulse $q(t)$ giving a stress wave propagation. The mesh is updated from time to time based on a post-processor error estimation. The time step is chosen small. So only the spatial discretization error is studied. Fig. 2b shows the predicted graded mesh for $t \in [3.1, 3.3]$. Fig. 2c shows that the solutions satisfy the given error tolerance in all stages. For comparison results from two conventional FE-analysis are also shown.

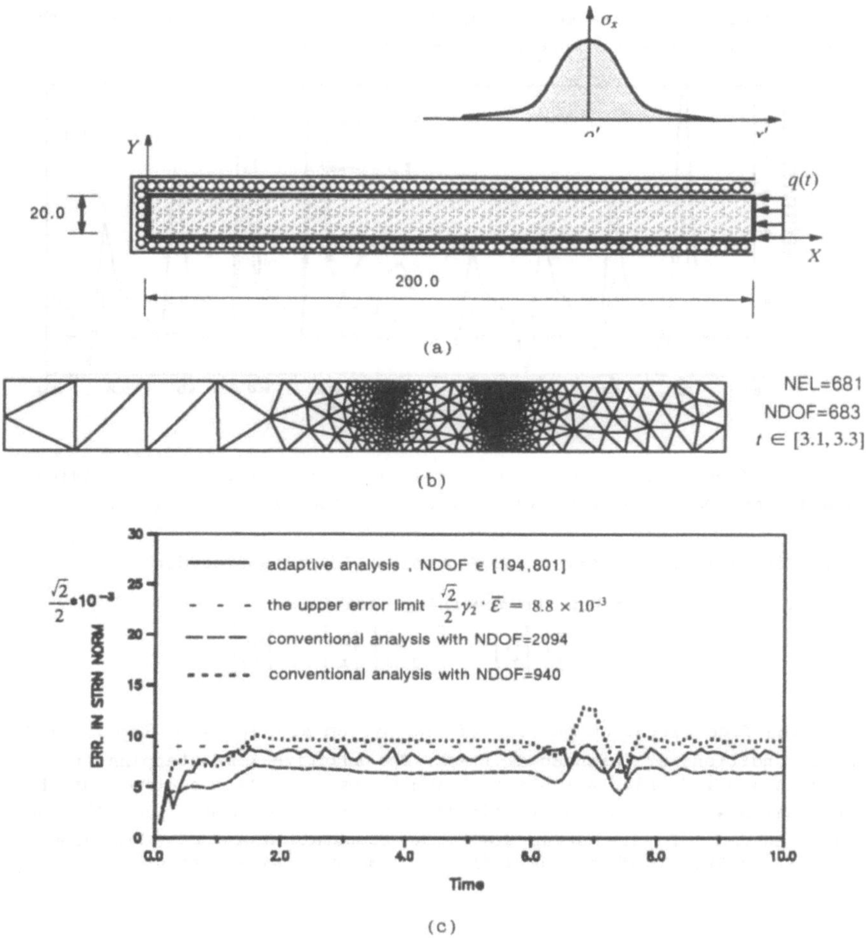

Fig.2 (a) Rectangular domain subjected to a half-sinusoidal pulse $q(t) = \sin(\pi t/4)$ and a sketch of its stress solution
(b) Predicted mesh at time station $t \in [3.9, 4.1]$
(c) Temporal error $\| e \|_E / \sqrt{2}$ for FE-analysis, [13]

(b) Time integration error

The equation of motion and the initial conditions for an undamped SDOF system has the form

$$\ddot{x} + \omega^2 x = 0 \ , \quad \omega = 1.571, \ x(0) = x_0 = 1.0, \ \dot{x}(0) = \dot{x}_0 = 0.0$$

As we know the exact solution, we want to assess the accuracy and reliability of the proposed error estimator (11). The effectivity index is plotted in Fig. 3 for $\Delta t/T = 0.1$, where T is the free vibration period. For the original Newmark method error estimators $\| e^{\infty} \|$ and $\| e^{o} \|$

do not give an accurate measure of the local error. The new local error estimator $\|e^N\|$ gives a more accurate and reliable estimation.

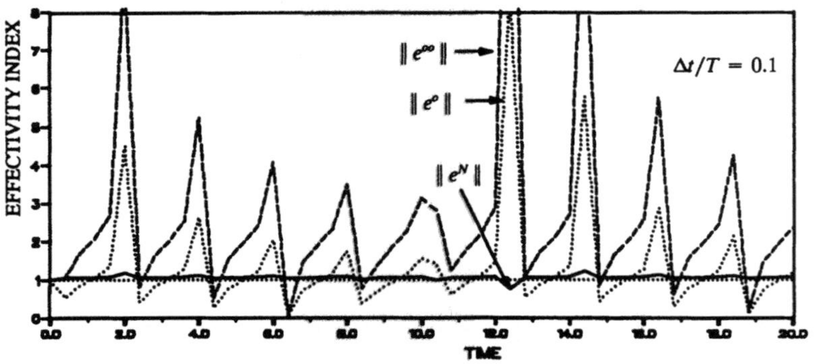

Fig. 3 Effectivity index of the local error estimator for the original Newmark method for SDOF undamped free vibration $\Delta t/T = 0.1$, [12].

(c) Time integration error, MDOF-problem, Adaptive procedure.

A two-DOF system has the governing equations

$$\begin{bmatrix} 2 & 0 \\ 0 & 1 \end{bmatrix} \begin{bmatrix} \ddot{x}_1 \\ \ddot{x}_2 \end{bmatrix} + \begin{bmatrix} 6 & -2 \\ -2 & 4 \end{bmatrix} \begin{bmatrix} x_1 \\ x_2 \end{bmatrix} = \begin{bmatrix} 0 \\ 10 \end{bmatrix}$$

The free vibration periods of the system are $T_1 = 4.45$, $T_2 = 2.8$ and rest conditions are assumed at $t = 0$. The adaptive time-stepping procedure is applied. Fig. 4 shows that the estimated relative error for the original Newmark method is within the prescribed accuracy $\eta \le \nu_2 \eta^{tol}$. The total number of time steps for the computation in a time domain [0,10s] is 590, in which 102 different time step sizes are used.

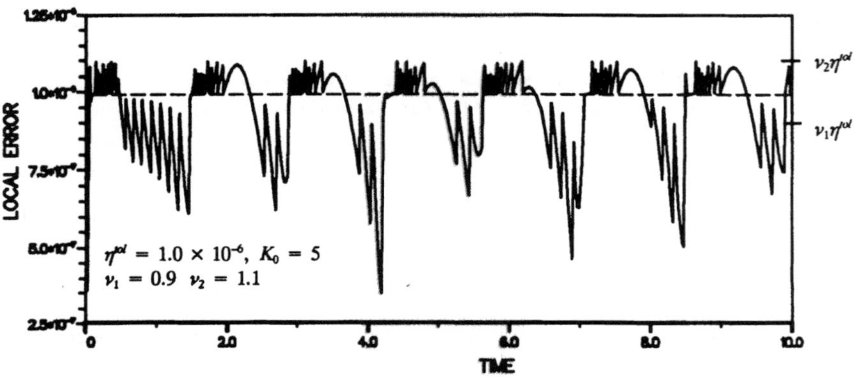

Fig. 4 The relative local error achieved by the adaptive time-stepping procedure for the two-DOF model with the original Newmark method, [12]

3. ADAPTIVE FE-PROCEDURES FOR ELASTOPLASTIC PROBLEMS

3.1. Governing equations

The constitutive equations in small deformation elastoplasticity with an associated flow rule can for v. Mises theory and isotropic hardening be written

$$\begin{aligned} \epsilon(\dot{\mathbf{u}}) &= \mathbf{D}^{-1}\cdot\dot{\sigma} + \dot{\epsilon}^P \\ (\sigma-\tau)\cdot\dot{\epsilon}^P &\geq 0, \quad \forall\tau \in B(\kappa) \\ \kappa &= c|\epsilon^P| \end{aligned} \tag{14}$$

where $\mathbf{D}$ is the tensor of elastic stiffness moduli, $\dot{\epsilon}^P$ is the rate of plastic strain, κ a plastic strain controlling isotropic hardening and

$$B(\kappa_v) = \{\sigma : F(\sigma,\kappa) \leq 0\} \tag{15a}$$

and

$$F(\sigma,\kappa) = \bar{\sigma} - \sigma_Y \tag{15b}$$

where $\bar{\sigma} = |\sigma_d|/c$ is the equivalent stress, σ_Y is the current reduced yield stress and $c = \sqrt{2/3}$. Subscript d denotes a deviator quantity.

With a backward Euler scheme for the time integration the following discrete equations should be satisfied at $t_{n+1} = t_n + \Delta t$

$$\begin{aligned} -\nabla\cdot\sigma &= \mathbf{f} \\ (\sigma-\tau)\cdot\mathbf{D}^{-1}\cdot(\sigma-\sigma^E) &\leq 0 \quad \forall\tau \in B \end{aligned} \tag{16}$$

where the fictitious elastic stress is defined as

$$\sigma^E = \sigma_n + \Delta t\mathbf{D}\cdot\epsilon(v), \qquad \mathbf{u}_{n-1} = \mathbf{u}_n + \Delta t v \tag{17}$$

The inequality in (16) has the solution

$$\sigma = \pi_c\sigma^E \tag{18}$$

where π_c is the projection operator onto B in complementary energy that is the solution of

$$\min_{\tau \in B^*} |\sigma^E - \tau|_c^2, \qquad |\tau|_c^2 = \tau\cdot\mathbf{D}^{-1}\cdot\tau \tag{19}$$

Explicit formulas for the solution can be obtained, see [14].

With a finite element space discretization the nodal velocities v are obtained from the variational formulation in the usual way as

$$\int_\Omega B^T\sigma\epsilon(v)dx = \int_\Omega \phi\mathbf{f}(t_{n+1})dx \tag{20}$$

There are several sources of error in these calculations. There are the discretization errors in time and space and there are numerical errors in the iterative solution of (20). Finally, there are round-off errors.

We will here discuss only the errors due to space discretization and procedures for an adaptive improvement of the finite element approximation. At the adaptive refinement a further source of errors is obtained at the mapping between finite element nets.

3.2. A posteriori error estimates

In a recent paper [15] Johnson and Hansbo give an a posteriori estimate for an antiplane shear problem. The material is assumed to be elastic-perfectly plastic. For the Hencky problem (or a time-step in a transient problem) and CST-elements they prove for zero volume load the following L_2-estimate in stresses

$$\|\sigma-\sigma_h\|\,|_{L_2(\Omega)} = C(\|hD_h\sigma_h\|_{L_2(\Omega^e)} + \|hD_h\sigma_h\|^{\frac{1}{2}}_{L_\infty(\Omega^p)}) \tag{21}$$

with for an element K

$$D_h\sigma_h = \max_{\tau\subset\partial K} \frac{|[\sigma_h\cdot n_\tau]|}{h_K} \tag{22}$$

where [ ] is the interelement jump, τ is a side of K with unit normal n_τ and the region is divided into an elastic and a plastic region Ω^e and Ω^p. A slower convergence can thus be expected in the plastic region than in the elastic one. We will in a later paper test the use of this estimate for adaptive control.

In [16] Bernspång has tried for CST elements the following error control measure for the consistent stress

$$\|\sigma-\sigma_h\|_{L_2} = C_1\|hD_h\sigma_h\|_{L_2} \tag{23}$$

with $D_h\sigma_h = |\nabla\bar{\sigma}_h^*|$ where $\bar{\sigma}_h^*$ is the effective stress obtained from ϵ_h in a postprocess by smoothening of ϵ_h to ϵ_h^*, calculation of σ_h^*, and iteration to equilibrium.

He also tried in the spirit of Zienkiewicz-Zhu

$$\|\bar{\sigma}-\bar{\sigma}_h\|_{L_2} \approx \|\bar{\sigma}_h^*-\sigma_h\|_{L_2} \tag{24}$$

We observe that (24) is an error in the conventional stresses while (23) is an error in the consistent stresses obtained after smoothening.

3.3. Numerical results with adaptive elastoplasticity

A elastic-perfectly plastic plate, Fig. 5a, has been calculated in [16] with an adaptive procedure based on an unstructured mesh generator, [7], based on isolines (ADMESH) followed by refinements. With (23) the final net, Fig. 5b, was obtained after 11 remeshings and 12 refinements. The distribution of the plastic zone after the adaptive procedure is shown in Fig. 5c. After a recalculation with the final net the plastic zone in Fig. 5a was obtained. The same problem solved with (24) gave after 11 remeshings and 10 refinements the mesh, Fig. 5e. The distribution of the plastic zone directly and after recalculation is shown in Fig 5f and 5g. The results are rather different. Looking at the displacements they appear to converge faster with the approach based on (23) than with (24). The extra equilibrium iterations involved in (23), however, reduces the gain considerably.

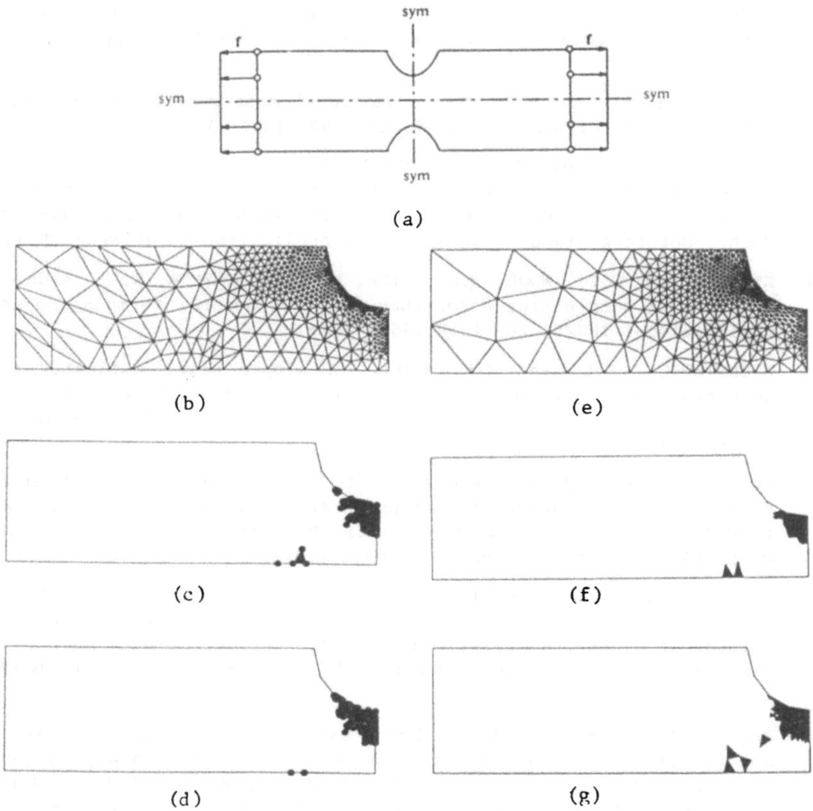

Fig 5. Adaptive procedure for elastic-perfectly plastic plate (from [16])

4. REFERENCES

1. PEANO, A., PASINI, A., RICCIONI, R. and SARDELLA, L. - Adaptive approximations in finite element analysis, Computers and Structures, 10, 333-342 (1979).

2. ZIENKIEWICZ, O.C., GAGO, J. and KELLY, D.W. - The hierarchical concept in finite element analysis, Computers and Structures, 16, 53-65 (1983).

3. WIBERG, N-E. and MÖLLER, P.- Formulation and Solution of hierarchical finite element equations, INJME, 26, 1213-1233 (1988).

4. BABUSKA, I; ZIENKIEWICZ, O.C., GAGO, J. and OLIVEIRA, E.R. de A. (Ed) - Accuracy estimates and adaptive refinements in finite element computations, Wiley & Sons, 1986.

5. ZIENKIEWICZ, O.C. and ZHU, J.Z.- A simple error estimator and adaptive procedure for practical engineering analysis, INJME, 24, 333-357 (1987).

6. ERIKSSON, K. and JOHNSON, C. - An adaptive finite element method for linear problems, Mathematics and Computations, 50, 361-383 (1988).

7. JIN, H. and WIBERG, N-E - Two dimensional mesh generation, adaptive remeshing and refinement. INJME, 29, 1501-1526 (1990).

8. ZENG, L.F., WIBERG, N-E. and BERNSPÅNG, L. - An adaptive finite element procedure for 2D-dynamic transient analysis using direct integration, Publ:90:8, Dep. of Struct. Mech., Chalmers Univ. of Tech., Göteborg, Sweden, accepted for publication in INJME (1990).

9. ZIENKIEWICZ, O.C., WOOD, W.L., HINE, N.W. and TAYLOR, R.L.- A unified set of single step algorithms, Part 1. General formulation and applications. INJME, 20, 1529-1552 (1984).

10. ZIENKIEWICZ, O.C. and XIE, Y.M. - A simple error estimator and adaptive time stepping procedure for dynamic analysis, Research Report, CR/663/91, Inst. for Num. Meth. in Eng., Univ. College of Swansea, (1991).

11. ZENG, L-F, WIBERG, N-E and LI, X.D. - A posteriori local error estimation and adaptive time-stepping for original Newmark Integration in dynamic analysis, Publ 91:11, Dep. of Struct. Mech., Chalmers Univ. of Tech., Göteborg, Sweden (1991).

12. LI, X.D., ZENG, L-F and WIBERG, N-E. - A simple local error estima tor and an adaptive time-stepping procedure for direct integration methods in dynamic analysis, Publ. 91:12, Dep. of Struct. Mech., Chalmers Univ. of Tech., Submitted to Comm. Appl. Num. Meth. (1991).

13. ZENG, L.Z. and WIBERG, N-E. - Error estimation and adaptivity of spatial discretization in semidiscrete finite element analysis for dynamic problems, Publ 91:10, Dep. of Struct. Mech., Chalmers Univ. of Techn., Göteborg, Sweden. Submitted to Computational Mechanics (1991).

14. RUNESSON, K., SAMUELSSON, A., BERNSPÅNG, L.- Numerical technique in plasticity including solution advancement control, IJNME, 22, 769-788, (1986).

15. JOHNSON, C. AND HANSBO, P. - 'Adaptive finite element methods for small strain elasto-plasticity', Preprint 1991:23, Dep. of Math., Chalmers Univ. of Techn., Göteborg, Sweden, (1991).

16. BERNSPÅNG, L. - Iterative and adaptive solution technique in computational plasticity, Thesis, Publ 91:8, Dep. of Struc. Mechanics, Chalmers Univ. of Technology, Göteborg, Sweden, (1991).

FINITE ELEMENTS IN THE 90's, E. Oñate, J. Periaux, A. Samuelsson (Eds.)
Springer-Verlag/CIMNE, Barcelona 1991

ON THE USE OF FINITE ELEMENTS IN FAST TRANSIENT DYNAMICS

J. Donea and **J.P. Halleux**
Commission of the European Communities,
Joint Research Centre, I-21020 Ispra (VA),
ITALY

Foreword

It is now more than twenty years that we have had the privilege of knowing Professor O. C. Zienkiewicz since participating in the first courses he organized at Swansea to promote the use of the then new method of finite element analysis by us, engineers involved in applied research. It is true to say that the eminent qualities of Professor Zienkiewicz and his enthusiasm in describing the potential of finite elements for modelling complex engineering problems were decisive factors for our early involvement in what would soon become the most powerful numerical method so far devised [1].

The present paper is dedicated to Professor Zienkiewicz on the occasion of his seventieth birthday and as an expression of our gratitude to him and his team at Swansea for many years of fruitful collaboration and friendship.

1 BRIEF HISTORICAL INTRODUCTION AND BASIC APPROACH

Computer codes able to predict the response of structures undergoing large elasto-plastic deformation were already in use in the early sixties. Their development started in the fifties and relied mainly upon finite difference techniques. For instance HEMP developed at U.S. Lawrence Livermore National Laboratory, was a two-dimensional explicit finite difference Lagrangian code making use of a Cauchy stress rate formulation [2]. This code belonged to a family of codes generally used for compressible fluid flow which are often referred to as "hydrocodes". While the first efforts aimed at the computation of high pressure shock wave propagation related to weapon effects, the codes were progressively applied to more conventional engineering problems among which were the safety related aspects of nuclear reactors.

It is in this field, at JRC Ispra, in the early seventies, that the senior author became interested in the numerical treatment of fast transient dynamic phenomena. At that time finite element methods (FEM) had already been firmly established as the best computational tools for linear structural analysis. These nowadays classical methods consisted essentially of two phases, namely the construction of a stiffness matrix followed by the solution of the associated linear system of equations. Their success induced many researchers to tackle nonlinear analysis by extending the apparently so promising existing methods. This led to the production of implicit computer codes. However, few of the codes developed along these lines appear to be able to deal realistically with fast transient dynamic finite deformation problems.

Precursor work [3, 4] was already indicating that the hydrocode methodology was likely to be largely and easily improved by substituting the spatial finite differencing by a finite element (FE) stencil directly derived from virtual work considerations. This led to the production at JRC Ispra of the EURDYN FE codes for nonlinear dynamic analysis of structures [5]. Following the successful application to structures, the idea came to apply FEM also to the description of the fluid domain present in most of the reactor safety problems (interest was then focused on pool type liquid metal fast breeder reactors).

The continuous following of the fluid motion in case of an abnormal event ensures an optimal evaluation of the pressure transmitted to the resisting structures and thereby provides very precise analysis of the plant safety. While structural deformation in large displacement-large rotation problems could be easily and well captured through a Lagrangian description of motion, it quickly appeared that Lagrangian fluid meshes often became excessively distorted. On the other hand a pure Eulerian description, as was often adopted in finite difference approaches, was inflexible with respect to the inevitable boundary motions. Therefore an arbitrary Lagrangian-Eulerian (ALE) [6] methodology was introduced which quickly became popular among the then emerging new generation of "FE hydrocodes". The ALE method also appears to be a good candidate for large strain solid mechanics (metal forming for instance).

While the EURDYN computer codes were originally produced for reactor safety problems, their geometrical flexibility in general and in particular their capacity to handle not only 2-D but also 3-D problems, thanks to the use of FEM, prompted their diffusion in many application fields. Today entire FE code systems with element and material libraries are being built up for the treatment of transient dynamic problems in general. One such system, PLEXIS-3C [7], is currently being built up at JRC Ispra in collaboration with CEA (F).

In FE hydrocodes, as already mentioned, the principle of virtual work (variational form of conservation of momentum) serves as a basis for a finite element discretization in space [1]. The following set of discrete differential equations in time can then be obtained:

$$\boldsymbol{M}\ddot{\boldsymbol{u}} = \boldsymbol{f}^{\text{ext}} - \sum_{e} \int_{V^e} \boldsymbol{B}^T \boldsymbol{\sigma}\, dV, \tag{1}$$

where $\boldsymbol{M}$ is a mass matrix; $\ddot{\boldsymbol{u}}$ is the nodal acceleration vector; $\boldsymbol{f}^{\text{ext}}$ is a vector of externally applied loads; V^e represents the element (e) volume in the current configuration; $\boldsymbol{B}$ is the matrix of shape functions derivatives; $\boldsymbol{\sigma}$ is a matrix representing the approximate Cauchy stresses acting in the media.

It may be noted that stiffness matrices do not appear explicitly in the above equations. In fact, the dynamic character of the problem will here be fully exploited. The equations are solved for $\ddot{\boldsymbol{u}}$ and a direct time marching algorithm is used to obtain suitable estimates for velocities and displacements. The solution of these equations is rendered almost trivial by an a priori diagonalization of the mass matrix $\boldsymbol{M}$. The main difficulties related to this methodology are concentrated in the evaluation of internal forces (last term in eq. (1). In practice once a new (current at time t^n) configuration is obtained ($\boldsymbol{u}^n$), the corresponding stresses ($\boldsymbol{\sigma}^n$) are evaluated (most often using an incremental technique) in order to get an estimate for $\ddot{\boldsymbol{u}}$. Then the time integration algorithm is applied (see next section) in order to obtain a new velocity which in turn will be used to update to the next configuration. In the common case of elasto-plastic material behaviour the stress update is performed via the radial return algorithm [2]. The underlying theoretical and numerical aspects can be found in [8].

2 THE EXPLICIT TIME INTEGRATION SCHEME

Velocities and displacements at time $t^{n+1} = t^n + \Delta t$ (Δt being a time increment) are obtained by a standard central difference integration scheme:

$$\dot{u}^{n+1} = \dot{u}^n + \frac{\Delta t}{2}(\ddot{u}^n + \ddot{u}^{n+1}), \tag{2}$$

$$u^{n+1} = u^n + \Delta t(\dot{u}^n + \frac{\Delta t}{2}\ddot{u}^n). \tag{3}$$

In fact motion is linearized between n and $n+1$. It is therefore useful to introduce an associated constant velocity:

$$v^{n+1/2} = v^{n-1/2} + \Delta t \ddot{u}^n. \tag{4}$$

Eqs. (2,3) are then re-written as:

$$v^{n+1/2} = v^{n-1/2} + \frac{\Delta t_- + \Delta t_+}{2}\ddot{u}^n, \tag{5}$$

$$u^{n+1} = u^n + \Delta t v^{n+1/2}, \tag{6}$$

where Δt_- and Δt_+ stand for successive time increments.

This last form (5,6) of the integration scheme is of course perfectly equivalent to the original scheme (2,3) but has the merit to put better into evidence the essence of the method used here. It should perhaps also be recalled that no damping is associated with this integrator. The implications related to the use of central difference time integration (stability, use of a diagonal mass matrix) can be found in [9].

Explicit time integration as presented here is only conditionally stable. This means that a very large number of time steps may be necessary in practical cases. It is, however, important to realize that implicit integrators do perform well in problems involving high frequencies only if time steps of the size required by explicit integration are used.

3 CONTINUUM ELEMENTS

Triangular shaped elements are well known and have of course often been used in FE analysis. Their ability in meshing geometries with cumbersome boundaries is a clear advantage with respect to competitors. Also their relative simplicity is appealing. On the other hand, too stiff behaviour can often be observed and some retain that optimum meshing of certain geometries (like thin members or regularly shaped regions) cannot be achieved.

Quadrilateral elements are very popular in FE hydrocodes. The 4 noded (8 noded in 3-D) bilinear quadrilateral element is likely to be the absolute best-seller here. This success is in great part explained because of the complexity often involved in the use of higher order elements for strongly nonlinear problems. In fact for linear elements, many simplifications due to the presence of constant terms are found with respect to the general FE formulation. The treatment of compressible fluid flow (see section 5) relies on simple quadrilateral elements while large strain solid mechanics modelling (briefly considered later in this section) is approached both with linear and quadratic elements.

The computation of internal nodal forces at element level is generally obtained via numerical quadrature (Gaussian). This poses a problem of choice: depending on the application, and of course on the FE of interest, what is the most suitable integration scheme? One may think of systematically trying to compute in order to come to, or close to an "exact" integration. This however is unconvincing in at least two aspects. Firstly the computational costs of higher order quadrature may become a burden; for example if the material behaviour is history dependent and many quadrature points are used it is easy to imagine the amount of CPU time which will go into stress update operations. Secondly, too precise integration leads to locked (stiff), often even unusable answers if the case of interest is prone to constraints.

Therefore element quadrature is frequently performed in a reduced manner. For instance 4 (9) node quadrilaterals are integrated with only 1 (4) quadrature point, leading to large CPU savings and to solutions which never suffer from any locking phenomena. Unfortunately under-integration is very likely to provoke rank deficiency in the matrices corresponding to the resulting numerical representations of the real problem tackled. As these matrices are never formed nor explicitly used in the present implementation the case will run in practice but the solution pattern may be spoiled by the propagation of so-called mechanisms ("hourglass" modes in the 4-node quadrilateral with 1 point quadrature). It is important to note though that, despite the anaesthetic distortion of the mesh due to the presence of the mechanism, the overall solution obtained is of good quality and becomes affected only for very large distortions. It is therefore sufficient in practice to impeach excessive propagation of these modes and not at all necessary to completely suppress in some (complicated and costly!) way the rank deficiency, as would result mandatory in most "standard" finite element approaches.

This can often be achieved by adequate setting of boundary conditions. If this is still insufficient to inhibit spurious propagation of mechanisms, a simple trick consists in picking a few elements in the mesh and integrating these with a quadrature scheme impervious to the mechanism (higher order). Experience shows that one such element can be sufficient in stabilizing an entire mesh. The solution obtained cannot lock if only a few elements are integrated at a higher order; furthermore it retains cost effectiveness and accuracy. Indeed in under-integrated elements, the integration points are often optimally located [1].

The incremental treatment of elasto-plastic finite deformation of solids requires the definition of a so-called objective rate of stress. While various rates have been advocated in the literature, it is found that the Jaumann rate of Cauchy stress is well adapted to a large class of problems [10] and that it is most readily implemented in the explicit solution scheme proposed here even for higher order elements [11].

In a subsequent section it will be shown that the ALE technique is of great help in order to cope with large fluid squeezing and distortion. In solid mechanics, the media are less prone to such excessive motion, but in metal forming processes for example similar situations may be encountered. If this forming action is to be handled as an elasto-plastic deformation process, a further difficulty appears in applying ALE techniques; history-dependent quantities need to be convected. Recent research indicates that this is indeed feasible and its implementation in a FE hydrocode does not require any major methodological change [12].

4 BEAM, PLATE AND SHELL ELEMENTS

It is well-known that thin members can be modelled by specialized elements derived from theories exploiting various hypotheses related to the definition of privileged directions. Reactor structures are typically rather thin and consequently such elements are widely used for reactor safety analysis. Plates and shells are used in the most general cases but very frequently the problem can be tackled assuming axisymmetric conditions. This led to the development of conical shell elements. The practical application which is shown in the last section (see Fig. 2) is an example of reactor containment safety analysis where the structural members are meshed with such an element. It is therefore chosen

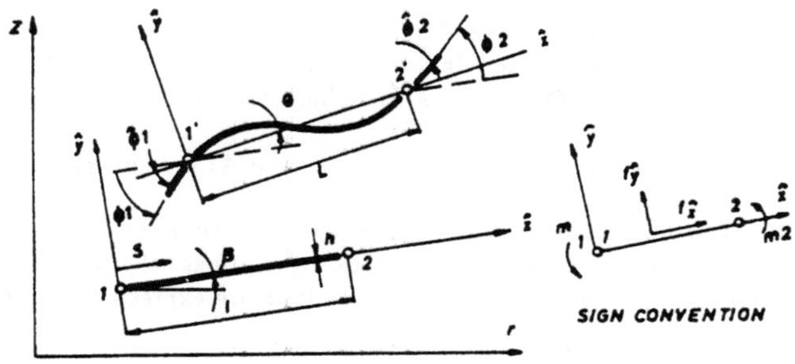

Figure 1: Conical shell element.

to describe the kinematics of this element in somewhat more detail here.

The Euler-Bernoulli theory is used in a corotational coordinates formulation (Fig. 1). The transverse displacement is assumed to be cubic in $\hat{x}$, the longitudinal displacement linear in $\hat{x}$. Angle θ is assumed to approximate the rigid body rotation of the element (small strain hypothesis). If ϕ_1 and ϕ_2 are the total rotations obtained from the integration procedure, the deforming rotations (rotations relative to the $\hat{x}$ axis) are:

$$\hat{\phi}_1 = \phi_1 - \theta, \qquad \hat{\phi}_2 = \phi_2 - \theta. \tag{7}$$

Using normalized coordinates ranging from -1 to 1, $\xi = (2s - \ell)/\ell$, $\eta = 2\hat{y}/h$, one obtains:

$$\epsilon_x = \frac{L-\ell}{\ell} - \frac{h}{2\ell}\eta\left[(3\xi - 1)\hat{\phi}_1 + (3\xi + 1)\hat{\phi}_2\right], \tag{8}$$

$$\epsilon_\theta = \frac{1-\xi}{2}\frac{u_{r1}}{r} + \frac{1+\xi}{2}\frac{u_{r2}}{r} - \frac{\eta h}{2r}\cos(\beta+\theta)\left[\theta + \frac{\hat{\phi}_1}{4}(3\xi^2 - 2\xi - 1) + \frac{\hat{\phi}_2}{4}(3\xi^2 + 2\xi - 1)\right], \tag{9}$$

with $r = \frac{1-\xi}{2}r_1 + \frac{1+\xi}{2}r_2$.

It is easily seen that when the element has a node on the axis (for instance $r_1 = 0$, then $u_{r1} = 0$ and $\hat{\phi}_1 = -\theta$), one consistently gets for $\xi \longrightarrow -1$:

$$\epsilon_r = \epsilon_\theta = \frac{u_{r2}}{r_2} + \frac{h\eta}{\ell}(2\hat{\phi}_1 + \hat{\phi}_2)\cos(\beta + \theta). \tag{10}$$

Expressions for the corresponding internal forces and for the lumped masses are reported in [13].

5 ALE DESCRIPTION OF COMPRESSIBLE FLUID FLOW

Arbitrary Lagrangian-Eulerian (ALE) finite element methods were introduced in the late 1970's in response to the need for very versatile modelling techniques for treating transient fluid-structure systems. By allowing the mesh points to move with a velocity $\boldsymbol{w}$ independent from the particle velocity $\boldsymbol{v}$, the ALE kinematic description provides the possibility to handle greater distortions in the fluid motion than would be allowed by a purely Lagrangian method, with more resolution of details of the flow and interfaces than is afforded by a purely Eulerian method. A summarized account of early developments of

ALE finite element methods for compressible flow problems is given in [14]. Subsequent studies described in the recent literature indicate the potential of this mixed formulation for solving large boundary-motion problems, including cases involving nonlinear, path-dependent materials.

Several difficulties have to be overcome to make the ALE finite element method practical and accurate.

A first difficulty arises because of the presence of convective terms in the mathematical formulation of flow problems based on kinematic descriptions other than Lagrangian. In the ALE description, the governing conservation laws are expressed in terms of the material derivative

$$\frac{Df}{Dt} = \frac{\partial f}{\partial t} + (v_j - w_j)\frac{\partial f}{\partial x_j} \tag{11}$$

where f denotes a typical physical property of the continuum. The first term in the right-hand side of Eq. (11) represents the variation per unit time felt by an observer moving with the velocity $\boldsymbol{w}$ of the mesh points. The second is the convective term expressed in terms of the relative velocity $(\boldsymbol{v} - \boldsymbol{w})$ of the material points with respect to the mesh points. Convection operators are nonsymmetric and, as a result, the best approximation property of the standard Galerkin finite element method, which is the basis for its success in symmetric cases (e.g. solid mechanics and diffusion-type problems), is lost when convection dominates the transport process. In practice, solutions to convection dominated transport problems by the Galerkin method are often corrupted by spurious node-to-node oscillations or "wiggles". This has provided motivation for the development of alternatives to the standard Galerkin formulation specifically adapted to the treatment of problems governed by convective transport. A brief overview of recent developments in this area is given in [15].

At Ispra, we favour the use of the Taylor-Galerkin method, a generalization to finite elements of the well-known Lax-Wendroff finite difference method. To ease its implementation in conjunction with the solution of the basic conservation equations for mass, momentum and energy, a splitting-up, or fractional step approach is employed. This consists of two distinct phases: Phase 1 is a Lagrangian phase in which convective contributions are ignored and Phase 2 is the convective phase. Explicit time integration is systematically used in the treatment of transient, compressible flow problems. To enhance its numerical efficiency, partitioning techniques are employed, in which different time steps are used in different parts of the mesh, so as to fully exploit local stability conditions.

Another difficulty in implementing ALE finite element methods is linked to the freedom in moving the element mesh offered by the ALE formulation. Though very attractive, such freedom can, in fact, be overshadowed by the burden of specifying grid velocities $\boldsymbol{w}$ well suited to a particular problem. As a consequence, the practical implementation of the ALE description requires that an automatic mesh displacement prescription algorithm be supplied. Such an algorithm has been implemented into the computer code PLEXIS-3C and successfully used in the numerical application described in the next section.

6 APPLICATION EXAMPLE

The example considered here is a benchmark problem illustrating the analysis of a hypothetical core disruptive accident in a liquid-metal cooled fast breeder reactor [16].

The geometry is shown in Fig. 2 and assumes symmetry around the reactor axis (2D axisymmeric calculation). The deformable reactor tank is clamped to the roof which is assumed to be rigid and fixed in space. The effects of internal structures are

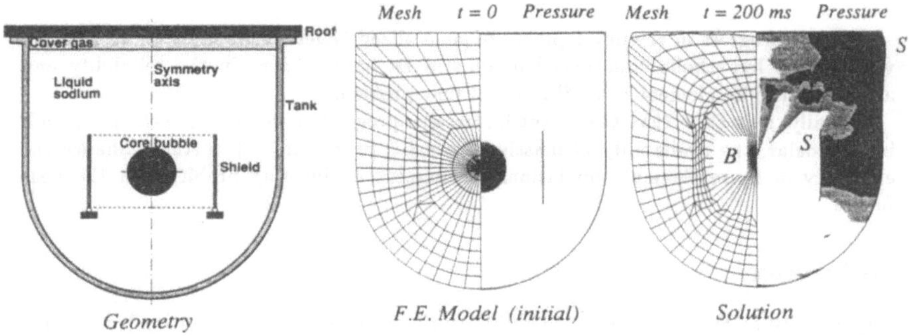

Figure 2: Reactor benchmark example

schematically taken into account by an internal stainless steel shield, hinged at its lower end. Three fluids, sodium as coolant, argon as cover gas and expanding fuel vapour as a core bubble are contained in the reactor tank. All material properties are defined in [16]. The finite element model (Fig. 2) uses conical shell elements to represent the deformable tank and shield, and 4-node quadrilaterals (with a few triangles) to mesh the bulk fluid. At first sight, the problem might seem quite simple, but a closer look reveals that correct modelling of various phenomena is required in order to get a reasonable answer. Critical points are, e.g., sodium sliding along the reactor tank, particularly near the sodium/argon interface, and the flow of sodium around the thin deformable internal shield.

In the PLEXIS-3C [7] solution illustrated in Fig. 2, ALE sliding is adopted around the internal shield and in the lower part of the reactor tank, while Lagrangian sliding is used close to the sodium/tank/argon interface. The fuel vapour bubble and the liquid sodium in its vicinity undergo very large displacements and distortions (see point "B" in Fig. 2), therefore the fluid domain rezoning algorithm is also severely tested. Finally, although strains are not extremely large in the structures, they attain values (nearly 4 % at "S" in Fig. 2) well beyond the elastic limit.

When compared with benchmark results [16], the solution obtained by PLEXIS-3C looks quite satisfactory. Permanent plastic strains in the tank and in the shield are within the scatterband of the reference solutions and so is the total roof impulse. The CPU time needed for this calculation was 30 minutes on an HP-9000 835s workstation, which appears to be rather low for such a problem.

7 CONCLUSIONS

A brief survey has been presented of finite element methods developed for simulating the nonlinear transient response of structures and fluid-structure systems exposed to fast dynamic loading.

Explicit time-stepping methods appear to be ideally suited for tracing the transient response in situations, such as those describing wave propagation, where high-frequency signals play an important rôle. They are of an extremely simple algorithmic structure, offer an easy treatment of material and geometric nonlinearities, and are computationally attractive in that few operations are required per time step, since no system solver is needed.

When used in conjunction with an automatic rezone algorithm, the ALE kinematic description allows an elegant treatment of the hydrodynamic domain and easily handles fluid-structure coupling.

Recent progress in the development of generalized Galerkin methods for solving advection and hyperbolic equations has allowed an improvement in the reliability and accuracy of finite element computations of flow problems.

Finally, it is clear that the recent impressive progress in computer technology and, in particular, the availability of massively parallel processors will further enhance the efficiency of finite element simulations of complex engineering problems in the near future.

References

[1] **ZIENKIEWICZ, O.C.** : "The Finite Element Method, 3rd Edition", McGraw-Hill, Maidenhead, 1977.

[2] **WILKINS, M.L.** : "Calculation of Elastic-Plastic Flow", Methods of Computational Physics, 3, Eds. B. Alder et al., Academic Press, New York, 1964.

[3] **KRIEG, R.D. and KEY, S.W.** : "Transient shell Response by Numerical Time Integration", Int. J. Num. Meths. Eng., 7, 273-286, 1973.

[4] **BELYTSCHKO, T. and HSIEH, B.J.** : "Nonlinear Transient Finite Element Analysis with Convected Coordinates", Int. J. Num. Meths. Eng., 7, 255-271, 1973.

[5] **DONEA, J. et al** : "Prediction of the Nonlinear Dynamic Response of Structural Components Using Finite Elements", Nucl. Eng. Des., 37, 95-114, 1976.

[6] **DONEA, J., FASOLI-STELLA, P. and GIULIANI, S.** : "Lagrangian and Eulerian Finite Element Techniques for Transient Fluid-Structure Interaction Problems", Paper B1/2, Trans. 4th SMIRT Conf., San Francisco, 15-19 August, 1977.

[7] **BUNG, H. et al.** : "PLEXIS-3C: a computer code for fast dynamic problems in structures and fluids". SMiRT 10, Anaheim, 1989.

[8] **HUGHES, T.J.R.** : "Numerical Implementation of Constitutive Models: Rate-Independent Deviatoric Plasticity", Theoretical Foundation for Large-Scale Computations for Nonlinear Material Behaviour, Eds. Nemat-Nasser et. al., Mechanics of Elastic and Inelastic Solids, 6, M. Nijhoff Publ., Dordrecht, 1984.

[9] **KEY, S.W.** : Transient Response by Time Integration: Review of Implicit and Explicit Operators, Advanced Structural Dynamics, Ed. J. Donea, Applied Science Publ., 1980.

[10] **HALLEUX, J.P. and DONEA, J.** : "A Discussion of Cauchy Stress Formulations for Large Strain Analysis", Finite Element Methods for Nonlinear Problems, Eds. Bergan et al., Springer Verlag, Berlin, 1986.

[11] **HALLEUX, J.P. and CASADEI, F.** : "Transient Large-Strain Finite Element Analysis of Solids", Computational Methods for Nonlinear Problems, Eds. Taylor, C. et al., Pineridge Press, Swansea, 1987.

[12] **HUERTA, A. and CASADEI, F.** : "New ALE Applications in Nonlinear Fast Transient Solid Dynamics", submitted for publication in Engineering Computations, 1991.

[13] **HALLEUX, J.P.** : "Non-Linear Transient Analysis of Thin Axisymmetric Structures", Numerical Methods for Non-Linear Problems, Vol. 1, Eds Taylor C. et al., Pineridge Press, Swansea, 1980.

[14] **DONEA, J.** : "Arbitrary Lagrangian-Eulerian Finite Element Methods", Chapter 10 in Computational Methods for Transient Analysis, Eds. Belytschko T. and Hughes T.J.R., North-Holland, 1983.

[15] **DONEA, J.** : "Generalized Galerkin Methods for Convection Dominated Transport Phenomena", Applied Mechanics Reviews, 44, 5, 205-214, 1991.

[16] **BENUZZI, A.** : "Comparison of different LMFBR primary containment codes applied to a benchmark problem", Nucl. Eng. Des., 100, 239-249, 1987.

FINITE ELEMENTS IN THE 90's, E. Oñate, J. Periaux, A. Samuelsson (Eds.)
Springer-Verlag/CIMNE, Barcelona 1991

INDUSTRIAL USE OF LIQUEFACTION ANALYSIS IN 1980s

T. Shiomi
Takenaka Technical Research Laboratory,
2-5-14, Minamisuna, Koto, Tokyo,
JAPAN

ABSTRACT

Liquefaction analysis is yet another research field of Professor O. C. Zienkiewicz. There are so many people who studied and researched on this subject in the University College of Swansea (UCS), that it is interesting to make a genealogical tree for this research field. We would like to summarize its history in this article. The history started in mid 1970s. In those days material nonlinearity problems had been commonly solved and was being studied with large deformation problem. But liquefaction analysis added more difficulty. The difficulty was due to phase changes of material (solid <-> fluid). This is related to the FE formulation and the algorithm of constitutive equations. To overcome the computational difficulties and to validate the approach, a project called Diana Project was started by the UCS and other research group including, Cambridge University(U.K.), Tokyo University (Japan) and Industries (Japan). This paper reports the works done related to the Diana Project. The reports includes many verification studies, e.g. problems of a shaking table test, the damage due to Niigata Earthquake, a land-slide, a soil-pile-structure interaction etc.

1. Brief review of liquefaction analysis

The failure of the Lower San Fernand Dam due to liquefaction solved by O. C. Zienkiewicz, C. T. Chang and E. Hinton in the late 1970s[1]. And it was the first numerical solution of a liquefaction failure problem. Adding K. H. Leung, permanent deformation was surveyed [2]. The analysis was based on two phase theory. The governing equation used in the theory was same as the one proposed by A. Biot in 1956 for the saturated porous media [3] but re-examined from the point of mixture theory. Following this work, extensive researches have been done by P. Bettess, E. Hinton, K.H. Leung, N. Bicanic, D. K. Paul, T. Shiomi and A. H. Chang during the late 1970s and the early 1980s [4-6]. These theories was re-organized and generalized in his lecture notes as coupled problems [7-9]. He was interested in Dam problem from the start of his academic carrier and he also solved liquefaction problems which took place in dams. These works were done with C. T. Chang and K. H. Leung [10-13].

Drainage was not included for liquefaction problems until 1980 although basic formulation was proposed in the series of paper mentioned above. But one conclusion lead by M. Akiba and H. Senda [31] triggered him to take drainage into account. They found that most of sandy dams failed either few hours or upto 24 hours *after* earthquake finished. It was considered that the reason might be re-distribution of exess pore pressure due to drainage of pore water. So drained and undrained condition were examined with those dam failure problems. But the analysis did not show the failure after earthquake [9, 13].

Concerning constitutive equation, he was working on a muli-surface type model for one phase problem in those time but the densification model (or Endochronic model) was mostly used for a liquefaction problem. The numerical formulation was the u-p formulation which means that unknown variables are u (displacement) for solid phase and p (pressure) for fluid phase. A staggered time integration scheme for fluid and solid phase was used by K. H. Leung et al [12]. Then the u-U formulation was implemented with the author in which U stand for fluid displacement [6]. The program written with this formulation was named as DIANA because it was completed in the University of Wales when Lady Diana become the princes of the Wales. DIANA also stands for Dynamic Interaction Approaches and Nonlinear Analysis. With DIANA program, a project to investigated the applicability of the approach to liquefaction problem of industrial size was formed as a consortium of the UCS, Cambridge University, the University of Tokyo and eighteen companies in Japan (lead by Takenaka Corp., Kajima Corp., Chihoda Chemical Engineering & Construction Co. Ltd., Software Science Laboratory). Since the project started Zienkiewicz and M. Pastor developed and improved a constitutive equation for the liquefaction phenomena [14]. The research has recently extended to partially saturated problems [15].

2. DIANA Project

On April 1984, a project to develop a computer code for liquefaction phenomena started. A development team was organized by four Japanese companies. The development team was supported by a steering committee and a technical committee. The steering committee consisted of four leader companies and was financially supported by fourteen companies. Technical committee headed by Professor Tadahiko Kawai (Tokyo University then) and consisted of Professor Kenji Ishihara, Professor O. C. Zienkiewicz et al. That was three year project and completed on July 1987.

It was the first project to develop a finite element code jointly between Japan and United Kingdom. It was also the first project that Japanese general contractors agreed with the joint development of a computer code. So it was good chance to learn each other for verification studies as a cooperation work . Professor Zienkiewicz also promoted the collaborated work with Professor Andrew Schofield in Cambridge University. So the soil mechanics group in Cambridge University had made Centrifuge test for earthquake problem. They also provided the tri-axial test to determine the soil properties.

3. Formulation of Governing Equation

We do not discuss the details of the governing equations but this was the place O. C. Zienkiewicz contributed most. "Formulation" is one of the words which can most often found titles of his papers. For both solid and fluid materials Lagrangian Formulation was adopted. It may be correct when the strain is small. But when the material liquefies solid part become fluid and deformation is not anymore small. So that we must know the limitation of the formulation. The equilibrium equation of the solid phase is

$$\sigma_{ji,j} + \rho b_i - \rho \ddot{u}_i - \rho_f\left(\frac{\partial \ddot{w}_i}{\partial t} + \dot{w}_k \dot{w}_{i,k}\right) = 0 \qquad \cdots(1)$$

where $\sigma_{ji,j}$ is stress tensor, ρ and ρ_f are mass density of average and fluid respectively. u_i is displacement of solid and w_i is relative displacement of fluid to solid displacement. b_i is the body force vector. The equilibrium equation of fluid is

$$-np_{,i} + n\rho_f b_i - \rho_f\left(n\ddot{u}_i + \left(\frac{\partial \dot{w}_i}{\partial t} + \dot{w}_k \dot{w}_{i,k}\right)\right) + R_i = 0 \qquad \cdots(2)$$

where p is pore water pressure and n is porosity. R_i is interaction force between solid and fluid phase. Using absolute displacement of fluid, w_i is written as $w_i = n(U_i - u_i)$. Substituting the relation to eqn. (1) and (2), equilibrium equation can be written with displacement of solid u_i and fluid U_i. Both formulations are equivalent. Assuming the convective term has no effect, normally we neglect it. For solid phase, strain or rate of deformation is defined as a function, displacement and stress are determined by the strain or rate of deformation. If fluid is imcompressible, the mass conservation law must be satisfied. It could directly be solved. This formulation is u-U formulation. On the other hand, penalty method can be used as in fluid mechanics. Using the penalty method, u_i and U_i become only unknown variable and a symmetric matrix formulation is derived. These formulations can reduced to consolidation equation and even to undrained single variable equation.

4. Fine Element Method

General Finite Element Code should have the following capability to solve problems of real projects.

(1) External force

Normally, static, cyclic, random earthquake motion can be easily treated as we have done in linear analysis. Recently, the problem to be solved become larger and larger so input motion can be different at any points. DIANA is not capable to treat the problem.

(2) Initial condition

Determine the initial condition is most difficult job. DIANA provides the stage analysis to determine the more correct initial stress. Fig. 1 shows the stages of (1) composition of flat ground, (2) excavation, (3) construction of piles and super-structure and earthquake response analysis.

(3) Boundary condition

Infinite boundary is problem. We utilized the smith type of the problem. But this boundary is only available for linear problems. When it is non-linear analysis superposition can not be used. Instead of it, we have repeated boundary.

(4) Element

To model soil, plane strain (2D) and solid element (3D) are implemented. The beam elements (2D and 3D) are used to model frame structures. Viscous damper is for an infinite boundary. For the practical use, spring element and joint element are also implemented.

(5) Constitutive equation

Several constitutive equations are implemented since there is no perfect constitutive model which simulate sandy soil. Details are mentioned in the next section.

(6) Nonlinear solver

Again, there is no almighty solver so the following three solver are the users choice. (1) Newton Raphson method, (2) Modified Newton Raphson method, (3) Initial stiffness method, (4) BFGS method. Scant method was also tested but did not gave good performance.

(7) Analysis

Eigenvalue and frequency analysis are available as well as non-linear time domain response. Eigenvalue analysis is necessary when Rayleigh type damping is used. Since eigenvalue analysis for two phase problem is not stable, frequency analysis give characteristics of vibration.

(8) Time integration

New time integration scheme (SSpj) has been developed by Zienkiewicz and Wood [16].

(9) pre- & post- processing

Commercial preprocessor is used such as IDEAS etc.

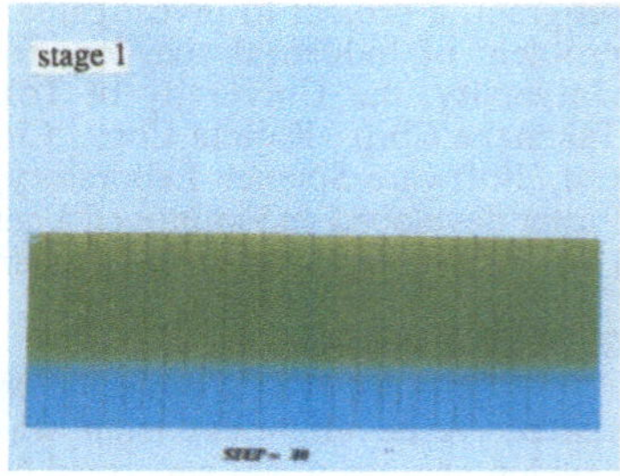

Applying Self-Weight (Effective Mean Stress)

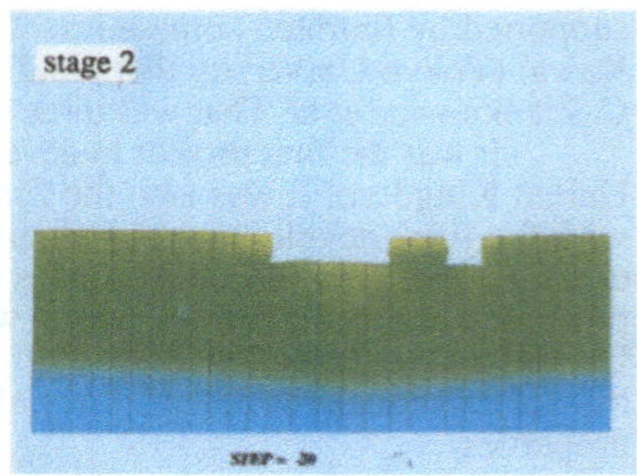

Excavation Process(Effective Mean Stress)

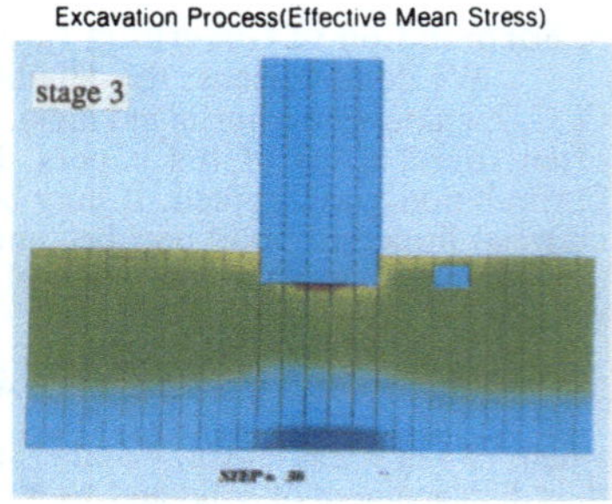

Construction Process (Effective Mean Stress)

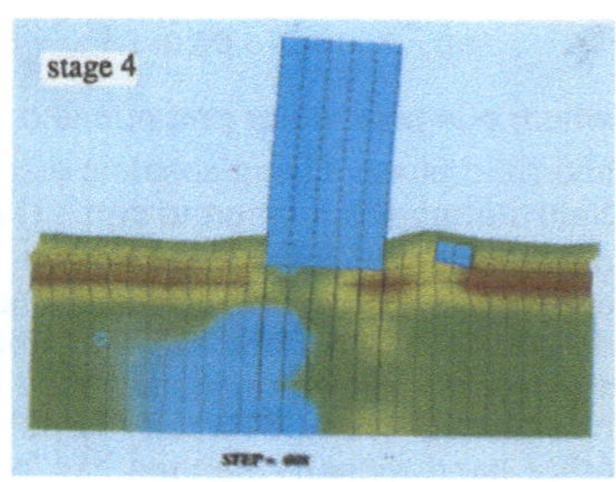

Liquefaction Behaviour at 8 sec.
(Excess Pore Water Pressure)

Fig. 1 Stage Analysis

5. Constitutive Equations

We chose four soil models which were developed for the liquefaction phenomena and verified them with triaxial tests. They may represent the following type of models.

1. The group of endochronic type model / Densification model (D-N): pore pressure is depend on the accumulation of shear strain. Failure criterion is defined by Mohr-Coulomb or Drucker-Prager criterion [12].
2. Single mechanism model with volumetric strain hardening (Reflecting Surface model: R-S) [17].
3. Single mechanism model with shear and volumetric hardening (Pastor Zienkiewicz model: P-Z) [14].
4. Multi-mechanism model where each mechanism is defined on three mobilized planes and volumetric change (Hujeux-Kabilamany-Ishihara model: M-M) [18, 19].

A series of triaxial tests have been performed with the Toyoura sand. A test result of Medium dense sand has been chosen for the validation test of the constitutive models. Fig. 2a and 2b shows pore pressure accumulation due to dilatancy and stress path of simulation of the tri-axial test.

Cyclic histories of the accumulation of excess pore pressures (EPP.) are shown for the four models in Fig 2a. The oscillation of the EPP observed in both experiment and simulation may be induced by the compression of the axial strain and is kept almost same until cyclic mobility starts but the center line shifted to the compression side. When cyclic mobility starts the significant plastic deformation is induced. The envelop of the measured EPP shows slight convex upward. The D-N and the M-M model shows the same tendency but the EPP envelop of the P-Z is concave. Since the P-Z model use more parameters it was difficult to choose the all the parameters correctly and some parameters might have been chosen inadequately.

Stress strain relationships are shown for axial stress in Fig. 2b, All simulation results are good at the average slope of the diagram but none of them reproduced good hysteresis loop. As the effective mean stress decrease during the cyclic loading the stiffness of the skeleton also decrease. These trend was seen in the both measurement and simulation. The M-M show the very good agreement with the measured data at the first cycle. As a whole the difference of the results between the models are much less than the difference between models and measured results. There must be further research to improve the models.

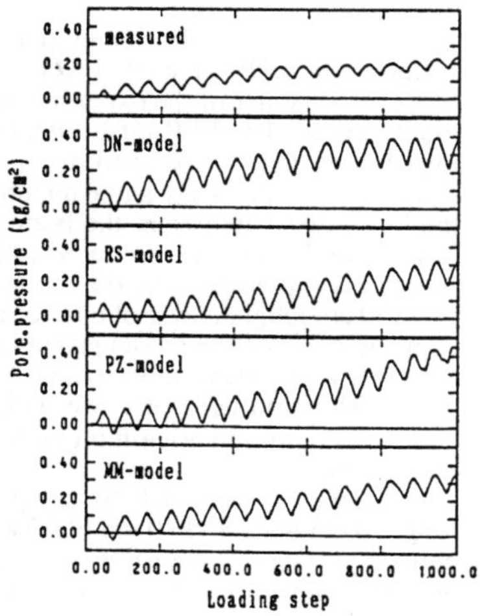

Fig. 2a Simulation of Undrained Triaxial Test

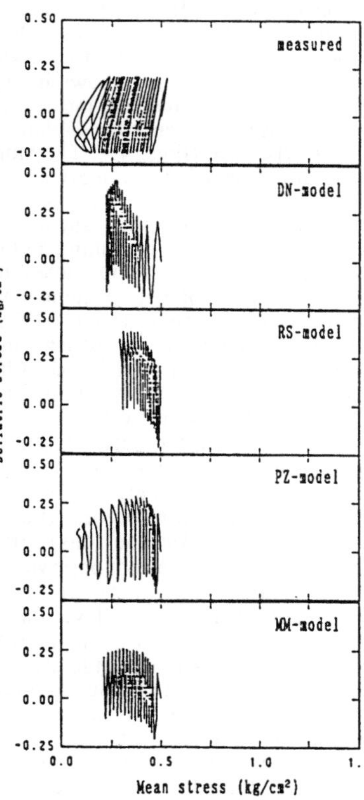

Fig. 2b Time History of Porewater Pressure

6. Verification Studies

The first verification model was a shaking table test of sandy ground with flat surface. The EPPs generated were almost same at the same depth so we compare the EPPs at the center of the experimental model. Fig. 3 shows the history of EPPs and the input motion. The EPPs of Experiment rise in very short period but those of analysis took more time to rise. And dissipation of EPPs started faster in the analysis. Both results did not precisely agree quantitatively but agreed qualitatively which means that the EPPs dissipated from the bottom to the top so dissipation did not start until the lower layer 's EPPs were reduced to the same pressure.

As shown Fig. 4.a the model ground was made of saturated sand 40 cm deep and 160 cm long. A layer of coarse sand was sticked to the bottom of the box to prevent slippage between the model ground and the box. Furthermore, 5 cm thick foam rubber cushions were placed at booth side boundaries as radiational boundaries.

The sand used here was a clean Toyoura sand. The specific gravity was 2.65, and the maximum and minimum dry densities of the sand are 1.648 and 1.356 g/cm^3 respectively.

Two experimental tests are performed. Both models have rigid structures (20 cm long, 28 cm wide and 10 cm high) at the center of the surface of the level ground. The contact pressure of the rigid structure was 21.5 gf/cm^2. In the test 1, the soil ground was homogeneous and its relative density was 47.6%. In the test 2, the area just under rigid structure was compacted as shown in Fig. 4c with shaded region. This area was 20 cm long, 15 cm deep. Relative density of the compacted area was 90 %. The initial water table was 3.3 cm deep for the non-compacted model and 2.5 cm deep for the compacted model. In both cases of the tests, the EPPs at 12 points, the horizontal acceleration at 3 points and the settlement of the rigid structure were measured. The points measuring pore water pressure are shown in Fig. 4a.

The EPPs reached the maximum value at almost same time (about 5 sec) for all the case (Figs. 4b). The slope of accumulation of the numerical results were nearly constant. On the other hand the experimental results showed slow accumulation at the beginning and sharp accumulation as they approached to the peak. In spite of the existence of the compacted area the maximum EPP under the rigid structure are almost same in both case of the non-compacted model (data of compacted model are shown in Fig. 4b). This may be due to water flow which transferred from the outer region of the compacted area in short time.

After EPP reached the peak, it started dissipation before the input motions stop in the case of the experiment. This was not observed in the numerical results except a little dissipation at the bottom layer under the rigid structure. This phenomena may be due to the strain dependency of the permeability during the earthquake excitation. In this study the permeability was assumed to constant.

The settlement of the rigid structure was 0.45 cm for the experiment and 0.65 cm for the calculation in the case of the compacted model. The error was about 31% and not negligible in the design purpose

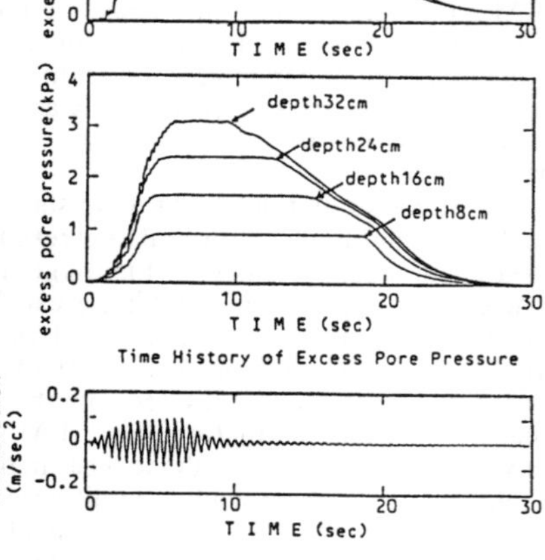

Fig. 3 Flat Susface Probrem

However, considering that research stage of liquefaction analysis was not well matured, the error would be in the acceptable range. On the other hand the settlement of the non-compacted model are 0.95 cm and 2.0 cm for the calculation and the experiment respectively. The error of these results was about 50% and the settlement was underestimated. The reason of this substantial discrepancy is due to full liquefaction of the surface layer where the sand particles are suspended in the water and the surface kept flat by the suspended water flow. This phenomena release a large amount of stress of the foundation area of the rigid structure, so a large settlement was expected that settlement might be reduced effectively if the compaction prevents the full liquefaction near the rigid structure.

Ground slide (or Permanent ground displacement) during or after earthquake due to liquefaction liquefaction has reported by Hamada et al for the Niigata Earthquake 1964, Nihonkai-Chubu Earthquake (1983), Kanto Earthquake (1923) and Fukui Earthquake (1944) [20]. Straight road was cut by shear defamation but it was not fault line. Permanent displacement was found in some square hundreds of square kilo meters. It is like land slide.

It must be noticed that it took place in city area which was flat field. There was neither hill or valley. No significant reason of the slid was found except liquefaction. But there was no clear idea how that took place mechanically. Fig.5a shows a map of the area around Ohgata School in Niigata City where ground slide was observed. The ground moved from Ohgata School to the Tsusen River about several meters. Area the phenomena took place was about 200m x 200m. There might be different condition for liquefied and non-liquefied zone.

The places the ground slide took place have mostly flat surface. The slope gradient are order of one several hundredth in some cases. So that lateral force due to the slope was one of hundreds of gravity force. It is difficult to imagine this force enforce the large permanent ground displacement of wide area. But it may be possible when stiffness of a layer underneath of a solid surface layer become zero.

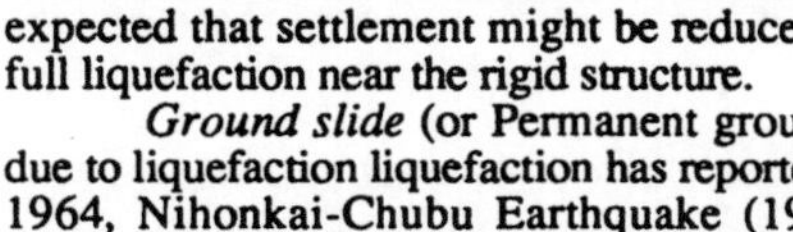

Fig. 4a Measuring Points

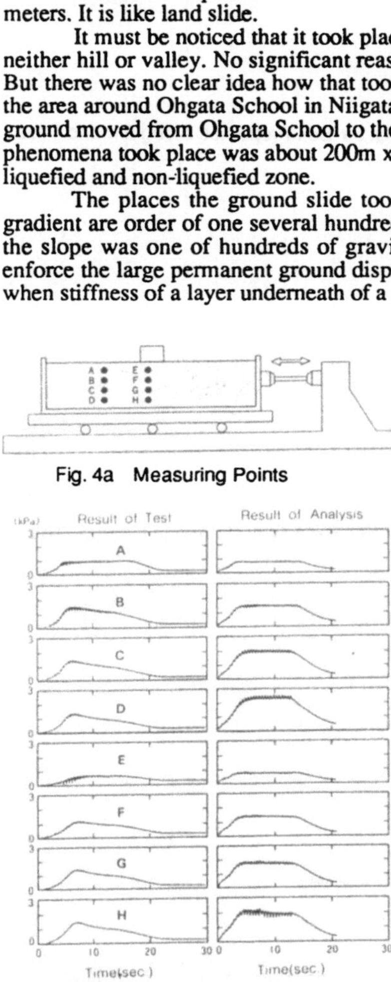

Fig. 4b Excess Pore Water Pressure

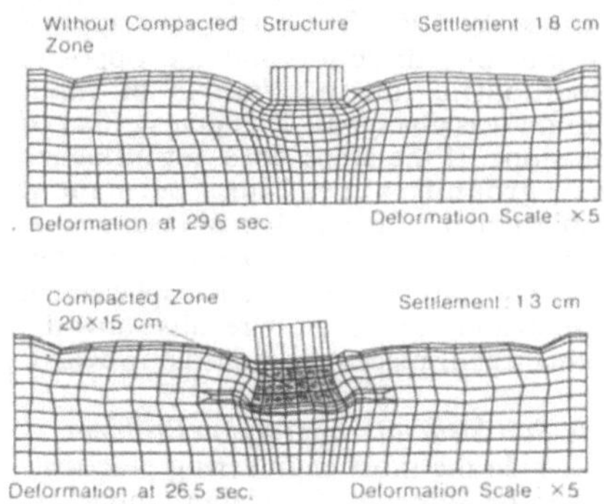

Fig. 4c Simulation of Settlement of Structure

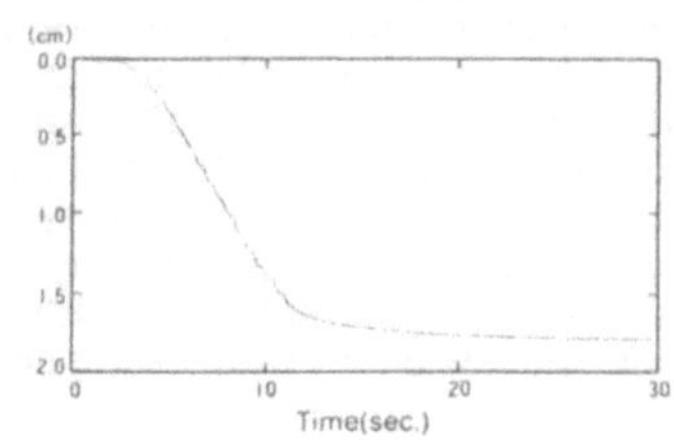

Fig. 4d Depth of Settlement

The situation may be like that the surface layer and a non-liquefied layer have a thin oil seats between them. In this case the surface layer always move toward downward of the slope. Or it may be like that upper layer floats on a thick liquefied layer. In this case the direction of the movement may be depend on a significant component of earthquake motions. Any small force is theoretically able to move the surface layer. Both hypotheses are might be valid in sequence of an earthquake response.

In any case the liquefied layer should be sandwiched and sealed by impermeable upper and lower layers. The direction of the slide is clearly understood in case of quay wall. Near Shinano river and Akita harbor the slide moved toward the river and sea side respectively. But it must be noticed the depth of the river is order of about 4 meters and liquefaction took place between 2 to 10 meters. The similar situation was found in case of Akita ground slide. In any case history of the ground slide due to liquefaction, there is a small slope. Our study starts from "What situation induce the ground slide?".

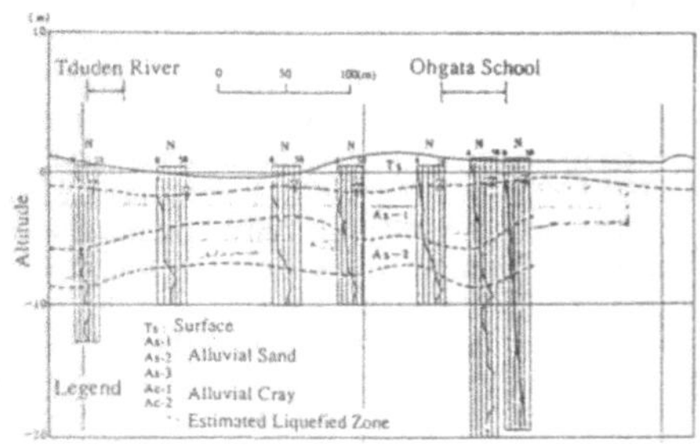

Fig. 5a Soil Properties Along Ground Slide

And "Is the very small slope acts a major role to determine the direction of ground slide?". Input motions was the component EW of an earthquake data recorded at Akita City in Niigata Earthquake 1964. The earthquake applied to both one direction and the opposite direction(i.e. the plus/minus of input motion was changed). This gives effect of the slope. If the slope does not influence the ground slide, the permanent displacement of the analytical results must be mirror image of the one of the other case. The result of permanent displacement is shown in Fig. 5c. Both results show the permanent displacement moved downward.

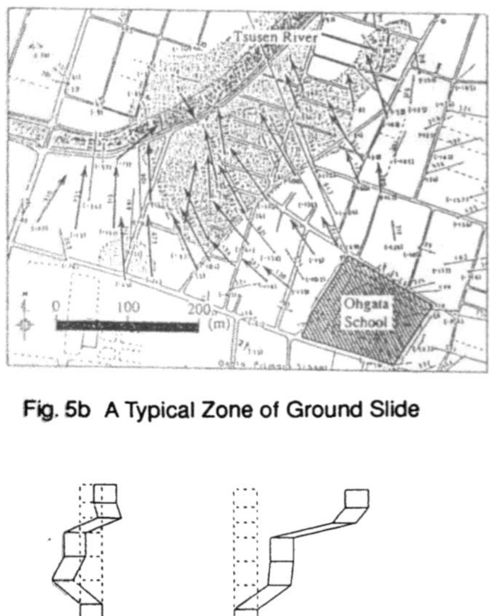

Fig. 5b A Typical Zone of Ground Slide

Pile-Soil-Structure Interaction Problem

In the most engineering practice, "beam-on-Winckler-foundation" model is commonly used, where non-linearity of soil is considered by nonlinear springs if liquefaction is ignored. Recently some researchers consider the weakening factor to soil stiffness can be calculated according to results of one dimensional analysis. But this may not correct procedure since one dimensional analysis can not correctly interpret EEPs built-up for interaction behavior of soils and pile-structure.

To estimate error of calculation, we performed a kind of equivalent linear analysis with the educed soil stiffness. In this example the shear modulus was reduced to on third of the initial value.

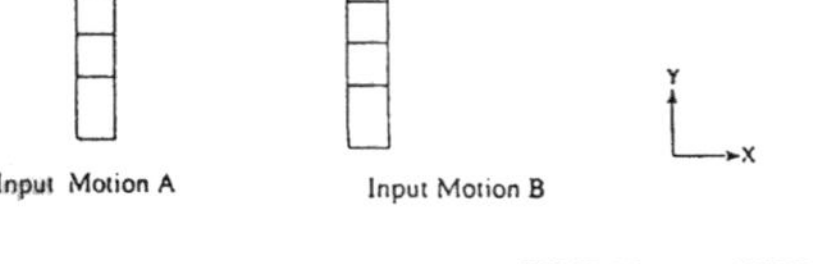

Fig. 5c Permanent Displacement

Since no formula has not been found yet to determine the reduction ratio of the shear modulus depending upon accumulated shear strain for this equivalent linear analysis. We take this ratio intuitively. When liquefaction take place, the shear modulus become zero. But zero modulus will not produce correct results. We must find an average value to represent the liquefaction behavior for whole period. The ratio should be studied in more details. Results with the reduced shear modulus are shown for the maximum bending moment and shear forces in Fig. 6b and 6c respectively. From the results it may be concluded that the maximum acceleration and displacement of the building and the piles show the similar profile with the results of liquefaction analysis (figure is not shown). But bending moment and shear force profile along the piles shows the different tendency. This is because the piles and the building are linear and bear horizontal resistance. But if we take into account non-linearity of piles, acceleration and displacement will also show difference. So this approach can be used to assess superstructure but not piles.

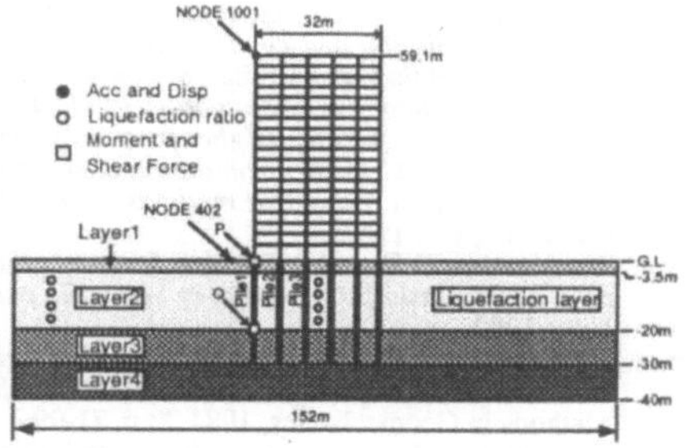

Fig. 6a Pile-soil-structure Probrem

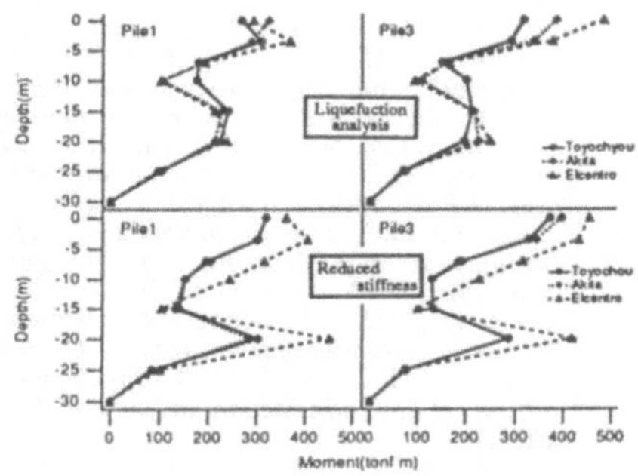

Fig. 6b Bending Moment of Piles

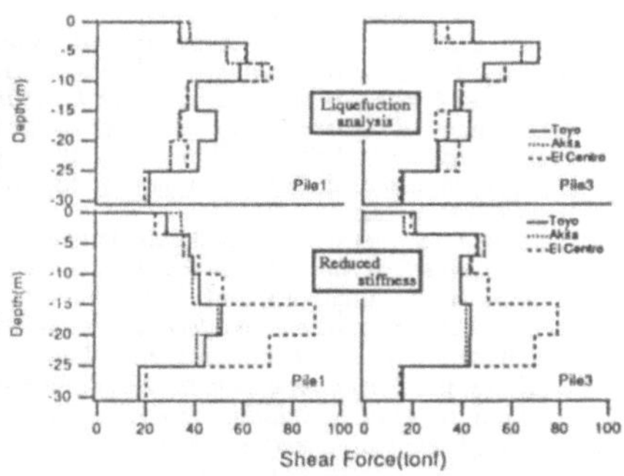

Fig. 6c Shear Force of Piles

7. Concluding Remarks

DIANA project started by strong necessity in Japanese industries and provided a good tools to understand the soil behavior and helped the water-front development. The two phase and nonlinear analysis, however, does not solve only liquefaction problem but also any geotechnical problems including soil-structure interaction problem such as sliding between two structures, uplift of building, seepage problem and those coupled problems. At the last we included the typical results which we solved [21-30]. These are the failure of Kawagishi-cho apartment during Niigata Earthquake (2D & 3D), uplift of utility tunnel, soil-structure interaction problem in Tokyo Bay area.

Acknowledgement

We would like to sincerely express my thanks to Professor O.C. Zienkiewicz for his guidance and advices during the DIANA project. It was not possible to reach to the current state without his help. DIANA program is now used more than 20 users who are all experts of soil dynamics.

References

1. Zienkiewicz, O.C., C.T. Chang, and E. Hinton, *Nonlinear seismic response and liquefaction.* Int. J. Num. and Anal. Meth. in Geomech., 1978. 2(4): p. 381-404.

2. Zienkiewicz, O.C., *et al. Liquefaction and permanent deformation under dynamic conditions. Numerical solution and constitutive relations.* in *Proc. Int. Conf. of Soils under Cyclic and Transient Loading, John Wiley & Sons.* 1980.
3. Biot, M.A., *Theory of propagation of elastic waves in a fluid-saturated porous solid, part I - Low-frequency range, part II - Higher frequency range.* J. Acoust. Soc. Am., 1956. **28**: p. 168-191. .
4. Zienkiewicz, O.C. *Basic formulation of static and dynamic behaviour of soil and other porous media.* in *Numerical methods in geomechanics,(Ed. Martins, J.B.), D. Reidl Publishing Co.* 1982.
5. Zienkiewicz, O.C. and P. Bettess. *Soil and other porous media under transient dynamic conditions. Basic formulation.* in *Scientific papers of Inst. Geotech. Engng., Wroclaw Tech. Univ.* 1980.
6. Zienkiewicz, O.C. and T. Shiomi, *Dynamic Behaviour of Saturated Porous Media; The Generalized BIOT formula and its Numerical Solution.* Int. J. for Numerical and Analytical Methods in Geomechanics, 1984. **8**: p. 71-99.
7. Zienkiewicz, O.C. *Coupled problems and their numerical solution.* in *Keynote lecture pres. at Int. Conf. Coupled Problems, Univ. Col. of Swansea. To be published by John Wiley & Sons.* 1981.
8. Zienkiewicz, O.C. *Finite element methods for coupled problems in engineering.* in *Proc. Int. Conf. Finite Element Methods, Shanghai.* 1982. Gorden and Breach Inc., New York. .
9. Zienkiewicz, O.C., C.T. Chang, and P. Bettess, *Drained, undrained, consolidation and dynamic behaviour assumptions in soils.* Geotech., 1980. **30**(4): p. 385-395.
10. Chang, C.-T. *Nonlinear response of earth dams and foundations in earthquakes.* in *Ph.D Thesis, Univ. Col. of Swansea, Swansea.* . 1979.
11. Zienkiewicz, O.C., *et al. Earthquake response of earth and concrete in the partial damage range.* in *Proc. 13th Int. Congress Large Dams, R.14.* 1979.
12. Zienkiewicz, O.C., *et al. Earth dam analysis for earthquakes: Numerical solution and constitutive relations for non-linear (damage) analysis.* in *Design of Dams to Resist Earthquakes.* 1980. London.: I.C.E.
13. Zienkiewicz, O.C., K.H. Leung, and E. Hinton. *Earthquake behaviours of soils with drainage.* in *4th Int. Conference of Num. Meth. in Geomech.* 1982. Canada:
14. Pastor, M., Z.O. C., and A.H.C. Chang. *Simple Models for Soil Behaviour and Application to Problems of Soil Liquefaction.* in *Numerical Methods in Geomechanics.* 1988. Innsbrug: A. A. Balkema.
15. Zienkiewicz, O.C., *et al., Static and dynamic behaviour of soils: a rational approach to* 11-321.
16. Zienkiewicz, O.C., W.L. Wood, and R.L. Taylor, *An alternative single step algorithm for dynamic problems.* Earthquake and Struct. Dynamics, 1980. **8**: p. 31-40.
17. Pande, G.N. and S. Pietruszczak. *Reflecting surface for soils.* in *Pro. of Int., Sym. on Num. Models in Geomech., Zurich.* . 1982.
18. Hujeux, J.C. and D. Aubry. *A critical state type stress-strain law for monotonous and cyclic loading: Getechincal and numerical consideration.* in *The Sym. on Implementation of the Computer Procedures and Stress-Strain Laws in Geotechnical Engng.,.* 1981. Chicago:
19. Kabilamany, K., *Modelling of cyclic behaviour of sands and its application to seismic response analysis.* 1986, The University of Tokyo:
20. Hamada, M. and e. al, *Research report of Jishinyochi-Sogo-Kenkyu-Kai.* 1990, Jishinyochi-Sogo-Kenkyu-Kai:
21. Shiomi, T., S. Pietruszczak, and G.N. Pande. *A liquefaction study of layers using the reflecting surface model.* in *Numerical Methods in Geomechanics.* 1982. Zurich:
22. Shiomi, T., S. amamoto, and T. Matsumoto. *Application of numerical method liquefaction problems.* in *Fifth international conference on numerical methods in geomechanics.* 1985. Nagoya:
23. Hatanaka, M., *et al. Simulation of sand liquefaction in shaking table test by two phase FE analyses.* in *International Symposium of numerical models in geomechanics.* 1986. Ghent:
24. Tsukuni, q., *et al. Simulation analysis of liquefaction of soil-structure system with compacted area (by the shaking table test).* in *The seventh Japan symposium of earthquake engineering.* 1986.
25. Shiomi, T., *et al., Simulation analysis of ground liquefaction induced by earthquake.* Computers and geotechnics, 1987. **4**.
26. Tanaka, Y., T. Shiomi, and T. Hirose. *Simulation of geotechnical centrifuge test by two phase FE analyses.* in *International conference on COMPUTATIONAL PLASTICITY MODELS, SOFTWARE & APPLICATIONS.* 1987. BARCELONA, SPAIN:
27. Shiomi, T., *et al. Simulation analysis of ground liquefaction in a soil-structures system of a shaking table test.* in *International conference series on advances in numerical methods in engineering: theory and applications.* 1987. Swansea:
28. Shiomi, T., *et al. Study of numerical constitutive for liquefaction problem.* in *Numerical methods in geomechanics.* 1988. Innsbruck:
29. Shiomi, T., S. Tsukuni, and Y. Tanaka. *A simulation study of ground liquefaction in a soil-structure interaction problem.* in *Proceedings of Ninth World Conference on Earthquake Engineering.* 1988. Tokyo - Kyoto:
30. shiomi, T., *Applications for dynamic problems,* in *Mechanics of granular materials,* R.o.I.T.C.O.M.O.G. MATERIALS, Editor. 1989, RIO DE JANEIRO. p. 135 - 144.
31. Akiba, M. and H. Senda, *The earthquake and its influence on reservoirs in Akita prefecture,* in *J. Agric. Engng. Soc., Japan.* 1941,

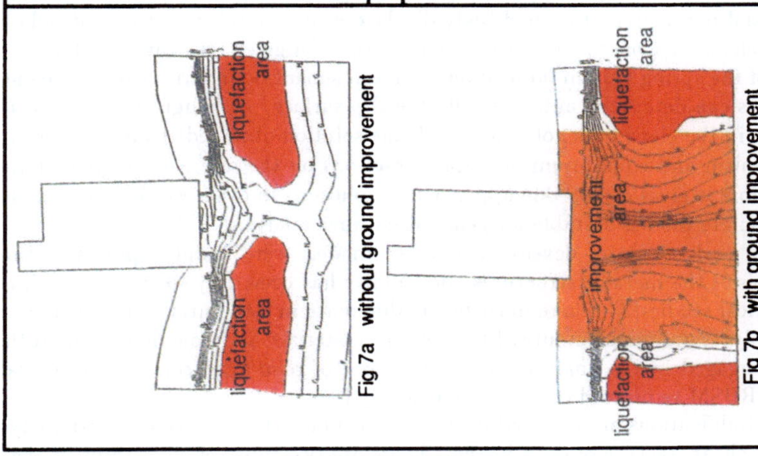

Fig 7a without ground improvement

Fig 7b with ground improvement

Fig 7 Kawagishi-cho apartment

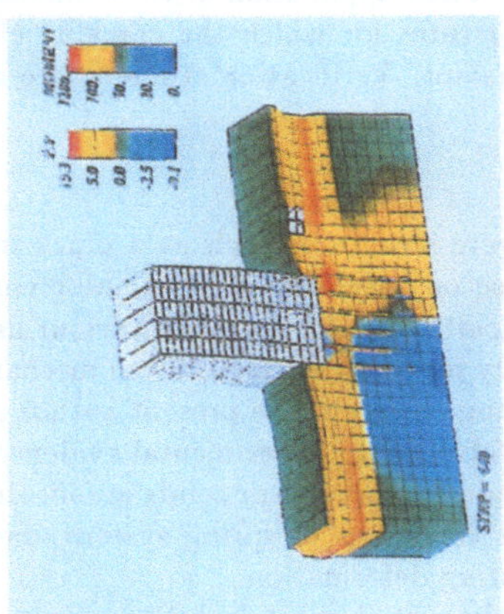

Fig 8 pile-soil-structure

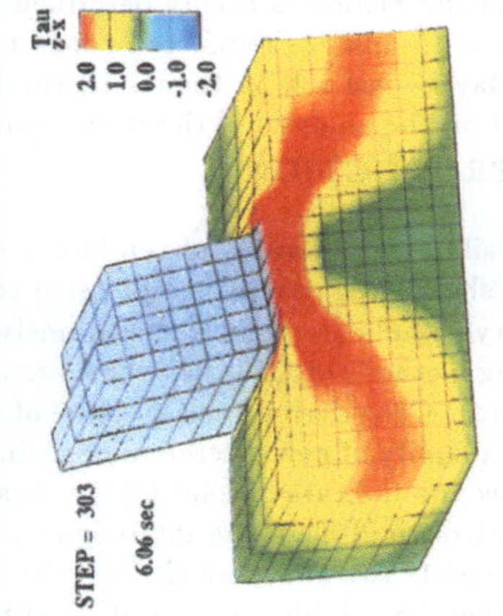

Fig 9 Kawagishi-cho (3-dimension)

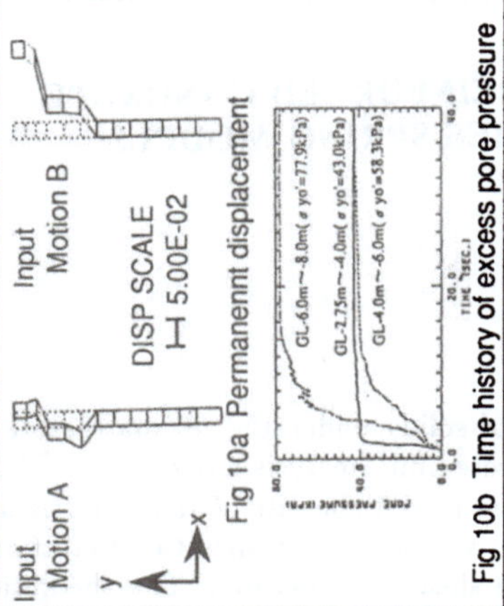

Fig 10a Permanennt displacement

Fig 10b Time history of excess pore pressure

Fig 10 landslide

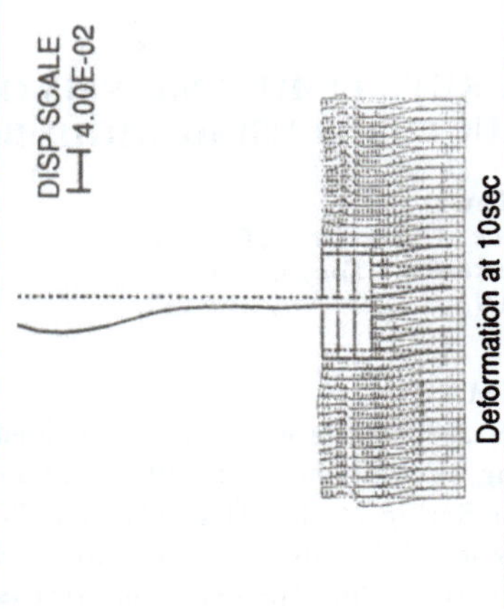

Deformation at 10sec

Fig 11 waterfront structure

FINITE ELEMENTS IN THE 90's, E. Oñate, J. Periaux, A. Samuelsson (Eds.)
Springer-Verlag/CIMNE, Barcelona 1991

DISCRETE LIMIT ANALYSIS OF REINFORCED CONCRETE STRUCTURES USING RIGID BODIES-SPRING MODELS

T. Kawai
Science University of Tokyo,
Shinjuku-Ku, Tokyo, 162,
JAPAN

SUMMARY

Based on the experimental evidence of solids under the ultimate state of loading, the present author proposed a discrete limit analysis named as the Rigid Bodies-Spring Models (RBSM) in 1976. The model consists of rigid bodies and two types of distributed connection springs, one of which resists the dilatational deformation while the other one resists the shear deformation. The theoretical basis of the models is briefly described first. Then application of these models is proposed to analyze reinforced concrete structures for which the cracking effect may play a vital role in their structural behavior. Verification studies were conducted on the analysis of the deep beams.

1.INTRODUCTION

Failure mechanisms of reinforced concrete structures such as a crack initiation, shear slip, partial compressive collapse or tensile failure of reinforcement are very complicated and it is extremely difficult to analyze their behaviour using existing methods of analysis, which are usually based on the continuum mechanics approach. Considering such a status of existing methods, the present author proposed a family of new discrete models in 1976 basing on experimental evidence on the flow and fracture of solids[1]. In these models structures or solids are idealized as a set of rigid elements interconnected by two types of spring system, one of which resists the dilatational, the other shearing deformation.

Therefore sliding or separation of two adjacent elements can be made easily. As nodal parameters, the rigid body displacements are taken at the centroid of a given element. Consequently the strain energy of a given element is evaluated by that of the spring system lumped on the interelement boundaries and incremental analysis can determine automatically the slip surface. Although the stress distribution in the element is not considered, the solution obtained always satisfies the mechanism and equilibrium of surface tractions on the interelement boundaries. Therefore it gives clearly one upper bound solution. In other words, these models are discrete models suitable for generalized limit analysis.

In early stage of development these models were mainly applied to limit analysis of the metallic structures and validity has been duly shown by a series of numerical studies[2]. There must be big difference in structural behavior between the polycrystalline structures like metallic materials and concrete. The author have engaged in the analysis of various types of reinforced concrete shear walls using RBSM reinforced concrete elements[3,4].

Modifications are required in the algorithm for the incremental load analysis on the shear fracture of deep beams. There are two types of algorithms in one of which unbalanced forces are relieved by the iteration, and the other is Yamada's method, in which the yielding, failure and unloading are judged for each integration point. Norio Takeuchi developed a new algorithm which may be applicable to analize coupled failure due to slip, tensile cracking and solid contact[5].

In this paper, a proposal is made for modified to accommodate cracking, recontact and compressive collapse; and analysis for a deep beam was made using the constitutive relation for reinforced concrete obtained with the previously developed RBSM and the results of analysis are compared with test results.

2.FORMULATION OF TWO DIMENSIONAL RBSM

For simplicity, consider two dimensional rigid triangular element of RBSM as shown in **Fig.1**. They are assumed to be equilibrium with external loads and reaction forces of the spring system which is distributed over contact surface of two adjacent bodies.

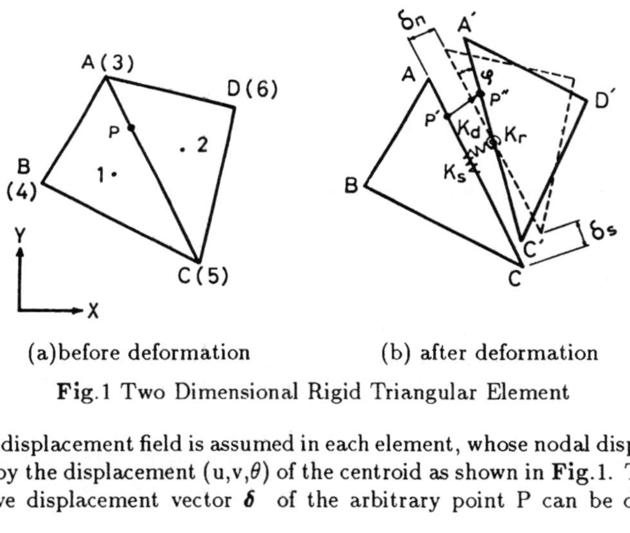

(a)before deformation (b) after deformation

Fig.1 Two Dimensional Rigid Triangular Element

Rigid displacement field is assumed in each element, whose nodal displacement are given by the displacement (u,v,θ) of the centroid as shown in **Fig.1**. Therefore, the relative displacement vector $\boldsymbol{\delta}$ of the arbitrary point P can be derived as follows:

$$\boldsymbol{\delta} = \sum_{i=1}^{2} \mathbf{B}_i \mathbf{u}_i \tag{2.1}$$

$$\mathbf{u}_1 = (u_1 \;\; v_1 \;\; \theta_1)^t \,, \quad \mathbf{u}_2 = (u_2 \;\; v_2 \;\; \theta_2)^t \,, \quad \boldsymbol{\delta} = (\delta_n \;\; \delta_s)^t$$

$$\mathbf{B}_1 = \begin{bmatrix} -l_1 & -m_1 & l_1(y-y_1) - m_1(x-x_1) \\ -l_2 & -m_2 & l_2(y-y_1) - m_2(x-x_1) \end{bmatrix}$$

$$\mathbf{B}_2 = \begin{bmatrix} l_1 & m_1 & -l_1(y-y_2) + m_1(x-x_2) \\ l_2 & m_2 & -l_2(y-y_2) + m_2(x-x_2) \end{bmatrix}$$

$$l_1 = \cos(\overline{x}, x) \,, \quad l_2 = \cos(\overline{x}, y) \,, \quad m_1 = \cos(\overline{y}, x) \,, \quad m_2 = \cos(\overline{y}, y)$$

On the other hand, the following relation are obtained from the definition of the spring constants:

$$\boldsymbol{\sigma} = \mathbf{D} \cdot \boldsymbol{\delta} \tag{2.2}$$

$$\boldsymbol{\sigma} = (\sigma_n \ \tau_s)^t \ , \quad \mathbf{D} = \begin{bmatrix} k_s & 0 \\ 0 & k_n \end{bmatrix}$$

$$k_n = \frac{(1-\nu)E_c}{(1+\nu)(1-2\nu)h} \ , \quad k_s = \frac{E_c}{(1+\nu)h}$$

where $h = h_1 + h_2$ is the projected length of a vector connecting centroids along the normal drawn, and τ_s and σ_n is tangential and normal stress respectively.

Basing on the above preliminaries, the strain energy expression of the in-plane element V can be obtained as the following:

$$V = \frac{1}{2}\int_{l_{35}} (\boldsymbol{\delta}^t \cdot \mathbf{D} \cdot \boldsymbol{\delta}) dS = \frac{1}{2}\mathbf{u}_i^t \int_{l_{35}} (\mathbf{B}^t \cdot \mathbf{D} \cdot \mathbf{B}) dS \mathbf{u}_i \tag{2.3}$$

Applying Castiglian's theorem, the following stiffness equation can be derived:

$$\mathbf{F} = \frac{\partial V}{\partial \boldsymbol{\delta}} \tag{2.4}$$

$$\mathbf{F} = (X_1, Y_1, M_1; X_2, Y_2, M_2) \ , \quad \mathbf{u} = (u_1, v_1, \theta_1; u_2, v_2, \theta_2)$$

where $\mathbf{K}$ is a (6×6) symmetric matrix and $\mathbf{F}$ is a nodal load vector.

3.REINFORCED CONCRETE MODEL

There are two discrete modeling of reinforced concrete structures. As shown in **Fig.**2 one is a composite model consisting of concrete and reinforcing bar elements, while the other is sandwich plate element composed of orthotropic materials.

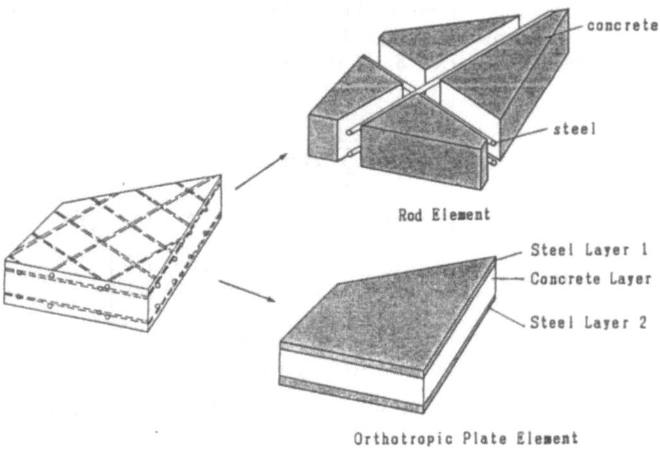

Fig.2 Modeling of a Reinforced Concrete Element

The latter model may be convenient for analysis of reinforced concrete structures, since such idealization is well established in modern design practice of heavyly reinforced concrete structures.

Therefore, the stress-strain relation along the reinforcing bar axis are assumed as follows:

$$\boldsymbol{\sigma} = \mathbf{D_s} \cdot \boldsymbol{\varepsilon} \tag{3.1}$$

$$\boldsymbol{\varepsilon} = (\varepsilon_{X'} \ \gamma_{XY'})^t \ , \quad \boldsymbol{\sigma} = (\sigma_{X'} \ \tau_{XY'})^t \ , \quad \mathbf{D_s} = \begin{bmatrix} E_s & 0 \\ 0 & \beta E_s \end{bmatrix}$$

where β is the reduction factor of dowel stiffness, and E_s is the Young's modulus for reinforcement.

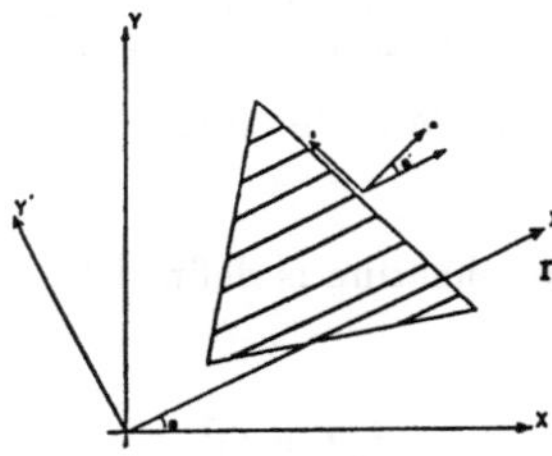

Fig.3 Reinforcing Bar Element Using Orthotropic Plate Element

4. MATERIAL PROPERTIES AND CONSTITUTIVE LAW

1) Stress-strain relations of concrete

The stress strain relations of concrete are shown in **Fig.**4. The stress at the first and second yielding level are $0.5F_c$ and $0.95F_c$, respectively. After first yielding, the tangent modulus is reduced to $0.5E_c$.

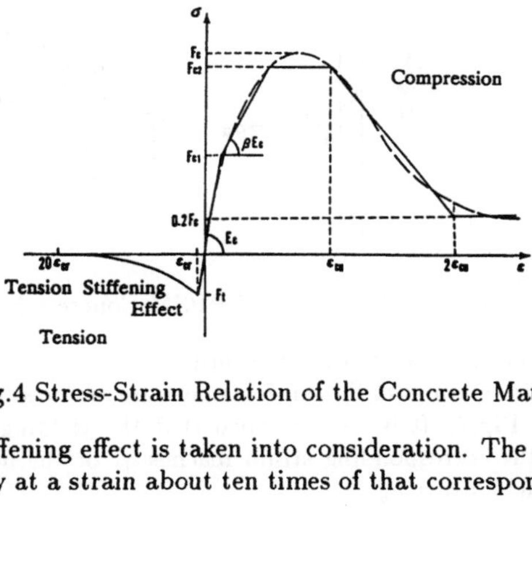

Fig.4 Stress-Strain Relation of the Concrete Material

Tension-Stiffening effect is taken into consideration. The tensile stress is released completely at a strain about ten times of that corresponding to the tensile strength.

2) Effective shear modulus

Cedolin suggested a linear relation for the effective shear modulus as a function of the principal tensile strain normal to the crack[6]. Shear stiffness reduction ratio of Cedolin and Dei Poli is adopted as shown in **Fig**.5.

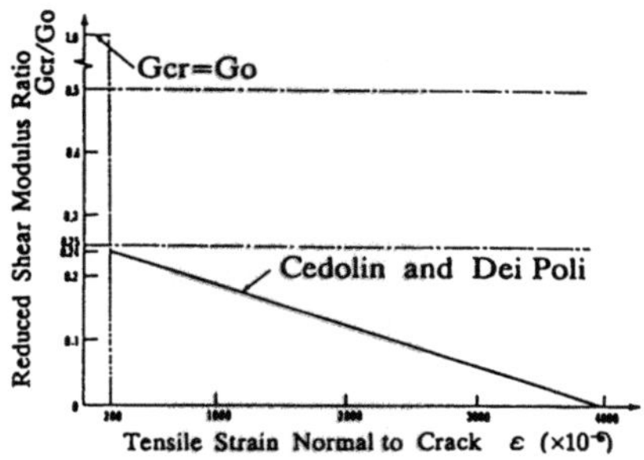

Fig.5 Reduced Shear Modulus Ratio G/Go

3) Failure surface of concrete

The concrete slip surface due to shearing is defined by the following Mohr-Coulomb's equation.

$$f^2 = \tau^2 - (c - \sigma_n \tan\phi)^2 \tag{4.1}$$

where $C = 0.138F_c$ (cohesion), $\phi = 37$ deg (internal shear friction angle)

According to the stress or strain conditions, the failure surface of concrete in the in-plane triangular plate are divided into seven regions as shown in **Fig**.6. After the shear strain at the element boundary reaches the constant value γ_u , the spring value of the boundary is set to zero. It is assumed that the γ_u is twice as much as the ultimate compressive strain ε_{cu} .

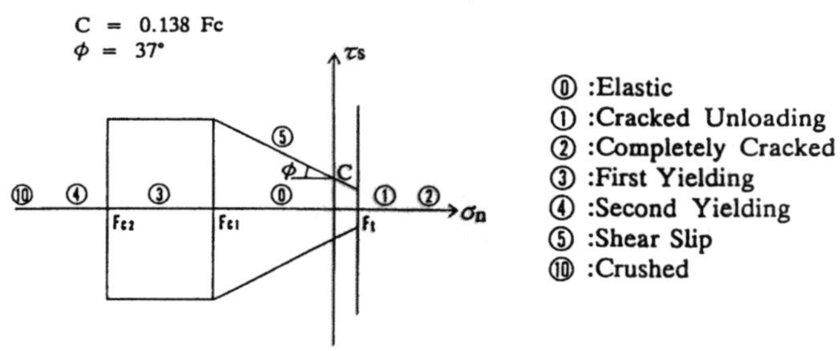

Fig.6 Adopted Yield Surface of RBSM Concrete Model

4) Stress strain relations of reinforcement

The stress-strain relations of the reinforcement are assumed to be ideal elasto-plastic as shown in **Fig**.7. It is also assumed that the yielding due to shearing force occurs when the corresponding strain reaches γ_u or the normal axial steel stress reaches the yield stress F_y .

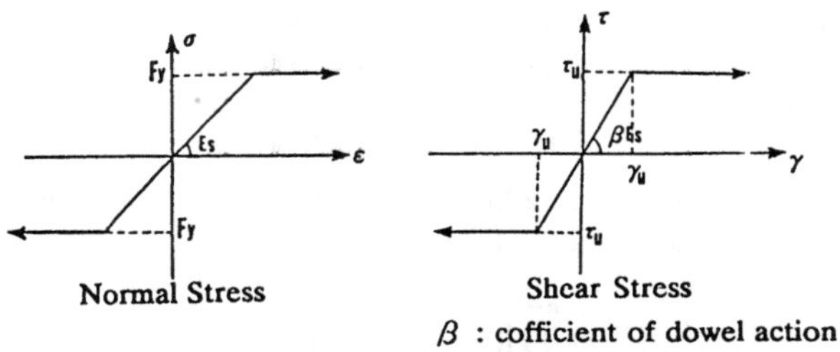

Fig.7 Stress-Strain Relation of Steel Reinforcement

The dowel shear stress is defined by the following equation.

$$\tau = \frac{3}{4}(\frac{d}{a_m})^2 \cdot E_s \cdot \gamma = \beta \cdot E_s \cdot \gamma \tag{4.2}$$

where d = diameter of the reinforcement, a_m = average distance between the cracks, E_s = Young's Modulus of the reinforcement, β = reduction factor of dowel stiffness ($\beta \approx 0.005$)

5.ANALYSIS ALGORITHM

A new algorithm is proposed by applying the incremental loading procedure developed by YAMADA[7]. In Yamada's method, necessary rate of the load increment to yield the most heavily stressed element can be calculated by stress distribution and load increment at the present stage as shown in **Fig.**8a. From this condition the required rate of load increment r can be calculated. Once the stress point lies on the failure curve, it may move according to the plastic flow rule until the unloading occurs.

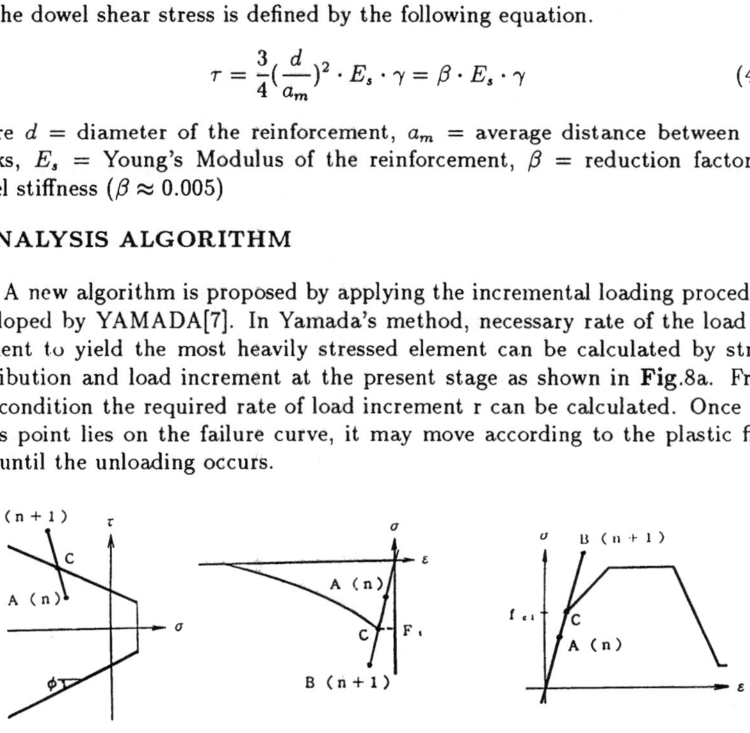

(a)shear failure (b) tensile failure (c) compressive yield
Fig.8 Rate of Load Increment

Similar calculation must be made in case of the tensile failure as shown in **Fig.**8b. All the possible rate of load increment corresponding to failure patterns should be calculated in all the elements and the minimum rate of load increment must be determined at each step as the load is increased to the next load level.

Stress relaxation is usually followed by the tensile failure. If Yamada's method is applied to this stress relaxation process exactly, endless calculation should be repeated corresponding to the tensile failure which may occur continuously.

The load $\mathbf{P}^{i+1}$ at the (i+1)th step can be calculated by using load $\mathbf{P}^i$ and rate of load increment r_i at the present step (i) as follows:

$$\mathbf{P}^{i+1} = (1 - r_i)\mathbf{P}^i \tag{5.1}$$

Therefore, in case of shearing failure, residual load $\mathbf{P}^n$ at the n_{th} step can be obtained by using initial load $\mathbf{P}$ as follows:

$$\mathbf{P}^n = \prod_{i=0}^{n-1}[(1 - r_i)]\mathbf{P} \quad (r_0 = 0) \tag{5.2}$$

This is the same result obtained Yamada's method.

On the other hand, if stress relaxation is caused by crack initiation, relief of forces are taken into account as follows:

$$\mathbf{P}^n = \prod_{i=0}^{n-1}[(1 - r_i)]\mathbf{P} + \sum_{k=1}^{n}(\prod_{i=k}^{n}[(1 - r_i)]\mathbf{F}^{k-1}) \tag{5.3}$$

where $\mathbf{F}^k$ is the force to be relieved at k_{th} step. Here r_{TOTAL} implies the cumulative rate of load increment and it can be defined as follows:

$$r_{TOTAL} = \sum_{k=1}^{n}(\prod_{i=0}^{k-1}[(1 - r_i)])r_k \tag{5.4}$$

The calculation must be repeated until $r_{TOTAL} = 1$ in each stage of loading.

6.NUMERICAL EXAMPLE

The deep beam test specimens, Model 1 and 2 (**Fig.**9) with and without stirrups were analyzed by two types of incremental load mthods, the normal Macal's method (Algo. 1) and the method proposed here(Algo.2).

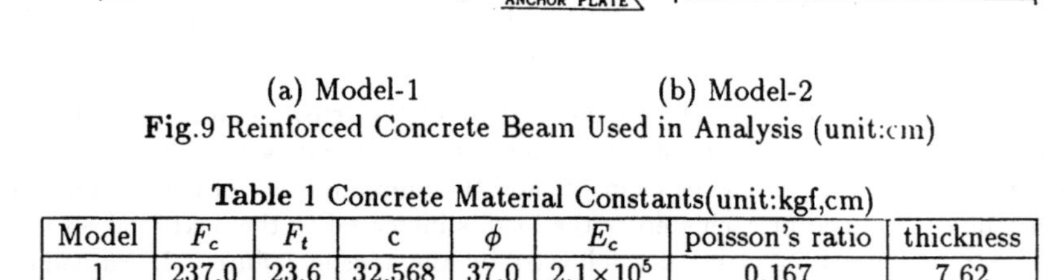

(a) Model-1 (b) Model-2

Fig.9 Reinforced Concrete Beam Used in Analysis (unit:cm)

Table 1 Concrete Material Constants(unit:kgf,cm)

Model	F_c	F_t	c	ϕ	E_c	poisson's ratio	thickness
1	237.0	23.6	32.568	37.0	2.1×10^5	0.167	7.62
2	203.0	20.3	28.014	37.0	2.1×10^5	0.167	7.62

Table 2 Reinforcement Material Constants(unit:kgf,cm)

Reinforcement	Modulus of Elasticity E_s	Yield Strength F_y	thickness t_s
Teinsile	2.1×10^6	3325.0	4.072
Conpressive	2.1×10^6	3620.0	2.243
Shear	2.1×10^6	2250.0	0.083

The material constants used in the analysis are given in **Tables** 1 and 2. The load-displacement relationships for Models 1 and 2 are shown in **Fig.**10. It can be seen from the figure that appreciable difference in the obtained load-deflection curves exists between two methods and the proposed algorithm(Algo.2) gives the better solution which are in good agreement with the failure load maxPexp obtained in the test.

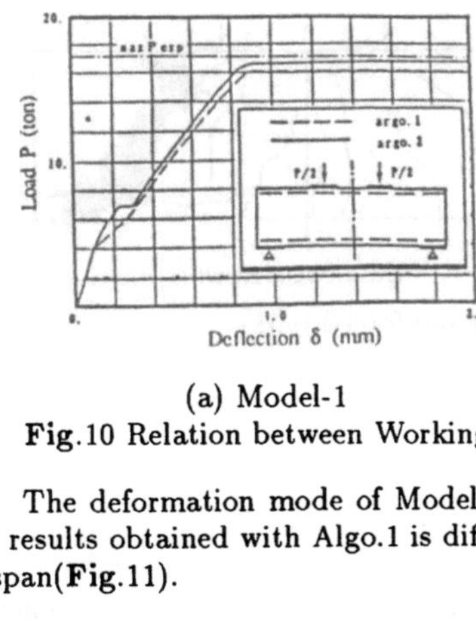

(a) Model-1

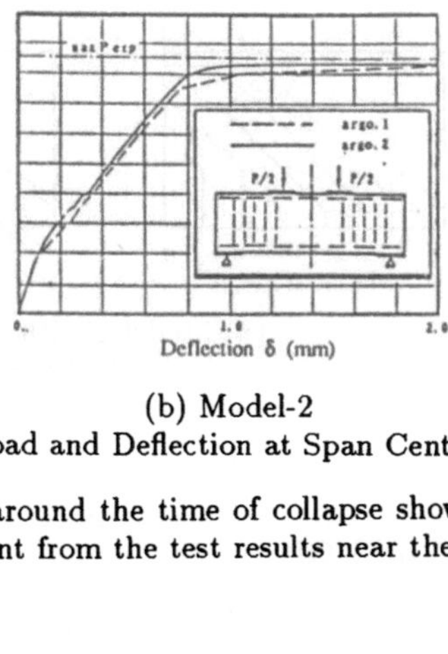

(b) Model-2

Fig.10 Relation between Working Load and Deflection at Span Centre

The deformation mode of Model 1 around the time of collapse shows that the results obtained with Algo.1 is different from the test results near the centre of span(**Fig.**11).

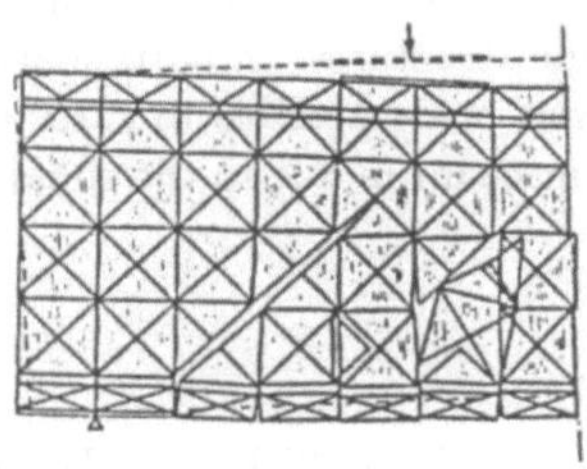

(a) Argo.1

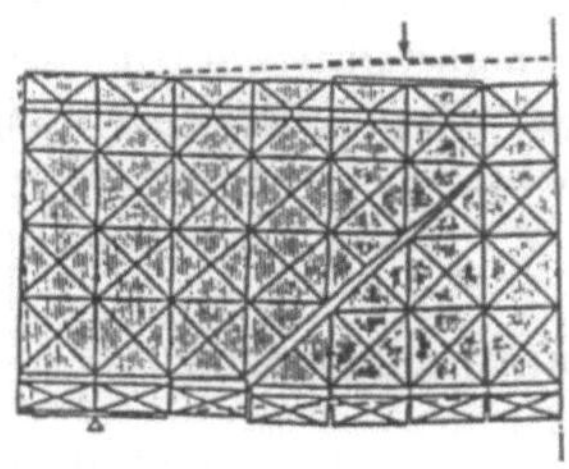

(b) Algo.2

Fig.11 Deformation Mode of Model 1 Around Time of Collapse (P-16 t,1/2Zone)

The failure characteristics at the time of collapse give by Algo.2 are shown in **Fig.**12. The failure mechanisms of the deep beams is well represented in both Model 1 and 2.

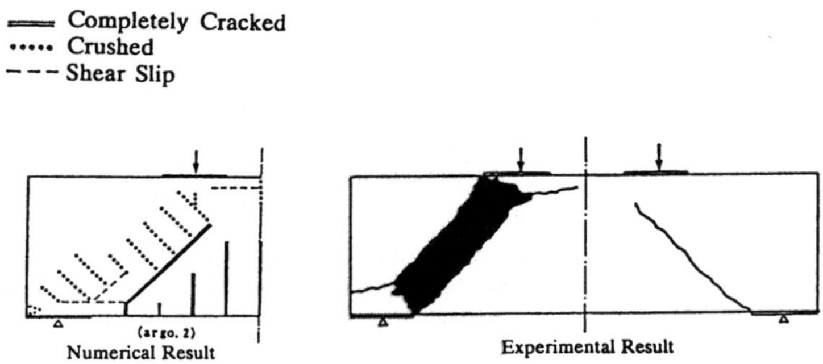

(a) Model-1(without shear reinforced bars)

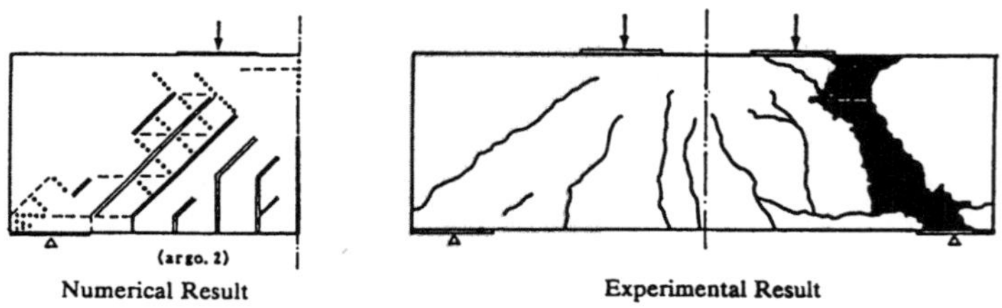

(b) Model-2(with shear reinforced bars)

Fig.12 Failure Modes

7.CONCLUSION

It was confirmed through analysis of the behavior of reinforced concrete beams using RBSM that they can precisely represent maximum shear strength and failure characteristics of reinforced concrete structures. In particular, adopting the algorithm proposed in this paper, it was made possible to carry out satisfactory evaluation of the progressive failure mechanism due to the simultaneous occurrence of cracking and recontact.

ACKNOWLEDGEMENTS

The present author would like to thank Dr. Norio Takeuchi, Department of Civil Engineering, Meisei University for his efforts made in preparation of this paper. He is also grateful to Mr. Masatoshi Ueda, Takenaka Corporation for his collaboration in this research project.

REFERENCES

[1] KAWAI,T. - New Discrete Structural Models and Generalization of the Method of Limit Analysis, Int. Conf. on Finite Elements in Nonlinear Solid and Struct. Mech., 2, 885-906 (1977)

[2] KAWAI,T. - New Discrete Models and Their application to Seismic Response Analysis of Structures, Int. Journal of Nuclear Eng. and Design, 48, 1, 207-229 (1978)

[3] UEDA,M., TAKEUCHI,N. and KAWAI,T. - A Discrete Limit Analysis of Reinforced Concrete Structures, Proc. of Int. Conf. on Computer Aided Analysis and Design, Design of Concrete Structures Part 2, 1369-1384 (1984)

[4] UEDA,M. and KAWAI,T. - Evaluation of a Discrete Limit Analysis to the Reinforced Concrete Shear Walls, Seminar on Finite Element Analysis of Reinforced Concrete Structures,JCI, 287-294 (1985)

[5] TAKEUCHI,N. and KAWAI,T. - A Discrete Limit Analysis of Granular Materials Including Effects of the Solid Contact, US-JAPAN Seminar,'MICROMECHANICS OF GRANULAR MATERIALS', Sendai, October 26-30, 116-125 (1987)

[6] CEDOLIN,L., and DEI POLI,S. - Dinite Element Non-Linear Plane Stress Analysis of Reinforced Concrete, Studi e struzioni in Cement Armato,13,3-33 (1976)

[7] YAMADA,Y.,YOSHIMURA,T. and SAKURAI.T - Plastic Stress-Strain Matrix and its Application for the Solution of Elastic-Plastic Problems by the Finite Element Method, Int. J. Mech. Sci.,10,343-354 (1968)

FINITE ELEMENTS IN THE 90's, E. Oñate, J. Periaux, A. Samuelsson (Eds.)
Springer-Verlag/CIMNE, Barcelona 1991

ERROR ANALYSIS OF STAGGERED PREDICTOR-CORRECTOR SCHEME FOR CONSOLIDATION OF POROUS MEDIA

K. Wisniewski and **E. Turska**
Polish Academy of Sciences, IFTR,
ul. Swietokrzyska 21,
00-049 Warszawa,
POLAND

L. Simoni and **B.A. Schrefler**
University of Padova,
Civil Engineering Department,
Via Marzolo 9, 35131 Padova,
ITALY

Summary

A systematic approach to the estimation of errors of a staggered predictor-corrector solution scheme for consolidation of porous media problems is presented. The total error is split into the temporal discretization error, the error of iterations and the stability error. The errors are investigated for linear and nonlinear problems and comparisons with the direct method are made.

1. INTRODUCTION

The mechanical behavior of porous media is determined by an interaction of several overlapping fields: the displacement field of the solid skeleton, and the flow fields of water and gas.

Usually, for the analysis of partially saturated porous media, with the gas occupying a part of the domain, a simplifying hypothesis of a static gas phase is introduced. This assumption can be removed as shown by Zienkiewicz *et al.* in [6, 12], where governing equations for transient (i.e. consolidation) and for dynamical behavior of such media were derived. Since these equations are very difficult to solve by analytical methods even for one-dimensional problems, numerical methods are used.

Two classes of numerical methods can be distinguished, direct solution schemes and staggered predictor-corrector (P-C) solution schemes. Although the latter scheme can be more efficient and easier to implement it also causes some disadvantages. Convergence and stability conditions are not always fulfilled, especially when dealing with strongly interacting fields, which usually yield nonlinear, nonsymmetric and coupled equations. Recently efforts have been made to modify the governing equations so that the staggered P-C scheme is unconditionally stable, see Zienkiewicz *et al.* in [11]. A good survey of methods for coupled problems can be found in Zienkiewicz *et al.* [10].

In the present paper we estimate errors of the staggered P-C scheme using a particular one, employed in [7, 8]. We assume that the finite element discretization in space is consistent and split the total error of the scheme into the temporal discretization error, the convergence and the stability error. These errors are examined for linear and nonlinear equations and comparisons with the direct scheme are made. Conditions sufficient to control the total error of the analyzed method are discussed.

Numerical tests are related to the determination of the lower bound of the time step Δt for fixed mesh size, the examination of the influence of the θ parameter on the spectral norm of the Jacobian matrix $\|f'\|$, and the comparison of $\|f'\|$ for the staggered P-C and direct schemes.

2. STAGGERED PREDICTOR-CORRECTOR SCHEME

As an example we consider consolidation problems of partially saturated porous media. The spatial discretization of these equations is performed using the Galerkin procedure of weighted residuals. These equations and the initial conditions are written as

$$\mathrm{B}\,\dot{\mathrm{X}} + \mathrm{C}\,\mathrm{X} = \mathrm{F} \quad , \quad \mathrm{X}(t_0) = \mathrm{X}_0 \tag{1}$$

where matrices B and C are nonsymmetric and depend on X but do not depend on time nor the time derivatives of X. The state vector is defined as $\mathrm{X} = (\mathrm{u}, \mathrm{p}_w, \mathrm{p}_c)^T$, where u is the displacement vector of the solid phase, p_w is the water (liquid) pressure, and p_c is the capillary pressure.

Eq.(1) is an ordinary differential equation in time and its discretization in the time domain is carried out by the generalized trapezoidal method

$$\dot{\mathrm{X}}_{n+\theta} = (\mathrm{X}_{n+1} - \mathrm{X}_n)/\Delta t \quad , \quad \mathrm{X}_{n+\theta} = (1-\theta)\,\mathrm{X}_n + \theta\,\mathrm{X}_{n+1} \tag{2}$$

where, X_n and X_{n+1} are state vectors at time instants t_n and t_{n+1}, Δt is the time step length, and θ is a parameter, usually $0 \le \theta \le 1$. Eq.(1) written for the time instant $t_{n+\theta}$ becomes

$$[\mathrm{B} + \theta\Delta t\,\mathrm{C}]\,\mathrm{X}_{n+1} = [\mathrm{B} - (1-\theta)\Delta t\,\mathrm{C}]\,\mathrm{X}_n + \Delta t\,\mathrm{F}_{n+\theta} \tag{3}$$

Let us consider an example of gas and liquid flow in deformable porous media. Then the basic matrices are of form (see [6])

$$\mathrm{B} = \begin{bmatrix} \mathrm{B}_{11} & \mathrm{B}_{12} & \mathrm{B}_{13} \\ \mathrm{B}_{21} & \mathrm{B}_{22} & \mathrm{B}_{23} \\ \mathrm{B}_{31} & \mathrm{B}_{32} & \mathrm{B}_{33} \end{bmatrix} \quad , \quad \mathrm{C} = \begin{bmatrix} 0 & 0 & 0 \\ 0 & \mathrm{C}_{22} & 0 \\ 0 & \mathrm{C}_{32} & \mathrm{C}_{33} \end{bmatrix} \quad , \quad \mathrm{F} = \begin{Bmatrix} \mathrm{F}_1 \\ \mathrm{F}_2 \\ \mathrm{F}_3 \end{Bmatrix} \tag{4}$$

The matrix $[\mathrm{B} + \theta\Delta t\,\mathrm{C}]$ from Eq.(3) can be partitioned as follows

$$[\mathrm{B} + \theta\Delta t\,\mathrm{C}] = \begin{bmatrix} \mathrm{B}_{11} & \mathrm{B}_{12} & 0 \\ \mathrm{B}_{21} & \mathrm{D}_{22} & 0 \\ 0 & 0 & \mathrm{D}_{33} \end{bmatrix} + \begin{bmatrix} 0 & 0 & \mathrm{B}_{13} \\ 0 & 0 & \mathrm{B}_{23} \\ \mathrm{B}_{31} & \mathrm{D}_{32} & 0 \end{bmatrix} \tag{5}$$

where $\mathrm{D}_{22} = \mathrm{B}_{22} + \theta\Delta t\,\mathrm{C}_{22}$, $\mathrm{D}_{33} = \mathrm{B}_{33} + \theta\Delta t\,\mathrm{C}_{33}$, $\mathrm{D}_{32} = \mathrm{B}_{32} + \theta\Delta t\,\mathrm{C}_{32}$. If Eq.(3) is rearranged in such a way that only the first component of sum (5) is left on the left-hand side and the second component is transferred to the right-hand side then the third and the first two equations can be solved independently. Thus we can define a staggered P-C solution scheme and solve these equations sequentially, i.e. we begin with solving the third equation for $\mathrm{p}_{c,n+1,k-1}$ and later solve the first two for $\mathrm{u}_{n+1,k}$ and $\mathrm{p}_{w,n+1,k}$. From the third equation we calculate

$$\mathrm{p}_{c,n+1,k-1} = -\mathrm{G}_3(\mathrm{x}_{n+\theta})\,\mathrm{x}_{n+1,k-1} + \mathrm{H}_3(\mathrm{X}_{n+\theta})\,\mathrm{X}_n + \Delta t\,\tilde{\mathrm{F}}_3 \tag{6}$$

where by G_3 and H_3 we denote matrices multiplied by $x_{n+1,k-1}$ and X_n, and $\tilde{F}_3$ is a vector obtained from F_3. Substituting Eq.(6) into the first two equations we obtain

$$x_{n+1,k} = G(x_{n+\theta})\, x_{n+1,k-1} - H(X_{n+\theta})\, X_n + \Delta t\, \tilde{F} \qquad (7)$$

where $x_{n+1} = \{u_{n+1}, p_{w,n+1}\}^T$, and k and $k-1$ are iteration indices. The symbols of the other matrices are chosen as in Eq.(6). The values of $u_{n+1,0}$ and $p_{n+1,0}$, which are necessary to initiate iterations, need to be predicted using some extrapolation formula. The iterations are repeated until subsequent corrected values of an error differ less than a required accuracy. The last iteration index will be denoted by K.

Other examples of partitioning and staggered solution schemes can be found in [2, 5, 10].

3. ERROR ANALYSIS

Let us consider errors of the staggered P-C procedure defined in the previous section. We distinguish the following solutions for time instant t_{n+1}:

$X(t_{n+1})$ the exact solution of the differential Eq.(1),

X_{n+1} the exact solution of the temporally discrete Eq.(2),

$X_{n+1,K}$ the solution of Eq.(6) and (7), where K denotes the last iteration,

$\tilde{X}_{n+1,K}$ the numerical solution of Eq.(6) and (7), affected by the round-off error.

The total error of the numerical solution can be expressed as follows

$$e_{n+1}^{total} = e_{n+1} + e_{n+1,K} + \tilde{e}_{n+1,K} \qquad (8)$$

where,

$e_{n+1}^{total} = X(t_{n+1}) - \tilde{X}_{n+1}$ is the total error,

$e_{n+1} = X(t_{n+1}) - X_{n+1}$ is the temporal discretization error,

$e_{n+1,K} = X_{n+1} - X_{n+1,K}$ is the iteration error at a fixed time instant,

$\tilde{e}_{n+1,K} = X_{n+1,K} - \tilde{X}_{n+1,K}$ is the stability error.

When the norms of e_{n+1}, $e_{n+1,K}$ and $\tilde{e}_{n+1,K}$ tend to zero, then the norm of e_{n+1}^{total} also diminishes to zero, i.e. the numerical solution $\tilde{X}_{n+1,K}$ converges to the exact solution $X(t_{n+1})$.

A. Temporal discretization error e_{n+1}

The discretization error e_{n+1} is estimated by examining Eq.(3) subsequently for $X(t_{n+1})$ and X_{n+1}. We obtain the following formula

$$e_{n+1} = A\, e_n + r_{n+1} \qquad (9)$$

where $A = [B + \theta\Delta t\ C]^{-1}\ [B - (1-\theta)\Delta t\ C]$. The residuum r_{n+1} is a measure of consistency of the temporally discrete Eq.(3) with the differential one (1) and is obtained by substitution of $X(t_{n+1})$ into Eq.(3). The error e_n on the right-hand side of (9) is evaluated recursively resulting in

$$e_{n+1} = A^{n+1}\, e_0 + \sum_{i=0}^{n} A^i\, r_{n+1-i} \qquad (10)$$

For initial values without round-off errors $e_0 = X(t_0) - X_0 = 0$ and only the second term of Eq.(10) remains. The evaluation obtained in [9] shows that the generalized trapezoidal method yields

$$r_{n+1} = \frac{1}{2}(1-2\theta)\,\ddot{X}(t_{n+\theta})\Delta t^2 + \frac{1}{2}\theta^2[B+\theta\Delta t C]^{-1}C\,\ddot{X}(t_{n+\theta})\Delta t^3 + 0(\Delta t^4) \quad (11)$$

The residuum $r_{n+1} \to 0$ for $\Delta t \to 0$ what means that Eq.(3) is consistent with Eq.(1). Finally, by the identity $A = I - \Delta t[B + \theta\Delta t\ C]^{-1}C$, the norm of error (10) can be expressed as

$$\|e_{n+1}\| = \frac{1}{2}(1-2\theta)\ 0(\Delta t^2) + \frac{1}{2}\theta^2\ 0(\Delta t^3) \quad (12)$$

where the symbol $0(\cdot)$ specifies the order of omitted terms. Thus, the error due to the use of the generalized trapezoidal method $\|e_{n+1}\| \to 0$ for $\Delta t \to 0$.

B. Iteration error $e_{n+1,K}$

The scheme implied by Eq.(7) can be concisely written as

$$x_{n+1,k} = f(x_{n+1,k-1}) \quad (13)$$

This scheme is convergent if the function f satisfies the assumptions of the Banach contraction mapping theorem (see [3], [4]), i.e.

1. $f : D \subseteq R^m \to R^m$ produces values belonging to its own domain $f(D) \subseteq D$,
2. f is a strict contraction on D, i.e. for arbitrary $y, z \in D$

$$\|f(y) - f(z)\| \leq L\,\|y - z\| \quad , \quad L < 1. \quad (14)$$

Then, for any starting value $y_0 \in D$ there exists a unique fixed point $y^* \in D$ that $y^* = f(y^*)$. If f is a differentiable function, then (14) can be replaced by

$$\|f'(y)\| \leq L \quad , \quad L < 1 \quad (15)$$

where $f'(y)$ is the Jacobian of $f(y)$.

In the numerical verification of assumptions 1 and 2 of the above theorem the elements y and z are replaced by arbitrary elements of $\{x_{n+1,0}, ..., x_{n+1,K}\}$. The fixed point $x_{n+1,K}$ of iterations (13) is reached when the relative error is limited by a required tolerance τ

$$\frac{\|e_{n+1,K}\|}{\|x_{n+1,K}\|} \leq \frac{L}{1-L}\,\frac{\|x_{n+1,K} - x_{n+1,K-1}\|}{\|x_{n+1,K}\|} \leq \tau. \quad (16)$$

where K denotes the last iteration.

To find the error of the staggered P-C method $e_{n+1,K}$ we define the following component errors

$$e^{s1}_{n+1,K} = x_{n+1} - x_{n+1,K} \quad , \quad e^{s2}_{n+1,K} = p_{c,n+1} - p_{c,n+1,K} \quad (17)$$

where x_{n+1} and $p_{c,n+1}$ are exact solutions of Eq.(7) and (6).

1. If G in Eq.(7) does not depend on x, i.e. G = *const*, then f' = G and $L = \|G\|$, thus for $\|G\| < 1$ the scheme (13) is convergent. Using definition (17) and writing Eq.(7) for the K-th iteration we get

$$e^{s1}_{n+1,K} = G\, e^{s1}_{n+1,K-1} = G^K\, e^{s1}_{n+1,0} \tag{18}$$

where $e^{s1}_{n+1,0} = x_{n+1} - x_{n+1,0}$, and $x_{n+1,0}$ is the predicted value. Let us write Eq.(6) for iteration K

$$p_{c,n+1,K} = -G_3(x_{n+\theta})\, x_{n+1,K} + H_3(X_{n+\theta})\, X_n + \Delta t\, \tilde{F}_3 \tag{19}$$

Using definition (17) and Eq.(6) the error of $p_{c,n+1,K}$ can be expressed by the error $e^{s1}_{n+1,K}$ as follows

$$e^{s2}_{n+1,K} = -G_3(x_{n+\theta})\, e^{s1}_{n+1,K} \tag{20}$$

By Eq.(18) and (20) we obtain

$$e_{n+1,K} = \begin{bmatrix} I \\ -G_3 \end{bmatrix} e^{s1}_{n+1,K} \tag{21}$$

Finally, using the recursive form of (18) we obtain

$$\|e_{n+1,K}\| \le c\, \|G\|^K\, \|e^{s1}_{n+1,0}\| \tag{22}$$

where c is a constant depending on the multiplier of $e^{s1}_{n+1,K}$ in Eq.(21). It is apparent that if $\|G\| < 1$ then $\|e_{n+1,K}\| \to 0$ for $K \to \infty$.

2. If G in Eq.(7) depends on x, i.e. $G = G(x_{n+\theta})$, then we have to make the usual assumption that it satisfies the Lipschitz condition with constant L. From the definition (17) and the Lipschitz condition we get

$$\|e^{s1}_{n+1,K}\| \le L\, \|e^{s1}_{n+1,K-1}\| \le L^K\, \|e^{s1}_{n+1,0}\| \tag{23}$$

Considering the error $e^{s2}_{n+1,K}$ related to $p_c(x_{n+1,K})$ for some bounded domain of f_c given by the right-hand side of Eq.(6) we have

$$\|e^{s2}_{n+1,K}\| = \|f_c(x_{n+1}) - f_c(x_{n+1,K})\| \le C\, \|x_{n+1} - x_{n+1,K}\| \le C\, L^K\, \|e^{s1}_{n+1,0}\| \tag{24}$$

Here we require that the function f_c satisfies the Lipschitz condition, with the Lipschitz constant C being arbitrary. Thus instead of Eq.(22) we obtain

$$\|e_{n+1,K}\| = \sqrt{\|e^{s1}_{n+1,K}\| + \|e^{s2}_{n+1,K}\|} \le \sqrt{1+C^2}\, L^K\, \|e^{s1}_{n+1,0}\| \tag{25}$$

It is apparent that if $L < 1$ then $\|e_{n+1,K}\| \to 0$ for $K \to \infty$.

C. Stability error $\tilde{e}_{n+1,K}$

To find the stability error of the staggered P-C method $\tilde{e}_{n+1,K}$ we define the following component errors

$$\tilde{e}^{s1}_{n+1,K} = x_{n+1,K} - \tilde{x}_{n+1,K}\,, \qquad \tilde{e}^{s2}_{n+1,K} = p_{c,n+1,K} - \tilde{p}_{c,n+1,K}\,. \tag{26}$$

where $\tilde{x}_{n+1,K}$ and $\tilde{p}_{c,n+1,K}$ are the solutions affected by the truncation error. Writing Eq.(7) first for $x_{n+1,K}$ and $X_{n,K}$, and then for $\tilde{x}_{n+1,K}$ and $\tilde{X}_{n,K}$, and subtracting the latter from the first one we have

$$\begin{aligned} x_{n+1,K} - \tilde{x}_{n+1,K} &= G(x_{n+\theta})x_{n+1,K-1} - G(\tilde{x}_{n+\theta})\tilde{x}_{n+1,K-1} - \\ & \quad H(X_{n+\theta})X_{n,K} + H(\tilde{X}_{n+\theta})\tilde{X}_{n,K} \end{aligned} \tag{27}$$

1. In the linear case the basic matrices do not depend on $X_{n+\theta}$. When $G(x_{n+\theta}) = G = const$ and $H(x_{n+\theta}) = H = const$ Eq.(27) becomes

$$\tilde{e}^{s1}_{n+1,K} = G\ \tilde{e}^{s1}_{n+1,K-1} - H\ \tilde{e}_{n,K} \tag{28}$$

where index K can assume different values depending on the number of iterations performed to obtain $X_{n,K}$.

Writing Eq.(6) first for $p_{c,n+1,K-1}$ and $X_{n,K}$ and later for $\tilde{p}_{c,n+1,K-1}$ and $\tilde{X}_{n,K}$, and subtracting the latter from the first one we get

$$\begin{aligned} p_{c,n+1,K-1} - \tilde{p}_{c,n+1,K-1} &= -G_3(x_{n+\theta})\ x_{n+1,K-1} + G_3(\tilde{x}_{n+\theta})\ \tilde{x}_{n+1,K-1} + \\ & \quad H_3(X_{n+\theta})\ X_{n,K} - H_3(\tilde{X}_{n+\theta})\ \tilde{X}_{n,K} \end{aligned} \tag{29}$$

For $G_3(x_{n+\theta}) = G_3 = const$ and $H_3(x_{n+\theta}) = H_3 = const$, on use of the definitions of component errors, Eq.(29) reads as follows

$$\tilde{e}^{s2}_{n+1,K-1} = -G_3\ \tilde{e}^{s1}_{n+1,K-1} + H_3\ \tilde{e}_{n,K} \tag{30}$$

Let us calculate $\tilde{e}^{s1}_{n+1,K-1}$ from this equation and substitute it into Eq.(28). Combining the so obtained equation and Eq.(30) together we find an equation which can be concisely written as

$$\tilde{e}_{n+1,K} = -P\ \tilde{e}_{n+1,K-1} + R\ \tilde{e}_{n,K} \tag{31}$$

where the determined matrix expressions are denoted respectively by by P and R. The stability error $\tilde{e}_{n+1,K}$ depends not only on the component error related to x but on the whole error of X. This fact substantially differs the above formula from the one obtained for the convergence error, Eq.(21). Using Eq.(31) recursively

$$\tilde{e}_{n+1,K} = [I - P + P^2 - P^3 \pm ... + (-P)^{K-1}]\ R\ \tilde{e}_{n,K} + (-P)^K\ \tilde{e}_{n+1,0} \tag{32}$$

The geometrical series in this equation has a partial sum given as

$$I - P + P^2 - P^3 \pm ... + (-P)^{K-1} = (I + P)^{-1}\ [I - (-P)^K] \tag{33}$$

If we assume as predictors the values from the last iteration of the previous time step i.e. $X_{n+1,0} = X_{n,K}$ and $\tilde{X}_{n+1,0} = \tilde{X}_{n,K}$, then $\tilde{e}_{n+1,0} = \tilde{e}_{n,K}$, and

$$\tilde{e}_{n+1,K} = Q_{n+1}\ \tilde{e}_{n,K} \tag{34}$$

where $Q_{n+1} = (I+P)^{-1} [I-(-P)^K] R + (-P)^K$. Using this equation recursively we obtain a formula which discribes the propagation of the initial round-off error in time

$$\tilde{e}_{n+1,K} = Q_{n+1} \cdot Q_n \cdot \ldots \cdot Q_1 \, \tilde{e}_{0,K} \tag{35}$$

A necessary condition for the decay of initial round-off errors is

$$\|Q_{n+1} \cdot Q_n \cdot \ldots \cdot Q_1\| < 1 \tag{36}$$

where Q's can be different for different time steps. If $\|P\| \leq 1$ then for $K \to \infty$

$$Q_{n+1} = Q = [B + \theta \Delta t \, C]^{-1} \, [B - (1-\theta)\Delta t \, C] \tag{37}$$

and condition (36) reduces to $\|Q\| \leq 1$, the one ensuring stability of the direct method. Because of the complexity of the above expressions usually stability is verified for $K = 1$, then

$$Q_{n+1} = Q = R - P \tag{38}$$

and Eq.(36) becomes

$$\|R - P\| < 1, \tag{39}$$

the condition verified for the last term predictor, e.g. in [8, 9].

2. In the nonlinear case, i.e. when matrices G, H, G_3 and H_3 depend on $X_{n+\theta}$ causing that matrices P and R also depend on $X_{n+\theta}$, condition (36) must be verified at each time instant.

4. NUMERICAL TESTS

Numerical tests are related to a 1-D problem in which the deformation of the solid skeleton is coupled with the flow of air and water caused by gravitational forces. The test is conceived on the basis of the Liakopoulos experiment [6] and is performed here to check different possible formulations and numerical procedures.

In our calculations a linear elastic model of the solid and a nonlinear constitutive equation for liquid [1] are assumed. A single finite element with a linear approximation function is used, [7].

4.1. Determination of the lower bound on the time step Δt

For the linear case the neccesary condition for convergence of iterations of the staggered P-C scheme gives for fixed mesh size a lower bound on the time step Δt, see [5, 8]. Here we examine this fact for nonlinear equations and the direct and staggered P-C scheme. The spectral norm $\|f'\|$ is checked for the first time step. Results are shown in Fig. 1.

4.2. Examination of the influence of the θ parameter on $\|f'\|$

The value of θ modifies the matrix G from Eq.(7) affecting also $\|f'\|$. In this test we check $\|f'\|$ for the first time step of the direct and staggered P-C scheme. Results are shown in Fig. 2. It is seen that the influence of θ is not very strong.

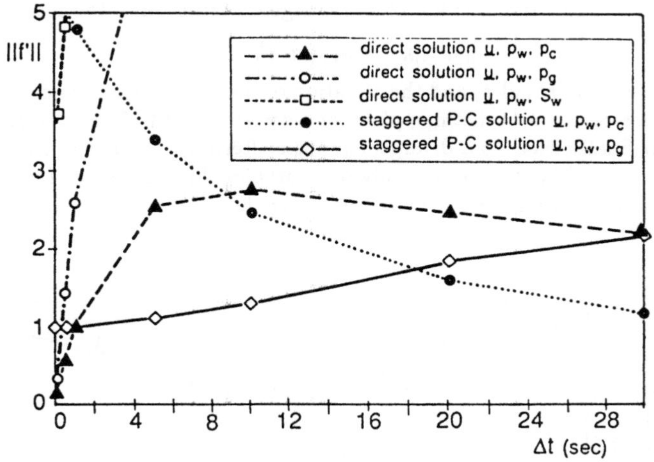

Fig. 1. Norm $\|f'\|$ for different Δt. Nonlinear equations.

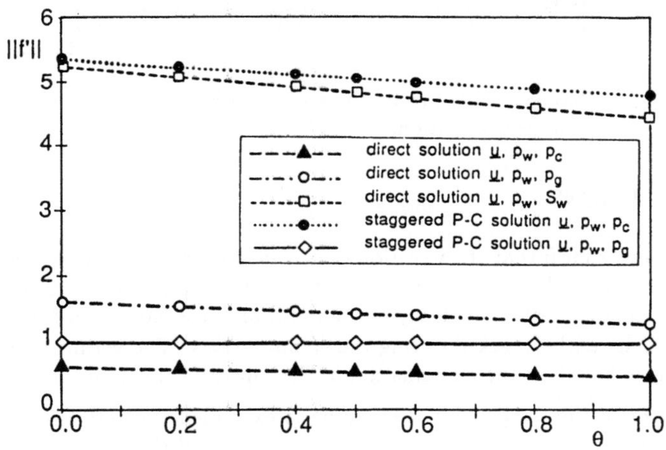

Fig. 2. Norm $\|f'\|$ for different θ. Nonlinear equations.

4.3. Comparison of the staggered P-C and the direct method

This test is to show that the use of a convergence criterion of the form

$$\|x_{n+1,K} - x_{n+1,K-1}\| \leq \tau$$

for not sufficiently small τ, and the violation of the convergence condition, which is 'only' necessary, can sometimes yield for a certain period of time almost correct results but finally leads to a divergent process. Results are shown in Fig. 3. The condition $\|f'\| < 1$ is satisfied for only one curve in this figure, i.e. for the direct solution method and $\Delta t = 1$. For all the others $\|f'\| \geq 1$.

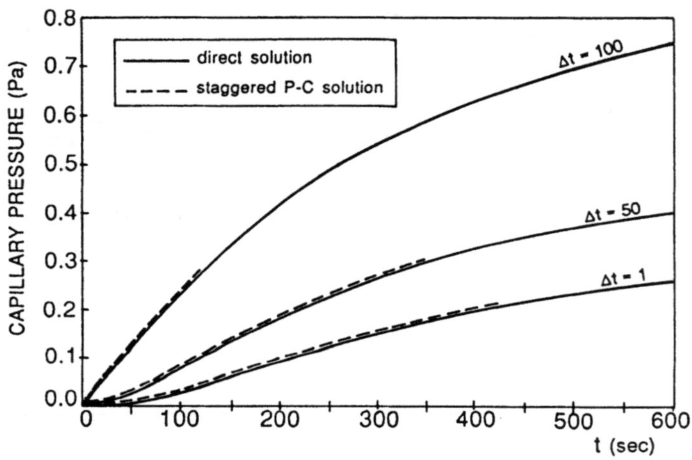

Fig. 3. Capillary pressure by the staggered P-C and direct method. Nonlinear equations.

5. CONCLUSIONS

1. It has been shown that the iteration error depends on the iteration error of the first two equations but not on the error of the third equation, Eq.(22) and (25). Necessary conditions for the decay of the iteration error have been formulated in a form which can be checked during the computations. The verification of these conditions is based on the examination of the Lipschitz constant.

2. The error of the temporal discretization is examined theoretically, and it has been shown that for the generalized trapezoidal method this error diminishes for $\Delta t \to 0$.

3. The stability error depends on the stability error of the first two equations as well as on the error of the third equation, Eq.(35). The formulas to control the stability error are very complex and have to be evaluated computationally. In our opinion further study is necessary to shed some light on the meaning of the derived formulas.

REFERENCES

1. BRUSTKERN, R.L., MOREL-SEYTOUX, H.J. - Analitical treatment of two-phase infiltration, J. of Hydraulics Div. ASCE, 96, 2535-2548, (1970)

2. LEWIS, R.W., SCHREFLER, B.A. - The finite element method in the deformation and consolidation of porous media, J. Wiley & Sons, Chichester, 1987.

3. ORTEGA, J.M., RHEINBOLDT, W.C. - Iterative solution of nonlinear equations in several variables, Academic Press, Inc., San Diego, 1970.

4. RALL, L.B. - Computational solution of nonlinear operator equations, J. Wiley & Sons, New York, 1969.

5. SAETTA, A., SCHREFLER, B.A., VITALIANI, R. - 'Partitioned solution procedure and convergence', Proc. New. Developments in Structural Mechanics, Catania 1990 (in print).

6. SCHREFLER, B.A., SIMONI, L., XIKUI, L., ZIENKIEWICZ, O.C. - 'Mechanics of partially saturated porous media', Numerical Methods and Constitutive Modelling in Geomechanics, Eds. Desai, C.S., Gioda, G., CISM Lecture Notes, Springer Verlag - Wien, 1990

7. SCHREFLER, B.A., SIMONI, L., 'Comparison between different finite element solutions for immiscible two phase flow in deforming porous media', Proc. 7th Conf. IACMAG, Balkema, 1215 -1220, 1991.

8. SIMONI, L., SCHREFLER, B.A. - A staggered F.E. solution for water and gas flow in deforming porous media, Commun. Appl. Num. Methods, 7, 213-223, (1991).

9. TURSKA, E., SCHREFLER, B.A. - On convergence conditions of partitioned solution procedures for consolidation problems, submitted for publication in Comp. Meth. in Appl. Mech. and Engng.

10. ZIENKIEWICZ, O.C., CHAN, A.H.C. - 'Coupled problems and their numerical solution', Advances in Computational Nonlinear Mechanics, Ed. Doltisnis, J.St., CISM Lecture Notes, Springer Verlag-Wien, 1989.

11. ZIENKIEWICZ, O.C., PAUL, D.K., CHAN, A.H.C. - Unconditionally stable staggered solution procedure for soil-pore fluid interaction problems, Int. J. Num. Meth. in Engng, 26, 1039-1055, (1988).

12. ZIENKIEWICZ, O.C., SCHREFLER, B.A., SIMONI, L., XIE Y.M., ZHAN, X.Y. - 'Two and three-phase behaviour in semi-saturated soil dynamics', Nonlinear Computational Mechanics - a State of the Art, Eds. Wriggers, P., Wagner, W., Springer Verlag, 1991

FINITE ELEMENTS IN THE 90's, E. Oñate, J. Periaux, A. Samuelsson (Eds.)
Springer-Verlag/CIMNE, Barcelona 1991

OPTIMAL SHAPE DESIGN UNDER TRANSIENT DYNAMIC LOADING

C.V. Ramakrishnan, A.C. Paul and D.K. Sehgal
Department of Applied Mechanics,
Indian Institute of Technology,
New Delhi-110016,
INDIA

Summary

This paper deals with the optimal design of structural shapes of member under the action of transient dynamic loads. A mathematical programming approach is proposed and the use of sequential linear programming is recommended for optimization. Transient dynamic analysis is carried out using finite element discretization and modal superposition approach. The paper mainly concentrates on the computation of design sensitivities for shape changes. First the design profile is modelled using a finite number of design ordinates using which the smooth shapes are generated by employing algebraic curves, cubic splines or B-splines. The FE mesh is automatically generated and the nodal coordinate derivatives with respect to design variables are computed and stored.By carrying out the numerical integration of the transformed equation of motion and the adjoint equations in terms of generalized coordinates, the design sensitivities for the time dependent constraints are computed.

1. INTRODUCTION

Structural shape design is an important field of research activity both from the point of view of economy and optimal functional behaviour. The interest in this area has grown enormously in recent years on account of the following developments.

- Ability to represent smooth shapes using appropriate surface/curve modelling procedures.
- Ability to generate automated meshes with good control on refinement.
- Ability to estimate the error measure in finite element analysis and design derivative computations
- Better understanding of design derivative computations and the use of a semianalytical material derivative or implicit differentiation approaches.

In some of the very early papers on the topic[1,2],the first author had emphasized the importance of profile definition and discretisation error control. While a large number of researchers are active in studies on various aspects of shape design(Bennett and Botkin[3]) the problem of optimal shape design under transient dynamic load has not been addressed at all. There are a few papers both old and recent which deal with the general optimum structural design under dynamic loads. Fox and Kapoor [4] used a bounding technique to obtain approximate peak response of planar truss frames under shock loading. Levy[5] presented fully stressed designs of rod and beam under dynamic loads. Feng [6],Feng et al [7] presented a general approach for optimal design of structures under general dynamic loads and used a generalized steepest descent method for optimization.

In this paper optimal shape design problem under transient loading is formulated and a systematic procedure for the computation of stress derivatives is described. While some of the steps for the analysis are similar to those for truss frame optimization,several salient points pertinent to shape optimization of continua are emphasized.

2. CONCEPTS IN DYNAMIC STRUCTURAL ANALYSIS USING MODAL SUPERPOSITION

The governing equations of motion for a structural system after discretising using finite element modelling are

$$\boldsymbol{M}(\boldsymbol{b})\, \ddot{\boldsymbol{z}} + \boldsymbol{K}(\boldsymbol{b})\, \boldsymbol{z} = \boldsymbol{Q}(t) \tag{1}$$

where **M** is the system mass matrix
K is the system stiffness matrix
z(t) is the vector of nodal displacements
Q(t) is the time dependant load vector
b is the vector of design variables representing shape
with initial conditions,

$$\boldsymbol{z}(0) = \boldsymbol{z}^{\circ} \quad \textit{and} \quad \dot{\boldsymbol{z}}(0) = \dot{\boldsymbol{z}}^{\circ} \tag{2}$$

and appropriate boundary conditions.

Here it should be noted that shape change is effected by changing the design vector **b** through an appropriate definition using spline functions.
Considering the relevant eigen value problem for equation (1)

$$\boldsymbol{K}\,\phi = \omega^2\, \boldsymbol{M}\,\phi \tag{3}$$

where (ω^2,ϕ) is an eigen pair. Assuming (ω_i^2,ϕ_i) $i=1,2,\ldots\ldots,n$ to be eigen pairs of the system. We can express **z** as a linear combination of eigen vectors through generalized coordinates.

$$\boldsymbol{z} = \Phi\,\zeta \tag{4}$$

It is usual to approximate the response of the system by considering only a few of the principal modes to reduce the size of the problem and where ζ is the vector of generalized coordinates.

Thus the original equation (1) reduces to

$$\boldsymbol{M^*} \ddot{\zeta} + \boldsymbol{K^*} \zeta = \boldsymbol{S}(t) \tag{5}$$

$$\text{where} \quad \boldsymbol{M^*} = \Phi^T \boldsymbol{M} \Phi$$

$$\boldsymbol{K^*} = \Phi^T \boldsymbol{K} \Phi$$

$$\boldsymbol{S} = \Phi^T \boldsymbol{Q} \tag{6}$$

The boundary conditions are

$$\dot{\zeta}(0) = \dot{\zeta}_o \quad \text{and} \quad \zeta(0) = \zeta_o \tag{7}$$

In a finite element context, the stiffness and consistent mass matrices can be written as

$$\boldsymbol{M} = \int_\Omega \boldsymbol{N}^T \rho \boldsymbol{N} \; d\Omega$$

$$\boldsymbol{K} = \int_\Omega \boldsymbol{B}^T \boldsymbol{D} \boldsymbol{B} \; d\Omega \tag{8}$$

where the field distribution of displacement in the interior of an element is expressed as

$$\boldsymbol{u}^{(e)} = \boldsymbol{N} \boldsymbol{z}^{(e)} \tag{9}$$

The strain and stress vector can be written as

$$\epsilon^{(e)} = \boldsymbol{B} \, \boldsymbol{z}^{(e)}$$

$$\sigma^{(e)} = \boldsymbol{D} \, \epsilon^{(e)} \tag{10}$$

where **D** is the elasticity matrix

3. PROBLEM FORMULATION FOR SHAPE DESIGN

The optimal shape design problem is concerned with the minimization of a suitable objective function subject to a set of constraints. The objective function could be minimization of area or a function of stress values at selected points. The constraints in a dynamic problem could be on deflections, frequencies or stresses. Thus the objective function is

$$J(\boldsymbol{b}) = F(\sigma_i(\boldsymbol{b})) \tag{11}$$

Where **b** represents the vector of design variable defining the shape. The manner in which shape is defined in the present investigation will be addressed later.

The constraint function for displacements and stresses can be written as follows

$$|z_i(\boldsymbol{b},t)| - z_i^p \leq 0 \qquad i=1,2,..,k_1 \tag{12}$$

$$|\sigma_j(\boldsymbol{z},\boldsymbol{b},t)| - \sigma_j^p \leq 0 \qquad j=1,2,....,k_2 \tag{13}$$

In general, these can be written in a vector form

$$\boldsymbol{h}(\boldsymbol{z},\boldsymbol{b},t) \leq 0 \quad \text{for all } 't \tag{14}$$

In the present formulation the above constraint is handled by considering the integral

$$\psi_i = \int_0^T h_i(\boldsymbol{z},\boldsymbol{b},t)\,dt \qquad \text{if } h_i<0 \;\; \text{in} \;\; t\in(0,T) \tag{15}$$

$$= \int_0^T \langle h_i(\boldsymbol{z},\boldsymbol{b},t) \rangle \,dt \qquad \text{if } h_i>0 \quad \text{in} \quad t \in (0,T)$$

Thus the quantity $\psi_i(\boldsymbol{z},\boldsymbol{b})$,if negative is a measure of the buffer domain available for optimization before violation of this constraint occurs. On the other hand if $\psi_i(\boldsymbol{z},\boldsymbol{b})$ is positive, it represents a measure of the constraint violation only over the entire time domain. In the present investigation we concentrate in obtaining the design derivatives viz., the derivatives of the objective function (F) and the constraint ψ_i 's with respect to the design variables $(\frac{\partial F}{\partial \boldsymbol{b}}, \frac{\partial \psi}{\partial \boldsymbol{b}})$. While the frequency constraints are important these are not considered here since the procedure to handle them is not different from frames and trusses [7]

4. COMPUTATION OF DESIGN DERIVATIVES

In any nonlinear optimization algorithm which is gradient based, the efficient computation of design derivatives is the key. In several earlier works the first author has advocated the use of an improved move limit method of SLP which is also called as the convex linearization algorithm by some researchers. However, just scanning the design derivatives yields a lot of information about desirable changes and in this section the necessary mathematical steps will be given for the computation of the derivatives in a finite element context under transient dynamic loading conditions.

The first variation of $\psi_i(\boldsymbol{z},\boldsymbol{b})$

$$\delta\psi_i = \frac{\partial \psi_i}{\partial \boldsymbol{z}}\,\delta \boldsymbol{z} + \frac{\partial \psi_i}{\partial \boldsymbol{b}}\,\delta \boldsymbol{b} \tag{16}$$

The second term represents the direct changes in the constraint function due to shape changes assuming the field displacements to remain unchanged while the first term corresponds to the constraint function changes due to changes in field displacement. For static problems,the computation procedure of these sensitivities have been well researched [1,8] but till date no systematic information is available for transient dynamic problems involving shape change. To express $\delta \mathbf{z}$ in terms of $\delta \mathbf{b}$ in equation (16) we consider a small neighbourhood of design space and taking variation of equation (1)

$$\boldsymbol{M}\,\delta\ddot{\boldsymbol{z}} + \boldsymbol{K}\,\delta\boldsymbol{z} = \left(\frac{\partial \boldsymbol{Q}}{\partial \boldsymbol{b}} - \frac{\partial \boldsymbol{M}}{\partial \boldsymbol{b}}\ddot{\boldsymbol{z}} - \frac{\partial \boldsymbol{K}}{\partial \boldsymbol{b}}\boldsymbol{z}\right)\delta\boldsymbol{b} = \frac{\partial \boldsymbol{R}}{\partial \boldsymbol{b}}\delta\boldsymbol{b} \qquad (17)$$

$$\delta\dot{\boldsymbol{z}}(0) = 0 \qquad \delta\boldsymbol{z}(0) = 0$$

Following the steps for formulation as an optimal control problem, we get an adjoint system

$$\boldsymbol{M}\,\ddot{\lambda}(t) + \boldsymbol{K}\,\lambda(t) = \left[\frac{\partial \boldsymbol{h}}{\partial \boldsymbol{z}}\right]^T.$$

$$\textit{subject to} \quad \dot{\lambda}(T) = 0 \qquad \lambda(T) = 0 \qquad (18)$$

The solution of the above system for λ using the same eigen system for the direct problem results in a simple procedure for the evaluation of

$$\int_0^T \frac{\partial \boldsymbol{h}}{\partial \boldsymbol{z}}(\boldsymbol{z},\boldsymbol{b},t)\,\delta\boldsymbol{z}\,dt \qquad \textit{as} \qquad \int_0^T \left(\lambda^T \frac{\partial \boldsymbol{R}}{\partial \boldsymbol{b}}(\boldsymbol{b},t)\,dt\right)\delta\boldsymbol{b} \qquad (19)$$

Thus we get

$$\delta\psi_i = \int_0^T \left\{ \left[\lambda^T \frac{\partial \boldsymbol{R}(\boldsymbol{b},t)}{\partial \boldsymbol{b}} + \frac{\partial h_i(\boldsymbol{z},\boldsymbol{b},t)}{\partial \boldsymbol{b}}\right] dt \right\} \delta\boldsymbol{b}$$

$$\textit{i.e.} \quad \frac{\partial \psi}{\partial \boldsymbol{b}} = \int_0^T \left[\lambda^T \frac{\partial \boldsymbol{R}(\boldsymbol{b},t)}{\partial \boldsymbol{b}} + \frac{\partial \boldsymbol{h}(\boldsymbol{z},\boldsymbol{b},t)}{\partial \boldsymbol{b}}\right] dt \qquad (20)$$

where λ represents the matrix of adjoint vector solutions for the set of constraints.

For shape optimization problems the evaluation of various quantities $\frac{\partial \boldsymbol{h}}{\partial \boldsymbol{z}}, \frac{\partial \boldsymbol{R}}{\partial \boldsymbol{b}}$ *and* $\frac{\partial \boldsymbol{h}}{\partial \boldsymbol{b}}$ require further mathematical steps. The interpolation of stress to the boundary is accomplished using

$$\sigma_1 = \boldsymbol{L}\,\{\sigma_{1g}\}^e \qquad (21)$$

where $\{\sigma_{1g}\}^{(e)}$ represent the vector of principal stresses at the gauss points of the element 'e'.

$$\{\sigma_{1g}\}^{e} = (\sigma_{1,1}, \sigma_{1,2}, \dots)$$

The principal stresses at gauss points are

$$\sigma_{1g} = \left\{ \frac{\sigma_x + \sigma_y}{2} + \sqrt{\left(\frac{\sigma_x - \sigma_y}{2}\right)^2 + \tau_{xy}^2} \right\}_g \qquad (22)$$

where $\mathbf{L}$ is a constant interpolation matrix,

and $\{\sigma_1\}^{(e)}$ is the vector gauss point principal stresses pertaining to the element 'e'.

Thus $\{\sigma_{1g}\}^{(e)}$ is a nonlinear function of the vector of element guess point stresses $\{\sigma_g\}^{(e)}$

$$\{\sigma_{1g}\}^{(e)} = f_1(\{\sigma_g^{(e)}\}) \qquad (23)$$

Thus differentiating equation no. (13) w.r.t a selected design variable b_k

$$\frac{\partial h_j}{\partial b_k} = \frac{\partial \sigma_j}{\partial b_k}\, sign(h_j)$$

Denoting the partial differentiation w.r.t a design variable with a prime on top

$$h_j' = sign(h_j)\mathbf{L}\{\sigma_{1g}'\}^{(e)} \qquad (24)$$

$$\sigma_{1g}' = \left(\frac{\sigma_x' + \sigma_y'}{2}\right)_g + \frac{\left(\frac{\sigma_x - \sigma_y}{2}\right)_g \left(\frac{\sigma_x' - \sigma_y'}{2}\right)_g + (\tau_{xy})_g\,(\tau_{xy}')_g}{\sqrt{\left(\frac{\sigma_x - \sigma_y}{2}\right)_g^2 + (\tau_{xy}^2)_g}} = f_2(\sigma_g, \sigma_g') \qquad (25)$$

The stress derivative vector at any gauss point is obtained by differentiating equation no. (10) w.r.t. the design variable

$$\sigma_g' = D^{(e)}\, B^{(e)}\, z^{(e)} \qquad (26)$$

To compute $\frac{\partial \boldsymbol{R}}{\partial \boldsymbol{b}}$ we consider equation no. (17)

$$\frac{\partial R}{\partial b_k} = \frac{\partial Q}{\partial b_k} - \frac{\partial M}{\partial b_k}\ddot{z} - \frac{\partial K}{\partial b_k} z$$

Using equation no. (8) and utilizing expressions for numerical quadrature

$$M' = \Sigma_e \Sigma_I \Sigma_J N^T \rho N \, |J|' \, w_I w_J$$

$$K' = \Sigma_e \Sigma_I \Sigma_J [(B^T D B' + B'^T D B)|J| + B^T D B \, |J|' \,] \, w_I w_J \quad (27)$$

Thus $M'\ddot{z}$ *and* $K'z$ can be written as

$$M'\ddot{z} = \Sigma_e \Sigma_I \Sigma_J N^T \rho \, |J|' \, \ddot{u}_g \, w_I w_J$$

$$K'z = \Sigma_e \Sigma_I \Sigma_J (B^T D B' u_g |J| + (B'^T |J| + B^T |J|') \{\sigma_g\}) \, w_I w_J \quad (28)$$

The significance of various terms of the second of equations(28)are well documented in the literature in the context of shape optimization problems.

The computation of $\frac{\partial \psi}{\partial z}$ is fairly straightforward

$$\frac{\partial \psi_I}{\partial z} = sign(h_I) \, L \, \{ \frac{\partial \sigma_{1g}}{\partial z} \}^{(e)} \quad (29)$$

$$\frac{\partial \sigma_{1g}}{\partial z} = \left(\frac{\overset{+}{\sigma}_x + \overset{+}{\sigma}_y}{2}\right) + \frac{\left(\frac{\sigma_x - \sigma_y}{2}\right)\left(\frac{\overset{+}{\sigma}_x - \overset{+}{\sigma}_y}{2}\right) + \overset{+}{\tau}_{xy}\tau_{xy}}{\sqrt{\left(\frac{\sigma_x - \sigma_y}{2}\right)^2 + \tau_{xy}^2}} \quad (30)$$

where

$$\overset{+}{\sigma}_g = \frac{\partial \sigma_g}{\partial z} = \frac{\partial}{\partial z}\begin{Bmatrix}\sigma_x \\ \sigma_y \\ \tau_{xy}\end{Bmatrix}_g$$

$$= DB_g \qquad \textit{from equation no. } (10)$$

where the superscript (+) represents the differentiation w.r.t z and the subscript 'g' represents computation at a gauss point 'g'.

All intergrations w.r.t time can be performed using a suitable Newton-Cotes formula.

5. COMPUTATIONAL ALGORITHM

It is obvious that the computational procedure has to be efficient for the computation of design derivatives.The algorithm is described briefly.

1. Read control data for analysis,automated mesh generation,identification of boundary definition points etc.
2. For the given set of design variables,generate mesh and also compute nodal coordinate derivatives($(x_i', y_i'), i=1,2,....$number of nodal points).
3. Compute a finite number of eigen vectors and eigen modes for the structural vibration problem.
4. Compute transient dynamic response using modal superposition technique and store nodal displacements at various time instants.
5. Carry out backward integration of the adjoint equation using the modal superposition technique since the eigen values and eigen vectors of the adjoint problem are the same.At each backward time step perform steps 6 to 12.
6. Identify all the elements sensitive to shape change.For each element, compute $\dfrac{\partial \boldsymbol{R}}{\partial \boldsymbol{b}}$ using appropriate expressions.
7. Check if the element under consideration is also a candidate for the evaluation of stress constraints.If so,compute $\dfrac{\partial h_i}{\partial \boldsymbol{b}}$ .If stress constraints are expressed for boundary stresses,suitable extrapolation smoothing techniques are employed.In the present context,simple extrapolation of principal stresses was used (fig.1).It may be noted that the steps 6 and 7 have to be carried out w.r.t each design variable.

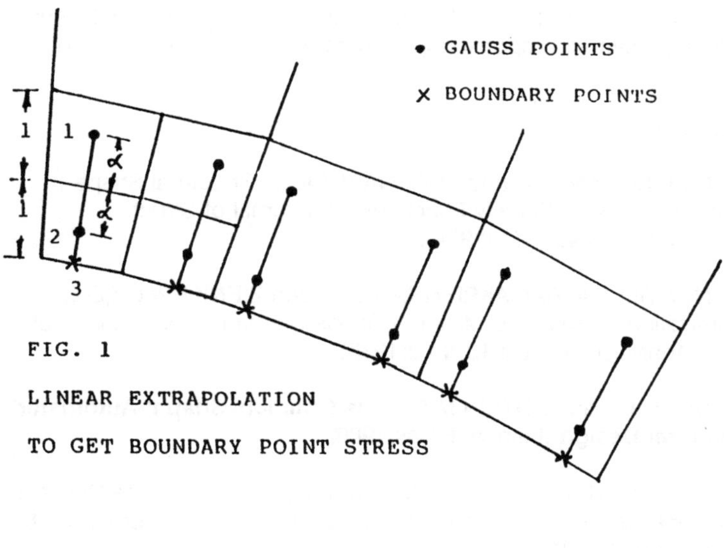

FIG. 1

LINEAR EXTRAPOLATION

TO GET BOUNDARY POINT STRESS

8. For each element identified for stress constraint evaluation,compute $\frac{\partial \psi_i}{\partial \boldsymbol{z}}$ and store

9. For each $\{\frac{\partial \psi_i}{\partial \boldsymbol{z}}\}$ vector carry out the computation for incremental computation of adjoint vector at the current backward time step.

10. Perform calculations at step 9 for each constraint derivative vector and obtain the matrix of adjoint vectors $\boldsymbol{\lambda}$.

11. Compute the inner product $\boldsymbol{\lambda}^T \frac{\partial \boldsymbol{R}}{\partial \boldsymbol{b}}$.

12. Use Newton-Cotes numerical integration coefficients and compute the contribution for $\frac{\partial \psi}{\partial \boldsymbol{b}}$ for the time interval 't' to 't+δt'.

13. The matrix of constraint vector sensitivities are now available $[\frac{\partial \psi_i}{\partial \boldsymbol{b}_k}]$.

6. NUMERICAL EXAMPLES

Presently computations have been performed on small test problems like a six element cantilever beam etc,.Detailed results will be presented separately.

7. ACKNOWLEDGEMENTS

The second author thanks the authorities of Regional Engineering college,Silchar,Assam for sponsoring him to I.I.T.,New Delhi for higher studies.

8. REFERENCES

1. RAMAKRISHNAN,C.V. and FRANCAVILLA,A.- Structural Shape Optimization using Penalty Functions,Int. Journal of Struct. Mechanics,vol.**3**,403-422(1974).

2. FRANCAVILLA,A.,RAMAKRISHNAN,C.V. and ZIENKIEWICZ,O.C.- Optimization of Shape to Minimize Stress Concentration, Journal of Strain Analysis(U.K), vol.**10**,No.2(1975).

3. BENNETT,J.A. and BOTKIN,M.E.- **'The Optimum Shape'-Automated Structural Design** ,Plenum Press,1986.

4. FOX,R.L. and KAPOOR,M.P.- 'Structural Optimization in the Dynamic Response Regime : A Computational Approach',AIAA Journal, vol.**8**, No.10, 1798-1804(1970).

5. LEVY,H.J.-Minimum Weight Design Under Dynamic Loading,Ph.D Dissertation,New York Univ,1972.

6. FENG,T.T. - Optimal Design of Elastic Structures Under Dynamic Loads, Ph.D Dissertation,Mechanics and Hydraulics Program,College of Engineering,Univ.of Iowa,May 1975.

7. FENG,T.T.,ARORA,J.S. and HANG Jr,E.J. - Optimal Structural Design Under Dynamic Loads,Int.Jour. of Num.Meth. Engg,vol.**11**,39-52(1977).

8. YANG,R.J. and BOTKIN,M.E. - The Relationship Between the Variational Approach and the Implicit Differentiation Approach to Shape Design Sensitivities - in **'The Optimum Shape'** by Bennet, J.A. and Botkin, M.E.,1986.

9. RAMAKRISHNAN,C.V. - "Development of a Two/Three Dimentional Shape Optimization Program with Solid Modelling,Semi-Automatic Mesh Generation and Adaptive Mesh Refinement" in **'Control of Boundaries and Stabilization'**- J.Simon(Ed.),Springer-Verlag,1989.

FINITE ELEMENTS IN THE 90's, E. Oñate, J. Periaux, A. Samuelsson (Eds.)
Springer-Verlag/CIMNE, Barcelona 1991

A FINITE ELEMENT METHOD FOR DYNAMIC CONTACT PROBLEMS

R.L. Taylor and **P. Papadopoulos**
Department of Civil Engineering,
University of California at Berkeley,
Berkeley, CA 94720,
USA

Abstract

This paper addresses the formulation and discrete approximation of dynamic contact initial-value problems. The continuous problem is presented in the context of non-linear kinematics. Standard semi-discrete time integrators are introduced and are shown to be unsuccessful in modeling the kinematic constraints imposed on the interacting bodies during persistent contact. A procedure that bypasses the aforementioned difficulties is proposed by means of a novel variational formulation. Numerical simulations are conducted and the results are reported and discussed.

1 Introduction

Mechanical contact is encountered whenever two or more bodies physically interact along their boundaries. Contact is of particular interest in numerous engineering applications ranging from metal forming and machine design to soil-structure and structure-structure interaction under dynamic excitation.

The non-linear character of contact boundary conditions allows for very few interesting problems to be solved analytically. Since the inception of the finite element method in the late 50s, numerical solutions to contact problems have been intensively investigated by various researchers. Conry and Seireg appear to be the first to treat contact as a quadratic programming problem, [CS71], while Chan and Tuba first introduced a general slideline methodology, [CT71]. Moreover, the works of Zienkiewicz on a flexibility approach, [FZ75], and Hughes on Lagrange multiplier methods, significantly contributed to the development of robust finite element approximations, applicable to the simulation of large-scale problems.

This work is concerned with the time dimension of contact problems. By way of background, continuous non-linear elastodynamics is briefly reviewed in Section 2. Section 3 addresses the two-body contact problem in the continuous setting, while Section 4 presents Lagrange multiplier and penalty formulations in space and semi-discrete integrators in time suitable for finite element approximations. Section 5 investigates standard second-order implicit time integrators of the Newmark family and their performance in modeling jump conditions in the kinematic fields (e.g., velocities and accelerations) of the interacting bodies. These integrators are shown to produce undesirable oscillatory solutions along the contact surface. The original Lagrange multiplier formulation of the two-body problem is appropriately extended so that it eliminate this inconsistency. One- and two-dimensional numerical simulations are offered in Section 6 and demonstrate the significance of the proposed formulation.

[†]This work has been supported by the Lawrence Livermore National Laboratory under contract Nr. W-7405-ENG-48.

2 Non-linear elastodynamics

In this section the initial-value problem of non-linear elastodynamics is briefly reviewed. To this end, a deformable body B is identified with an open set Ω in the R^3 linear space equipped with the standard basis $(\mathbf{E}_1, \mathbf{E}_2, \mathbf{E}_3)$. It follows that a typical material point of B is algebraically represented by vector coordinates $\mathbf{X} = (X_1, X_2, X_3)$ in the undeformed (reference) configuration. The boundary of the body, $\partial\Omega$, possesses a unique outer normal $\mathbf{N}$ at each of its points, provided it is sufficiently smooth (at least pointwise differentiable). Furthermore, assume that there exists an invertible mapping $\mathcal{X}$ defined as

$$\mathcal{X} \; : \; \Omega \times R_0^+ \mapsto R^3 \quad | \quad \mathbf{x} = \mathcal{X}(\mathbf{X}, t) ,$$

where $\mathbf{x}$ denotes the position of particle $\mathbf{X}$ in the deformed (current) configuration at a generic time t. The displacement field, $\mathbf{u}$, associated with the motion, is introduced according to

$$\mathbf{u}(\mathbf{X}, t) = \mathcal{X} - \mathbf{X}$$

and the deformation gradient, $\mathbf{F}$, as

$$\mathbf{F} = \frac{\partial \mathbf{x}}{\partial \mathbf{X}} .$$

The strong form of the initial-value problem of non-linear elastodynamics is described by the following set of equations:

$$\begin{aligned}
\nabla \cdot \mathbf{P} + \rho_0 \mathbf{b} &= \rho_0 \ddot{\mathbf{u}} \quad on \quad \Omega \times R_0^+ , && (1)\\
\mathbf{u} &= \bar{\mathbf{u}} \quad on \quad \Gamma_u \times R_0^+ , && (2)\\
\mathbf{N} \cdot \mathbf{P} &= \bar{\mathbf{T}} \quad on \quad \Gamma_t \times R_0^+ , && (3)\\
\mathbf{u}(\mathbf{X}, 0) &= \mathbf{u}_0(\mathbf{X}) \quad on \quad \Omega , && (4)\\
\dot{\mathbf{u}}(\mathbf{X}, 0) &= \dot{\mathbf{u}}_0(\mathbf{X}) \quad on \quad \Omega . && (5)
\end{aligned}$$

In the above, $\mathbf{u}_0$ and $\dot{\mathbf{u}}_0$ denote the initial displacement and velocity fields of the body, respectively, while $\mathbf{P}$ is the 1st Piola-Kirchhoff stress tensor. Also, $\rho_0 = \rho_0(\mathbf{X})$ is the mass density in the reference state and Γ_u, Γ_t are the Dirichlet and Neumann portions of the boundary, on which boundary displacements, $\bar{\mathbf{u}}$, and surface tractions, $\bar{\mathbf{T}}$, are specified, respectively. A hyperelastic constitutive assumption is made in the present context. Thus

$$\mathbf{P}(\mathbf{X}, t) = \rho_0 \frac{\partial W}{\partial \mathbf{F}} , \qquad (6)$$

where W is a strain energy functional, per unit mass, expressed in terms of some invariant measure of deformation. Equations (1-6) constitute *Problem (P)*.

An integral counterpart of *Problem (P)* with reference to both the temporal and the spatial dimension can be stated by means of Hamilton's law of Varying Action, [Ham34, Ham35] as follows:

Problem (V)

For any given time interval $(t_i \, , \, t_f] \subset R^+$, find $\mathbf{u} \in \mathcal{U}$ such that

$$D_{\tilde{\mathbf{u}}} \int_{t_i}^{t_f} \int_\Omega [\frac{1}{2} \rho_0 \dot{\mathbf{u}} \cdot \dot{\mathbf{u}} - \rho_0 W(\mathbf{u})] \, dV \, dt + \int_{t_i}^{t_f} \int_\Omega \rho_0 \mathbf{b} \cdot \tilde{\mathbf{u}} \, dV \, dt$$

$$+ \int_{t_i}^{t_f} \int_{\Gamma_t} \bar{\mathbf{T}} \cdot \tilde{\mathbf{u}} \, d\Gamma \, dt + \rho_0 \dot{\mathbf{u}} \cdot \tilde{\mathbf{u}} \Big]_{t_i}^{t_f} = 0 \quad \forall \, \tilde{\mathbf{u}} \in \tilde{\mathcal{U}} .$$

In the above equation, $D_{\tilde{\mathbf{u}}}$ denotes the Gâteaux derivative in the direction of $\tilde{\mathbf{u}}$. Also, $\mathcal{U}$ is the space of kinematically admissible displacements for the given problem given by

$$\mathcal{U} = \Big\{\mathbf{u} \in H^1(\Omega \times R_0^+) \;\; | \\ \mathbf{u} = \bar{\mathbf{u}} \;\; on \;\; \Gamma_u \times R_0^+ \;\; , \;\; \mathbf{u}(\mathbf{X}\,,\,0) = \mathbf{u}_0 \;\; , \;\; \dot{\mathbf{u}}(\mathbf{X}\,,\,0) = \dot{\mathbf{u}}_0 \;\; on \;\; \Omega\Big\} \; ,$$

while $\tilde{\mathcal{U}}$ is the associated space of admissible variations defined as

$$\tilde{\mathcal{U}} = \Big\{\tilde{\mathbf{u}} \in H^1(\Omega \times R_0^+) \;\; | \\ \tilde{\mathbf{u}} = \mathbf{0} \;\; on \;\; \Gamma_u \times R_0^+ \;\; , \;\; \tilde{\mathbf{u}}(\mathbf{X}\,,\,0) = \mathbf{0} \;\; , \;\; \dot{\tilde{\mathbf{u}}}(\mathbf{X}\,,\,0) = \mathbf{0} \;\; on \;\; \Omega\Big\} \; .$$

After integration by parts on the kinetic energy, Hamilton's law of Varying Action gives rise to a statement of virtual work for the initial-value problem according to

$$\int_{t_i}^{t_f} \Big\{ \int_\Omega [\rho_0 \ddot{\mathbf{u}} \cdot \tilde{\mathbf{u}} + \rho_0 D_{\tilde{\mathbf{u}}} W(\mathbf{u}) - \rho_0 \mathbf{b} \cdot \tilde{\mathbf{u}}] \, dV \\ - \int_{\Gamma_t} \bar{\mathbf{T}} \cdot \tilde{\mathbf{u}} \, d\Gamma \Big\} \, dt = 0 \quad \forall \, \tilde{\mathbf{u}} \in \tilde{\mathcal{U}} \; , \; (t_i \, , \, t_f] \subset R^+ \; . \tag{7}$$

Equation (7) constitutes the basis for the space-time discretization of the problem and will be subsequently utilized in the development of finite elements for dynamic contact.

3 Dynamic contact problem

In this section a local formulation of the dynamic contact problem is furnished. Considerations are restricted to the two-body problem, with generalization to multi-body contact omitted for simplicity. Observing the notation of the previous section, the boundary of bodies B^1 and B^2 can be uniquely decomposed into three distinct regions according to

$$\partial\Omega^\alpha = \partial\Gamma_u^\alpha \bigcup \partial\Gamma_t^\alpha \bigcup C \quad , \quad \alpha = 1,2 \; ,$$

where C is the common contact surface. Contact is said to occur if $meas(C) > 0$. Gap and pressure functions can be defined along C, by means of projection of boundary points from one body onto the other, see [Pap91], as

$$g \; : \; (\partial\Omega^1 - \Gamma_u^1) \times (\partial\Omega^2 - \Gamma_u^2) \times R_0^+ \mapsto R_0^+ \quad | \quad g = g_{1\hookrightarrow 2}(t) \; ,$$

$$p \; : \; (\partial\Omega^1 - \Gamma_u^1) \times (\partial\Omega^2 - \Gamma_u^2) \times R_0^+ \mapsto R_0^- \quad | \quad p = p_{1\hookrightarrow 2}(t) \; .$$

The dynamic contact problem distinguishes itself from the static in that inequality constraint conditions hold not only for the displacements along the contacting surfaces, but also for their rates. The impenetrability constraint may be expressed as

$$[(\mathbf{X}_{P^2} + \mathbf{u}(\mathbf{X}_{P^2}\,,\,t)) - (\mathbf{X}_{P^1} + \mathbf{u}(\mathbf{X}_{P^1}\,,\,t))] \cdot \mathbf{n}_{P^1}(\mathbf{X}\,,\,t) = 0 \; , \tag{8}$$

where $\mathbf{n}_{P^1} \equiv \mathbf{n}$ is the outer unit normal from body B^1 in the current configuration. Time differentiation of (8) results in

$$[\dot{\mathbf{u}}(\mathbf{X}_{P^2}\,,\,t) - \dot{\mathbf{u}}(\mathbf{X}_{P^1}\,,\,t)] \cdot \mathbf{n}_{P^1}(\mathbf{X}\,,\,t) \\ + [(\mathbf{X}_{P^2} + \mathbf{u}(\mathbf{X}_{P^2}\,,\,t)) - (\mathbf{X}_{P^1} + \mathbf{u}(\mathbf{X}_{P^1}\,,\,t))] \cdot \dot{\mathbf{n}}_{P^1}(\mathbf{X}\,,\,t) = 0 \; ,$$

which, taking into account that the unit normal, $\mathbf{n}_{P^1}$ is, by definition, parallel to the distance between the contacting points, can be readily shown to reduce to

$$[\dot{\mathbf{u}}(\mathbf{X}_{P^2}\,,\,t) - \dot{\mathbf{u}}(\mathbf{X}_{P^1}\,,\,t)] \cdot \mathbf{n}_{P^1}(\mathbf{X}\,,\,t) = 0 \; . \tag{9}$$

Likewise, a second time differentiation on (8) yields

$$[\ddot{\mathbf{u}}(\mathbf{X}_{P^2}\,,\,t) \;-\; \ddot{\mathbf{u}}(\mathbf{X}_{P^1}\,,\,t)]\cdot\mathbf{n}_{P^1}(\mathbf{X}\,,\,t) \;=\; 0\,. \tag{10}$$

Similar results may be obtained for higher time derivatives of the boundary displacements. As it has been just illustrated in the continuous setting, equations (9) and (10) are obtained directly from the impenetrability condition and, therefore, do not represent additional constraints. They have been derived here because of their significance in discrete solutions.

In the case of persistent contact, the two body problem is summarized as follows:

Problem (P_c)

Given state at time t_i for Ω^α, $\alpha = 1,2$,
solve (1-6) at interval $(t_i\,,\,t_f]$ subject to
$pg = 0 \;,\; p \leq 0 \;,\; g \geq 0 \;\; on \;\; C\,.$

Integral formulations of the problem can be obtained by exploiting the statement of Hamilton's law (or, equivalently, the statement of virtual work) presented earlier in this section. A formulation based on Lagrange multipliers reads:

Problem (V_L)

Find $(\mathbf{u}^1\,,\,\mathbf{u}^2\,,\,p) \in \mathcal{U}^1 \times \mathcal{U}^2 \times \mathcal{P}$ such that

$$\int_{t_i}^{t_f} \sum_{\alpha=1}^{2} \Big\{ \int_{\Omega^\alpha} [\rho_0^\alpha \ddot{\mathbf{u}}^\alpha \cdot \tilde{\mathbf{u}}^\alpha + \rho_0^\alpha D_{\tilde{\mathbf{u}}^\alpha} W^\alpha(\mathbf{u}^\alpha) - \rho_0^\alpha \mathbf{b}^\alpha \cdot \tilde{\mathbf{u}}^\alpha]\, dV$$
$$- \int_{\Gamma_t^\alpha} \bar{\mathbf{T}}^\alpha \cdot \tilde{\mathbf{u}}^\alpha\, d\Gamma \Big\}\, dt + \int_{t_i}^{t_f} \int_C (\tilde{p} g + p\delta g)\, d\Gamma\, dt = 0$$
$$\forall \;\; (\tilde{\mathbf{u}}^1\,,\,\tilde{\mathbf{u}}^2\,,\,\tilde{p}) \in \tilde{\mathcal{U}}^1 \times \tilde{\mathcal{U}}^2 \times \tilde{\mathcal{P}}\,,\; (t_i\,,\,t_f] \subset R^+\,.$$

The spaces of admissible and virtual displacements are as in the previous section, while the respective pressure fields is defined as

$$\mathcal{P} = \left\{ p \in H^{-1/2}(C \times R_0^+) \;\;|\;\; p \leq 0 \right\}$$

and

$$\tilde{\mathcal{P}} = \left\{ \tilde{p} \in H^{-1/2}(C \times R_0^+) \;\;|\;\; \tilde{p} \leq 0 \right\}\,,$$

Integral statements such as *Problem (V_L)* generally form the basis of finite element approximations to problems in elastodynamics. Formulations based on various penalty regularizations (in the classical penalty, perturbed Lagrangian or augmented Lagrangian sense) can also be obtained in a straightforward manner.

4 Spatial and temporal discretization

For a given time, the displacement field, $\mathbf{u}$, and its rates are approximated according to

$$\left\{ \begin{array}{c} \mathbf{u} \\ \dot{\mathbf{u}} \\ \ddot{\mathbf{u}} \end{array} \right\} \approx \left\{ \begin{array}{c} \mathbf{u}_h \\ \dot{\mathbf{u}}_h \\ \ddot{\mathbf{u}}_h \end{array} \right\} = \mathbf{A}_{(I)}(\mathbf{X}) \left\{ \begin{array}{c} \mathbf{u}_{(I)} \\ \mathbf{v}_{(I)} \\ \mathbf{a}_{(I)} \end{array} \right\}\,,$$

where $\mathbf{A}_{(I)}$ is a matrix of basis functions for the entire domain, normally composed by standard interpolation functions within each element (no summation is implied on I). Further, vectors $\mathbf{u}_{(I)}$, $\mathbf{v}_{(I)}$ and $\mathbf{a}_{(I)}$ consist of all nodal displacements, velocities and accelerations, respectively. In addition, the pressure field is spatially approximated according to

$$p \approx p_h = \mathbf{A}^p_{(I)}\mathbf{p}_{(I)} .$$

Moreover, subscript h characterizes spatially discretized fields. It follows that *Problem (V_L)* may be recast in the form:

Find $(\mathbf{u}^1_h , \mathbf{u}^2_h , p_h) \in \mathcal{U}^1_h \times \mathcal{U}^2_h \times \mathcal{P}_h$ such that

$$\int_{t_n}^{t_{n+1}} \sum_{\alpha=1}^{2} \left\{ [\mathbf{M}^\alpha \mathbf{a}^\alpha_{(I)} + \mathbf{Q}^\alpha(\mathbf{u}^\alpha_{(I)}) - \mathbf{W}^\alpha + \mathbf{W}^\alpha_L(\mathbf{u}^\alpha_{(I)})] \cdot \tilde{\mathbf{u}}^\alpha_{(I)} \right\} dt$$

$$+ \int_{t_n}^{t_{n+1}} \mathbf{W}_p \tilde{\mathbf{p}}_{(I)}\, dt = 0$$

$$\forall \; (\tilde{\mathbf{u}}^1_h , \tilde{\mathbf{u}}^2_h , \tilde{p}_h) \in \tilde{\mathcal{U}}^1_h \times \tilde{\mathcal{U}}^2_h \times \tilde{\mathcal{P}}_h \quad , \quad (t_n , t_{n+1}] \in R^+ ,$$

where

$$\mathbf{M}^\alpha = \sum_{\Omega^\alpha_h} \rho^\alpha_0 \mathbf{A}^T_{(I)} \mathbf{A}_{(I)} \quad , \quad \mathbf{Q}^\alpha = \sum_{\Omega^\alpha_h} \nabla \mathbf{A}^T_{(I)} \mathbf{P}(\mathbf{u}^\alpha_{(I)}) ,$$

$$\mathbf{W}^\alpha = \sum_{\Omega^\alpha_h} \rho^\alpha_0 \mathbf{A}^T_{(I)} \mathbf{b}^\alpha + \sum_{\Gamma^\alpha_{th}} \mathbf{A}^T_{(I)} \bar{\mathbf{T}}^\alpha \quad , \quad \mathbf{W}^\alpha_L = (-1)^\alpha \sum_{C_h} p_h \mathbf{A}^T_{(I)} \mathbf{n} ,$$

and

$$\mathbf{W}_p = \sum_{C_h} g(\mathbf{U}_h) \mathbf{A}^{pT}_{(I)} \quad , \quad \alpha = 1,2 .$$

In the above definitions, summation symbols over the domain or parts of its boundary refer to the usual assembly operation.

A Newmark scheme is employed for the time integration of the spatially discretized equations of motion resulting from the Lagrange multiplier or penalty formulation of the contact problem. The family of Newmark integrators, as originally suggested in [New59], imposes dynamic equilibrium at the two end points of the finite time interval $\Delta t_n \equiv t_{n+1} - t_n$. The general form of the integrator is

$$\mathbf{u}_{n+1} = \mathbf{u}_n + \mathbf{v}_n \Delta t_n + \frac{1}{2}[(1 - 2\beta)\mathbf{a}_n + 2\beta \mathbf{a}_{n+1}]\Delta t^2_n , \tag{11}$$

$$\mathbf{v}_{n+1} = \mathbf{v}_n + [(1 - \gamma)\mathbf{a}_n + \gamma \mathbf{a}_{n+1}]\Delta t_n , \tag{12}$$

where $(\bullet)_n \equiv (\bullet)(t_n)$. Parameters β and γ (called the *Newmark parameters*) may vary according to

$$0 \leq \beta \leq 0.5 \quad , \quad 0 \leq \gamma \leq 1 .$$

The overall characteristics of any particular Newmark scheme governed by the choice of the above two parameters. In particular, unconditional stability is guaranteed for $2\beta \geq \gamma \geq 0.5$, as shown in [Woo90]. Furthermore, integration is globally first-order accurate (error $O(\Delta t^2)$) provided that $\gamma = 0.5$.

For the remainder of this section, all subscripts associated with spatial discretization of the various fields will be dropped in favor of those pertaining to the temporal description. The Lagrange multiplier formulation of the previous section leads to equations of motion written at time t_{n+1} as

$$\sum_{\alpha=1}^{2} [\mathbf{M}^\alpha \mathbf{a}^\alpha_{n+1} + \mathbf{Q}^\alpha(\mathbf{u}_{n+1}) - \mathbf{W}^\alpha_{n+1} + \mathbf{W}^\alpha_L(\mathbf{u}^\alpha_{n+1})] = \mathbf{0} , \tag{13}$$

$$g(\mathbf{u}^1_{n+1} , \mathbf{u}^2_{n+1}) = 0 . \tag{14}$$

Use of equations (11) and (12) in (13) results in elimination of the unknown velocity and acceleration fields, thus yielding a set of equations with the displacement vector at time t_{n+1} as the only unknown, according to

$$\sum_{\alpha=1}^{2} \left\{ \frac{1}{\beta \Delta t_n^2} \mathbf{M}^\alpha \mathbf{u}_{n+1}^\alpha + \mathbf{Q}^\alpha(\mathbf{u}_{n+1}^\alpha) + \mathbf{W}_L^\alpha(\mathbf{u}_{n+1}^\alpha) - \mathbf{W}_{n+1}^\alpha \right. \tag{15}$$

$$\left. - \mathbf{M}^\alpha \frac{1}{\beta} \left[\frac{1}{\Delta t_n^2} \mathbf{u}_n^\alpha + \frac{1}{\Delta t_n} \mathbf{v}_n^\alpha + \frac{(1 - 2\beta)}{2} \mathbf{a}_n^\alpha \right] \right\} = \mathbf{0} .$$

Equations (14-15) are generally non-linear and have to be solved by an iterative algorithm such as Newton's method or any of its variants.

5 Dynamic contact/release conditions

In the preceding section, Newmark integrators were employed in the solution of the two-body contact problem, where contact/release constraints were imposed on the displacement fields, in accordance with (8). Unfortunately, in contrast with the continuous case, the velocity and acceleration fields recovered from such an integration scheme do not satisfy the respective rate impenetrability constraints realized by (9) and (10). This can be easily seen with the help of a simple example; consider contact between two nodal points, each belonging to one of the contacting bodies, as in Figure 1. The points come into initial contact at time t_n and the following conditions hold:

$$\mathbf{X}^1 + \mathbf{u}_n^1 = \mathbf{X}^2 + \mathbf{u}_n^2 = \mathbf{0} ,$$

$$\mathbf{v}_n^1 = \mathbf{v}_0 \quad , \quad \mathbf{v}_n^2 = \mathbf{0} ,$$

$$\mathbf{a}_n^1 = \mathbf{a}_n^2 = \mathbf{0} .$$

Enforcement of the impenetrability condition on the discrete displacement fields, as

Figure 1: Dynamic contact; failure of Newmark algorithms

defined in (11), along the normal, n, to the contact surface at time t_{n+1}, yields

$$(\mathbf{X}^1 + \mathbf{u}_n^1) \cdot \mathbf{n} + \Delta t_n \mathbf{v}_n^1 \cdot \mathbf{n} + \frac{1}{2} \Delta t_n^2 [(1 - 2\beta)\mathbf{a}_n^1 + 2\beta \mathbf{a}_{n+1}^1] \cdot \mathbf{n}$$

$$= (\mathbf{X}^2 + \mathbf{u}_n^2) \cdot \mathbf{n} + \Delta t_n \mathbf{v}_n^2 \cdot \mathbf{n} + \frac{1}{2} \Delta t_n^2 [(1 - 2\beta)\mathbf{a}_n^2 + 2\beta \mathbf{a}_{n+1}^2] \cdot \mathbf{n} ,$$

or, taking into account the conditions at t_n,

$$(\mathbf{a}_{n+1}^2 - \mathbf{a}_{n+1}^1) \cdot \mathbf{n} = \frac{1}{\beta \Delta t_n} \mathbf{v}_0 \cdot \mathbf{n} , \tag{16}$$

provided that $\beta > 0$. Likewise, subtracting the discrete velocities of the two nodes at t_{n+1} from one another and using again the initial conditions at t_n gives

$$(\mathbf{v}^2_{n+1} - \mathbf{v}^1_{n+1}) \cdot \mathbf{n} \doteq \Delta t_n(\mathbf{a}_{n+1} - \mathbf{a}^1_{n+1}) \cdot \mathbf{n} - \mathbf{v}_0 \cdot \mathbf{n} ,$$

which, with the aid of (16), results in

$$(\mathbf{v}^2_{n+1} - \mathbf{v}^1_{n+1}) \cdot \mathbf{n} = (\frac{\gamma}{\beta} - 1)\mathbf{v}_0 \cdot \mathbf{n} . \tag{17}$$

Equations (16) and (17) illustrate the problematic behavior of Newmark integrators, when directly used in simulating dynamic contact/release conditions. It can be readily concluded that the discontinuity in the post-contact velocities is independent of the time step, while the discontinuity in the accelerations is inversely proportional to the time step, thus consistency is not attained as $\Delta t \rightarrow 0$. Finally, notice that equations (16) and (17) were obtained without any reference to dynamic equilibrium (which, in this case would only determine the exact position of the nodes at time t_{n+1}).

The above analysis indicates that a special treatment is necessary for an accurate simulation of contact/release conditions. Indeed, the discrepancy in the computed rate quantities, although immaterial within linear elasticity, can be potentially devastating in non-linear problems employing rate-dependent constitutive assumptions. In an earlier attempt, Hughes and coworkers have used the Newmark solution as a predictor to be subsequently followed by a corrector step, in which velocities were matched on the contact surface by means of a local wave propagation analysis. Moreover, accelerations were weighted by the (lumped) masses of the interacting nodes, so that dynamic equilibrium be observed after the treatment, see [HTS+76]. Here a slightly different methodology is proposed based on *a priori* satisfaction of the impenetrability constraint and its two rate forms, corresponding to equations (8) and (9-10), respectively. To this end, the Lagrange multiplier formulation in Section 3 is augmented by the variational terms

$$\int_{C_h} \sum_{\alpha=1}^{2} \Big\{ w^\alpha_v [\mathbf{u}^\alpha_{n+1} - \mathbf{u}^\alpha_n - \Delta t_n \mathbf{v}^\alpha_n - \Delta t^2_n (\frac{1}{2} - \beta)\mathbf{a}^\alpha_n \tag{18}$$
$$- \Delta t^2_n \beta \mathbf{a}^\alpha_{n+1}] \cdot \mathbf{n} \, (\bar{\mathbf{v}}^\alpha_{n+1} \cdot \mathbf{n}) + \lambda_v (-1)^\alpha \bar{\mathbf{v}}^\alpha_{n+1} \cdot \mathbf{n} + \bar{\lambda}_v (-1)^\alpha \mathbf{v}^\alpha_{n+1} \cdot \mathbf{n} \Big\} d\Gamma = 0$$

and

$$\int_{C_h} \sum_{\alpha=1}^{2} \Big\{ w^\alpha_a [\mathbf{v}^\alpha_{n+1} - \mathbf{v}^\alpha_n - \Delta t_n (1 - \gamma)\mathbf{a}^\alpha_n \tag{19}$$
$$- \Delta t_n \gamma \mathbf{a}^\alpha_{n+1}] \cdot \mathbf{n} \, (\bar{\mathbf{a}}^\alpha_{n+1} \cdot \mathbf{n}) + \lambda_a (-1)^\alpha \bar{\mathbf{a}}^\alpha_{n+1} \cdot \mathbf{n} + \bar{\lambda}_a (-1)^\alpha \mathbf{a}^\alpha_{n+1} \cdot \mathbf{n} \Big\} d\Gamma = 0 ,$$

where w^α_v, $w^\alpha_a > 0$ are weighting functions. Equations (18) and (19) reveal that the Newmark integrator is modified along the contact boundary. Particularly, the admissible contact velocity and acceleration fields are, in principle, assumed independent of the displacements, while the Lagrange multipliers λ_v and λ_a enforce the conditions

$$[\mathbf{v}_{n+1}]_c = 0 \quad , \quad [\mathbf{a}_{n+1}]_c = 0 ,$$

where

$$[(\bullet)]_c \equiv [(\bullet)^2 - (\bullet)^1] \cdot \mathbf{n} .$$

The multipliers λ_v and λ_a are viewed as generalized momenta, energy conjugate to the contact velocities and accelerations, respectively. Equation (18) requires that

$$w^1_v [\mathbf{u}^1_{n+1} - \mathbf{u}^1_n - \Delta t_n \mathbf{v}^1_n - \Delta t^2_n (\frac{1}{2} - \beta)\mathbf{a}^1_n - \Delta t^2_n \beta \mathbf{a}^1_{n+1}] \cdot \mathbf{n} - \lambda_v = 0$$

and

$$w_v^2[\mathbf{u}_{n+1}^2 - \mathbf{u}_n^2 - \Delta t_n \mathbf{v}_n^2 - \Delta t_n^2(\frac{1}{2} - \beta)\mathbf{a}_n^2 - \Delta t_n^2 \beta \mathbf{a}_{n+1}^2] \cdot \mathbf{n} + \lambda_v = 0$$

along C_h. Premultiplying each of the above with the weighting function of the other, recalling that

$$[\mathbf{X} + \mathbf{u}_{n+1}]_c = 0 ,$$

and, finally, differencing, gives

$$\lambda_v = \frac{[\mathbf{X} + \mathbf{u}_n]_c + \Delta t_n[\mathbf{v}_n]_c + \Delta t_n^2(\frac{1}{2} - \beta)[\mathbf{a}_n]_c}{\frac{1}{w_v^1} + \frac{1}{w_v^2}} . \tag{20}$$

Similarly, equation (19) dictates that

$$w_a^1[\mathbf{v}_{n+1}^1 - \mathbf{v}_n^1 - \Delta t_n(1 - \gamma)\mathbf{a}_n^1 - \Delta t_n \gamma \mathbf{a}_{n+1}^1] \cdot \mathbf{n} - \lambda_a = 0$$

and

$$w_a^2[\mathbf{v}_{n+1}^2 - \mathbf{v}_n^2 - \Delta t_n(1 - \gamma)\mathbf{a}_n^2 - \Delta t_n \gamma \mathbf{a}_{n+1}^2] \cdot \mathbf{n} + \lambda_a = 0 .$$

Repeating the same process as before yields

$$\lambda_a = \frac{[\mathbf{v}_n]_c + \Delta t_n(1 - \gamma)[\mathbf{a}_n]_c}{\frac{1}{w_a^1} + \frac{1}{w_a^2}} . \tag{21}$$

In case one of the contacting bodies is rigid, then it is, by assumption, associated with infinite weighting functions, so that (20) and (21) are modified accordingly. Velocity and acceleration fields normal to the contact surface are recovered with the aid of the above constraint equations as

$$\mathbf{v}_{n+1}^\alpha \cdot \mathbf{n} = [\mathbf{v}_n^\alpha + \Delta t_n(1 - \gamma)\mathbf{a}_n^\alpha + \Delta t_n \gamma \mathbf{a}_{n+1}^\alpha] \cdot \mathbf{n} - (-1)^\alpha \frac{\lambda_a}{w_a^\alpha}$$

and

$$\mathbf{a}_{n+1}^\alpha \cdot \mathbf{n} = \frac{1}{\beta \Delta t_n^2} \left\{ [\mathbf{u}_{n+1}^\alpha - \mathbf{u}_n^\alpha - \Delta t_n \mathbf{v}_n^\alpha - \Delta t_n^2(\frac{1}{2} - \beta)\mathbf{a}_n^\alpha] \cdot \mathbf{n} + (-1)^\alpha \frac{\lambda_v}{w_v^\alpha} \right\} .$$

In the absence of contact, as well as during persistent contact, the proposed treatment reduces naturally to the associated Newmark method. Indeed, in the former case, as $meas(C_h) = 0$, all Lagrange multipliers disappear at the outset, while in the latter, equations (20) and (21) require the multipliers to be again identically equal to zero.

The weighting functions w_v^α and w_a^α, are chosen so that the integrands in (18) and (19) be (virtual) work quantities. Here, they are set to be

$$w_v^\alpha = \frac{m^\alpha}{\Delta t_n} \quad , \quad w_a^\alpha = m^\alpha \Delta t_n ,$$

where m^α denotes a characteristic mass quantity of body α.

Whenever release is detected (as defined in [HTS+76]), equations (18) and (19) yield, in direct analogy to initiation of contact

$$[\mathbf{u}_{n+1}]_c - \Delta t_n^2 \beta [\mathbf{a}_{n+1}]_c + \lambda_v(\frac{1}{w_v^1} + \frac{1}{w_v^2}) = 0 \tag{22}$$

and

$$[\mathbf{v}_{n+1}]_c - \Delta t_n \gamma [\mathbf{a}_{n+1}]_c + \lambda_a(\frac{1}{w_a^1} + \frac{1}{w_a^2}) = 0 . \tag{23}$$

In this work, both Lagrange multipliers in (22) and (23) are set to zero at release, as their appearance was originally identified with the (currently inactive) constraint of equal velocities and accelerations in the direction normal to the contact surface.

6 Numerical simulations

Numerical solutions for two dynamic contact/impact problems are presented in this section. Computations are conducted within the environment of the fully non-linear general purpose Finite Element Analysis Program (FEAP), briefly documented in Chapter 15 of [ZT89] and Chapter 16 of [ZT91].

Although implicit time integration schemes are used throughout, an attempt is made to keep the time step as close to optimal, with regard to the explicit stability limit, as possible. Diagonal (lumped) mass matrices are used for all types of elements, as their behavior is considered superior to that of corresponding consistent matrices whenever, as in given class of problems, the associated kinematic fields exhibit discontinuities (or near-discontinuities) in time. Lumped masses are also used as weighting functions of the modified Newmark impact treatment.

6.1 Impact of dissimilar bars

Two bars of lengths and initial velocities as in Figure 2 come into contact at time $t = 0$. With reference to the same figure, the properties of the two bars are

$$\begin{array}{llll} \text{Bar 1:} & E = 0.49 & \rho = 1.0 & A = 1.0 \\ \text{Bar 2:} & E = 1.0 & \rho = 1.0 & A = 1.0\,. \end{array}$$

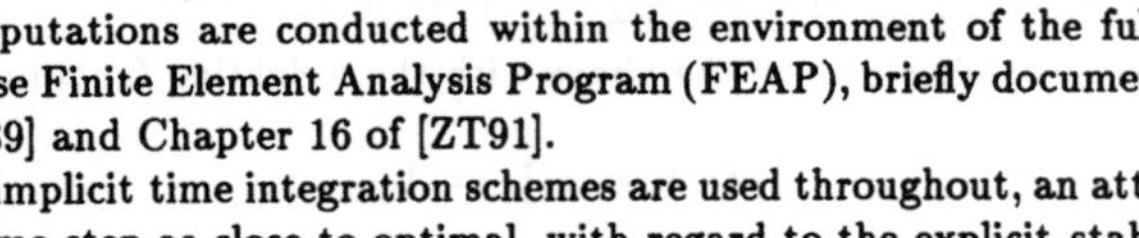

Figure 2: Impact of dissimilar bars

Bar 1 is discretized uniformly by 100 elements and Bar 2 by 70 elements, so that, for time step $\Delta t = 0.142857$, both are integrated in time optimally. Plots of the displacements, velocities and contact force at the contacting tip are shown in Figure 3. Excellent agreement with the exact solutions is exhibited for all three fields.

6.2 Cylinder on rigid surface

An elastic cylinder in plane strain is dropped with initial velocity $\mathbf{v} = -1.0\mathbf{E}_2$ towards a rigid surface, see Figure 4. No gravity effects are included in the analysis. The cylinder is modeled by two-dimensional 9-node bi-quadratic elements and the mesh is shown in Figure 5. The elastic constants for the cylinder are

$$E_{cyl} = 5. \times 10^2 \quad , \quad \nu_{cyl} = 0.3$$

and the mass density is $\rho = 1.0$. The Newmark parameters are set to $\beta = 0.25$ and $\gamma = 0.5$. The deformed meshes corresponding to selected stages of the analysis conducted with time step $\Delta t = 0.05$, are illustrated in Figure 6. The history of total force developed due to contact is plotted in Figure 7. Three choices of time step are considered, namely $\Delta t = 0.02, 0.05, 0.15$. It is evident that $\Delta t = 0.05$ produces slightly smoother results than the smaller time step. This can be explained by observing that the stability limit is, in an average sense, closer to one for the larger time step than for the smaller. The larger time step yields somewhat inaccurate resultant forces, as it fails to capture the abrupt changes in the response.

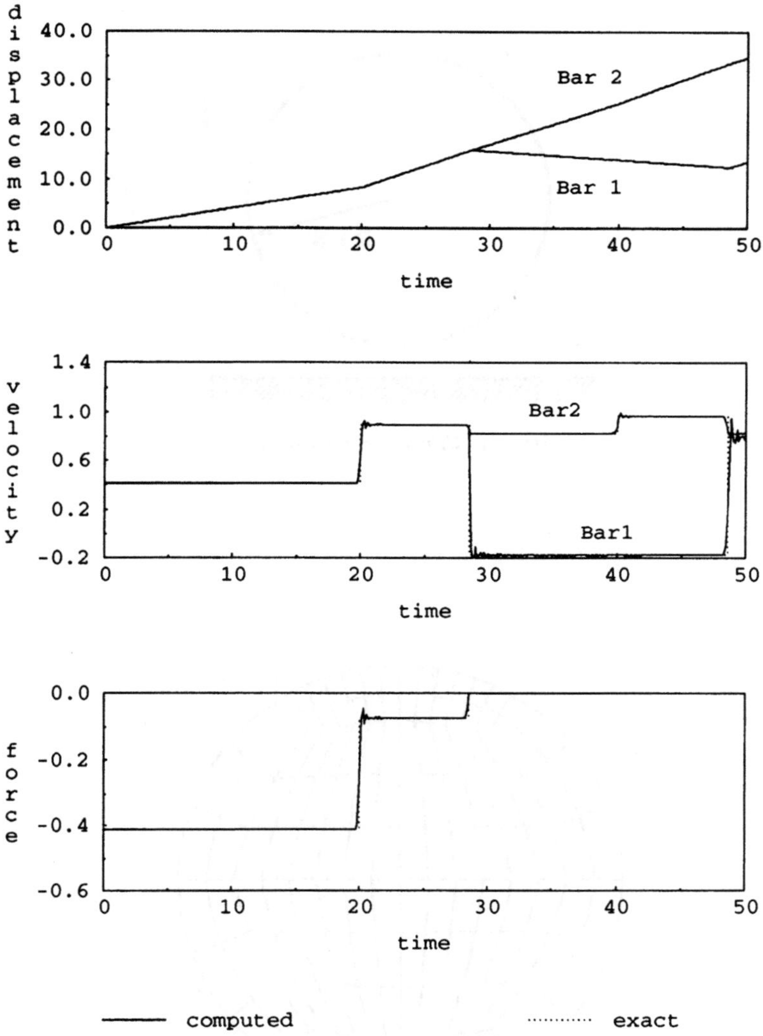

Figure 3: Impact of dissimilar bars; history of tip displacements, velocities and contact force

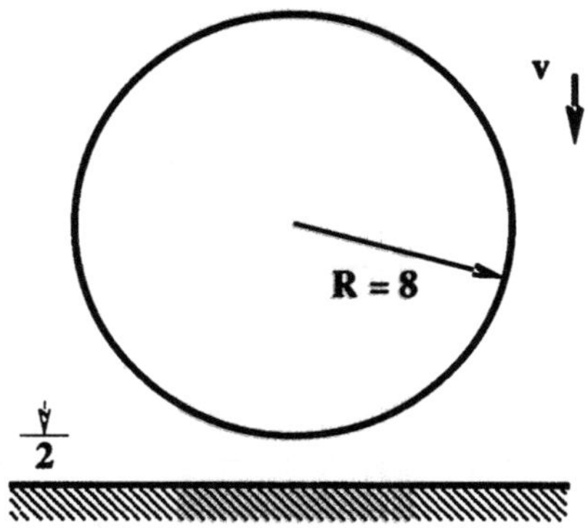

Figure 4: Cylinder on rigid surface

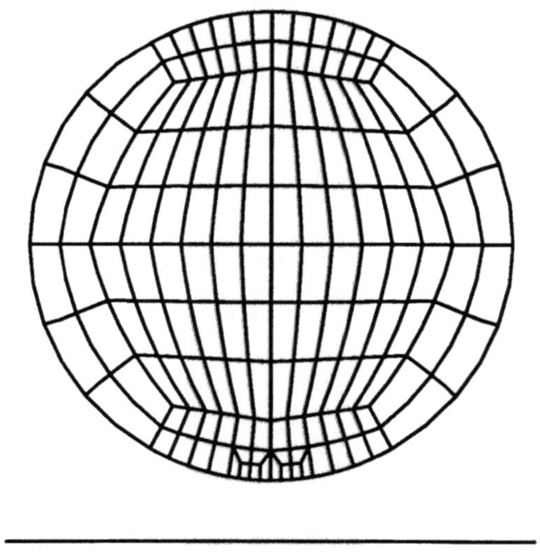

Figure 5: Cylinder on rigid surface; discretization

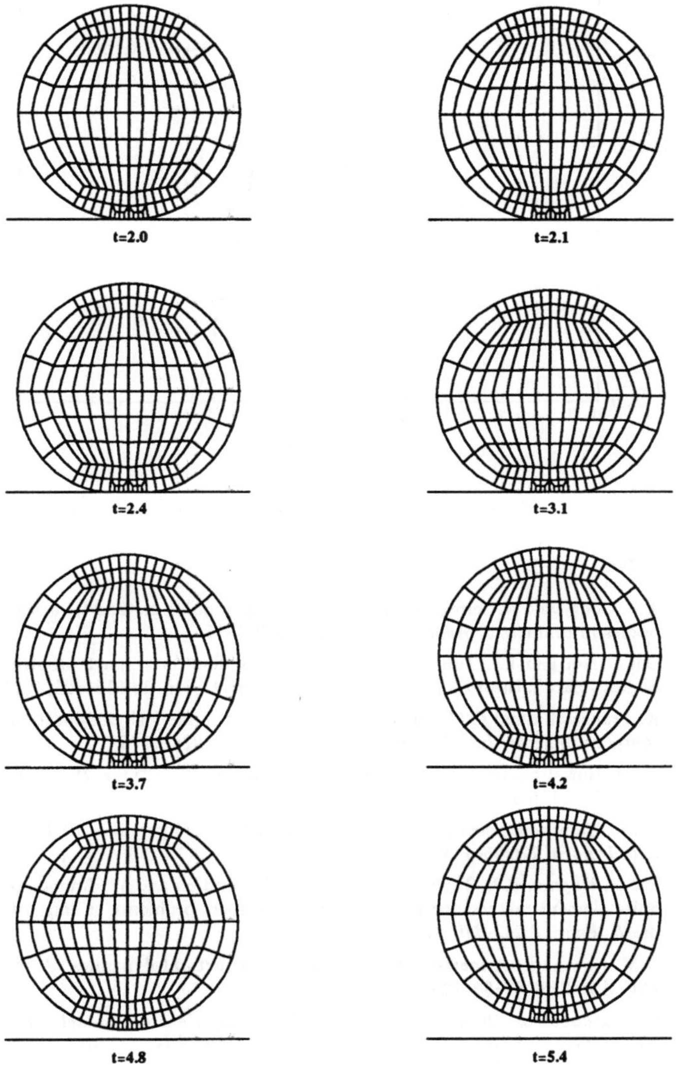

Figure 6: Cylinder on rigid surface; deformed meshes

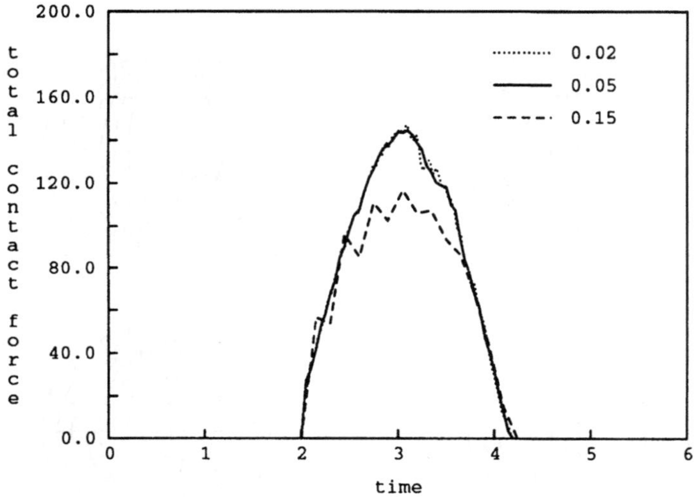

Figure 7: Cylinder on rigid surface; history of total contact force

References

[CS71] T.F. Conry and A. Seireg. A mathematical programming method for design of elastic bodies in contact. *J. Appl. Mech.*, 38:387–392, 1971.

[CT71] S.K. Chan and I.S. Tuba. A finite element method for contact problems of solid bodies - part I. theory and validation. *Int. J. Mech. Sci.*, 13:615–625, 1971.

[FZ75] A. Francavilla and O.C. Zienkiewicz. A note on numerical computation of elastic contact problems. *Int. J. Num. Meth. Engr.*, 9:913–924, 1975.

[Ham34] W. R. Hamilton. On a general method in dynamics. *Phil. Trans. Royal Soc. London*, 124:247–308, 1834.

[Ham35] W. R. Hamilton. Second essay on a general method in dynamics. *Phil. Trans. Royal Soc. London*, 125:95–144, 1835.

[HTS+76] T.J.R. Hughes, R.L. Taylor, J.L. Sackman, A. Curnier, and W. Kanoknukulchai. A finite element method for a class of contact-impact problems. *Comp. Meth. Appl. Mech. Engr.*, 8:249–276, 1976.

[New59] N.M. Newmark. A method of computation in structural dynamics. *J. Engr. Mech. Div. ASCE*, 85:67–94, 1959.

[Pap91] P. Papadopoulos. *On the Finite Element Solution of General Contact Problems.* PhD dissertation, Department of Civil Engineering, University of California at Berkeley, Berkeley, USA, 1991.

[Woo90] W.L. Wood. *Practical Time-stepping Schemes.* Oxford University Press, Oxford, 1990.

[ZT89] O.C. Zienkiewicz and R.L. Taylor. *The Finite Element Method; Basic Formulation and Linear Problems*, volume 1. McGraw-Hill, London, 4th edition, 1989.

[ZT91] O.C. Zienkiewicz and R.L. Taylor. *The Finite Element Method; Solid and Fluid Mechanics, Dynamics and Non-linearity*, volume 2. McGraw-Hill, London, 4th edition, 1991.

FINITE ELEMENTS IN THE 90's, E. Oñate, J. Periaux, A. Samuelsson (Eds.)
Springer-Verlag/CIMNE, Barcelona 1991

THE TAYLOR IMPACT TEST PROBLEM REVISITING THE STANDARD 2-NODE ROD FINITE ELEMENT

I. Cormeau
Cisi Ingénierie,
CEN Saclay, 91192 Gif sur Yvette Cedex,
FRANCE

The Taylor impact test is a standard benchmark problem in high speed computational mechanics. Sophisticated Euler or Lagrange methods exist to solve it while it still poses challenges to applied mathematicians. It is shown that some qualitatively acceptable results can be achieved by bringing limited changes to the standard 2 node rod finite element. While not very accurate, the resulting 1D model is attractive for educational purposes.

1. INTRODUCTION

Material experimenters use the Taylor impact test to investigate the behaviour of metals at very high rates of strain.

A circular cross section rod is fired at a fixed,supposedly rigid, target; the residual plastic deformation, reduction in length and diameter increase at the impacted end of the test specimen are subsequently used in an attempt to identify dynamic stress/strain properties as well as equation of state parameters for the material under investigation.

The accurate modelling of the Taylor impact is still considered a challenge by many applied mathematicians, physicists and engineers.
This is due to the complex nonlinear shock wave patterns and thermomechanical interactions present in the rod, together with the existence of a moving stress free lateral surface on which correct boundary conditions must be obeyed.

Finite difference (FD) algorithms, mostly based on the Eulerian approach typical of CFD applications, have been reported (see e.g. [1]), in which special care is taken of the derivation and accurate numerical treatment of velocity, stress and density discontinuities.
They involve the solution of a Riemann problem at each grid points of a single space coordinate hyperbolic system (x,t) grid.
Extension to 2D (x,y,t) or axisymmetric (r,z,t) systems is possible through a coordinate splitting technique [2], which unfortunately results in some further loss of accuracy.
The Eulerian approach faces a difficult task in dealing with the moving free suface; the correct boundary condition is poorly represented even if fine grids are used.

Finite element (FE) software, based on the Lagrange description of the motion, has long been available, almost routinely, to tackle this problem.
No difficulty arises in connection with the moving free surface since the computational mesh exactly fits the boundary at any time. Of course, trouble will occur if excessive local distorsions appear in highly stressed regions, as the nodal coordinated are continuously updated.
Mesh rezoning may then be needed, with all the associated interpolation problems that follow.
No explicit care is taken of the discontinuities in this approach.

Applied to the analysis of a long, thin rod, both FD and FE methods just mentioned require an excessive number of grid points or elements, if regular meshes with an acceptable aspect ratio are needed to cover the entire computational domain.

The question then naturally arises:

could one use the conventional 1D, 2-node rod finite element to simulate the Taylor impact test?

The answer is yes, to a limited extent, as we shall see later, provided some changes are made to this element.

Indeed, if the rod element as it is currently implemented in most commercial FE programs is used in an attempt to model this problem, it is most likely that weird results will come out of the computer.

It is our purpose to see why.

2. SMALL STRAIN, LINEAR ELASTIC IMPACT

As a first step, we applied a standard FE code to the following benchmark case:

rod length	L = 1 m
rod cross section	S = 0.0001 m2
density	ρ = 7800 kg/m3
Young's modulus	E = 210000 MPa
Poisson's ratio	ν = 0.3
speed of sound	c = 5188.75 m/s
initial velocity	Vo = 0.1 c
uniform 1D mesh	100 elements

The exact analytical solution to this problem of linear elastodynamics is straightforward; the magnitude of the stress discontinuity which propagates at sound speed along the rod is given by

$$\Delta\sigma = E\ Vo / c \tag{1}$$

Figure 1 depicts the time history of the displacement at half length and at the free end of the rod, in nondimensional units; these results are apparently acceptable.

Figure 2 shows the time history of the velocity at the same points; huge oscillations about the theoretical solution are observed; the same is true for the stress history.

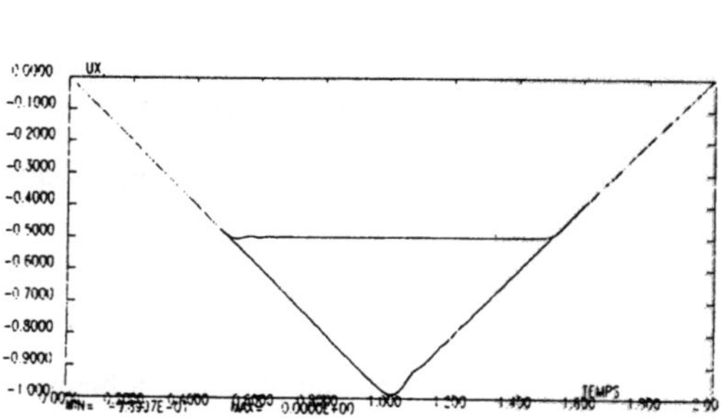

Figure 1

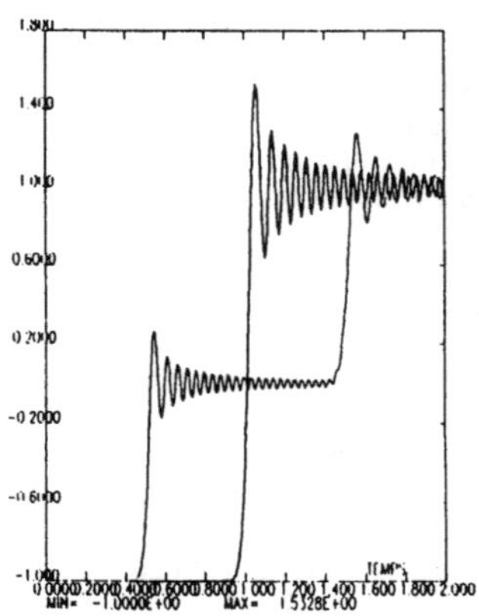

Figure 2

No improvement whatsoever is obtained when reducing the time step; neither do the oscillations disappear when the mesh is refined.

This may come as a surprise to readers unfamiliar with transient linear dynamics.

Without elaborating further, let us simply recall that the FE method essentially calls for a piecewise continuous approximation of the field variables; while discontinuities may exist (think about the stresses in linear displacement-type finite elements) at the interelement boundaries, the discretisation method does not support, by nature, traveling discontinuities.

The oscillations just observed are similar to the well known Gibbs phenomenon encountered in Fourier series of discontinuous functions.

If, for aesthetic reasons, artificial damping is introduced in the model, which means cheating really, the oscillations might be mostly eliminated, but this trick changes the nature of the system which ceases to be conservative.

We have just shown, with this trivial one dimensional example, one of the difficulties inherent to transient dynamics, which are present as well in 2D or 3D:

Finite elements do not like moving discontinuities!

3. PLASTIC TAYLOR IMPACT

It has become almost impossible, nowadays, to record all existing commercial, inhouse or research-oriented FE software that would tolerate as input data:

- a linear, uniformly spaced mesh of 2 node rod elements of uniform cross section
- a single fixed node at one end of the rod
- zero initial displacements
- uniform nodal initial velocity, except at the fixed node where it is zero
- constant Young's modulus and Poisson's ratio
- uniform initial density
- a perfectly plastic von Mises material associated with a constant, uniform yield stress Y

In order to bring the rod into the plastic range the only condition to be met is to specify a large enough initial velocity

$$Vo \quad > \quad Y / \sqrt{\rho E} \tag{2}$$

To the author's guess, most readers will have access to such a FE program, and we actually encourage them to run this apparently trivial benchmark .

For accurate results, a fine mesh will tentatively be used.

Consider an internal node close to the impact point, say at distance d from the end of the rod.

Most probably, if the computer run has not been aborted before, this node will reach, and subsequently penetrate, the target at time

$$t = \frac{(1 - Y/E).(d/L)}{(Vo/c \ - \ Y/E)} \ (L/c) \tag{3}$$

With the data mentioned in Section 2 and a yield stress Y = 0.05 E, we get t = 0.19 (L/c) for the first node next to the impact point.

This unacceptable result follows from the underlying assumption, present in most 2 node rod element implementations, that the cross section S remains constant in time.

This assumption is completely inadequate here: the total impulse developed over some small, but finite, time interval Δt for a perfectly plastic rod of constant cross section undergoing stress σ is equal to $\sigma\, S\, \Delta t$; the total loss of momentum of the portion of the rod between the impact point and the propagating compression wave, necessary to bring this portion to a standstill, is equal to $\rho o\, So\, c\, Vo\, \Delta t$.

Hence $$\sigma S = \rho o\, So\, c\, Vo \qquad (4)$$

and this equation cannot be satisfied if $\sigma <= Y$ and $S = So$.

One needs either $S > So$ or $\sigma > Y$, or both.

The required changes to the standard 2-node rod element are thus three-fold:

a) allow for a variation of current cross section as a function of current axial stress and strain

b) consider large, radial and hoop as well as axial, strains

c) allow for $\sigma > Y$, due to the presence of non zero radial and hoop stresses near the impact point in the rod which, being compressive too, contribute to the von Mises equivalent stress.

4. A SIMPLE LARGE STRAIN, PERFECTLY PLASTIC 1D IMPACT MODEL

Since space does not permit the full theoretical description of the model, we shall limit ourselves to the statement of the essential assumptions, and present a few results.

4.1 Simplifying assumptions.

In the following, we refer to cylindrical coordinates (r,z):

a) 2 node, linear displacement rod element

b) cross section S, Young's modulus E, Poisson's ratio ν and density ρ are uniform over each individual element
(i.e. piecewise constant functions of z)

c) S and ρ may vary in time

d) axial displacement uz and velocity vz are functions of z and t only
(i.e. piecewise linear in z)

e) radial displacement ur and velocity vr are continuous functions of r and t, and are considered elementwise; no reference is made to strains γrz which need not be defined

f) use is made of logarithmic strains

$$\varepsilon z = \ln(\Delta z/\Delta zo) \qquad \varepsilon r = \ln(\Delta R/\Delta Ro) \qquad \varepsilon\theta = \ln(R/Ro) \qquad (5)$$

where R,Ro denotes the current and initial external radius and $\Delta z,\Delta zo$ the current and initial length of each individual element.

g) the stress boundary condition on the rod free surface at r = R is simply $\sigma r = 0$

h) the axial stress σz does not vary with r

i) stresses σr and $\sigma\theta$ are continuous functions of r in the range $0 < r < R$, and are considered elementwise

j) shear stresses τrz are ignored

k) the stress/strain law is elastic/perfectly plastic von Mises in terms of the three stress components mentioned above, which are considered true, Cauchy, stresses

l) plastic flow is incompressible

4.2 Main theoretical results of the 1d model

With the assumptions of Section 4.1, the variation of S is given by the simple equation

$$\pi\ R2 = S = So\ \exp((1-2\nu)\ \sigma z/E - \varepsilon z) \tag{6}$$

From the solution of the radial equation of motion, element by element, it is possible to express σz as a function of the element outer radius acceleration, as

$$\sigma z/E = \mu(\ \rho R\ddot{R}/Y\ ) \tag{7}$$

Function $\mu(x)$, which is determined by numerical integration of this radial equation of motion, is illustrated at Figure 3 and depends on whether we are in compression or in tension.

For compression, it can be approximated with good accuracy as a 5th degree least squares fit polynomial

$$\mu(x) = -1 - (2.48992E\text{-}1)x + (3.41599E\text{-}2)x{*}{*}2 + (1.18362E\text{-}2)x{*}{*}3 + (3.27586E\text{-}3)x{*}{*}4 + (3.64373E\text{-}4)x{*}{*}5 \tag{8}$$

while for tension we simply have $\mu(x) = -\mu(-x)$ (9)

As shown in Figure 3, the range of the independent variable is limited:

$$-3.24 < x < 0.72 \text{ (compression) and } -0.72 < x < 3.24 \text{ (tension)} \tag{10}$$

Consequently, this limited range precludes the use of a very fine mesh; inequalities (10) may be violated when the mesh size Δz becomes smaller than the current radius R, and then the above approximations cease to be valid.

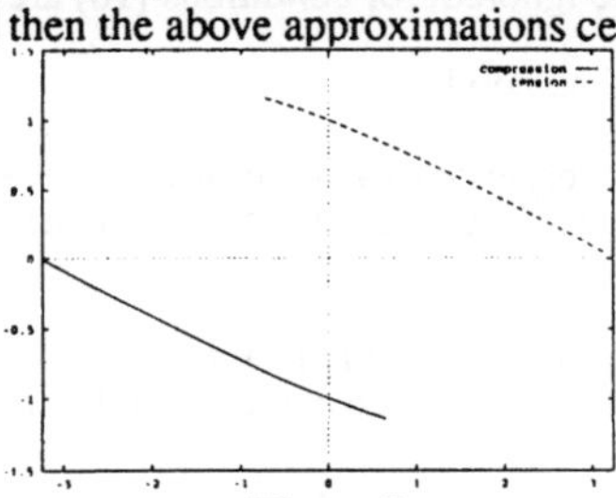

Figure 3

4.3 Implementation of the model

The 1D modified rod element was implemented in the high speed dynamics finite element code PLEXUS [3].

The axial (z,t) equation of motion is integrated using an updated Lagrange representation, a diagonal mass matrix and an explicit, numerical damping free, time integrator.

The coupling between R and σz (equations 6 and 7) is solved using a simple predictor-corrector scheme and finite difference approximations for $\dot{R}$ and $\ddot{R}$.

Equation 7 is considered only as plastic flow takes place; upon elastic unloading, we simply have $\Delta\sigma z = E\,\Delta\varepsilon z$.

4.4 A few numerical results.

The test problem data are those of Section 2 except for the yield stress, Y = 10500 MPa, i.e. an elastic strain limit of 5% .

We present time history plots in the early phases of the impact (though the actual calculations were extended to a time interval allowing the stress wave to travel twice along the length of the rod):

- axial velocity of the node next to the impact point (Figure 4)
- axial stress in the first three elements on the impact side (Figure 5)
- outer radius of the first element (Figure 6)
- radial acceleration at the outer radius of the first element (Figure 7)
- axial dynamic stress factor $\sigma z/Y$ (Figure 8)
- density in the first element (Figure 9)

The dominant feature of these results is the increase in radius (15%) and cross section (34%).

The same problem was solved with a thicker rod (R = 40 mm, S = 0.00503 m2) for a rod of identical length, material and initial conditions, in order to make a comparison with a straightforward large strain, axisymmetric Lagrange model using 4*100 4-node quadrilateral elements with 4 Gauss points as shown in Figure 10.

The radial expansion of the 1D model at the impacted end (Figure 11) is substantially smaller than the one predicted by the axisymmetric model in Figure 12, at least in the end "slice" of the rod of length 40 mm adjacent to the impact plane; at a distance of the impact point greater than the rod radius, both models predict the same final radius R = 45.5 mm.

Since the mesh size in z is only 10 mm, i.e. a quarter of the radius, the dynamic factor μ of equation (7) had to be ignored, for conditions (10) are not satisfied.

The 1D mesh is actually too fine now!

Purists will object that no convergence to a yet to be discovered theoretical solution can be obtained since the mesh and time step sizes cannot be arbitrarily reduced to zero.

Inspection of the transient stress histories in the axisymmetric model shows that both σz (Figure 13) and σr (Figure 14) are considerably higher than Y during a short duration following the impact, with σz as large as 4Y and σr about 3Y, 5 mm away from the impact plane. Moreover, the radial stress can reach such high levels even very close to the outer boundary where a stress free boundary condition should normally be observed.

Here lies in fact the basic difference between the 1D and axisymmetric models: a radial, as well as an axial wave propagation takes place at the tip of the rod, irrespective of how thin the rod might be.

The radial stress discontinuity associated with this wave propagation is the mechanism that permits the simultaneous high compressive radial stress inside the rod, and the fulfilment of the boundary condition at the skin.

The 1D model theory postulates that σz is uniform in the rod cross section, and that the radial stress σr is continuous across the thickness, in effect precluding the radial wave propagation just mentioned above.

Since stresses in the 1D model are about only 25% of those predicted by the axisymmetric model, it is no surprise that plastic flow at the end of the rod will be substantially smaller in the 1D model.

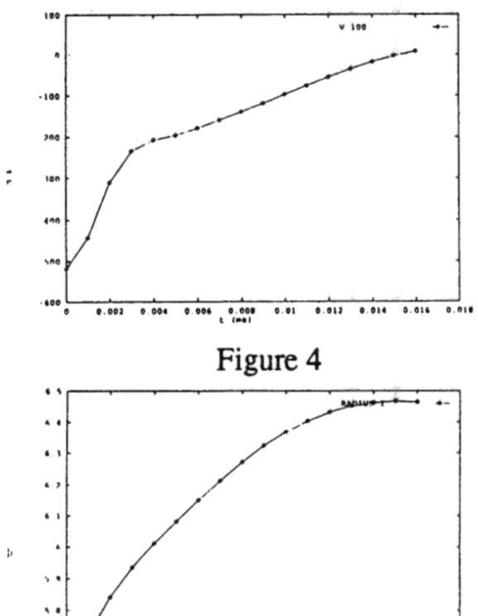

Figure 4

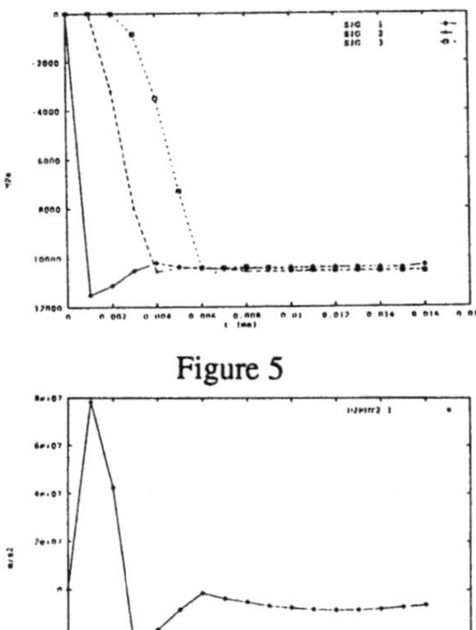

Figure 5

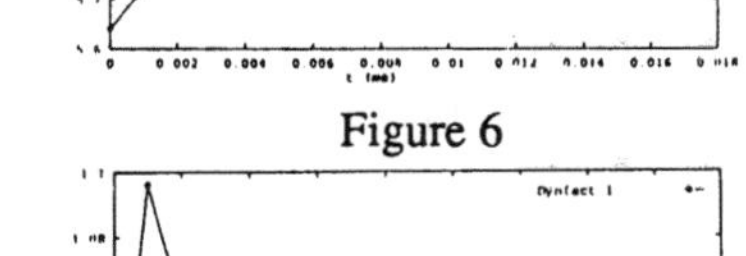

Figure 6

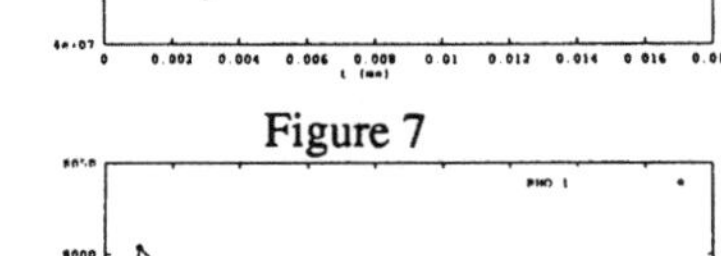

Figure 7

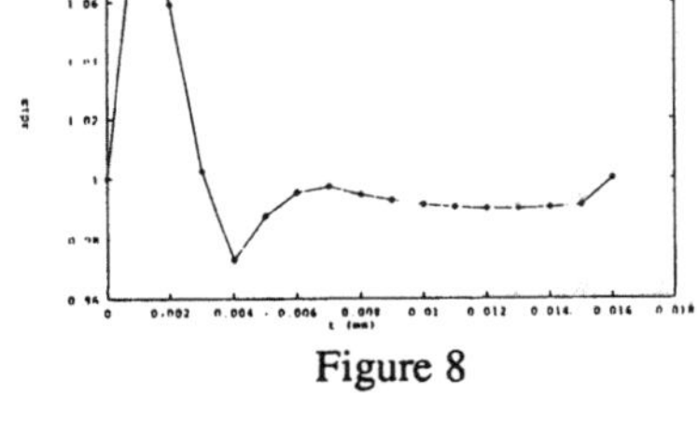

Figure 8

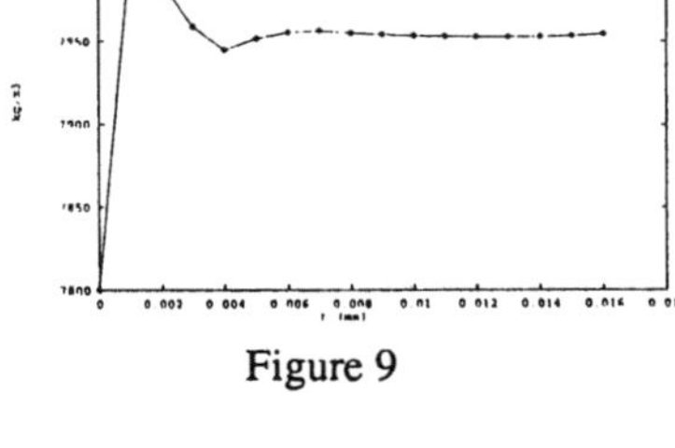

Figure 9

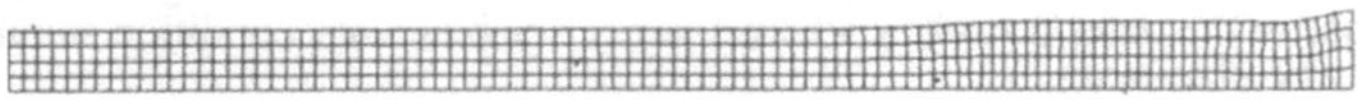

Figure 10

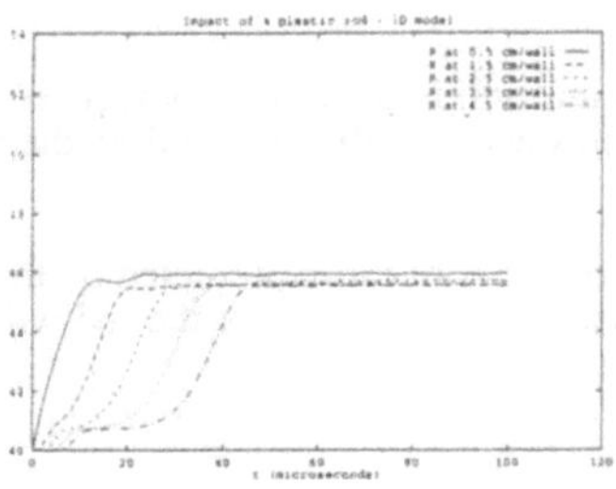

Figure 11

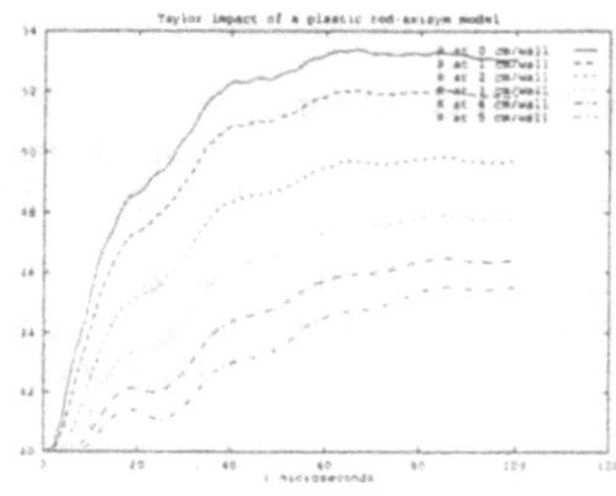

Figure 12

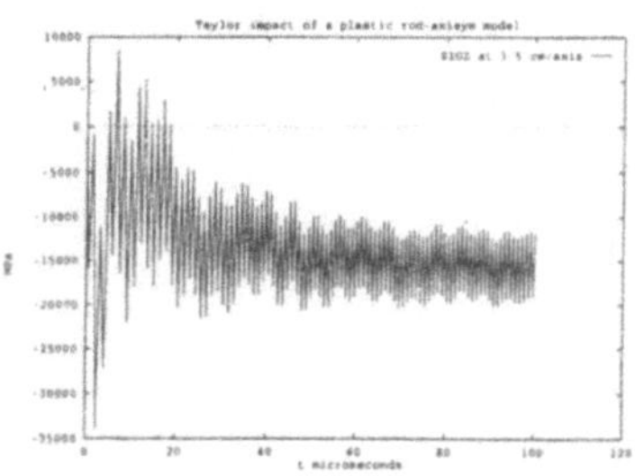

Figure 13

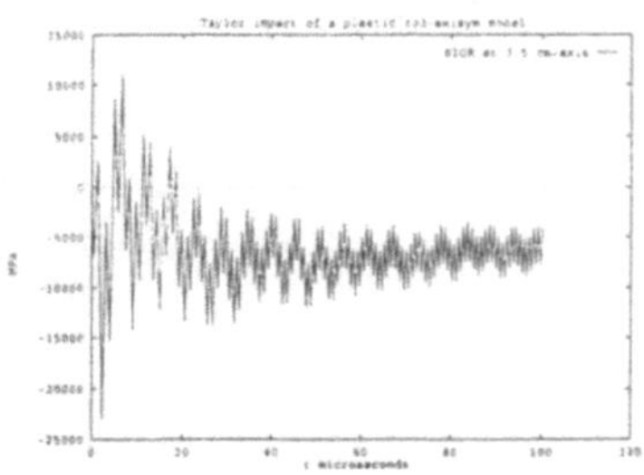

Figure 14

5. CONCLUSIONS

The Taylor impact test, while it might appear at first glance , to inexperienced as well as some experienced FE users, as a problem almost too trivial to deserve attention, still poses considerable challenges to physicists and applied mathematicians.

A detailed analysis will reveal such complexity that on the contrary, it might seem foolish to the experienced specialist to even contemplate using such a crude tool as a 2 node rod element to address the problem.

Yet some qualitatively acceptable results may be obtained with a suitably modified rod element which, thanks to its simplicity as compared to full 2D Euler or Lagrange nonlinear approaches, provides the non specialist with an easier first introduction to this topic typical of high speed dynamics.

Not to be considered a production tool, it nevertheless might be used as an aid to all those who, either in Academia or in Industry, face the never ending task to introduce new people to computational solid mechanics.

REFERENCES

1. NOURI, A. - Problèmes hyperboliques en élastoplasticité dynamique.
Thesis submitted to the University of Nice, presented on January 17, 1991

2. ARNAUD, L. - Quelques schémas numériques adaptés à l'élastodynamique en configuration axisymétrique.
Thesis submitted to the University of Bordeaux I, presented on June 28, 1990

3. CHAVANT,C. PLEXUS : A general computer code for explicit Lagrangian computation.
SMIRT 5 B2/8 Berlin 1979

FINITE ELEMENTS IN THE 90's, E. Oñate, J. Periaux, A. Samuelsson (Eds.)
Springer-Verlag/CIMNE, Barcelona 1991

ONE DIMENSIONAL TRANSIENT DYNAMIC ANALYSIS WITH ADAPTIVE MESH REFINEMENT

A. Selman and E. Hinton
Department of Civil Engineering,
University College of Swansea,
UNITED KINGDOM

Abstract

An adaptive finite element (FE) method for the solution of one dimensional transient dynamic problems is described. The mesh adaption is achieved by regenerating the grid into a number of regions using an error estimation procedure coupled with an automatic mesh generator. A numerical example of this mesh adaption technique is presented for the problem of a bar subjected to a suddenly applied end load.

1. Introduction

In the FE analysis of steady state problems there has been a growing interest in the use of adaptive procedures [1-3]. The incorporation of error estimators and mesh refinement techniques in FE codes has enabled user-specified accuracy to be achieved with a near optimal rate of convergence. In optimal meshes, the error in energy norm is equally distributed within each element; however, from a practical standpoint, it is not necessary to optimise the meshes exactly, instead we seek only adaptive mesh refinement (AMR) strategies that are computationally efficient. Among the h-adaptive procedures, the most commonly used are (a) mesh enrichment and (b) mesh regeneration. Mesh regeneration is becoming more popular because of its inherent simplicity and also because it is easy to incorporate into existing FE codes. Most adaptive studies on steady state problems have dealt with static stress analysis [4,5].

In transient problems, particularly those involving hypersonic fluid flow and multiphase heat flow, there have been major advances in the application of adaptive remeshing procedures [6-8]. However, there has been relatively little work done in applying such techniques in structural dynamics [9].

In this report we introduce an AMR procedure for the FE analysis of one dimensional transient dynamic problems. An example of a compression stress wave propagating down a rod is presented to illustrate the procedure.

2. Adaptive Mesh Refinement

Typically, for transient problems involving wave propagation, there are regions of the domain in which the solution is smooth and small regions where the solution changes rapidly. A fine mesh is needed in those areas of high gradient in order to successfully model the behaviour of the structure.

As a fine mesh over the whole domain would be impractical, it is necessary to use an error indicator to identify the regions of high gradient which must be refined.

At any time station we may write the error in the energy norm (associated with strain energy) as

$$e = \int E^{-1}(\sigma - \sigma^{*})^2 \, dl \qquad (2.1)$$

where E is the elastic modulus, σ is the FE stress and σ^{*} is the exact value of the stress. If we do not know the exact solution we may estimate the energy norm as

$$e \approx \hat{e} = \int E^{-1}(\sigma - \hat{\sigma})^2 \, dl \qquad (2.2)$$

in which $\hat{\sigma}$ is an improved estimate of the stress obtained using some stress recovery technique.

We find that it is best to normalise the error, in which case the error indicator will be dimensionless and will lie in the range 0 – 1.

Now we can sort the elements into a number of (say four) regions[1] as follows :

Region 1 :	Elements i for which	$0.75 < e \leq 1.00$
Region 2 :	Elements i for which	$0.50 < e \leq 0.75$
Region 3 :	Elements i for which	$0.25 < e \leq 0.50$
Region 4 :	Elements i for which	$0.00 \leq e \leq 0.25$

For each region we determine the boundary nodes, and remesh using a specified maximum size for region 4, and minimum size for region 1 (which, in fact, sets the maximum allowable time step to be used in connection with any conditionally stable integration algorithm). For the remaining regions set an intermediary size.

Figure 1 illustrates the procedure just described for the example in Section 4.

3. Algorithm description

The complete description of the algorithm for the advancement of the solution in time may be written as follows [6,8] :

1. Generate an initial grid to represent the computational domain and to allow an adequate initial solution.

2. Advance the solution for a prescribed number of time steps.

3. Use the error indicator to compute a new mesh distribution.

[1] More generally into m regions – obviously the more regions used the better mesh gradation obtained. However, this will be more expensive. We find, in our one dimensional problem, that a round (or two) of mesh smoothing will give that desired effect.

4. Obtain the nodal values of the solution on the new grid by direct interpolation from the previous grid.

5. If the desired time interval has elapsed stop, otherwise go to step 2.

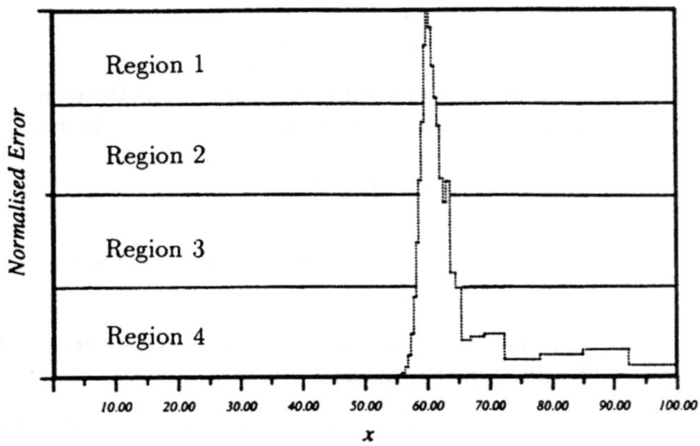

Figure 1 – Regions Definition.

4. Numerical Example

Consider the uniform steel bar and tip loading shown in Figure 2. The bar is undamped and initially at rest.

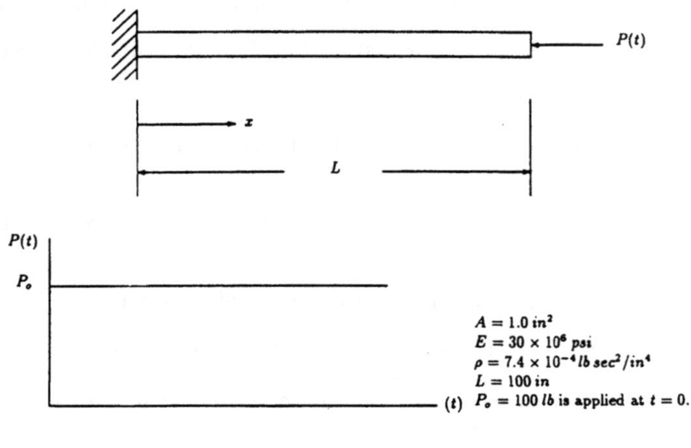

Figure 2 – One dimensional uniform bar with instantaneous tip loading.

It should be noted that only a relatively short time step enables accurate investigation in the case of shock waves on a structure.[2]

In this case, if a diagonal mass matrix is employed, then the central difference method is certainly the most attractive scheme to be used, as the system of equations can be solved without factorising a matrix. Consequently, it is not necessary to assemble the stiffness or mass matrices. Thereby reducing the computational cost.

Figure 3 compares the solution (at some specific positions in time) when using the refinement procedure described above and a uniform mesh of 250 elements, that is a mesh of element size equal to the minimum size set in Section 2. The exact solution is also shown. The time step is taken as $\Delta t = 1. \times 10^{-6}$ *sec.*

Figure 4 shows the temporal degrees of freedom used in the adaptive and uniform mesh computation, and demonstrates the potential of adaptive mesh refinement in transient dynamic problems.

5. Conclusions

It has been demonstrated that the use of adaptive mesh refinement is a promising approach for the solution of (one dimensional) transient dynamic problems. It no only reduces the computational cost by reducing the number of elements, but, as a bonus, it filters high energy modes which cause oscillations in the solution.

In two dimensional problems, the use of error estimation procedures coupled with an automatic mesh generator capable of creating meshes with elements varying considerably in size throughout the domain will certainly prove to be extremely useful. The savings in computational cost will be more dramatic, when the cost of a computation on a (uniform) fine mesh is prohibitive.

6. Acknowledgements

The authors wish to thank Dr. O. Hassan for helpful discussions.

---

[2] The use of large time steps will smear the solution cross the wave front.

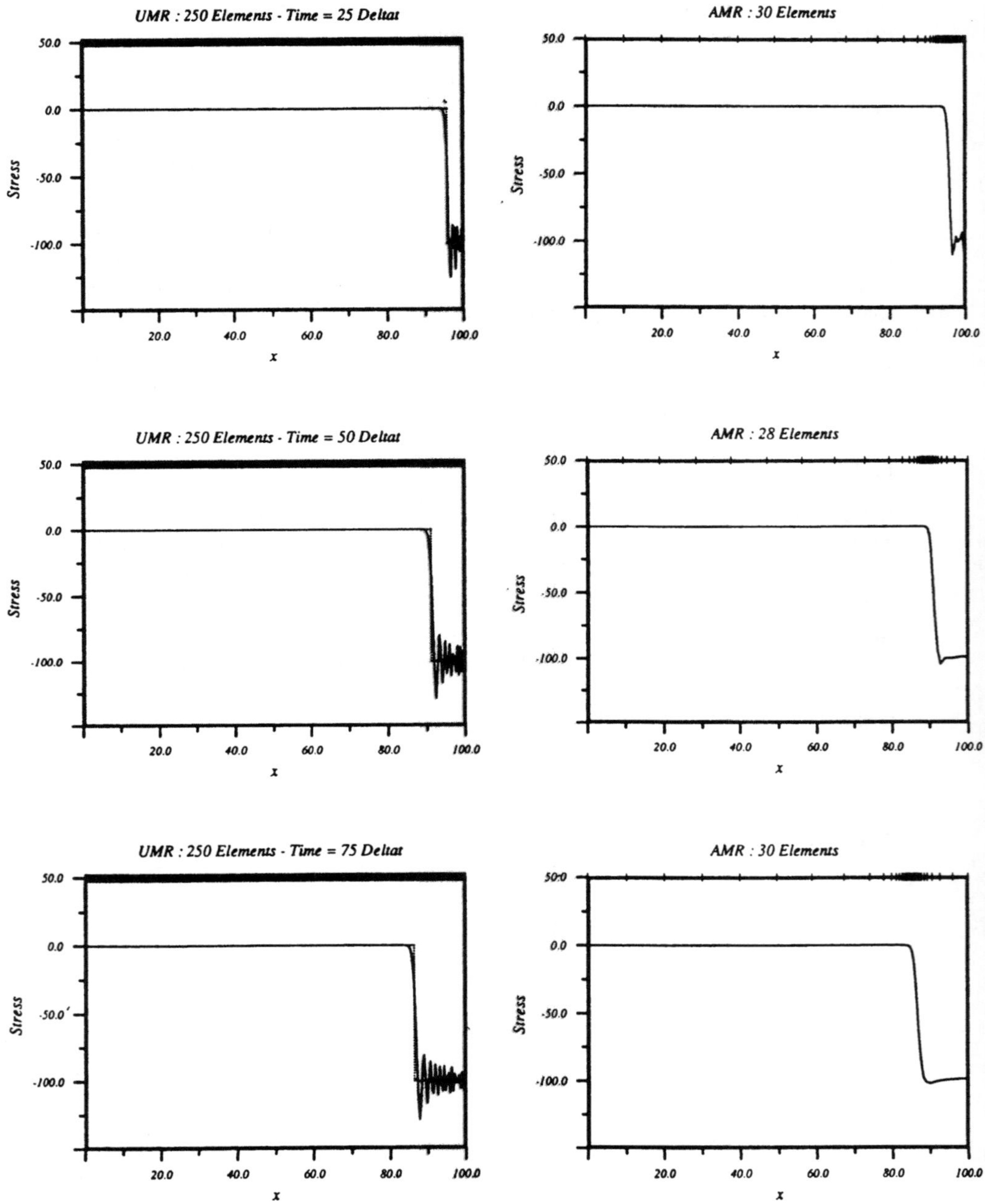

Figure 3 – Stress distribution (at some specific positions in time) when using uniform and adaptive meshes.

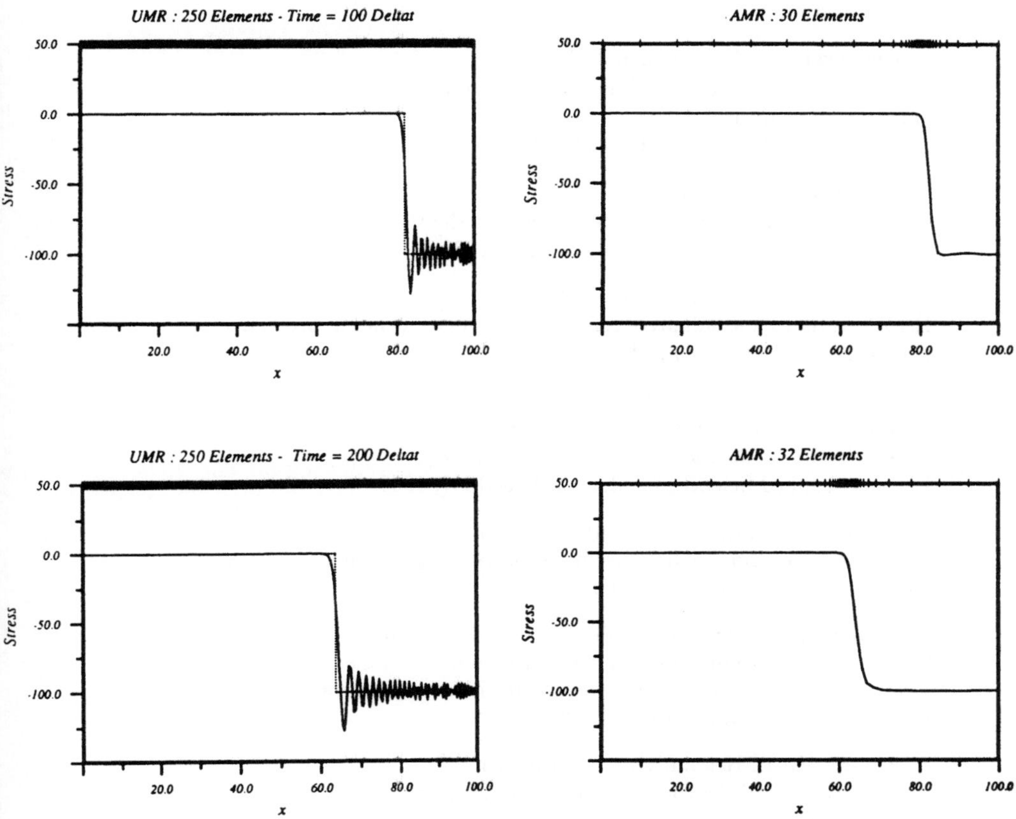

Figure 3 – Continued.

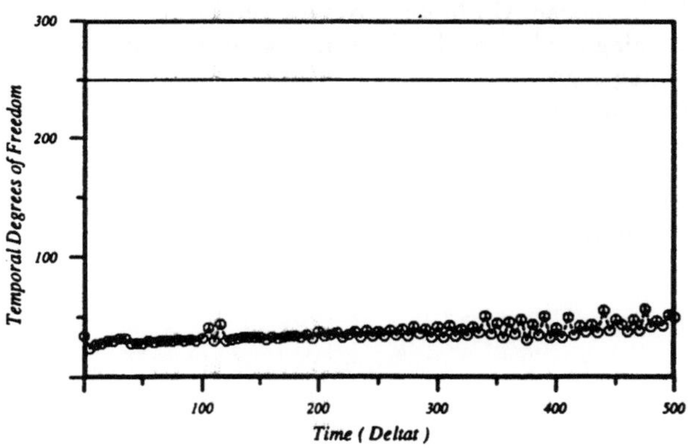

Figure 4 – Temporal degrees of freedom used in uniform and adaptive meshes.

7. References

1. Zienkiewicz O. C. and Zhu, J. Z.
A simple error estimator and adaptive procedure for practical engineering analysis,
Int. J. Num. Meth. Eng., **24**, 337-357, 1987.

2. Zhu, J. Z. and Zienkiewicz O. C.
Adaptive techniques in the finite element method,
Commun. Appl. Num. Meth., **4**, 197-204, 1988.

3. Zienkiewicz O. C. and Zhu, J. Z.
Error estimates and adaptive refinement for plate bending problems,
Int. J. Num. Meth. Eng., **28**, 2839-2853, 1989.

4. Atamaz Sibai, W. and Hinton, E.
Adaptive mesh refinement with Morley plate element
Proc. NUMETA 90 Conf., Swansea (Wales, UK),
Elsevier Applied Science, London, **2**, 1044-1057, 1990.

5. Selman, A., Hinton, E. and Atamaz Sibai, W.
Edge effects in Mindlin-Reissner plates using adaptive mesh refinement
Eng. Comput., **7**, 217-226, 1990.

6. Probert, E. J., Hassan, O., Peraire J. and Morgan K.
Transient adaptive methods for moving boundary problems
Proc. 5th Int. Symp. on Numerical Methods in Engineering,
Lausanne, Switzerland, 801-808, 1989.

7. Probert, E. J., Hassan, O. and Morgan K.
An adaptive finite element method for transient compressible flows with moving boundaries
Submitted to the special issue of the *Int. J. Num. Meth. Eng.* on Adaptive Mesh Methods.

8. Probert, E. J., Hassan, O., Morgan K. and Peraire J.
Adaptive remeshing applied to the thermal analysis of convectively cooled cylindrical leading edge
Proc. 7th Int. Conf. on Numerical Methods in Thermal problems, Stanford, USA, 811-823, 1991.

9. Samuelsson, A.
Private communication.

FINITE ELEMENTS IN THE 90's, E. Oñate, J. Periaux, A. Samuelsson (Eds.)
Springer-Verlag/CIMNE, Barcelona 1991

INCREMENTAL FINITE ELEMENT ANALYSIS OF NONLINEAR STRUCTURAL DESIGN SENSITIVITY PROBLEMS

M. Kleiber
RCAST, Univ. of Tokyo, on leave of absence from the Polish Academy of Sciences, Warsaw, POLAND

T.D. Hien and E. Postek
Polish Academy of Sciences, Warsaw, POLAND

ABSTRACT: An incremental finite element formulation for structural sensitivity analysis of nonlinear (possibly path dependent) systems with fixed overall shape is proposed. Both the direct differentiation and adjoint variable methods, as well as a mixed differentiation-adjoint technique are employed to evaluate 1st- and 2nd-order sensitivity increments during the load step. Numerical illustration is given.

1 INTRODUCTION

Structural sensitivity concerns the relationship between design parameters (design variables) and structural behaviour characterized by a response functional. What is most frequently sought is the gradient of the response functional with respect to the design variables. A general overview of problems and techniques typical of sensitivity analysis of nonlinear structural response are given elsewhere, [1]–[5]. The purpose of this note is to provide a consistently incremental formulation for such problems. The analysis is restricted to multi-degree-of-freedom systems with fixed overall configuration. On the basis of the Taylor expansion technique both the direct differentiation and adjoint variable algorithms, as well as a mixed differentiation-adjoint procedure are employed to deal with 1st- and 2nd-order sensitivity problems.

2 PROBLEM STATEMENT

Consider response of a nonlinear system with N degrees of freedom defined by the following functional

$$^{(t+\Delta t)}\mathcal{G} = {}^{(t+\Delta t)}G[\,^{(t+\Delta t)}\mathbf{q}(\mathbf{h}), \mathbf{h}] \tag{1}$$

where $^{(t+\Delta t)}G$ is a given function of its arguments at time $t + \Delta t$, $\mathbf{h} = \{h^e\}$, $e = 1, 2, \ldots, E$, is the vector of design variables and $^{(t+\Delta t)}\mathbf{q}(\mathbf{h}) = \{\,^{(t+\Delta t)}q_\alpha(h^e)\}$, $\alpha = 1, 2, \ldots, N$, is the vector of nodal displacement-type parameters. The finite element equilibrium equations can be presented in the incremental form as

$$K_{\alpha\beta}[\,^{(t+\Delta t)}\mathbf{q}(\mathbf{h}), \mathbf{h}]\, \Delta q_\beta(\mathbf{h}) = \Delta Q_\alpha(\mathbf{h}) \tag{2}$$

where $K_{\alpha\beta}$ is the system tangent stiffness matrix and

$$\Delta q_\alpha = {}^{(t+\Delta t)}q_\alpha - {}^{(t)}q_\alpha \tag{3}$$

is the vector of nodal incremental displacements due to the load increment ΔQ_α applied in the (pseudo-time) interval $[t, t{+}\Delta t]$. The simple functional dependence assumed for the stiffness matrix in Eq. (2) is just for algebraic convenience; the final result is easily shown to be valid for a more general class of problems including inelastic material response.

Assume now that $K_{\alpha\beta}$ and ΔQ_α are S times continuously differentiable with respect to h^e. According to the implicit function theorem, [6,7], ${}^{(t+\Delta t)}q_\alpha$ is then also S times continuously differentiable. Assume further that ${}^{(t+\Delta t)}\mathcal{G}$ has at least S continuous derivatives with respect to h^e and $S{+}1$ continuous derivatives with respect to q_α. Considering $S = 2$ we shall be concerned below with the computation of sensitivity functions ${}^{(t+\Delta t)}\mathcal{G}^{\cdot e}$ and ${}^{(t+\Delta t)}\mathcal{G}^{\cdot ef}$, $e, f = 1, \ldots, E$, where $(.)^{\cdot e}$ stands for the partial differentiation with respect to h^e and the following incremental relations hold true:

$$\begin{aligned} {}^{(t+\Delta t)}\mathcal{G} &= {}^{(t)}G + \Delta\mathcal{G} \\ {}^{(t+\Delta t)}\mathcal{G}^{\cdot e} &= {}^{(t)}\mathcal{G}^{\cdot e} + \Delta\mathcal{G}^{\cdot e} \\ {}^{(t+\Delta t)}\mathcal{G}^{\cdot ef} &= {}^{(t)}\mathcal{G}^{\cdot ef} + \Delta\mathcal{G}^{\cdot ef} \end{aligned} \tag{4}$$

For the sake of notation simplicity the left upper index (t) indicating the function evaluation at the beginning of the step will be omitted from now on.

Assuming $\mathcal{G}^{\cdot e}$ and $\mathcal{G}^{\cdot ef}$ known the objective of the incremental structural design sensitivity analysis is to evaluate $\Delta\mathcal{G}^{\cdot e}$ and $\Delta\mathcal{G}^{\cdot ef}$ on the basis of the incremental finite element solution Δq_α.

By using the Leibnitz rule, differentiation of Eq. (1) gives

$${}^{(t+\Delta t)}\mathcal{G}^{\cdot e} = {}^{(t+\Delta t)}G^{\cdot e} + {}^{(t+\Delta t)}G_{.\alpha}\, {}^{(t+\Delta t)}q_\alpha^{\cdot e} \tag{5}$$

where $(.)_{.\alpha}$ denotes the partial derivative with respect to q_α. The 1st-order Taylor expansion for ${}^{(t+\Delta t)}G$ about q_α evaluated at time t reads

$${}^{(t+\Delta t)}G = G + G_{.\alpha}\Delta q_\alpha \tag{6}$$

By substituting Eq. (6) and its derivatives with respect to h^e and q_α into Eq. (5) we arrive at

$$\Delta\mathcal{G}^{\cdot e} = \left(G_{.\alpha}^{\cdot e} + G_{.\alpha\beta}q_\beta^{\cdot e}\right)\Delta q_\alpha + \left(G_{.\alpha} + G_{.\alpha\beta}\Delta q_\beta\right)\Delta q_\alpha^{\cdot e} \tag{7}$$

in which only $\Delta q_\alpha^{\cdot e}$ remains to be calculated. This can be done alternatively by direct differentiation method (DDM) or by the adjoint variable method (AVM).

The former is based on differentiating Eq. (2) with respect to h^e to get

$$K_{\alpha\beta}\Delta q_\beta^{\cdot e} = \Delta Q_\alpha^{\cdot e} - (K_{\alpha\beta}^{\cdot e} + K_{\alpha\beta.\gamma}q_\gamma^{\cdot e})\Delta q_\beta \qquad e = 1, 2, \ldots, E \tag{8}$$

Having solved Eq. (8) for $\Delta q_\alpha^{\cdot e}$ the incremental sensitivity gradient $\Delta\mathcal{G}^{\cdot e}$ follows by the explicit use of Eq. (7) with the number of equations to be solved equal to $E{+}1$.

The main idea behind AVM is to define an adjoint vector $\boldsymbol{\lambda} = \{\lambda_\alpha\}$, $\alpha = 1, 2, \ldots, N$ such that

$$K_{\alpha\beta}\lambda_\beta = G_{.\alpha} + G_{.\alpha\beta}\Delta q_\beta \tag{9}$$

which implies that

$$\Delta \mathcal{G}^{\cdot e} = \Delta Q^{\cdot e}_{\alpha} \lambda_{\alpha} + \left[G^{\cdot e}_{.\alpha} + G_{.\alpha\beta} q^{\cdot e}_{\beta} - (K^{\cdot e}_{\beta\alpha} + K_{\beta\alpha.\gamma} q^{\cdot e}_{\gamma}) \lambda_{\beta} \right] \Delta q_{\alpha} \tag{10}$$

It is readily seen that both the formulations are fully equivalent in terms of the final result but will generally require a different number of algebraic operations.

The second derivatives of ${}^{(t+\Delta t)}\mathcal{G}$ with respect to the variables h^e can be expressed by Eq. (5) as

$$\begin{aligned} {}^{(t+\Delta t)}\mathcal{G}^{\cdot ef} &= \left({}^{(t+\Delta t)}G^{\cdot e} + {}^{(t+\Delta t)}G_{.\alpha} \, {}^{(t+\Delta t)}q^{\cdot e}_{\alpha} \right)^{\cdot f} \\ &= {}^{(t+\Delta t)}G^{\cdot ef} + {}^{(t+\Delta t)}G^{\cdot e}_{.\alpha} \, {}^{(t+\Delta t)}q^{\cdot f}_{\alpha} + {}^{(t+\Delta t)}G^{\cdot f}_{.\alpha} \, {}^{(t+\Delta t)}q^{\cdot e}_{\alpha} \\ &+ {}^{(t+\Delta t)}G_{.\alpha\beta} \, {}^{(t+\Delta t)}q^{\cdot f}_{\beta} \, {}^{(t+\Delta t)}q^{\cdot e}_{\alpha} + {}^{(t+\Delta t)}G_{.\alpha} \, {}^{(t+\Delta t)}q^{\cdot ef}_{\alpha} \end{aligned} \tag{11}$$

which, by noting Eqs. (11),(6),(3),$(4)_3$ yields the incremental expression

$$\begin{aligned} \Delta \mathcal{G}^{\cdot ef} &= \left(G^{\cdot ef}_{.\alpha} + G^{\cdot e}_{.\beta\alpha} q^{\cdot f}_{\beta} + G^{\cdot f}_{.\beta\alpha} q^{\cdot e}_{\beta} + G_{.\beta\gamma\alpha} q^{\cdot e}_{\beta} q^{\cdot f}_{\gamma} + G_{.\beta\alpha} q^{\cdot ef}_{\beta} \right) \Delta q_{\alpha} \\ &+ \left[G^{\cdot e}_{.\alpha} + G_{.\beta\alpha} q^{\cdot e}_{\beta} + \left(G_{.\alpha\beta} + G_{.\gamma\alpha\beta} q^{\cdot e}_{\gamma} \right) \Delta q_{\beta} \right] \Delta q^{\cdot f}_{\alpha} \\ &+ \left[G^{\cdot f}_{.\alpha} + G_{.\alpha\beta} q^{\cdot f}_{\beta} + \left(G_{.\alpha\beta} + G_{.\alpha\gamma\beta} q^{\cdot f}_{\gamma} \right) \Delta q_{\beta} \right] \Delta q^{\cdot e}_{\alpha} \\ &+ \left(G_{.\alpha\beta} + G_{.\alpha\beta\gamma} \Delta q_{\gamma} \right) \Delta q^{\cdot f}_{\beta} \Delta q^{\cdot e}_{\alpha} + \left(G_{.\alpha} + G_{.\alpha\beta} \Delta q_{\beta} \right) \Delta q^{\cdot ef}_{\alpha} \end{aligned} \tag{12}$$

To find $\Delta \mathcal{G}^{\cdot ef}$ only $\Delta q^{\cdot e}_{\alpha}$ and $\Delta q^{\cdot ef}_{\alpha}$ have to be evaluated. This can be done by using the DDM or AVM algorithms, or by developing a mixed differentiation-adjoint scheme.

First we shall have a look at the DDM algorithm. By differentiating Eq. (8) with respect to h^e we arrive at

$$\begin{aligned} K_{\alpha\beta} \Delta q^{\cdot ef}_{\beta} &= \Delta Q^{\cdot ef}_{\alpha} - \left[K^{\cdot ef}_{\alpha\beta} + K^{\cdot e}_{\alpha\beta.\gamma} q^{\cdot f}_{\gamma} + \left(K^{\cdot f}_{\alpha\beta.\gamma} + K_{\alpha\beta.\gamma\zeta} q^{\cdot f}_{\zeta} \right) q^{\cdot e}_{\beta} \right. \\ &+ \left. K_{\alpha\beta.\gamma} q^{\cdot ef}_{\gamma} \right] \Delta q_{\beta} + \left(K^{\cdot e}_{\alpha\beta} + K_{\alpha\beta.\gamma} q^{\cdot e}_{\gamma} \right) \Delta q^{\cdot f}_{\beta} \\ &- \left(K^{\cdot f}_{\alpha\beta} + K_{\alpha\beta.\gamma} q^{\cdot f}_{\gamma} \right) \Delta q^{\cdot e}_{\beta} \qquad e, f = 1, 2, \ldots, E \end{aligned} \tag{13}$$

The complete solution to our problem is provided by Eqs. (2),(8),(13),(12). Since Eq. (13) is symmetric with respect to e and f the formulation requires $1+E(E+3)/2$ equations to be solved.

Employing the AVM algorithm we differentiate Eq. (10) with respect to h^e to obtain

$$\begin{aligned} \Delta \mathcal{G}^{\cdot ef} &= \Delta Q^{\cdot ef}_{\alpha} \lambda_{\alpha} + \left\{ G^{\cdot ef}_{.\alpha} + G^{\cdot e}_{.\alpha\beta} q^{\cdot f}_{\beta} + \left(G^{\cdot f}_{.\alpha\beta} + G_{.\alpha\beta\gamma} q^{\cdot f}_{\gamma} \right) q^{\cdot e}_{\beta} + G_{.\alpha\beta} q^{\cdot ef}_{\beta} \right. \\ &- \left. \left[K^{\cdot ef}_{\beta\alpha} + K^{\cdot e}_{\beta\alpha.\gamma} q^{\cdot f}_{\gamma} + \left(K^{\cdot f}_{\beta\alpha.\gamma} + K_{\beta\alpha.\gamma\zeta} q^{\cdot f}_{\zeta} \right) q^{\cdot e}_{\gamma} + K_{\beta\alpha.\gamma} q^{\cdot ef}_{\gamma} \right] \lambda_{\beta} \right\} \Delta q_{\alpha} \\ &+ \left[G^{\cdot e}_{.\alpha} + G_{.\alpha\beta} q^{\cdot e}_{\beta} - \left(K^{\cdot e}_{\beta\alpha} + K_{\beta\alpha.\gamma} q^{\cdot e}_{\gamma} \right) \lambda_{\beta} \right] \Delta q^{\cdot f}_{\alpha} \\ &+ \left[\Delta Q^{\cdot e}_{\alpha} - \left(K^{\cdot e}_{\alpha\beta} + K_{\alpha\beta.\gamma} q^{\cdot e}_{\gamma} \right) \Delta q_{\beta} \right] \lambda^{\cdot f}_{\alpha} \end{aligned} \tag{14}$$

with $\Delta q^{\cdot e}_{\alpha}$ and $\lambda^{\cdot e}_{\alpha}$ to be computed. In order to do this we first differentiate Eq. (9) with respect to h^e, and then define two sets of E adjoint variable vectors $\boldsymbol{\vartheta}^e = \{\vartheta^e_{\alpha}\}$ and $\boldsymbol{\xi}^e = \{\xi^e_{\alpha}\}$, $e = 1, 2, \ldots, E$, $\alpha = 1, 2, \ldots, N$ such that

$$\begin{aligned}
K_{\alpha\beta}\vartheta^e_\beta &= \Delta Q^{\cdot e}_\alpha - \left(K^{\cdot e}_{\alpha\beta} + K_{\alpha\beta.\gamma}q^{\cdot e}_\gamma\right)\Delta q_\beta \\
K_{\alpha\beta}\xi^e_\beta &= G^{\cdot e}_{.\alpha} + G_{.\alpha\beta}q^{\cdot e}_\beta - \left(K^{\cdot e}_{\alpha\beta} + K_{\alpha\beta.\gamma}q^{\cdot e}_\gamma\right)\lambda_\alpha + G_{.\alpha\beta}\vartheta^e_\alpha \\
& \qquad e = 1, 2, \ldots, E
\end{aligned} \tag{15}$$

Having solved $2(E+1)$ equations (2),(9),(15) for λ_α, ϑ^e_α and ξ^e_α we arrive at

$$\begin{aligned}
\Delta\mathcal{G}^{\cdot ef} = {} & \Delta Q^{\cdot ef}_\alpha\lambda_\alpha + \Big\{G^{\cdot ef}_{.\alpha} + G^{\cdot e}_{.\alpha\beta}q^{\cdot f}_\beta + \left(G^{\cdot f}_{.\alpha\beta} + G_{.\alpha\beta\gamma}q^{\cdot f}_\gamma\right)q^{\cdot e}_\beta + G_{.\alpha\beta}q^{\cdot ef}_\beta \\
& - \left[K^{\cdot ef}_{\beta\alpha} + K^{\cdot e}_{\beta\alpha.\gamma}q^{\cdot f}_\gamma + \left(K^{\cdot f}_{\beta\alpha.\gamma} + K_{\beta\alpha.\gamma\zeta}q^{\cdot f}_\zeta\right)q^{\cdot e}_\gamma + K_{\beta\alpha.\gamma}q^{\cdot ef}_\gamma\right]\lambda_\beta\Big\}\Delta q_\alpha \\
& + \Big[G^{\cdot f}_{.\alpha} + G_{.\alpha\beta}q^{\cdot f}_\beta + \left(G^{\cdot f}_{.\alpha\beta} + G_{.\alpha\beta\gamma}q^{\cdot f}_\gamma\right)\Delta q_\beta \\
& - \left(K^{\cdot f}_{\alpha\beta} + K_{\alpha\beta.\gamma}q^{\cdot f}_\gamma\right)\lambda_\beta\Big]\vartheta^e_\alpha \\
& + \left[\Delta Q^{\cdot e}_\alpha - \left(K^{\cdot e}_{\alpha\beta} + K_{\alpha\beta.\gamma}q^{\cdot e}_\gamma\right)\Delta q_\beta\right]\xi^f_\alpha
\end{aligned} \tag{16}$$

A mixed algorithm can be proposed by making the observation that the last term on the right-hand side of Eq. (12) is the product of the first adjoint load (Eq. (9)) multiplied by the 2nd-order sensitivity of the displacement increments (Eq. (13)) so that

$$\begin{aligned}
\Delta\mathcal{G}^{\cdot ef} = {} & \left(G^{\cdot ef}_{.\alpha} + G^{\cdot e}_{.\beta\alpha}q^{\cdot f}_\beta + G^{\cdot f}_{.\beta\alpha}q^{\cdot e}_\beta + G_{.\beta\gamma\alpha}q^{\cdot e}_\beta q^{\cdot f}_\gamma + G_{.\beta\alpha}q^{\cdot ef}_\beta\right)\Delta q_\alpha \\
& + \left[G^{\cdot e}_{.\alpha} + G_{.\beta\alpha}q^{\cdot e}_\beta + \left(G_{.\alpha\beta} + G_{.\gamma\alpha\beta}q^{\cdot e}_\gamma\right)\Delta q_\beta\right]\Delta q^{\cdot f}_\alpha \\
& + \left[G^{\cdot f}_{.\alpha} + G_{.\alpha\beta}q^{\cdot f}_\beta + \left(G_{.\alpha\beta} + G_{.\alpha\gamma\beta}q^{\cdot f}_\gamma\right)\Delta q_\beta\right]\Delta q^{\cdot e}_\alpha \\
& + \left(G_{.\alpha\beta} + G_{.\alpha\beta\gamma}\Delta q_\gamma\right)\Delta q^{\cdot f}_\beta\Delta q^{\cdot e}_\alpha + \Big\{\Delta Q^{\cdot ef}_\alpha - \Big[K^{\cdot ef}_{\alpha\beta} + K^{\cdot e}_{\alpha\beta.\gamma}q^{\cdot f}_\gamma \\
& + \left(K^{\cdot f}_{\alpha\beta.\gamma} + K_{\alpha\beta.\gamma\zeta}q^{\cdot f}_\zeta\right)q^{\cdot e}_\gamma + K_{\alpha\beta.\gamma}q^{\cdot ef}_\gamma\Big]\Delta q_\beta \\
& + \left(K^{\cdot e}_{\alpha\beta} + K_{\alpha\beta.\gamma}q^{\cdot e}_\gamma\right)\Delta q^{\cdot f}_\beta - \left(K^{\cdot f}_{\alpha\beta} + K_{\alpha\beta.\gamma}q^{\cdot f}_\gamma\right)\Delta q^{\cdot e}_\beta\Big\}\lambda_\alpha
\end{aligned} \tag{17}$$

The complete solution is provided by Eqs. (2),(8),(9),(17) with the total number of solutions reduced to $E+2$.

3 NUMERICAL ILLUSTRATION

To illustrate the formulation the 1st-order sensitivity response of an offshore drilling steel structure sited in a shallow continental shelf is considered in the range of large elastic displacements, Fig. 1. (A number of simple academic text examples are discussed in [8]). A detailed description of geometric, material and load data for the structure of Fig. 1 is given in [5]. The loading consists of a component caused by sea currents and waves calculated according to the Morisons' rule, [9]); the platform is also loaded by wind and operational loads, Fig. 1.

The objective of the analysis is to evaluate sensitivity of the response with respect to variations of some cross-sectional areas with the response functional taken as

$$\mathcal{G} = \frac{|\,q_A\,|}{\hat{q}} - 1 \leq 0,$$

where $\hat{q}$ is an admissible value and q_A is the z-component of the displacement vector at the point A, Fig. 1. The structure is modelled as the 3D truss with 692 finite elements. Two cases are considered: (i) structure with the ideal geometry and (ii) structure with some geometry imperfection in the form of a slight overall twist of the truss, [5].

The results of the nonlinear incremental analysis are shown in Fig. 2 for two components of the displacement vector at the point A. The imperfection pattern turns out largely insignificant in the case considered.

The nonlinear sensitivity results are documented in Fig. 3. The sensitivity of the z-displacement at A has been considered with respect to variations of cross-sectional areas in four corner columns.

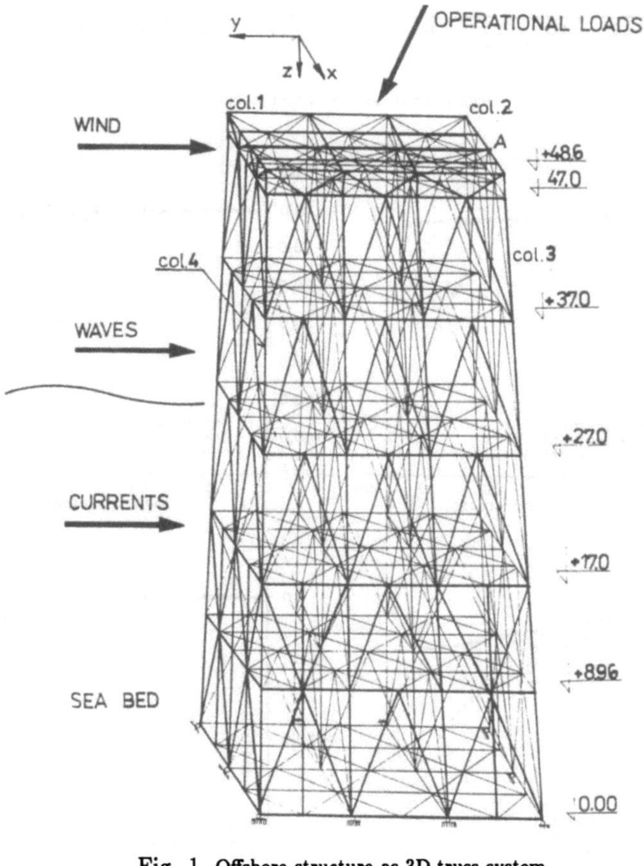

Fig. 1 Offshore structure as 3D truss system.

Each column has been considered as consisting of 6 segments as indicated in Fig. 1. The response sensitivity with respect to cross-sectional variations of the segment I (level 0.0 to 8.96), segment III (level 17.0 to 27.0) and segment VI (level 47.0 to 48.6) has been plotted in Fig. 3 for each of the four columns.

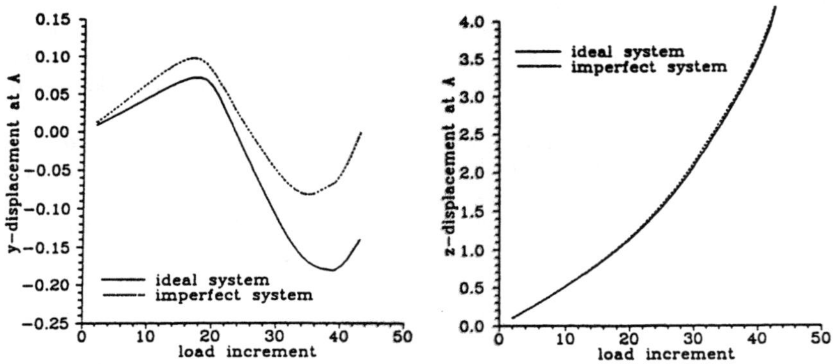

Fig. 2 Nonlinear response of displacements at A.

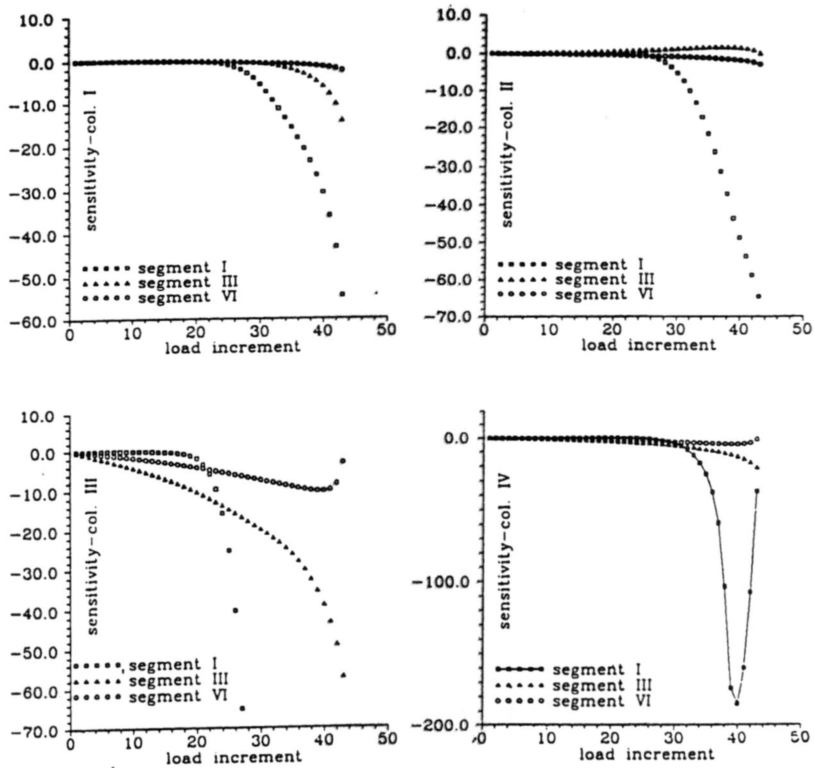

Fig. 3 Nonlinear structural sensitivity response.

It is seen that the response is the most sensitive with respect to the segment I cross-sectional variation in each column, which is intuitively rather obvious. It is also seen that the most crucial design parameters (i.e. cross-sectional areas in this case) for the loading considered are those characterizing the column No. 3 as the corresponding sensitivity gradients grow unacceptably fast after some 20 incremental steps. As before, no significant sensitivity with respect to the overall geometry imperfection has been observed.

References

1 Cardoso J.B. and Arora J.S. Variational method for design sensitivity analysis in nonlinear structural mechanics. *AIAA J.*, 26(5):595–603, 1988.

2 Cardoso J.E.B. and Arora J.S. Design sensitivity analysis of nonlinear structural response. *AIAA J.*, 26(5):595–603, 1988.

3 Tsay J.J. and Arora J.S. Nonlinear structural design sensitivity analysis for path dependent problems. Part 1: General theory. *Comput. Meth. Appl. Mech. Eng.*, 81:183–208, 1990.

4 Tsay J.J., Cardoso J.E.B., and Arora J.S. Nonlinear structural design sensitivity analysis for path dependent problems. Part 2: Analytical examples. *Comput. Meth. Appl. Mech. Eng.*, 81:209–228, 1990.

5 Kleiber M., Hien T.D., and Postek E. Incremental finite element formulation for first- and second-order structural sensitivity problems, *in preparation*.

6 Coddington E.A. and Levinson N. *Theory of Ordinary Differential Equations*. McGraw-Hill, 1955.

7 Goffman C. *Calculus of Several Variables*. Harper-Row, 1965.

8 Kleiber M., Hien T.D., and Postek E. Incremental finite element formulation for nonlinear structural design sensitivity analysis. In *Proc. Asian Pacific Conf. Comput. Mech.*, Dec. 11–13 1991.

9 Zienkiewicz O.C., Lewis R.W., and Stagg K.G. (Ed.). *Numerical Methods in Offshore Engineering*. Wiley, 1978.

FINITE ELEMENTS IN THE 90's, E. Oñate, J. Periaux, A. Samuelsson (Eds.)
Springer-Verlag/CIMNE, Barcelona 1991

A COMPARISON OF TWO VARIATIONAL FORMULATIONS FOR RUBBER-LIKE MATERIALS

C.H. Liu, G. Hofstetter and H.A. Mang
Technical University of Vienna,
A-1040 Vienna,
AUSTRIA

Abstract

Two different variational formulations for rubber-like materials within the framework of the finite element method are briefly reviewed. The first formulation deals with the constraint of incompressible material behavior by means of Lagrange multipliers. The second one can be viewed as a Hu-Washizu variational principle valid for both compressible and nearly incompressible material behavior. Since for three-dimensional nonlinear analyses the efficiency of a finite element formulation strongly depends on the amount of cpu-time it is worthwhile to compare these two methods on the basis of a numerical study to show their advantages and disadvantages.

1 Introduction

Three-dimensional finite element analysis of structures made of rubber-like materials undergoing large displacements and large strains is a task which is very computer-time consuming. For example, for three-dimensional analysis of tires both the large number of degrees of freedom and the geometrically and materially nonlinear behavior contribute to the large amount of cpu-time. In addition the high ratio of bulk to shearing stiffness, yielding a nearly incompressible behavior of rubber-like materials, may lead to numerical problems within the context of a pure displacement formulation.

The classical method to incorporate the condition of incompressibility to a displacement formulation is to add a constraint by means of the Lagrange multiplier method (LMM) [1,2,3,4]. Such a formulation for 'rubber elements' is implemented, for example, in the multi-purpose finite element program MARC [3]. Compared with a pure displacement formulation by means of the LMM, the number of degrees of freedom is increased by the number of Lagrange multipliers and the banded structure of the resulting coefficient matrix is destroyed. This results in a substantial increase of cpu-time for large scale problems.

As an alternative to LMM, hybrid finite element formulations for nearly incompressible materials [1,5,6,11] have been proposed. These formulations are usually valid for both, compressible and nearly incompressible materials. In [5] a formulation, based on a mixed interpolation of the displacements and the hydrostatic pressure is presented. Making use of discontinuous approximations for the pressure, the pressure related degrees of freedom can be eliminated on the element level. In [6] and [11] a mixed finite element formulation on the basis of a three-field Hu-Washizu variational principle is proposed, characterized by the independent approximation of the displacements, the dilatation and the hydrostatic pressure. In addition, in [11] the incompressible limit is enforced by means of an augmented Lagrangian procedure. Employing discontinuous approximations for the dilatation and the pressure between the elements, a generalized displacement model is obtained. As compared with a pure displacement formulation, hybrid methods are characterized by an unchanged number of degrees of freedom in the global system of equations and by the preservation of the banded structure of the coefficient matrix. These methods avoid numerical problems which may occur with a pure displacement formulation for nearly incompressible materials.

For three-dimensional nonlinear problems the amount of computer time frequently still is an important factor for the efficiency of a finite element formulation. This was the reason for implementing a hybrid finite element formulation for the three-dimensional analysis of rubber-like materials, valid for compressible and nearly incompressible behavior into the finite element program MARC by adding a user-defined element.

The paper gives a brief review of the two different formulations and points out the advantages and disadvantages of the two methods by means of a comparison of numerical results.

2 Constitutive Equations for Rubber-Like Materials

2.1 Incompressible Formulation

Assuming rubber-like materials to be hyperelastic materials, the stresses can be derived from a free energy function Ψ. For the case of an incompressible material, characterized by

$$J = Det\,\mathbf{F} = 1, \tag{1}$$

where $\mathbf{F}$ denotes the deformation gradient, Ψ only consists of a volume-preserving part W. The Mooney-Rivlin form of W is given as

$$W = A_1(I_1 - 3) + A_2(I_2 - 3), \qquad I_3 = 1, \tag{2}$$

where A_1 and A_2 are material constants and

$$I_1 = tr\,\mathbf{C}, \qquad I_2 = \frac{1}{2}(tr\,\mathbf{C}\ tr\,\mathbf{C} - tr\,\mathbf{C}^2), \qquad I_3 = Det\,\mathbf{C} \tag{3}$$

are the invariants of the right Cauchy-Green deformation tensor $\mathbf{C} = \mathbf{F}^T\mathbf{F}$. The Mooney-Rivlin form (2) can be viewed as a special case of the more general constitutive model proposed by Ogden [7]. The latter gives a considerably better correspondence between experimental and numerical results at strains exceeding 50 %. Since in our applications the strains usually do not exceed this range, we use the Mooney-Rivlin form for W.

For incompressible materials the free energy function W only determines the Cauchy stress $\boldsymbol{\sigma}$ up to a hydrostatic pressure λ. λ can be obtained from a variational principle using Lagrange multipliers to formulate the incompressibility condition.

2.2 Compressible Formulation

A motivation for extending the free energy function for incompressible rubber-like materials to a compressible formulation [5,11] is that such materials are not strictly incompressible. They are characterized rather by a high ratio of the bulk to the shear modulus. There are cases, for which a finite compressibility is important insofar as it has some influence on the resulting stresses and strains in boundary value problems [8]. Apart from this, within the context of a Hu-Washizu three-field variational principle [6,11], such an extension proves to be computationally advantageous.

A suitable formulation for the volumetric part U of the free energy is [10]

$$U(J) = K(J \, lnJ - J + 1), \tag{4}$$

where K denotes the bulk modulus. A synopsis of different formulations for the volumetric part of the free energy can be found in [9]. The volume-preserving part of $\mathbf{F}$ is given as

$$\bar{\mathbf{F}} = J^{-1/3}\mathbf{F} \quad \rightarrow \quad Det\,\bar{\mathbf{F}} = 1. \tag{5}$$

Consequently the volume-preserving part $\bar{\mathbf{C}}$ of $\mathbf{C}$ is obtained as

$$\bar{\mathbf{C}} = \bar{\mathbf{F}}^T\bar{\mathbf{F}} = J^{-2/3}\mathbf{C} \quad \rightarrow Det\,\mathbf{C} = 1. \tag{6}$$

Replacing $\mathbf{C}$ by $\bar{\mathbf{C}}$ in (3) and consequently I_1 and I_2 by $\bar{I}_1$ and $\bar{I}_2$ in (2) and (3) the free energy for a compressible Mooney-Rivlin material is obtained as

$$\Psi = U(J) + W(\bar{\mathbf{C}}), \tag{7}$$

where

$$W(\bar{\mathbf{C}}) = A_1(\bar{I}_1 - 3) + A_2(\bar{I}_2 - 3). \tag{8}$$

It follows from (5) and (6) that the first and second invariant of $\bar{\mathbf{C}}$ in (8) are

$$\bar{I}_1 = J^{-2/3}I_1 \qquad \bar{I}_2 = J^{-4/3}I_2. \tag{9}$$

The Cauchy stress is obtained from the free energy function (7) as

$$\boldsymbol{\sigma} = p\,\mathbf{1} + dev\,\boldsymbol{\sigma}, \tag{10}$$

where the hydrostatic pressure p and the deviatoric part of $\boldsymbol{\sigma}$ are given as

$$p = \frac{\partial U}{\partial J} = K \;\, ln\, J, \tag{11}$$

$$J\, dev\,\boldsymbol{\sigma} = 2\mathbf{F}\frac{\partial W}{\partial \mathbf{C}}\mathbf{F}^T = 2A_1 dev\,\bar{\mathbf{b}} + 2A_2(tr\,\bar{\mathbf{b}}\;\; dev\,\bar{\mathbf{b}} - dev\,\bar{\mathbf{b}}^2). \tag{12}$$

In (12), $\mathbf{b} = \mathbf{F}\mathbf{F}^T$ denotes the left Cauchy-Green deformation tensor and $\bar{\mathbf{b}} = J^{-2/3}\mathbf{b}$ is the volume preserving part of $\mathbf{b}$.

3 Variational Formulations

3.1 Variational formulation using Lagrange multipliers

The classical method to incorporate the condition of incompressible material behavior (1) into a variational principle is to add the constraint by means of a Lagrange multiplier λ. To this end a potential function Π is formulated

$$\Pi(\mathbf{u},\lambda) = \int_V [W(\mathbf{C}) + \lambda(J-1)]dV + \Pi^{ext}(\mathbf{u}), \tag{13}$$

where $\mathbf{u}$ is the displacement vector, V denotes the volume in the reference configuration and Π^{ext} is the potential of the external forces, which for simplicity are assumed to be conservative. Variation of Π with respect to $\mathbf{u}$ and λ gives

$$\delta_u \Pi = \int_V \left(\frac{\partial W}{\partial \mathbf{C}} : \delta\mathbf{C} + \lambda\,\delta J \right) dV + \delta_u \Pi^{ext}(\mathbf{u}), \tag{14}$$

$$\delta_\lambda \Pi = \int_V \delta\lambda\,(J-1)\,dV. \tag{15}$$

Making use of the relations

$$\delta\mathbf{F} = grad\,\delta\mathbf{u}\,\mathbf{F}, \tag{16}$$

$$\delta\mathbf{C} = \mathbf{F}^T grad^T \delta\mathbf{u}\,\mathbf{F} + \mathbf{F}^T grad\,\delta\mathbf{u}\,\mathbf{F}, \tag{17}$$

$$\delta J = J\,grad\,\delta\mathbf{u} : \mathbf{1}, \tag{18}$$

the transformation of (14) to the current configuration yields

$$\delta_u \Pi = \int_v (\tilde{\boldsymbol{\sigma}} + \lambda\,\mathbf{1}) : \delta\boldsymbol{\varepsilon}\,dv + \delta_u \Pi^{ext}(\mathbf{u}), \tag{19}$$

where v is the volume in the current configuration, $\delta\boldsymbol{\varepsilon}$ denotes the variation of the linearized Eulerian strain tensor and

$$\tilde{\boldsymbol{\sigma}} = 2\frac{1}{J}\mathbf{F}\frac{\partial W}{\partial \mathbf{C}}\mathbf{F}^T \tag{20}$$

is the part of the Cauchy-stress, which is determined by W. Thus, the Lagrange multiplier λ can be interpreted as a corrector of the hydrostatic pressure, ensuring satisfaction of the constraint condition.

In order to solve the nonlinear equations (15) and (19) by means of a Netwon iteration technique, these equations have to be linearized. Finally, discretization of the linearized equations yields a system of algebraic equations for the unknown nodal values of $\mathbf{u}$ and λ.

Thus, compared with an unconstrained displacement formulation, the number of unknowns is increased by the number of the Lagrange multipliers. In addition, it follows from the structure of the equations that the banded structure of the coefficient matrix is destroyed.

3.2 Hu-Washizu variational principle

In order to prevent the well-known 'locking' phenomenon of displacement formulations within the framework of finite elements and the context of nearly incompressible material behavior, Simo [6] has proposed a three-field Hu-Washizu variational principle for elastoplastic problems at finite strains, treating the displacements $\mathbf{u}$, the dilatation θ and the hydrostatic pressure p as independent variables. In this paper this approach is adopted for rubber-like materials. Simo and Taylor [11] have applied this variational principle to quasi-incompressible problems of finite elasticity, formulated in terms of principal stretches.

Replacing J by θ in (4), the potential function

$$\Pi(\mathbf{u},\theta,p) = \int_V [U(\theta) + W(\bar{\mathbf{C}}) + p(J-\theta)]dV + \Pi^{ext}(\mathbf{u}), \tag{21}$$

can be stated [6]. Variation of (21) with respect to $\mathbf{u}$, θ and p yields

$$\delta_u\Pi = \int_V \left(\frac{\partial W}{\partial \mathbf{C}} : \delta\mathbf{C} + p\,\delta J\right) dV + \delta_u\Pi^{ext}(\mathbf{u}) \tag{22}$$

$$\delta_\theta\Pi = \int_V \left(\frac{\partial U}{\partial \theta} - p\right) \delta\theta\, dV, \tag{23}$$

$$\delta_p\Pi = \int_V (J-\theta)\,\delta p\, dV. \tag{24}$$

Linearization of (22) and transformation to the current configuration gives

$$\begin{aligned}&\int_v grad^T\delta\mathbf{u} : \left[\mathbf{a}^{dev} + p(\mathbf{1}\otimes\mathbf{1} - 2\mathbf{I})\right] : grad\Delta\mathbf{u}\, dv \, + \\ &+ \int_v \boldsymbol{\sigma} : (grad^T\delta\mathbf{u}\ \ grad\Delta\mathbf{u})\, dv + \int_v \Delta p\,\mathbf{1} : grad\delta\mathbf{u}\, dv + \delta\Delta\Pi^{ext}(\mathbf{u}) = 0.\end{aligned} \tag{25}$$

In (25) $\mathbf{a}^{dev}$ denotes the deviatoric part of the spatial tangent moduli and $\mathbf{1}$ and $\mathbf{I}$ are the second and fourth order unit tensor, respectively. $\mathbf{a}^{dev}$ is obtained by inserting (8) into the equation relating the spatial tangent moduli a_{ijkl} to the material tangent moduli A_{IJKL}

$$a_{ijkl} = \frac{1}{J} F_{iI}F_{jJ}F_{kK}F_{lL}A_{IJKL} \qquad A_{IJKL} = 4\frac{\partial\Psi}{\partial C_{IJ}\partial C_{KL}} \tag{26}$$

as

$$\begin{aligned} J\,\mathbf{a}^{dev} = &\ \frac{4}{3}A_1\left[\bar{I}_1\left(\mathbf{I} - \frac{1}{3}\mathbf{1}\otimes\mathbf{1}\right) - \left(dev\bar{\mathbf{b}}\otimes\mathbf{1} + \mathbf{1}\otimes dev\bar{\mathbf{b}}\right)\right] \\ &+4\,A_2\left\{\frac{2}{3}\left(\frac{2}{3}\mathbf{1}\otimes\mathbf{1} + \mathbf{I}\right)\bar{I}_2 - \frac{2}{3}\left[\mathbf{1}\otimes\left(\bar{\mathbf{b}}\bar{I}_1 - \bar{\mathbf{b}}^2\right) + \left(\bar{\mathbf{b}}\bar{I}_1 - \bar{\mathbf{b}}^2\right)\otimes\mathbf{1}\right]\right. \\ &\left.+\bar{\mathbf{b}}\otimes\bar{\mathbf{b}} - \bar{\mathbf{H}}\right\}, \end{aligned} \tag{27}$$

where

$$\bar{H}_{ijkl} = \frac{1}{2}(\bar{b}_{ik}\bar{b}_{jl} + \bar{b}_{il}\bar{b}_{jk}). \tag{28}$$

Linearization of (23) and (24) gives

$$\int_V \left(\frac{\partial^2 U}{\partial\theta^2}\Delta\theta - \Delta p\right)\delta\theta\, dV = 0, \tag{29}$$

$$\int_V (\Delta J - \Delta\theta)\,\delta p\, dV = 0. \tag{30}$$

Assuming a discontinuous approximation of θ and p between the elements, (25), (29) and (30) can be reduced to a generalized displacement formulation. For a constant value of θ and p within one finite element, it follows from (29) and (30) that

$$\Delta p_e = \frac{\partial^2 U}{\partial \theta_e^2} \frac{1}{V_e} \int_{V_e} \mathbf{1} : grad \Delta \mathbf{u} \, dV_e, \tag{31}$$

(32)

where e denotes a finite element and $\partial^2 U / \partial \theta_e^2 = K / \theta_e$. Substitution of (31) into the last but one term on the left hand side of (25) and discretization of the so obtained equation finally yields a system of algebraic equations for the unknown nodal displacements only. Thus, compared with a pure displacement formulation, the number of unknowns remains unchanged and the banded structure of the coefficient matrix is preserved.

4 Numerical Investigation

The subsequent numerical investigation consists of a comparison of the formulation available in the multi-purpose finite element program MARC, where the constraint is taken into account by means of Lagrange multipliers, with the hybrid method, derived from a three-field Hu-Washizu variational principle. The latter has been implemented into MARC by adding a user defined element. The aim of the numerical study is to point out the advantages and disadvantages of these two formulations.

4.1 FE-Analysis of a Rubber Bushing

Fig.1 contains the cross-section of a rubber bushing. The rubber is assumed to be perfectly bonded to the rigid frame and the rigid internal shaft. This example was analysed in [5] and had previously been solved in [12].

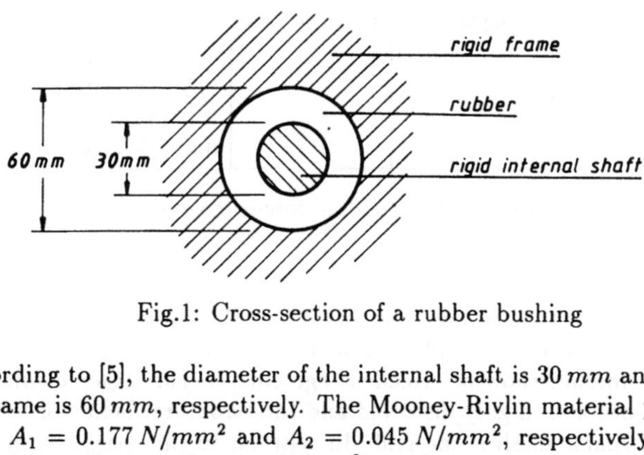

Fig.1: Cross-section of a rubber bushing

According to [5], the diameter of the internal shaft is 30 mm and the diameter of the frame is 60 mm, respectively. The Mooney-Rivlin material parameters are given as $A_1 = 0.177\, N/mm^2$ and $A_2 = 0.045\, N/mm^2$, respectively, and the bulk modulus is specified as $K = 666\, N/mm^2$, yielding a ratio of $K/G = 1500$, where $G = 2(A_1 + A_2)$ denotes the initial shear modulus. The FE-analysis is carried out under the assumption of plane strain conditions, prescribing the displacement of the internal shaft. Because of symmetry reasons only one half of the cross-section is discretized. The coarse mesh (Fig.2a) consists of 72 4-node elements with constant dilatation and pressure. The fine mesh (Fig.2b), consisting of 288 4-node elements with constant dilatation and pressure, is obtained by a systematic refinement of the coarse mesh.

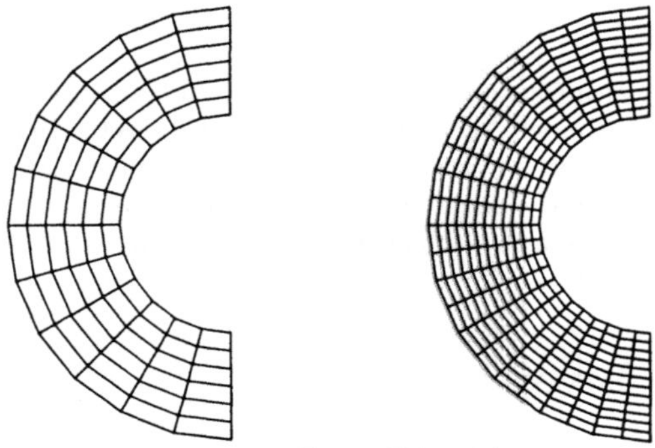

Fig.2: FE-meshes

A total displacement of the shaft of $8\,mm$ is applied in 16 equal incremental steps. In Table 1, for a few selected incremental steps for each iteration step within the Newton solution procedure, the norm of the residual forces is summarized.

iteration	increment 1	increment 8	increment 16
1	0.1418E+04	0.1519E+04	0.1999E+04
2	0.5887E+02	0.8462E+02	0.4675E+03
3	0.2009E+00	0.6551E+00	0.1792E+03
4	0.3366E−01	0.2357E+00	0.3352E+02
5	0.3470E−06	0.4671E−04	0.8340E+02
6	0.5645E−11	0.2138E−08	0.2413E+01
7			0.7105E+01
8			0.2392E−01
9			0.1844E−02
10			0.8952E−08

Table 1 - Norm of the residual forces (coarse mesh)

The deformed structure is shown in Fig.3 and the load-displacement curves are plotted in Fig.4. Up to a displacement of $4\,mm$ the results obtained on the basis of the two meshes show an excellent agreement. For larger displacements the coarse mesh becomes highly distorted and the structural behavior, computed by means of the coarse mesh is too stiff. In contrast to the coarse mesh the fine mesh does not exhibit this pronounced 'locking' behavior. In addition in Fig.4 the results, obtained with MARC on the basis of the LMM are plotted. With MARC for displacements beyond 5 mm even for incremental steps as small as 0.1 mm convergence problems occur.

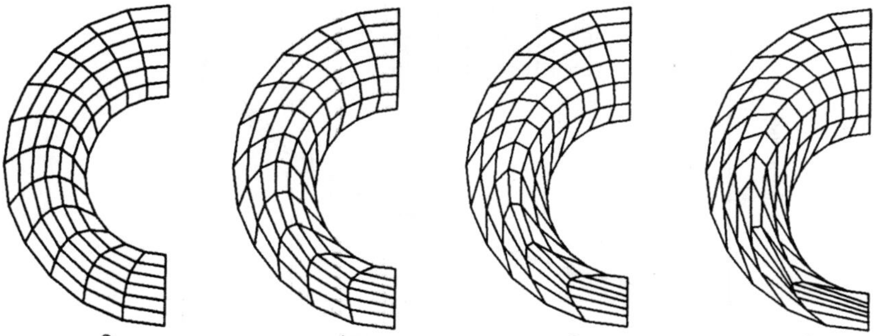

Fig.3: Deformed structure (coarse mesh)

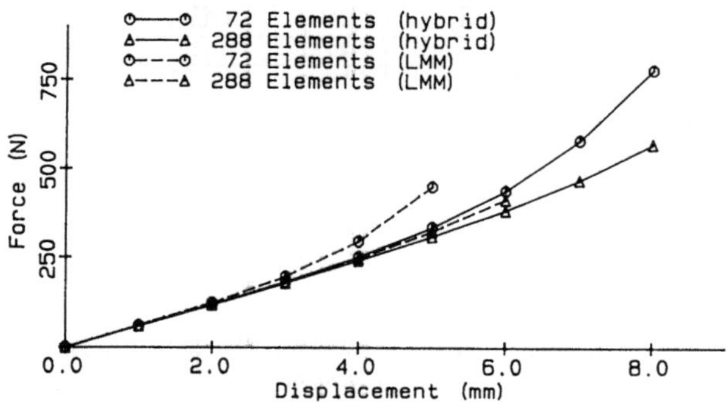

Fig.4: Load-displacement diagrams for the two meshes

4.2 Compression of a Rubber Cube

In order to demonstrate the computational advantages of the hybrid formulation compared with the Lagrange multiplier method within the framework of a 3D finite element analysis, a rubber cube with the dimensions $20 \times 20 \times 20$ mm is subjected to a uniformly vertical displacement of the top surface. The horizontal displacements of both the top and the bottom surface are constrained. The material parameters have the same values as in the first example of this paper. The cube is discretized by means of 8-node 3D-elements with constant dilatation and pressure, which have been implemented into the FE-program MARC by adding a user-defined element. Four meshes, consisting of 27, 125, 512 and 1000 elements are employed for the FE-analysis. The results, computed on the basis of these elements, are compared with results, obtained by the LMM, which is available in MARC. Fig.5 shows the load-displacement diagram computed on the basis of the first two meshes. Table 2 contains a comparison of the number of degrees of freedom (DOF) occurring in the global system of equations, the dimension of the workspace (WS), the number of iteration steps (IS) within one increment and the cpu-time (CPU) in seconds for the solution of the global system of equations within one iteration step on the basis of a HP9000/835 mini-computer with 80 MB memory.

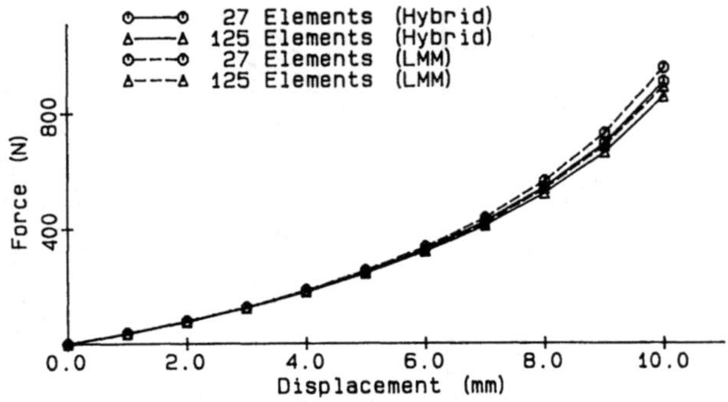

Fig.5: Load-displacement diagrams

	27 elements		125 elements		512 elements		1 000 elements	
	LMM	Hybrid	LMM	Hybrid	LMM	Hybrid	LMM	Hybrid
DOF	131	104	569	444	2 231	1 719	4 289	3 289
WS	93 907	32 019	772 639	178 591	8 048 151	1 196 071	26 786 027	3 148 759
IS	3	3	3	3	3	3	3	3
CPU	1.45	0.53	31.4	8.0	888.	250.	4 796.	1 188.

Table 2 - Comparison of the two formulations

5 Conclusions

Compared with the Lagrange multiplier method, the hybrid formulation, based on a three-field Hu-Washizu principle, proves to be computationally advantageous, provided the ratio of the bulk to the shearing stiffness is not too large, i.e. within the range of our examples. In this case, the hybrid formulation leads to a significant reduction of computer time, which especially for three-dimensional analyses is an important factor for the efficiency of a finite element formulation. A drawback of the hybrid formulation is encountered for very high ratios of the bulk to the shearing stiffness, leading to numerical problems. In this case, either the Lagrange multiplier method, enforcing strictly incompressible behavior, can be used or the formulation presented in [11] can be employed, consisting of a combination of the three-field Hu-Washizu variational principle with an augmented Lagrangian procedure, preserving the advantages of the hybrid formulation even for the incompressible limit.

6 References

1. O.C. Zienkiewicz and R.L. Taylor, The Finite Element Method, 4th ed., Vol.1, McGraw-Hill, UK, 1989.

2. L.R. Herrmann, 'Elasticity equations for incompressible and nearly incompressible materials by a variational theorem', *JAIAA*, **3**, 1986-1900, 1965.

3. MARC Analysis Research Cooperation, User Information Manual, Rev. K.4, Palo Alto, CA94306, USA, Jan. 1990.

4. T. Scharnhorst and T.H.H. Pian, 'Finite element analysis of rubber-like materials by a mixed model', *Int.J.Num.Meth.Eng.*, **12**, 665-676, 1978.

5. T. Sussman and K.J. Bathe, 'A finite element formulation for nonlinear incompressible elastic and inelastic analysis', *Comp.Struct.*, **26**, No.1/2, 357-409, 1987.

6. Simo, J.C., "A framework for finite strain elastoplasticity based on maximum plastic dissipation and the multiplicative decomposition, part II. computational aspects", *Comp. Meth. Appl. Mech. Eng.*, **68**, pp.1-31, 1988.

7. R.W. Ogden, 'Elastic deformations of rubber-like solids', *Mechanics of Solidsr*, The Rodney Hill 60th Anniversary Volume, Eds. H.G. Hopkins and M.J. Sewell, Pergamon Press, 499-537, 1981.

8. S.R. Swanson, L.W. Christensen and M. Ensign, 'Large deformation finite element calculations for slightly compressible hyperelastic materials', *Comp.Struct.*, **21**, 81-88, 1985.

9. P.A.J. van den Bogert and R. de Borst, 'Constitutive aspects and finite element analysis of 3D rubber specimens in compression and shear', Proceedings of the NUMETA 90, Swansea, Wales, UK, 1990.

10. J.C. Simo and T.J.R. Hughes, *Elastoplasticity and Viscoplasticity. Computational Aspects*, in press.

11. J.C. Simo and R.L. Taylor, 'Quasi-incompressible finite elasticity in principal stretches. Continuum basis and numerical algorithms', *Comp.Meth.Appl. Mech.Engr.*, **85**, 273-310, 1991.

12. A.B. Zdunek and M. Bercovier, 'Numerical evaluation of finite element methods for rubber parts', *Proceedings of the 6th International Conference on Vehicle Structural Mechanics*, Society of Automotive Engineers, Detroit, 1986.

FINITE ELEMENTS IN THE 90's, E. Oñate, J. Periaux, A. Samuelsson (Eds.)
Springer-Verlag/CIMNE, Barcelona 1991

CONVERGENCE OF THE NEWTON-RAPHSON METHOD IN ELASTIC-PLASTIC-SOFTENING STRUCTURAL DYNAMICS

C. Comi, A. Corigliano and G. Maier
Department of Structural Engineering,
Technical University of Milan,
Piazza L. da Vinci 32, 20133 Milano,
ITALY

SUMMARY - Elastoplastic associative constitutive laws with linear kinematic hardening or *softening* are adopted. The boundary-initial-value problem of dynamic elastoplastic analysis is discretized in space by finite elements (using Prager's notion of *generalized variables*) and in time by various implicit procedures (including mid-point and backward-difference schemes). It is proved that the Newton-Raphson iterative predictor-corrector algorithm for the solution of the nonlinear finite-step problem, converges to the solution, subject to conditions which bound from above, in a correlated way, the *amount of constitutive softening* and the time step. These conditions imply also the uniqueness of the step solution.

NOTATION - All symbols are defined where they are used for the first time. Superscript T means transpose; dots mark time derivatives. Vector inequalities apply componentwise. A factor in parentheses is preceded by a dot denoting scalar product, whenever it may be mistaken for an argument of functional dependence.

1. INTRODUCTION

Many computational aspects of incremental static and dynamic analysis in small-deformation plasticity *stable* in Drucker's sense have been investigated since several years in an abundant literature surveyed in the Zienkiewicz-Taylor book [1]. Constitutive instabilities or softening is also a timely subject of growing research efforts, many of which condensed in another recent treatise [2]. In the intersection of these two areas, much work has been done on strain localization phenomena (representative contributions in dynamics are those by Needleman [3] and Loret-Prevost [4]). Less attention has been paid to computational aspects of structural dynamic analysis in the presence of softening: seismic responses of frames with softening plastic hinges have been studied by Sanjayan and Darvall [5]; bifurcations and chaotic phenomena due to softening by Maier-Perego [6]; implicit time-integration for pseudodynamics by Shing-Vannan [7]; algorithmic nonlinear stability of stepping algorithms and convergence of iterative algorithms for solving the step problem have recently been studied by the present writers [8], [9]. The last of the above topics is further pursued here with specific reference to the Newton-Raphson method.

2. GOVERNING RELATIONS OF THE DISCRETIZED PROBLEM

A space-discretization by finite elements of a dynamic analysis problem in the small-deformation range and in associative elastoplasticity with linear hardening or softening, leads to the following set of (nonlinear differential) equations:

$$\underline{q} = \underline{C}\,\underline{u}, \qquad \underline{M}\,\ddot{\underline{u}} + \underline{V}\,\dot{\underline{u}} + \underline{C}^T\underline{Q} = \underline{F}(t) \tag{1}$$

$$\underline{q} = \underline{e} + \underline{p}, \qquad \underline{Q} = \underline{E}\,\underline{e}\,, \quad \underline{\chi} = \underline{A}\,\underline{\eta} \tag{2}$$

$$\dot{\underline{p}} = \frac{\partial \underline{\phi}^T}{\partial \underline{Q}}(\underline{Q},\underline{\chi})\,\dot{\underline{\lambda}}, \quad \dot{\underline{\eta}} = -\frac{\partial \underline{\phi}^T}{\partial \underline{\chi}}(\underline{Q},\underline{\chi})\,\dot{\underline{\lambda}} \tag{3}$$

$$\underline{\phi} \equiv \underline{\Phi}(\underline{Q},\underline{\chi}) - \underline{Y} \le \underline{0}, \qquad \dot{\underline{\lambda}} \ge \underline{0}, \qquad \underline{\phi}^T\dot{\underline{\lambda}} = 0 \tag{4}$$

$$\underline{u}(0) = \underline{u}_0, \quad \dot{\underline{u}}(0) = \dot{\underline{u}}_0 \tag{5}$$

The compatibility equation (1a) relates the m-vector of generalized strains $\underline{q}$ to the n-vector of free nodal displacements (d.o.f.) $\underline{u}$; eq. (1b) expresses dynamic equilibrium, $\underline{M}$ and $\underline{V}$ are symmetric positive definite and semidefinite mass and viscous damping matrices, respectively, $\underline{Q}$ generalized stress vector, $\underline{F}$ nodal load data (for brevity no kinematic external actions are envisaged). Equation (2a) represents the additivity of elastic $\underline{e}$ and plastic $\underline{p}$ generalized strains; eq. (2b) relates elastic strains to stresses through positive definite element stiffness matrix $\underline{E}$ and eq. (2c) relates kinematic $\underline{\eta}$ to static $\underline{\chi}$ internal variables through the constant symmetric matrix $\underline{A}$. Equations (3) express the normality rule. Equation (4a) defines the yield functions (assumed convex) as the sum of effective stresses $\underline{\Phi}$ (assumed as order-one homogeneus in $\underline{Q},\underline{\chi}$) and (constant, positive) original yield limits $\underline{Y}$; eqs. (4) govern the loading-unloading processes and confer to the system dissipative, nonholonomic property. The initial conditions (5) complete the governing set.

The following remarks further elucidate and specify the problem in point.

(a) The constitutive relations concern, for notation convenience, all elements simultaneously, but are decoupled (i.e. $\underline{E}$, $\underline{A}$, $\partial\underline{\phi}^T/\partial\underline{Q}$, $\partial\underline{\phi}^T/\partial\underline{\chi}$ are block-diagonal matrices and each diagonal block pertains to an element, say i). Three pairs of conjugate generalized variable vectors are involved: $\underline{q}^i$ (or $\underline{e}^i$, $\underline{p}^i$) and $\underline{Q}^i$; $\underline{\eta}^i$ and $\underline{\chi}^i$; $\underline{\lambda}^i$ and $\underline{\phi}^i$. In each pair the former, kinematic variables, are e.g. local values in Gauss points governing the field through chosen interpolations; the latter, static variables, are modelled consequently in order to preserve the dot product (e.g. $\underline{Q}_i^T\underline{q}_i = \int_{\Omega_i} \underline{\sigma}^T\underline{\varepsilon}\; d\Omega$). This requirement characterizes *generalized* (in Prager's sense) variables; the relevant FE formulation is conceptually (but not necessarily computationally) different from the conventional ones in local (Gauss points) values. The generalized variable concept is discussed in detail in [10] and may be illustrated here by the elementary examples of trusses ($\underline{q}$ = member elongations; $\underline{Q}$ = axial forces) and constant strain tetrahedral FE in Argyris *natural* variables.

(b) Matrix $\underline{A}$ plays in the present context a central role due to the following relations:

$$\frac{1}{2}\,\dot{\underline{Q}}^T\dot{\underline{p}} = \frac{1}{2}\,\dot{\underline{\lambda}}^T\underline{H}\,\dot{\underline{\lambda}}, \quad \text{where} \quad \underline{H} \equiv \frac{\partial\underline{\phi}}{\partial\underline{\chi}^T}\,\underline{A}\,\frac{\partial\underline{\phi}^T}{\partial\underline{\chi}} \tag{6}$$

$$\frac{1}{2}\,\underline{e}^T\underline{E}\,\underline{e} + \frac{1}{2}\,\underline{\eta}^T\underline{A}\,\underline{\eta} = \mathcal{E} \tag{7}$$

$$\underline{Q}^T\dot{\underline{p}} - \underline{\eta}^T\underline{A}\,\dot{\underline{\eta}} = \dot{D} \geq 0 \tag{8}$$

Equations (6), a consequence of $\dot{\underline{\phi}}^T\dot{\underline{\lambda}} = 0$ which in turn flows from eqs. (4), shows that if we do not require the sign definition of $\underline{A}$ and, hence, of the hardening matrix $\underline{H}$, the plastic and the total second-order work may be negative: this means *softening*, i.e. unstable constitutive behaviour (while $\underline{A}$ and $\underline{H}$ positive semidefinite means hardening or perfectly plastic behaviour in full accordance with Drucker's stability postulate). Equation (7) states that the quadratic form associated to $\underline{A}$ represents the contribution to Helmholtz *free* energy $\mathcal{E}$ due to the material rearrangements described by the internal variables $\underline{\eta}$. Equations (8) show the presence of $\underline{A}$ in the expression for dissipated power $\dot{D}$, which is nonnegative for thermodynamical reasons.

(c) In the following we will specialize the constitutive laws (2)-(4) to kinematic hardening by assuming:

$$\underline{\Phi} = \underline{\Phi}(\underline{Q} - \underline{\chi}) \quad \text{and, hence:} \quad \dot{\underline{\eta}} = \dot{\underline{p}} \tag{9}$$

A time-discretization by finite differences can be achieved through the following provisions, [11]. Focussing on a time step Δt starting from instant t_n at which all variables (subscript n) are known, we will denote by subscript ρ their values at instant $t_\rho = t_n + \rho\Delta t$ with $0 < \rho \leq 1$ (ρ = α, β, γ, δ, 1). The constitutive laws, eqs. (2)-(4) and (9), will be made path-independent (holonomic) in increments (marked by Δ) over Δt by enforcing them at instant t_α:

$$\Delta\underline{Q} = \underline{E}\;\underline{C}\Delta\underline{u} - \underline{E}\Delta\underline{p}\,, \quad \Delta\underline{p} = \frac{\partial\underline{\phi}^T}{\partial\underline{Q}}(\underline{Q}_\alpha - \underline{A}\;\underline{p}_\alpha)\;\Delta\underline{\lambda} \tag{10}$$

$$\underline{\phi}_\alpha = \underline{\Phi}(\underline{Q}_\alpha - \underline{A}\;\underline{p}_\alpha) - \underline{Y} \leq \underline{0}, \qquad \underline{\phi}_\alpha^T\Delta\underline{\lambda} = 0, \qquad \Delta\underline{\lambda} \geq \underline{0}, \tag{11}$$

Assuming linear interpolations over Δt, we may write for any variable vector $\underline{X}$:

$$\underline{X}_\rho \cong (1 - \rho)\;\underline{X}_n + \rho\;\underline{X}_1 = \underline{X}_n + \rho\;\Delta\underline{X}, \qquad 0 < \rho \leq 1. \tag{12}$$

For displacements and velocities a truncated series expansion, β and γ being chosen in the interval (0, 1], reads:

$$\underline{u}_1 \cong \underline{u}_n + \dot{\underline{u}}_\beta\;\Delta t\,, \qquad \dot{\underline{u}}_1 \cong \dot{\underline{u}}_n + \ddot{\underline{u}}_\gamma\;\Delta t \tag{13}$$

Using eq. (12) for $\rho = \delta$, β, γ and eqs.(13) one can eliminate $\dot{\underline{u}}_\delta$ and $\ddot{\underline{u}}_\delta$ from the equilibrium equation (1b) imposed at instant δ, thus transforming it into an equation in displacement and stress increments only:

$$\underline{K}_D\Delta\underline{u} + \underline{C}^T\Delta\underline{Q} = \Delta\underline{\tilde{F}} \tag{14}$$

having set:

$$\underline{K}_D \equiv \frac{1}{\beta\gamma\Delta t^2}\underline{M} + \frac{1}{\beta\,\Delta t}\underline{V}\,; \quad \Delta\underline{\tilde{F}} \equiv \frac{1}{\delta}\cdot\left(\underline{\tilde{F}}_\delta - \underline{C}^T\underline{Q}_n\right) \tag{15}$$

$$\underline{\tilde{F}}_\delta \equiv \underline{F}_\delta + \underline{M}\cdot\left(\frac{\delta-\gamma}{\gamma}\underline{\ddot{u}}_n + \frac{\delta}{\beta\gamma\,\Delta t}\underline{\dot{u}}_n\right) + \underline{V}\cdot\left(\frac{\delta-\beta}{\beta}\underline{\dot{u}}_n\right) \tag{16}$$

The set of eqs. (10), (11) and (14) governs the problem discretized in both space and time, i.e. the *finite step dynamic problem* referred to as FSDP in what follows.

Substituting eq. (10a) into the equilibrium eq. (14), this becomes:

$$\underline{K}\,\Delta\underline{u} - \underline{C}^T\underline{E}\,\Delta\underline{p} = \Delta\underline{\tilde{F}} \tag{17}$$

where:

$$\underline{K} \equiv \underline{K}_D + \underline{K}_E, \qquad \underline{K}_E \equiv \underline{C}^T\underline{E}\underline{C} \tag{18}$$

The (assembled) elastic stiffness and the *dynamic stiffness* matrices $\underline{K}_E$ and $\underline{K}_D$ are positive definite, as $\underline{C}$ is supposed full-rank (no rigid body motions) and $\underline{M}$ and $\underline{V}$ are assumed positive definite.

3. A CLASS OF PREDICTOR-CORRECTOR ALGORITHMS AND A GENERAL CONVERGENCE CONDITION

Various iterative procedures potentially apt to solve numerically the nonlinear algebraic problem over the finite step (FSDP) formulated in what precedes, can be based on the following customary strategy.

(a) Prediction phase within the r-th iteration gives updated displacement increments $\Delta\underline{u}^r$ as solution to the *global*, linear equation:

$$\underline{S}^r\cdot(\Delta\underline{u}^r - \Delta\underline{u}^{r-1}) = \Delta\underline{\tilde{F}} - \underline{K}\,\Delta\underline{u}^{r-1} + \underline{C}^T\underline{E}\,\Delta\underline{p}^{r-1} \tag{19}$$

where $\Delta\underline{u}^{r-1}$, $\Delta\underline{p}^{r-1}$ are known from the preceding iterations and $\underline{S}^r$ represents a suitably chosen symmetric matrix.

(b) Correction phase for given incremental strains $\Delta\underline{q}^r = \underline{C}\Delta\underline{u}^r$ (provided by phase *(a)* through compatibility) computes new incremental strains $\Delta\underline{p}^r$ by means of the constitutive equations, i.e. by solving locally (in each element Gauss point, separately) a number of nonlinear small-size problems (10), (11).

The outline of the basic algorithm is completed by: *(c) initialization phase* for r = 1 with $\underline{S}^o = \underline{K}_E + \underline{K}_D$ and $\Delta\underline{u}^o$ and $\Delta\underline{p}^o$ assumed by guess; *(d) termination phase* which truncates the iteration sequence when a suitable convergence test is passed.

For the present context of dynamic softening plasticity, we have established in [9] the following convergence theorem.

Proposition 1 - The above predictor-corrector iterative sequence leads to the solution of the FSDP under the following two sufficient conditions:

(a) $\underline{R}^r \equiv \underline{S}^r - \frac{1}{2}(\underline{K}_D + \underline{K}_E)$ symmetric positive definite (20)

(b) a scalar μ can be found in the interval [0,1) such that:

(b') $\underline{A} + \mu\underline{E}$ positive semidefinite (21)

(b'') $\underline{K}_D - \zeta\underline{K}_E$ with $\zeta = \frac{\mu}{1-\mu}$, positive definite (22)

The proof of this statement, not duplicated here for brevity, rests on an extremum property of the FSDP solution. This property, together with solution uniqueness, was proved in [9] subject to condition (b). It is worth noting that material stability in Drucker's sense, i.e. matrix $\underline{A}$ positive semi-definite (perfect or hardening plasticity), makes condition (b) always satisfied (for $\zeta = 0$); then Proposition 1 basically reduces to the convergence theorem proved in [12] for statical ($\underline{K}_D = \underline{0}$) and in [11] for dynamical situations.

The present purpose is to derive, specifically for the Newton-Raphson algorithm, convergence conditions stronger and of more direct applicability than those stated in Proposition 1.

4. THE NEWTON-RAPHSON METHOD WITH SOFTENING: SPECIFIC CONVERGENCE CONDITIONS

Within the symbology of the preceding Section, Newton-Raphson (NR) procedure is characterized by adopting in the predictor eq. (19) the matrix:

$$\underline{S}^r = \underline{S}_N^r \equiv \underline{K}_D + \underline{K}_T^r \quad \text{for } r = 2, 3... \tag{23}$$

where $\underline{K}_T^r$ denotes a tangent stiffness matrix computed at the current r-th iteration. It is well known that, due to faster convergence, NR strategy may be often preferred to other strategies with $\underline{S}$ unrelated to iteration index r, despite the computing cost implied by changing $\underline{S}^r$. Clearly, if the so-called *modified NR* strategy is adopted, i.e. $\underline{S}^r = \underline{K}_E + \underline{K}_D$, condition (a), eq. (20), is complied with and only the latter condition (b) has to be checked (see [9]). Instead, when NR assumption (23) is adopted, condition (20) is not a priori satisfied even in the case of stable materials [11]. Moreover, in the presence of softening $\underline{K}_T^r$ may be nondefinite and this circumstance may, in turn, entail a violation of condition (20), which therefore becomes meaningful and, hence, is investigated below.

In view of eq. (23), matrix $\underline{R}^r$ in eq. (20) can be written in the forms:

$$\underline{R}^r = \underline{R}_N^r \equiv \frac{1}{2}\underline{K}_D + \underline{K}_T^r - \frac{1}{2}\underline{K}_E = \underline{K}_E + \underline{K}_T^r + \frac{1}{2}(\underline{K}_D - 3\underline{K}_E) \tag{24}$$

Therefore, condition (a) of Proposition 1 is fulfilled if the two following conditions are complied with:

(a') $\underline{u}^{*T}(\underline{K}_E + \underline{K}_T^r)\,\underline{u}^* \geq 0 \quad \forall\, \underline{u}^*$; (25)

(a'') $\underline{u}^{*T}(\underline{K}_D - 3\underline{K}_E^r)\,\underline{u}^* > 0 \quad \forall\, \underline{u}^* \neq \underline{0}$; (26)

where $\underline{u}^*$ is a d.o.f. vector without specific physical meaning or role in the problem.

Considering first condition (a′), eq. (25), and denoting by $\underline{T}^r$ the constitutive block diagonal matrix of current local (at element or Gauss point level) tangent stiffnesses, we can write through geometric compatibility eq. (1a):

$$\underline{u}^{*T} (\underline{K}_E + \underline{K}_T^r) \underline{u}^* = \underline{q}^{*T} (\underline{E} + \underline{T}^r) \underline{q}^* \tag{27}$$

The constitutive tangent matrix $\underline{T}^r$ may be computed as *consistent* matrix from the mapping $\Delta\underline{q} \rightarrow \Delta\underline{Q}$ represented by the time-integrated, stepwise holonomic constitutive law. However, this matrix is not necessarily symmetric, while symmetry is required by the convergence conditions stated in Proposition 1 (as by the extremum theorem from which it was derived in [9]). The condition:

$$\underline{E} + \underline{T}^r \text{ positive semidefinite} \tag{28}$$

is sufficient for (a′) to hold and entails a restriction on the *amount of softening* admitted in the material model with reference to the elastic stiffness.

Consider now the second condition (a″), eq. (26), let us adopt the usual assumption that the viscous matrix $\underline{V}$ is a linear combination of $\underline{K}_E$ and $\underline{M}$ (Rayleigh damping). Then expressing u* in terms of principal coordinates $\underline{P}$ through the modal matrix $\underline{\Psi}$ (normalized with respect to $\underline{M}$) of the system regarded as linear elastic, account taken of the expression of the dynamic stiffness $\underline{K}_D$, we may write in full equivalence to (26):

$$\frac{1}{\beta\gamma\Delta t^2} \underline{P}^{*T} \operatorname*{diag}_j \left[1 + 2\gamma\nu_j\omega_j\Delta t - 3\beta\gamma\omega_j^2\Delta t^2 \right] \underline{P}^* > 0 \quad \forall\, \underline{P}^* \neq \underline{0} \tag{29}$$

where: $\omega_j = (k_j/m_j)^{1/2}$, $\nu_j = 1/2\ v_j\ (k_j m_j)^{-1/2}$, k_j, m_j and v_j represent eigenfrequency, damping ratio, equivalent stiffness, mass and viscous damping, respectively, pertaining to the j-th eigenmode (j = 1 ...n).

Condition (a″) in form (29) can be readily translated into an equivalent upper bound on the time step:

$$\Delta t < \min_j \left\{ \frac{\nu_j + (\nu_j^2 + 3\beta/\gamma)^{1/2}}{3\beta\omega_j} \right\} \tag{30}$$

Turning now to condition (b), eqs. (21)-(22), note that, like condition (a), it encompasses a restriction (b′) on the constitutive softening and a restriction (b″) on the dynamic properties and their representation in the time-discrete model. However, unlike condition (a), the two restrictions are coupled by the equation $\zeta = \mu/(1-\mu)$ in (22) relating to each other the scalars μ and ζ. Let the number μ_m be defined on the basis of the constitutive laws, as:

$$\mu_m \equiv \min \{ \mu \in [0,1), \text{ such that: } \underline{A} + \mu\, \underline{E} \text{ positive semidefinite} \} \tag{31}$$

Let $\zeta_m \in [0,\infty)$ denote the value of ζ corresponding to μ_m through (22). Then condition (b″) leads to the following restriction on the time-step:

$$\min_j \left\{ \frac{1 + 2\, \nu_j\, \omega_j \gamma\, \Delta t}{\omega_j^2\, \Delta t^2} \right\} > \beta\gamma\, \zeta_m \tag{32}$$

If we can set $\underline{V} = \underline{0}$ and hence $\nu_j = 0$ because plastic dissipation prevails over structural damping, inequality (32) can be expressed as another upper bound on the time step, ω_M being the highest eigenfrequency of the system:

$$\Delta t < (\beta\gamma\, \zeta_m)^{-1/2} \omega_M^{-1} \tag{33}$$

The developments leading from formulation (22) to (32) of condition (b″) are analogous to those leading from formulation (26) to (30) for condition (a″) and have been presented in [9]. Therefore a formal derivation of eq. (32) is not given here for brevity.

The conclusions achieved in what precedes can now be condensed in the following statement.

Proposition 2 - The NR algorithm converges to the solution of the FSDP if the following two sufficient (not necessary) conditions are fulfilled:

(a′) the sum of the elastic and tangent constitutive (local) stiffness matrices is symmetric and positive semidefinite everywhere at any iterations r;

(a″) the time step Δt is less than the critical threshold (30);

(b) the time step satisfies inequality (32) where $\zeta_m \equiv \mu_m/(1-\mu_m)$ and μ_m is the minimum scalar μ in the interval [0,1) which makes positive semidefinite the matrix $\underline{A} + \mu\underline{E}$.

Mechanical interpretations of this result, its numerical tests, assessment of computational use and possible extensions to more general hardening-softening models are pursued in a separate paper.

ACKNOWLEDGEMENTS - Part of this work was done while author G. Maier was visiting University of Colorado. He wants to thank his colleagues of UC, Boulder, and especially Professor K. Willam for friendly hospitality and stimulating discussions. A grant from CNR-GNDT is also gratefully acknowledged.

REFERENCES

1. ZIENKIEWICZ, O.C. and TAYLOR, R.L. - The finite element method, 2[d] volume, 4-th edition, Mc Graw-Hill, London, 1991.
2. BAŽANT, Z.L., and CEDOLIN, L. - Stability of structures, Oxford Univ. Press., Oxford, 1991.
3. NEEDLEMAN, A. - Dynamic shear band development in plain strain, J.. Appl. Mech. ASME, 56, 1-9, (1989).
4. LORET, B. and PREVOST J.H. - Dynamic strain localization in elasto-(visco)-plastic solids, Part 1. General formulation and one -dimensional examples, Comp. Meth. Appl. Mech. Engng., 83, 247-273, (1990).
5. SANJAYAN, G. and DARVALL, P.L. - Dynamic response of softening structures, J. Struct. Eng. Div. ASCE, 113, 1203-1220, (1987).
6. MAIER, G. and PEREGO, U. - Effects of softening in elastic-plastic structural dynamics, Int. J. Num. Meth. Eng., (to appear).
7. SHING, P.S.B. and VANNAN, M.T. - Implicit time integration for pseudodynamic tests: convergence and energy dissipation, Earthquake Eng. & Struct. Dyn., (to appear).

8. COMI, C., CORIGLIANO, A. and MAIER, G. - Nonlinear stability conditions for stepping solutions of elastoplastic dynamic analysis in the presence of softening, (forthcoming).
9. COMI, C., CORIGLIANO, A. and MAIER, G. - Dynamic analysis of elastoplastic-softening discretized structures, (submitted for publication).
10. COMI, C., MAIER, G., and PEREGO, U. – Generalized variables finite element modelling and extremum theorems in stepwise holonomic elastoplasticity with internal variables,, Int. J. Computer Meth. Appl. Mech. Eng., (to appear).
11. CORIGLIANO, A. and PEREGO, U. – Generalized mid-point finite element dynamic analysis of elastoplastic structures, (submitted for publication).
12. COMI, C. and MAIER, G. - Extremum theorem and convergence criterion for an iterative solution to the finite-step problem in elastoplasticity with mixed nonlinear hardening, Eur. J. Mech./A, 9, 563-585 (1990).

FINITE ELEMENTS IN THE 90's, E. Oñate, J. Periaux, A. Samuelsson (Eds.)
Springer-Verlag/CIMNE, Barcelona 1991

SOME RECENT RESEARCH ON THE NON-LINEAR ANALYSIS OF SHELLS

M.A. Crisfield, X. Peng and **J. Shi**
Dept. of Aeronautics, Imperial College, London, UNITED KINGDOM

Summary

The paper describes some recent research on the non–linear analysis of shells using a co–rotational formulation with the constant strain/ constant moment facet–triangular shell element.

INTRODUCTION

The first use of facet elements for shells was due to Clough et al [1] and Zienkiewicz et al [2]. This concept was extended to non–linear analysis by Horrigmoe and Bergan [3] using an "up–dated Lagrangian" technique that was very similar to the "co–rotational approach". The latter was introduced by Wempner [4] and Belytschko et al [5] and has much in common with the "natural approach" of Argyris [6].

In the present paper, we describe the application of the co–rotational approach to the simplest possible facet formulation which combines the constant strain membrane triangle with the constant curvature bending triangle of Morley [7]. This combination of elements has previously been applied to non–linear analysis by Backlund [8] and Chen [9]. The main emphasis of the present work is on the generation of an "increment–independent formulation" with an associated "consistent tangent". Numerical examples include a post–bifurcation analysis with "branch–switching".

THEORY

Fig. 1 shows the element which has three translation variables at each corner and a single "normal rotation" at each mid–side (Fig. 1b). The main problem for the creation of an "increment–independent formulation" is the definition of the global mid–side rotation using a form that can later be differentiated for the virtual work and the tangent stiffness matrix.

As a first stage, an **R** triad is constructed at each mid–side in the initial configuration (Fig. 2a). To this end, we require the initial co–ordinates of a "dummy node" at each mid–side, which allows the computation of the unit mid–side vector $\mathbf{R}_3$ (Fig. 2a). For smooth–shells, the unit vector, $\mathbf{R}_3$, can be simply computed as the average normal from the adjoining elements, using a pre–processor.

In the "current configuration", the unit vector, $\mathbf{r}_2$, which is tangential to the side (Fig. 2c) can easily be computed. In order to move the initial mid–side **R** triad to the final **r** triad, a two–stage process is adopted, with the **R** triad being firstly rotated in the plane formed by $\mathbf{r}_2$ and $\mathbf{R}_2$ through the vector angle:

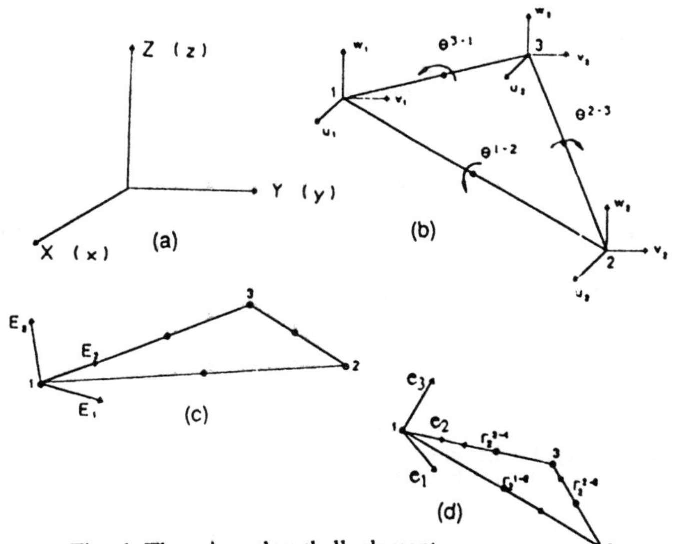

Fig. 1 The triangular shell element

a) Axes and co-ordinates b) "Global displacement" variables
c) Local element frame in initial configuration
d) Local element frame in final config. and side $\mathbf{r}_2$ vectors

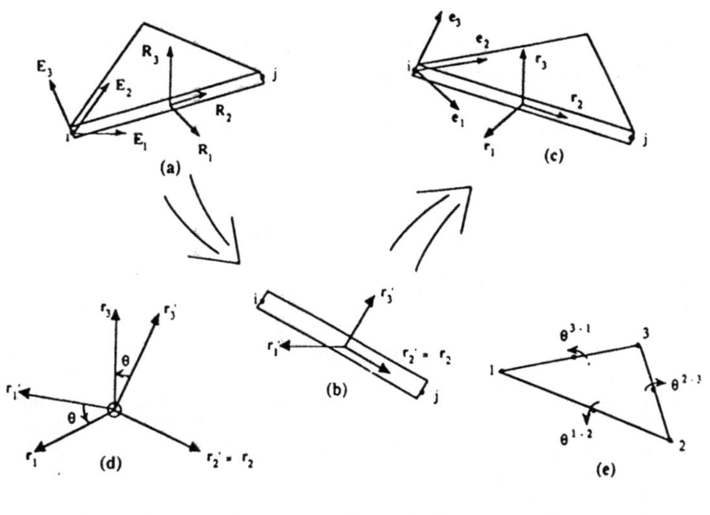

Fig. 2. Mid-side triads in initial, intermediate and final configuration and local slopes

a) Initial b) Intermediate c) Final
d) Global θ's e) Local θ's

$$\gamma = \cos^{-1}(R_2^T r_2) \frac{R_2 x r_2}{\|R_2 x r_2\|} \qquad (1)$$

This process leads to the intermediate triad (Fig. 2b) with unit vectors given by:

$$r_1' = R_1 - \frac{b_1}{1 + b_2}(R_2 + r_2) \qquad (2a)$$

$$r_2' = r_2 \qquad (2b)$$

$$r_3' = R_3 - \frac{b_3}{1 + b_2}(R_2 + r_2) \qquad (2c)$$

where:

$$b_k = R_k^T r_2 \qquad (3)$$

In order to reach the final configuration with the mid-side triad of r_1–r_3 (Fig. 2c), we apply the global rotation, θ, about the $r_2 = r_2'$ axis (Figs. 2c and d). This results in the relationships:

$$r_1 = r_1'\cos\theta + r_3'\sin\theta \qquad (4a)$$

$$r_2 = r_2' \qquad (4b)$$

$$r_3 = -r_1'\sin\theta + r_3'\cos\theta \qquad (4c)$$

Knowing the total (global) finite element variables, p_u and p_θ, the current configuration of the element can be fully defined with the aid of the auxiliary mid-side triads r_1–r_3 and the current local element frame given by e_1–e_3 (Fig. 1d). The latter is given by:

$$e_2 = \frac{x_{31}}{\|x_{31}\|}; \; e_3 = \frac{x_{31} x x_{12}}{\|x_{31} x x_{12}\|}; \; e_1 = e_2 x e_3 \qquad (5)$$

where, for example,

$$x_{31} = X_3 + u_3 - X_1 - u_1 = X_{31} - u_{31} \qquad (6)$$

with X_i being the initial position vector of node i and u_i containing the three translational displacements at node i. Because we also know the initial values R_1–R_3 and E_1–E_3, we can compute the total local "strains".

For the in-plane response, this is achieved by super-imposing the initial element configuration on the current configuration with coincidence at node 1 and with coincidence along side 1–3. As a result, the rigid body rotation is removed and we are left with three "strain-producing" local in-plane displacements. Further details are given in [10] with full details in [11]. In the limited space that is available here, we will concentrate on the local bending "strains".

Because the local element frame passes through the three corner-nodes, with regard to bending, the only "strain-producing displacements" are the three local mid-side rotations (Fig. 2e). At a typical mid-side node, this quantity can be computed (Figs. 2c and 2e) from:

$$\sin\theta_\ell = e_3^T r_1 \qquad (7)$$

where e_3 is a function of the current translational displacements and, via the two stage process in equations (1)–(4), r_1 is a function of both the translational displacements and the mid–side global rotation θ.

With θ_ℓ known from (6), one can simply use the standard expressions in [7] (but with the local w_ℓ's as zero) to obtain the local constant curvatures and hence the local bending moments, $\mathbf{M}_\ell$. The key element in the derivation of a consistent co–rotational formulation is the differentiation of (7) with a view to the virtual work and the internal force vector with further differentiation leading to the tangent stiffness matrix.

From (7), we obtain:

$$\delta\theta_\ell = \frac{1}{\cos\theta_\ell}(e_3{}^T\delta r_1 + \delta e_3{}^T r_1) \qquad (8)$$

The change δe_3 can be directly related to the change in the translational variables, δp_u, via differentiation of (5) while by differentiating (2)–(4), δr_1 can be related to both the changes in translational variables, δp_u, and the changes in the "global mid–side rotations", δp_θ. The latter results [11] in the relationship:

$$\delta r_1 = A\delta u_{ij} + r_3\delta\theta \qquad (9)$$

where:

$$A = \frac{1}{(1+b_2)s}[(R_2 + r_2)r_1{}^T - (R_2{}^T r_1)(I - r_2 r_2{}^T)] \qquad (10)$$

with b_2 from (3) and s as the current length of the side under consideration. The complete process leads to a matrix and vector relationship of the form:

$$\delta\theta_\ell = T_{\theta u}\delta p_u + T_{\theta\theta}\delta p_\theta \qquad (11)$$

(where $T_{\theta\theta}$ turns out to be the identity matrix [11]).

Considering only the bending action, virtual work now leads to the global internal force vectors as:

$$q_u = A_o T_{\theta u}{}^T B_{\theta\ell}{}^T M_\ell \qquad (12a)$$

$$q_\theta = A_o B_{\theta\ell}{}^T M_\ell \qquad (12b)$$

where A_o is the intial area of the element, $\mathbf{M}_\ell$ the three local bending moments and $\mathbf{B}_{\theta\ell}$ the standard matrix relating the the three local curvatures and the three local mid–side rotations.

The tangent stiffness matrix follows from the differentiation of (12) with the "initial stress" contribution arising from the $\delta T_{\theta u}$ term from (12a).

APPLICATIONS

The first example involves the application of an end–moment to a cantilever (Fig. 3). Many finite element formulations fail or give erroneous solutions for this problem once M^* (Fig. 3) exceed 0.5 and they may also require a significant number of increments to reach this stage. For example, using a to total Lagrangian implementation of the semi–loof element [12], the displacements effectively locked beyond $M^* = 0.5$. For the present co–rotational formulation, the exact solution (within the limits of a facet modelling – Fig. 3) was obtained for $M^* = 1.0$ using only two steps.

The second example involves the analysis of the deep arch of Fig. 4 for which the length L was 80 and the radius, R, was 100. An analytical solution has been given by Huddleston [13] and is shown in Fig. 4 in conjunction with the finite element results. In plan, the mesh configuration was of the same form as that used for the cantilever of Fig. 3. However, forty element divisions were used along the length of the arch.

For the symmetric, unstable solution, one negative pivot would be encountered beyond the bifurcation point and two beyond the limit point. In order to obtain this solution with a conventional spherical arc–length method [14], a horizontal restraint was applied at the central node so that the bifurcation was removed.

It is possible to devise path–following techniques that will converge directly onto a singular point with the aid of the derivative of the tangent stiffness matrix [15]. As a by–product, these techniques compute the eignemode associated with the near singular $\mathbf{K}_t$ so that branch–switching can easily follow.

Because the derivative of $\mathbf{K}_t$ is not available within most finite element systems, the authors have introduce a simple "bracketed interpolation" with the aim that a solution should be obtained where the minimum pivot is nearly zero. With the additional aid of the current stiffness parameter [14], the system will either indicate a bifurcation or a limit point. In the former case, the user is advised to re–start with an "eigenmode injection". For the latter, the authors simply use the conventional cylindrical arc–length method [14], with a predictor given by:

$$\Delta p_p = \Delta \ell z \; ; \qquad \Delta\lambda_p = 0 \tag{13}$$

where z is the lowest eigenmode at (or close to) the bifurcation point.

Figure 14 shows that this technique has been successfully used, in conjunction with the co–rotational facet element, to obtain a post–bifurcation path that is in close agreement with the analytical solution of Huddleston [13].

Simo et al [16] have analysed a symmetric octant of a sphere as illustrated in Fig. 5a which also shows the present mesh. The adopted dimensions and properties, which were taken from experiments by Taber [16], are:

$$E = 4000\text{kPa}, \; \nu = 0.5, \; R = 26.3\text{mm}, \; t = 4.4\text{mm}$$

Fig. 4c compares the present load/deflection response with both that obtained by Simo et al [16] and the experimental solution due to Taber [17]. The figure also shows the results obtained using a total Lagrangian implementation of the semi–loof triangle [12] . It can be seen from the figure that the response for the latter soon became too stiff and that no effective progress was achieved beyond $d_c/R = 0.5$. The deformed shape that was computed using the present formulation is shown in Fig. 5b.

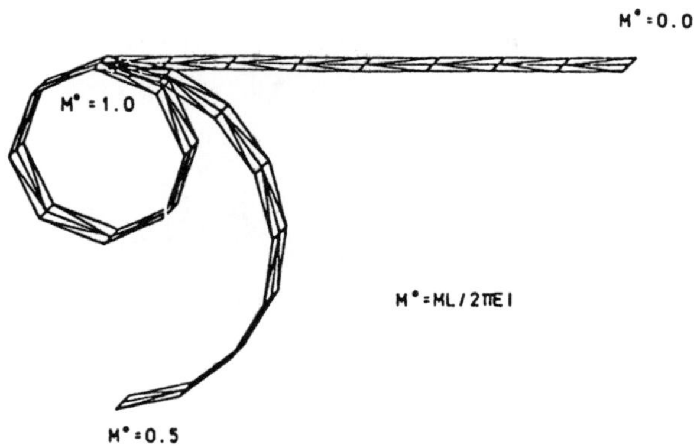

Fig. 3. Response of cantilever subject to tip–moment

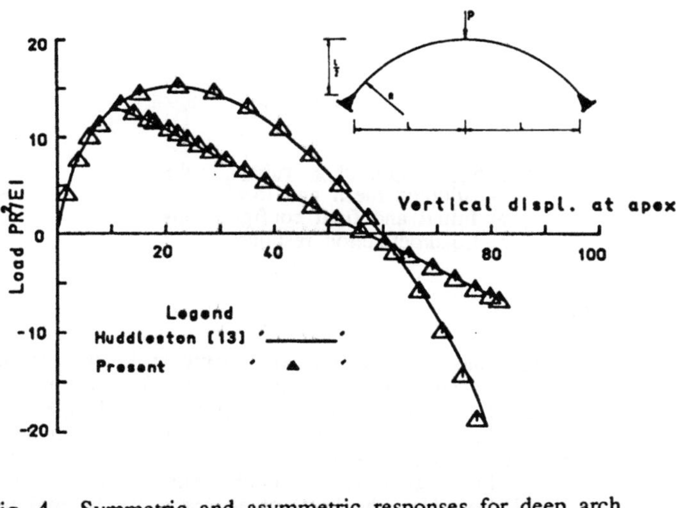

Fig. 4. Symmetric and asymmetric responses for deep arch

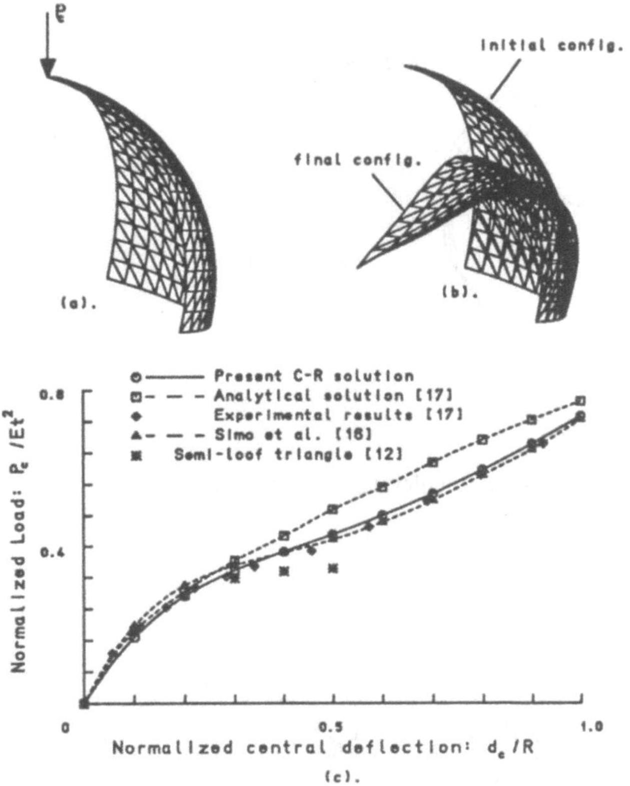

Fig. 5. Collapse of a "rubber" sphere
a) Adopted mesh and loading
b) Initial and final configurations
c) Load/deflection response

CONCLUSIONS

This brief paper has outlined the key elements in the derivation of consistent co-rotational formulation based on a simple facet formulation using a combination of the constant-strain membrane triangle and the constant-curvature bending triangle. The results which are given here and in [10,11] are very encouraging. While the element does not give very accurate results for a linear analysis, it would seem that by introducing an effective co-rotational procedure, the level of accuracy can be maintained for a non-linear analysis. This performance contrasts with that of many more sophisticated elements which may give very accurate linear results but for which the non-linear performance can degrade badly.

It is intended to introduce plasticity and a simple eccentric stiffener and to test the formulation for crash analysis.

REFERENCES

1. Clough, R.W. & Tocher, J.L., Analysis of thin arch dams by the finite element method, Proc. Symp. Theory of Arch Dams, Southampton Univ., 1964.
2. Zienkiewicz, O.C., Parekh, C.J. & King, I.P., Arch dams analysed by a linear finite element shell solution program, Proc. Symp. Arch Dams, Inst. Civ. Engrs., London, 1968.
3. Horrigmoe, G. & Bergan, P.G., Nonlinear analysis of free-form shells by flat finite elements, Comp. Meth. in Appl. Mech. & Engng., 16, pp.11–35, 1978.
4. Wempner, G., Finite elements, finite rotations and small strains of flexible shells, Int. J. Solids & Structs., 5, pp. 117–153, 1969.
5. Belytschko, T. & Hseih, J., Nonlinear transient analysis with convected co-ordinates, Int. J. Num. Meth. in Engng., 7, pp. 255–271, 1973.
6. Argyris, J.H., Balmer, H., Doltsinis, J. St., Dunne, P.C., Haase, M., Klieber, M., Malejannakis, G.A., Mlejenek, J.P., Muller, M. & Scharpf, D.W., Finite element method – the natural approach, Comp. Meth. in Appl. Mech. & Engng., 17/18, 1–106, 1979.
7. Morley, L.S.D., The constant-moment plate-bending element, J. Strain Analysis, 6, 1, pp.20–24, 1971.
8. Backlund, J., Finite element analysis of nonlinear structures, Ph. D. Thesis, Chalmers Univ., Gotenburg, Sweden, 1973.
9. Chen, K.K., A triangular plate finite element for large-displacement elastic-plastic analysis of automobile structural components, Comp. & Struct., 10, pp.203–215, 1979.
10. Crisfield, M.A. & Peng, X., Recent research on co-rotational beam and shell elements, Proc. Int. Conf. on Nonlinear Engineering Computations, Swansea, Sept. 1991.
11. Peng, X. & Crisfield, M.A., A consistent co-rotational formulation for shells using the constant stress/ constant moment triangle, submitted to Int. J. for Num. Meth. in Engng., 1991
12. LUSAS user's manual, FEA Ltd., Kingston-upon-Thames, Surrey KT11HN, England, 1990.
13. Huddleston, V.J., Finite deflection and snap-through of high circular arches, J. Appl. Mech., ASME, 35, 763–769, 1968.
14. Crisfield, M.A., Non-linear finite element analysis of solids and structures, Vol. 1, J. Wiley & Sons., Chichester, 1991.
15. Wriggers, P. & Simo, J.C., A general procedure for direct computation of turning and bifurcation points, Int. J. for Num. Meth. in Engng., 30, pp.155–176, 1990.
16. Simo, J.C., Fox, D.D. & Rafai, M.S., On a stress resultant geometrically exact model, Part 3: Computational aspects of the non-linear theory, Comp. Meth. in Appl. Mech. & Engng., 79, pp.21–70, 1990.
17. Taber, L.A., Large deflection of a fluid filled spherical shell under a point load, J. of Appl. Mech., 49, 121–128, 1982.

FINITE ELEMENTS IN THE 90's, E. Oñate, J. Periaux, A. Samuelsson (Eds.)
Springer-Verlag/CIMNE, Barcelona 1991

CONSTITUTIVE MODELLING OF GEOMATERIALS WITH ACCOUNT FOR DEFORMATIONAL ANISOTROPY

Z. Mróz
Institute of Fundamental Technological Research, POLAND

S. Jemiolo
Warsaw Technical University, Warsaw, POLAND

ABSTRACT

In modelling inelastic response of soils, it is usually assumed that initial overconsolidation generates an anisotropic structure described by the yield surface with its center shifted in the prestrain direction. In multisurface models, the external consolidation surface is usually assumed to be isotropic with all internal surfaces evolving during unloading or reloading events to become eventually tangential to the external surface. In this note the description of structural anisotropy represented by the external surface is coupled with the "Bauschinger anisotropy" represented by the interior surfaces, thus providing a generalization of previous models.

1. INTRODUCTION

The present note follows previous research conducted jointly with Professor O.C.Zienkiewicz and Dr. V.A.Norris in 1979-86 on constitutive modelling of inelastic response of soils subjected to complex loading histories, in particular to cyclic loading [1-6]. It was assumed that besides the isotropic hardening due to porosity changes, the material acquires also an anisotropy represented by the "back stress" represented by the center of the yield surface. In multisurface description, the evolution of the back stress followed from the condition of relative motion of consecutive loading surfaces without intersection. In the two-surface description the back stress was assumed to evolve toward the image or "conjugate" stress on the external surface. However, this surface was assumed to preserve isotropy and evolve only due to variation of scalar state variables. This assumption was abandoned in the recent work by Mróz and Niemunis [7] where the deformation anisotropy was separated into two phenomena: i) the fabric or texture anisotropy associated with preferred structure orientation due to initial consolidation (clay platelet alignment, grain contact arrangement, directional porosity, etc.) and ii) the "Bauschinger or back stress" anisotropy associated with the oriented structure of microstresses varying rapidly with contact slip and grain rearrangement. The Bauschinger anisotropy is assumed to be represented by the evolving centers of the yield and internal surfaces. On the other hand, the fabric anisotropy is represented by the varying shape of external consolidation surface and the associated shape variation of internal surfaces.

In this paper, we shall discuss further this problem and provide more general description of fabric anisotropy by assuming this anisotropy to preserve orthotropic symmetry coaxial with the principal axes of the plastic strain tensor. The general representation theorem is used and some specific forms are considered in detail.

2. DESCRIPTION OF TEXTURE OR FABRIC ANISOTROPY

Referring to Fig. 1 which presents a typical shear or compression stress-strain curve for soil during first loading and unloading, we may specify the yield point by the departure from linearity or by a small offset value (point 1"). On the other hand, by extrapolating backward the portion of the stress-strain curve obtained for advanced plastic deformation, and interesting with the elastic line or stress-axis, we obtain the yield stress represented by 1' or 1. Similar definitions can be used for the unloading portion thus specifying points 3", 3' and 3. The yield surface generated by the extrapolation method is the external surface to surfaces specified by increasing offset values. The model with a set of nested surfaces provides a proper visualization of varying material stiffness for loading in different directions with respect to initial prestrain direction.

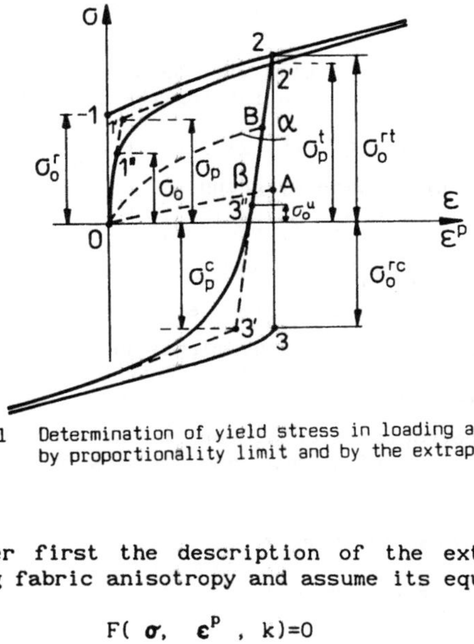

Fig.1 Determination of yield stress in loading and unloading by proportionality limit and by the extrapolation method.

Consider first the description of the external yield surface representing fabric anisotropy and assume its equation in the form

$$F(\boldsymbol{\sigma}, \ \boldsymbol{\varepsilon}^P, \ k)=0 \tag{1}$$

such that under rigid body rotation specified by the orthogonal matrix $\mathbf{Q}$, there is

$$F(\boldsymbol{\sigma}, \boldsymbol{\varepsilon}^P, k) = F(\mathbf{Q}\, \boldsymbol{\sigma}\, \mathbf{Q}^T, \ \mathbf{Q}\, \boldsymbol{\varepsilon}^P \mathbf{Q}^T, \ k) = 0 \tag{2}$$

and $\mathbf{Q}\,\mathbf{Q}^T = \mathbf{Q}^T\mathbf{Q} = \mathbf{1}$. Here $\boldsymbol{\sigma}$ denotes the Cauchy stress, $\boldsymbol{\varepsilon}^P$ is the plastic strain and k denotes a scalar state variable. The condition (2) requires F to be an isotropic function of $\boldsymbol{\sigma}$ and $\boldsymbol{\varepsilon}^P$, expressed in terms of basic invariants of two tensors, thus

$$J_1 = \mathrm{tr}\,\boldsymbol{\sigma} = \sigma_{ii}, \quad J_2 = \mathrm{tr}\,\boldsymbol{\sigma}^2 = \sigma_{ij}\sigma_{ij}, \quad J_3 = \mathrm{tr}\,\boldsymbol{\sigma}^3 = \sigma_{ik}\sigma_{kj}\sigma_{ji}$$
$$J_4 = \mathrm{tr}\,\boldsymbol{\varepsilon}^p = \varepsilon^p_{ii}, \quad J_5 = \mathrm{tr}(\boldsymbol{\varepsilon}^p)^2 = \varepsilon^p_{ij}\varepsilon^p_{ij}, \quad J_6 = \mathrm{tr}(\boldsymbol{\varepsilon}^p)^3 = \varepsilon^p_{ik}\varepsilon^p_{kj}\varepsilon^p_{ji},$$
$$J_7 = \mathrm{tr}(\boldsymbol{\sigma\varepsilon}^p) = \sigma_{ij}\varepsilon^p_{ij}, \quad J_8 = \mathrm{tr}(\boldsymbol{\sigma}^2\boldsymbol{\varepsilon}^p) = \sigma_{ik}\sigma_{kj}\varepsilon^p_{ij}, \tag{3}$$
$$J_9 = \mathrm{tr}(\boldsymbol{\sigma\varepsilon}^{p2}) = \sigma_{ij}\varepsilon^p_{jk}\varepsilon^p_{ki}, \quad J_{10} = \mathrm{tr}(\boldsymbol{\sigma}^2\boldsymbol{\varepsilon}^{p2}) = \sigma_{ik}\sigma_{kj}\varepsilon^p_{js}\varepsilon^p_{si},$$

and

$$F = F(J_1, J_2, J_3, \dots J_{10}, k) = 0 \tag{4}$$

Particular forms of the representation (4) were discussed by Shrivastava, Mróz and Dubey [9], for the case of plastically incompressible materials. However, such representation does not assure in general an orthotropic structure of the prestrained material. In fact, assuming existence of three mutually perpendicular orthotropy axes specified by an orthonormal triad $\mathbf{t}_i$, such that $\mathbf{t}_i \cdot \mathbf{t}_j = \delta_{ij}$, $i,j=1,2,3$, there must be a symmetry of material response with respect to three orthogonal orthotropy planes. Considering the transformed tensor ε^p with positive eigenvalues, so that

$$\tilde{\boldsymbol{\varepsilon}}^p = |\varepsilon^p_1|\mathbf{e}_1 \otimes \mathbf{e}_1 + |\varepsilon^p_2|\mathbf{e}_2 \otimes \mathbf{e}_2 + |\varepsilon^p_3|\mathbf{e}_3 \otimes \mathbf{e}_3 \tag{5}$$

then the function (1) $F=F(\boldsymbol{\sigma}, \tilde{\boldsymbol{\varepsilon}}^p, k)=0$ will be an orthotropic function with respect to stress tensor $\boldsymbol{\sigma}$, satisfying the condition

$$F(\boldsymbol{\sigma}, \boldsymbol{\varepsilon}^p, k) = F(Q\sigma Q^T, \tilde{\boldsymbol{\varepsilon}}^p, k), \qquad \forall Q \in S_o \tag{6}$$

where S_o is the orthotropic symmetry group

$$S_o = \{\mathbf{I}, -\mathbf{I}, \mathbf{S}_1, \mathbf{S}_2, \mathbf{S}_3\} \tag{7}$$

where $\mathbf{S}_k = [S^{(k)}_{ij}]$, $S^{(k)}_{ij} = \delta_{ij} - 2e^{(k)}_i e^{(k)}_j$, $k=1,2,3$, and $e^{(k)}_i$ is the k-th component of unit orthonormal vector $\mathbf{e}_i$. Here $\mathbf{S}_k$ specify the reflections with respect to orthotropy planes and $\mathbf{I}$ is the identity operator cf. Vakulenco and Markov [12]. Introduce the parametric tensors $\mathbf{M}_{\alpha\alpha} = \mathbf{e}_\alpha \otimes \mathbf{e}_\alpha$ (no summation with respect to α). The yield condition for an orthotropic material can now be expressed in terms of seven joint invariants of stress and tensors $\mathbf{M}_{\alpha\alpha}$, namely, cf. Boehler [11]

$$\begin{aligned} &F=F[\mathrm{tr}(\mathbf{M}_{11}\boldsymbol{\sigma}), \mathrm{tr}(\mathbf{M}_{22}\boldsymbol{\sigma}), \mathrm{tr}(\mathbf{M}_{33}\boldsymbol{\sigma}), \\ &\mathrm{tr}(\mathbf{M}_{11}\boldsymbol{\sigma}^2), \mathrm{tr}(\mathbf{M}_{22}\boldsymbol{\sigma}^2), \mathrm{tr}(\mathbf{M}_{33}\boldsymbol{\sigma}^2), \mathrm{tr}\boldsymbol{\sigma}^3, k] = 0 \end{aligned} \tag{8}$$

where

$$\begin{aligned} &\mathrm{tr}(\mathbf{M}_{11}\boldsymbol{\sigma}) = \sigma_{11}, \quad \mathrm{tr}(\mathbf{M}_{22}\boldsymbol{\sigma}) = \sigma_{22}, \quad \mathrm{tr}(\mathbf{M}_{33}\boldsymbol{\sigma}) = \sigma_{33}, \\ &\mathrm{tr}(\mathbf{M}_{11}\boldsymbol{\sigma}^2) = \sigma^2_{11} + \sigma^2_{12} + \sigma^2_{13}, \quad \mathrm{tr}(\mathbf{M}_{22}\boldsymbol{\sigma}^2) = \sigma^2_{12} + \sigma^2_{22} + \sigma^2_{23}, \\ &\mathrm{tr}(\mathbf{M}_{33}\boldsymbol{\sigma}^2) = \sigma^2_{13} + \sigma^2_{23} + \sigma^2_{33}, \\ &\mathrm{tr}\,\boldsymbol{\sigma}^3 = \sigma^3_{11} + \sigma^3_{22} + \sigma^3_{33} + 3[\sigma_{11}(\sigma^2_{12} + \sigma^2_{13}) + \sigma_{22}(\sigma^2_{12} + \sigma^2_{23}) + \\ &\qquad + \sigma_{33}(\sigma^2_{13} + \sigma^2_{23}) + 6\sigma_{12}\sigma_{13}\sigma_{23} \end{aligned} \tag{9}$$

From (8) and (9) it follows that the yield condition can be expressed in terms of the following new set of invariants

$$F = F\,(\sigma_{11},\sigma_{22},\sigma_{33},\ \sigma_{12}^2,\sigma_{13}^2,\sigma_{23}^2,\ \sigma_{12}\sigma_{13}\sigma_{23},\ k) = 0 \tag{10}$$

or

$$F = F\,(\sigma_{11},\sigma_{22},\sigma_{33},\ |\sigma_{12}|,|\sigma_{13}|,|\sigma_{23}|,\ \sigma_{12}\sigma_{13}\sigma_{23},\ k) = 0 \tag{11}$$

that is in terms of normal stress components σ_{11}, σ_{22}, σ_{33} with respect to orthotropy axes and absolute values or squares of shear stress components σ_{12}, σ_{23}, σ_{31}. For instance, for the plane stress state with orthotropy axes lying within the plane, the representation (10) becomes

$$F(\sigma_{11},\sigma_{22},\sigma_{12}^2,\ k) = 0, \qquad \text{or} \qquad F(\sigma_{11},\sigma_{22},|\sigma_{12}|,\ k) = 0 \tag{12}$$

In particular, any quadratic yield condition, proposed by Hill [14]

$$a_1\sigma_{11}^2 + 2\,a_2\sigma_{11}\sigma_{22} + a_3\sigma_{22}^2 + a_4\sigma_{12}^2 - 1 = 0 \tag{13}$$

follows from (12). The material parameters can be identified by measuring yield limits in tension and shear along orthotropy axes and in tension along the line inclined at 45° to the orthotropy axes, thus

$$a_1 = \frac{1}{\sigma_0^2}, \qquad a_3 = -\frac{1}{\sigma_{90}^2}, \qquad a_4 = \frac{1}{\tau_0^2},$$
$$a_2 = \left(\frac{2}{\sigma_{45}^2} - \frac{1}{2\sigma_0^2} - \frac{1}{2\sigma_{90}^2} - \frac{1}{2\tau_0^2}\right) \tag{14}$$

where σ_0, σ_{45}, σ_{90} and τ_0 are the yield limits along orthotropy axes and along the bisector line. As there is no symmetry with respect to σ_{11}, σ_{22}, the principal axes of (13) within the plane σ_{11}, σ_{22} are specified by

$$\cot 2\alpha = \frac{a_1 - a_3}{2a_2}, \qquad a_2 = 0 \tag{15}$$

and in the case of elliptical yield condition Eq.(13) can be rewritten in the coordinate system coinciding with principal axes

$$F = \left(\frac{\sigma'_{11}}{a}\right)^2 + \left(\frac{\sigma'_{22}}{b}\right)^2 + \left(\frac{\sigma'_{12}}{c}\right)^2 - 1 = 0 \tag{16}$$

where σ'_{11}, σ'_{22} and σ'_{12} are stress components (16) referred to principal axes 1, 2 of the yield ellipsoid, and a, b, c, are the principal orthotropy coefficients.

Our discussion was so for concerned with the representation of an orthotropic yield condition. However, more important is formulation of proper evolution rules. Several important cases of increasing complexity can be distinguished in formulating such rules, namely

i) proportional loading with orthotropy axes fixed with respect to the material

ii) non-proportional loading with orthotropy axes fixed

iii) non-proportional loading with rotating orthotropy axes

Consider first the case i) and assume that the Lagrangian triad is fixed with respect to the material, Fig.1. Apply the polar decomposition of the deformation gradient

$$\mathbf{F} = \mathbf{RU} = \mathbf{VR} \tag{17}$$

where $\mathbf{R}$ is a proper orthogonal tensor, $\mathbf{RR}^T=1$, $\det(\mathbf{R})=1$, and $\mathbf{U}$, $\mathbf{V}$ are symmetric deformation tensors. Both $\mathbf{U}$ and $\mathbf{V}$ can be represented in terms of principal stretches $\Lambda_\alpha=\lambda_\alpha$, so that

$$\mathbf{U} = \sum_{\alpha=1}^{3} \Lambda_\alpha \mathbf{N}_\alpha \otimes \mathbf{N}_\alpha, \qquad \mathbf{V} = \sum_{\alpha=1}^{3} \lambda_\alpha \mathbf{n}_\alpha \otimes \mathbf{n}_\alpha, \qquad \mathbf{n}_\alpha = \mathbf{RN}_\alpha, \tag{18}$$

where $\mathbf{N}_\alpha$ and $\mathbf{n}_\alpha$ (α=1,2,3) are unit vectors of Lagrangian and Eulerian triads. The principal plastic strains can be defined in terms of principal plastic stretches, for instance

$$\varepsilon_1^p = \ln \lambda_1^p, \qquad \varepsilon_2^p = \ln \lambda_2^p, \qquad \varepsilon_3^p = \ln \lambda_3^p,$$
$$\dot{\varepsilon}_1^p = \frac{\dot{\lambda}_1^p}{\lambda_1^p}, \qquad \dot{\varepsilon}_1^p = \frac{\dot{\lambda}_2^p}{\lambda_2^p}, \qquad \dot{\varepsilon}_1^p = \frac{\dot{\lambda}_3^p}{\lambda_3^p} \tag{19}$$

or as the Green strain measure

$$\varepsilon_1^p = \tfrac{1}{2}[(\lambda_1^p)^2 - 1], \qquad \varepsilon_2^p = \tfrac{1}{2}[(\lambda_2^p)^2 - 1], \qquad \varepsilon_3^p = \tfrac{1}{2}[(\lambda_3^p)^2 - 1], \tag{20}$$

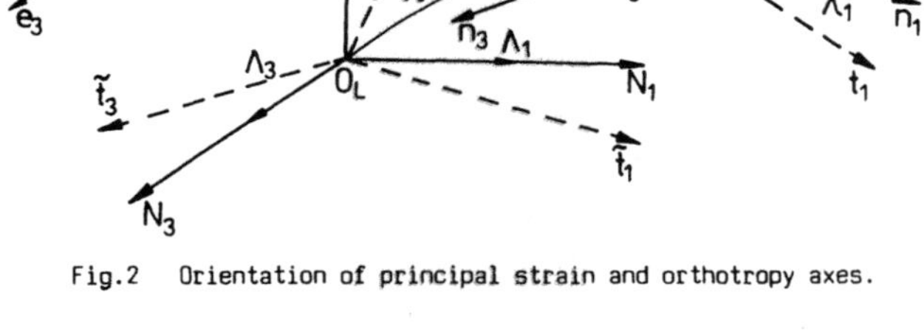

Fig.2 Orientation of principal strain and orthotropy axes.

Proportional straining is now represented by a straight line in the plastic strain space with the principal strain components increasing proportionally. It is natural to assume that the orthotropy axes now coincide with the principal plastic strain directions which are fixed with respect to material. Neglecting the effect of elastic strain and referring to Fig.2, it can be assumed that the orthotropy triad $\tilde{t}_1$, $\tilde{t}_2$, $\tilde{t}_3$ in the unloaded configuration or triad t_1, t_2, t_3 in the deformed configuration coincide with the

Lagrangian or Eulerian triads, thus $\tilde{t}_\alpha = N_\alpha$, $t_\alpha = n_\alpha$ and $t_\alpha = R\,\tilde{t}_\alpha$. Apply the Voigt notation for stress and strain regarding them as vectors in the 6-dimensional space, so that

$$\begin{aligned} &\sigma_1=\sigma_{11},\quad \sigma_2=\sigma_{22},\quad \sigma_3=\sigma_{33},\quad \sigma_4=\tau_{12},\quad \sigma_5=\tau_{23},\quad \sigma_6=\tau_{13},\\ &\varepsilon_1=\varepsilon_{11},\quad \varepsilon_2=\varepsilon_{22},\quad \varepsilon_3=\varepsilon_{33},\quad \varepsilon_4=2\varepsilon_{12},\quad \varepsilon_5=2\varepsilon_{23},\quad \varepsilon_6=2\varepsilon_{13} \end{aligned} \tag{21}$$

The texture (or consolidation) yield condition can now be expressed as a quadratic function of "effective" stress $\sigma_i - \beta_i$

$$F_c = [A_{ij}(\sigma_i - \beta_i)(\sigma_j - \beta_j)]^{\frac{1}{2}} - \sigma_p = \sigma_e - \sigma_p = 0 \tag{22}$$

or

$$\begin{aligned} F_c = {} & [A_{11}(\sigma_1 - \beta_1)^2 + A_{22}(\sigma_2 - \beta_2)^2 + A_{33}(\sigma_3 - \beta_3)^2 + A_{44}(\sigma_4 - \beta_4)^2 \\ & + A_{55}(\sigma_5 - \beta_5)^2 + A_{66}(\sigma_6 - \beta_6)^2 + 2A_{12}(\sigma_1 - \beta_1)(\sigma_2 - \beta_2) + \\ & + 2A_{23}(\sigma_2 - \beta_2)(\sigma_3 - \beta_3) + 2A_{13}(\sigma_1 - \beta_1)(\sigma_3 - \beta_3)]^{\frac{1}{2}} - \sigma_p = 0 \end{aligned} \tag{23}$$

where the orthotropy matrix

$$\mathbf{A} = \begin{bmatrix} A_{11} & A_{12} & A_{13} & 0 & 0 & 0 \\ A_{21} & A_{22} & A_{23} & 0 & 0 & 0 \\ A_{31} & A_{32} & A_{33} & 0 & 0 & 0 \\ 0 & 0 & 0 & A_{44} & 0 & 0 \\ 0 & 0 & 0 & 0 & A_{55} & 0 \\ 0 & 0 & 0 & 0 & 0 & A_{66} \end{bmatrix} \tag{24}$$

depends on the principal plastic strains. The yield surface specified by (22) is symmetric with respect to its center at β_1, β_2, β_3, whose position evolves during plastic straining.

If the texture surface is the active loading surface reached by the stress point, the associated flow rule provides

$$\dot{\varepsilon}^p_i = \frac{1}{H}\frac{\partial F_c}{\partial \sigma_i}\dot{F}_c = \frac{1}{H}\frac{A_{ij}(\sigma_j - \beta_j)}{\sigma_e}\left[\frac{A_{kl}(\sigma_l - \beta_l)\,\dot{\sigma}_k}{\sigma_e}\right] \tag{25}$$

where H is the hardening modulus. For a plastically incompressible material the condition $\dot{\varepsilon}^p_1 + \dot{\varepsilon}^p_2 + \dot{\varepsilon}^p_3 = 0$ imposes constraints on A_{ij}, namely

$$A_{11}+A_{12}+A_{13}=0,\qquad A_{12}+A_{22}+A_{23}=0,\qquad A_{13}+A_{23}+A_{33}=0 \tag{26}$$

from which the off-diagonal terms A_{12}, A_{13}, A_{23} can be determined in terms of the diagonal coefficients and there are 6 independent orthotropy coefficients. However for a compressible material all 9 coefficients should be specified and their evolution rules formulated. The coefficients A_{11}, A_{22}, A_{33} represent the yield limits under uniaxial tension or compression along the orthotropy directions and will be termed the extension or compression orthotropy coefficients. On the other hand, A_{44}, A_{55}, A_{66} are specified by shear limits along orthotropy directions and will be called the shear orthotropy coefficients. Finally, the components A_{12}, A_{23} and A_{13} specify the orientation of principal axes of the ellipsoid (23) in the principal stress space and could be identified by tension or compression along diagonal directions with respect to orthotropy axes. We shall call them the diagonal orthotropy coefficients.

The evolution rules may be considerably simplified by the following assumption: *proportional prestrain induces alignment of principal directions of the texture ellipsoid (23) with the plastic strain vector*. Since the plastic strain and strain rate vectors are coaxial, this condition, in view of (25), can be expressed as follows

$$A_{ij}(\sigma_j-\beta_j)=\lambda(\sigma_i-\beta_i)=\mu\varepsilon^p_i, \qquad i=1,2,3 \tag{27}$$

where λ and μ are the proportionality factors. Note that (27) provides eigenvectors and eigenvalues of the orthotropy matrix A_{ij}. In general there are six eigenvalues and eigenvectors specifying the ellipsoid orientation and principal orthotropy coefficients. As the principal directions are specified by plastic prestrain, Eqs. (27) provide additional relations between orthotropy components and the canonical form of (23) written in the principal components $\bar{\sigma}_i-\bar{\beta}_i$ takes the form

$$F_c=\left[\,\bar{A}_{11}(\bar{\sigma}_1-\bar{\beta}_1)^2+\bar{A}_{22}(\bar{\sigma}_2-\bar{\beta}_2)^2+\bar{A}_{33}(\bar{\sigma}_3-\bar{\beta}_3)^2+\bar{A}_{44}(\bar{\sigma}_4-\bar{\beta}_4)^2+\bar{A}_{55}(\bar{\sigma}_5-\bar{\beta}_5)^2+\bar{A}_{66}(\bar{\sigma}_6-\bar{\beta}_6)^2\right]^{\frac{1}{2}}-\sigma_p=0 \tag{28}$$

where $\bar{A}_{11}$, $\bar{A}_{22}$, $\bar{A}_{23}$ and $\bar{A}_{44}=A_{44}$, $\bar{A}_{55}=A_{55}$, $\bar{A}_{66}=A_{66}$ are the principal orthotropy coefficients. It is now natural to assume that

$$\bar{A}_{11}=f_1(\varepsilon^p_v,\ \varepsilon^p_q)\ ,\quad \bar{A}_{22}=\bar{A}_{23}=f_2(\varepsilon^p_v,\ \varepsilon^p_q)\ ,\quad A_{44}=A_{55}=A_{66}=f_3(\varepsilon^p_v,\ \varepsilon^p_q) \tag{29}$$

where ε^p_v is the volumetric plastic strain and ε^p_q is the integrated deviatoric plastic strain. Here $f_1(\varepsilon^p_v,\ \varepsilon^p_q)$ corresponds to main hardening along prestrain directions and $f_2(\varepsilon^p_v,\ \varepsilon^p_q)$, $f_3(\varepsilon^p_v,\ \varepsilon^p_q)$ represent the cross-hardening in tension or compression and shear along principal directions. Equations (29) imply transverse isotropy with respect to proportional prestrain. A more general assumption with respect to that expressed by (27), can be formulated as follows: deviatoric prestrain induces alignment of texture ellipsoid axes with the transformed plastic strain vector $\tilde{\varepsilon}^p_i=e^p_i+\eta\frac{1}{3}\varepsilon^p_v\delta_{ij}$, where e^p_i is the strain deviator, ε^p_v is the volumetric plastic strain and η is the transformation factor.

As an example, consider the consolidation of clays in the uniaxial strain condition (K_0-consolidation). The inherent anisotropy developed in naturally deposited soils is usually manifested by directional variation of the undrained shear strength, the largest strength being exhibited by vertical specimens, coaxial with the consolidation stress. Since the consolidation stress is axisymmetric, $\sigma_2=\sigma_3$, we can use the usual notation

$$p= -\frac{1}{3}J'_1 = -\frac{1}{3}(\sigma_1+2\varepsilon_2) \quad , \qquad q=(3J_2)^{\frac{1}{2}}=(\frac{3}{2}S_{ij}S_{ij})^{\frac{1}{2}}=\sigma_2-\sigma_1 \tag{30}$$

where J'_1 and J_2 are the first invariant of effective stress and the second invariant of the stress deviator S_{ij}. The corresponding strains are

$$\varepsilon_v = -\varepsilon_{ii} = -(\varepsilon_1+2\varepsilon_2) \; , \; \varepsilon_q=(\frac{2}{3}e_{ij}e_{ij})^{\frac{1}{2}} = \frac{2}{3}(\varepsilon_1-\varepsilon_2) \tag{31}$$

where e_{ij} is the strain deviator. Referring to Fig.3a, let us now determine the yield surface during uniaxial consolidation. In the uniaxial consolidation there is $\varepsilon_q/\varepsilon_v=\frac{2}{3}$, so according to our generalized assumption the yield ellipse should have the major axis AA' making a specified angle $\delta-\beta$ with plastic strain vector, where $\tan\beta=\frac{2}{3}$. Assuming the tensile strength as zero, the ellipse should be tangential at 0 to the p-axis. At the symmetrically situated point B there is $\dot{\varepsilon}^p_q=0$. The stress state S^o_{ij}, p^o at B specifies the center position 0_c at $\frac{1}{2}S^o_{ij}$, $\frac{1}{2}p^o$. Thus both q^o and p^o could be regarded as state variables specifying size and rotation of the consolidation surface. Neglecting elastic strain, we can assume that during the K_o-consolidation the stress state evolves along the line OA, the yield surface center moves along OB and the yield surface axis AA' is inclined at constant angle δ to the p-axis. Following the previous treatment by Mróz and Niemunis [13], the consolidation surface can be assumed in the form

$$F_c = p^2 + \beta_{ij}S_{ij}p + \beta S_{ij}S_{ij} + \gamma p = 0 \tag{32}$$

where

$$\beta_{ij} = -\frac{m^o_{ij}}{M^2}, \quad \beta = \frac{1}{2M^2}, \quad \gamma = p^o[\frac{m^o_{ij}m^o_{ij}}{2M^2} - 1], \quad m^o_{ij} = \frac{s^o_{ij}}{p^o} \tag{33}$$

and M is the critical state parameter, so that $f_c=q-Mp=0$ along the critical state line. Let us now specify the evolution equations for p^o and S^o_{ij}. Assume that

$$\dot{m}^o_{ij} = \frac{m^o_m(\dot{e}^p_{ij}\varepsilon^p - e^p_{ij}\dot{\varepsilon}^p)}{(\varepsilon^p)^2} \qquad \text{or} \qquad m^o_{ij} = m^o_m\frac{e^p_{ij}}{\varepsilon^p} \tag{34}$$

where

$$\varepsilon^p=\{\frac{4}{3}(e^p_q)^2+(\varepsilon^p_v)^2\}^{\frac{1}{2}}, \qquad e^p_q=(\frac{2}{3}e^p_{kl}e^p_{kl})^{\frac{1}{2}} \tag{35}$$

The evolution of p^o is associated with density and shear hardening. We have

$$\dot{p}^o = \dot{p}^o_1 + \dot{p}^o_2, \qquad \dot{p}^o_1 = \dot{\varepsilon}^p_v\frac{(1+e)}{\lambda-k}p^o, \qquad \dot{p}^o_2 = \dot{m}^o_{ij}\frac{p^o m^o_{ij}-S_{ij}}{\frac{1}{2}m_{ij}m_{ij}-M^2} \tag{36}$$

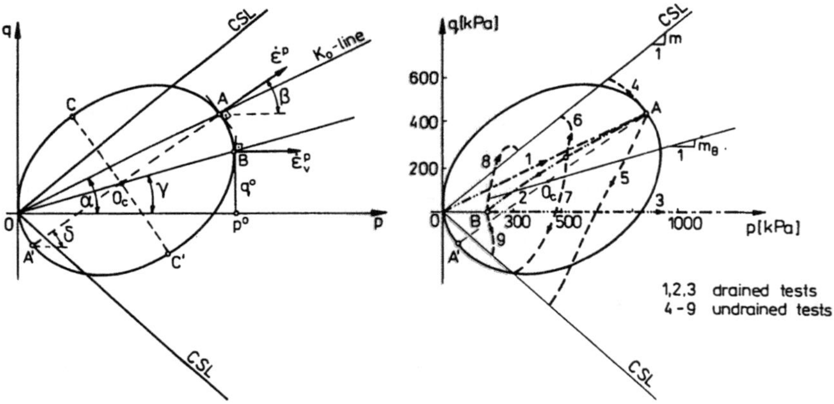

Fig.3 Uniaxial consolidation
a) orientation of K_0-line,(α) diameter line AA'(δ), strain rate line $\dot{\varepsilon}^p$, $^0(\beta)$, and pure consolidation line OB, (γ).
b) strain paths for drained and undrained tests after K_0-consolidation

The first term is usual expression relating rate of pressure with volumetric strain rate (λ,k and m being soil constants, and e the void ratio). The second term of (36) is associated with the evolution of soil anisotropy. Fig.3b presents the predicted stress paths on the p-q plane for various loading programs in triaxial testing after initial consolidation OA. For program 2 the specimen is unloaded to vanishing shear stress q and for program 3 it is subsequently isotropically consolidated. It is seen that for subsequent isotropic consolidation the surface $F_c=0$ rotates backwards towards the pressure axis and the material restores is structure corresponding to isotropic consolidation. More details on this anisotropy description can be found in [13].

The model discussed could be applied both for proportional and non-proportional loading with orthotropy axes fixed. Generally, for any deformation path, the plastic strain rate is known at each instant within the plastic regime. The principal orthotropy axes will then undergo rotation toward new orientation corresponding to new plastic strain vector. Associate with the plastic strain rate a virtual strain path with which the limiting values of orthotropy coefficients A^1_{ij} and β^1_{ij} are associated. The evolution rules can then be formulated as follows

$$\dot{A}_{ij} = k_1(A^1_{ij} - A_{ij}) , \qquad \dot{\beta}_i = k_2(\beta^1_i - \beta_i) \tag{37}$$

where k_1 and k_2 are constant parameters.

Finally, let us briefly discuss to case of rotating orthotropy axes t_1, t_2, t_3. For simplicity, assume that these axes coincide with the Eulerian triad in a deformed configuration and with the Lagrangian triad in a material configuration, thus

$$\mathbf{t}_\alpha = \mathbf{n}_\alpha, \qquad \tilde{\mathbf{t}}_\alpha = \mathbf{N}_\alpha \tag{38}$$

In other words, the orthotropy axes are assumed to follow the principal axes of deformation tensors **U** or **V**. Referring to material reference system $\mathbf{e}_i$, we can write, cf. Fig.2

$$\tilde{\mathbf{t}}_i = \mathbf{N}_i = \mathbf{R}_L\mathbf{e}_i \ , \qquad \mathbf{n}_\alpha = \mathbf{R}\,\mathbf{N}_\alpha \ , \qquad \mathbf{t}_\alpha = \mathbf{R}\,\tilde{\mathbf{t}}_\alpha \tag{39}$$

and differentiating (39) we obtain

$$\dot{\mathbf{t}}_\alpha = \dot{\mathbf{n}}_\alpha = \boldsymbol{\Omega}\mathbf{n}_\alpha + \mathbf{R}\,\boldsymbol{\Omega}_L\,\mathbf{R}^T\mathbf{n}_\alpha = \boldsymbol{\Omega}\,\mathbf{t}_\alpha + \mathbf{R}\,\boldsymbol{\Omega}_L\,\mathbf{R}^T\,\mathbf{t}_\alpha \tag{40}$$

where

$$\boldsymbol{\Omega}_L = \dot{\mathbf{R}}_L\mathbf{R}_L^T \ , \qquad \boldsymbol{\Omega} = \dot{\mathbf{R}}\,\mathbf{R}^T \tag{41}$$

Here $\boldsymbol{\Omega}_L$ denotes the rate of rotation of Lagrangian triad with respect to material and $\boldsymbol{\Omega}$ is the relative spin of Eulerian and Lagrangian triads. Writing the yield condition in the form

$$F_c\,(\,\boldsymbol{\sigma},\ \mathbf{A},\ \mathbf{t}_\alpha,\ \sigma_p\,) = 0 \tag{42}$$

it is seen that the rotation of orthotropy axes affects the plastic behaviour. As the triad $\mathbf{t}_\alpha$ was used as the reference system for orthotropic yield condition, introduce the corotational rates of stress $\overset{\nabla}{\boldsymbol{\sigma}}$ and of orthotropy tensor $\overset{\nabla}{\mathbf{A}}$ with respect to $\mathbf{t}_\alpha$. The consistency condition now is

$$\frac{\partial F_c}{\partial \boldsymbol{\sigma}}\cdot\overset{\nabla}{\boldsymbol{\sigma}} + \frac{\partial F_c}{\partial \mathbf{A}}\cdot\overset{\nabla}{\mathbf{A}} + \frac{\partial F_c}{\partial \sigma_p}\dot{\sigma}_p = 0 \tag{43}$$

Note that $\overset{\nabla}{\mathbf{t}}_\alpha=0$ and $\overset{\nabla}{\mathbf{A}}=\dot{\mathbf{A}}$ since the orthotropy coefficients were specified in terms of the principal strain components. A more detailed study of rate equations (43) and of rotational hardening was presented in [15].

Once the textural anisotropy is described, the Bauschinger anisotropy can easily be modelled by assuming a set of nested surfaces with the same shape and orientation, translating inside the texture surface, cf [1-6]. This part of modelling is not discussed in detail in this paper.

REFERENCES

1. MRÓZ, Z., NORRIS, V.A., and ZIENKIEWICZ, O.C., An anisotropic hardening model for soils and its application to cyclic loading, Int. Num. Anal. Geom. 2, 203-221, (1978).

2. MRÓZ, Z., NORRIS, V.A., and ZIENKIEWICZ, O.C., Application of an anisotropic hardening model in the analysis of elastoplastic deformation of soils, Géotechnique, 29, 1-34, (1979).

3. MRÓZ, Z., NORRIS, V.A., and ZIENKIEWICZ, O.C., An anisotropic critical state model for soils subject to cyclic loading, Géotechnique, 31, 451-469, (1981).

4. MRÓZ, Z., NORRIS, V.A., Elastoplastic and viscoplastic models for soils with application to cyclic loads, in "Soil Mechanics-Transient and Cyclic Loads", Ed. G.N. Pande and O.C.Zienkiewicz J.Wiley and Sons, Chapter 8, 173-217, (1982).

5. MRÓZ, Z., and ZIENKIEWICZ, O.C., Uniform formulation of constitutive equations for clays and sands, in "Mechanics of Eng. Mat.", Ed. C.S.Desai and R.H.Gallagher, J.Wiley and Sons, pp.415-449, (1984).

6. ZIENKIEWICZ, O.C., and MRÓZ, Z., Generalized plasticity formulation and application to geomechanics, in "Mechanics of Eng. Mat.", Ed. C.S.Desai and R.H.Galagher, J.Wiley and Sons, pp. 655-679, (1984).

7. MRÓZ, Z. and PIETRUSZCZAK, S., A constitutive model for sand with anisotropic hardening rule, Int. J. Num. Anal. Meth. Geomechanics, 7, 19-38, (1983).

8. JARZĘBOWSKI, A. and MRÓZ, Z., A constitutive model for sands and its application to monotonic and cyclic loading, in "Const. Equations for Granular Non-Cohesive Soils", Ed. Saada, Bianchini, Balkema Publ. pp.307-323, (1988).

9. SHRIVASTAVA, H.P., MRÓZ, Z. and DUBEY, R.N., Yield criterion and the hardening rule for a plastic solid, ZAMM, 53, 625-633, (1973).

10. WANG, C.C., A new representation theorems for isotropic functions, Part I and II, Arch. Rat. Mech. An., 36, 166-223, (1970).

11. BOEHLER, J.P., A simple derivation of representations for non - polynominal constitutive equations in some cases of anisotropy, ZAMM, 59, 157-167, (1979).

12. VAKULENKO, A. and MARKOV, K., On the bases in the space of second order tensors, Bull. Acad. Pol. Sci. Tech., 24, 441-450, (1976).

13. MRÓZ, Z. and NIEMUNIS, A., On the description of deformation anisotropy of materials, Proc. IUTAM Symp. "Yielding, Damage and Failure of Anisotropic Solids", Ed. J.P.Boehler, Mech. Eng. Pub. Ltd., pp.171-187, (1990).

14. HILL, R., The Mathematical Theory of Plasticity, Clarendon Press, Oxford, (1950).

15. RANIECKI, B., MRÓZ, Z., On the strain-included anisotropy and texture in rigid-plastic solids, "Inelastic Solids and Structures", Ed. M.Kleiber and A. Konig, Pineridge Press, Swansea, (1989).

FINITE ELEMENTS IN THE 90's, E. Oñate, J. Periaux, A. Samuelsson (Eds.)
Springer-Verlag/CIMNE, Barcelona 1991

THE CO-ROTATIONAL APPROACH TO SUBSTRUCTURING FOR FLEXIBLE MULTIBODY ANALYSIS

M. Géradin
University of Liège,
BELGIUM

A. Cardona
Conicet,
ARGENTINA

Abstract

A new formulation of a component mode method for flexible multibody system analysis is presented. It allows for a full independence between the vibration analysis module – where the component mode model of the elastic member is built – and the mechanism analysis program. It is then easy to link the mechanisms program to *any* existing linear finite element code of structural vibrations analysis. The formulation permits to treat cases in which only translation and/or rotation degrees of freedom are retained at certain nodes on the boundary. The sole limitation to the user, when building the component mode reduced model, is to select the three components of translation (or rotation) at each boundary node. Therefore, the algorithm gains flexibility and ease of application with respect to previous approaches.

1. Introduction

Multibody dynamics problems are highly nonlinear, the nonlinearities being mainly due to the large relative rotations between bodies. In fact, many cases occur where the deformation effects inside each body are small enough to consider that its elastic behavior remains linear in a local frame. Then, it may be said in some sense that the nonlinearities are concentrated at the joints. This fact allows the development of methods for modeling complex elastic mechanism members based on the linear expansion of the elastic displacements field in a basis of deformation modes of the body.

It is however important to stress out the limitations of the linearized approach. Situations may be identified where geometric stiffness effects are of paramount importance in at least some of the members of the system, in which case the linearized approach presented hereafter remains no longer valid and must be replaced by a fully nonlinear description of elastic body deformation.

Several authors have proposed forms of the component mode method for analyzing multibody systems [1-12]. Most of them used vibration modes to model the dynamics of flexible multibodies with the limitation that the bodies are modeled by finite elements embedded in the mechanism analysis program. Ref. [4-6] also included bodies whose flexibility effects are represented by modal superposition, but they were oriented to the modeling of multibody systems in tree topology. Refs. [9,10] presented a multibody formalism in which also static correction modes are included to account for effects of local deformation due to kinematics constraints. In ref. [11], geometric non-linear effects are accounted for by partitioning the elastic member into several linear substructures (or superelements).

This paper presents a component mode method for multibody analysis. In it, we seek to obtain a full independence between the vibration analysis module and the mechanism analysis one. The objective is to be able to use *any* existing linear finite element code of structural vibrations analysis to build the component mode model of the elastic member. In this way, we take advantage of the already developed capabilities of modeling complex structural members of many well known finite element programs for vibration analysis (think, for instance, in the problem of modeling mechanism members of composite materials).

In our formulation, flexible bodies are represented by a collection of fixed-boundary free vibration modes plus some "static correction" or "constraint" modes which account for local effects at the boundaries. The approach of fixed-boundary mode was chosen because it gives a perfect compatibility between bodies, a fact considered necessary to place appropriately the

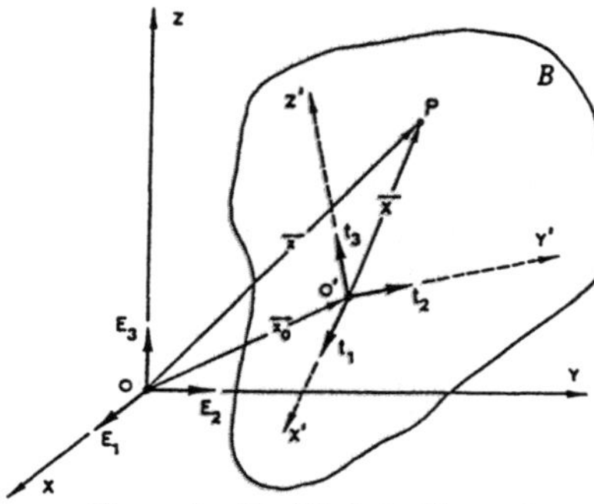

Figure 1 : Flexible body kinematics

joints between them. A standard finite-element structural analysis code [13] performs the modal analysis determining the modes of the superelement. Then, the mechanism analysis package [14] makes the dynamic analysis, linking the body to the rest of the system by the selected joints. The degrees of freedom of the superelement are the translations and the rotations at boundaries, plus a given number of internal mode amplitudes.

Three different alternatives for treating the inertia terms have been discussed and compared in an earlier paper [22] :

- *the consistent approach*, in which all couplings present in the consistent mass matrix are retained to compute the kinetic energy of the flexible body;
- *the lumped mass approach*, in which lumped masses at each node of the body are used to compute the kinetic energy;
- *the co-rotational approach*, in which the consistent mass matrix provided by the linear analysis is used, but the velocities interpolation is not completely consistent with the displacements one.

The third way of computing the inertia, based on a co-rotational approximation of kinetic energy, proved to be the best of all for the simplicity of formulation and easy interfacing of both modules. The sole information used from the vibration analysis module to build the superelement are the reduced stiffness and mass matrices. The inconsistency of interpolations of displacements and velocities did not introduce any appreciable errors in most practical cases. In order to construct the reduced model, the user should specify the nodes at the boundary in which all degrees of freedom are retained, and the number of modes to employ as internal degrees of freedom.

In this paper, we present a slight improvement over the latter technique. The formulation of the inertia forces is reviewed, so as to allow to treat cases in which only translation and/or rotation degrees of freedom are retained at certain nodes on the boundary. The sole limitation to the user, when building a component modes reduced model, is to retain the three components of translation (or rotation) at each boundary node. In this way, the algorithm gains flexibility and ease of application. At the same time, the new method of computing the inertia terms is more clear and simple than the previous one.

The paper is organized as follows. The kinematic hypotheses of component mode synthesis are briefly recalled first, and the discretization of strain energy is described next. The fourth section presents the new formulation to discretize the kinetic energy following the co-rotational approach. The last section is devoted to an example applying the new method and comparing results with those of previous approaches.

2. Kinematics hypotheses

Let $\mathbf{x}$ be the position of an arbitrary point P of the flexible body; we write it in terms of variables in a local reference frame of the body:

$$\mathbf{x} = \mathbf{x}_0 + \mathbf{R}_0(\mathbf{X} + \mathbf{u}) \tag{1}$$

where $\mathbf{x}_0$ is the position of the local reference frame, $\mathbf{R}_0$ is the rotation of the local frame about the global one, $\mathbf{X}$ is the position of point P in the local frame and $\mathbf{u}$ is the elastic displacement of P measured in the local frame (see figure 1).

After time-differentiating equation (1), the virtual displacement, the velocity and the acceleration at point P result:

$$\begin{aligned}\delta\mathbf{x} &= \delta\mathbf{x}_0 + \mathbf{R}_0\delta\widetilde{\mathbf{\Theta}}_0(\mathbf{X}+\mathbf{u}) + \mathbf{R}_0\delta\mathbf{u}\\ \dot{\mathbf{x}} &= \dot{\mathbf{x}}_0 + \mathbf{R}_0\widetilde{\mathbf{\Omega}}_0(\mathbf{X}+\mathbf{u}) + \mathbf{R}_0\dot{\mathbf{u}}\,.\\ \ddot{\mathbf{x}} &= \ddot{\mathbf{x}}_0 + \mathbf{R}_0(\widetilde{\mathbf{\Omega}}_0^2 + \widetilde{\mathbf{A}}_0)(\mathbf{X}+\mathbf{u}) + 2\mathbf{R}_0\widetilde{\mathbf{\Omega}}_0\dot{\mathbf{u}} + \mathbf{R}_0\ddot{\mathbf{u}}\end{aligned} \tag{2}$$

with $\mathbf{\Omega}_0, \mathbf{A}_0$ and $\delta\mathbf{\Theta}_0$ being the material angular velocities, accelerations and the variation of angular displacements of the local frame [12,16]. Rotations can also be expressed as increments with respect to a reference value, giving

$$\mathbf{\Psi} = \mathbf{\Psi}_0 \circ \boldsymbol{\psi} \tag{3}$$

where the operation $\circ$ symbolizes the composition of rotations [12], and $\mathbf{\Psi}_0$ are the parameters of the current rotation of the local frame.

From equations (1,3), we can then compute the relative displacements and slopes inside the elastic body in terms of absolute positions and rotations:

$$\begin{bmatrix}\mathbf{u}\\ \boldsymbol{\psi}\end{bmatrix} = \begin{bmatrix}\mathbf{R}_0^T(\mathbf{x}-\mathbf{x}_0)-\mathbf{X}\\ (-\mathbf{\Psi}_0)\circ\mathbf{\Psi}\end{bmatrix} \tag{4}$$

Let us assume that the elastic displacements and slopes in the local reference frame of each body are small compared to the unity:

$$\frac{\|\mathbf{u}\|}{\|\mathbf{X}\|}\,,\ \|\boldsymbol{\psi}\| \ll 1 \tag{5}$$

These requirements imply a geometric linearity condition in the local frame; that is to say, although the superelement as a whole undergoes finite rotations in the three-dimensional space, the displacements in a local frame remain small enough to assure the linearity of relations between local values of forces and displacements.

Let us also assume that the dynamic loading conditions are such that the local displacements and slopes can be accurately expanded in terms of a few global shape functions:

$$\begin{bmatrix}\mathbf{u}\\ \boldsymbol{\psi}\end{bmatrix} = \mathbf{\Phi}\mathbf{y} \tag{6}$$

Here, $\mathbf{\Phi}$ is the set of global shape functions and $\mathbf{y}$ are the generalized displacement amplitudes. The equations above express then a relation between local relative displacements of the body and global absolute positions, which can be conveniently used to formulate a discrete model of the body.

The relation between local and global variables can be employed to make a reduced model from a discrete one. The latter model, usually having a large number of degrees of freedom, could have been built by using, for instance, the finite element method. In this case, equations (1,3) can be rewritten at each node of the discretization in the following form:

$$\begin{bmatrix}\mathbf{x}_i\\ \mathbf{\Psi}_i\end{bmatrix} = \begin{bmatrix}\mathbf{x}_0+\mathbf{R}_0(\mathbf{X}_i+\mathbf{u}_i)\\ \mathbf{\Psi}_0\circ\boldsymbol{\psi}_i\end{bmatrix} = \begin{bmatrix}\mathbf{x}_0+\mathbf{R}_0(\mathbf{X}_i+\mathbf{\Phi}_i\mathbf{y})\\ \mathbf{\Psi}_0\circ(\mathbf{\Phi}_i\mathbf{y})\end{bmatrix} \tag{7}$$

where subindex i refers to magnitudes at node i, and index 0 denotes the same quantities computed at the origin of the reference frame.

If the component-mode method is followed, the global shape functions are of two kinds:

$$\begin{bmatrix}\mathbf{u}_i\\ \boldsymbol{\psi}_i\end{bmatrix} = \mathbf{\Phi}_{B\,i}\mathbf{y}_B + \mathbf{\Phi}_{I\,i}\mathbf{y}_I \tag{8}$$

equation in which we distinguish between the boundary modes $\mathbf{\Phi}_B$ –obtained by the static condensation procedure– and the internal vibration modes $\mathbf{\Phi}_I$ –computed by fixing the boundaries.

This particular choice of global shape functions permits to represent both the local deformation effects induced by the joints acting on the boundary degrees of freedom, and the global deformation effects induced by the dynamic behavior of the body itself [17].

The boundary generalized displacements $\mathbf{y}_B$ can be computed in terms of position and rotation values at the boundary:

$$\mathbf{y}_B = \begin{bmatrix} \mathbf{u}_B \\ \boldsymbol{\psi}_B \end{bmatrix} = \begin{bmatrix} \mathbf{R}_0^T(\mathbf{x}_B - \mathbf{x}_0) - \mathbf{X}_B \\ (-\boldsymbol{\Psi}_0) \circ (\boldsymbol{\Psi}_B) \end{bmatrix} \tag{9}$$

Then, equations (8, 9) express a kinematics relation between the local relative displacements at the nodes of the discretized body, and global absolute positions and rotations. These global variables constitute the set of generalized displacements $\mathbf{q}$ of the superelement. It is formed by 6 degrees of freedom expressing position and orientation of the local frame, $3 \times (N_{B\ tras} + N_{B\ rot})$ degrees of freedom expressing positions (at $N_{B\ tras}$ nodes of the boundary) and orientations (at $N_{B\ rot}$ nodes of the boundary), and a certain number of internal modal amplitudes $\mathbf{y}_I$:

$$\mathbf{q}^T = [\mathbf{x}_0^T \quad \boldsymbol{\Psi}_0^T \quad \mathbf{x}_B^T \quad \boldsymbol{\Psi}_B^T \quad \mathbf{y}_I^T] \tag{10}$$

Positions and rotational vectors at the boundary connect the body to the rest of the multibody system.

In what follows, we treat $\mathbf{x}_B$ and $\boldsymbol{\Psi}_B$ as vectors in $\Re^3$, but the reader should keep in mind that their actual dimension depends on the number of nodes (and degrees of freedom) retained at the boundary.

3. Computation of the strain energy

The energy of deformation of the body can be directly obtained by making the double discretization process on the continuum expression of the strain energy; i.e., if $\pi = \frac{1}{2}\int_V \sigma\epsilon \, dV$, the finite element method gives a first discrete equation as follows

$$\pi = \frac{1}{2}\mathbf{d}^T\mathbf{S}\mathbf{d} \tag{11}$$

with $\mathbf{d}$ the nodal displacements vector and $\mathbf{S}$ the stiffness matrix. The second discretization (expansion into the modal basis $\mathbf{d} = \boldsymbol{\Phi}\mathbf{y}$) gives the expression:

$$\pi = \frac{1}{2}\,\mathbf{y}^T\overline{\mathbf{S}}\mathbf{y} \tag{12}$$

where $\overline{\mathbf{S}}$ is the reduced stiffness matrix of the body:

$$\overline{\mathbf{S}} = \boldsymbol{\Phi}^T\mathbf{S}\boldsymbol{\Phi} = \begin{bmatrix} \overline{\mathbf{S}}_{BB} & 0 \\ 0 & \overline{\mathbf{S}}_{II} \end{bmatrix} \tag{13}$$

We note that in the latter equation we used the orthogonality relation which characterizes the constraint modes ($\boldsymbol{\Phi}_B^T\mathbf{S}\boldsymbol{\Phi}_I = 0$).

The variation of generalized displacements can be computed in terms of the variation of the superelement degrees of freedom $\mathbf{q}$ as follows:

$$\delta\mathbf{y} = \begin{bmatrix} \delta\mathbf{u}_B \\ \delta\boldsymbol{\psi}_B \\ \delta\mathbf{y}_I \end{bmatrix} = \begin{bmatrix} \mathbf{R}_0^T(\delta\mathbf{x}_B - \delta\mathbf{x}_0) + (\mathbf{X}_B + \mathbf{u}_B) \times \delta\boldsymbol{\Theta}_0 \\ \delta\boldsymbol{\Theta}_B - \delta\boldsymbol{\Theta}_0 \\ \delta\mathbf{y}_I \end{bmatrix} \tag{14}$$

By taking into account equation (4) which expresses the condition of small displacements and rotations in the local frame, the variation of displacements can be simplified to give:

$$\delta\mathbf{y} \simeq \begin{bmatrix} \mathbf{R}_0^T(\delta\mathbf{x}_B - \delta\mathbf{x}_0) + \widetilde{\mathbf{X}}_B\delta\boldsymbol{\Theta}_0 \\ \delta\boldsymbol{\Theta}_B - \delta\boldsymbol{\Theta}_0 \\ \delta\mathbf{y}_I \end{bmatrix} = \mathbf{Y}\delta\mathbf{q} \tag{15}$$

with the definition of the configuration dependent matrix

$$\mathbf{Y} = \begin{bmatrix} -\mathbf{R}_0^T & \widetilde{\mathbf{X}}_B & \mathbf{R}_0^T & \mathbf{0} & \mathbf{0} \\ \mathbf{0} & -\mathbf{I} & \mathbf{0} & \mathbf{I} & \mathbf{0} \\ \mathbf{0} & \mathbf{0} & \mathbf{0} & \mathbf{0} & \mathbf{I} \end{bmatrix} \tag{16}$$

and the vector of generalized coordinates

$$\delta\mathbf{q}^T = [\delta\mathbf{x}_0^T \quad \delta\mathbf{\Theta}_0^T \quad \delta\mathbf{x}_B^T \quad \delta\mathbf{\Theta}_B^T \quad \delta\mathbf{y}_I^T] \tag{17}$$

where $\delta\mathbf{\Theta}_0, \delta\mathbf{\Theta}_B$ are respectively the angular displacements variations at the reference frame and at the boundary nodes.

The internal forces vector of the superelement is then calculated as follows:

$$\delta\pi = \delta\mathbf{q}^T\, \mathbf{Y}^T\overline{\mathbf{S}}\mathbf{y} = \delta\mathbf{q}^T\, \mathbf{G}_{int} \tag{18}$$

By differentiating the internal forces and by neglecting the derivatives of $\mathbf{Y}$, we arrive at the expression of the stiffness matrix of the superelement:

$$\left[\frac{\partial\mathbf{G}_{int}}{\partial\mathbf{q}}\right]\Delta\mathbf{q} \simeq \mathbf{Y}^T\overline{\mathbf{S}}\mathbf{Y}\Delta\mathbf{q} = \overline{\mathbf{S}}_{sup}\Delta\mathbf{q} \tag{19}$$

with

$$\overline{\mathbf{S}}_{sup} = \mathbf{Y}^T\overline{\mathbf{S}}\mathbf{Y} \tag{20}$$

4. Co-rotational evaluation of the kinetic energy

As said in the introduction, the most convenient way to evaluate the kinetic energy of the superelement is to compute it from the velocities rotated to the material frame, in a corotational approach:

$$T = \frac{1}{2}\int_V \dot{\mathbf{x}}\cdot\dot{\mathbf{x}}\,\rho\, dV = \frac{1}{2}\int_V (\mathbf{R}_0^T\dot{\mathbf{x}})\cdot(\mathbf{R}_0^T\dot{\mathbf{x}})\,\rho\, dV \tag{21}$$

Here, $\mathbf{R}_0$ gives the rotation at the reference frame, attached to the elastic body at node 0.

Let us denote the material velocities by

$$\mathbf{v}(\mathbf{X}) = \mathbf{R}_0^T\dot{\mathbf{x}} \tag{22}$$

and interpolate them in terms of nodal velocities in the form

$$\mathbf{v}(\mathbf{X}) = \sum_{i=1}^{n} N_i(\mathbf{X})\,\mathbf{v}_i \tag{23}$$

where the summation extends to all nodes of the flexible member. Note that this interpolation is not consistent with the displacement interpolation used to build the strain energy expression and also that, for finite element models using rotational degrees of freedom, the same interpolation has to be made on the material angular velocities.

After performing the volume integral, the kinetic energy of the superelement can be written:

$$T = \frac{1}{2}\sum_i\sum_j [\mathbf{v}_i^T \quad \mathbf{\Omega}_i^T]\int_V \mathbf{N}_i^T\mathbf{N}_j\,\rho\, dV \begin{bmatrix}\mathbf{v}_i\\ \mathbf{\Omega}_i\end{bmatrix} = \frac{1}{2}\sum_i\sum_j [\mathbf{v}_i^T \quad \mathbf{\Omega}_i^T]\,\mathbf{M}_{ij}\begin{bmatrix}\mathbf{v}_i\\ \mathbf{\Omega}_i\end{bmatrix} \tag{24}$$

where $\mathbf{M}_{ij}$ denotes the block of the mass matrix coupling nodes i and j.

The next step is to form a reduced model by following the component modes approach. A second stage discretization is made by assuming that the material velocities can be expressed in terms of a few global shape functions

$$\begin{bmatrix}\mathbf{v}_i\\ \mathbf{\Omega}_i\end{bmatrix} = \mathbf{\Phi}_i\dot{\mathbf{y}} = \Big[[\mathbf{\Phi}_{0u} \quad \mathbf{\Phi}_{0\Psi}]_i \quad [\mathbf{\Phi}_{Bu} \quad \mathbf{\Phi}_{B\Psi}]_i \quad \mathbf{\Phi}_{Ii}\Big]\begin{bmatrix}\mathbf{v}_0\\ \mathbf{\Omega}_0\\ \mathbf{v}_B\\ \mathbf{\Omega}_B\\ \dot{\mathbf{y}}_I\end{bmatrix} \tag{25}$$

where the part $[\mathbf{v}_0^T \quad \boldsymbol{\Omega}_0^T]$ corresponds to (material) velocities at the reference frame, $\dot{\mathbf{y}}_I$ are the time derivatives of the internal mode amplitudes, and the contribution of (material) velocities at the boundary nodes of the superelement can be written in the form:

$$[\mathbf{v}_B^T \quad \boldsymbol{\Omega}_B^T] = \left[\left((\mathbf{v}_1^T\,\boldsymbol{\Omega}_1^T)\ \ldots\ (\mathbf{v}_k^T\,\boldsymbol{\Omega}_k^T)\right)\left(\mathbf{v}_{k+1}^T\ \ldots\ \mathbf{v}_{k+\ell}^T\right)\left(\boldsymbol{\Omega}_{k+\ell+1}^T\ \ldots\ \boldsymbol{\Omega}_{k+\ell+m}^T\right)\right] \tag{26}$$

Note that at nodes $k+1,\ \ldots\ k+\ell$ of the boundary, only the translation degrees of freedom have been retained to form the superelement, while at nodes $k+\ell+1,\ \ldots\ k+\ell+m$ only the rotation terms are conserved. The sole restriction is that the three components of translation and/or rotation (if any) should be incorporated to the model. The translation material velocities are computed by projection over the reference frame, while material velocities are the true material velocities at the considered node.

Even when the flexible body suffers large rotations, the material velocities pattern does not change; then, this second stage discretization continues to be valid and we can still apply the same modal expansion. The total contribution of the superelement to the kinetic energy of the structure T can then be written as follows:

$$T = \frac{1}{2}\,\dot{\mathbf{y}}^T\overline{\mathbf{M}}\dot{\mathbf{y}} = \frac{1}{2}\,[\mathbf{v}_0^T \quad \boldsymbol{\Omega}_0^T \quad \mathbf{v}_B^T \quad \boldsymbol{\Omega}_B^T \quad \dot{\mathbf{y}}_I^T]\begin{bmatrix}\overline{\mathbf{M}}^{(00)} & \overline{\mathbf{M}}^{(0B)} & \overline{\mathbf{M}}^{(0y)}\\ \overline{\mathbf{M}}^{(B0)} & \overline{\mathbf{M}}^{(BB)} & \overline{\mathbf{M}}^{(By)}\\ \overline{\mathbf{M}}^{(y0)} & \overline{\mathbf{M}}^{(yB)} & \mathbf{I}\end{bmatrix}\begin{bmatrix}\mathbf{v}_0\\ \boldsymbol{\Omega}_0\\ \mathbf{v}_B\\ \boldsymbol{\Omega}_B\\ \dot{\mathbf{y}}_I\end{bmatrix} \tag{27}$$

The constant mass matrix $\overline{\mathbf{M}}$ in (27) results from the projection of the element mass matrices over the modal basis

$$\overline{\mathbf{M}} = \sum_i\sum_j \boldsymbol{\Phi}_i^T\mathbf{M}_{ij}\boldsymbol{\Phi}_j \tag{28}$$

Inertia forces are computed by differentiating the kinetic energy. Its first variation is :

$$\delta T = \delta\dot{\mathbf{y}}^T\,\overline{\mathbf{M}}\,\dot{\mathbf{y}} \tag{29}$$

The vector of variations of generalized velocities reads :

$$\delta\dot{\mathbf{y}}^T = [\,\delta\mathbf{v}_0^T \quad \delta\boldsymbol{\Omega}_0^T \quad \delta\mathbf{v}_B^T \quad \delta\boldsymbol{\Omega}_B^T \quad \delta\dot{\mathbf{y}}_I^T\,] \tag{30}$$

with the variations of material and angular velocities at nodes 0 and B computed from

$$\delta\mathbf{v} = \mathbf{R}_0^T\delta\dot{\mathbf{x}} - \delta\boldsymbol{\Theta}_0\times(\mathbf{R}_0^T\dot{\mathbf{x}}) \qquad \text{and} \qquad \delta\boldsymbol{\Omega} = \delta\dot{\boldsymbol{\Theta}} + \boldsymbol{\Omega}_0\times\delta\boldsymbol{\Theta} \tag{31}$$

By introducing the latter expressions into (29) and by integrating by parts, we get

$$\begin{aligned}\delta T &= -\delta\mathbf{q}^T\,\mathbf{G}_{iner}\\ &= -\delta\mathbf{q}^T\left(\mathbf{P}\,\overline{\mathbf{M}}\,\mathbf{P}^T\,\ddot{\mathbf{q}} - \mathbf{P}\left(\overline{\mathbf{M}}\mathbf{W} + \mathbf{W}^T\overline{\mathbf{M}} + \mathbf{U}^T\overline{\mathbf{M}}\right)\mathbf{P}^T\,\dot{\mathbf{q}}\right)\end{aligned} \tag{32}$$

where variations of the generalized displacements at the global frame are given by (17) and where

$$\dot{\mathbf{q}}^T = [\dot{\mathbf{x}}_0^T \quad \boldsymbol{\Omega}_0^T \quad \dot{\mathbf{x}}_B^T \quad \boldsymbol{\Omega}_B^T \quad \dot{\mathbf{y}}_I^T] \tag{33}$$

$$\mathbf{P} = \begin{bmatrix}\mathbf{R}_0 &&&&\\ & \mathbf{I} &&&\\ && \mathbf{R}_0 &&\\ &&& \mathbf{I} &\\ &&&& \mathbf{I}\end{bmatrix} \quad \mathbf{W} = \begin{bmatrix}\widetilde{\boldsymbol{\Omega}}_0 &&&&\\ & \mathbf{0} &&&\\ && \widetilde{\boldsymbol{\Omega}}_0 &&\\ &&& \mathbf{0} &\\ &&&& \mathbf{0}\end{bmatrix} \quad \mathbf{U} = \begin{bmatrix}\mathbf{0} & \widetilde{\mathbf{v}}_0 & \mathbf{0} &&\\ \mathbf{0} & \widetilde{\boldsymbol{\Omega}}_0 & \mathbf{0} &&\\ \mathbf{0} & \widetilde{\mathbf{v}}_B & \mathbf{0} &&\\ &&& \widetilde{\boldsymbol{\Omega}}_B &\\ &&&& \mathbf{0}\end{bmatrix} \tag{34}$$

Then, by differentiating the inertia forces with respect to the generalized accelerations in the global frame $\ddot{\mathbf{q}}$ we get the superelement tangent mass matrix $\overline{\mathbf{M}}_{sup}$:

$$\overline{\mathbf{M}}_{sup} = \mathbf{P}\,\overline{\mathbf{M}}\,\mathbf{P}^T \tag{35}$$

The inertia forces also depend on the velocities $\dot{\mathbf{q}}$. In order to get full quadratic convergence rate, it will be necessary in some cases to compute the gyroscopic matrix of derivatives of the inertia forces with respect to velocities. This is a non symmetric matrix, which proved to be of value for improving convergence in several examples.

$$\overline{\mathbf{C}}_{,up} = \mathbf{P}\Big(\underbrace{\left[\overline{\mathbf{M}}\mathbf{U} - \mathbf{U}^T\overline{\mathbf{M}} - \mathbf{V}\right]}_{\text{antisym.}} - \underbrace{\left[\overline{\mathbf{M}}\mathbf{W} + \mathbf{W}^T\overline{\mathbf{M}}\right]}_{\text{sym.}} \Big)\mathbf{P}^T \tag{36}$$

The antisymmetric matrix $\mathbf{V}$ is defined by

$$\mathbf{V} = \begin{bmatrix} 0 & \tilde{\mathbf{g}}_{u\,0} & & & \\ \tilde{\mathbf{g}}_{u\,0} & \tilde{\mathbf{g}}_{\Psi\,0} & \tilde{\mathbf{g}}_{u\,B} & & \\ & \tilde{\mathbf{g}}_{u\,B} & 0 & & \\ & & & \tilde{\mathbf{g}}_{\Psi\,B} & \\ & & & & 0 \end{bmatrix} \tag{37}$$

where vectors $\mathbf{g}_{u\,i}, \mathbf{g}_{\Psi\,i}$ are computed as follows:

$$[\mathbf{g}_{u\,0}^T \quad \mathbf{g}_{\Psi\,0}^T \quad \mathbf{g}_{u\,B}^T \quad \mathbf{g}_{\Psi\,B}^T \quad \mathbf{g}_I^T] = \dot{\mathbf{q}}^T\mathbf{P}\overline{\mathbf{M}} \tag{38}$$

We note that the pseudo-damping matrix $\overline{\mathbf{C}}_{,up}$ is formed by adding up two terms: a symmetric and an antisymmetric matrix which are clearly identified in equation (36). We finally remark that in this formulation, all contributions to the inertia terms (inertia forces $\mathbf{G}_{iner}$, mass matrix $\overline{\mathbf{M}}_{,up}$ and pseudo-damping matrix $\overline{\mathbf{C}}_{,up}$) are evaluated directly from the reduced mass matrix $\overline{\mathbf{M}}$, the projection over the modal basis of the finite element mass matrix. In this way, we can very easily interface the vibration analysis and the mechanism analysis modules. It should also be noted that, contrary to what happened with the technique we proposed in ref. [15], it is not necessary now to include all degrees of freedom at each boundary node but only the triplets of positions and/or rotations.

5. Example

The equations of motion that we get by following the techniques described in the preceding paragraphs are time-integrated using a particular implementation of the Hilbert-Hughes-Taylor algorithm to solve mechanism equations. This implementation of the HHT integrator has special provisions to treat large rotations and the equations of constraint, and is fully described in references [12,18].

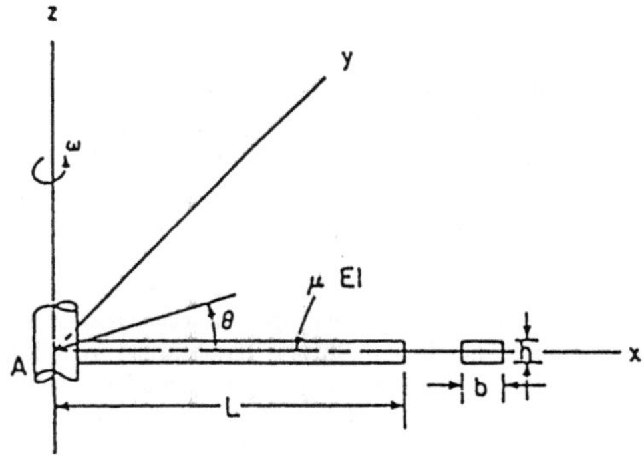

Figure 2 : rotating beam problem.

The example to be analyzed in this paper is a rotating beam, taken from ref. [11]. Figure 2 displays the physical system: a beam built into a rigid shaft which is given an angular

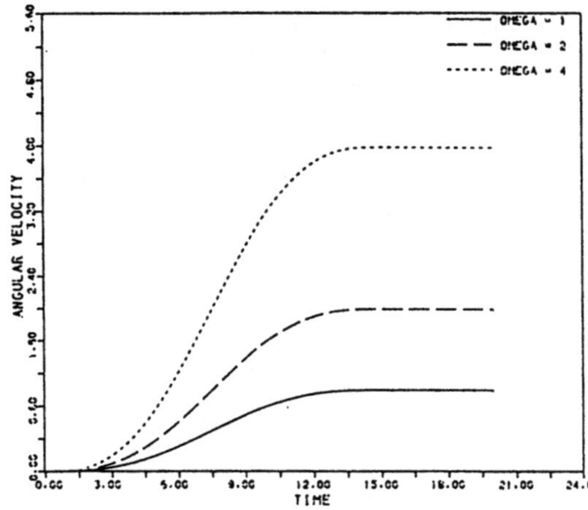

Figure 3 : imposed angular velocity at shaft.

displacement θ about the z−axis defined as

$$\theta = \begin{cases} \dfrac{\omega_s}{T_s}\left[\dfrac{1}{2}t^2 + \left(\dfrac{T_s}{2\pi}\right)^2\left(\cos\left(\dfrac{2\pi t}{T_s}\right) - 1\right)\right] & t < T_s \\ \omega_s\left(t - \dfrac{T_s}{2}\right) & t \geq T_s \end{cases} \tag{39}$$

(see figure 3). The shaft reaches steady state motion after T_s sec and then rotates at constant angular velocity ω_s rad/sec. Beam properties are length 8.0, mass density 2766.67, elasticity modulus 6.895×10^{10}, inertia modulus 8.2181×10^{-9} and cross sectional area 7.299×10^{-5}. Transient responses were computed for $T_s = 15$, and for final angular velocities $\omega_s = 1, 2$ and 4.

First, we solved the problem by using 8 equally spaced nonlinear beam elements [19] giving a master solution for comparison.

Figure 4 displays the evolution of the elastic displacement at the tip of the beam − the displacement of the free end, measured with respect to the undeformed state of the beam. We computed it using the full system model and the reduced system one, when $\omega_s = 1$. The reduced model was obtained by retaining the node at the base and including four vibration modes to build a five degrees of freedom model (the z−rotation at the base plus the four vibration modes). We see that both solutions are in almost complete agreement. We remark also that the results of the reduced system coincide completely with those obtained using the co-rotational inertia formulation presented in ref. [15].

When increasing the speed, the geometric stiffening effect due to the centrifugal forces come into play. Coupling between centrifugal forces and bending moments make the spinning beam stiffer than predicted by linear theory. Figures 5 and 6 display the evolution of the elastic displacement for $\omega_s = 2$ and for $\omega_s = 4$, respectively. We can appreciate that slight differences appear between the solutions given by the full system model and the reduced system one. These discrepancies become more obvious when $\omega_s = 4$.

Figure 5 also shows two solutions computed in ref. [11] : these ones are labeled respectively as "linear solution" and "6 substructures" in the graph, in accordance to the notation employed by Wu and Haug in their paper. The "linear solution" is obtained by following in some sense a similar procedure to that employed in this work. A reduced model is formulated by employing static correction modes (those ones computed by retaining certain degrees of freedom at the boundaries) and internal vibration modes. We can note that our solution is much closer to the true solution than their linear solution, showing that our model takes good account of the geometric stiffening effect. Wu and Haug present in their paper the formulation of a reduced

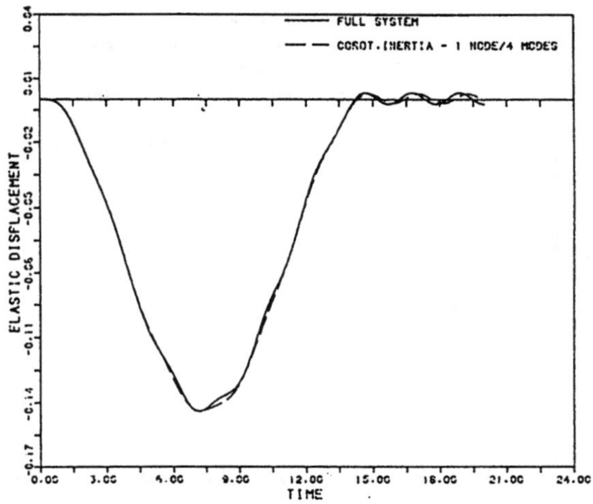

Figure 4 : elastic displacement at the tip of the beam. Angular velocity $\omega = 1$

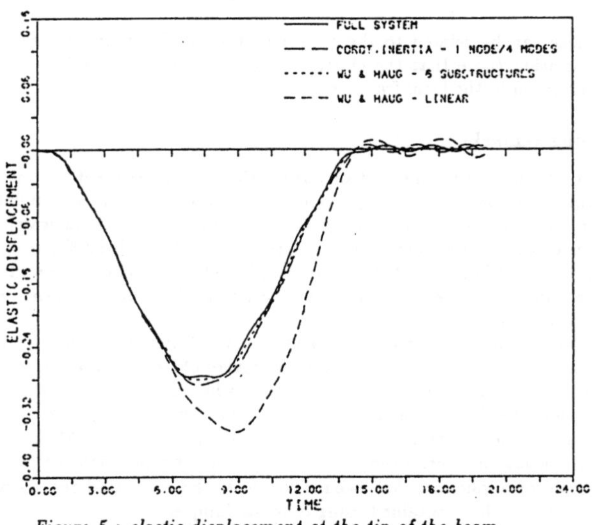

Figure 5 : elastic displacement at the tip of the beam. Angular velocity $\omega = 2$

model nonlinear solution, which consists into dividing the structure to be analyzed into several substructures. In this way, they obtain the "6 substructures" solution which is displayed in figure 5 and which is more accurate than the previous one substructure "linear solution" (it is also more accurate that our solution, but at a higher computational cost).

In figure 6, which displays the results of computations for $\omega_s = 4$, the results obtained by Wu and Haug are also shown. In this case, the discrepancies between their "6 substructures" and "linear" solutions became more drastic, since the "linear" solution presents a divergent behavior. The "6 substructures" solution is instead very close to the full system response computed using the beam nonlinear elements. Our reduced system solution "A", which was computed by retaining the node at the base plus four vibration modes, evidences some discrepancies with the

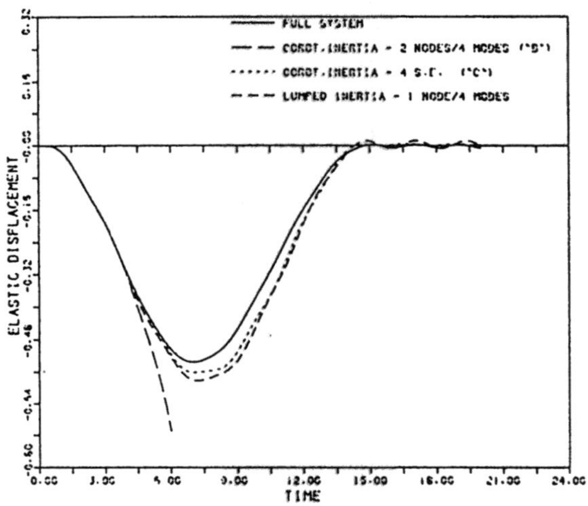

Figure 6 : elastic displacement at the tip of the beam.
Angular velocity $\omega = 4$

solution. These can be related to the approximation of the inertia terms made by the corotational formulation (note that the elastic displacement is close to $L/10$, indicating that the linearity assumption in the local frame is no longer valid).

6. Concluding remarks

The corotational inertia method of formulating a superelement for modeling mechanism members, which was presented in a previous paper [15], has been reformulated. The method allows for an easy interfacing to existing vibration analysis codes, in which the stiffness and mass matrices of the superelement are computed, without requiring the computation of extra-terms within them. It is best suited for easy implementation, since it permits to build reduced models of mechanism members constructed with *any* kind of finite element, without modifying the vibration analysis code.

This paper presents a slight improvement over the latter technique. It allows now to treat cases in which only translation and/or rotation degrees of freedom are retained at certain nodes on the boundary. The sole limitation to the user, when building a component modes reduced model, is to select the three components of translation (or rotation) at each boundary node. In this way, the algorithm has gained flexibility and ease of application. At the same time, the new method of computing the inertia terms is more clear and simple than the previous one.

An example of application has been developed; the computed results using this method compare favorably to those obtained using other techniques.

References

1. SHABANA A.A. and WEHAGE R.A., *A coordinate reduction technique for dynamic analysis of spatial substructures with large angular rotations* , J. Struct. Mech., Vol. 11 , pp. 401-431 (1983).

2. SHABANA A.A., *Substructure synthesis methods for dynamic analysis of multi-body systems* , Computers and Structures, Vol. 20 , pp. 737-744 (1985).

3. SHABANA A.A., *Dynamics of multibody systems*, J. Wiley & Sons (1989).

4. GERADIN M., ROBERT G. and BERNARDIN C., *Dynamic modeling of manipulators with flexible members*, in *Advanced Software in Robotics*, ed. by A. Danthine and M. Géradin, North-Holland (1984).

5. SINGH R.P., Vander VOORT R.J., and LIKINS P.W., *Dynamics of flexible bodies in tree-topology – A computer-oriented approach* , J. Guidance and Control, Vol. 8 , pp. 584-590 (1985).

6. GERADIN M., BERNARDIN C. and ROBERT G., *A computer approach to dynamic modeling of robot manipulators with flexible members*, Report RA-3, LTAS, University of Liege (1986).

7. AGRAWAL O.P. and SHABANA A.A., *Application of deformable-body mean axis to flexible multibody system dynamics*, Comp. Meth. Appl. Mech. and Engng. , Vol. 56 , pp. 217-245 (1986).

8. SHABANA A.A., *Thermal analysis of viscoelastic multibody systems*, Mechanism and Machine Theory, Vol. 21 , pp. 231-242 (1986).

9. YOO W.S. and HAUG E.J., *Dynamics of articulated structures. Part I. Theory*, J. Struct. Mech., Vol. 14, pp. 105-126 (1986).

10. YOO W.S. and HAUG E.J., *Dynamics of articulated structures. Part II. Computer implementation and applications*, J. Struct. Mech., Vol. 14, pp. 177-189 (1986).

11. WU S.C. and HAUG E.J., *Geometric non-linear substructuring for dynamics of flexible mechanical systems*, Int. Jour. Num. Meth. Engng., Vol. 26, pp. 2211-2226 (1988).

12. CARDONA A. and GERADIN M., *A superelement formulation for mechanism analysis* (1989), to be published in Comp. Meth. Appl. Mech. Engng.

13. SAMCEF, *Manuel d'utilisation*, LTAS report, University of Liège, Belgium (1987).

14. CARDONA A. and GERADIN M., *SAMCEF: module d'analyse de mécanismes MECANO (manuel d'utilisation)*, LTAS report, University of Liège, Belgium (1987).

15. CARDONA A. and GERADIN M., *Modeling of superelements in mechanism analysis* (1990), to be published in Int. Jour. Num. Meth. Engng.

16. GERADIN M., PARK K.C. and CARDONA A., *On the representation of finite rotations in spatial kinematics*, LTAS report, University of Liège, Belgium (1988).

17. MEIROVITCH L., *Substructure synthesis*, Chapter 11 of *Computational methods in structural dynamics*, pp. 384-409, Sijthoff & Noordhoff (1980).

18. CARDONA A. and GERADIN M., *Time integration of the equations of motion in mechanism analysis*, Computers and Structures, Vol. 33, pp. 801-820 (1989).

19. CARDONA A. and GERADIN M., *A beam finite element nonlinear theory with finite rotations*, Int. Jour. Num. Meth. Engng., Vol. 26, pp. 2403-2438 (1988).

FINITE ELEMENTS IN THE 90's, E. Oñate, J. Periaux, A. Samuelsson (Eds.)
Springer-Verlag/CIMNE, Barcelona 1991

REDUCED BASIS TECHNIQUE FOR CALCULATING SENSITIVITY DERIVATIVES OF THE NONLINEAR STRUCTURAL RESPONSE

A.K. Noor and **J.M. Peters**
Center for Computational Structures Technology,
University of Virginia,
NASA Langley Research Center, Hamton,
Virginia 23665,
U.S.A.

1. SUMMARY

An efficient reduced basis technique is presented for calculating the sensitivity of nonlinear structural response to variations in the design variables. The structure is discretized by using two-field mixed finite element models. The vector of structural response and its sensitivity derivatives, with respect to design variables, are each expressed as a linear combination of a small number of global approximation (or basis) vectors. The Bubnov-Galerkin technique is then used to approximate each of the finite element equations governing the response, and the sensitivity derivatives, by a small number of algebraic equations in the amplitudes of these vectors. The path derivatives (derivatives of the response vector with respect to path parameters - e.g., load parameter) are used as basis vectors for approximating the response. A combination of the path derivatives and their derivatives with respect to the design variables is used for approximating the sensitivity derivatives. The potential of the proposed technique is discussed and its effectiveness is demonstrated by means of a numerical examples of laminated composite plates subjected to uniform transverse loading.

2. NOMENCLATURE

d_ι	material and lamination parameters of the plate
E_L, E_T	elastic moduli of the individual layers in the direction of fibers and normal to it, respectively
G_{LT}, G_{TT}	shear moduli in the plane of fibers and normal to it, respectively
$[K]$	global linear structure matrix which includes the flexibility and the linear strain-displacement matrices, see Eqs. (1) and A.2 (Appendix A)
$\{G(Z)\}$	vector of nonlinear terms, see Eqs. (1)
$\{\tilde{G}(\psi)\}, \{\bar{G}(\psi)\}$	vectors of nonlinear terms of the reduced systems, see Eqs. (5) and (6)
h	thickness of the plate
$[\tilde{K}], [\bar{K}]$	linear matrices of the reduced systems, see Eqs. (5) and (6)
L	side length of the plate
p_o	intensity of uniform transverse loading
q	load parameter
$\{Q\}$	normalized load vector, see Eqs. (1)

$\{\tilde{Q}\}, \{\bar{Q}\}$	load vectors of the reduced systems, see Eqs. (5) and (6)
r	number of basis vectors used in evaluating the response
U	total strain energy of the plate
u_1, u_2, w	displacement components in the coordinate directions
x_1, x_2, x_3	Cartesian coordinate system (x_3 normal to the middle plane of the plate)
$\{Z\}$	response vector which includes both the nodal displacements and the stress parameters
$[\Gamma], [\bar{\Gamma}_\iota]$	matrices of basis vectors, see Eqs. (3) and (4)
θ	fiber orientation angle
ν_{LT}	major Poisson's ratio of the individual layers
ϕ_1, ϕ_2	rotation components of the middle plane of the plate, see Fig. 1
$\{\psi\}, \{\bar{\psi}_\iota\}$	vectors of undetermined coefficients of the reduced equations, see Eqs. (5) and (6)

Superscripts:
t denotes matrix transposition
Range of Indices:
I,J,L : 1 to the total number of degrees of freedom (nodal displacements and stress parameters)
i,j : 1 to the total number of reduced degrees of freedom used in evaluating the sensitivity derivatives
ι : 1 to the total number of lamination and material parameters

3. INTRODUCTION

The continuing active development of nonlinear analysis procedures and the increasing use of these procedures in automated optimum design of structural systems has stimulated interest in the development of efficient techniques for calculating the sensitivity of nonlinear structural response to variations in the design variables (see, for example, [1] to [10], and the survey papers, [11] to [14]). The sensitivity derivatives are used to:

a) determine a search direction in the direct application of nonlinear mathematical programming algorithms. When approximation concepts are used for the optimum design of structures, sensitivity derivatives are used to construct explicit approximations for the critical and potentially critical, behavior constraints;

b) generate an approximation for the response of a modified structure (in conjunction with a reanalysis technique);

c) assess the effects of uncertainties in the material and geometric parameters of the computational model on the structural response; and

d) predict the changes in the structural response due to changes in the parameters.

Two general procedures are currently being used for calculating the sensitivity derivatives of the nonlinear structural response. The two approaches are: direct differentiation method, and adjoint variable method (see, for example, [3], [8], [9], [10] and [11]). The first procedure is based on the implicit differentiation of the equations that describe the nonlinear structural response with respect to the desired parameters, and the solution of the resulting sensitivity equations. In the adjoint variable method an adjoint physical system is introduced whose solution permits rapid evaluation of the desired sensitivity derivatives. Both procedures can be applied to either the governing discrete equations or to the functional of the variational formulation of the structure (with a consequent change in the order of discretization and implicit differentiation).

The discrete models for complex structures typically have large number of degrees of freedom, and the calculation of the sensitivity derivatives of the nonlinear structural response can become quite expensive. Although efficient reduced basis techniques have been developed for substantially reducing the number of degrees of freedom in the initial discretization and the computational effort involved in nonlinear structural analysis (see, for example, [15], [16] and [17]), no attempt has been made to adapt these techniques to the calculation of sensitivity derivatives. The present study is an attempt to fill this void. Specifically, the objective of this paper is to present a reduced basis technique and a computational procedure for the efficient calculation of the sensitivity of the nonlinear structural response to variations in the design variables. The crux of the technique is the effective choice of basis vectors for approximating the sensitivity derivatives.

To sharpen the focus of the study, discussion is limited to multilayered composite plates with geometric nonlinearities. The design variables consist of lamination and material properties.

4. MATHEMATICAL FORMULATION

4.1 Governing Finite Element Equations

The analytical formulation is based on a form of the moderate rotation, geometrically nonlinear theory of the structure. A total Lagrangian formulation is used for describing the deformation, and the structure is discretized by using two-field mixed finite element models. The governing finite element equations for the response vector and its sensitivity derivatives can be written in the following compact form [10],[15],[17]:

$$[K]\{Z\} + \{G(Z)\} - q\{Q\} = 0 \tag{1}$$

and

$$\left[[K] + \left[\frac{\partial G_I}{\partial Z_J}\right]\right]\left\{\frac{\partial Z}{\partial d_\iota}\right\} = -\left[\frac{\partial K}{\partial d_\iota}\right]\{Z\} - \left\{\frac{\partial G}{\partial d_\iota}\right\} + q\left\{\frac{\partial Q}{\partial d_\iota}\right\} \tag{2}$$

where $[K]$ is a global linear structure matrix which includes the flexibility, and the linear strain-displacement matrices; $\{Z\}$ is the response vector which includes both nodal displacements and stress parameters; $\{G(Z)\}$ is the vector of nonlinear terms; q_I

is a load parameter; {Q} is normalized load vector; and d_ι are design variables, selected in the present study to be lamination and material parameters. The form of the arrays [K], {G(Z)}, {Q} is described in Appendix A. Note that Eqs. (1) are nonlinear in {Z} but Eqs. (2) are linear in $\left\{\frac{\partial Z}{\partial d_\iota}\right\}$.

4.2 Basis Reduction and Reduced System of Equations

The response vector, {Z}, and its derivatives with respect to d_ι, $\left\{\frac{\partial Z}{\partial d_\iota}\right\}$, are expressed as linear combinations of a few preselected global approximations (basis) vectors. The approximation can be expressed by the following transformation:

$$\{Z\} = [\Gamma]\{\psi\} \tag{3}$$

and

$$\left\{\frac{\partial Z}{\partial d_\iota}\right\} = [\bar{\Gamma}_\iota]\{\bar{\psi}_\iota\} \tag{4}$$

The columns of the matrices $[\Gamma]$ and $[\bar{\Gamma}_\iota]$ in Eqs. (3) and (4) are the approximation (basis) vectors, and the elements of the vectors $\{\psi\}$ and $\{\bar{\psi}_\iota\}$ are the amplitudes of the approximation vectors which are, as yet, unknowns. Note that the number of basis vectors in Eqs. (3) and (4) is considerably smaller than the total number of degrees of freedom (components of the vectors {Z} and $\left\{\frac{\partial Z}{\partial d_\iota}\right\}$).

A Bubnov-Galerkin technique is now used to replace the governing equations for the structural response and its sensitivity derivatives, Eqs. (1) and (2), by the following reduced equations in the unknowns $\{\psi\}$ and $\{\bar{\psi}_\iota\}$:

$$[\tilde{K}]\{\psi\} + \{\tilde{G}(\psi)\} - q\{\tilde{Q}\} = 0 \tag{5}$$

and

$$\left[[\bar{K}] + \left[\frac{\partial \bar{G}_i}{\partial \psi_j}\right]\right]\{\bar{\psi}_\iota\} = -\left[\frac{\partial \bar{K}}{\partial d_\iota}\right]\{\psi\} + \left\{\frac{\partial \bar{G}}{\partial d_\iota}\right\} + q\left\{\frac{\partial \bar{Q}}{\partial d_\iota}\right\} \tag{6}$$

where

$$[\tilde{K}] = [\Gamma]^t[K][\Gamma] \tag{7}$$

$$\{\tilde{G}(\psi)\} = [\Gamma]^t\{G(\psi)\} \tag{8}$$

$$\{\tilde{Q}\} = [\Gamma]^t\{Q\} \tag{9}$$

$$[\bar{K}] = [\bar{\Gamma}_\iota]^t[K][\bar{\Gamma}_\iota] \tag{10}$$

$$\left[\frac{\partial \bar{G}_i}{\partial \psi_j}\right] = [\bar{\Gamma}_\iota]^t\left[\frac{\partial G_I}{\partial Z_J}\right][\bar{\Gamma}_\iota] \tag{11}$$

$$\left[\frac{\partial \bar{K}}{\partial d_\iota}\right] = [\bar{\Gamma}_\iota]^t \left[\frac{\partial K}{\partial d_\iota}\right] [\Gamma] \tag{12}$$

$$\left\{\frac{\partial \bar{G}}{\partial d_\iota}\right\} = [\bar{\Gamma}_\iota]^t \left\{\frac{\partial G}{\partial d_\iota}\right\} \tag{13}$$

$$\left\{\frac{\partial \bar{Q}}{\partial d_\iota}\right\} = [\bar{\Gamma}_\iota]^t \left\{\frac{\partial Q}{\partial d_\iota}\right\} \tag{14}$$

Note that Eqs. (5) are nonlinear in the unknowns $\{\psi\}$ and Eqs. (6) are linear in $\{\bar{\psi}_\iota\}$. The vector $\{G(\psi)\}$ in Eqs. 8 is obtained by replacing $\{Z\}$ in the vector $\{G(Z)\}$ by its expression in terms of $\{\psi\}$, Eqs. (3).

4.4 Selection and Generation of Basis Vectors

The effectiveness of the proposed technique for calculating the sensitivity derivatives depends, to a great extent, on the proper choice of the basis vectors, the columns of the matrices $[\Gamma]$ and $[\bar{\Gamma}_\iota]$. An effective choice for the basis vectors used in approximating the response vector $\{Z\}$, Eqs. (3), was found to be the various order path derivatives (derivatives with respect to the control parameter q - see, for example, [16],[17]); i.e., the matrix $[\Gamma]$ used in approximating $\{Z\}$, over a range of values of q, includes the response vector corresponding to a particular value of q (viz, q^0) and its various-order derivatives with respect to q, evaluated at the same value of q^0, or

$$[\Gamma] = \left[\{Z\} \left\{\frac{\partial Z}{\partial q}\right\} \left\{\frac{\partial^2 Z}{\partial q^2}\right\} \left\{\frac{\partial^3 Z}{\partial q^3}\right\} \cdots \right]_{q^0} \tag{15}$$

The basis vectors for approximating the sensitivity derivatives $\left\{\frac{\partial Z}{\partial d_\iota}\right\}$ were selected to be a combination of:

a) the various order path derivatives, i.e., the columns of the matrix $[\Gamma]$; and

b) the derivatives of $[\Gamma]$ with respect to d_ι, i.e., $\left[\frac{\partial \Gamma}{\partial d_\iota}\right]$. Therefore

$$[\bar{\Gamma}_\iota] = \left[[\Gamma] \quad \left[\frac{\partial \Gamma}{\partial d_\iota}\right] \right] \tag{16}$$

and the number of basis vectors in $[\bar{\Gamma}_\iota]$, for each d_ι, is twice the number of basis vectors in $[\Gamma]$.

The rationale for the particular choice of $[\bar{\Gamma}_\iota]$ is based on the following two facts:

a) Differentiating Eqs. (3) with respect to d_ι leads to:

$$\left\{\frac{\partial Z}{\partial d_\iota}\right\} = [\Gamma]\left\{\frac{\partial \psi}{\partial d_\iota}\right\} + \left[\frac{\partial \Gamma}{\partial d_\iota}\right]\{\psi\} \tag{17}$$

i.e., the expression for $\left\{\frac{\partial Z}{\partial d_\iota}\right\}$ includes *both* $[\Gamma]$ and $\left[\frac{\partial \Gamma}{\partial d_\iota}\right]$; and

b) the use of free parameters $\{\Psi_\iota\}$, instead of the fixed amplitudes $\left\{\frac{\partial \psi}{\partial d_\iota}\right\}$ and $\{\psi\}$, is expected to improve the accuracy of the approximation for $\left\{\frac{\partial Z}{\partial d_\iota}\right\}$, over a wide range of values of q. The free parameters $\{\Psi_\iota\}$ are obtained by applying the Bubnov-Galerkin technique to Eqs. (2), resulting in Eqs. (6).

The path derivatives (columns of the matrix $[\Gamma]$) are obtained by successive differentiation of the governing finite element equations of the structure, Eqs. 1, with respect to the parameter q. The recursion relations for evaluating the path derivatives are given in [18]. The first three recursion relations are summarized in Appendix B. Note that *only one matrix factorization is needed for generating all the path derivatives.*

The derivatives of $[\Gamma]$ with respect to d_ι, $\left[\frac{\partial \Gamma}{\partial d_\iota}\right]$, are obtained by differentiating each of the recursion relations for evaluating the path derivatives with respect to d_ι. The resulting equations have the same left-hand side as that of the recursion relations, and therefore, *no additional matrix factorizations are needed for generating* $\left[\frac{\partial \Gamma}{\partial d_\iota}\right]$ (see Appendix B).

5. COMPUTATIONAL PROCEDURE

For each value of q, the vector of reduced unknowns, $\{\psi\}$, is obtained by solving the reduced nonlinear equations, Eqs. (5). Then the vectors $\{\Psi_\iota\}$, associated with the same value of $\{\psi\}$, are evaluated by solving the reduced linear equations, Eqs. (6). The response vector and its sensitivity derivatives, $\{Z\}$ and $\left\{\frac{\partial Z}{\partial d_\iota}\right\}$, are obtained by using Eqs. (3) and (4). The process is repeated for different combinations of q.

The computational procedure proposed in [16], [17] and [18] for sensing and controlling the error of the response vector $\{Z\}$, predicted by the reduced system of equations, is used for error sensing and control of the sensitivity derivatives $\left\{\frac{\partial Z}{\partial d_\iota}\right\}$.

6. NUMERICAL STUDIES

To test and evaluate the effectiveness of the proposed reduced basis technique, the sensitivity derivatives of the nonlinear and postbuckling responses of several structures were generated by this technique. Comparisons were made with the sensitivity derivatives obtained by using the full system of equations of the finite element model. Typical results are presented herein for the eight-layer quasi-isotropic composite plate shown in Fig. 1. The sensitivity derivatives are calculated for the nonlinear response due to uniform transverse loading, ($q=p_o$).

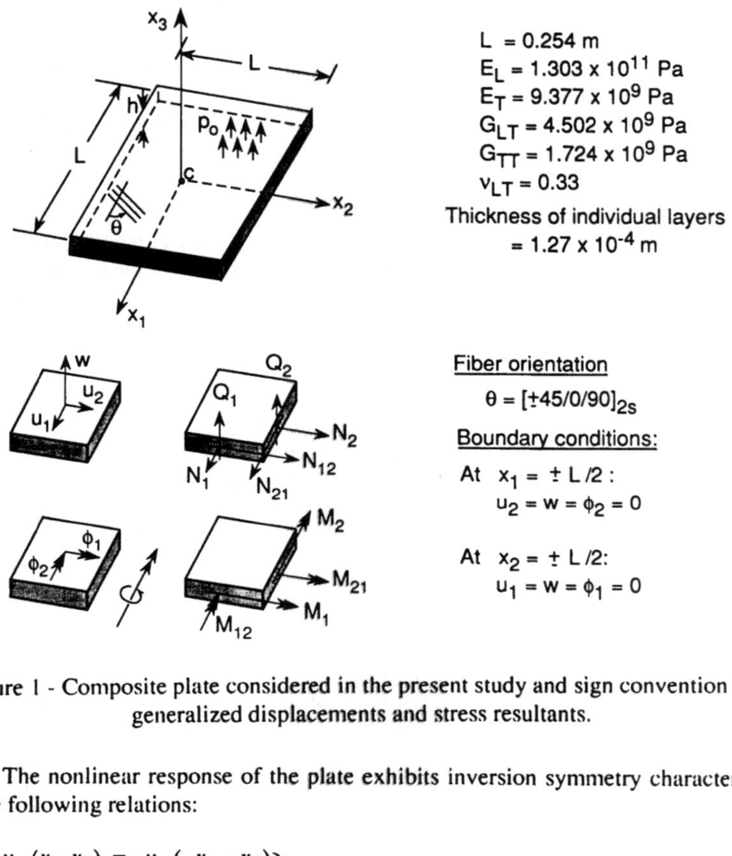

Figure 1 - Composite plate considered in the present study and sign convention for generalized displacements and stress resultants.

The nonlinear response of the plate exhibits inversion symmetry characterized by the following relations:

$$\left.\begin{aligned} u_\alpha(x_1,x_2) &= -u_\alpha(-x_1,-x_2) \\ w(x_1,x_2) &= w(-x_1,-x_2) \\ \phi_\alpha(x_1,x_2) &= -\phi_\alpha(-x_1,-x_2) \end{aligned}\right\} \quad \alpha=1,2 \tag{18}$$

The above symmetry relations were used in conjunction with the procedure outlined in [19] to reduce the size of the analysis model to one-half of the plate. An 8×16 grid of mixed finite elements was used for the discretization of the plate. Biquadratic shape functions were used for approximating each of the generalized displacements, and bilinear shape functions were used for approximating each of the stress resultants. The total number of stress resultant parameters was 4096, and the total number of generalized displacement degrees of freedom was 2525.

The characteristics of the finite element model are given in [20]. The nonlinear and postbuckling responses as well as their sensitivity derivatives were calculated by using the reduced basis technique. Typical results are presented in Figs. 2, 3 and 4, and are discussed subsequently.

Figure 2 shows plots of the loading versus the maximum transverse displacement, w_c, and the total strain energy, U, as well as their sensitivity derivatives with respect to the material parameters, E_L, E_T, G_{LT}, ν_{LT} and G_{TT}. As can be seen from Fig. 2, both w_c and U are very sensitive to variations in E_L; they are slightly sensitive to variations in E_T and G_{LT} and fairly insensitive to variations in ν_{LT} and G_{TT}.

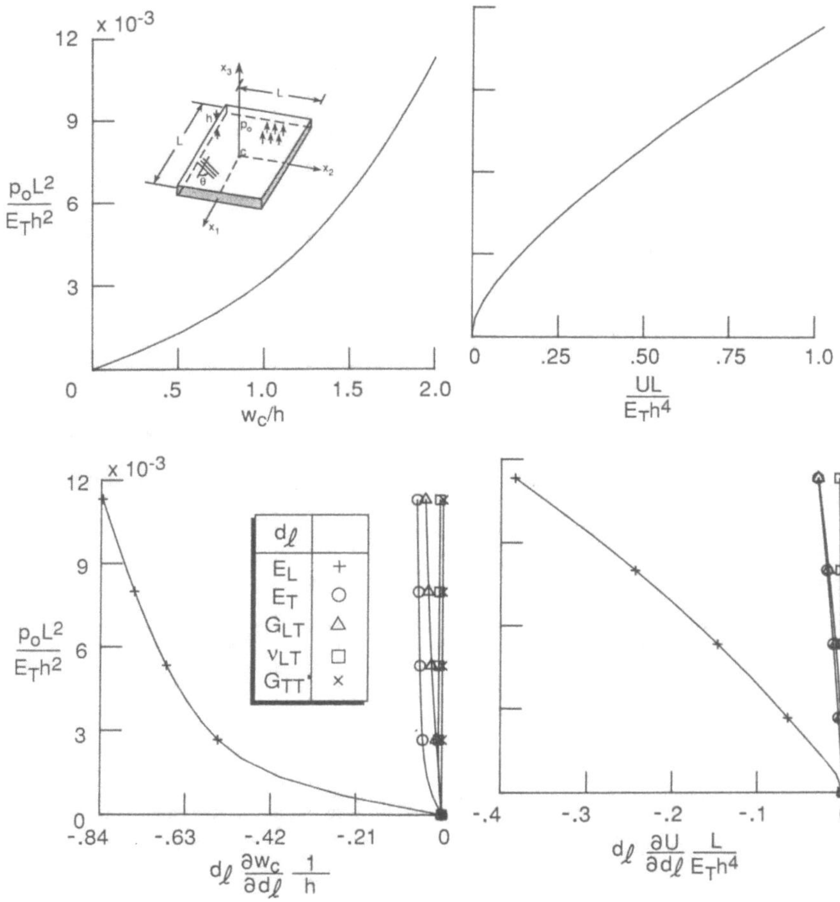

Figure 2 - Nonlinear response and sensitivity derivatives for the composite plate subjected to uniform transverse loading (see Fig. 1).

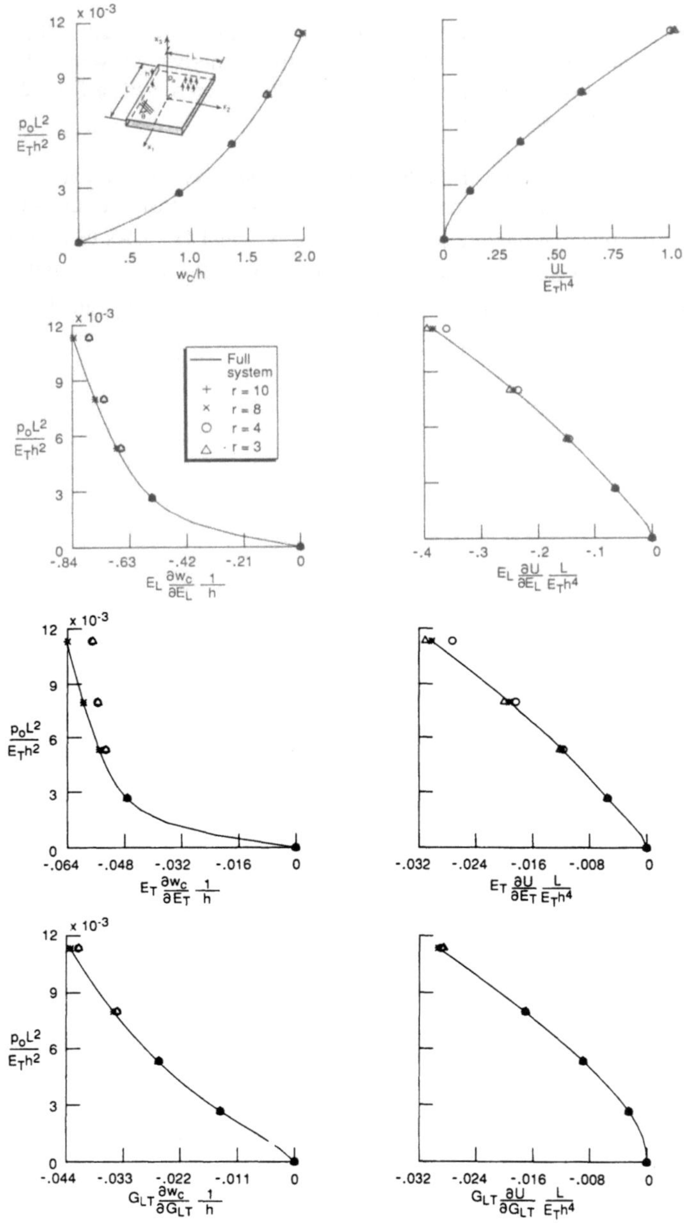

Figure 3 - Accuracy of the nonlinear response and sensitivity derivatives obtained by the reduced basis technique. Composite plate subjected to uniform transverse loading (see Fig. 1).

The basis vectors for both the response and the sensitivity derivatives were generated for the unloaded plate, p_o=0, and were thus obtained by solving a linear system of finite element equations. Then the reduced equations for evaluating the response and the sensitivity derivatives were generated. The basis vectors were not updated throughout the range of loading considered. An indication of the accuracy of w_c and U, and their sensitivity derivatives, obtained by the reduced basis technique is given in Fig. 3.

For the range of loading considered, w_c and U obtained by using ten basis vectors are almost indistinguishable from those obtained by using the full system of finite element equations. The sensitivity derivatives of w_c and U with respect to E_L, E_T and G_{LT} obtained by using twenty basis vectors (r=10, i.e., first ten derivatives of

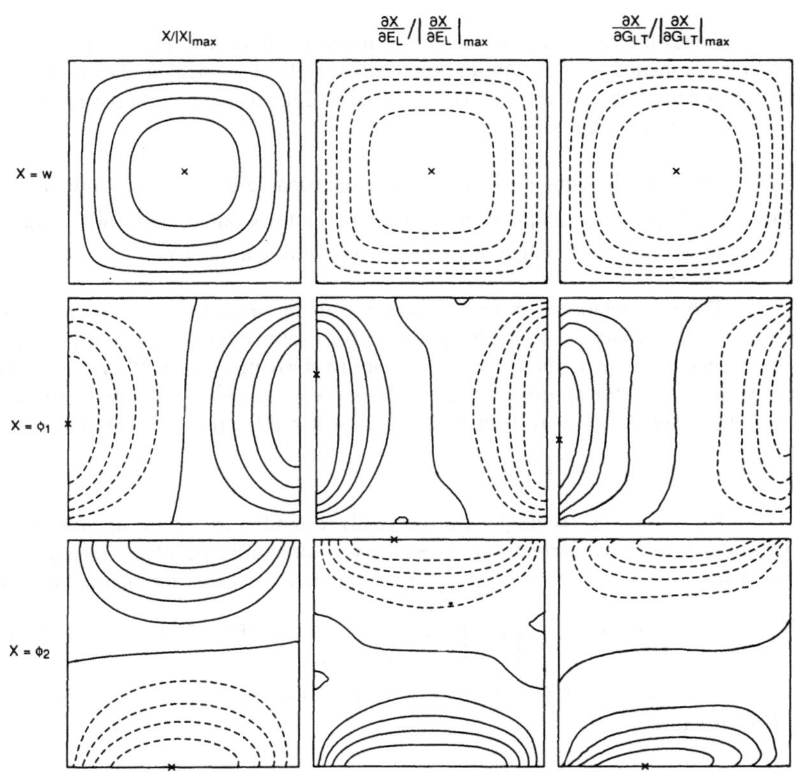

Figure 4 - Normalized contour plots for the generalized displacement components and their sensitivity derivatives. Composite plate subjected to uniform transverse loading (see Fig. 1). Spacing of contour lines is 0.2 and dashed lines denote negative contours. An × denotes location of maximum absolute value of the component. $\frac{p_o L^2}{E_T h^2} = 11.33\times10^{-3}$

{Z} with respect to p_o, and their first derivatives with respect to d_ι) are as accurate as w_c and U obtained by using ten basis vectors (r=10, i.e., first ten derivatives of {Z} with respect to p_o). Normalized contour plots for w, ϕ_1, ϕ_2 and their sensitivity derivatives with respect to E_L and G_{LT}, at $\frac{p_o L^2}{E_T h^2} = 11.33 \times 10^{-3}$, are shown in Fig. 4. Note that the contour plots of the generalized displacements bear some resemblance to the contour plots of their sensitivity derivatives. Also note that the contour lines for the sensitivity derivatives are somewhat closer to each other than the corresponding ones for the response functions. This supports the observation, which has been mentioned in [21], that accurate sensitivity calculations may require a more refined grid than accurate response calculations.

5. POTENTIAL OF THE FOREGOING REDUCED BASIS TECHNIQUE

The foregoing reduced basis technique appears to have high potential for use in large-scale automated analysis and design systems. The numerical studies conducted clearly demonstrate the accuracy and effectiveness of the technique. In particular, the following comments seem to be in order:

1. The particular choice of the basis vectors used herein for approximating the sensitivity derivatives (columns of the matrix $[\bar{\Gamma}_\iota]$) allows the accurate prediction of these derivatives for a wide range of values of the control parameters.

2. The computational procedure developed for predicting the nonlinear response using the reduced basis technique (see, for example, [16], [17] and [18]) can now be extended to the prediction of the sensitivity derivatives as well. When a new set of basis vectors, for approximating the response, is generated; the derivatives of each of these vectors with respect to the design variables (columns of the matrix $\left[\frac{\partial\Gamma}{\partial d_\iota}\right]$) are generated as well.

3. The foregoing technique can be easily extended to the evaluation of the second-order sensitivity derivatives, as well as to structural reanalysis (analysis of modified structures). For the second-order sensitivity derivatives the matrix of basis vectors is selected to be $[\bar{\Gamma}_\iota] = \left[\,[\Gamma]\,\left[\frac{\partial\Gamma}{\partial d_\iota}\right]\left[\frac{\partial^2\Gamma}{\partial d_\iota^2}\right]\right]$. In structural reanalysis, the set of basis vectors included in $[\bar{\Gamma}_\iota]$ are used for predicting the nonlinear response of the modified structure.

6. CONCLUDING REMARKS

An efficient reduced basis technique is presented for calculating the sensitivity of the nonlinear structural response to variations in the design variables. The analytical formulation is based on a form of the moderate rotation, geometrically nonlinear theory of the structure. A total Lagrangian description is used and the structure is discretized by using two-field mixed finite element models. The equations governing the sensitivity derivatives are obtained by implicit differentiation of the discrete finite element equations of the structure. The vector of structural response and its sensitivity derivatives, with respect to design variables, are each expressed as a linear combination of a small number of global approximation (basis) vectors and the Bubnov-Galerkin technique is used to approximate each of the finite element equations governing the response and the sensitivity derivatives by a small number of algebraic equations in the amplitudes of these vectors. The path derivatives (derivatives of the response vector with respect to a path parameter) are used as basis vectors for approximating the response. A combination of the path derivatives and their derivatives with respect to the design variables is used for approximating the sensitivity derivatives.

The potential of the proposed technique is discussed and the effectiveness of the basis vectors used in approximating the sensitivity derivatives is demonstrated by means of a numerical example of the nonlinear response of an eight-layer quasi-isotropic plate subjected to uniform transverse loading.

7. ACKNOWLEDGEMENT

The present research is partially supported by NASA Cooperative Agreement NCCW-0011 and by an Air Force Office of Scientific Research Grant AFOSR-90-0369. The numerical studies were performed on the CRAY Y-MP8/864 at the San Diego Supercomputer Center. The authors appreciate the encouragement of Samuel L. Venneri of NASA Headquarters and Spencer Wu of AFOSR.

8. REFERENCES

1. RYU, Y. S., HARIRIAN, M., WU, C. C. and ARORA, J. S., Structural Design Sensitivity Analysis of Nonlinear Response, Computers and Struct., 21(1/2), 245-55 (1985).
2. MROZ, Z., KAMAT, M. P. and PLAUT, R. H., Sensitivity Analysis and Optimal Design of Nonlinear Beams and Plates, Journal of Struct. Mech., 13(3/4), 245-66 (1985).
3. HAFTKA, R. T. and MROZ, Z., First- and Second-Order Sensitivity Analysis of Linear and Nonlinear Structures, AIAA J., 24(7), 1187-92 (1986).
4. WU, C. C. and ARORA, J. S., Design Sensitivity Analysis and Optimization of Nonlinear Structural Response Using Incremental Procedure, AIAA J., 25(8), 1118-25 (1987).
5. CARDOSO, J. B. and ARORA, J. S., Variational Method for Design Sensitivity Analysis in Nonlinear Structural Mechanics, AIAA J., 26(5), 595-603 (1988).
6. TORTORELLI, D. A., HABER, R. B. and LU, S. C.-Y., Design Sensitivity Analysis for Nonlinear Thermal Systems, Comp. Meth. in Appl. Mech. and Eng., 77, 61-77 (1989).
7. ARORA, J. S. and CARDOSO, J. E. B., A Design Sensitivity Analysis Principle and Its Implementation into ADINA, Computers and Struct., 32(3/4), 691-705 (1989).

8. TORTORELLI, D. A., HABER, R. B. and LU, S. C.-Y., Adjoint Sensitivity Analysis for Nonlinear Dynamic Thermoelastic Systems, AIAA J., 29(2), 253-263 (1991).
9. PHELAN, D. G., VIDAL, C. and HABER, R. B., An Adjoint Variable Method for Sensitivity Analysis of Nonlinear Elastic Systems, Int. Journal for Num. Meth. in Eng., 31, 1649-1667 (1991).
10. NOOR, A. K., TANNER, J. A. and PETERS, J. M., Sensitivity of Tire Response to Variations in Material and Geometric Parameters, Finite Elements in Design and Analysis (to appear).
11. ADELMAN, H. M. and HAFTKA, R. T., Sensitivity Analysis of Discrete Structural Systems, AIAA J., 24(5), 823-832 (1986).
12. HAFTKA, R. T. and ADELMAN, H.M., 'Recent Developments in Structural Sensitivity Analysis,' Structural Optimization I, Springer-Verlag, 1989, pp. 137-151.
13. HABER, R. B., TORTORELLI, D. A., VIDAL, C. A. and PHELAN, D. G. - 'Design Sensitivity Analysis of Nonlinear Structures - I: Large-Deformation Hyperelasticity and History-Dependent Material Response,' Structural Optimization: Status and Promise, Ed. Kamat, M. P., AIAA Series: Progress in Astronautics and Aeronautics (to appear).
14. ARORA, J. S., LEE, T. H. and KUMAR, V. - 'Design Sensitivity Analysis of Nonlinear Structures - III: Shape Variation of Viscoplastic Structures,' Structural Optimization: Status and Promise, AIAA Series: Progress in Astronautics and Aeronautics (to appear).
15. NOOR, A. K., Recent Advances in Reduction Methods for Nonlinear Problems, Computers and Struct., 13(1/2), 31-44 (1981).
16. NOOR, A. K., On Making Large Nonlinear Problems Small, Comp. Meth. in Appl. Mech. and Eng., 34, 955-985 (1982).
17. NOOR, A. K. and PETERS, J. M., Recent Advances in Reduction Methods for Instability Analysis of Structures, Computers and Struct., 16(1-4), 67-80 (1983).
18. NOOR, A. K. and PETERS, J. M., Multiple-Parameter Reduced Basis Technique for Bifurcation and Postbuckling Analyses of Composite Plates, Int. Journal for Num. Meth. in Eng., 19, 1783-1803 (1983).
19. NOOR, A. K. and CAMIN, R. A., Symmetry Considerations for Anisotropic Shells, Comp. Meth. in Appl. Mech. and Eng., 9, 317-335 (1976).
20. NOOR, A. K. and ANDERSEN, C. M., Mixed Models and Reduced/Selective Integration Displacement Models for Nonlinear Shell Analysis, Int. Journal for Num. Meth. in Eng., 18, 1429-1454 (1982).
21. CHOI, K. K. and TWU, S. L., On Eigenvalue of Continuum and Discrete Methods of Shape Design Sensitivity Analysis, AIAA J., 27(10), 1418-1424 (1989).

Appendix A - Form of the Arrays in the Governing Discrete Equations of the Structure

The governing discrete equations of the structure, Eqs. (1), consist of both the constitutive relations and the equilibrium equations for the structure. The response vector, $\{Z\}$, can be partitioned into the subvectors of stress parameters, $\{H\}$, and the nodal displacements, $\{X\}$, as follows:

$$\{Z\} = \begin{Bmatrix} H \\ X \end{Bmatrix} \tag{A.1}$$

The different arrays in Eqs. (1) can be partitioned as follows:

$$[K] = \begin{bmatrix} -F & S \\ S^t & 0 \end{bmatrix} \tag{A.2}$$

$$\{G(Z)\} = \begin{Bmatrix} M(X) \\ N(H,X) \end{Bmatrix} \tag{A.3}$$

$$\{Q\} = \begin{Bmatrix} 0 \\ P \end{Bmatrix} \tag{A.4}$$

where $[F]$ is the flexibility matrix; $[S]$ is the linear strain-displacement matrix; $\{M(X)\}$ and $\{N(H,X)\}$ are the subvectors of nonlinear terms; and $\{P\}$ is the vector of normalized applied mechanical loads.

Appendix B - Evaluation of Path Derivatives and Their First Derivatives with Respect to d_l

The path derivatives in Eqs. (15) are obtained by successive differentiation of the governing finite element equations, Eqs. (1), with respect to the load parameter q, and solving the resulting system of linear algebraic equations. The recursion relations for the path derivatives can be written in the following compact form:

$$\left[K_{IJ} + \frac{\partial G_I}{\partial Z_J} \right] \frac{\partial^s Z_J}{\partial q^s} = R_I^{(s)} \tag{B.1}$$

where I,J range from 1 to the total number of degrees of freedom in the model. The range of s is 1 to r, where r is the number of basis vectors, and the explicit forms of the components of the right-hand sides, $R_I^{(s)}$, are given in Table 1.

Table 1

s	$R_I^{(s)}$
1	Q
2	$-\frac{\partial^2 G_I}{\partial Z_J \partial Z_L} \frac{\partial Z_J}{\partial q} \frac{\partial Z_L}{\partial q}$
3	$-3 \frac{\partial^2 G_I}{\partial Z_J \partial Z_L} \frac{\partial Z_J}{\partial q} \frac{\partial^2 Z_L}{\partial q^2}$

In Table 1, a repeated uppercase Latin index denotes summation over its full range and G_I are bilinear (or quadratic) functions of Z_I.

The sensitivity derivatives of the path derivatives are obtained by differentiating Eqs. (B.1) with respect to the design variables d_ι. The resulting equations can be written in the following compact form:

$$\left[K_{IJ} + \frac{\partial G_I}{\partial Z_J}\right] \frac{\partial^{s+1} Z_J}{\partial q^s \partial d_\iota} = \frac{\partial R_I^{(s+1)}}{\partial d_\iota} - \left[\frac{\partial K_{IJ}}{\partial d_\iota} + \frac{\partial^2 G_I}{\partial Z_J \partial d_\iota} + \frac{\partial^2 G_I}{\partial Z_J \partial Z_K} \frac{\partial Z_K}{\partial d_\iota}\right] \frac{\partial^s Z_J}{\partial q^s} \tag{B.2}$$

where d_ι refers to any of the design variables and $s > 0$. For the case when $s=0$, the sensitivity derivatives are given by Eqs. (2).

FINITE ELEMENTS IN THE 90's, E. Oñate, J. Periaux, A. Samuelsson (Eds.)
Springer-Verlag/CIMNE, Barcelona 1991

A NUMERICAL APPROACH TO THE EXACT FINITE ELEMENT LINEAR BUCKLING ANALYSIS OF STRUCTURES

Z. Waszczyszyn and **M. Janus**
Cracow University of Technology,
31-155 Kraków, ul. Warszawska 24,
POLAND

J. Pieczara
Cracow Academy of Agriculture,
30-059 Kraków, ul. Mickiewicza 24,
POLAND

Summary

Instead of *a priori* assumed shape functions a numerical solution of all field equations can be used to compute the finite element matrices. Finite difference scheme with a midpoint is used to integrate the canonic set of ordinary differential equations. The equations describe the 2-point boundary value problem for unidimentional structural problems. The matrices of such exact finite elements are obtained as a result of solution of the appropriate set of linear algebraic equations. An algorithm for the analysis of a modified eigenproblem is presented. Numerical examples deal with simple bar structures for which analytical solutions exist. In addition to [13] a new example deals with buckling of cooling tower models under combined loads and a comparsion of results with those obtained experimentally is made. Canonic set of incremental equations is given in Appendix for a circular, slightly curved bar.

1. Introduction

The main idea of the finite element method for solving boundary value problems lies in subdivision of the domain into subdomains (finite elements) over which appropriate field functions are approximated by so-called shape functions. Corresponding finite elements can be called Standard Finite Elements (SFE).

If the shape functions satisfy all field equations then the appropriate FE can be called Exact Finite Element (EFE). Such EFEs are well known in the simple problems of the stiffness method. Exact stiffness matrices in linear statics and buckling can be analitically computed for uniform beams and columns. In case of tapered beam-columns instead of Bessel functions (cf.[1]) power series can be used (cf.[2]) in order to compute the stiffness matrices for EFE.

EFE can be formulated not only for bar structures but also for a class of Unidimensional Structural Problems (USP) which are governed by ordinary differential equations. Plates and shells of revolution are good examples of structures for which such USP can be formulated. Exact solution of an appropriate two-point boundary value problem leads to the exact siffness matrices. This way was

[1] Now at the Cracow Academy of Agriculture,
30-059 Kraków, ul.Mickiewicza 24, Poland

followed for conical shells [3,4], cylindrical shells [5] and circular plates [6,7]. More general formulae were given in [8].

The above mentioned solutions have been obtained for rather small number of special cases of USPs. Difficulties with integration of differential equations have influenced development of the semi-analitycal FE approach to the analysis of shells of revolution [9] and strip-type structures [10]. Instead of FEM numerical integration has been successfully applied to the analysis of highly nonlinear USPs - cf. [11].

In paper [12] the idea of EFE is considered from numerical point of view. The canonic set of linear ordinary differential equations was solved by means of forward numerical integration. Simple matrix manipulation led to the components of FE matrices. A more efficient approach was suggested in [13] where finite differences were used for integrating the differential equations.

The lattest approach in explored in the present paper. The finite differences with a midpoint are used for solving the 2-point BVP which can be formulated for a separate finite element. The finite difference model is shortly discussed for the fundamental (prebuckling) state as well for the adjacent configuration associated with the bifurcation of equilibrium states.

Linear elastic material, small strains and moderate rotations are assumed, as it is usually done in the linear buckling analysis. Appropriate sets of differential equations for a slightly curved bar is given in Appendix.

2. Finite-Difference Analysis of 2-Point BVP

In case of USP the full set of equations (geometrical, equlibrium and constitutive equations) can be transformed into the canonic set of ordinary differential equations:

$$\mathbf{y}' = \mathbf{A}(\mathbf{y},\xi)\mathbf{y} + \mathbf{p}(\xi) \equiv \mathbf{f}(\mathbf{y},\xi) \ . \tag{2.1}$$

These equations are completed by boundary conditions

$$\mathbf{B}_b \ \mathbf{y}_b = \mathbf{g}_b \qquad \text{for} \qquad b = 0, L \ . \tag{2.2}$$

In Eqs. (2.1) and (2.2) the state vector is used:

$$\mathbf{y} = \{\mathbf{q} \ \vdots \ \mathbf{r}\} \in \mathbb{R}^n \ , \tag{2.3}$$

where: $\mathbf{q}$, $\mathbf{r} \in \mathbb{R}^{n/2}$ are vectors of generalised displacements and forces correspondingly. The boundary conditions (2.1) are formulated at the end points ξ_b for $\xi = 0, L$ and $\mathbf{g}_b \in \mathbb{R}^{n/2}$, $\mathbf{B}_b \in \mathbb{R}^{n/2} \times \mathbb{R}^n$.

In what follows it is assumed that the function $\mathbf{f}(\mathbf{y},\xi)$ in Eq. (2.1) is continuous and obeys Lipshitz's conditions in the domain $(\xi,\ \mathbf{y}(\xi)) \subset D$ for $\xi \in [\xi_0, \xi_L]$.

Eqs (2.1) and (2.2) formulate the symmetric 2-Point BVP of range r = n/2. Various methods for solving such a problem have been sugested in literature (cf. e.g. [14]). A very simple method is that related to the following finite - difference scheme with a midpoint [15]:

$$y'_{i+1/2} \approx \frac{1}{\Delta\xi_i} (y_{i+1} - y_i) = \frac{1}{2} (f_{i+1} + f_i) + O\left(\frac{1}{24} y'''_{i+1/2} \ \Delta\xi_i^2\right), \tag{2.4}$$

where the subscript $\imath = 0,1,\ldots,L$ numerates the FD stations.

Applying the scheme (2.4) to (2.1) and adding (2.2) the following set of algebraic equations is obtained:

$$\mathbf{B}_0\mathbf{y}_0 = \mathbf{g}_0$$

$$\mathbf{C}_\imath\mathbf{y}_\imath + \mathbf{D}_{\imath+1}\mathbf{y}_{\imath+1} = \mathbf{g}_{\imath+1/2} \qquad \text{for} \quad \imath = 0,1,\ldots,L-1 \tag{2.5}$$

$$\mathbf{B}_l\mathbf{y}_l = \mathbf{g}_l \,,$$

where:

$$\underset{(n\times n)}{\mathbf{C}_\imath} = -\mathbf{I} - (\Delta\xi_\imath/2)\mathbf{A}_{\imath+1/2} \,, \qquad \underset{(n\times n)}{\mathbf{D}_\imath} = \mathbf{I} - (\Delta\xi_\imath/2)\mathbf{A}_{\imath+1/2} \,, \tag{2.6}$$

$$\mathbf{g}_{\imath+1/2} = (\Delta\xi_\imath/2)\,(\mathbf{p}_\imath + \mathbf{p}_{\imath+1}) \,.$$

Eqs (2.5) can be written in the following shortened form:

$$\mathbf{H}\,\mathbf{y} = \mathbf{g} \,, \tag{2.7}$$

where the matrix **H** and the vectors **y**, **g** are composed of the blocks (2.6) as it shown in Fig. 1.

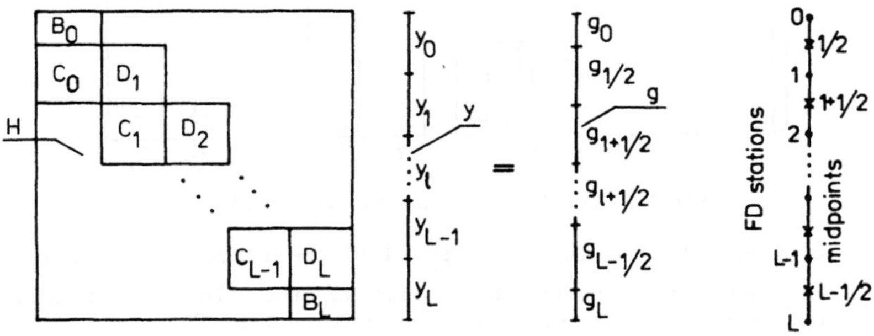

Fig. 1. Structure of the FD equation (2.7)

3. FD Equations for Linear Buckling Analysis

The fundamental and adjacent equilibrium are considered by means of the state vectors:

$$\mathbf{y} = \mathring{\mathbf{y}} + \Delta\mathbf{y} \,, \tag{3.1}$$

which correspond to the following equations:

$$\mathbf{H}_0\mathring{\mathbf{y}} = \mathring{\mathbf{g}} \equiv \Lambda\,\mathring{\mathbf{g}}^* \,, \tag{3.2}$$

$$\mathbf{H}_T\,\Delta\mathbf{y} \equiv [\mathbf{H}_0 + \mathbf{H}_G(\Lambda\,\mathring{\mathbf{y}}^*)]\,\Delta\mathbf{y} = 0 \,. \tag{3.3}$$

For linear prebuckling state the matrix $\mathbf{H}_0$ is to be formulated under asumption that the matrix $\mathbf{A}(\xi)$ in (2.1) is independent of the state vector $\mathbf{y}(\xi)$ and the load factor Λ can be used in (3.2).

4. Computation of Stifness Matrices

Let us consider the matrix form of Eq.(3.2) in the FE element domain and ξ_i for $i = 0,1,\dots,L$:

$$\mathbf{H}_0\mathbf{Y} = \mathbf{G}\ , \tag{4.1}$$

where the matrices **Y** and **G** are split into blocks:

$$\mathbf{Y} = [\mathbf{Y}_I \mid \mathbf{Y}_P]\ , \quad \mathbf{G} = [\mathbf{G}_I \mid \mathbf{G}_P]\ . \tag{4.2}$$

The subscript I is related to unit displacements of the FE ends b = 0,L respectively. The subscript P is associated with the homogeneous boundary conditions but non-zero loads. In Fig.2 the structure of Eq.(4.1) corresponds to that in Fig.1 but it is extended to n unit displacements and m load configurations.

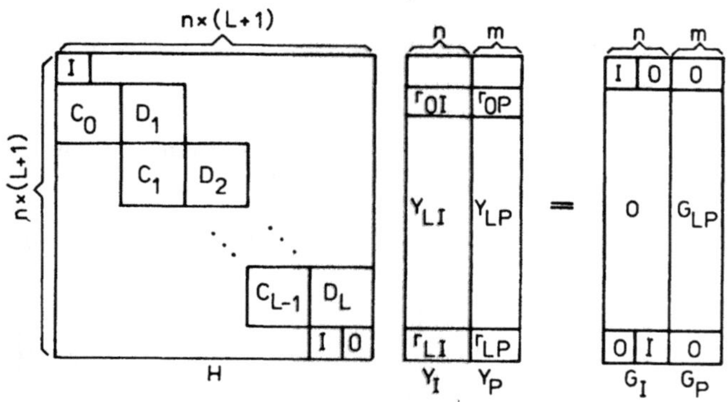

Fig.2. Structure of the matrix Eq.(4.1)

As a result of solution of Eq.(4.1) the reactions $\mathbf{r}_{bI}$, $\mathbf{r}_{bP}$ are obtained. They create the blocks of the linear stiffness matrix $\mathbf{k}_0^e$ and the vector of equivalent nodal forces $\mathbf{f}^e$:

$$\mathbf{k}_0^e = \begin{bmatrix} -\mathbf{r}_{0I} \\ \mathbf{r}_{LI} \end{bmatrix}, \quad \mathbf{f}^e = \begin{bmatrix} -\mathbf{r}_{0P} \\ \mathbf{r}_{LP} \end{bmatrix}. \tag{4.3}$$

The sign minus is put at $\mathbf{r}_{0I}$ and $\mathbf{r}_{0P}$ because of relations of the sense of those reactions to the FE system of coordinates.

The solution of the matrix equation

$$\mathbf{H}_T\mathbf{Y} = \mathbf{G}_I \tag{4.4}$$

can be used to compute the geometrical stiffness matrix $\mathbf{k}_G^e$:

$$\mathbf{k}_G^e = \mathbf{k}_T^e - \mathbf{k}_0^e \tag{4.5}$$

The geometrical stiffness matrix $\mathbf{k}_G^e$ can be computed also in an approximate way usually used in Stardard FEs. The matrix equals (notation is taken after [16]):

$$\mathbf{k}_G^e = \int_{(l)} \hat{\Gamma}^T \mathbf{S} \hat{\Gamma}\, d\xi \approx \sum_l W_l \cdot (\hat{\Gamma}^T \mathbf{S} \hat{\Gamma})_l\, \Delta\xi_l \tag{4.6}$$

where: $\hat{\Gamma}$ - matrix of displacement gradients in nonlinear part of strains, $\mathbf{S}$ - matrix of generalized stresses in prebuckling state of equilibrium.

In approximate approaches the matrix $\hat{\Gamma}$ is defined by polynominal shape functions [9] or by exact solutions which can by obtained analitycally for linear statics [17]. In the quadrature formula, shown in (4.6), appropriate columns of the solution $\mathbf{Y}_{lI}$ from Fig.2 can be used to create $\hat{\Gamma}$. The components of the matrix $\mathbf{S}$ are associated with the prebuckling state of equilibrium, usually as a result of the linear static analysis.

In case of plates and shells of revolution all functions can be extended into the trigonometric series of the circumferential variable In such a way for shells of revolution the following formulae are used [13]:

$$\overset{j}{\mathbf{k}}{}^e = c_j \begin{bmatrix} -\rho_0 \overset{j}{\mathbf{r}}_{0I} \\ \rho_L \overset{j}{\mathbf{r}}_{LI} \end{bmatrix}, \quad \overset{j}{\mathbf{f}}{}^e = c_j \begin{bmatrix} -\rho_0 \overset{j}{\mathbf{r}}_{0P} \\ \rho_L \overset{j}{\mathbf{r}}_{LP} \end{bmatrix}. \tag{4.7}$$

where $c_j = 2\pi$ for $j = 0$ and $c_j = \pi$ for $j \geq 1$; ρ_0, ρ_L are radii of the edge contours. The formula $(4.7)_1$ is valid for both the linear stiffness matrix $\overset{j}{\mathbf{k}}{}^e_0$ and the geometric stiffness matrix $\overset{j}{\mathbf{k}}{}^e_G$.

5. Algorithm of Eigenanalysis

Let us assume that nodal systems of coordinates are adopted. The transformation, assembling and fullfilling the kinematic boundary conditions lead to the standard FE set of algebraic equations with the global stiffness matrices $\mathbf{K}_0$, $\mathbf{K}_G$ and the global vector od nodal displacements $\mathbf{Q}$ respectively.

The linear buckling analysis is carried out in two stages. The linear set of equations is firstly solved for the load factor $\Lambda = 1$ and the reference load $\overset{o}{\mathbf{F}}{}^*$:

$$\overset{o}{\mathbf{K}}_0 \overset{o}{\mathbf{Q}}{}^* = \overset{o}{\mathbf{F}}{}^*, \tag{5.1}$$

where overscript $\circ$ corresponds to the fundamental (prebuckling) state.

A modified eigenvalue problem is formulated in the following form:

$$[\mathbf{K}_0 + \lambda_{(i)} \mathbf{K}_G(\Lambda_{(i)} \overset{o}{\mathbf{Q}}{}^*)]\, \mathbf{v}_{(i)} = \mathbf{0}\,. \tag{5.2}$$

The critical value of the load factor is found in an iterative way by means of the following formula:

$$\Lambda_{(i)} = \Lambda_{(i-1)}\lambda_{(i-1)} \quad \text{and} \quad \Lambda_{(1)} = 1 \quad \text{for } i = 1,\dots,n . \tag{5.3}$$

The iteration is ended if the following convergence condition is fullfilled:

$$|\lambda_n - 1| \le \varepsilon_\lambda . \tag{5.4}$$

Numerical tests, carried out for various structures, have shown that the number of iterations n = 3 - 4 for $\varepsilon_\lambda = 0.001$.

In case of axisymmetric plates and shells the stiffness matrices in (5.2) should be computed for a selected harmonics j (number of circumferential buckling waves), i.e. $\mathbf{K}_0^j$ and $\mathbf{K}_G^j(\Lambda_{(i)}\overset{o}{Q}{}^{*})$ are to be substituted in Eq.(5.2). The critical value of load factor is defined as:

$$\Lambda_{cr} = \min_j (\Lambda_{cr}^j) \tag{5.5}$$

and the corresponding buckling mode is $\mathbf{v}_{cr}^j$.

6. Numerical examples

Specific structure of the matrix **H** makes it ill-conditoned at $\Delta\xi \rightarrow 0$. That is why special attention ought to be paid to solvers both for the analysis of linear algebraic equations as well for the eigenanalysis. In comparison to [13] more satisfying solvers have been tested and used to compute the following numerical examples. In all examples the geometric stiffness matrix $\mathbf{k}_G^e$ is computed according to (4.5).

The examples are related rather to bar structures since for them many analytical solutions can be found in existing literature. The main goal of the examples is evaluation of the number of FD stations and their relation to the number of FE enough for the assumed level of accuracy.

6.1. Euler's column. The critical load for axially compressed column buckling in the plane (x,y) - cf.Fig.3a - is defined by the formula $P_{cr} = 2P_E/(1+(1-4P_E/(EA))^{1/2})$ where $P_E = \pi^2 EI_z/(4a^2)$

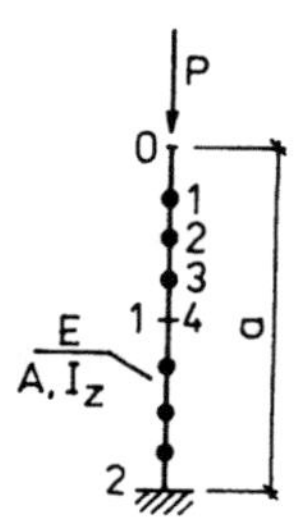

Fig.3. Euler's column with NFD=4 and NFE=2

NFD	NFE		
	1	2	4
2		2.63	0.65
4	2.63	0.65	0.16
8	0.65	0.16	0.04
16	0.16	0.04	0.01

Table 1. Error for different NFD and NFE

Computations have been carried out for $I_2/Aa^2 = 10^{-6}$. For such data the critical load $P_{cr} = 1.0005\ P_z$ and in Table 1 the following error is shown:

$$e = \frac{P^{num}_{cr} - P^{anal}_{cr}}{P^{anal}_{cr}}\ 100\% \quad , \tag{6.1}$$

where the numerical values have been obtained for a different number of finite difference stations NFD per one FE and a different number of 'exact' finite elements NFE.

It is visible in Table 1 that the error e is the same for the fixed value of product N = NFD•NFE. Another remark is that the error decreasesquadratically with increasing number of N.

6.2. Plane buckling of various columns. In Fig. 4 there are shown three columns under various loads and of various shapes. Analitycal results are computed according to formulae given in [18].

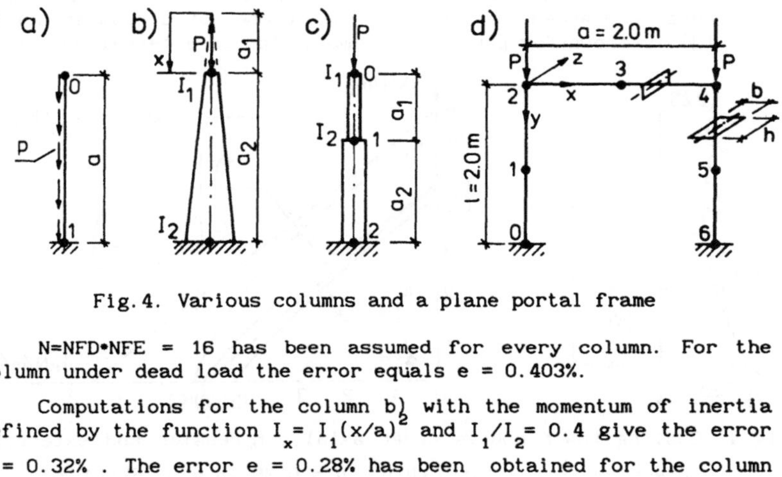

Fig. 4. Various columns and a plane portal frame

N=NFD•NFE = 16 has been assumed for every column. For the column under dead load the error equals e = 0.403%.

Computations for the column b) with the momentum of inertia defined by the function $I_x = I_1(x/a)^2$ and $I_1/I_2 = 0.4$ give the error e = 0.32% . The error e = 0.28% has been obtained for the column c) assuming $I_1/I_2 = 0.25$, $a_1/a_2 = 1.0$.

6.3. In- and out-of-plane buckling of a portal frame. A constant, rectangular cross-section hxb has been assumed. For h = 0.03m and b = 0.01m - cf. Fif. 4d - the in-plane, antisymmetric mode of buckling occurs and e = 0.47% for NFD•NFE = 8•6. The analytical value of critical loads has been computed by the formula $kl/tg(kl) = -3/(1+24I_z/(Aa^2))$ taken from [18], where $k^2 = P/(EI_z)$.

If h = 0.01m and b = 0.03m then out-of-plane buckling takes place and for other data as above the error is e = 0.16% as in the case of Euler's column at N = 16.

6.4. Buckling of circular rings under dead and follower external pressure. Critical value of plane buckling loads depends on the behaviour of external pressure. In case of dead load the formula $p_{cr} = 4\ EI_z/R^3$ is valid but for follower load the formula $p_{cr} = 3\ EI_z/R^3$ is to be used.

One quarter of the ring has been considered for NFD•NFE = 8•4. For both types of loads the errors are nearly the same: e = 1.37% for dead load and e = 1.42% for the follower load. Computations have been carried out for thin rings since the ratio h/R = 1/250 has been assumed.

6.5. Cooling tower models under combined loads. In 1983 a number of epoxy models was tested against buckling under compression and external pressure [19]. The hyperboidal shells were smooth, with one or two ring-stiffness as well - Fig.5a. Experimental stability boundaries are nearly the same for the smorth and one-ring shells.

The computations have been carried out according to shell equations taken from [13] for NFD = 10 and NFE = 4-12 . A greater number of FD stations or finite elements does not improve results more than 1-2%.

In Fig.5b an increase of buckling loads is visible if ring-stiffenerss are added. In brackets the number of circumferential waves (j) is shown for pure axial compression or external pressure.

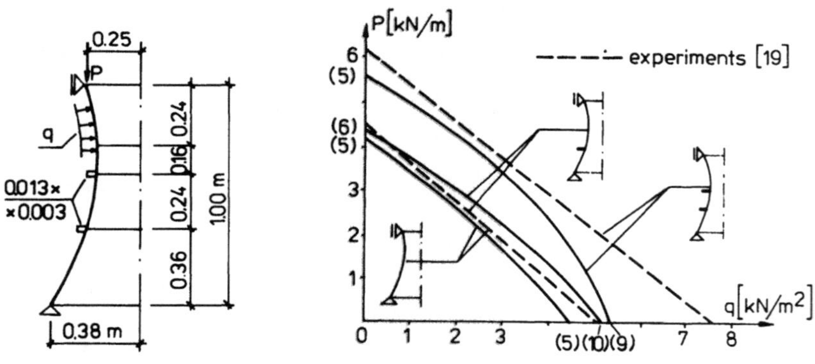

Fig.4. Cooling tower models under axial compression and external pressure ($E = 3.15 \cdot 10^6 kN/m^2$, $\nu = 0.38$, $h = 0.158 \cdot 10^{-2} m$)

7. Conclusion

The main advantage of the presented approach is that no approximate shape functions are necessary and only physical-type degress of freedom are used. In comparision to SFEs langer finite elements can be obtained. The presented FE can be called 'exact' in numerical sense, i.e. the accuracy depends only on the number of FD stations which are applied to the solution of an appropriate 2-point BVP.

The approach is under development from the view point of the analysis of thin-walled beams and box girders.

References

1. BANARJEE, J.R. and WILLIAMS, F.W. - Exact Bernoulli-Euler static stiffness matrix for a range of tapered beam-columns, Int. J.Num. Meth.Eng., **23**, 1615-1628 (1986).
2. EISENBERGER, M. - Buckling loads for variable cross-section members with variable axial forces, Int.J.Solids Struct., **27**, 135-143 (1991).
3. MEYER, G.H. and HARMON, M.B., - Conical segment method for analyzing open crown shells of revolution for edge loading, AIAA J., **4**, 886-891 (1963).
4. POPOV, E.P., PENZIEN, J., and ZUNG-AU LU - Finite element solution for axisymmetrical shells, Proc. ASCE, EM5, **90**, 119-145 (1964).
5. LAKIS, A.A. - Cylindrical finite element for analysis of cylindrical structures, Dept.Mech.Eng., McGill Univ., R. No 70-2, Montreal (1969).
6. PARDOEN, G.C. - Asymmetric vibration and stability of circular plates, Comput. Struct., **9**, 89-95 (1978).
7. SHIZHONG HAN and PILKEY, W.D. - Stiffness matrices for the static, dynamic, and buckling analysis of circular plates, Finite Elem Analysis/Design, **7**, 27-5- (1990).
8. PILKEY, W.D. - Analytically based finite elements for rods, plates and shells, Finite Elem. Analysis/Design, **7**, 123-127 (1990).
9. ZIENKIEWICZ, O.C. -The Finite Element Method in Engineering Science, McGraw-Hill, 1971.
10. CHEUNG, Y.K. - Finite Strip Method in Structural Analysis, Pergamon Press, 1976.
11. WUNDERLICH, W. et al. - A semi-analytical approach to the nonlinear analysis of shells of revolution, 'Analytical and Computational Models of Shells', Eds. Noor, A.K. et al., ASME, CED - Vol.3, 1989, pp.509-536.
12. WASZCZYSZYN, Z. - Exact finite elements in the analysis of shells of revolutions, Theor.Appl.Mech. Bulgarian Acad.Sci.,**9**, 101-104 (1978).
13. WASZCZYSZYN, Z. and PIECZARA, J. - Exact finite elements for the linear buckling analysis of unidimensional structures, 'Discretisation Methods in Structural Mechanics', Eds.Kuhn, G. and Mang,H., Springer-Verlag, 1990, pp.307-316.
14. MEYER, G.H. - Initial Value Methods for Boundary Value Problems, Academic Press, New York - Toronto, 1973.
15. BEAFAIT, W. and REDDIEN, G.W. - Midpoint difference method for analyzing beam structures, Comput.Struct., **8**, 745-751 (1978).
16. WASZCZYSZYN, Z., CICHON, Cz. and RADWAŃSKA, M. - Finite Element Method and Stability of Structures (in Polish), Arkady, Warsaw, 1990.
17. KARABALIS, D.L. and BESKOS, D.E. - Static, dynamic and stability analysis of structures composed of tapered beams, Comput. Struct., **16**, 731-748 (1983).
18. TIMOSHENKO, S.P. and GERE, J.M. - Theory of Elastic Stability, McGraw-Hill, 1961.
19. KAŁUŻA,R. - Complete buckling tests for cooling tower models of smooth and ring- stiffened shells (in Polish), Opole Hight Eng. School Internal Report, 1983.

Appendix: Stability Equations for a Slightly Curved Circular Bar

The following assumptions are adopted:

i) Bernoulli-Euler hypotheses are valid,

ii) theory of small strains but moderate rotations is used (2nd-order theory),

iii) bar is thin and its axis x is circular in the plane of symmetry (x,y) at the undeformed configuration,

iv) material is linear elastic, isotropic and homogeneous.

The state vector has the following components:

$$\mathbf{y} = \{u\ v\ w\ \varphi_x\ \varphi_y\ \varphi_z \ \vdots\ N\ Q_y\ Q_z\ M_x\ M_y\ M_z\} \in \mathbb{R}^{12} \qquad \text{(A1)}$$

Strain - displacement relations are:

$$\begin{aligned} u' &= \varepsilon_0 + v/R - (\varphi_y^2 + \varphi_z^2)/2, \\ v' &= \varphi_z - u/R, \qquad w' = -\varphi_y, \qquad \text{(A2)} \\ \varphi_x' &= \kappa_x - \kappa_y\varphi_z + (\kappa_z + k)\varphi_y, \quad \varphi_y' = \kappa_y + \kappa_x\varphi_z + (\kappa_z + k)\varphi_x, \\ \varphi_z' &= \kappa_z - \kappa_x\varphi_y + \kappa_y\varphi_x, \end{aligned}$$

where $k = 1/R$ is the curvature of bar axis and $()'=d()/dx$.

According to the above listed assumptions the physical relations are:

$$\varepsilon_0 = N/(AE), \quad \kappa_x = M_x/(GI_x), \quad \kappa_y = M_y/(EI_y), \quad \kappa_z = M_z/(EI_z). \qquad \text{(A3)}$$

On the base of formulae (A2) and (A3) appropriate equilibrium equations can be derived applying the principle of virtual work.

Stability equations are formulated from the above mentioned set of equations. After conservation of the linear terms with respect to increments of the components of state vector (A1) the following canonic form od differential equations can be obtained:

$$\begin{aligned} \Delta u' &= \Delta\varepsilon_0 + k\Delta v - \left[\mathring{\varphi}_y\,\Delta\varphi_y + \mathring{\varphi}_z\,\Delta\varphi_z\right], \quad \Delta v' = \Delta\varphi_z - k\Delta u, \quad \Delta w' = -\Delta\varphi_y, \\ \Delta\varphi_x' &= \Delta\kappa_x + k\Delta\varphi_y - \left[\mathring{\kappa}_y\Delta\varphi_z + \mathring{\varphi}_z\Delta\kappa_y - \mathring{\kappa}_z\Delta\varphi_y - \mathring{\varphi}_y\Delta\kappa_z\right], \\ \Delta\varphi_y' &= \Delta\kappa_y + k\Delta\varphi_x - \left[\mathring{\kappa}_x\Delta\varphi_z + \mathring{\varphi}_z\Delta\kappa_x - \mathring{\kappa}_z\Delta\varphi_x - \mathring{\varphi}_x\Delta\kappa_z\right], \\ \Delta\varphi_z' &= \Delta\kappa_z - \left[\mathring{\kappa}_x\Delta\varphi_y + \mathring{\varphi}_y\Delta\kappa_x - \mathring{\kappa}_y\Delta\varphi_x - \mathring{\varphi}_x\Delta\kappa_y\right], \qquad \text{(A4)} \\ \Delta N' &= k\Delta Q_y, \quad \Delta Q_y' = -k\Delta N, \quad \Delta Q_z' = 0, \\ \Delta M_x' &= k\Delta M_y + \left[\mathring{N}\,\Delta\varphi_x + \mathring{\varphi}_x\Delta N\right], \quad \Delta M_y' = \Delta Q_z - k\Delta M_x + \left[\mathring{N}\,\Delta\varphi_y + \varphi_y\Delta N\right], \\ \Delta M_z' &= -\Delta Q_y + \left[\mathring{N}\,\Delta\varphi_z + \mathring{\varphi}_z\,\Delta N\right], \end{aligned}$$

where overscript o denotes the prebuckling state and relations (A3) are used also for the increments of generalized strains in order to shorten the notation.

FINITE ELEMENTS IN THE 90'S, E. Oñate, J. Periaux, A. Samuelsson (Eds.)
Springer-Verlag/CIMNE, Barcelona 1991

AN ASSESSMENT OF THE COSSERAT CONTINUUM THROUGH THE FINITE ELEMENT SIMULATION OF A STRAIN LOCALISATION PROBLEM

J. Yu, D. Peric and D.R.J. Owen
Department of Civil Engineering,
University College of Swansea,
Swansea SA2 8PP, Wales,
UNITED KINGDOM

Abstract — *As a* panacea *for serious limitations exhibited by classical continuum models in the post-instability region, the generalization of standard J_2-elastoplasticity within the Cosserat continuum is introduced. Constitutive equations are briefly reviewed and a computational framework is modified to accomodate for additional degrees of freedom. Consistent linearization of the field equations is performed which leads to a fully implicit scheme which preserves asymptotic quadratic convergence of the Newton-Raphson method. Through the finite element simulation of the slope localisation problem for an elasto-plastic softening material the Cosserat continuum model is critically assessed against the classical continuum model.*

1. Introduction

Renewed interest in generalized continuum theories is emerging after serious limitations of the classical continuum theory have been exhibited in the analysis of strain localisation problems. The notion of a classical continuum, as understood hereafter, is based on the assumption of a *simple material* which relies on dependence of the constitutive function only on the first derivative of deformation (see Noll [1]) and Cauchy's second law which restricts the stress tensor to be symmetric.

Although significant progress in modelling strain localisation has been achieved by incorporating, for example, various constitutive descriptions (yield criterion, rate dependence), kinematic conditions (strain range, strain discontinuity), special numerical solution schemes (criterion for localisation, solution algorithms), dicretization techniques (element type, shape function, mesh dependence) (see [2-11] for details), the classical continuum may be generally regarded as describing satisfactorily the material response prior to the onset of material instability. The key problem encountered in the numerical analysis which is continued through to the post localisation regime, is the mesh size

§ On leave from *University of Science and Technology Beijing, Beijing, China* effect. Well documented numerical experience (see again [2-11]) shows that different mesh dimensions can lead to different localisation zones and with continuously reducing mesh sizes the localisation band merely narrows as much as the mesh permits. As a result, the energy dissipated inside such a localisation zone tends to zero in the absence of residual stresses and the load-displacement curve is not uniquely determined. Such a behaviour is physically unacceptable which is confirmed by experimental evidence (see [12]).

Defects experienced in the numerical analysis of strain localisation problems are fundamentally related to the physical fact that in the post localisation regime a proper description of deformation whose wave length aproaches to the scale of the characteristic size of the material becomes increasingly important. This suggests extension of the classical continuum model to generalized continuum models which would intrinsically introduce an *internal length* scale to the constitutive model. Among various theories we mention the introduction of *higher-order* deformation gradients in the constitutive equation (*gradient continuum theories*) (see [13,14]) or introduction of additional degrees of freedom within a *polar continuum* (see [15]). The mathematical structure of such generalized continuum models is equipped with the necessary ingredients to not permit change of the type of governing differential equations in the post localisation regime. This makes the mathematical problem well-posed and removes spurious mesh size dependence. The width of the shear band and spacing emerges here as a natural result of the analysis in the post localisation regime.

The Cosserat theory has been recently introduced into the finite element analysis of localisation problems by de Borst [11] in the study of localisation of an infinitely long layer under pure shear stress state. In the Cosserat continuum model an additional material parameter is required to represent the relationship between microcurvature and couple stress which contains the internal length scale. Although this constitutive model needs more theoretical and experimental investigations it is a prospective method for localisation analysis.

In the present paper the Cosserat continuum is employed in the finite element analysis of the slope localisation for an elasto-plastic strain softening material, under plane strain conditions. An elastic-predictor plastic-corrector integration scheme is developed within the operator split methodology and the consistent elasto-plastic tangent modulus for the generalized von Mises yield criterion is obtained which preserves a quadratic rate of asymptotic convergence of the Newton-Raphson method. Through numerical results, it is shown that by introducing the internal length parameter within the Cosserat continuum a definite width of shear band can be attained and that in the post localisation regime the solution converges uniformly with reduction of the characteristic element size.

2. Constitutive model for elasto-plastic Cosserat continuum

In what follows, the *von Mises* yield criterion with nonlinear strain hardening is considered, with attention restricted to plane strain conditions. A classical model of J_2-elastoplasticity may be readily generalized by introducing additional deegres of freedom within the Cosserat continuum to obtain a standard set of constitutive equations (see de Borst [11])

$$
\begin{aligned}
\dot{\boldsymbol{\varepsilon}} &= \dot{\boldsymbol{\varepsilon}}^e + \dot{\boldsymbol{\varepsilon}}^p \\
\dot{\boldsymbol{\sigma}} &= \boldsymbol{D}\dot{\boldsymbol{\varepsilon}}^e \\
\dot{\boldsymbol{\varepsilon}}^p &= \dot{\lambda}\frac{\partial f}{\partial \boldsymbol{\sigma}} \\
f &:= \sqrt{J_2} - k(\bar{\varepsilon}^p) = 0
\end{aligned}
\tag{2.1}
$$

where strain and stress may now be represented respectively in vector form as

$$
\boldsymbol{\varepsilon} = (\varepsilon_{xx}, \varepsilon_{yy}, \varepsilon_{xy}, \varepsilon_{yx}, \kappa_{zx}l, \kappa_{yz}l, \varepsilon_{zz})^T
$$

$$
\boldsymbol{\sigma} = (\sigma_{xx}, \sigma_{yy}, \tau_{xy}, \tau_{yx}, m_{zx}/l, m_{yz}/l, \sigma_{zz})^T \tag{2.2}
$$

and l is the internal-length. The yield function needs to be changed appropriately to accomodate nonsymmetry of the standard stress tensor and, in addition, to introduce the effects of the couple-stress tensor. For plane strain J_2-elastoplasticity within a Cosserat continuum the yield criterion generates a hypersurface in seven-dimensional stress space, which may be expressed as

$$
\begin{aligned}
J_2 = {} & \frac{1}{2}(s_{xx}^2 + s_{yy}^2 + s_{zz}^2) + \frac{1}{4}(\tau_{xy}^2 + \tau_{yx}^2 + 2\tau_{xy}\tau_{yx}) \\
& + \frac{1}{2}\{(m_{zx}/l)^2 + (m_{yz}/l)^2\}
\end{aligned}
\tag{2.3}
$$

or in matrix format

$$
J_2 = (\frac{1}{2}\boldsymbol{\sigma}^T \boldsymbol{P}\boldsymbol{\sigma})^{1/2} \tag{2.4}
$$

where

$$
\boldsymbol{P} = \begin{bmatrix}
\frac{2}{3} & -\frac{1}{3} & 0 & 0 & 0 & 0 & -\frac{1}{3} \\
-\frac{1}{3} & \frac{2}{3} & 0 & 0 & 0 & 0 & -\frac{1}{3} \\
0 & 0 & \frac{1}{2} & \frac{1}{2} & 0 & 0 & 0 \\
0 & 0 & \frac{1}{2} & \frac{1}{2} & 0 & 0 & 0 \\
0 & 0 & 0 & 0 & 1 & 0 & 0 \\
0 & 0 & 0 & 0 & 0 & 1 & 0 \\
-\frac{1}{3} & -\frac{1}{3} & 0 & 0 & 0 & 0 & \frac{2}{3}
\end{bmatrix} \tag{2.5}
$$

Equation (2.1d) defines the hardening law $k(\bar{\varepsilon}^p)$ in terms of the equivalent plastic strain $\bar{\varepsilon}^p$ which in rate form is given by

$$
\dot{\bar{\varepsilon}} = (\frac{3}{2}\dot{\boldsymbol{\varepsilon}}^{pT} \boldsymbol{P}\dot{\boldsymbol{\varepsilon}}^p)^{1/2} \tag{2.6}
$$

It should be noted that choice of the yield criterion for the elastic-plastic Cosserat continuum is not unique since the influence of the additional terms arising from introduction of new degrees of freedom is not easily assessed. The mathematical representation of the yield surface given by (2.3) is one possible choice which in the limit of vanishing components of the couple-stress tensor recovers the conventional J_2 yield criterion.

3. Numerical integration algorithm

3.1 Stress update algorithm

Starting from the associated flow rule (2.1c) which may be expressed in matrix format as †

$$\dot{\boldsymbol{\varepsilon}}^p = \lambda \frac{\partial f}{\partial \boldsymbol{\sigma}} = \frac{\lambda}{2\sqrt{J_2}} \boldsymbol{P}\boldsymbol{\sigma} \tag{3.1}$$

and using the standard Euler backward integation algorithm, the stresses at step n may be written as

$$\boldsymbol{\sigma}_n = \boldsymbol{D}\boldsymbol{\varepsilon}_n - \frac{\lambda}{2\sqrt{J_{2n}}} \boldsymbol{D}\boldsymbol{P}\boldsymbol{\sigma}_n \tag{3.2}$$

Rearranging this equation we obtain

$$\boldsymbol{\sigma}_n = (\boldsymbol{I} + \frac{\lambda G}{\sqrt{J_{2n}}} \boldsymbol{P})^{-1} \boldsymbol{\sigma}_n^{\text{trial}} \tag{3.3}$$

where $\boldsymbol{\sigma}_n^{\text{trial}}$ is a trial stress which can be easily obtained from the elastic constitutive law. Multiplying both sides of equation (3.3) by $\boldsymbol{\sigma}^T \boldsymbol{P}$ and substituting into equation (2.1d), the yield function has the form

$$f_n(\lambda) = \sqrt{J_{2n}^{\text{trial}}} - G\lambda - (\sigma_Y + H\bar{\varepsilon}^P)/\sqrt{3} = 0 \tag{3.4}$$

To solve (3.4) for consistency parameter λ this expression is linearized which results in

$$\lambda^{(k+1)} = \lambda^{(k)} + \frac{\sqrt{J_{2n}^{(k)}} - \sigma_Y/\sqrt{3}}{G + H/3} \tag{3.5}$$

and then a local Netwon-Raphson iteration is performed in a standard manner.

3.2 Consistent tangent operator

By differentiating equation (3.2) and rearranging one obtains

$$d\boldsymbol{\sigma}_n = \boldsymbol{D}_n^*(d\boldsymbol{\varepsilon}_n - \frac{d\lambda}{2\sqrt{J_{2n}}} \boldsymbol{P}\boldsymbol{\sigma}_n + \frac{\lambda}{2J_{2n}} \boldsymbol{P}\boldsymbol{\sigma}_n d\sqrt{J_{2n}}) \tag{3.6}$$

where

$$\boldsymbol{D}_n^* = (\boldsymbol{D}^{-1} + \frac{\lambda}{2\sqrt{J_{2n}}} \boldsymbol{P})^{-1} \tag{3.7}$$

† In the following, to simplify notation, $\dot{\lambda}$ is replaced by λ.

Analogously, differentiation of the yield function (2.1) gives

$$d\sqrt{J_{2n}} = \frac{\sigma_n^T}{2\sqrt{J_{2n}}} P d\sigma_n = \frac{H}{3} d\lambda \tag{3.8}$$

Substituting (3.8) into (3.6) gives

$$d\lambda = \frac{\sigma_n^T P D_n^* d\varepsilon_n}{\{\frac{2\sqrt{J_{2n}}H}{3} + (\frac{1}{2\sqrt{J_{2n}}} - \frac{\lambda H}{6J_{2n}})\sigma_n^T P D_n^* P \sigma_n\}} \tag{3.9}$$

which together with equation (3.6) and (3.9) allows the incremental elastic-plastic constitutive relation to be recast in matrix format as

$$d\sigma_n = D_n^{**} d\varepsilon_n \tag{3.10}$$

where

$$D^{**} = D_n^* - \frac{D_n^* P \sigma_n \sigma_n^T P D_n^*}{K_1 + \sigma_n^T P D_n^* P \sigma_n} \tag{3.11}$$

is the *consistent tangent modulus* for a Cosserat continuum model and

$$K_1 = \frac{4 J_{2n} H}{3(1 - \frac{\lambda H}{3\sqrt{J_{2n}}})} \tag{3.12}$$

4. Remarks on the formulation of the boundary value problem and finite element discretization

Introducing the additional degrees of freedom within the Cosserat continuum the displacement vector now takes the form

$$\boldsymbol{u} = (u, v, \omega)^T \tag{4.1}$$

which upon discretization may be written

$$\boldsymbol{u}^e = \boldsymbol{N}\boldsymbol{a}^e \tag{4.2}$$

where $\boldsymbol{a}^e$ denotes the collection of nodal displacements and $\boldsymbol{N}$ are standard shape functions. The rotational dof ω is here interpolated by the same shape functions as the translational dof u and v.

The strain vector then may be expressed as

$$\boldsymbol{\varepsilon}^e = \boldsymbol{B}\boldsymbol{a}^e \tag{4.3}$$

where

$$\boldsymbol{B} = \boldsymbol{L}\boldsymbol{N} \tag{4.4}$$

is the strain-displacement matrix and $\boldsymbol{L}$ is the differential operator which may be represented as

$$\boldsymbol{L} = \begin{bmatrix} \frac{\partial}{\partial x} & 0 & 0 \\ 0 & \frac{\partial}{\partial y} & 0 \\ 0 & \frac{\partial}{\partial x} & -1 \\ \frac{\partial}{\partial y} & 0 & 1 \\ 0 & 0 & l\frac{\partial}{\partial x} \\ 0 & 0 & l\frac{\partial}{\partial y} \end{bmatrix} \tag{4.5}$$

Upon linearization of the weak form of the equilibrium of the boundary value problem we arrive at the system of algebraic equations

$$\Sigma_e\{\int_{\Omega^e} \boldsymbol{B}^T \boldsymbol{D}^{**} \boldsymbol{B}\ dv\ \boldsymbol{a}^e\} = \Sigma_e \boldsymbol{f}^e_{out} \tag{4.6}$$

where

$$\boldsymbol{f}^e_{out} = \int_{\Omega^e} \boldsymbol{N}\boldsymbol{b}\ dv + \int_{\partial\Omega^e} \boldsymbol{N}\boldsymbol{t}\ ds \tag{4.7}$$

is a vector of the out-of-balance forces.

Since zero order derivatives are included in the $\boldsymbol{B}$ matrix, the integration rule corresponding to a particular set of shape functions should be selected with care. For the Cosserat continuum model which includes second order of the deformation gradient quadratic shape functions are generally needed. To obtain an accurate integration in equation (4.6) a nine gauss point integration rule, for example, must be used for eight-noded isoparametric quadrilateral finite elements.

5. Numerical example

The finite element simulation of the slope localisation of the elastic-plastic softening material is presented in this section to assess the Cosserat continuum model against the classical approach. A full Newton-Raphson method is employed, with a tangent stiffness arising from the linerization described in Section 3. Converegence of the finite element solution is established on the basis of the standard Euclidian norm of the out-of-balance forces. The use of consistent tangent moduli (3.16-17) is shown to be of utmost importance for preserving the quadratic rate of convergence typical for the Newton-Raphson method.

EXAMPLE 5.1 Slope localisation problem for an elastic-plastic softening material. The geometric and material characteristics for this example are shown in Figure 5.1. The footing is chosen to be composed of a 100 times stiffer material then the rest of the structure. Loading is enforced by prescribing displacement increments along the Y direction at point A while the other two degrees of freedom are set free. Displacements U, V on the bottom boundary and U along the vertical boundary are constrained to zero. Computation is performed for three different finite element meshes, denoted A, B and C, consisting of 120, 420 and 920 eight-noded isoparametric finite elements, respectively. Full integration is applied based on nine integration points.

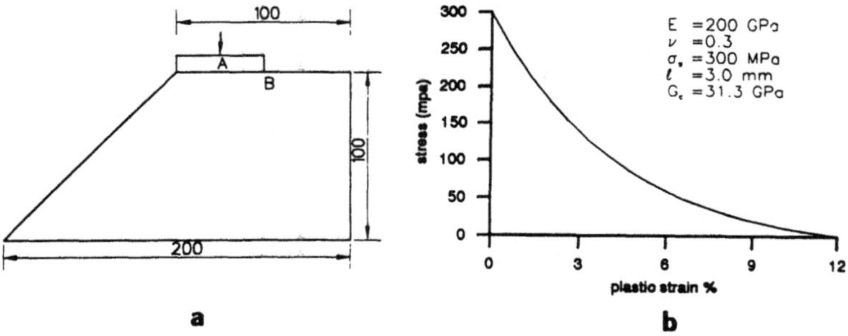

Fig.5.1: Slope localisation for an elastic-plastic softening material: Geometry and material characteristics.

In [10], Larsson considered this problem in a plastic localisation analysis with an adaptive technique. The results showed a strong mesh dependence. The localisation patterns rely on the mesh, that is, with a biased mesh in which elements were arranged along the possible sliding direction in which localisation occured; otherwise a diffuse pattern was obtained. Furthermore, in an adaptive mesh in which the element size was very small in the assumed localisation zone the width of the shear band was limited to one element size (it should be noted that in such a case standard adaptivity techniques break down due to the loss of ellipticity of the underlying differential equations).

To provide a basis for comparison, the analysis is performed with a conventional elastoplastic model and the results obtained are illustrated in Figure 5.2 for various meshes employed. Although localisation has been achieved for all three meshes, the results show a strong mesh dependence. As is clearly illustrated in Fig.5.2(d-f), not only does the width ot the shear band differ significantly for different meshes, but the localisation pattern changes as the number of finite elements increases. In Fig.5.2(c,f) which presents deformed mesh and contour plots of equivalent plastic strain for mesh C the shear band branches in two distinct shear bands with the finite elements exhibiting randomly organized pattern inside the band. This is complemented by a sudden drop in the force-displacement diagram shown in Fig.5.5. Such behaviour may be explained by the loss of ellipticity of the underlying differential equations which indicates localisation and is closely related to the non-uniqueness of solution in the post-localisation regime.

Results obtained within the framework of a Cosserat continuum are presented in Fig.5.3 for the final deformed configurations and various finite element meshes. Although some dependence of the width of shear band on the finite

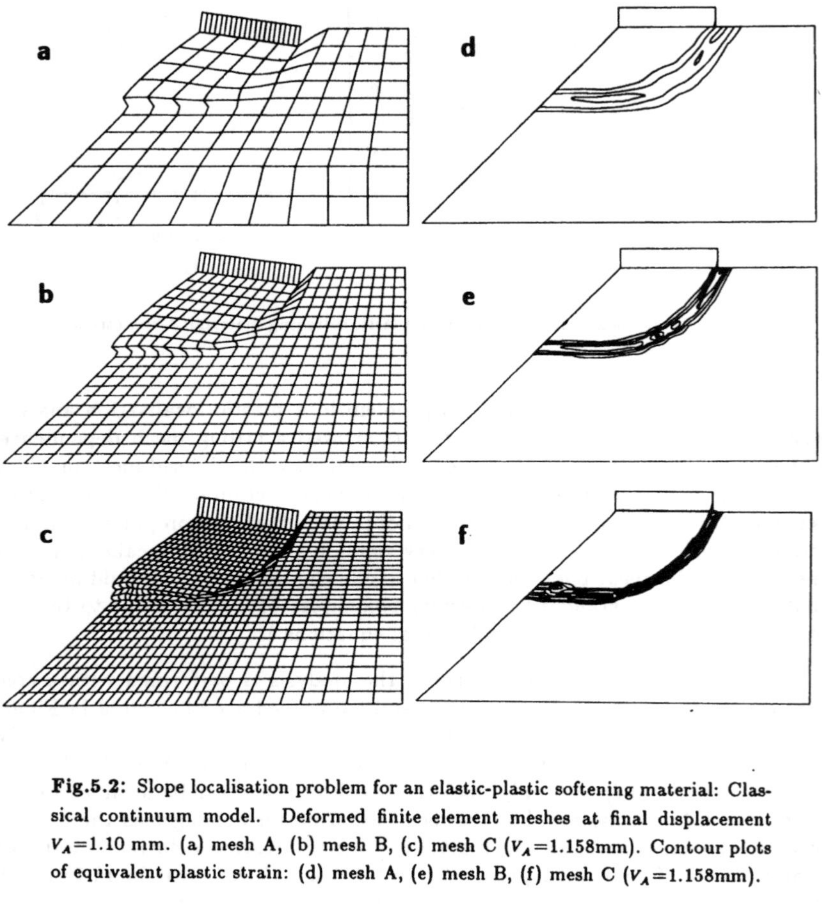

Fig.5.2: Slope localisation problem for an elastic-plastic softening material: Classical continuum model. Deformed finite element meshes at final displacement v_A=1.10 mm. (a) mesh A, (b) mesh B, (c) mesh C (v_A=1.158mm). Contour plots of equivalent plastic strain: (d) mesh A, (e) mesh B, (f) mesh C (v_A=1.158mm).

element mesh is visible in Fig.5.3(d-f) and may be explained by difficulties of representing the localisation pattern as a short deformation wave within the spatial finite element discretization on the macroscale, a trend towards a definite width of the shear band is observed. Contrary to results within the classical continuum, the corresponding finite element meshes show regular patterns as depicted in Fig.5.3(a-c) and Fig.5.4.

A uniform convergence of the finite element solution for this boundary value problem exhibited by the Cosserat continuum model as opposed to the solution by the classical continuum approach is illustrated in Fig.5.5 which presents force-displacement diagrams for both models and for the various meshes employed in

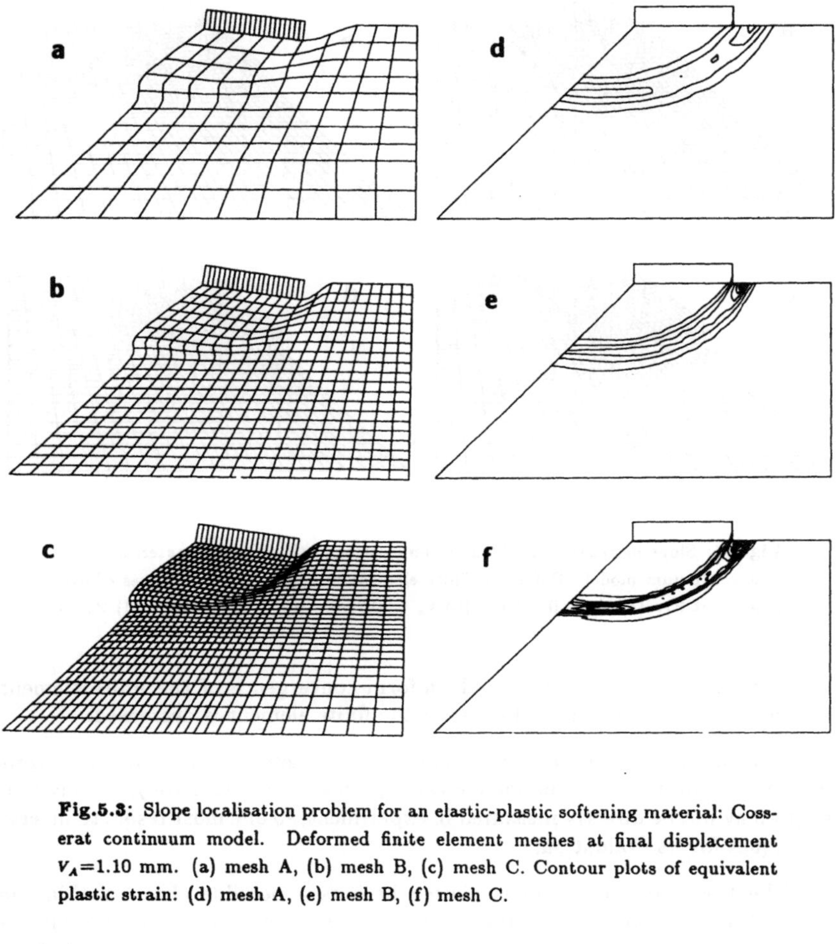

Fig.5.3: Slope localisation problem for an elastic-plastic softening material: Cosserat continuum model. Deformed finite element meshes at final displacement v_A=1.10 mm. (a) mesh A, (b) mesh B, (c) mesh C. Contour plots of equivalent plastic strain: (d) mesh A, (e) mesh B, (f) mesh C.

analysis.

6. Summary and conclusions

Following the pioneering work of de Borst [11], the computational framework for the elastic-plastic Cosserat continuum has been outlined and applied to the finite element analysis of the slope localisation of an elastic-plastic strain softening material. As a consequence of introduction of an internal length within the constitutive model, well posedness of the underlying mathematical problem in the post-localisation regime is attained, which:

(i) effectively removes the spurious mesh dependence of the solution.

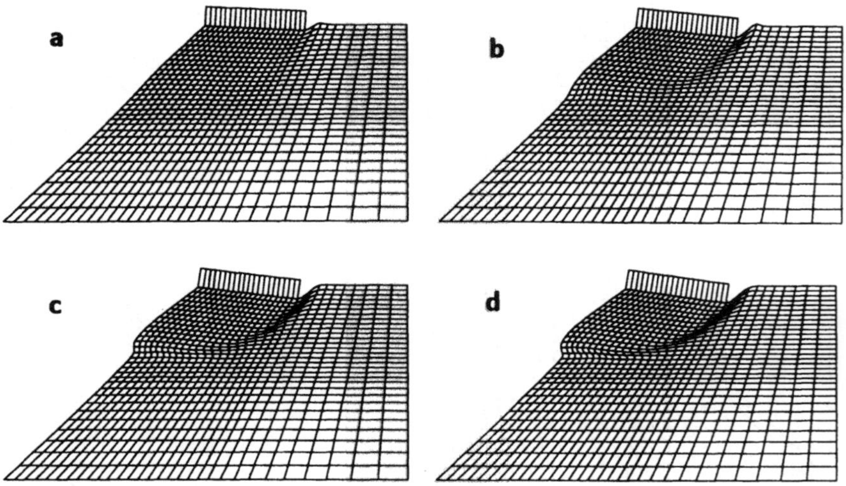

Fig.5.4: Slope localisation problem for an elastic-plastic softening material: Cosserat continuum model. Deformed finite element meshes at various stages of displacement v_A. (a) $v_A = 0.395$mm (b) $v_A = 0.70$mm (c) $v_A = 1.90$mm (d) $v_A = 1.10$mm.

(ii) generates unique solution and uniform convergence of the finite element results as the characteristic size of the finite element is decreased.

The introduced internal length scale is a parameter which needs to be *apriori* specified for every new boundary value problem. Unfortunately, the internal length may not be directly measured in experiment, so one must resort to inverse techniques for its evaluation.

The Cosserat continuum is viewed here as a possible choice within the whole class of generalized continuum models, as a remedy for the limitations of the conventional continuum approach. The gradient dependent theories, for example, clearly exhibit advantages when applied to localisation in mode I (see [19]).

References

[1] W.NOLL, 'A mathematical theory of the mechanical behavior of continuous media', *Arch. Rat. Mech. Anal.*, **2**, 198–226 (1958)

[2] V.TVERGAARD, A.NEEDLEMAN and K.K.LO, 'Flow localization in the plane strain tensile test', *J. Mech. Phys. Solids*, **29**, 115-142 (1981)

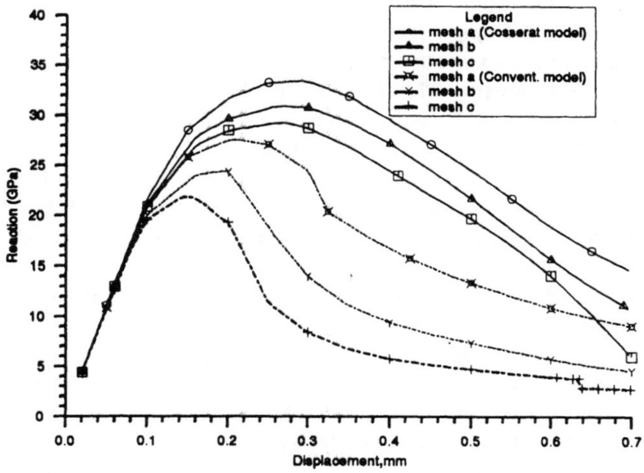

Fig.5.5: Slope localisation problem for an elastic-plastic softening material: Force versus displacement V_A diagram.

[3] R.DE BORST, 'Bifurcations in finite element models with a non-associated flow law' , *Int. J. Num. Anal. Meth. Geomech.*, **12**, 99–116 (1988)

[4] A.NEEDLEMAN and V.TVERGAARD, 'Finite element analysis of localization in plasticity', In *Finite Elements V.- Special Problems in Solid Mechanics*, (Edited by J.T. Oden and G.F. Carey), 94–157, Prentice-Hall, Englewood Cliff, New Jersey, 1984

[5] Y.LEROY and M.ORTIZ, 'Finite element analysis of strain localization in frictional materials' , *Int. J. Num. Anal. Meth. Geomech.*, **13**, 53–74(1989)

[6] M.ORTIZ, Y.LEROY and A.NEEDLEMAN, 'A finite element method for localized failure analysis' , *Comp. Meth. Appl. Mech. Eng.*, **61**, 189–214 (1987)

[7] D.PERIĆ, D.R.J.OWEN and M.E.HONNOR, 'A model for large strain elastoplasticity based on logarithmic strain: computational issues' , *Comp. Meth. Appl. Mech. Engng.*, (accepted for publication), (1991)

[8] A.NEEDLEMAN, 'Non-normality and bifurcation in plane strain tension and compression', *J. Mech. Phys. Solids,* **27**, 231–254 (1979)

[9] O.C.ZIENKIEWICZ and G.C.HUANG, 'A note on localization phenomena and adaptive finite-element analysis in forming processes' , *Comm. Appl. Num. Meth.*, **6**, 71–76 (1990)

[10] R.LARSSON, *Numerical Simulation of Plastic Localization*, Publication 90:5, Chalmers University of Technology, Gōteborg, (1990)

[11] R.de Borst, 'Simulation of strain localization: a reappraisal of Cosserat continuum', *Eng. Comput.*, (accepted for publication) (1990)

[12] H.B.Mülhaus and I.Vardoulakis, 'The thickness of shear band in granular materials', *Géotechnique*, **37**, 271–283 (1987)

[13] B.D.Coleman and M.L.Hodgdon, 'On shear band in ductile materials', *Arch. Rat. Mech. Anal.*, **90**, 219–247(1985)

[14] N.Triantafyllidis and E.C.Aifantis, 'A gradient approach to localization of deformation: I. Hyperelastic materials', *J. Elasticity*, **16**, 225–237(1986)

[15] R.D.Mindlin, 'Influence of couple-stresses on stress concentrations', *Exper. Mech.*, **2**, 1–8 (1962)

[16] D.R.J.Owen and E.Hinton, *Finite Elements in Plasticity*, Pineridge Press, Swansea (1980)

[17] R.D.Wood, 'Finite element analysis of plane couple-stresses problems using first order stress functions', *Int. J. Num. Meth. Eng.*, **26**, 489–509 (1988)

[18] O.C.Zienkiewicz and R.L.Taylor, *The Finite Element Method. Volume 2*, McGraw-Hill, Maidenhead, 4-th ed. (1991)

[19] R.de Borst and H.-B.Mülhaus, 'Computational strategies for gradient continuum models with a view to localization of deformation', In *Nonlinear Engineering Computations*, (Edited by N.Bićanić et al.), 239–260, Pineridge Press, Swansea (1991)

FINITE ELEMENTS IN THE 90's, E. Oñate, J. Periaux, A. Samuelsson (Eds.)
Springer-Verlag/CIMNE, Barcelona 1991

SIMULATED HEAD IMPACTS WITH UPPER INTERIOR STRUCTURES USING RIGID AND ANATOMIC BRAIN MODELS

F. DiMasi
RSPA/Volpe National Transportation Systems Center, USA

P. Tong
Hong Kong University of Science and Technology, HONG KONG

J.H. Marcus, H.C. Gabler III and R.H. Eppinger
NHTSA/Office of Crashworthiness Research, USA

ABSTRACT:

This article first describes modeling and analysis of relatively rigid head impacts with upper interior structures and compares results with laboratory test measurements. In addition to demonstrating the capacity of a well defined model to accurately replicate dynamic response and to predict levels of injury mitigation obtainable from structural modification, such models are also capable of estimating physical quantities (e.g. detailed stress/strain patterns) known to affect the functional capacity of brain tissue. Measuring these quantities in a test program would be virtually impossible. To provide estimates of such quantities, a first generation anatomic brain model was developed to estimate strains induced in the brain as a result of typical head impacts sustained in automotive collisions. The latter model is viewed as the first step in a complicated process, leading towards the ability to assess soft tissue injuries, and to the development of an expanded head-injury criteria which addresses specific forms of brain injuries known to result from automobile crashes.

INTRODUCTION AND OVERVIEW:

In addition to automobile crash testing, NHTSA conducts a wide range of structural and biomechanical studies aimed at mitigating injuries sustained in automobile collisions. Although finite element analysis has been available for many years, recent advances in computing speed and the development of robust explicit integration computer codes have made the simulation of transient impact events typical of the automobile collision environment very practicable. In particular, the finite-element codes INGRID, DYNA3D and TAURUS [1,2,3] developed by the Department of Energy's Lawrence Livermore National Laboratory (LLNL), provide for analyzing the large deformation response of inelastic solids and structures in three dimensions. Contact-impact algorithms within the explicit integration code DYNA3D make this code particularly useful for simulating transient impact events involving large displacements and sliding contact between impacting surfaces as well as for computing detailed physical quantities within the continuum.

These public domain codes are currently being used by the aerospace industry, foreign and domestic auto manufactures [4], and by agencies of the U.S. government. In the mid-eighties, NHTSA and VNTSC began to apply these codes to structural design problems relating to head injuries sustained in automobile collisions, since this class of injuries is a major cause of death and serious trauma. Modeling of thoracic impact and chest deflections in response to impact

[1] Currently on leave from RSPA/Volpe National Transportation Systems Center

loading from contact with steering columns and wheels, shoulder harness, belts and air bags, is also in process.

Initial head-impact simulations modeled the headpart as a relatively simple, homogeneous elastic body with inertial properties similar to a Hybrid III crash dummy head. Results of simulated impacts with upper interior structures were very successful and demonstrated the feasibility of utilizing these tools to predict head-impact response [5]. These simulations also provide computations of detailed stress/strain distributions, within the continuum, some of which would be virtually impossible to measure in a laboratory setting. Prediction of such quantities will be required to develop a better understanding of collision mechanics and injuries to the brain which are responsible for so many of the devastating injuries sustained in automotive collisions.

The current head injury criteria (HIC) as specified in Federal Motor Vehicle Safety Standard (FMVSS) 208 evolved from the Wayne State Tolerance Curve [6] and provides a quantitative indication of the potential for serious head injury based on translational acceleration measurements at the center of gravity of a dummy head. The Wayne State Tolerance Curve, which relates the probability of a head injury to a combination of time and acceleration, was developed based on a number of tests in which the heads of human cadaveric subjects were dropped onto rigid and padded plates, and post test autopsies revealed the resulting injuries. Because cadaveric subjects were used it was not possible to evaluate functional impairments such as memory loss, paralysis, emotion changes, or intellectual impairments, nor could vascular injuries such as hematoma be evaluated because the vascular system was not perfused. Thus, while HIC has been shown to be indicative of many forms of head injury, questions have been raised about its sensitivity to certain classes of brain injuries resulting from automobile crashes. These brain injuries, called Diffuse Axonal Injuries (DAI), do not have an identifiable lesion but result in neurological impairments. In addition, vascular injuries such as sub-dural hematomas may not be adequately predicted by HIC.

Results of recent experimental and pathophysiological studies performed by the University of Pennsylvania suggest that the functional impairment of a single nerve cell is directly related to the magnitude of strain imposed on the cell, and that the extent of injury to brain tissue is strongly related to both the magnitude and direction of applied loading [7]. These findings also indicate that strain levels developed in brain tissue and in bridging veins (the blood vessels between the top of the brain and the inside of the skull), are particularly sensitive to rotational acceleration. Bridging vein disruption is believed to be the cause of many brain hematoma. One can immediately imagine the difficulties involved in measuring and analyzing these quantities, even in laboratory experiments using relatively simple geometric shapes to emulate the complex geometry of the brain [8].

For these reasons, subsequent activities were aimed at developing a first generation anatomic brain model as a predictive tool for estimating strains induced in the soft cortical tissues of the brain in response to dynamic forces and accelerations representative of the automobile collision environment. The more detailed anatomic model was subjected to impacts with upper interior structures and/or surfaces (such as the rails and posts supporting the roof of the vehicle) previously modeled. The following paragraphs first describe impact analyses conducted with relatively rigid headparts and subsequently, the development of an anatomic brain model for estimating strains induced in brain tissue in response to similar impacts. Closing thoughts outline the direction of future work and how these activities relate to development of improved head injury criteria.

RIGID BODY HEADPART/A-PILLAR IMPACT SIMULATIONS:

The A-pillar is the post supporting the roof of a car at the point where the front windshield and the side of the car meet, and is frequently struck by the driver's or passenger's head during frontal or lateral crashes. To evaluate potential injury mitigation efforts associated with head/A-pillar impacts, component-level laboratory tests were conducted at the Vehicle Research and Test Center (VRTC) wherein various padding materials were added to 1981 Chevrolet Citation A-pillars. Analytical studies were also conducted at VNTSC using finite element models to support hardware development efforts at VRTC and to evaluate additional countermeasure options for experimental evaluation.

The tests at VRTC utilized a Hybrid III crash dummy head, with a modified posterior surface to permit the headform to be launched into free flight with controlled speed and direction using a hydraulically actuated impact accelerator [9]. The Free Motion Head-Form (FMHF) was instrumented with a nine-accelerometer package (NAP) capable of measuring three rotational as well as three translational head accelerations, and was launched into baseline (i.e. unpadded) or padded Citation A-pillars at speeds of 20 and 25 mph. The headpart was positioned at the height of a normally seated 50th percentile occupant facing the A-pillar. This resulted in a velocity vector of 45 degrees relative to the vehicle longitudinal axis, and an impact location near the top of the pillar. A complete test description is provided in [9,10].

A-PILLAR STRUCTURAL CONFIGURATION AND MODELING - The Citation A-pillar used in the above tests is a hollow curved structure, fabricated from three formed sheet stock sections. To provide descriptions of the A-pillar's material, geometry, and construction, required for developing a computer model with structural fidelity, several General Motors X-body (the GM platform designation for the Citation) A-pillars were purchased from a recycled auto parts dealer and used for geometry, cross-section and material property measurements.

Figure 1a illustrates a representative pillar cross-section used for generating the mesh model, and a schematic illustration of the A-pillar/door/windshield interface. Although some minor variations in thickness and cross section were noted along the length of the A-pillar, these variations were considered insignificant and the pillar's cross-section was modeled as uniform along its length.

The principal contact surface and FMHF velocity vector (projected into the plane of the cross-section) is also shown. The windshield is attached to the pillar using a bonded rubber seal/support and the door assembly abuts a thick rubber molding attached to the A-pillar flange. Considering this construction, the windshield and door assembly were considered to have a minimal influence on A-pillar stiffness and were not modeled. A thin plastic finishing cover (not shown) was also disregarded for similar reasons.

Figure 1b contains an "exploded-view" mesh plot of the three formed sheet metal sections comprising the A-pillar model to illustrate construction and modeling features. The sections are assembled using a combination of spot welds and tack welds along the lengths of adjoining sections. These welds are modeled using combinations of tied and sliding slide-surfaces options. (Later versions of INGRID support a spot-weld slide-line option.) The three-segment A-pillar is modeled with 550 shell elements.

Material testing of A-pillar coupons was conducted by MIT's Department of Material Science and Engineering. Measurement accuracy was 5% or better on yield strength and 10% or better for small stains beyond yield. Past yield, material behavior was characterized as elastic with linear strain hardening or elastic-perfectly plastic. Material properties of the three sections comprising the Citation A-pillar are shown below.

Section (ref Fig 1b)	Thickness (in.)	Yield Stress (psi)	Tangent Modulus (psi)
A	.063	33,000	1.5 x 10E6
B	.040	48,000	perf. plastic
C	.049	28,000	2.2 x 10E6

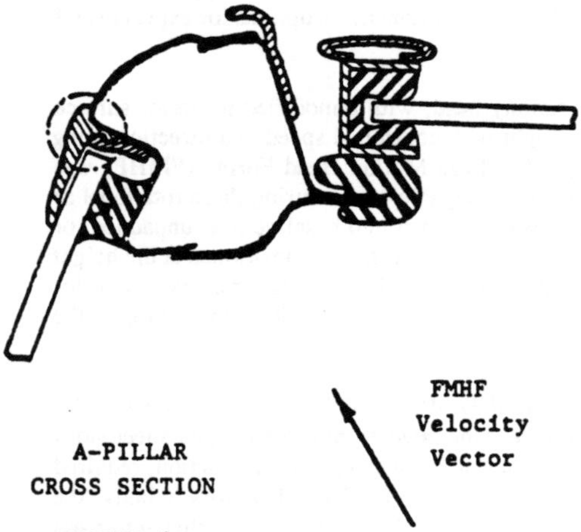

A-PILLAR
CROSS SECTION

FIGURE 1a

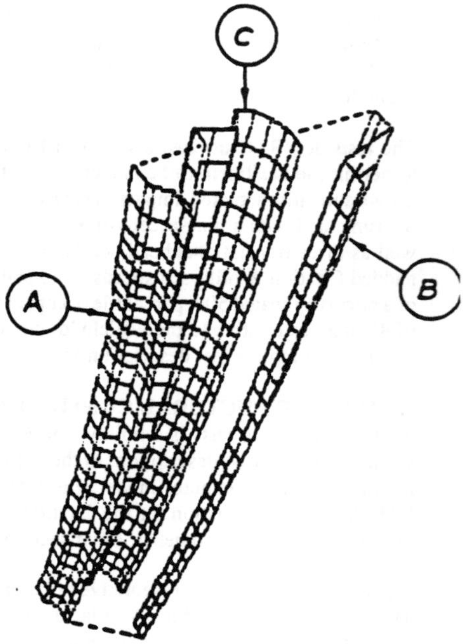

A-PILLAR EXPLODED VIEW
FIGURE 1b

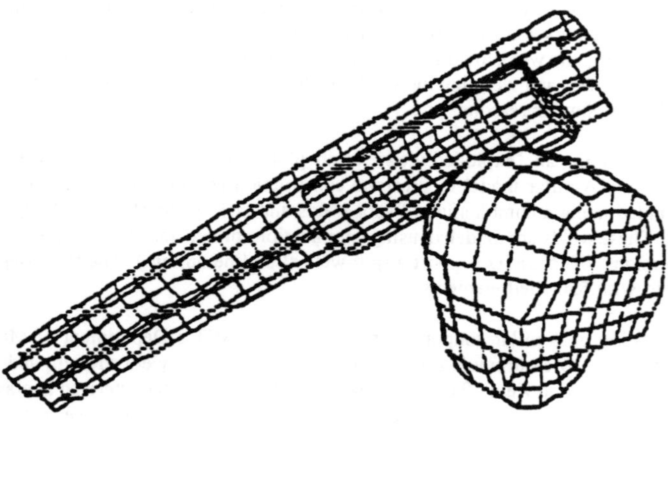

ASSEMBLED FMHF/A-PILLAR/CUSHION MODEL

FIGURE 2

Nodes located at the top and bottom extremities of the A-pillar were constrained to simulate structural stiffness effects. This is an important approximation since it has potential influence on the impact response.

THE FREE-MOTION HEADFORM was modeled using solid elements, to approximate geometric profile, mass and inertial properties. An "equivalent" elastic modulus was defined by simulating a drop-test used to calibrate FMHF accelerometers prior to impact testing. This procedure involves dropping the FMHF from a specified height onto a smooth steel surface and requires that the resultant acceleration measured at the head center of gravity (CG) to be between 225 and 275 g's. An elastic modulus of 150,000 psi for the simulated FMHF was found to closely replicate the impact response measured at the CG of the modified HYBRID III headpart during calibration tests.

SIMULATION OF A-PILLAR PADDING - Force-deflection data was developed for a range of candidate padding materials by firing the FMHF into cushions comprised of various padding materials, thicknesses and combinations. This data was used to develop relationships between stroke and stiffness parameters and probable HIC values expected when padding was added to the A-pillar. Material constants were again estimated (first for visco-elastic and later, crushable foam material models) by simulating acceleration-based force-deflection data measured during FMHF impacts with padding samples. This provided elastic constants for the simulated padding material which replicate the impulse response characteristics of the actual padding material. The simulated A-pillar padding component was modeled to conform with the general proportions of the cushion used in laboratory testing, using 311 solid elements.

ASSEMBLED A-PILLAR MESH MODEL - The assembled head/A-pillar/cushion model comprised of the various parts described above is illustrated in Figure 2. The baseline configuration uses the identical mesh (with the headpart moved closer to the pillar). The assembled model consists of approximately 900 elements for the baseline A-pillar and 1,100 elements with the padding. Typical simulation times were 10 to 12 msec. Calculations were performed on NHTSA's DEC VAX/8350 computer and these simulations typically required 15 to 24 CPU hours of computer time per simulation.

RESULTS/COMPARISON WITH TEST DATA - The mesh model described above was used to simulate 20 and 25 mph impacts for each configuration, with all restraint conditions and material properties held invariant. However, since the exact impact location on the pillar could not be exactly defined, and since it was necessary to reposition the FMHF propulsion apparatus during testing, some small variations in headpart initial location were permitted in order to match impact severities measured during testing.

Principal evaluation criteria used to compare test vs simulated impacts included resultant accelerations at the head CG and computed HIC values. Table 1 contains a summary of measured vs computed peak resultant accelerations, computed HIC values and impact severity for the matrix of four test conditions simulated. Resultant acceleration time histories compared very favorably with test data in terms of amplitude and duration, overall pulse shape and net change in velocity. (The one exception was the 20 mph baseline condition where the simulated impact severity was about 8 percent higher than measured.)

Predicted vs measured reductions in HIC resulting from the addition of padding are:

Speed	Percent Reduction in HIC	
(mph)	Via Testing	Via Simulation
20	53%	56%
25	64%	55%

TABLE 1: COMPARISON OF TEST VS SIMULATION DATA FOR A-PILLAR IMPACTS

	TEST			SIMULATION		
Configuration	Peak G's	HIC No.	Delta V (mph)	Peak G's	HIC No.	Delta V (mph)
20 mph Baseline	194	1058	14.8	245	1428	16.0
25 mph Baseline	312	2666	18.0	310	2475	19.4
20 mph Padded	130	496	17.5	150	617	17.5
25 mph Padded	165	973	21.5	160	1113	20.0

These results are particularly satisfying considering the relatively crude headpart model used, approximated A-pillar end restraints, and uncertainties regarding cushion material properties, and contact point location.

The results presented above clearly demonstrate the feasibility of modeling head/A-pillar impact mechanics using the three dimensional finite-element tools discussed above, and the ease with which the model may be changed to evaluate different speeds and/or A-pillar configurations. Furthermore, a well calibrated model is also capable of predicting detailed physical quantities within the continuum, which would be virtually impossible to measure in a test program.

An instance where the latter capability has great potential application, is in predicting strains developed in soft cortical brain tissue in response to a head impact. Since many of the most traumatic head-injuries involve the permanent loss of function resulting from microscopic damage to soft tissue in which there is no observable lesion, and since the research has shown this microscopic damage is related to and caused by the strain imposed on brain tissue, the application of finite element techniques to this problem has great potential. We describe these efforts next.

FIRST GENERATION ANATOMICAL BRAIN MODEL:

Recent experimental and pathophysiological results suggest excessive strain experienced by soft cortical tissues is a leading cause of brain injury and that the extent of injury to soft cortical tissue is strongly related to both the magnitude and direction of loading applied to the head. These findings also indicate that strain levels developed in soft tissues and bridging veins are particularly important in assessing the potential for injuries such as Diffuse Axonal Injury (DAI), and Sub Dural Hematoma (SDH). In DAI, axons (i.e. single nerve cells) experience excessive stretching causing the axon's membrane to loose its ability to maintain the internal external calcium ion concentration ratio (approximately 50,000 times more calcium outside the cell). As the internal calcium level climbs the axon looses its ability to function. Functional impairment occurs before mechanical destruction of the axon is seen.

The difficulties involved with measuring and analyzing detailed stress/strain distributions even in laboratory experiments using simple geometric shapes [8] to emulate the complex geometry of the human brain is readily apparent. To provide an understanding of these strain fields, a first generation anatomic brain model was developed to calculate strains developed in brain tissue. Such a model will also be useful in studying the effects of directional loading, scaling or extrapolating injury data from primates to humans, and developing a better understanding of relationships between impact kinematics and injuries to soft cortical tissues. These injuries are responsible for many devastating brain injuries sustained in automotive collisions.

Before describing the model and simulated responses, it is important to note that additional research will be required before results from such a model can be related to injuries sustained by humans in the collision environment. Most importantly, basic research is needed to provide (1) a definition of damage tolerance limits in terms of maximum mechanical strains, strain rates and/or other physical quantities which can be related to various tissue failure, trauma and/or loss of function, and (2) improved constitutive relationships for various cortical tissues for modeling studies.

An analytical model is needed to relate mechanical distress experienced by the brain in response to the forces and accelerations found in an automobile collision. For various loads, appropriate indicators of physical damage may be computed from predicted nodal displacement data. In order to relate these computations with what occurs to a dummy within the occupant compartment during a collision, it should be noted that head kinematics are easily quantified using an NAP installed in the dummy head. These accelerations may be processed to define translational and rotational kinematics, relative to a fixed coordinate system, and used to "drive" anatomic models capable of computing internal strains or other physical quantities, induced in the brain. This hypothetical approach may eventually provide a link between dummy testing and analytical predictions of soft tissue strains which would be very difficult to measure directly. These same quantities may also be computed in response to direct impact simulations. This work is in progress and is described below.

ANATOMIC FEATURES - The first step in developing a model was to consider anatomic features which are likely to affect strain fields throughout the brain. Principal anatomic features and basic geometry of the cranial cavity are illustrated in Figure 3. Features of known or suspected importance to anatomical modeling include the following:

1. Anatomic partitions such as the falx cerebri, tentorium and falx cerebelli,
2. Basic cortex geometry including the longitudinal fissure and distinctive sagittal and coronal geometry,
3. Constraint conditions at the base of the cerebral cortex where the brain stem and surrounding dura (the tough outer cover of the brain) enter into the upper cranial cavity after passing through the foramen magnum (the opening of the skull through which the spinal cord enters the brain) and tentorium,
4. Constraint (or slip) condition between the brain tissue and the dura as effected by pia and arachnoid membranes on the surface of the cortex, and
5. Potential directional stiffness effects due what appears to be anisotropic material properties associated with the corpus callosum and corona radiata within the internal capsule.

Other factors which may be important include effects of the mennix (i.e. pia and arachnoid membranes) on local tissue stiffness at the outer perimeter of the cortex, vascular

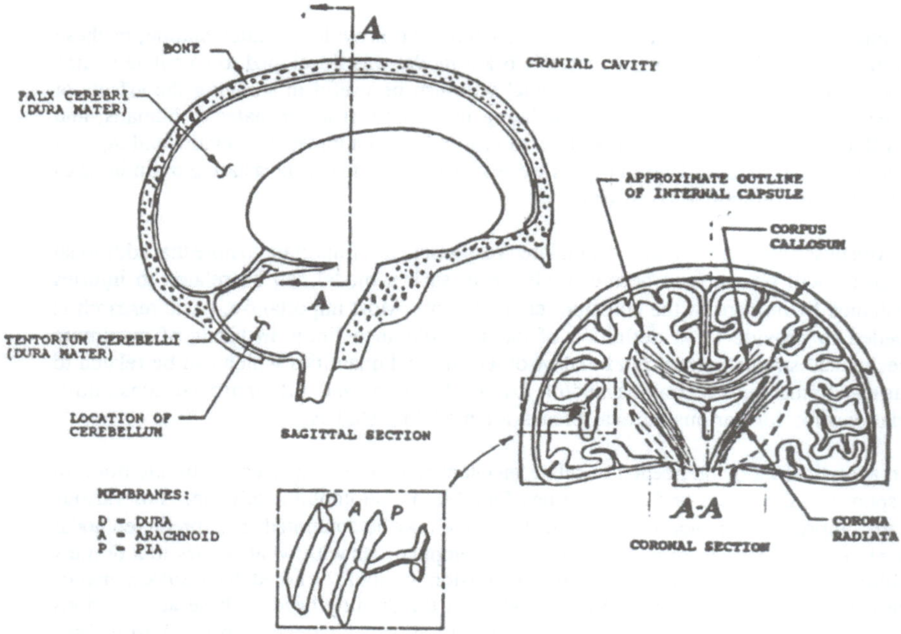

PRINCIPAL ANATOMIC FEATURES

FIGURE 3

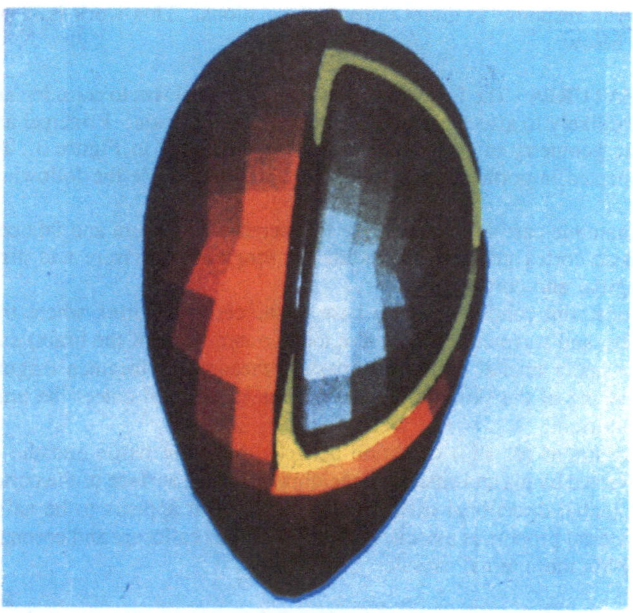

SECTIONED VIEW OF CORTEX/DURA/HEADPART MODEL

FIGURE 4

pressurization and cerebral spinal fluid (CSF) found in ventricles in interior regions and between mennix on the perimeter and in subdural and subarachnoid spaces.

MODELING APPROXIMATIONS - Since the main region of the brain is located in the upper cranium, the upper cerebral cortex was modeled using approximate but distinctive sagittal and coronal geometries including the longitudinal fissure which partitions right and left hemispheres, and interfaces with the relatively thin, hard falx cerebri. The inferior surface was simplified, and modeled with a cylindrical surface. The cerebellum was not included nor were details of brain stem egression, although an interior inferior surface of the cortex was constrained to emulate brain stem egression. (The cerebellum, the falx tentorium and details of the internal capsule and brain stem egression will be considered in subsequent modeling.)

The dura, which makes a double fold in the sagittal plane (the plane in the center of the body dividing right and left) to form the falx anatomic partition, is a thick tough membrane completely surrounding the cortex and brain stem. This was modeled using solid elements, and light friction was applied along the surface defined by the interior dural surfaces and the mating external surface of the cortex, to approximate slip conditions along this interface. (The arachnoid and pia mater also lie along this interface, and the friction modeled in this region is intended to approximate the global behavior of a more complex interaction of tissues and membranes. This is tantamount to defining a slip condition between the external surface of the brain as it mates with the interior surface of the skull.) Cerebral spinal fluid, vascular pressurization and effects of mennix at the perimeter of the cortex/dura interface could not be estimated with any accuracy and were not fully considered.

Little is known about the tissue properties of the corona radiata and corpus callosum which constitute the internal capsule within the upper cerebral cortex. Since these are described anatomically as bundles of nerve fibers, it would not be surprising if these tissues exhibited directional stiffness effects. Representative material properties for dura and cortex were selected based on published data, with consideration given to effects of mennix and potential directional stiffness effects. A visco-elastic material (Flugge) model with static and dynamic shear modulii of 2.5 and 5. psi, was used for the cortex and the dura was modeled as linear elastic.

In order to model contact events, it was necessary to model a skull with a cranial cavity to enclose the dura and cortex components. "Bone-like" elastic properties were assigned to provide a relatively rigid enclosure for simulating direct contact events. A sectioned view of the assembled cortex/dura/skull-headpart model is shown in Figure 4. The skull properties are approximated, and many details of the skull have been ignored. Thus, while the model is believed a good simulation of a head impact and the resulting strain response of the brain tissue, the model is not appropriate for studying skull fracture and other details of the skull's response (as opposed to the brain's response).

SIMULATED IMPACTS - To approximate strains induced in soft cortical tissue, the cortex/dura/headpart model described above was used to simulate a 25 mph impact with the padded A-pillar described above and a 30 mph oblique impact with a simulated windshield. In the latter simulation, the headpart velocity vector was oriented at an angle of about 26 degrees with a normal to the contact surface projected into the horizontal plane, and light friction was used between the head and windshield components to produce head rotations about multiple axes. The A-pillar model consisted of about 5,500 elements and 7700 nodes, while the windshield model consisted of approximately 5,000 elements and 7,150 nodes. Both models were run for simulation intervals of 20 msec on a CRAY2 computer. The A-pillar simulation required approximately 2 CPU hours to complete (time step was controlled by relatively stiff A-pillar materials). The windshield simulation was completed in a fraction of this time.

Figure 5 illustrates sequences of head kinematic response for the 25 mph A-pillar impact simulation and contours of shear strain for selected states. The kinematic response data is useful for visualizing the extent of headpart rotation. The first shear strain contour plot illustrates the development of strains at the contact point location (top figure). Regions of high shear strain are shown in red while lower strain levels are shown in bluish tones. Although these contours illustrate surface strains only, and are averages throughout the volume of the surface element, information on maximum strain levels throughout the volume of the cortex is also provided. Internal strain distributions may also be produced by "slicing" through the volume during post processing.

Review of the complete strain time-history data set for the A-pillar impact illustrates the development of a stress wave and its propagation and reflection through the volume of the brain. (The second and third strain contours of Figure 5 are selected from this subset.) In fact, as the stress wave travels across the cortex, maximum strains occur directly opposite the contact location (refereed to as the "contra-coup" location) at approximately 8 to 10 msec, and high strain regions at the interface with the falx are also obvious as the stress wave traverses the anatomic partition.

Although the cortex is subjected to combined rotational and translational excitations in the A-pillar impact, the principal axis of rotation is primarily about the pitch axis of the headpart (or in the sagittal plane). On the other hand, the oblique windshield impact was simulated to expose the headpart to additional rotational kinematics with a somewhat longer impact duration. A review of headpart kinematics for the latter simulation indicates rotations were induced about all three axes. Contact durations were approximately 10 msec in the A-pillar impact simulation and 20 msec for the windshield impact simulation.

Detailed review of the cortical strain data for each simulation indicates a maximum shear strain of about 40 percent for the A-pillar impact and 120 percent for the windshield impact. The significantly higher strains induced in the latter simulation demonstrates the importance of rotational kinematics on the strains developed. These findings are consistent with the distinctive sagittal and coronal geometries, and anatomic partitions which are thought to have a significant effect on induced strains. They also suggest that vastly different strain fields may result from different combinations of headpart rotational kinematics.

FUTURE CONSIDERATIONS:

The model described above has been useful for preliminary studies to relate various dynamic excitations to strain fields induced in brain tissue, and illustrates the importance of rotational kinematics and directional loading effects. Changing the model orientation and/or scaling the model size or dynamic excitation are relatively easy modifications. This (or similar) models may be used to develop scaling relationships for extrapolating dynamic excitation vs. injury data developed from primate experiments to humans head impact tolerance values. The models may also be used to develop pediatric head injury tolerance values based on adult tolerances, but allowing for the altered geometry and mass

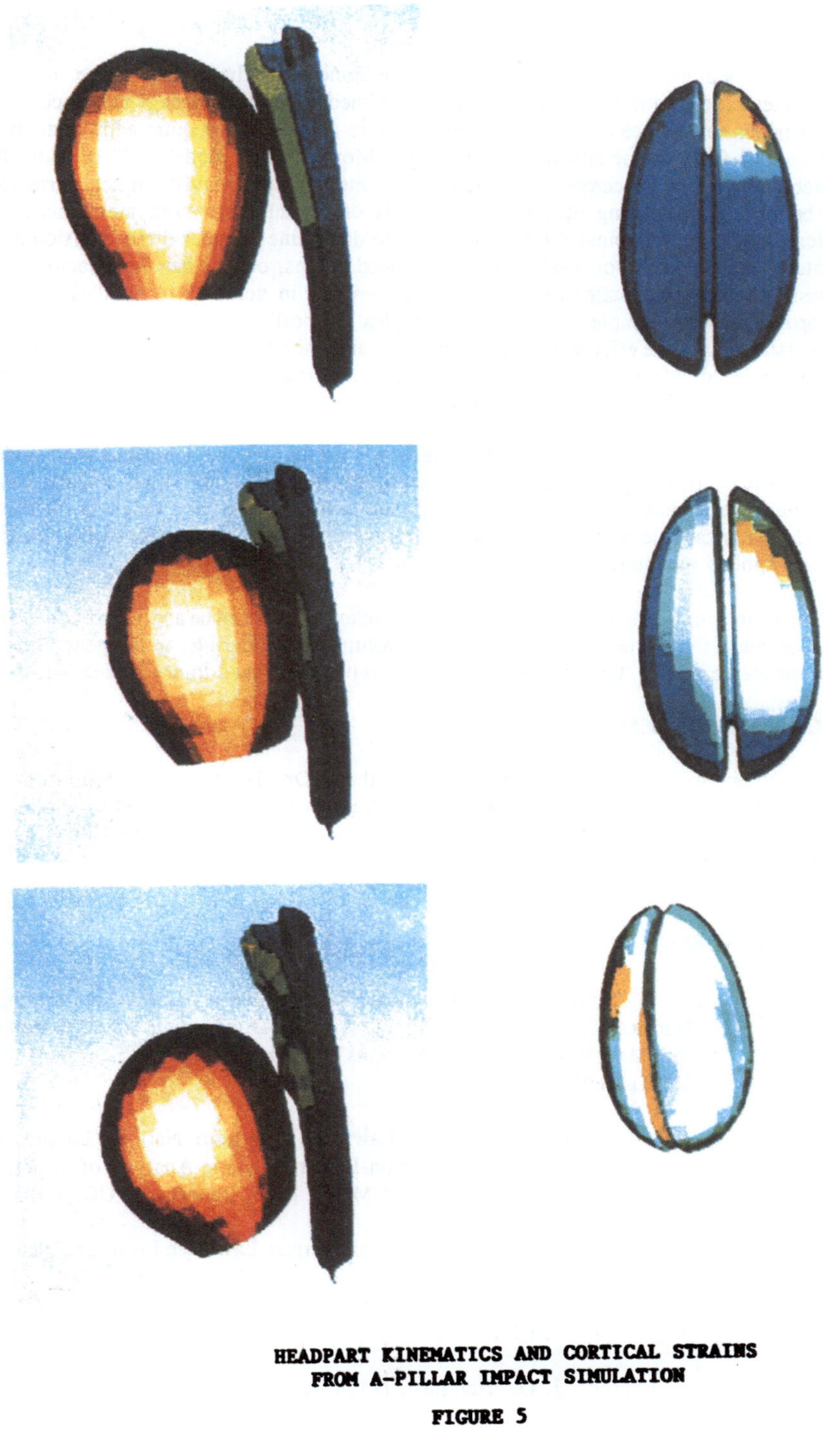

HEADPART KINEMATICS AND CORTICAL STRAINS FROM A-PILLAR IMPACT SIMULATION

FIGURE 5

ratios of children.

Future modeling efforts will require development of improved tissue material characterizations, and biomechanical research aimed at defining damage tolerance limits for various soft tissues and/or fibers. Also, future models will require additional "real-geometry" features for enhanced biofidelity. More detailed models will consider the cerebellum, tentorium cerebelli and details of the internal capsule and brain stem egression. A better understanding of the global effects of pia and arachnoid membranes and intervening cerebral spinal fluid is important to determine whether these physiological features have a strong or weak effect on induced strains, or whether these factors, and possibly anisotropic materials which may be present in nerve fibre bundles, may be approximated by simple "global" vs complex "local" representations. A better understanding of the effects of vascular pressurization and the presence of fluid ventricles is also needed. Moreover, this fragile, complex, delicate organ, filled with detail and intricacy, must also be understood as a structure.

The use of MRI/CAT scan data may be particularly useful for developing anatomic models with more detailed physiological features, especially in the complex inner capsule. Since different tissues have different resonant signatures when placed in a magnetic field, the MRI data is capable of discerning these differences at each planar cross-section excited by an applied magnetic field.

Some promising work is in progress [11]. If these techniques can be applied to converting the geometric information contained in 3-D volumetric renderings to the generation of a finite element mesh, this could be of tremendous benefit to head-injury modeling studies.

ACKNOWLEDGEMENTS:

The authors would like to acknowledge and thank Dr. Thomas Gennarelli from the University of Pennsylvania's Department of Neurosurgery, for his corroboration in discussing and defining anatomic features potentially affecting strains developed in soft cortical tissue.

REFERENCES:

1. D. W. Stillman, J. O. Hallquist, R. Rainsberger, Lawrence Livermore National Laboratory, "INGRID: A Three-Dimensional Mesh Generator for Modeling Nonlinear Systems", Revised July 1985, Report No. UCID-20506.

2. J. O. Hallquist, D.J. Benson, Lawrence Livermore National Laboratory, "DYNA3D User's Manual (Non-Linear Dynamic Analysis of Structures in Three-Dimensions), Revised March 1986, Report No. UCID-19592.

3. B. E. Brown, J.O. Hallquist, R. Rainsberger, Lawrence Livermore National Laboratory, "TAURUS: An Interactive Post Processor For The Analysis Codes NIKE3D, DYNA3D, TACO3D, and GEMINI", Revised May 1984, Report No. UCID-19392.

4. D. Benson, J. O.. Hallquist, (LLNL), M. Igrarshi, K. Shimomaki, M.

Mizuno, Suzuki Motor Co. Ltd, Japan, "The Application of DYNA3D to Large Scale Crashworthiness Calculation", 1986 ASME International Computers in Engineering Conference, Chicago, July 20-26, 1986.

5. DiMasi, Frank P., US/DOT/RSPA/VNTSC "Three-Dimensional Finite Element Modeling of Head Impacts with Vehicle Upper Interiors" Technical Memorandum No. TSC-HS-876-4, September, 1988.

6. H. R. Lissner, M. Lebow, and F. G. Evans, "Experimental Studies on the Relation Between Acceleration and Intracranial Pressure Changes in Man," Surgery, Gynecology, and Obstetrics, Volume 113, 1960, pp. 329-338

7. L. Thibault, T. Generelli, and S. Margolis, "Animal, Physical and Analytical Models for Use in the Development of Improved Head Injury Criteria" University of Pennsylvania, Department of Bioengineering and Department of Neurosurgery, DOT Report No. HS 807 481, Volume 1, March, 1989.

8. Susan S. Margolis, "Biomechanics of Traumatic Coma in the Primate", Doctoral Thesis, Department of Bioengineering, University of Pennsylvania, 1987.

9. R.A. Saul, M. Farson, D.A. Guenther "Development of a Component Level Head Impact Test Device", Thirtieth Stapp Car Crash Conference Proceedings, San Diego, California, Oct 27-29, 1986, SAE P-189. (Paper Number 861889)

10. M. W. Monk, L. K. Sullivan, NHTSA/Vehicle Research and Test Center, "Energy Absorption Material Selection For Head/A-pillar", Thirtieth Stapp Car Crash Conference Proceedings, San Diego, California, Oct 27-29, 1986, SAE P-189. (Paper Number 861887)

11. P. Finnigan, A. Hathaway and W. Lorenson, "Merging CAT and FEM", Mechanical Engineering, Computers in Mechanical Engineering Series, July, 1990.

FINITE ELEMENTS IN THE 90's, E. Oñate, J. Periaux, A. Samuelsson (Eds.)
Springer-Verlag/CIMNE, Barcelona 1991

MASS LUMPING BY SPECTRAL MATCHING

N. Bicanic
Department of Civil Engineering, University College Swansea, UNITED KINGDOM

J. Máca
Czechoslovak Academy of Sciences, Praha, CZECHOSLOVAKIA

INTRODUCTION

In direct integration of dynamic equilibrium equations it may be very beneficial to diagonalise a mass matrix. Despite the fact that the consistent mass matrix leads to optimal error estimates, there is a strong motivation for breaking the basic rules. In practice, diagonal or "lumped" mass matrices are often employed due to their general economy and because they lead to some especially attractive time-integration schemes (explicit methods).

Any mass lumping aims to produce a mass matrix, where the total mass of the element is distributed along the diagonal of the matrix and whose behaviour simulates that of the consistent matrix. There are several ways of constructing the lumped mass matrix (see e.g. Hughes [1]):

a) **Equal mass lumping, E**
(the same proportion of the mass is associated with each node);

b) **Mass lumping by nodal quadrature, NQ**
(e.g. a Lobatto rule used for the biquadratic Lagrangian element);

c) **Row-sum technique, RS**
(terms in each row of the consistent mass matrix are summed and lumped on the diagonal);

d) **Special mass lumping, S**
(Hinton-Rock-Zienkiewicz algorithm, [2], where the idea is to set the terms of the lumped mass matrix proportional to the diagonal terms of the consistent mass matrix with the constant of proportionality selected to conserve the total element mass).

However, is there a better way of lumping the mass to give more accurate approximation? Here, the **Spectral Matching Mass Lumping Scheme, SM** is proposed. Such an improved lumping scheme will be constructed by a comparative eigenvalue analysis of a single unconstrained element, formulated using a lumped and a consistent mass matrix. The eigenvalues and eigenvectors of the element formulated with a consistent mass matrix will be used as a target. The improved lumped mass matrix will be the one that matches best the target spectral characteristics, *eigenvalues and eigenmodes*, obtained when using the consistent mass matrix formulation.

Various improved lumping schemes will be obtained with respect to the number of eigenvectors included in the optimisation procedure and the error measure employed. Finally, the effect of the improved lumping scheme on the free and forced vibration analysis of the real finite element mesh will be investigated.

IMPROVED MASS LUMPING

As previously stated the improved mass lumping scheme by ***spectral matching*** is to be found by performing a comparative eigenvalue analysis. The spectral characteristics of the single unconstrained element with a consistent mass matrix (called ***"consistent"*** eigenvalues and eigenvectors) are used as a target. The eigenvalues and eigenmodes calculated for a single unconstrained element with a lumped mass matrix (***"lumped"*** eigenvalues and eigenvectors) are then compared with a target solution. Finally the minimum value of a certain error measure **E** is used to indicate the improved lumping scheme.

Such a scheme depends on the error measure employed (as it is possible to compare ***only eigenvalues*** or ***both eigenvalues and eigenmodes***) and on/or how many of the "consistent" eigenvectors are aimed to be approximated (e.g. only the fundamental "consistent" eigenvector can be approximated or, at the other extreme, the improved lumping scheme is aiming to approximate all eigenvectors.

Three types of error measures are considered

Error type 1

(error based on comparing eigenvalues only)

$$E1 = \left(\sum_{i=1}^{n} (\lambda_i - \bar{\lambda}_i)^2 \right)^{1/2}$$

where λ_i is the "consistent" eigenvalue
$\bar{\lambda}_i$ is the "lumped" eigenvalue
n is the number of eigenvalues considered.

Error type 2

(standard error measure used in the practical finite element analysis, see e.g. Bathe [3])

$$E2 = \frac{\|(\mathbf{K} - \bar{\lambda}\mathbf{M})\bar{\mathbf{\Phi}}\|}{\|\mathbf{K}\bar{\mathbf{\Phi}}\|}$$

where **K** is the stiffness matrix
M is the "consistent" mass matrix
$\bar{\lambda}$ and $\bar{\mathbf{\Phi}}$ are calculated "lumped" eigenvalue and mode approximations to the "consistent" eigenpair λ_i and $\mathbf{\Phi}_i$

Error type 3
(an energy based error measure, as the term $\bar{\lambda} = \bar{\mathbf{\Phi}}^T\mathbf{K}\bar{\mathbf{\Phi}}$ is very closely connected with the strain energy of a certain eigenmode)

$$E3 = \frac{\|(\mathbf{K} - \bar{\lambda}\mathbf{M})\bar{\mathbf{\Phi}}\|}{\bar{\lambda}}$$

Physically, the term $\mathbf{K}\bar{\mathbf{\Phi}}$ represents the elastic nodal forces and the term $\bar{\lambda}\mathbf{M}\bar{\mathbf{\Phi}}$ represents the inertia nodal forces related to an eigenpair $\bar{\lambda}$ and $\bar{\mathbf{\Phi}}$. The norm of these out-of-balance forces should be small for the consistent solution.

All three types of error measures are used for calculations based on comparing only the fundamental eigenvector (case **F**) and on comparing all non-zero eigenvectors (case **A**). In total, six error combinations (type + case) are studied and identified by a suitable symbol - e.g. **E1A** is an **E**rror measure type **1** comparing **A**ll eigenvalues produced. For the case **A**, the mean error measures type 2 and type 3 had to be calculated (sum of all errors is divided by the number of eigenvectors considered).

When mass lumping is performed, certain eigenmodes appear to "jump the queue" [4]. There may not be necessarily a direct relationship between the ordered eigenvalues λ_i and $\bar{\lambda}_i$, therefore the eigenvalues corresponding to the same eigenmode have to be compared.

TWO-DIMENSIONAL 8-NODED ELEMENT

Improved Lumping Scheme by Spectral Matching

The improved mass lumping procedure by spectral matching will be illustrated on the regular shape two-dimensional 8-noded Serendipity plane stress/strain element. The nodes of the element are divided into two types: mid-side nodes and corner nodes and the total mass M_t was distributed between them (Fig 1).

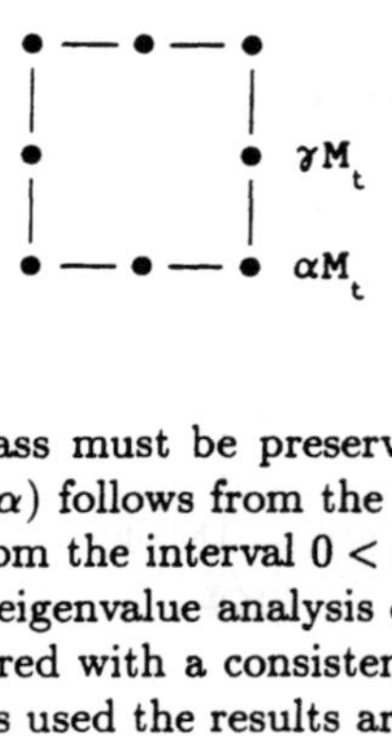

Fig. 1

As the total element mass must be preserved, only parameter α needs to be varied, as $\gamma = 1/4(1 - 4\alpha)$ follows from the above constraint condition. For all values of parameter α from the interval $0 < \alpha < 0.25$ the separate lumped mass matrices are assembled, eigenvalue analysis of the single unconstrained element is conducted and compared with a consistent mass solution. It is obvious that for all the error measures used the results are dependent on the parameter α.

All calculations were carried out for the following input values

element size 2.0X2.0
Young's modulus $E = 1000$
Poisson's ratio $\nu = 0.2$
element thickness $t = 1.0$
mass density $\rho = 10.$

Element matrices for plane strain conditions were calculated using the Gaussian 3X3 integration rule.

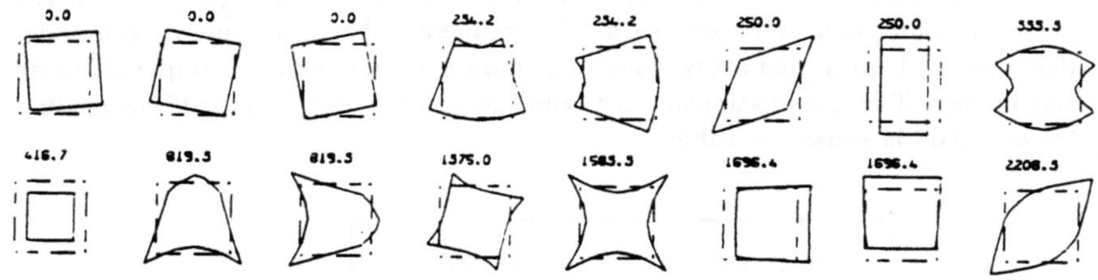

CONSISTENT MASS PROBLEM

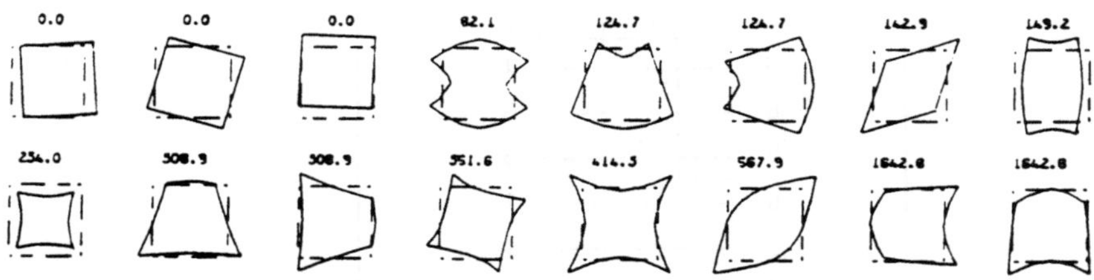

LUMPED MASS PROBLEM - $\alpha = 0.04$

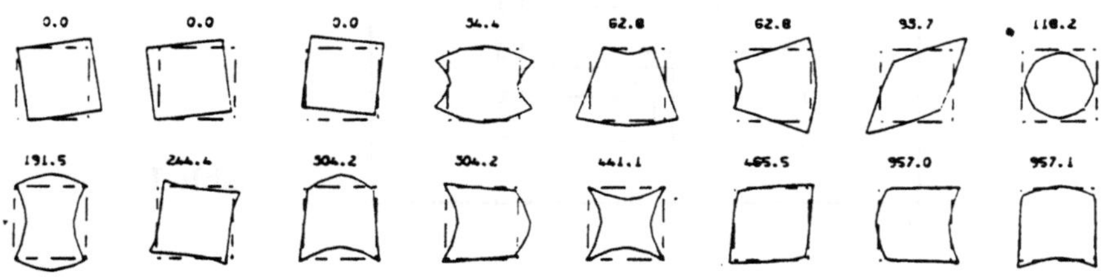

LUMPED MASS PROBLEM - $\alpha = 0.125$

Fig. 2

Fig 2 shows 16 mode shapes and corresponding eigenvalues for a consistent mass solution and the lumped mass solutions for different values of parameter α.

Comparing "consistent" and "lumped" eigenmodes, the "jump the queue" effect [4], where the corresponding eigenvectors change their possition in the eigenspectrum, can be clearly seen. The sequence of lumped mass eigenmodes depends on the lumping mass scheme (i.e.parameter α). As an example of "jump the queue" effect the relationship between "consistent" and "lumped" eigenvalues for $\alpha = 0.04$ is shown in Table I.

"consistent" modes	"lumped" modes /α = 0.04/
4	8
5	4
6	5
7	6
8	7
9	9
10	10
11	11
12	12
13	13
14	16
15	14
16	15

Table I

It is interesting to note that for all values of parameter α the first non-zero "lumped" eigenvector does not correspond to the fundamental non-zero "consistent" eigenvector.

The spectral matching scheme which corresponds to the fundamental non-zero "consistent" eigenvalue (error **E1F**) leads to the best approximation when α tends to zero. For other error combinations, to locate the minimum value of any error measure considered, values of errors are plotted against parameter α. Fig 3 shows two dominant peaks for error measure $E1A$, and an optimal value $\alpha = 0.025$ is adopted.
Table II shows all error measures used and corresponding "optimal" values of lumping parameter α.

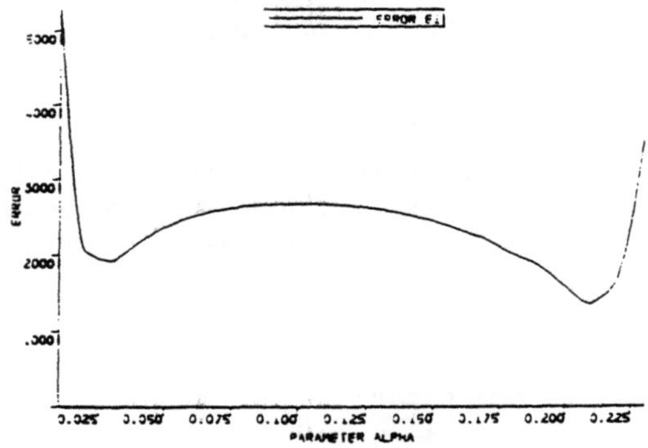

ERROR MEASURES E1A

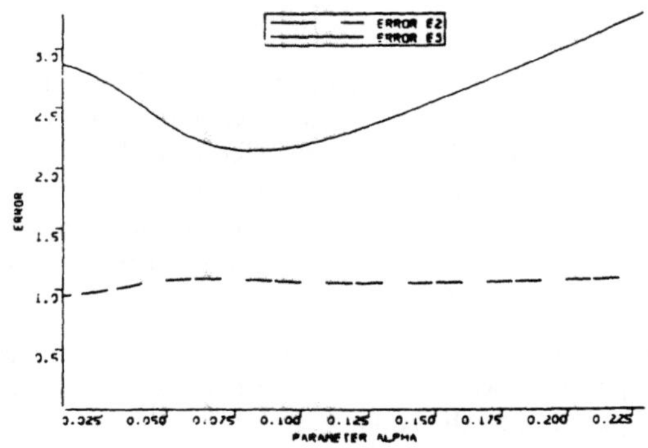

ERROR MEASURES E2F, E3F

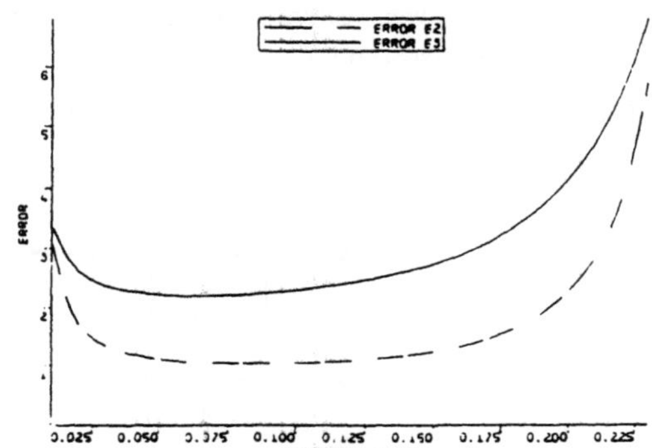

ERROR MEASURES E2A, E3A

Fig. 3

Essentially identical results were obtained for different values of Poisson's ratio, for element matrices calculated in 2X2 integration points as well as for a plane stress analysis.

3. DISTORTED ELEMENTS

In order to utilise the data from regular elements for an arbitrary distorted element shape the coordinates of the *transition points* can be calculated. These points are used to divide the element into areas associated with the mass at each node.

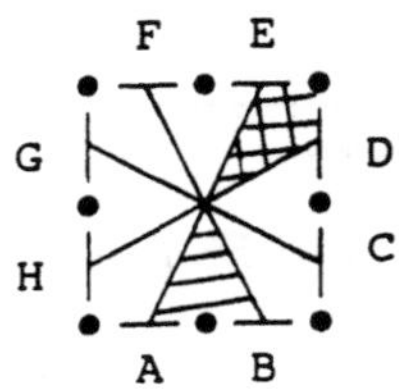

Fig 4 shows the regular 8-noded element, the shaded area is proportional to a mid-side node mass and the cross-hatched area is proportional to a corner node mass. Since the regular elements have effectively been studied within the natural coordinate system, ξ and η , the coordinates of the transition points are given as ξ_t and η_t coordinates, hence they can easily be mapped to a deformed element shape. This enables the equivalent mass at each node to be calculated proportional to a certain area of the element.

Clearly, the transition point coordinates ξ_t and η_t depend on the parameter α, i.e., the transition point between the mid side and corner node is positioned at the distance corresponding to the ratio α/γ.

4. EXAMPLES

4.1. Eigenvalue analysis

The effect of the *spectral matching* **SM** *lumping scheme* on the eigenvalue analysis of plane stress problems was tested. Two different rectangular meshes of isoparametric elements were investigated:

(a) the square mesh of 2X2 elements with minimum constraints

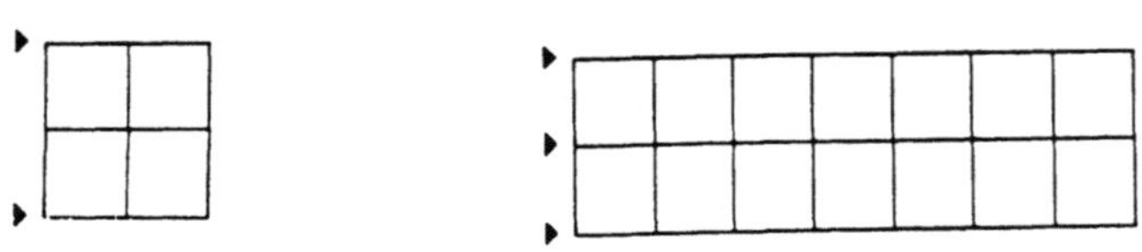

(b) the cantilever mesh with only one end fixed

The results are shown in Tables III and IV. It is obvious that the best results (using the ccnsistent mass matrix solution as a target) are for an **SM** lumping scheme with parameter $\alpha = 0.0$.

4.2. Forced vibration analysis

Wave propagation problem for the $2-D$ plane stress finite element mesh was solved. For the cantilever mesh where only one end is fixed, the displacement input on the other end was introduced. The following harmonic displacement excitation in the longitudinal direction was used in all nodal points at section $A-A$

$$u(t) = u_0 \sin \omega t \text{ for } 0 \leq \omega t < \pi$$

$$u(t) = 0 \text{ for } \quad \omega t \geq \pi$$

With respect to the theoretical wave velocity of the problem $c = \sqrt{E/\rho} = 50$, the input excitation frequency $\omega = 10\pi$ was used and the theoretical wave length $\lambda = 2\pi c/\omega$ was thenapproximately equal to 10.

A displacement response of the structure is shown on Fig 5 (a)-(c) for various times t during the solution interval 0 - 0.2 sec.

The "flatness" of the propagating wave was used as a measure of the effect of various mass matrix lumping schemes. Two types of error measure were considered :

Error type a

$$\mathrm{E-a} = \frac{|\max u(x_B, y, t_M) - \min u(x_B, y, t_M)|}{\int_0^b u(x_B, y, t_M)\, dy}$$

where x_B is the x-coordinate of the nodal points in the section $B-B, t_M$ is the time during the solution period when the integral used has the maximum value (when the propagating wave is just passing through the given cross section), b is the transversal dimension of the cantilever. This error was used at 1/3 of the cantilever length (section $B-B, x = x_B$), at 1/2 (section $C-C, x = x_C$) and at 2/3 of its length (section $D-D, x = x_D$).

Error type b

$$\mathrm{E-b} = \frac{\int_0^T \int_0^b u(x_C, y, t)\, dy\, dt}{\int_0^T \int_0^b u(x_A, y, t)\, dy\, dt}$$

where T is the solution period $/T = 0.2\,\mathrm{sec}/$.

In this case the total "volume" of the propagating wave pased through the section $C-C$ during the solution interval is divided by the same value calculated in the section $A-A$ (where the displacement input was prescribed).

The results are shown in Table V for several values of lumping parameter α. The best results - i.e. error $E-a$ tending to zero and error $E-b$ approaching one - were again obtained for the lumping scheme with parameter $\alpha = 0.0$ (here approximated by the value 0.005).

Error measure	α - optimal
E1F	0.0
E1A	0.025
E2F	0.0
E2A	0.08
E3F	0.08
E3A	0.06

Table II

Lumping scheme	Eigenvalue		Error measure	
	1st bending	2nd axial	E2A	E3A
consistent m.	3.02	4.21	-	-
α = 0.005	2.97	4.25	0.49	0.09
α = 0.040	2.93	4.23	0.89	0.14
α = 0.060	2.91	4.22	1.05	0.16
α = 0.080	2.89	4.21	1.24	0.18
α = 0.125	2.85	4.18	1.56	0.23
α = 0.250	2.73	4.09	1.59	0.37

Table III

Lumping scheme	Eigenvalue			Error measure	
	1st bending	2nd bending	3rd axial	E2A	E3A
consistent m.	0.39	1.94	2.25	-	-
α = 0.005	0.39	1.91	2.25	0.51	0.10
α = 0.040	0.39	1.90	2.25	0.89	0.15
α = 0.060	0.39	1.89	2.24	1.09	0.18
α = 0.080	0.39	1.88	2.24	1.29	0.21
α = 0.125	0.39	1.87	2.24	1.62	0.26
α = 0.250	0.39	1.83	2.23	1.65	0.42

Table IV

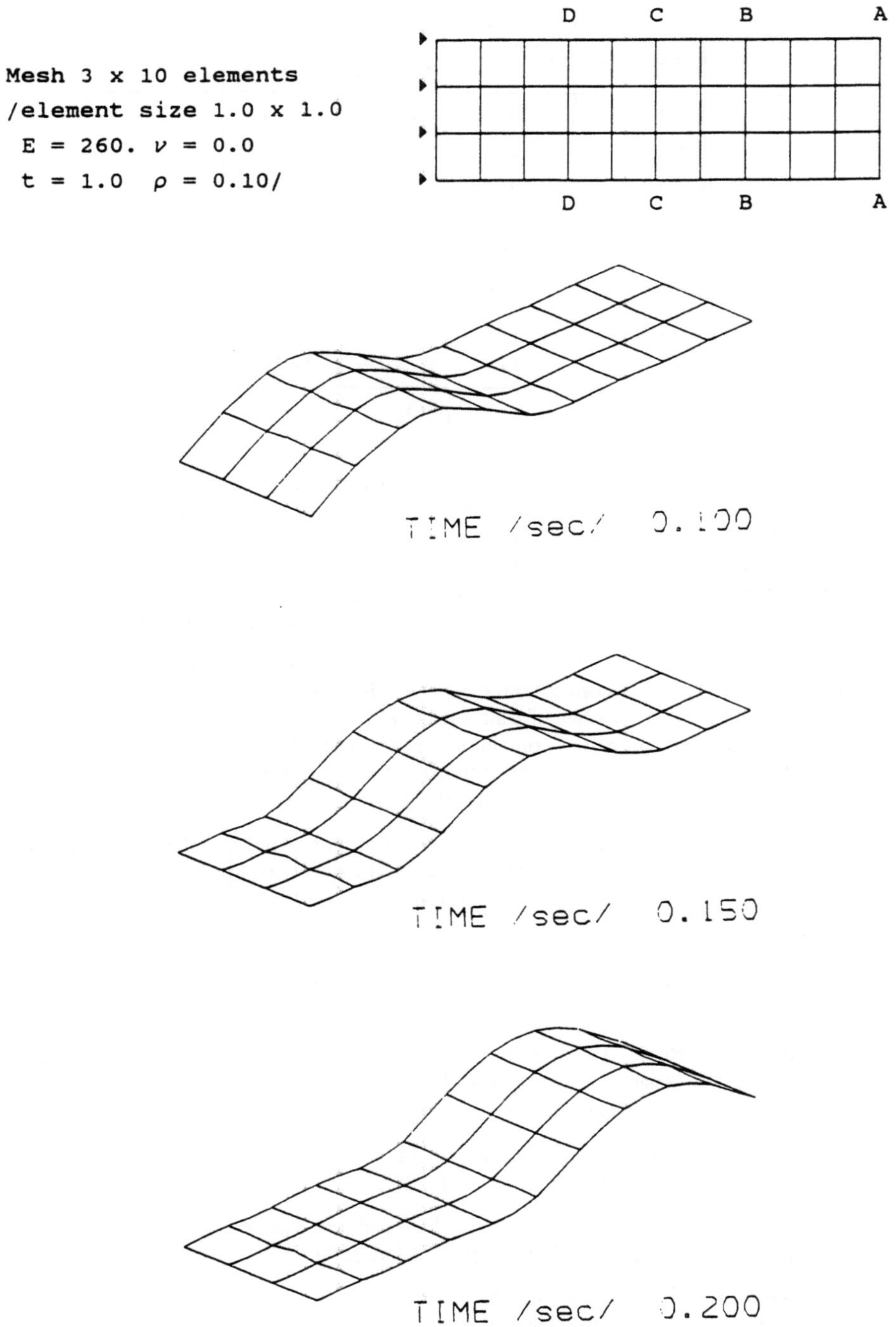
Mesh 3 x 10 elements
/element size 1.0 x 1.0
E = 260. ν = 0.0
t = 1.0 ρ = 0.10/
D C B A
D C B A
TIME /sec/ 0.100
TIME /sec/ 0.150
TIME /sec/ 0.200

Fig. 5

Parameter α	Error type E-a			E-b
	$x = x_B$	$x = x_C$	$x = x_D$	$x = x_C$
0.005	0.046	0.047	0.047	0.978
0.020	0.061	0.058	0.057	0.973
0.040	0.063	0.055	0.040	0.963
0.060	0.056	0.064	0.058	0.953
0.080	0.077	0.090	0.080	0.944
0.100	0.105	0.100	0.074	0.941
0.125	0.117	0.088	0.096	0.937
0.245	0.120	0.105	0.099	0.897

Table V

CONCLUSION

A novel way of lumping the element mass matrix by matching element spectral characteristicsis, **SM** scheme, proposed. Such an improved lumped mass matrix has approximately - depending on the error measure used - the best spectral characteristics comparing with a consistent mass matrix formulation. It is shown that a more commonly used "special" mass lumping scheme gives often very good results too - but the same cannot be said about the "equal" mass lumping scheme.

Moreover, **SM** *lumping scheme* with the parameter $\alpha = 0$ - when the mass of the 8-noded element is concentrated in the mid-side nodes only - may be very beneficial. The total number of unknowns can be considerably reduced using a static condensation procedure on the structure level. Such lumping scheme can be used in all problems when the fundamental eigenvector contribution to the overall response of the structure is significant.

The improved mass lumping procedure was demonstrated only for the case of two-dimensional plane stress/strain elements. However, the proposed method is more general and it can be easily extended for other types of finite elements.

REFERENCES

[1] Hughes, T.J.R., The Finite Element Method, Prentice-Hall International, Inc., 1987.

[2] Hinton, E., Rock, T., Zienkiewicz, O.C., A Note on Mass Lumping and Related Processes in the Finite Element Method, Journal Earthquake Engineering and Structural Dynamics, Vol 4(1976), $pp245 - 279$.

[3] Bathe, K.J., Wilson, E.L., Numerical Methods in Finite Element Analysis, Prentice-Hall, Englewood Cliffs, N.J. 1976.

[4] Bićanić, N., Seizmika i numerika, Proceedings 5th Yugoslav Earthquake Engineering Congress (ed. P. Fajfar), Bled 1990.

[5] Bićanić N., Máca J., An improved mass lumping technique by matching element spectral properties, Internal Report $CR/685/91$. Department of Civil Engineering, University College Swansea

FINITE ELEMENTS IN THE 90's, E. Oñate, J. Periaux, A. Samuelsson (Eds.)
Springer-Verlag/CIMNE, Barcelona 1991

VIBRATIONS OF FLUID-STRUCTURÉ COUPLED SYSTEMS

R. Ohayon
Office National d'Etudes et de Recherches Aérospatiales (ONERA),
F-92320 Châtilon,
FRANCE

SUMMARY

It is proposed to investigate alternative symmetric formulations valid for static and modal analysis based either on the representation of the liquid by a pressure field and a displacement potential field, either on the representation of the structure by a displacement field and a reaction force (or acceleration) field.

INTRODUCTION

Transient and modal analysis of fluid-structure problems have been the subject of many investigations [**1**]. For such problems, the choice of field variables for the variational statements is of prime importance. The description of the fluid, <u>in the compressible case</u>, by a stress (pressure) scalar variable p, or by a displacement potential φ introduces non symmetric matrices after a finite element discretization, we refer to the pioneering work of [**32, 33, 34**] followed then by [**2, 7, 12, 22, 28, 29**]. An attempt to symmetrize the non symmetric matrix system by using matrix manipulation algebra, has been investigated in [**4, 21**]. In this respect, let us also mention the analysis of [**9, 10**] based on matrix scaling concepts.

Another idea lies in the utilization of the velocity potential instead of the displacement potential or instead of the pressure. The resulting symmetric matrix system involves a complex $j\omega$ term, if ω denotes the circular frequency [**8**]. The corresponding formulation has been improved in [**27**] in order to take into account the static case. The analysis leads to a numerical procedure which involves a special purpose eigenvalue solver.

In reference [**19**] and in the subsequent references [**20, 25**], a three-field symmetric variational formulation has been introduced using p and φ as fluid unknowns, while the displacement field u is used to describe the structure. The formulation involves coupling $p - \varphi$ and $u - \varphi$ terms in the mass matrix. Its utilization in nuclear engineering problems is described in [**3**]. A complete re-investigation of this formulation in order to get rid of the "spurious zero-frequency fluid modes" generally encountered, is the subject of reference [**18**]. Some attempt in order to eliminate those spurious modes has also been analyzed in [**11**].

Let us also recall that a description of the fluid through a displacement variable, introduces difficulties related to the appropriate discretization of the irrotationality condition for the fluid. A displacement formulation has been presented in an appendix of [**19**], the corresponding finite element analysis is presented in [**13**].

In references [**23, 26**], a three-field (u; p, φ) variational symmetric formulation has been introduced. It involves coupling terms in the stiffness matrix. Applications with different interpolation functions for p and φ can be found in reference [**31**].

Two three-field (u, R; p) and (u, R; φ) symmetric formulations, have been introduced in references [**24, 26**]. Both involve an additional unknown in the structure, namely the dynamic reaction force R or the acceleration γ. We then obtain either a "mass-coupling" matrix eigenvalue system, either a "stiffness-coupling" matrix system.

Those formulations have been used through particular modal synthesis procedures in [**6**] and [**14**]. Finally, let us mention the introduction of sloshing simple model in [**25, 15**].

A complete modal interaction scheme taking into account both compressibility and gravity effects is analyzed in [**17**]. Modal synthesis procedure have already been investigated for elastoacoustic vibrations using structural modes in [**19**], and hydroelastic incompressible modes in [**20**], together with acoustic modes as Ritz vectors, in order to study the vibrations of a structure containing a gas [**19**] or a compressible liquid with a free surface while neglecting gravity effects [**20**]. Fluid-structure interface damping effects may be found in [**15**].

The subject of the present paper is to reinvestigate some of the previous formulations in order to take into account static cases. This means that the formulations should not contain, in the modal situations, non physical zero eigenvalues.

The variational methodology is based on a direct utilization of the local equations of the coupled system. Let us mention an asymptotic mathematical approach to coupled problems in [**30**].

In a first step, we shall recall the basic local equations of the coupled system [**18**].

In a second step, we shall extend the three-field (u; p, φ) symmetric variational formulation, of stiffness-coupling type, which has been established for a compressible liquid with a zero pressure free surface condition [**23, 26**], to the case of a structure completely filled with a compressible gas. We shall show that the static case can be easily handled.

Finally, in a third step, we shall investigate, for a partially clamped structure containing either a compressible gas, either a compressible liquid with a free-surface (neglecting gravity effects), a three-field symmetric variational formulation of "mass-coupling" type [**24**] extended in order to take into account static cases.

1. LOCAL EQUATIONS OF THE BOUNDARY VALUE PROBLEM

1.1 Notations

Ω_s (resp. Ω)	structure (resp. fluid) domain at equilibrium
Γ	(horizontal) equilibrium free surface
n	unit normal (external to the fluid boundary)
n_s	unit normal (external to the structure boundary)
Σ	fluid/structure contact surface
ρ_s (resp. ρ_0)	mass density of the structure (resp. of the fluid)
$\sigma_{ij}(u)$	stress tensor considered as a function of the displacement u according to the following definition: $\sigma_{ij} = a_{ijkl}\varepsilon_{kl}(u)$
$\varepsilon_{kl} = \frac{1}{2}(u_{k,l} + u_{l,k})$	strain tensor
a_{ijkl}	elastic coefficients
$\lambda = \omega^2$	square of the circular frequency
c	velocity of sound in the compressible fluid

The equations of the boundary eigenvalue problem are [**18**]:

$$-\nabla p + \lambda\rho_0 \nabla\varphi = 0 \qquad |_{\Omega} \tag{1}$$

$$p + \rho_0 c^2 \Delta\varphi = 0 \qquad |_{\Omega} \tag{2}$$

$$\frac{\partial\varphi}{\partial n} = u.n \qquad |_{\Sigma} \tag{3}$$

$$\sigma_{ij}(u)\, n_j{}^s = p\, n_i \qquad |_{\Sigma} \tag{4}$$

$$\sigma_{ij,j}(u) + \lambda\, \rho_s\, u_i = 0 \qquad |_{\Omega_s} \tag{5}$$

$$\sigma_{ij}(u)\, n_j{}^s = 0 \qquad |_{\partial\Omega_s \backslash \Sigma} \tag{6}$$

$$l(\varphi) = 0, \quad l(1) \neq 0 \tag{7}$$

Remark 1

Interpretation of the above equations

(1) Euler equation of the inviscid compressible irrotationnal fluid.
(2) "constitutive law" of the linear fluid.
(3) (kinematical) contact condition.
(4) force equilibrium condition at fluid-structure interface.
(5) (linear) dynamic equation for an elastic structure.
(6) free boundary condition on the structure.
(7) uniqueness condition for φ (defined through an additive constant).

2. (U; P, φ) SYMMETRIC FORMULATION (STIFFNESS-COUPLING)

2.1 The local boundary value equations

Let us recast the equation (1) as follows:

$$-\Delta p + \lambda\rho_0 \Delta\varphi = 0 \qquad |_{\Omega} \tag{8}$$

$$-\frac{\partial p}{\partial n} + \lambda\rho_0 \frac{\partial\varphi}{\partial n} = 0 \qquad |_{\partial\Omega=\Sigma} \tag{9}$$

The system (8), (9) is equivalent to (1). The new local boundary value equations are then (8), (9), (2)-(7). We make then use of the usual test function method [5] as follows:

We multiply (8) by a (regular) test function $\delta p \in V_p$ and we use the Green formula to take into account the boundary condition (9).

We multiply equation (2) by a (regular) test function $\delta\varphi \in V_\varphi$ and we use the Green formula to take into account (3).

At last, we multiply equation (5) by a (regular) test function $\delta u \in V_u$ and we use the Green formula to take into account the boundary conditions (4) and (6).

~~V_p, V_φ and V_u are admissible classes of regular functions.~~

We then obtain the following variational formulation.

2.2 The symmetric variational formulation

Find $(u; p \text{ and } \varphi) \in V_u \times V_p \times V_\varphi$ and $\lambda \geq 0$ such that:

$$\int_{\Omega_s} \sigma_{ij}(u)\varepsilon_{ij}(\delta u) - \lambda \int_{\Omega_s} \rho_s\, u.\delta u - \int_{\Sigma} \delta u.n\, p = 0 \tag{10}$$

$$\int_{\Omega} \nabla\varphi.\nabla\delta p - \frac{1}{\rho_0 c^2}\int_{\Omega} p\,\delta p - \int_{\Sigma} u.n\,\delta p = 0 \tag{11}$$

$$\int_{\Omega} \nabla p.\nabla\delta\varphi - \lambda\int_{\Omega} \rho_0 \nabla\varphi.\nabla\delta\varphi = 0 \tag{12}$$

$$\forall(\delta u; \delta p, \delta\varphi) \in V_u \times V_p \times V_\varphi \tag{13}$$

2.3 Matrix equations

After a finite element discretization, and using u, p and φ as the vector of nodal values of the continuum fields, we obtain the following matrix equation:

$$\begin{bmatrix} K & D & 0 \\ D^T & K_F & E \\ 0 & E^T & 0 \end{bmatrix} \begin{Bmatrix} u \\ p \\ \varphi \end{Bmatrix} = \lambda \begin{bmatrix} M & 0 & 0 \\ 0 & 0 & 0 \\ 0 & 0 & F \end{bmatrix} \begin{Bmatrix} u \\ p \\ \varphi \end{Bmatrix} \tag{14}$$

2.4 The case $\lambda = 0$

Let us set $\lambda = 0$ in the formulation (10)–(13). The equation (12) shows that p is constant over the domain Ω:

$$p = p_s \tag{15}$$

Therefore, the equation (11) gives the value of the constant pressure field p_s as a function of the normal displacement $u.n$ of Σ:

$$p_s = -\frac{\rho_0 c^2}{\text{Vol}(\Omega)}\int_{\Sigma} u.n \tag{16}$$

The equation (10) then becomes:

$$\int_{\Omega_s} \sigma_{ij}(u)\varepsilon_{ij}(\delta u) + \frac{\rho_0 c^2}{\mathrm{Vol}(\Omega)}\left(\int_\Sigma u.n\right)\left(\int_\Sigma \delta u.n\right) = 0 \tag{17}$$

One can easily see that the only zero-frequency movements are the six physical rigid body motions u_R:

$$u = u_R \tag{18}$$

for wich, we have:

$$\int_\Sigma u_R.n = 0 \tag{19}$$

Therefore, for those movements, we obtain as expected:

$$p_s(u_R) = 0 \tag{20}$$

This proves that the static case can be easily handled.

From a practical point of view, it is only necessary to use the stiffness matrix of (14) and to impose to all p-nodal values to be equal.

This is a way to overcome the difficulty concerning the non standard finite element discretization of the second bilinear form of (17).

2.5 The "condensed" (u, φ) symmetric variational formulation

p can be eliminated from equation (14), leading to a two-field symmetric stiffness-coupling formulation.

2.6 A four-field $(u_s; p_s, P, \varphi)$ formulation

In order to exhibit the constant (over the domain Ω) so-called pneumatic pressure p_s [19], it could be convenient to recast the formulation (10)–(14) through the following change of representation:

$$p = p_s + P \tag{21}$$

with:

$$p_s = -\frac{\rho_0 c^2}{\mathrm{Vol}(\Omega)}\int_\Sigma u.n \tag{22}$$

and

$$\int_\Omega P = 0 \tag{23}$$

For sake of brevity, we do not enter into the details of such modified formulation wich will be discussed elsewhere.

3. $(U, \gamma; P)$ SYMMETRIC FORMULATION (MASS-COUPLING)

3.1 The local boundary value equations

Referring to reference [24, 26], introducing a new unknown variable for the structure enables us to obtain two three-field symmetric formulations involving respectively p or φ for the fluid (resp. mass- or stiffness-coupling).

In a first step, let us recast the (u, p, φ) local equations as follows:

$$\Delta p + \frac{\lambda}{c^2} p = 0 \qquad |_\Omega \tag{24}$$

$$\frac{\partial p}{\partial n} = \lambda \rho_0 u.n \qquad |_{\partial\Omega=\Sigma} \tag{25}$$

$$\sigma_{ij}(u) n_j^s = p n_i \qquad |_\Sigma \tag{26}$$

$$\sigma_{ij,j}(u) + \lambda \rho_s u_i = 0 \qquad |_{\Omega_s} \tag{27}$$

$$\sigma_{ij}(u) n_j^s = 0 \qquad |_{\Sigma_1} \tag{28}$$

$$u = 0 \qquad |_{\Sigma_2=(\partial\Omega_s\setminus\Sigma)\setminus\Sigma_1} \tag{29}$$

and

$$p + \rho_0 c^2 \Delta\varphi = 0 \qquad |_\Omega \tag{30}$$

$$\frac{\partial\varphi}{\partial n} = u.n \qquad |_\Sigma \tag{31}$$

$$l(\varphi) = 0 \tag{32}$$

if we ignore the equations (30)–(32), one must add the following constraint equation in order to obtain $p = 0$ in the $\lambda = 0$ case:

$$\int_\Omega p \; + \rho_0 c^2 \int_\Sigma u.n = 0 \tag{33}$$

<u>Remark 2</u>

(33) is obtained by integrating (30) over Ω and by taking into account (31).

If we consider only the system defined by (24)–(29), one must add (33) in order to eliminate the constant zero-frequency mode ($p = constant$). With this respect, it is convenient to recast the system (24)-(29) by introducing two new variables p_s and P instead p, namely those defined by (21), (22) and (23).

The local equations then are as follows:

$$\Delta P + \frac{\lambda}{c^2} P + \frac{\lambda}{c^2} p_s = 0 \qquad |_\Omega \tag{34}$$

$$\frac{\partial P}{\partial n} = \lambda \rho_0 u.n \qquad |_\Sigma \tag{35}$$

$$p_s + \frac{\rho_0 c^2}{\mathrm{Vol}(\Omega)} \int_\Sigma u.n = 0 \, , \int_\Omega P = 0 \tag{36}$$

$$\sigma_{ij}(u) n_j^s = P n_i + p_s n_i \qquad |_\Sigma \tag{37}$$

$$\sigma_{ij,j}(u) + \lambda \rho_s u_i = 0 \qquad |_{\Omega_s} \tag{38}$$

$$\sigma_{ij}(u) n_j^s = 0 \qquad |_{\Sigma_1} \tag{39}$$

$$u = 0 \qquad |_{\Sigma_2} \tag{40}$$

In order to obtain a symmetric variational formulation, we shall use as previously [**24**, **26**], a new structural variable, namely the dynamic reaction force:

$$R = \rho_0 \lambda u \tag{41}$$

or the acceleration:

$$\gamma = \lambda u \tag{42}$$

The local equation (38) is then recasted as follows:

$$\sigma_{ij,j}(u) + \rho_s \gamma = 0 \tag{43}$$

$$\gamma = \lambda u \tag{44}$$

The local equations which lead to a variational formulation implying the unknowns $(u, \gamma; p_s, P)$ are then (34), (35), (36), (37), (39), (40), (43) and (44).

The variable P (and the corresponding test function δP) must satisfy the constraint:

$$\int_\Omega P = 0 \tag{45}$$

Remark 3

For sake of brevity, we do not enter here into the detailed analysis of (45).

3.2 The "condensed" symmetric variational formulation $(\gamma; p_s, P)$

The variational formulation corresponding to $(u, \gamma; p_s; P)$ can be established by using standard test function procedure. One obtains:

$$\int_{\Omega_s} \sigma_{ij}(u)\varepsilon_{ij}(\delta u) - \int_{\Omega_s} \rho_0 \gamma.\delta u$$
$$- \int_\Sigma P n.\delta u \quad - p_s \int_\Sigma n.\delta u = 0 .$$
$$\int_{\Omega_s} \gamma.\delta\gamma \quad - \lambda \int_{\Omega_s} u.\delta\gamma = 0 ,$$
$$- \int_\Omega \nabla P.\nabla\delta P + \lambda \int_\Sigma \rho_0\, u.n\, \delta P$$
$$+ \frac{\lambda}{c^2} \int_\Omega P\delta P + \frac{\lambda}{c^2} p_s \int_\Omega \delta P = 0 ;$$
$$p_s\, \delta p_s + \frac{\rho_0\, c^2}{\mathrm{Vol}(\Omega)} \delta p_s \int_\Sigma u.n = 0 ,$$
$$\forall(\delta u, \delta\gamma, \delta P, \delta p_s) \tag{46}$$

with the constraints:

$$\int_\Omega P = 0$$
$$u = 0 \qquad |_{\Sigma_2}$$

In (46), for clamped structure, *one can eliminate u* and it can be shown that the resulting three-field formulation $(\gamma; p_s, P)$ *is symmetric* and is valid for the static case.

Remark 4.

In the case of a free surface Γ, one must set the following equation (reference [**18**]).

$$p = 0 \qquad |_\Gamma \tag{47}$$

It is easy to see that this case does not require the constraint (33). Consequently, the variable p, does not appear.

CONCLUSION

We have introduced two alternative formulations valid for the static case, which mean that the only zero-frequency modes are the physical six rigid body structural motions.

The numerical applications of such formulations depends on the geometry of the fluid-structure system. For instance, for a partially clamped structure of shell type which contains a huge amount of compressible fluid, the formulation which introduces an additional variable for the structure is of interest.

REFERENCES

1. Belytschko T.B., Kennedy J.M., *A fluid-structure finite element method for the analysis of reactor safety problems*, Nucl. Eng. Des. **38** (1976), 71-81.
2. Craggs A., *The transient response of a coupled plate-acoustic system using plate and acoustic finite elements*, J. of Sound and Vib. **15** (1971), 509-528.
3. Combescure A., Gibert R.J., Jeanpierre F., Hoffmann A., Livolant.M., *Fluid-structure interaction in the CEASEMT computer program*, Comp. and Struct. J. **12** (1980), 459-469.
4. Daniel W.J.T., *Modal methods in finite element fluid-structure eigenvalue problems*, Int. J. Num. Meth. Eng. **15** (1980), 1161-1175.
5. Dautray R., Lions J.L., "Analyse mathématique et calcul numérique pour les sciences et les techniques," Tomes 1 à 9, Ed. Masson, Paris, 1988.
6. Deneuvy A.C., *Optimization of a coupled fluid-structure problem for maximum difference betwteen adjacent eigenfrequencies*, in "Proc. Innov. Num. Meth. in Eng.," Springer-Verlag, 1986.
7. Dowell E.H., Gorman G.F., Smith D.A., *Acoustoelasticity: general theory, acoustic natural modes and forced response to sinusoidal excitation, including comparison with experiments*, J. of Sound and Vib. **52** (1977), 519-542.
8. Everstine G.C., *A symmetrical potential formulation for fluid-structure interactions*, J. of Sound and Vib. **79, 1** (1981), 157-160.
9. Felippa C.A., *Symmetrization of the contained fluid vibration eigenproblem*, Comm. in Appl. Num. Meth. **1** (1985), 241-247.
10. Felippa C.A., Ohayon R. yr 1990, *Mixed variational formulation of finite elements analysis of acousto-elasto-elastic slosh fluid-structure interaction*, Int. J. Fluids and Structures **4**, 35-57.
11. Geradin M., Roberts G., Huck J., *Eigenvalue analysis and transient response of fluid-structure interaction problems*, Eng. Comp. J. **1** (1984), 152-160.

12. Gladwell G.M.L., *A variational formulation of damped acousto-structural vibration problems*, J. of Sound and Vib. **4** (1966), 172–186.
13. Hamdi M.A., Ousset Y., Verchery G., *A displacement method for the analysis of vibrations of coupled fluid-structure systems*, Int. J. Num. Meth. Eng. **19** (1978), 1657–1668.
14. Kanarachos A., Antoniadis I., *Symmetric variational principles and modal methods in fluid-structure interaction problems*, J. of Sound and Vib. **121, 1** (1988), 77–104.
15. Kehr-Candille V., Ohayon R., *Elasto-acoustic vibrations of a structure containing a gas*, in "Proc. of the European Conf. on New Advances in Computational Structural Mechanics," Giens, France, 1991.
16. Liu W.K., Uras R.A., *Variational approach to fluid-structure interaction with sloshing*, Nucl. Eng. Des. **106** (1988), 69–85.
17. Morand H.J-P., *Modal methods for the analysis of the vibrations of structures coupled with fluids*, in "ASME PVP Conference, Hawaii," 1989.
18. Morand H.J-P., Ohayon R., *Finite element method applied to the prediction of the vibrations of liquid-propelled launch vehicles*, in "ASME PVP Conference, Hawaii," 1989.
19. Morand H.J-P., Ohayon R., *Investigation of variational formulations for the elasto-acoustic problem*, in "Proc. of the Int. Symp. on FEM meth. in flow problems," Rappallo, Italy, 1976.
20. Morand H.J-P. Ohayon R., *Substructure variational analysis for the vibrations of coupled fluid-structure systems*, Int. J. Num. Meth. Eng. **14,5** (1979), 741–755.
21. Müller W.C., *Simplified analysis of linear fluid-structure interaction*, Int. J. Num. Meth. Eng. **13** (1981), 109–122.
22. Nefske D.J., Wolf Jr., Howell L.J., *Structural-acoustic finite element analysis of the automobile passenger compartment : a review of current practice*, J. of Sound and Vib. **80** (1982), 247–266.
23. Nicolas-Vullierme B., Ohayon R., *Fluid-structure interaction for bounded and infinite medium vibration problems. Some recent advances at ONERA*, in "Advances in Fluid-Structure Interaction," vol. 64, ASME/AMD, 1984.
24. Ohayon R., *Transient and modal analysis of bounded medium fluid-structure problems*, in "Proc. Int. Conf. on Num. Meth. for Transient and Coupled Problems," (Ed. R.W. Lewis, P. Bettess, E. Hinton), Pineridge Press, Swansea. 1984. TP-ONERA 1984-56.
25. Ohayon R., Valid R., *True symmetric variational formulation for fluid-structure interaction in bounded media. Finite element results*, in "Numerical Methods in Coupled Systems," Chapt.10, Ed. RW Lewis and al., John Wiley, 1984.
26. Ohayon R., *Interaction fluide-structure en milieu borné. Nouvelles formulations symétriques*, in "Tendances Actuelles en Calcul de Structures," Ed. Pluralis, 1985, pp. 435–448.
27. Olson L.G., Bathe K.-J., *Analysis of fluid-structure interactions. A direct symmetric coupled formulation based on the fluid velocity potential*, Comp. and Struct. **J. 21, 1/2** (1985), 21–32.

28. Petyt M., Lim S.P., *Finite element analysis of the noise inside a mechanically excited cylinder*, Int. J. Num. Meth. Eng. **13** (1978), 109–122.
29. Ramakrishnan J.V., Koval L.R., *A finite element model for sound transmission through laminated composite plates*, J. of Sound and Vib. **112, 3** (1987), 433–446.
30. Sanchez-Hubert J., Sanchez-Palencia E., "Vibration and Coupling of Continuous Systems," Springer-Verlag, 1989.
31. Sandberg G., Göransson P., *A symmetric finite element element formulation for acoustic fluid-structure interaction analysis*, J. of Sound and Vib. **123, 3** (1988), 507–515.
32. Zienkiewicz O.C., Newton R.E., *Coupled vibrations of a structure submerged in a compressible fluid*, in "Proc. of the Int. Symp. Finite Elements Techniques," Stuttgart, May 1969.
33. Zienkiewicz O.C. yr 1984, *Coupled problems and their numerical solution*, in "Numerical Methods in Coupled Systems," Chapt.1, Ed. R.W Lewis and al., John Wiley.
34. Zienkiewicz O.C., Taylor R.L., "The Finite Element Method," Mc Graw Hill, 1988.

FINITE ELEMENTS IN THE 90's, E. Oñate, J. Periaux, A. Samuelsson (Eds.)
Springer-Verlag/CIMNE, Barcelona 1991

BIOMECHANICS NEEDS FEM

P.M. Calderale and C. Bignardi
Technical University of Torino,
Corso duca degli Abruzzi 24, 10129 Torino,
ITALY

SUMMARY

Having identified biomechanics as of interest to clinical application, in particular structural analysis, the theoretical and experimental analysis difficulties were then looked into.
It may be concluded that presently only the finite element method can furnish significant guideline bases towards clinical practice and implant design.
Following are some examples of applications that we carried out in interdisciplinary work to study problems of orthopaedics and dentistry.

1.INTRODUCTION

We feel we do not err in thinking that each and every one of us, about to start up in the various research areas, at the thought of using finite element method must have momentarily wondered what on earth Oleg Zienkiewicz had done in the field to get inspiration and guidance [1].
This was also our case, over twenty years ago, when we began research in biomechanics.

2.DEFINITION OF BIOMECHANICS

Biomechanics is a part of bioengineering that can be defined as the application of ideas and methods of engineering and physical sciences to the investigation and solution of problems in medicine and biology.
Whilst the biomechanics field of interest is clear, its ways are quite various.
The subject abounds with abstrusely theoretical basic research, which we respect since research must not be limited in any way, but there is a paucity of practical application. There is also research auguring an application follow-up, yet to be clearly worked out.
There is some research (albeit little) to solve clinical problems; it is invariably the fruit of strictly interdisciplinary work and evolves alongside methodological improvements.
We elected to study biomechanics as a support to research and clinical application.

On this basis biomechanics may make a huge contribution to surgery.

3.BIOMECHANICAL STRUCTURAL METHODOLOGIES

The turning point in determining the potential role of engineering in studying biological and clinical matters occured towards the end of the 60's. Then it was realised that biological structure generally, and bone particularly, could be studied using engineering theory and experimentation methods [2][3].
Thus a biological structure could potentially be treated as a normal engineering one. It therefore became necessary to calculate acting forces within the various system sectors, characterise materials and make structural analysis.
All three phases gave rise to considerable difficulties. The forces acting on bone segments and joints are body mass forces resulting from static and dynamic actions during body movement and forces due to muscle contractions.
Determining these forces correctly is extremely difficult, due to the fact that mainly the locomotor apparatus is highly hyperstatic.
Two approaches may thus be taken:
a) Simplifying the system by ignoring the action of most of the muscles so as to make the system statically determinate. With this approach, only the joint reactions can be calculated.
b) Considering the action of many or all muscles, and solving indeterminacy by introducing other conditions e.g. minimizing a function of the unknowns [4][5].
With hyperstatic systems, though the degree of approximation is limited by the restricted number of conditions available, we can also produce an overall view of the forces acting in the muscles, and which thus stress the bone segments.
The validity of these approaches is doubtful. The optimization of muscular actions strategy perhaps may be accepted to model sport activity, but we think it less acceptable in patology uses.
Thus it was time to decide on how to cope with the problem.
Current practice prefers to utilize simplified load systems, considering external forces and loads on joints. However, by this system the structural calculations do not take into account the local actions in the various sectors of the system.
Bone is a non-homogeneous, anisotropic material with non-linear behaviour patterns. In addition, its characteristics vary from one individual and one skeletal area to another and according to sex, age, particular load history as well as due to measurement methods and test sample preservation conditions [6].
Material properties have been studied for many years and we now know their orders of magnitude. On the other hand, until we develop methods able to characterize biomaterials "in vivo", more precise values are of little use, given that they vary, not only from individual to individual, but also with time and pathology. Bone characteristics also vary according to physical activity. Tibia and patella mechanical characteristic variations in Boston maraton runners were recently calculated using ultrasonic techniques; before and after figures showed variations suggesting that an adaptation mechanism to withstand cyclical loads exists within the bone [7].
What greatly distinguishes a biological from an engineering problem is the fact that bone is a living material and prone to remodel; currently the stress levels leading to bone reabsorption or necrosis

are not known.
Remodelling makes bone a self-optimizing structure, in the sense that its trabeculae change orientation with stress direction variations and are aligned with the main tension trajectories.
Clearly, whenever bone is injured, subjected to a patology or surgery, this process is reactivated and the cancellous and compact bone are remodelled to enable the organ to bear the load. In other words, a bone structure always tends to optimization, in the sense that it changes as stress field changes.
In the presence of a prosthesis there are additional problems: it impairs structural optimization so severely that most surgical failures stem from structural origin. The main problem lies in the bone/prosthetic stem or cement/bone interface conditions and, more generally, in the structure of the mechanical system constituted by the prosthesis and bone, with or without cement [8][9][10][11].
The structural problems involving prosthetic implants derive from the fact that the bone/prosthesis system couples elements having different stiffness and always produces complex structural conditions as a result of bone resection and stiffness discontinuity.
It would seem clear that this connection is loaded in fatigue and that aseptic loosening, that is the most frequent (and substantial) cause of implant failure, has fatigue failure characteristics; loosening is progressive and the time spans reduce as stresses increase. Consequently, stress patterns are closely linked to prosthesis life.
We also believe that high stresses, and stress peak in particular, are the principal cause of pain experienced by most patients, especially during the first post-operative year. These stresses, which are greater than and differ from those of the original bone, lead to bone remodelling.
Bone tends to self-optimize and succedes if an exaggerated stress level is avoided.
Clinical problems are linked to functionality (to be re-acquired) and pain (to be eliminated or diminished). According to our criteria, based on our feeling that high stresses and pain are strongly associated, structural analysis can give a decisive contribution to assessing the acceptability of a clinical situation whether or not in connection with a surgical operation.
Closed form solutions are not suitable to solving problems of this type both due to irregular bone form plus strong mechanical characteristics variations from one point to another [12][13][14][15][16][17][18].
The introduction of FEM has therefore made it possible to carry out structural calculations otherwise not feasible, as well sophisticated local analysis.
Not even experimental methods can give a contribution since the only valid experimentation should be carried out in vivo. Dry bone experimentation is unreliable since it does not take into account remodelling nor bone-implant connection [19][20][21][22].
Developments in theoretical stress analysis have made it clear that three dimensional discretization through the finite element method is the best means of obtaining meaningful results; with FEM it is not possible to obtain absolute values, but either parametric analyses or comparisons between different situations are made possible.
However, the method is limited due to the prosthesis-bone interface conditions which are difficult to simulate since prosthesis stem-bone connection is not yet fully understood. We can hypothesize an initial bone-to-prosthesis anchorage state (during which bone-stem relative

micromotions may take place) followed by subsequent spontaneous structural optimization states wherein stresses cause adequate bone remodelling if strength limits are not exceeded.
As to the system of loads acting on bone segments it is difficult to determine the entity of the muscular forces acting and to simulate the load transfer across the bone attachment areas.
Oleg Zienkiewicz was one of the first to produce a three-dimensional human femur model (fig.1) and found himself, like all researchers in this field, up against these problems.
We feel that FEM will have further important applications also in biomechanics when it can reliably construct models directly i.e. as a matter of course from x rays at acceptable cost, personalize the structural calculation and design surgery.

4.METHOD APPLICATION EXAMPLES

One of the most important applications of the finite element method in biomechanics is the study of the load bearings arthroprosthesis-bone coupling in orthopaedics with or without cement. During the mid 70's, to have a useful deductions basis for a hip prosthesis design of optimized stress distribution, we produced a two-dimensional femur model with various prostheses inserted and subjected it to different simplified load conditions (fig.2). An initial finding from these calculations was the need to increase the cement-prosthesis contact surface in the highest stress areas (fig.3) [23][24].
A recent three-dimensional study on the influence of stem length on interface stress distribution showed that, contrary to past belief, a long stem confers no advantages of any kind [25][26].

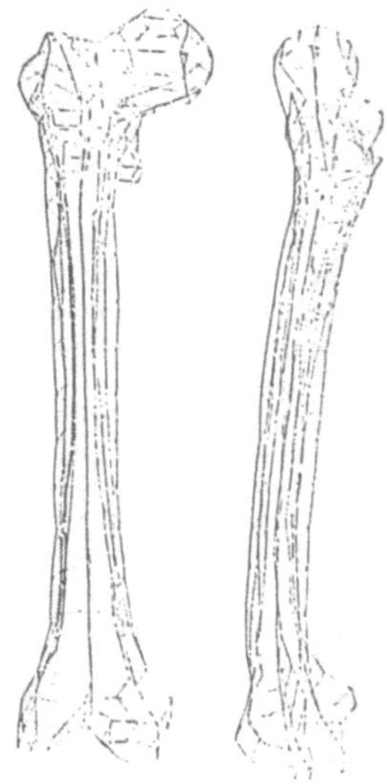

Fig.1 - Femur mesh by Oleg Zienkiewicz

This research shows the clinical advisability of using a short stem that make it possible to employ a high resection, immediately below the femoral head. This preserves the original strength of the proximal femur shell structure, and provides prosthesis support in an essentially cortical bone area. In addition, in the event of revision, a high resection would facilitate a later conventional resection at the femoral neck base for a second prosthesis.
Performing a second resection also make it easier to remove the stem, since it frees part of its surface.
At the end of the 70's we analysed a dynamic type biomechanics problem using the finite element method. A driver-vehicle system model was made to analyse mutual interactions during a head-on barrier collision in the case of a driver not wearing a seat or vehicle-attached safety belt [27]. The model was later improved upon and a lateral collision also considered (fig.4) [28].
The model is of general application; in other words, it can be used to determine the forces exchanged between pelvis and femur, between femur and patella and in the muscular structure for any hypothesis or initial condition whatsoever. These results make up the entry data

for the stress distribution analysis in the femur and pelvis bone structures (figs.5,6).
The values of the mechanical characteristics of the muscle groups introduced in the calculations were arrived at by process of later correction until dynamic answers comparable with the results of experimental trials on dummies and corpses were achieved.
The finite element method was further used to calculate temperatures reached by components of an arthroprosthesis during the exothermic acrylic cement polymerization process (fig.7) [24][29].
Using this method to analyse this type of problem, whilst requiring preliminary experimental inquiry to set the base figures, has an advantage over experimental trials in that it can quickly and easily calculate the influence of the various parameters involved.
The results of a calculation for a total ankle prosthesis with a talar component covering the talus and a tibial component inserted in the tibia by means of a stem highlighted the need for well-calculated cement use especially as regards quantity and the importance of using the lowest possible heat polymerization cements to avoid biological tissue necrosis.

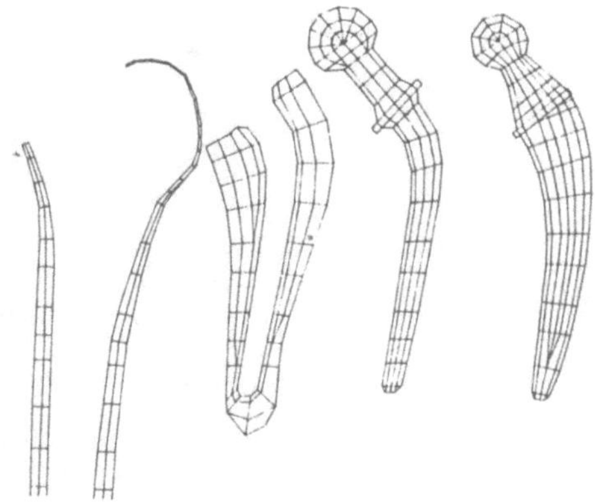

Fig.2 - Femur, acrylic cement and two types of hip prosthesis models

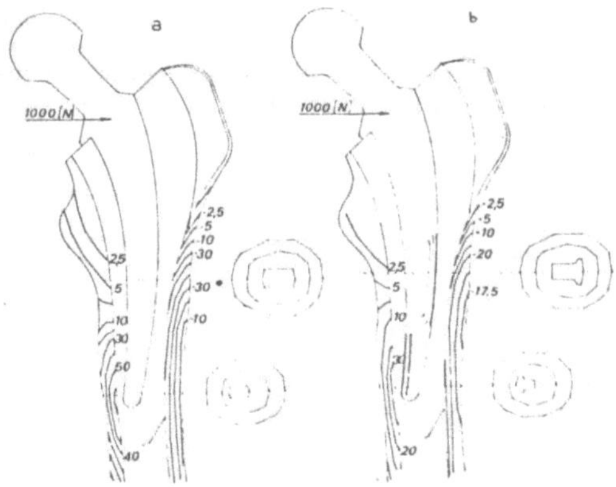

Fig.3 - Isostress lines [N/mm^2] in cortical bone with a) Müller prosthesis, b) modified Müller prosthesis

In addition, a three-dimensional talus model was made comparing stress distributions in the whole and severed bone (fig.8) [30].
Still within orthopaedics, a pelvis stress distribution analysis following various types of screwed-in acetabular component is underway, to parametrically calculate the form of acetabular component and screwthread least liable to affect natural stress flow (fig.9).
We also developed a human mandible model (fig.10) validated by experimental analyses on a photoelastic model [31][32][33]. Early results show the model to provide satisfactory experimental reproduction under centric occlusion conditions and is sufficiently accurate to study the biomechanical performance of partial removable prostheses and implants involving monolateral morsel; a microanalysis

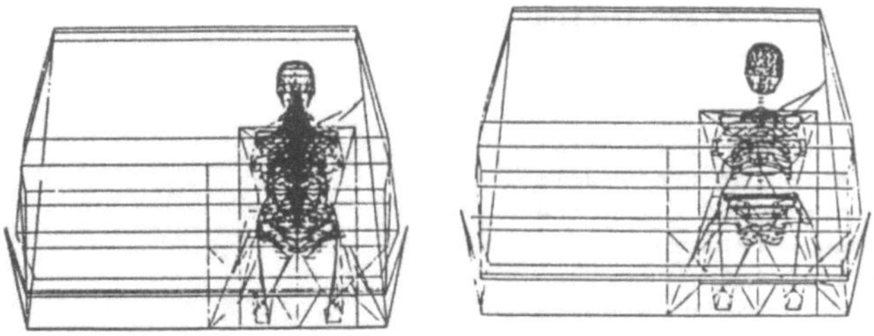

Fig.4 - Vehicle-driver mesh

of the tooth retaining system in the alveolus was also carried out (fig.11) [31].
Cancellous bone properties are vastly more complex than cortical bone ones and, as far as we know, have never been studied sufficiently in depth.
In general, researchers in this field attributed to cancellous bone properties values obtained on the basis of experimental trials on samples that, whilst small in size, follow the average behaviour of a structure similar to that of a honeycomb. The results of the finite element analyses, if the cancellous bone mean characteristics are suitably selected and distributed within the whole structure, are correct as to strains, but reveal stress values which do not reflect their complex distribution in the trabecular bone.

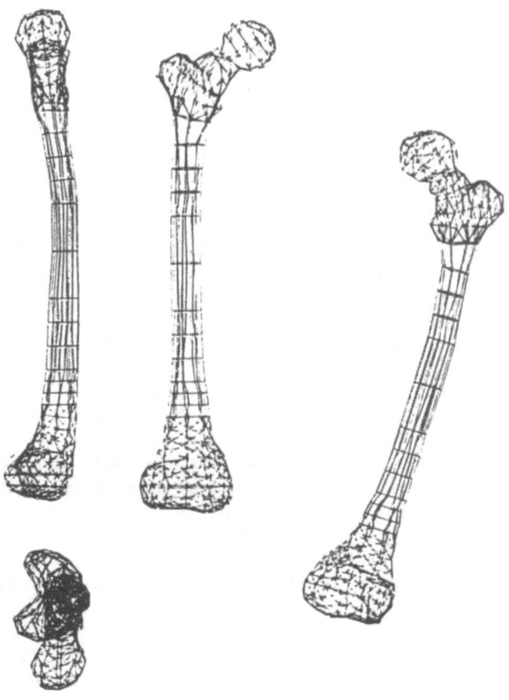

Fig.5 - Femur mesh

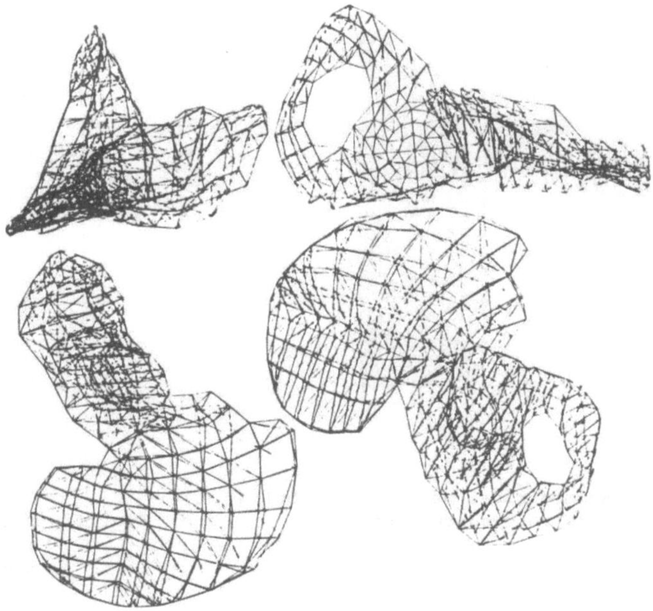

Fig.6 - Pelvis mesh

Thus we chose to carry out a local strain and stress microanalysis and came up with a cancellous bone model based on experimental observations which, while limited, allowed us to determine the order of magnitude of trabecular element stresses; as expected, stresses in the individual trabecula proved to be much higher than those resulting from the homogeneous elements of spongy bone and showed up the need to carry out local microanalysis in high stress area (fig.12) [34].

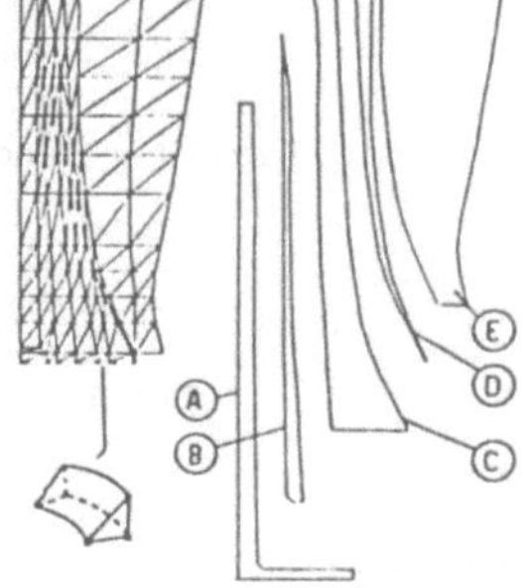

Fig.7 - FEM model for temperature behaviour analysis: A) prosthesis, B) cement C) spongy bone, D) cortical bone, E) muscle

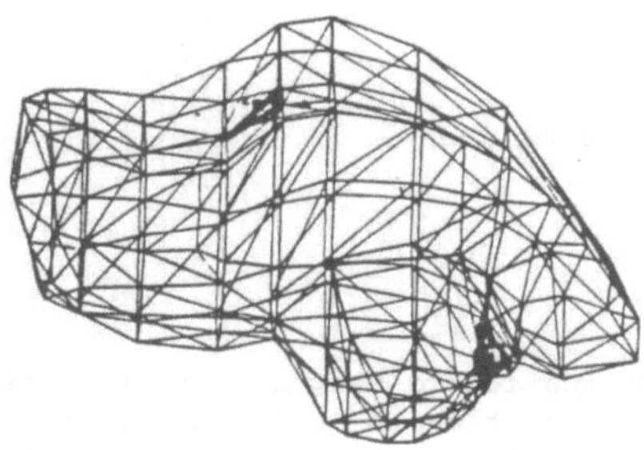

Fig.8 - Talus mesh

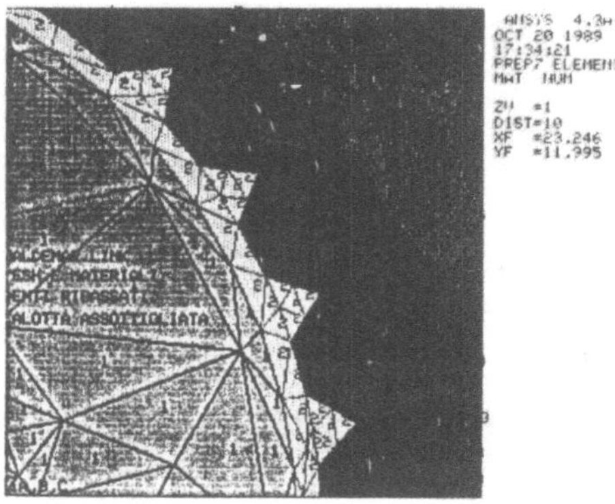

Fig.9 - Aximmetrical mesh of acetabulum with screwed cup inserted

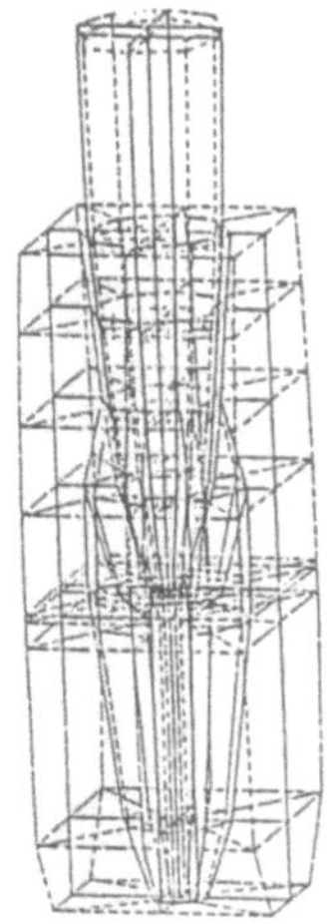

Fig.11 - Tooth-alveolus mesh

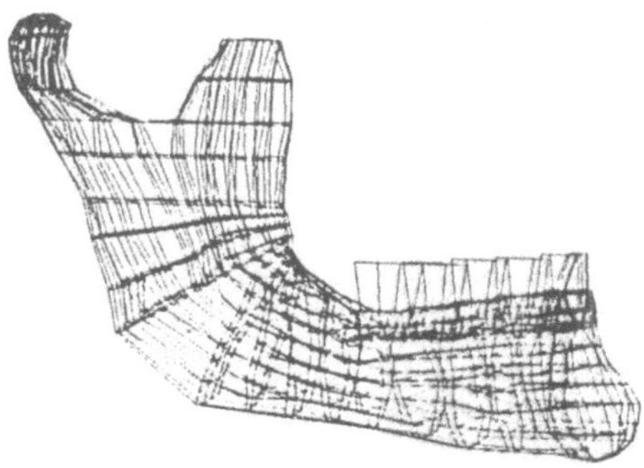

Fig.10 - Mandible mesh

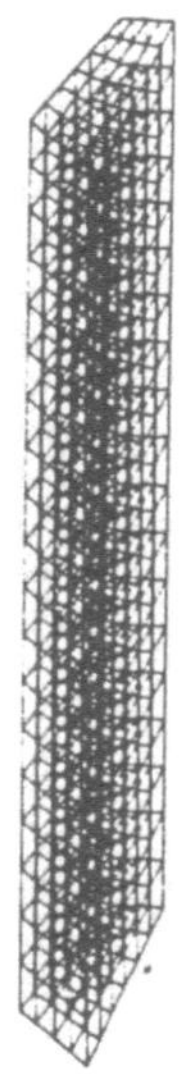

Fig.12 - Cancellous bone mesh

5.CONCLUSIONS

We have outlined some FEM applicationsto show how this method can make a far from negligible contribution to clinical practice. This is the path we have been following for some years, during which time our periodical friendly contacts with Oleg Zienkiewicz were most stimulating and enabled us to profit from his wisdom and great ability.

REFERENCES

1. ZIENKIEWICZ, O.C. - Finite element methods in continuum mechanics, Nato Advanced Study Institute, Lisbona, 1971.
2. CALDERALE, P.M. and LORENZI, G.L. - Problemi meccanici delle artroprotesi, LIX Congr. SIOT, Cagliari, Sett. 1974, 1-107.
3. CALDERALE, P.M. - Biomechanics of load-bearing human joints and arthroprostheses, 2nd Italian-Polish Symp. Bioengineering, Udine, Oct. 1975, 215-229.
4. CALDERALE, P.M. and SCELFO, G. - Mathematical models of musculo-skeletal systems - Eng. in Medicine, **16**, 3, 131-146 (1987).
5. CALDERALE, P.M. and SCELFO, G. - A matematical model of the locomotor apparatus, Eng. in Medicine, **16**, 3, 147-161 (1987).
6. CALDERALE, P.M. and BIGNARDI, C. - L'osso come materiale nall'interpretazione del tecnologo", Riunione Congiunta SIB-SIBOT, Trieste, Luglio 1988.
7. RUBIN, C.T., PRATT, G.W., PORETER, A.L., LANYON L.E., POSS, R. - The use of ultrasound in vivo to determine acute change in the mechanical properties of bone following intense physical activity, J. Biomechanics, **20**, 723-727 (1987).
8. CALDERALE, P.M. - Artroprotesi portanti, Conv AMMA, Torino, Nov.1980; Notiziario Tecnico AMMA, Torino,**XXXVI-4**, 66-77 (1981).
9. CALDERALE, P.M. - Esperienze di progettazione in biomeccanica, Proc. Inter. Conf. Teach. Eng. Des., ICED, Roma, Mar. 1981; Tecniche Nuove, **6**, W/2-1, 245-256 (1981).
10. CALDERALE, P.M. - Progettazione biomeccanica, Il Progettista Industriale, II, **1**, 30 -37 (1982) 1st part; II, **2**, 82-91 (1982), 2nd part.
11. BIGNARDI, C., CALDERALE, P.M., GOLA, M.M. and GUGLIOTTA, A. - Progettazione e valutazione di artroprotesi, in Bioingegneria della riabilitazione, T. Leo e G. Rizzolatti, Patron Editore, 295-309 (1987).
12. CALDERALE, P.M., GOLA, M.M.and GUGLIOTTA, A. - Analytical estimate of bone-prosthesis coupling, VI International Congr. Biomech., Copenhagen, Jul. 1977, 1-22.
13. CALDERALE, P.M., GOLA, M.M. and GUGLIOTTA, A. - An approximate analytical method for the design of arthroprosthesis stems, 1st Mediterranean Conf. Medical Biological Engineering, Sorrento, Sept. 1977, II/1.77-1.79.
14. BARBERI, G., GOLA, M.M., GUGLIOTTA, A. and CALDERALE, P.M. - Analytical considerations of the mechanical stem-femur coupling, Congr. Europ. Soc. Biomech., Bruxelles, May 1978, 216-222; Acta Orthopaedica Belgica, **46**, 6, 728-734 (1980).
15. BARBERI, G., CALDERALE, P.M., GOLA, M.M. and GUGLIOTTA, A. - Flexural coupling of hip prosthesis stem and femur: a simple method of theoretical analysis, Eng. in Med., **3**, 172-178 (1978).
16. CALDERALE, P.M., GOLA, M.M. and GUGLIOTTA, A. - Theoretical and experimental analysis of the stem-femur coupling: new improved procedures, 2nd Meeting Europ. Soc. Biomech., Strasbourg, Sept. 1979, 1-14.
17. CALDERALE, P.M., GOLA, M.M. and GUGLIOTTA, A. - Mechanical stress distribution in stems of implants, Biomechanics VII, 3A, Warsawa, Sept. 1979, 373-379.
18. CALDERALE, P.M., GOLA, M.M. and GUGLIOTTA, A. - Systematic approach to the experimental and theoretical study of hip prosthesis coupling, I World Biomat. Congr., Baden near Vienna, Apr. 1980; Advan. Biomat.,**3**, J.Wiley & Sons, chap. 17, 113-119.

19. GOLA, M.M. and GUGLIOTTA, A. - Some experimental results concerning the mechanical coupling in endoprosthetics, Eng. in Medicine, **16**, 1, 23-27 (1987).
20. MONGINI, F., PRETI, G., CALDERALE, P.M. and BARBERI, G. - Experimental strain analysis on the mandibular condyle under various conditions, Med. Biol. Eng., **19**, 521-523 (1981).
21. PEZZOLI, M., ROSSETTO, M. and CALDERALE, P.M., - Evaluation of load transmission by distal extension removable partial dentures by using reflection photoelasticity, 63rd I.A.D.R., March 1985, Las Vegas, NE, U.S.A.; J. Prosth. Dent., **56**, 3, 329-337 (1986).
22. CALDERALE, P.M. and ROSSETTO, M. - Biomechanical analysis of coupling between mandible and removable partial dentures, 5th ESB Meeting, Berlin (RFT), Sept. 1986; in Bergmann G. et al.- Biomechanics: Basic and Applied Research, Martinus Nijhoff Publishers, Dordrecht, 1987, 745-750.
23. CALDERALE, P.M., FASOLIO, G. and PACCA, G. - Problemi di biomeccanica studiati con il metodo degli elementi finiti", VIII Conv. Naz. AIAS, Firenze, Sett. 1980, 317-326.
24. CALDERALE, P.M. and FASOLIO, G. - Applicazioni del metodo degli elementi finiti in biomeccanica, Minerva Ortopedica, **33**, 3, 239-250 (1982).
25. CALDERALE, P.M. and BIGNARDI, C. - Mechanical analysis of stem length in hip prostheses, Progress in Bioengineering Seminar, Glasgow (GB), September 1988; in J.P.Paul, Progress in Bioengineering, Adam Higler, Bristol, 1989.
26. CALDERALE, P.M. and PIPINO, F. - Biomechanical study of load bearing arthroprostheses: our experience concerning with hip and ankle", Int. Conf. Appl. Mech., Beijing, 1989.
27. CALDERALE, P.M., GARRO A. and LORENZI, G.L. - Car accident mathematical model: femur and pelvis stress analysis, IVth Congr. IRCOBI, Goeteborg, Sept. 1979, 305-316; ATA 32-10, Ott. 1979, 418-428.
28. CALDERALE, P.M. - Interaction between man and car during accidents, Polish-Italian Seminar, Warsaw Univ. of Technology, oct. 1990, 13-28.
29. CALDERALE, P.M. and PIPINO F. - Effect of temperature changes in acrylic cement on prosthetic implant surgey, Italian Journal Of Orthopedics and Traumatology, **IX**, I, 57-65 (1983).
30. CALDERALE, P.M., GARRO, A., BARBIERO, R., FASOLIO G. and PIPINO F. - Biomechanical design of the total ankle prosthesis, ASME, Winter Meeting, Phoenix-AZ, 82-DET-48, Nov. 1982; Engineering in Medicine, **12**, 2, 69-80 (1983).
31. CALDERALE, P.M., ROSSETTO, M., TRITTONI, L. and GARRO, A. - Studio biomeccanico della mandibola mediante tecniche numeriche e sperimentali, Atti del XV Convegno Nazionale AIAS, Pisa, Settembre 1987, 265-278.
32. CALDERALE, P.M., MODICA, R., PEZZOLI, M. and ROSSETTO, M. - Modello matematico della mandibola umana, Atti del XXI Congresso Nazionale SIOCMF, Firenze, Novembre 1987, 1251-1256.
33. CALDERALE, P.M. and ROSSETTO, M. - Our activity in dentistry biomechanics by experimental and theoretical methods", Progress in Bioengineering Seminar, Glasgow (GB), September 1988; in J.P.Paul, Progress in Bioengineering, Adam Higler, Bristol, 1989, 179-184.
34. CALDERALE, P.M. and GARRO, A. - Theoretical analysis of the coupling between hip prosthesis stem and femur, Proc. Int. Conf. on Experimental Mechanics, Beijing (Rep. Pop. Cinese), Science Press, 1985,

SECTION 3

FLUID MECHANICS

FINITE ELEMENTS IN THE 90's, E. Oñate, J. Periaux, A. Samuelsson (Eds.)
Springer-Verlag/CIMNE, Barcelona 1991

NUMERICAL METHODS FOR THE HELMHOLTZ EQUATION WITH NON-REFLECTING BOUNDARY CONDITIONS IN EXTERIOR DOMAINS

I. Harari and **T. J. R. Hughes**
Division of Applied Mechanics,
Durand Building,
Stanford University,
Stanford, California 94305-4040,
U.S.A.

Take care of the sense, and the sounds will take care of themselves.
— LEWIS CARROLL, *Alice's Adventures in Wonderland* (1865)

SUMMARY

Finite element methods are constructed for the reduced wave equation with special attention to unbounded domains. Exterior boundary conditions for a computational problem are derived by the DtN method, giving rise to exact (and thereby unique) solutions. Simple expressions that determine a sufficient number of terms in the truncated DtN operator (which is employed in practice) for unique solutions at any given wave number are derived. A local approximation of the boundary conditions restores uniqueness for all wave numbers. Galerkin and Galerkin/least-squares finite element methods are presented. The Galerkin/least-squares method with DtN boundary conditions is designed to exhibit superior behavior, providing accurate solutions with relatively low mesh resolution and allowing damping of unresolved waves. Convergence results guarantee the good performance of Galerkin/least-squares methods on configurations of practical interest. Numerical tests validate these conclusions.

1. INTRODUCTION

An important building block of structural acoustics and fluid-structure interaction is the solution of problems of acoustics, governed by the Helmholtz equation (or reduced wave equation) in the time-harmonic case. These solutions, describing propagating and evanescent waves, are related to a variety of physical phenomena. Numerical solutions in exterior domains have been sought primarily via techniques that are based on Helmholtz integral representations, relating quantities on the *physical boundary*. These equations lead to boundary element methods (see the survey in [14]), with the benefit of *a priori* satisfaction of the radiation condition at infinity. On the other hand, methods of this kind are known to encounter difficulties, such as potential non-uniqueness of the *continuous* solution at certain wave numbers. This is *not* a property of *exact* solutions of the boundary-value problem and may lead to ill-conditioned discrete equations. Of the numerous remedies that have been attempted, the Burton-Miller approach [2] appears to offer the most promising basis for boundary-based computation in that non-unique solutions are rigorously excluded.

In this paper we develop finite element methods for the Helmholtz equation. These methods are not restricted to homogeneous, isotropic, linear problems, which is often the case for boundary elements. When applicable, boundary elements do have an ostensible advantage. For example, in three dimensions they require only surface discretization as compared to volumetric in the case of finite elements, thereby reducing the number of equations to be solved. However, the resulting systems of equations are not favorably structured, rendering the two methods economically competitive for large-scale computation [9]. Nevertheless the task of discretization is substantially simpler for boundary element methods, an important consideration in model preparation. Finite elements have a rich mathematical background which can be used to prove convergence of numerical solutions and at times aid in method design. In the field of structural acoustics, finite element methods are widely used to model elastic solids, coupled with (and on occasion supplanting) boundary elements for the acoustic fluid, e.g., [3]. Numerous schemes to circumvent the difficulty in applying finite elements to unbounded domains are surveyed in the exhaustive reviews [1], pp. 95–116, and [5]. In the following, a computational formulation is derived by the DtN method proposed by Givoli and Keller [6] for converting problems defined over large or unbounded domains to equivalent formulations that are suitable for domain-based computation. We address the degradation of stability that arises in problems of acoustics by employing a general technique called 'Galerkin/least-squares' that was developed to counteract such difficulties. This methodology is related to a profusion of new classes of methods, see survey in [11]. In such cases, additional quantities are introduced into the formulation in order to demonstrate convergence of numerical solutions to the exact solution. By endowing the method in this manner with a sound mathematical foundation stability properties are enhanced while maintaining higher-order accuracy.

In Section 2 boundary-value problems for the reduced wave equation in exterior domains are introduced. Problems that are suitable for domain-based computation are obtained via the DtN method, and properties of resulting continuous formulations are outlined. In Section 3, Galerkin and Galerkin/least-squares finite element methods are presented. Model problems are employed to examine the numerical formulations and design the Galerkin/least-squares operator. The general convergence of solutions obtained by Galerkin/least-squares finite element methods is stated. The performance of the methods proposed is validated by numerical examples. Conclusions are drawn in Section 4.

2. CONTINUOUS FORMULATIONS UNDERLYING THE COMPUTATION OF TIME-HARMONIC ACOUSTICS

2.1 An exterior problem for the Helmholtz equation

Let $\mathcal{R} \subset \mathbf{R}^d$ be an unbounded region, where $d = 2$ or 3 is the number of space dimensions. The boundary of $\mathcal{R}$, denoted by Γ, is internal and assumed piecewise smooth (Fig. 1), with an outward normal unit vector $\boldsymbol{n}$. A boundary-value problem for the Helmholtz equation is: Find $\phi: \overline{\mathcal{R}} \rightarrow \mathbf{C}$, the spatial component of the acoustic pressure or velocity potential, such that

$$-\mathcal{L}\phi \quad = \quad f \qquad \text{in } \mathcal{R} \tag{1}$$

satisfying boundary conditions on the interior boundary Γ and the *Sommerfeld radiation condition* at infinity ($\mathcal{L}\phi := \Delta\phi + k^2\phi$ is the Helmholtz operator, Δ is the Laplace operator and $k \geq 0$ is the wave number, and $f: \mathcal{R} \rightarrow \mathbf{C}$ is a given source distribution). The radiation condition admits only outgoing waves at infinity, guaranteeing that the solution is unique (see [15]). Appropriate representation of this condition is crucial to the reliability of numerical formulations.

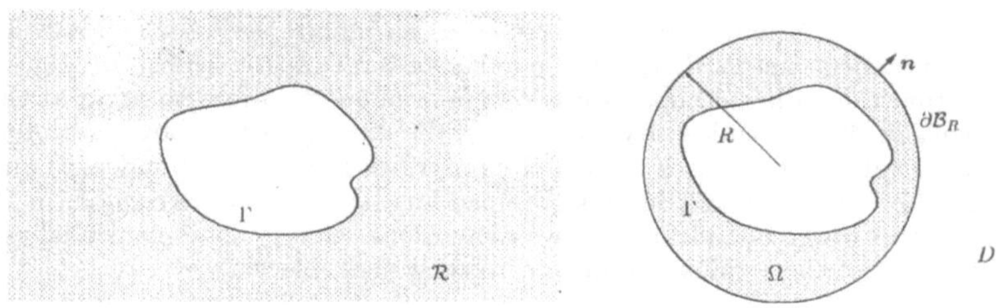

Fig. 1. An unbounded region converted to a computational domain.

2.2 The computational problem with a DtN boundary

Boundary-value problems in unbounded regions are not suitable for traditional computational techniques. In order to overcome this difficulty the DtN method [6 and 13] is employed. The original domain is decomposed into two parts by an artificial boundary. In the external region the problem is to be solved analytically. For this purpose, the artificial boundary is constructed in a simple shape, often a d-dimensional sphere of radius R, $\partial\mathcal{B}_R$, and located so that all geometric irregularities, nonlinearities and source terms, as well as material anisotropy and inhomogeneity, are contained within it. The analytical solution obtained for the external region $D := \mathbb{R}^d - \mathcal{B}_R$ (Fig. 1) is used to form the DtN map M, a relation between the function and its derivatives on the artificial boundary. This relation, which is the *exact* impedance (see [10]) is then imposed as the boundary condition on the artificial boundary

$$\phi_{,n} = -M\phi \qquad \text{on } \partial\mathcal{B}_R \tag{2}$$

completing the definition of the computational problem in $\Omega := \mathcal{R} - D$ ($\phi_{,n} := \nabla\phi \cdot \boldsymbol{n}$ is the normal derivative and ∇ is the gradient operator). Several implementational aspects of DtN formulations were discussed in [6 and 13].

The objective of the variational form of the problem is to find $\phi \in \mathcal{S}$, an admissible trial solution, such that $\forall w \in \mathcal{V}$, admissible weighting functions

$$A_k(w, \phi) = L(w) \tag{3}$$

where

$$A_k(w, \phi) := (\nabla w, \nabla\phi) - k^2(w, \phi) + (w, M\phi)_{\partial\mathcal{B}_R} \tag{4}$$

$L(w)$ contains terms emanating from the source distribution and Neumann boundary conditions on the interior boundary, and $(\cdot,\cdot)\colon \mathcal{V} \times \mathcal{S} \to \mathbb{C}$ is the $L_2(\Omega)$ inner product (with the first argument conjugated). Subscripts on inner products denote domains of integration other than Ω.

2.3 Properties of DtN formulations

Potential non-uniqueness of boundary integral representations of the Helmholtz equation prompted the investigation in [10] of properties of *continuous* DtN formulations that had not been addressed previously. It was established therein that the solution obtained by the DtN method is the restriction of the exact solution to the computational domain, and, as such, is unique for all wave numbers. The DtN map is often expressed as an infinite series, which is truncated in practice. Let N denote the number of terms in the series. In [10] we showed that the truncated operator allows non-unique solutions to occur at the harmonics of higher modes. By considering the annular region bounded internally by a sphere of radius a (which circumscribes the interior boundary) and externally by $\partial\mathcal{B}_R$ a lower bound is obtained for the critical wave number in Ω. In Fig. 2 this lower bound for the two-dimensional case is nondimensionalized by the radius of the artificial boundary, to show that selecting

$$N \geq kR \tag{5}$$

guarantees uniqueness of solutions to the two-dimensional problem (which is also valid for three dimensions [10]). Givoli and Keller derived a local approximation to their boundary conditions in two dimensions [7], in terms of higher-order tangential derivatives, preserving the local nature of finite element data. In [10] we demonstrated that employing the local boundary conditions retains uniqueness of solutions for all wave numbers, and derived their three-dimensional form.

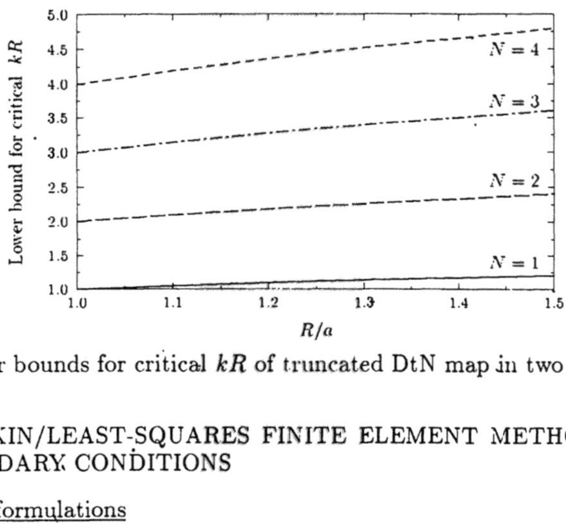

Fig. 2. Lower bounds for critical kR of truncated DtN map in two dimensions.

3. GALERKIN/LEAST-SQUARES FINITE ELEMENT METHODS WITH DTN BOUNDARY CONDITIONS

3.1 Discrete formulations

Galerkin (this is the classical method): Consider a partition of the domain into finite elements such that $\tilde{\Omega}$ is the union of element interiors. Let the function spaces $\mathcal{S}^h \subset \mathcal{S}$, $\mathcal{V}^h \subset \mathcal{V}$ consist of continuous piecewise polynomials of degree l. Find $\phi^h \in \mathcal{S}^h$ such that $\forall w^h \in \mathcal{V}^h$

$$A_k(w^h, \phi^h) = L(w^h) \tag{6}$$

where h is the mesh parameter. The coefficient matrix is a linear combination of standard finite element stiffness and mass matrices, and the DtN contribution.

Galerkin/least-squares (GLS): The variational form (3) is adjusted to include a residual of the governing differential equation

$$A_{GLS}(w^h, \phi^h) = L(w^h) - \tau\left(\mathcal{L}w^h, f\right)_{\tilde{\Omega}} \tag{7}$$

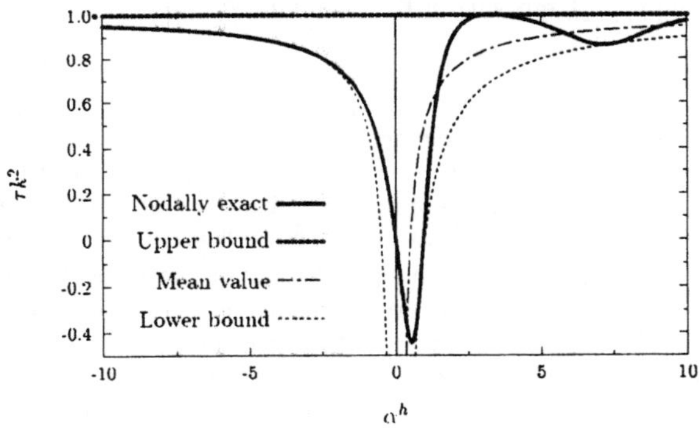

Fig. 3. Nodal exactness via the Galerkin/least-squares method, see (9).

where

$$A_{GLS}(w^h, \phi^h) := A_k(w^h, \phi^h) + \tau \left(\mathcal{L}w^h, \mathcal{L}\phi^h\right)_{\tilde{\Omega}} \tag{8}$$

and τ is a yet undetermined parameter. For linear finite elements the coefficient matrix of this formulation is still made up of a linear combination of standard stiffness and mass finite element matrices (and the DtN contribution, which is unchanged), adding virtually no complexity beyond the Galerkin method.

3.2 A model for the inhomogeneous Neumann problem

In order to assess the performance of the numerical methods proposed we constructed model problems that exhibit many of the physical features of acoustic phenomena, yet still possess readily obtainable analytical solutions [8]. These models yield the one-dimensional problem of an infinite plane radiator with a *modified wave number*, which can be either real or imaginary, corresponding to propagating and evanescent waves, respectively. First we wish to characterize the behavior of solutions in the mesh interior, by considering an infinite mesh.

Galerkin: Galerkin solutions are conveniently characterized by the numerical parameter $\alpha^h := (kh)^2/12$. When the exact solution is propagating ($\alpha^h > 0$) there are two cases: For $\alpha^h \leq 1$ the magnitude of the solution is correct, but there is a *phase lag*. When α^h is larger than one the solution decays. The case of decay ($\alpha^h < 0$) has been studied in [4] and is reviewed here for completeness. As long as $\alpha^h > -1/2$ there is no conceptual problem. However, for $\alpha^h < -1/2$ there are spurious *oscillations* localized to the neighborhood of boundary layers.

Galerkin/least-squares: We employ $\alpha_{GLS} := \alpha^h(1 - \tau k^2)$. The choice of

$$\alpha_{GLS} = \frac{1}{2}\frac{1 - \cos(kh)}{2 + \cos(kh)} \tag{9}$$

leads to a method for which the discrete solution is ***nodally exact*** in the ***entire*** range of phenomena under consideration, regardless of mesh refinement. The behavior of τk^2 is depicted in Fig. 3. The bounds suggest simple approximations which may be employed in practical implementations (see [12]). Other design criteria lead to alternative definitions of Galerkin/least-squares methods [8].

3.3 The Neumann problem with a DtN boundary

Plane waves: In numerical testing the Galerkin method was found to require wave resolution of about eight elements per wave whereas Galerkin/least-squares solutions were accurate to four elements per wave [8]. As an example we show results in the propagation region over the unit interval discretized by a uniform mesh of ten linear elements. A Neumann condition is enforced on the wet surface ($z = 0$) and the DtN boundary is at $z = 1$. For $\alpha^h = 1/10$ there are six elements per wave (Fig. 4). The Galerkin solution exhibits error in both magnitude, and phase, which causes the loss of accuracy to become more noticeable towards the DtN boundary. This is an indication of insufficient wave resolution. There is no apparent error in the Galerkin/least-squares results. Since far-field solutions are obtained by the DtN method from values on the artificial boundary, the accuracy of Galerkin/least-squares on the artificial boundary is significant.

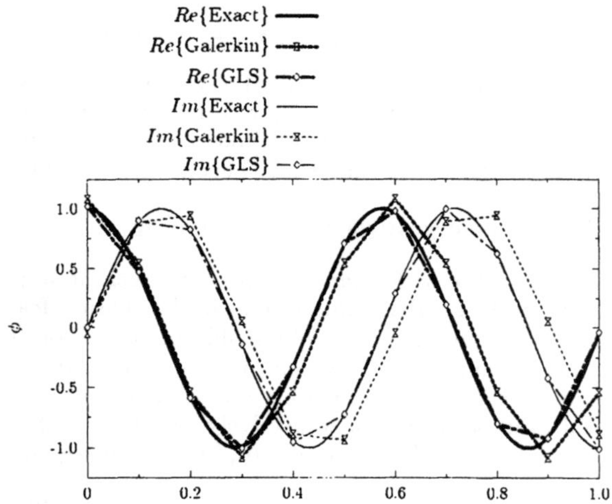

Fig. 4. Neumann model problem with a DtN boundary condition: propagating exact solution (six elements per wave).

Damping of unresolved waves: By fixing $\alpha_{GLS} = -1/2$ we obtain

$$\tau k^2 = 1 + \frac{1}{2\alpha^h} \tag{10}$$

For *any* uniform mesh the resulting solution vanishes over the entire domain except for the first element, in which it approximates the flux on the wet boundary

$$\phi^h_{,n} = \phi_{,n}/3 \qquad \text{on } \Gamma \tag{11}$$

Solutions thus obtained maintain this character as the nondimensional wave number is increased indefinitely. The example depicted in Fig. 5 demonstrates damping of unresolved waves ($\alpha^h = 1/2$, less than three elements per wave).

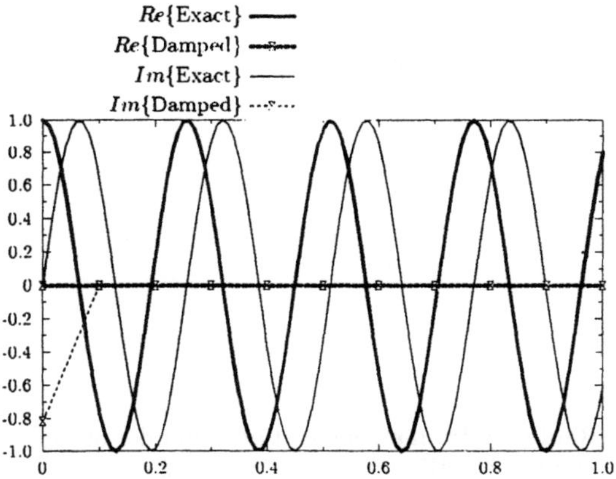

Fig. 5. Neumann model problem with a DtN boundary condition: unresolved propagating solution ($\phi^h_{,n} = \phi_{,n}/3$ on Γ).

3.4 Convergence of Galerkin/least-squares formulations

A detailed error analysis of the Galerkin/least-squares method for the entire range of propagation and decay was performed in [12]. The convergence result with optimal rates for all values of α^h provides an estimate of the error $e = \phi^h - \phi$ for resolved waves, and otherwise bounds the error, namely

$$\|\nabla e\|^2 + C_{kR}\|e\|^2_{\partial\mathcal{B}_R} \leq \begin{cases} C_\phi h^{2l}, & |\alpha^h| < 1 \\ |\alpha^h|\, C_\phi h^{2l}, & |\alpha^h| \geq 1 \end{cases} \tag{12}$$

where the constants C_{kR} and C_ϕ depend on nondimensionalized wave number and the exact solution, respectively.

3.5 Numerical results

Radiation from an element of a cylinder: Consider an infinite circular cylinder of radius a. The artificial boundary is positioned at $R = 2a$, and the domain is discretized with 3×32 linear quadrilateral elements. We consider non-uniform radiation from the rigid cylinder with a constant inhomogeneous value on an arc of $5\pi/16$ and vanishing elsewhere. The wave length is equal to the diameter of the cylinder ($ka = \pi$) and 8 terms are employed in the DtN kernel. In general, both the Galerkin and Galerkin/least-squares solutions capture the essential physics of the problem. The methods are compared in Fig. 6, which depicts the imaginary part of the solutions evaluated along the artificial boundary. The improvement of the Galerkin/least-squares results over Galerkin is evident.

Scattering of a plane wave from a cylinder: In this example we compute the scattering, by a cylinder of radius a, of a plane wave traveling along the positive x axis ($\theta = 0$) in a direction perpendicular to the cylinder's axis. The properties and discretization of the previous problem are retained. The real parts of the computed solutions are presented in Fig. 7, in comparison to an analytical solution denoted by the dotted contours. Both numerical solutions represent the expected behavior, but the Galerkin/least-squares results are clearly superior.

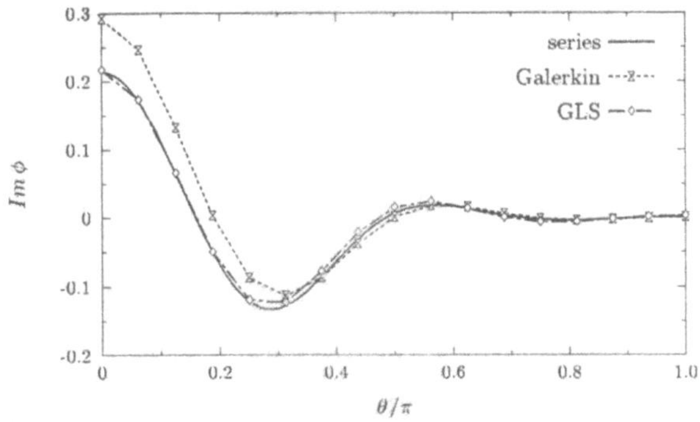

Fig. 6. Radiation from an element of a cylinder of radius a, $ka = \pi$, along the artificial boundary $R = 2a$.

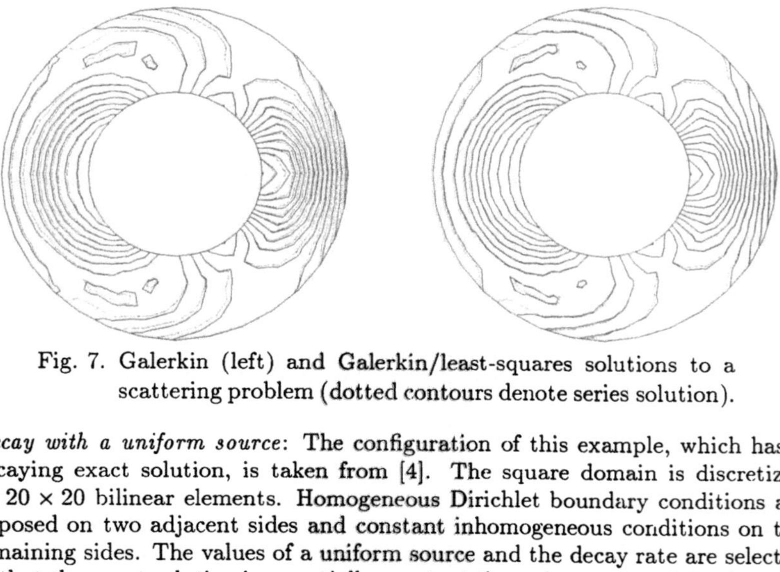

Fig. 7. Galerkin (left) and Galerkin/least-squares solutions to a scattering problem (dotted contours denote series solution).

Decay with a uniform source: The configuration of this example, which has a decaying exact solution, is taken from [4]. The square domain is discretized by 20×20 bilinear elements. Homogeneous Dirichlet boundary conditions are imposed on two adjacent sides and constant inhomogeneous conditions on the remaining sides. The values of a uniform source and the decay rate are selected so that the exact solution is essentially constant throughout most of the domain, decaying sharply in thin boundary layers near the boundaries with homogeneous conditions. Figure 8 shows elevation plots of the numerical solutions. The Galerkin solution exhibits the expected spurious oscillations in the vicinity of the boundary layer. These oscillations are absent in Galerkin/least-squares.

4. CONCLUSIONS

In this paper we have presented finite element methods for problems governed by the Helmholtz equation. The DtN approach provides a suitable basis for computing solutions to exterior problems by defining a bounded domain with appropriate boundary conditions, precluding singular behavior in finite element models. Numerical techniques were constructed by the Galerkin and Galerkin/least-squares methods. This variant of the Galerkin method was designed to enhance certain properties and yet retain advantageous features of

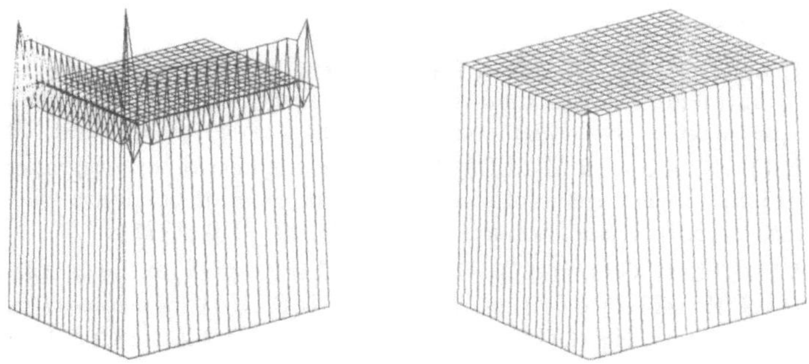

Fig. 8. Galerkin (left) and Galerkin/least-squares solutions to a Dirichlet problem with a decaying exact solution.

Galerkin formulations, while adding virtually no complexity in implementation. Solutions obtained by the Galerkin method were acceptable only with a relatively high degree of mesh refinement. *We found the Galerkin/least-squares finite element method with DtN boundary conditions (7) suitable for obtaining robust numerical solutions to the Helmholtz equation in unbounded domains over the entire range of interest*, achieving high accuracy without requiring excessive resolution. Galerkin/least-squares solutions are exceptionally accurate on the artificial boundary which is crucial to the representation of the far field by the DtN method. When physical phenomena are not sufficiently resolved by the mesh, the Galerkin/least-squares method is designed to damp out the numerical solution. *The method we advocate is defined via (9) for all decaying phenomena and to the limit of four elements per wave in propagation, and by (10) beyond that limit.* A global convergence analysis of the Galerkin/least-squares method guarantees good performance of the method on configurations of practical interest. Numerical calculations confirm the performance of these formulations for multi-dimensional configurations.

This paper has concentrated on problems of time-harmonic acoustics. Galerkin/least-squares finite element methods with DtN boundary conditions should provide a suitable framework for computing solutions to related exterior problems. A natural extension of the work presented herein is to investigate problems of interaction of acoustic fluids with elastic structures in the frequency domain, as well as in the transient case. The coupled problem is similar in nature to the abstract mixed problem, in which Galerkin/least-squares finite element methods have proven successful. Initial work on local DtN boundary conditions for time-dependent problems indicates that Galerkin/least-squares technology may play an important role in stabilizing related time-integration schemes.

ACKNOWLEDGEMENT

This research was supported by the U.S. Office of Naval Research under Contracts N00014-89-K-0027 and N00014-88-K-0446.

REFERENCES

1. ABBOUD, N.N. - A mixed finite element formulation for the transient and harmonic exterior fluid-structure interaction problem, Ph.D. Thesis, Department of Civil Engineering. Stanford University, Stanford, 1990.
2. BURTON, A.J. and MILLER, G.F. - The application of integral equation methods to the numerical solution of some exterior boundary-value problems, Proc. Roy. Soc. London, Ser. A 323, 201–210 (1971).
3. EVERSTINE, G.C. and QUEZON, A. - User's guide to the coupled NASTRAN/Helmholtz equation capability (NASHUA) for acoustic radiation and scattering, NSRDC Technical Report DTRC CMLD-88/03, David Taylor Research Center, Bethesda, Maryland (1988).
4. FRANCA, L.P. and DUTRA DO CARMO, E.G. - The Galerkin gradient least-squares method, Comput. Methods in Appl. Mech. Engrg., 74, 41–54 (1989).
5. GIVOLI, D. - Non-reflecting boundary conditions—a review, J. of Comp. Phys., 94(1), 1–29 (1991).
6. GIVOLI, D. and KELLER, J.B. - A finite element method for large domains, Comput. Methods in Appl. Mech. Engrg., 76, 41–66 (1989).
7. GIVOLI, D. and KELLER, J.B. - Non-reflecting boundary conditions for elastic waves, Wave Motion, 12, 261–279 (1990).
8. HARARI, I. and HUGHES, T.J.R. - Finite element methods for the Helmholtz equation in an exterior domain: model problems, Comput. Methods in Appl. Mech. Engrg., 87(1), 59–96 (1991).
9. HARARI, I. and HUGHES, T.J.R. - A cost comparison of boundary element and finite element methods for problems of time-harmonic acoustics, Comput. Methods in Appl. Mech. Engrg., to appear.
10. HARARI, I. and HUGHES, T.J.R. - Analysis of continuous formulations underlying the computation of time-harmonic acoustics in exterior domains, Comput. Methods in Appl. Mech. Engrg., to appear.
11. HARARI, I. and HUGHES, T.J.R. - What are C and h?: Inequalities for the analysis and design of finite element methods, Comput. Methods in Appl. Mech. Engrg., to appear.
12. HARARI, I. and HUGHES, T.J.R. - Galerkin/least-squares finite element methods for the reduced wave equation with non-reflecting boundary conditions in unbounded domains, in preparation.
13. KELLER, J.B. and GIVOLI, D. - Exact non-reflecting boundary conditions, J. of Comp. Phys., 82(1) (1989) 172–192.
14. SHAW, R.P. - Integral equation methods in acoustics, Boundary Elements X, Vol. 4, Ed. Brebbia, C.A., Springer-Verlag, Berlin, 1988, 221–244.
15. WILCOX, C.H. - Scattering Theory for the d'Alembert Equation in Exterior Domains, Springer-Verlag, Berlin, 1975.

FINITE ELEMENTS IN THE 90'S, E. Oñate, J. Periaux, A. Samuelsson (Eds.)
Springer-Verlag/CIMNE, Barcelona 1991

FINITE ELEMENT STUDY OF SHOCK INDUCED CAVITATION

R. E. Newton
U.S. Naval Postgraduate School,
Monterey, California, 93943,
U.S.A.

Explosively generated underwater shock waves may interact with a compliant submerged structure or with the free surface to induce cavitation. A finite element analysis of the former case is described. Discretised equations are derived using a displacement potential as the dependent variable. Formulations for a variety of boundary conditions are described. Computational strategy is discussed. A one-dimensional example is presented and results are shown to be in good agreement with those obtained by the method of characteristics.

1. INTRODUCTION

It is well known that fluids cannot transmit static tension. On the contrary, it has been demonstrated that high frequency sound waves may produce cyclic pressure excursions into the negative (absolute) regime [5].(1) Analytic investigation has shown that the capacity of a fluid to sustain transient negative pressure is dependent upon the concentration and size of gas bubbles present [12]. Despite these complications, when pressure excursions of many atmospheres are involved, negative pressures persist for only a few microseconds. The cavitation which follows involves appearance of visible bubbles and an accompanying decrease of mean fluid density.

The work reported here concerns structural responses to underwater shock. Peak shock pressures are typically from 10 to 100 atmospheres and it is assumed that cavitation begins without delay at zero absolute pressure. Vapor pressure is thus neglected and it is further assumed that the cavitated subregion is macroscopically homogeneous. Analysis will show that the mean density is negligibly smaller than that of the uncavitated fluid.

Cavitation induced by underwater shock has been under active investigation for at least 40 years. Two separate mechanisms are of interest. First of these is the scattered negative pressure wave produced by incidence of a shock wave on a compliant submerged structure. In the absence of high hydrostatic pressure, such an encounter can induce cavitation in the subregion that was traversed by the incident shock before reaching the structure. Early investigations were done by Temperley [19,20] and later work was reported by Bleich and Sandler [2,3].

A second means for inducing cavitation involves scattering of the shock wave from the free surface of the fluid. Again the scattered wave has a negative pressure and, in almost all encounters, cavitation occurs throughout an extensive subregion near the free surface. This behavior, called bulk cavitation, has been studied by Cushing [6,7], Clark [4], and Marcus [13].

(1) Numbers in square brackets refer to the list of References.

2. CHOICE OF DEPENDENT VARIABLE

The first treatment of cavitation by the finite element method was reported by the writer [14]. That paper includes an examination of candidate choices of dependent variable. It is shown that fluid pressure, introduced as the dependent variable in [22] and used for noncavitating underwater shock problems in [1,16, and 11], is not a suitable choice for cavitation problems. Although fluid particle displacement is a viable choice, it has the disadvantage of being a vector quantity. Two attractive choices are velocity potential and displacement potential. Both are scalar and both yield irrotational velocity fields. The displacement potential was chosen because a pressure discontinuity results only in a discontinuous second spatial derivative, whereas it produces a slope discontinuity in the velocity potential.

The defining equation for the displacement potential ψ is taken to be

$$\nabla\psi = \rho\underline{r} \tag{1}$$

where ∇ is the gradient operator, $\underline{r}$ is the fluid particle displacement, a vector, and ρ is the fluid density. Inclusion of the factor ρ is convenient, but not necessary. If e is the cubical dilatation (or volumetric strain), then the condensation s is defined to be

$$s = -\rho e \tag{2}$$

It is easily established that

$$s = -\nabla^2\psi \tag{3}$$

The constitutive law is taken as

$$p = c^2 s,\ s > 0; \quad p = 0,\ s \le 0 \tag{4}$$

where p is the absolute fluid pressure and c is the acoustic velocity in the fluid. The equation of motion $\nabla\ddot{\psi} = \nabla(p_h - p)$ may be spatially integrated to yield the scalar form

$$\ddot{\psi} = p_h - p \tag{5}$$

where the superior dots denote temporal derivatives and p_h is the hydrostatic (equilibrium) pressure.

3. FINITE ELEMENT EQUATIONS

The technique for discretising Eq.3 using the method of weighted residuals and Galerkin weighting is well known and details are omitted here. The resulting equation takes the form

$$\underline{Q}\underline{\ddot{s}} = \underline{H}\underline{\psi} - \underline{b} \tag{6}$$

where $\underline{s}$ and $\underline{\psi}$ are column vectors of nodal values of s and ψ, $\underline{Q}$ and $\underline{H}$ are analogs of mass and stiffness matrices in structural problems, and $\underline{b}$ is a vector determined by boundary conditions. A discussion of these conditions follows.

4. BOUNDARY CONDITIONS

4.1 Free Boundary. If there is a free surface where the fluid is in contact with air at atmospheric pressure p_a, the boundary condition takes the form

$$p = p_a \tag{7}$$

4.2 Fluid-Structure Interface. If the fluid remains in contact with the submerged structure, Eq.1 may be used to deduce that the normal derivative of ψ is equal to the component of the structural wet surface deflection in the direction of the interface normal.

4.3 Entry Boundary. The boundary condition needed to allow entry of a shock wave may be deduced from the displacement potential description of the wave. As an example, consider a plane wave travelling in the negative x direction with velocity c and entering through a boundary segment $x = x_1$. The wave may be described by $\psi = f(x + ct)$, where the function f depends on the wave shape and t represents time. At boundary points the normal derivative of ψ is a determinate function of time $f'(x_1 + ct)$, where the prime denotes differentiation of f with respect to its argument. This function may be used to evaluate $\underline{b}$ along the entry boundary.

Although the simplest entry condition (normal incidence of a plane wave) has been treated, it is not difficult to extend these ideas to situations involving oblique incidence or the entry of cylindric or spherical waves.

4.4 Nonreflecting Boundary. Many applications concern fluid regions too large to model. In such instances only a portion of the region is represented, introducing arbitrarily selected boundaries. A shock wave travelling inside the solution space and meeting such a boundary will undergo a spurious reflection unless appropriate conditions are imposed. One such condition, which completely eliminates reflection of plane waves normally incident on plane boundary segments, was described in [22]. This condition utilizes the relation that the normal derivative of ψ is equal to $-\dot{\psi}/c$ for such outbound waves. It has been successfully used in cavitation problems in cases where the cavitated region does not meet such a boundary. Further details are omitted here.

4.5 Combined Boundary Conditions. On an individual boundary segment it is possible to impose simultaneously both the entry and nonreflecting conditions described above.

5. FLUID - STRUCTURE INTERACTION

In an underwater shock problem the response of a submerged or boundary structure is usually the principal concern. A separate finite element code may be used to model the structure, giving rise to a host of operating problems. Attention here is confined to the strategy of the interacting scheme.

The scheme employed involves iterative use of a computation cycle which advances by one time step for the fluid, followed by an equal time step for the structure. This strategy was employed in [1,16, and 11] with a noncavitating fluid model. It is also used by Park, Geers, Felippa, and DeRuntz in [8,9,10 and 17] where it is called a "staggered" solution procedure.

Each forward step for the fluid requires evaluation of a subvector $\underline{b}_s$ corresponding to the interface nodes. The structural information utilized may be transmitted in a vector $\underline{\delta}$. A forward step for the structure requires pressure loads on the wet surface. These may take the form of a vector of fluid pressures at the interface nodes.

The scheme described is based on a common time step for fluid and structure. Computational efficiency requirements dictate the use of an explicit time integration algorithm for the fluid. The resulting stability limit fixes an upper bound on the time step. The time step thus selected will usually be too large for the structure if an explicit time integration algorithm is also employed there. A common remedy for this is to employ an unconditionally stable implicit algorithm for the structure. An alternate course is afforded by making a modal analysis for the structure (without fluid coupling) and omitting the high frequency modes beyond the stability limit of the time integration. A structural model based on the latter strategem was developed by Waller [21] and used by Newton in [15]. It is observed that either alternative is equivalent to using a "low-pass" filter on the structural modes. The two choices differ only in the gain vs. frequency characteristic of the filter.

A recent study by Stiehl and Haberman [18] looks at the effects of elastic coating on a submerged beam loaded by underwater shock. They use finite element modelling and include the effects of cavitation.

6. COMPUTATIONAL STRATEGY

Initial applications of the displacement potential formulation in [14] dealt with one-dimensional fluid regions. In such problems, if the spatial subdivision is uniform and the consistent "mass" matrix $\underline{Q}$ (Eq.6) is replaced by a "lumped" (diagonal) matrix, no coefficient matrices need be formed and there is no distinction between finite element and finite difference equations.

Nontrivial two-dimensional problems may require thousands of fluid nodes and a thousand time steps. For problems of this size it is desirable to control closely the storage requirements and the number of calculations per time step. Experience with wave propagation problems demonstrates that linear shape functions give excellent results [1]. Four-noded quadrilateral elements are well-suited and it is frequently possible to let most of the fluid elements be identical squares. If the structure is not rectangular a transition region adjacent to the structure will require non-square fluid elements. It may be possible to use symmetry properties to minimize the number of differently shaped fluid elements required.

In the fluid calculation the term $\underline{H}$ in Eq.6 may be evaluated by using $\underline{H}$ matrices at the element level, thus avoiding the need to assemble and store the system $\underline{H}$ matrix. Similarly, the coefficient matrix $\underline{Q}$ is reduced to a diagonal matrix $\underline{D}$ for which d_{ii} is the sum of the elements in the ith row of $\underline{Q}$. Although $\underline{D}$ needs to be assembled, the required storage space is that for a vector of length equal to the number of fluid nodes. Evaluation of vector $\underline{b}$ in Eq.6 can be performed employing operations at the element level. The other working equations (Eqs.4 and 5) involve only one node at a time.

One additional consideration arises when cavitation occurs. This is the requirement to damp spurious pressure oscillations near a cavity boundary. Although some time integration algorithms can provide the needed damping, experience has established that neither an explicit central difference scheme nor the Wilson algorithm provides sufficient damping. To remedy this an additional term $-\beta s$ is added to the right hand side of Eq.5. The coefficient β, which has the dimensions of [velocity x length], is chosen by trial. The minimum value of β needed to suppress "frothing" is used. The term "frothing", introduced in [14], describes either of two anomalous behaviors in the neighborhood of a cavity boundary. These are: (1) the appearance of isolated small cavities in the pressurized region; (2) the appearance of small pressurized islands within the cavitated region.

7. EXAMPLE

In order to establish the validity of the technique described above it is useful to compare results with those obtained by a different method. A problem described and solved by Bleich and Sandler in [3] is suitable. For its solution they used the method of characteristics, introducing additional relations to connect state variables on opposite sides of a shock front. They state that "[the problem concerns] the response of a horizontal layer of mass on the surface of a half-space of fluid, Fig.1. A plane pressure wave with a sudden rise and an exponential decay moves toward the surface, reaching the mass at the time t = 0. The system is subject to gravity and atmospheric pressure, all particles being at rest prior to the arrival of the shock."

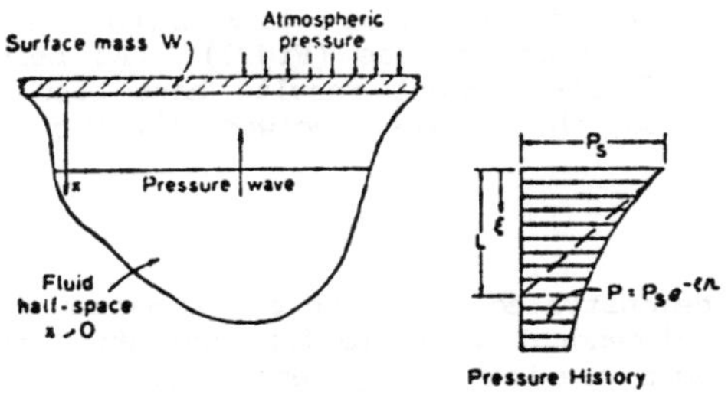

Fig.1. Particulars of Bleich-Sandler example.

For calculation and presentation of results Bleich and Sandler use the time constant of the pressure wave decay (≈ 1 ms) as the time unit. The acoustic velocity c is given the convenient value unity by choosing unit length to be the distance travelled by the pressure wave in unit time (≈ 1.4 m).

The history of the growth and decay of the cavitated region as found by Bleich and Sandler is shown in Fig.2. Discrete points superposed on their curves show results of a finite element analysis using linear elements with a node spacing of .01 over a region extending from the surface to a radiation boundary at x = 4. Time integration is performed using the Wilson algorithm and a time step .0025. A value of β = .0025 was sufficient to suppress frothing. It is clear that there is good agreement between finite element results and those of Bleich

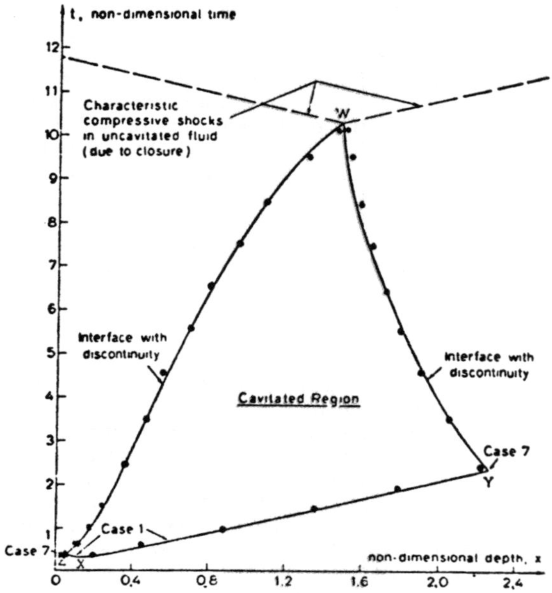

Fig.2. Time-history of the cavitated region.

Bleich and Sandler also presented a time-history of the nondimensional velocity of the surface mass (Fig.3). As before, discrete points representing finite element results have been superposed on their curve. Here, too, the agreement between the two sets of results is good.

8. CONCLUSION

It has been demonstrated that the displacement potential is a workable choice of dependent variable for cavitation problems. Two significant advantages are that the potential is scalar and that the discontinuity associated with a pressure step first appears in the second spatial derivative of the potential.

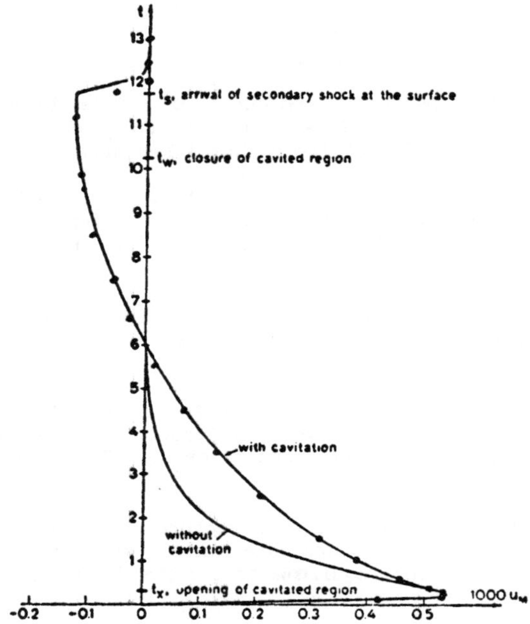

Fig.3. Nondimensional velocity of surface mass.

9. REFERENCES

1. ATCHISON, D.L. - Finite Element Study of the Interaction of a Plane Acoustic Blast Wave and a Cylindric Structure, thesis presented to the U.S. Naval Postgraduate School, at Monterey, CA, in partial fulfillment of the requirements for the degree Mechanical Engineer (1974).

2. BLEICH, H.H., and SANDLER, I.S. - Dynamic Interaction between Structures and Bilinear Fluids," *Technical Report No.* 43, Columbia University, New York, N.Y. (1968).

3. BLEICH, H.H., and SANDLER, I.S. - Interaction between Structures and Bilinear Fluids, Int. Journal of Solids and Structures, Great Britain, 6, 617-39 (1970).

4. CLARK, J.A. - Numerical Prediction of Supersonic Bulk-Cavitation Closure Pulses, *Technical Report* A-O-LAB-1-81, Catholic University of America, Washington, D.C. (1981).

5. CLAY, C.S., and MEDWIN, H. - Acoustical Oceanography; Principles and Applications, 1st ed., John Wiley and Sons, Inc., New York, N.Y., 1977, pp.170-5.

6. CUSHING, V., - Study of Bulk Cavitation and Consequent Water Hammer, *Final Report*, Engineering Physics Company, Rockville, Md., for Office of Naval Research (1961).

7. CUSHING, V., BOWDEN, G., and REILY, D. - Three-Dimensional Analysis of Bulk Cavitation, *Interim Report*, Engineering Physics Company, Rockville, Md., for Office of Naval Research (1962).

8. DeRUNTZ, J.A., GEERS, T.L., and FELIPPA, C.A. - The Underwater Shock Analysis Code (USA-Version 3): A Reference Manual, *Final Report No. LMSC-0777843*, Lockheed Missiles and Space Co., Inc., Palo Alto, CA (1980).

9. FELIPPA, C.A., and DeRUNTZ, J.A. - Finite Element Analysis of Shock-Induced Hull Cavitation, Computer Methods in Appl. Mech. and Engrg, 11, No.3, 297-337 (1981).

10. FELIPPA, C.A., and GEERS, T.L. - Partitioned Analysis for Multidisciplinary Problems Involving Structures, *Report No. LMSC/F-213651*, Lockheed Missiles and Space Co., Inc., Palo Alto, CA (1986).

11. GALLO, J.G. - Shock Induced Structural Response, thesis presented to the U.S. Naval Postgraduate School, at Monterey, CA in partial fulfillment of the requirements for the degree Mechanical Engineer (1978).

12. KEDRINSKII, V.K. - Negative Pressure Profile in Cavitation Zone at Underwater Explosion near Free Surface, Acta Astronautica, 3, 623-32 (1976).

13. MARCUS, M.H. - The Response of a Cylindrical Shell to Bulk Cavitation Loading, *Final Report NSWC/TR-81-295*, Naval Surface Weapons Center, White Oak, MD (1983).

14. NEWTON, R.E. - Effects of Cavitation on Underwater Shock Loading - Part I, *NPS69-78-013*, U.S. Naval Postgraduate School, Monterey, CA, for Defense Nuclear Agency (1978).

15. NEWTON, R.E. - Effects of Cavitation on Underwater Shock Loading - Plane Problem, *NPS69-81-001A*, U.S. Naval Postgraduate School, Monterey, CA, for Defense Nuclear Agency, (1981).

16. NEWTON, R.E., and ATCHISON, D.L. - Response of a Ring-stiffened Cylinder to an Acoustic Blast Wave, Proceedings, Second Int. Symposium on Finite Element Methods in Flow Problems, Santa Margherita Ligure, Italy, 1976, pp.703-13.

17. PARK, K.C., FELIPPA, C.A., and DeRUNTZ, J.A. - Stabilization of Staggered Solution Procedures for Fluid-Structure Interaction Analysis, Computational Methods for Fluid-Structure Interaction Problems, American Society of Mechanical Engineers, AMD-Vol.26, 1977, pp.95-124.

18. STIEHL, A.L., and HABERMAN, R.C. - Computer Simulation of the Shock Induced Response of a Coated Beam in Water with Cavitation, Stress Wave, Hysteresis and Strain Rate Effects, Proc. of 1990 ASME Int. Computers in Engrg. Conf. and Expo., ASCE, Boston, 1990, pp.67-72.

19. TEMPERLEY, H.N.V. - Theoretical Investigation of Cavitation Phenomena Occurring when an Underwater Pressure Pulse is Incident on a Yielding Surface - I, Underwater Explosion Research, 3, 255 (1950).

20. TEMPERLEY, H.N.V. - Theoretical Investigation of Cavitation Phenomena Occurring when an Underwater Pressure Pulse in Incident on a Yielding Surface - III, Underwater Explosion Research, 3, 283 (1950)

21. WALLER, J.T. - Dynamic Structural Model of a Submerged Ring, thesis presented to the U.S. Naval Postgraduate School, at Monterey, CA, in partial fulfillment of the requirements for the degree Master of Science in Mechanical Engineering (1979).

22. ZIENKIEWICZ, O.C., and NEWTON, R.E. - Coupled Vibrations of a Structure Submerged in a Compressible Fluid, Proc. of Int. Symposium on Finite Element Techniques, Ed. Sorensen, M., University of Stuttgart, W. Germany, 1969, pp.359-79.

FINITE ELEMENTS IN THE 90's, E. Oñate, J. Periaux, A. Samuelsson (Eds.)
Springer-Verlag/CIMNE, Barcelona 1991

GENERALISED MESH AND ADAPTIVITY TECHNIQUES FOR COMPUTATIONAL FLUID DYNAMICS

N. P. Weatherill
Institute for Numerical Methods in Engineering,
University College of Swansea,
Singleton Park, Swansea, SA2 8PP,
UNITED KINGDOM

1. Introduction

After more than a decade of interest and research effort there is still considerable attention given to techniques for the construction of meshes. This is because the majority of people involved with computational methods have to face the problem of mesh generation. For some applications the process is relatively simple, for others mesh generation represents the major problem in the numerical simulation procedure. Techniques applicable to the former class of problem are now widely understood. Methods for use in problems defined in complicated domains have been developed by a relatively small group of workers in both industry and academia. Today, automatic grid generation generally follows the body aligned approach in the form of structured multiblock, utilising interpolation or elliptic solvers, or unstructured meshes of tetrahedra constructed using the advancing front or Delaunay approaches. Much success has been achieved using the overlapping mesh approach, which is now synonymous with the 'Chimera' method, although this is not widely used outside a few notable groups.

A topic always likely to ensure a lively discussion is the argument of the advantages of structured vs. unstructured meshes. With the past efforts and present commitment of many groups to these approaches it is unlikely that the argument is likely to be definitively resolved. However, from experience gained in developing and applying both approaches it is clear, and we believe now widely recognised, that each have their own natural advantages and disadvantages and both will continue to be used. More recently there have been efforts to study the potential benefits of combining the structured and unstructured approaches in the form of hybrid meshes. *The purpose of this short paper is to present an overview of an approach which represents a possible template for a fully integrated, general mesh and computational fluid dynamics computer system. The work and applications are for two dimensions, although many modules for the extension to three dimensions have been developed and are available.*

2. System Overview

Figure 1 outlines the components of a fully integrated highly flexible CFD system. A new feature is the capability to combine structured and unstructured meshes to form hybrid meshes. Due discussion and motivation for such an approach has been widely covered[1]. The figure indicates the various pathways that can be achieved. It should be noted that in addition to mesh generation techniques emphasis is given to adaptivity methods which now and in the future will play an important role in CFD. Adaptivity can be achieved using point enrichment and derefinement, node movement, remeshing and combinations of these approaches in all grid types. Here, emphasis is given to how such a system can be applied and this is demonstrated by the use of examples, firstly for complicated geometries.

3. Applications - Complex Geometries

3.1 Two Component Aerofoil

Figure 2 shows a multiblock mesh for a 2 component aerofoil. This is constructed using 20 blocks and continuity of the grid lines across block boundaries has been achieved within the elliptic mesh generation equations and local boundary orthogonality has been implemented using an iterative modification to the control functions. This is a classic approach and in this case the flow simulation has been performed using a spatially centred cell vertex based flow algorithm for the Navier-Stokes equations. In this case local point enrichment has been applied to further resolve the flowfield.

3.2 Three Component Aerofoil

In the case of a 3 component aerofoil, the system affords several options. The Delaunay algorithm together with automatic point creation can be utilised to readily construct a mesh of triangles. Figure 3 shows such a mesh and the pressure coefficient on the aerofoils. However, on a previous occasion perhaps this high lift configuration had been studied utilising a structured multiblock approach, with a slat and no flap. Now it is required to insert the flap and reassess the aerodynamic performance with this additional component. A new multiblock mesh could be generated but this may be time consuming. An alternative approach which the system offers is to utilise the hybrid approach. Here, the flap with its own component mesh, can be inserted over the multiblock mesh. The underlying mesh is deleted and the local component structured mesh connected to the multiblock mesh with the use of the Delaunay triangulation. Figure 4 shows the initial hybrid mesh and the adapted mesh after node movement. The flowfield was obtained using a Jameson cell centred algorithm.

3.3 Moving Body

Figure 5 shows, in pictorial form, how a moving boundary problem can be resolved with the use of the hybrid approach. In the example shown a structured grid has been generated around the main aerofoil and the moving flap carries its own component grid which overlays, and is connected to, the underlying main grid. This requires a sequence of reconnections to the main mesh and tests the robustness of the approach. In this case the flap was moved with a prescribed motion requiring 250 reconnections which were performed automatically without user intervention. Present effort is applied in constructing a time accurate flow algorithm for moving boundaries.

4. Practical Implementation

4.1 Data structure

To implement the approach to CFD which is outlined in Figure 1 it is necessary to ensure that there is a minimum level of complexity in software development. To achieve this it is important to use a flexible data structure and in this case, for mesh generation, a quadtree data structure has been employed. This proves to be suitable for integrating structured and unstructured meshes, for the implementation of mesh/flow adaptivity and providing the mesh in a format for the flow algorithms. Further details of this data structure can be found elsewhere[6].

4.2 Mesh Construction

Two basic mesh generation approaches have been used. The structured meshes have been constructed using elliptic equations. For geometrical flexibility, this approach has been implemented within a multiblock environment[2]. The unstructured meshes have been constructed from a Delaunay triangulation algorithm making full use of an automatic point insertion routine[3]. Hybrid grids have been constructed using these basic algorithms where the implementation of boundary integrity routines within the Delaunay approach has been given particular attention. In this respect an edge swapping routine has been applied to ensure the correct connections between the structured and unstructured regions.

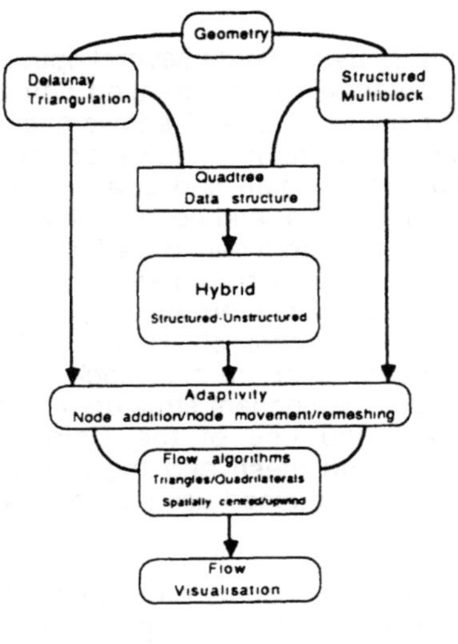

Figure 1. System Overview

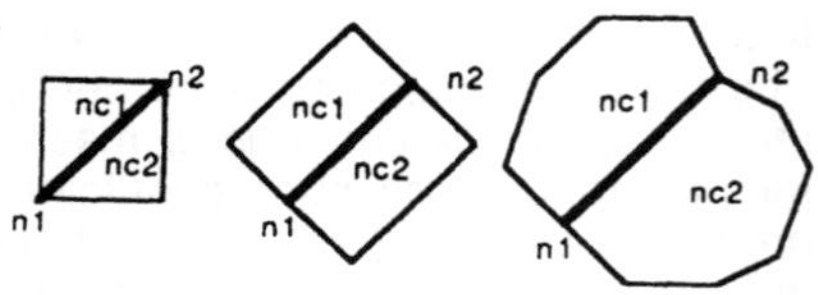

Figure 6 Edge/cell combinations

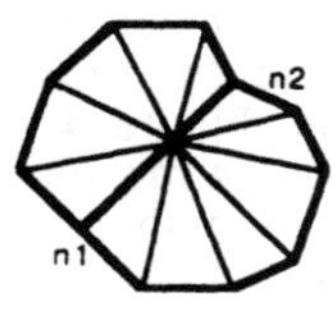

Figure 7 Possible nested grids for multigrid

4.3 Flow Algorithms

It is impractical to utilise different flow codes for the different grid types. Hence, a unified approach has been adopted. This can be achieved by again the use of appropriate data structures. In all cases finite volume flow codes have been used. For the Euler and Navier-Stokes equations it is necessary to evaluate a contour integral around each finite volume. For example, the evaluation of the flux in a domain with a contour c and flux vectors **f** and **g**, can be expressed as

$$\int_C f\,dy - g\,dx = \sum_{\text{sides}} f\Delta y - g\Delta x.$$

For a cell vertex implementation, the contribution from each edge is sent to the appropriate nodes on either side of the edge. For a cell centre formulation, the contribution from each edge is sent to the two cells on either side of the edge. Clearly, in the latter case there are only two additional memory requirements per edge, but for the cell vertex approach the number is different for quadrilateral/quadrilateral (4 required), triangle/quadrilateral (3 required) and triangle/triangle(2 required) combinations. This approach, first described by Jameson[4] for use on unstructured meshes, has been used here in the development of flow codes applicable to all mesh types. Figure 6 shows this in schematic form for a cell centre approach.
A typical computation for each edge, within a grid of triangles, can be implemented as:

```
do  1 to number of sides
        n1              = ncon(I,1)
        n2              = ncon(I,2)
        nc1             = ncon(I,3)
        nc2             = ncon(I,4)
        Computation = .............
        sum(nc1)        = sum(nc1) + computation
        sum(nc2)        = sum(nc2) + computation
End do loop
```

Figure 8b shows how, using point enrichment for adaptivity, interfaces can arise where the flexibility of the cell centre approach and the data structure and implementation previously outlined can be fully utilised. Flow codes for the Euler and Navier-Stokes equations based upon spatially centred, cell vertex and cell centre algorithms and upwind cell centre algorithms for inviscid flow have been developed in this way[5]. They can then be used on all flow types described in Figure 1.
Another advantage of this approach is in the implementation of multigrid and the requirement for nested grids. Figure 7 shows how the cell centre approach may be of value.

4.3 Mesh/Flow Adaptivity

In general, an initial mesh will not be capable of accurately resolving features of the flowfield. It is hence necessary to adapt the mesh by utilising node movement, point enrichment or remeshing techniques.

4.3.1 Point enrichment

Local point enrichment can be achieved on any grid type. It is an effective way to ensure greater resolution of the flowfield. It is most naturally employed on unstructured meshes, since on the addition of a point and the subsequent connection to the mesh the data format of the mesh does not change. Hence, flow solvers do not require any modifications. This is not true for structured meshes where the addition of points breaks the (i,j) data format and hanging nodes are created. In general, flow solvers have to be modified accordingly. In our implementation a structured mesh is converted to a quadtree data format[6] and with flow codes written in a general manner, as already indicated, the treatment of local point insertion on structured meshes does not cause any problems. Different types of embedding can be used, dependent upon the type of flow feature identified. An example of this approach is shown in Figure 8b where local point insertion on a structured mesh has been used in the simulation of the Mach 3 flow over a cylinder. The multiple hanging nodes do not cause any problems since a cell centre formulation, described in the previous section, was used. As a further example of point enrichment, Figure 9b shows this approach applied in the hybrid environment. Again with the data structures and flow algorithms previously described the apparent complication of mixed cell types does not cause any difficulties. Other examples of point enrichment are shown in Figures 10b and 11a, where it has been applied in a multiblock structured grid environment for a tandem aerofoil configuration and a single block grid for a channel, respectively.

4.3.2 Node movement

As demonstrated in the previous section, node enrichment successfully enhances the resolution of the flowfield. However, it can become computationally expensive and provides a diminishing return on successive enrichments. After ensuring that there is sufficient mesh point resolution, node movement can provide the required mechanism to achieve high resolution at negligible additional computational cost. Accepting this, it is necessary to devise a node movement technique which can be applied on all mesh types and can be easily implemented into flow codes. An approach which proves to be very successful is described as,

$$\mathbf{r}_0^{n+1} = \mathbf{r}_0^{n} + \omega \frac{\sum_{i=1}^{M} C_{i0} \left(\mathbf{r}_i^{n} - \mathbf{r}_0^{n}\right)}{\sum_{i=1}^{M} C_{i0}} ,$$

where $\mathbf{r}=(x,y)$, $\mathbf{r}_0^{n+1}$ is the position of node 0 at relaxation level n+1, C_{i0} is the adaptive weight function between nodes i and 0 and ω is the relaxation parameter. The summation is taken over all edges connecting point 0 to i, where it is taken that there are M surrounding nodes. In practice, this relaxation is typically applied over 50 cycles with a relaxation parameter of 0.1. The weight function C_{i0} can be taken as a measure of any flow parameter such as pressure, density or a measure of local error. This approach proves trivial to implement on all mesh types, but yet its effects are impressive. Figures 8a and d demonstrate node movement for the Mach 3 flow around a cylinder for both a structured and unstructured mesh, respectively. Figure 9c shows the effects of node movement on a hybrid grid, and in Figure 10a mesh movement for a Mach 1.2 flow over a tandem aerofoil configuration in which a multiblock mesh has been used is shown.

4.3.3 Combinations of node movement, point enrichment and derefinement

An optimum approach to adaptation is to combine node movement and point enrichment with derefinement. These procedures should be implemented in a dynamic way, i.e. applied at regular intervals within the flow simulation. Such an approach also provides the possibility of using movement and enrichment to independently capture different features of the flow. Examples of combinations of these adaptive mechanisms are given in Figure 8c, 10c and 11b.

4.4 Flowfield Visualisation

Visualisation of the flowfield is a fundamentally important aspect of choosing a general approach. Within the system general data cells made from polygonal regions have been employed. Output from the flow codes indicates the type of polygon and the visualisation software then acts accordingly.

5. Operational Aspects

From experience it is clear that the construction of a structured multiblock mesh for particular configurations proves difficult. For such a case use can be made of the unstructured approach or a structured mesh can be used on a simplified version of the configuration and the problematic regions can be handled with a region of unstructured mesh, thereby making the mesh hybrid. High quality multiblock meshes for complex geometries can take longer to generate than the unstructured or hybrid meshes.

6. Extension to 3 Dimensions

Some of the necessary modules to develop an equivalent system for 3-dimensional flowfield simulation have already been developed. Some new problems arise particularly with the connection of structured to unstructured meshes using the Delaunay approach[7]. A typical example is shown in Figure 12. A flow algorithm has been developed to solve the Euler equations on hybrid 3-dimensional grids and results have been obtained without adaptivity for several configurations[8] .

7. Summary

This short paper has outlined a possible approach to mesh generation and flow simulation which will be ultimately required to provide the necessary flexibility for a comprehensive CFD system. The approach integrates flexible approaches to mesh generation for complex shapes and fully integrates mesh/flow adaptivity. The next few years will see the development of fully generalised mesh and adaptivity techniques in 3 dimensions.

8. Acknowledgements

I would like to thank M. J. Marchant, D. K. Marchant, A. Evans and J. Szmelter for their assistance and efforts in producing results for this paper.

References

1. Weatherill N. P., 'Mixed structured-unstructured meshes for aerodynamic flow simulation', The Aeronautical Journal, Vol. 94, No. 934, p111-123, April 1990.

2. Weatherill N. P. and Forsey C. R. 'Grid generation and flow calculations for aircraft geometries', J. of Aircraft, Vol.22, No. 10, pp855-860, October 1985.

3 Weatherill N. P. 'A methods for generating irregular computational grids in multiply connected planar domains', Int. J. Num. Mthds in Fluids, Vol 8. pp181-197, 1988.

4 Jameson A., Baker T. J. and Weatherill, N. P. "Calculation of inviscid transonic flow over a complete aircraft", 24th Aerospace Sciences Meeting, Reno, AIAA Paper No. 86-0103, 1986.

5. Weatherill N. P., Mathur J. S., Natakusumah D.K and Marchant M. J. 'An upwind kinetic flux vector splitting method on general mesh topologies'. Submitted to Int. J. Num. Methods in Engineering.

6. Evans A., Marchant M.J., Szmelter J. and Weatherill N.P.' Adaptivity for compressible flow computations using point embedding on 2-D structured multiblock meshes', To appear, Int. J. Num. Mthds. in Engineering, 1991.

7. Childs P.N. and Weatherill N. P. 'Generation of unstructured grids within a hybrid multiblock environment' Proceedings of the Third Int. Conf. on Num. Grid Generation. Pub. Elsevier, 1991.

8. Shaw J.A., Georgala J.M., Peace A.J. and Childs P.N. 'The construction,application and interpretation of three-dimensional hybrid grids' Proceedings of the Third Int. Conf. on Num. Grid Generation. Pub. Elsevier, 1991.

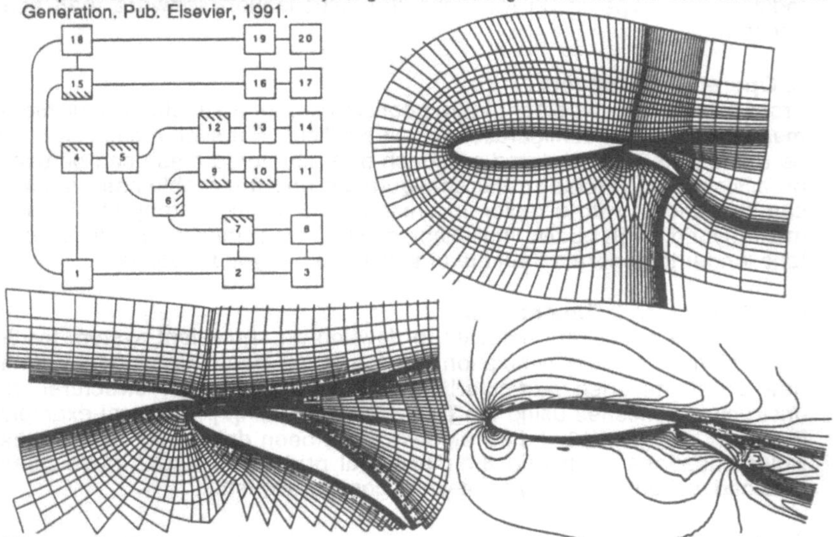

Figure 2 Embedded Navier-Stokes multiblock mesh and Mach no. contours

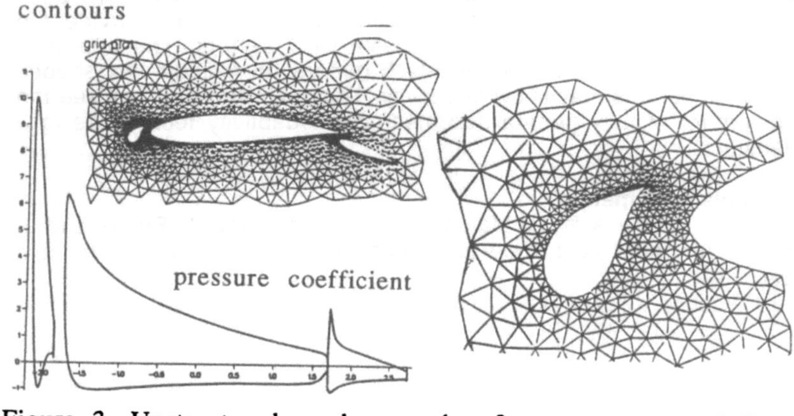

Figure 3. Unstructured mesh around a 3 component aerofoil

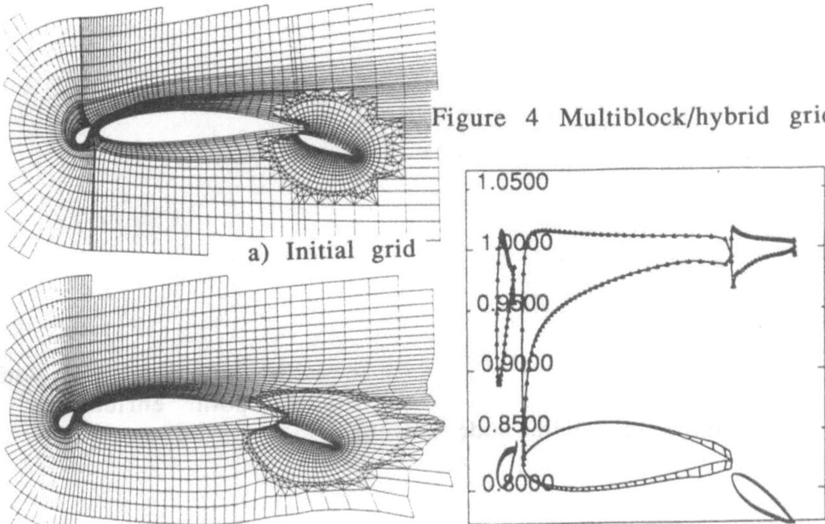

Figure 4 Multiblock/hybrid grid

a) Initial grid

b) Node movement

c) Pressure on aerofoils

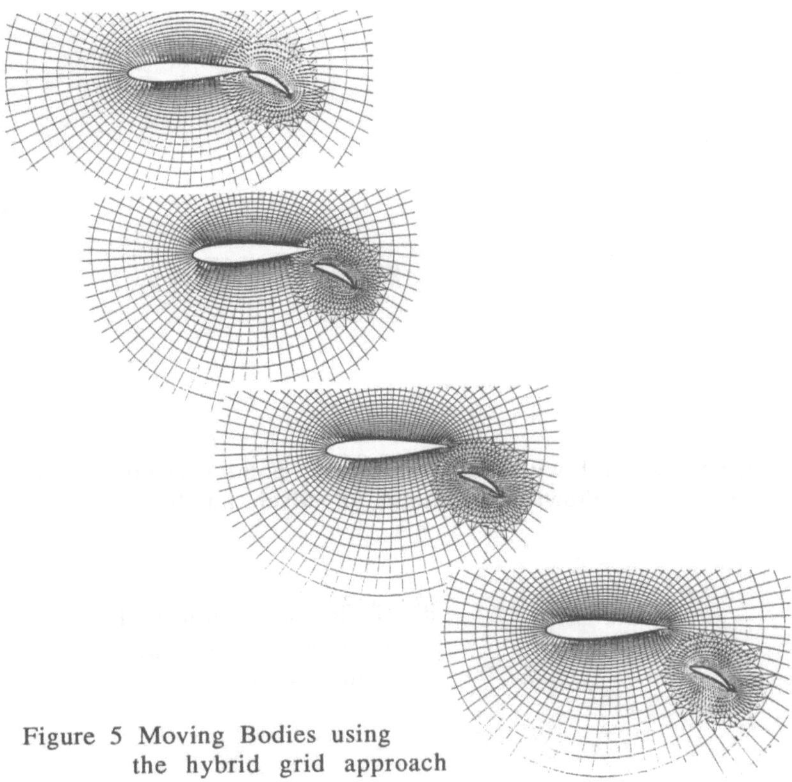

Figure 5 Moving Bodies using the hybrid grid approach

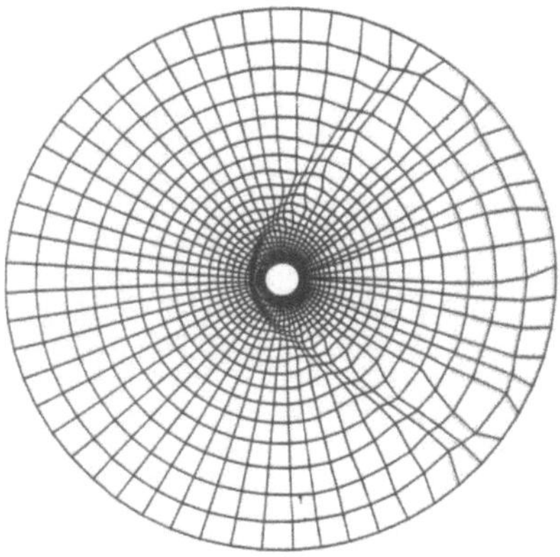

a) Structured grid with node movement

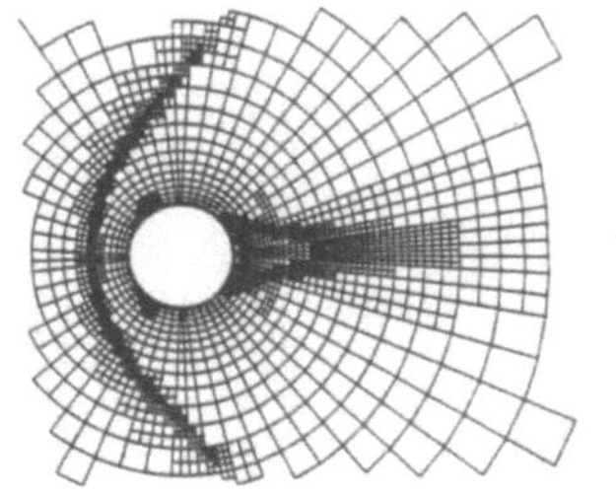

b) Structured grid with point enrichment

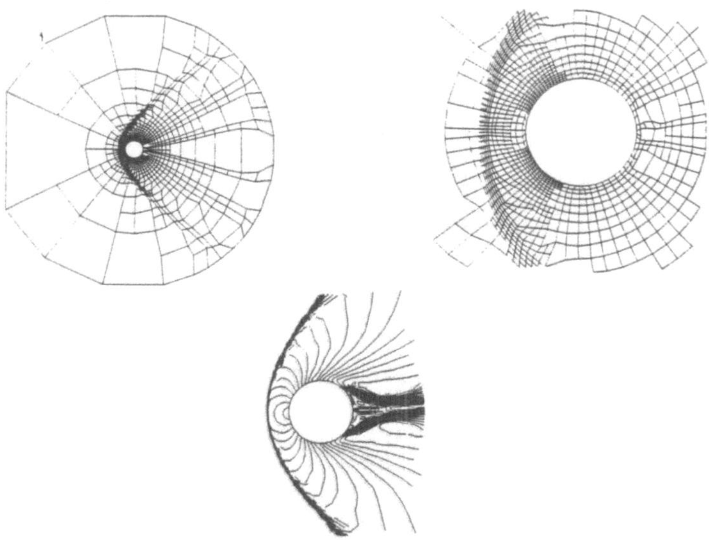

c) Structured grid with node movement, point enrichment and derefinement. Flowfield contours of Mach number.

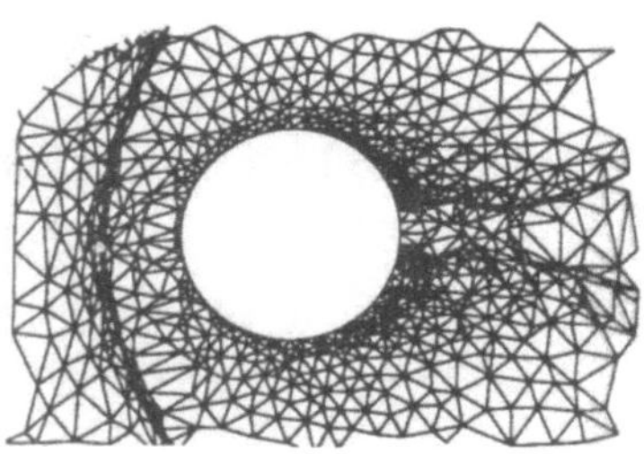

d) Unstructured grid with node movement and point enrichment

Figure 8. Mach 3 inviscid flow around a cylinder.

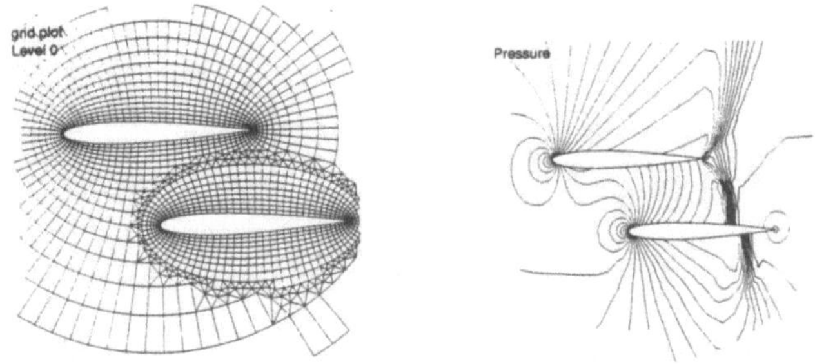

a) Initial hybrid grid and flowfield contours of pressure
(Mach No. = 0.8, $\alpha = 0.0^{\circ}$)

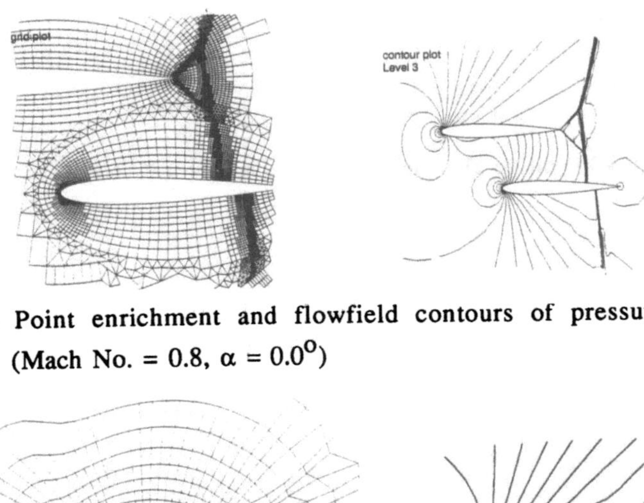

b) Point enrichment and flowfield contours of pressure
(Mach No. = 0.8, $\alpha = 0.0^{\circ}$)

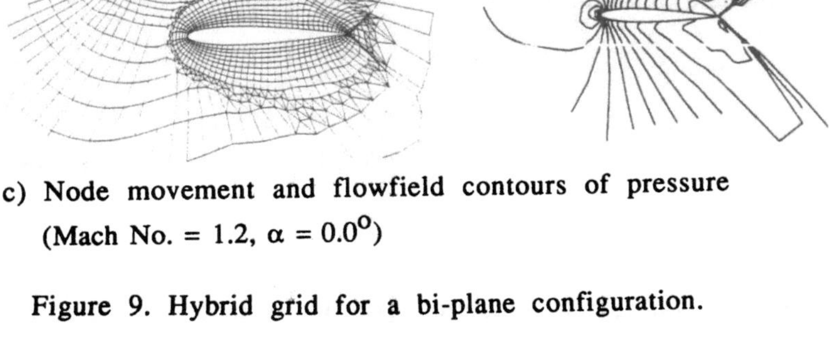

c) Node movement and flowfield contours of pressure
(Mach No. = 1.2, $\alpha = 0.0^{\circ}$)

Figure 9. Hybrid grid for a bi-plane configuration.

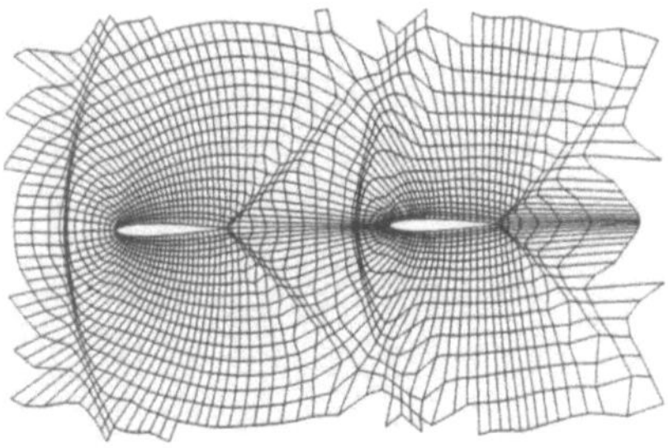

a) Node movement to flowfield of Mach No. = 1.2, $\alpha = 0.0^{o}$

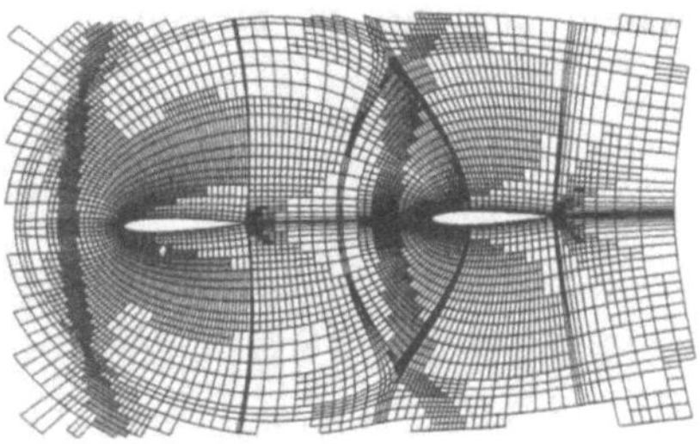

b) Point enrichment to flowfield of Mach No. = 1.2, $\alpha = 0.0^{o}$

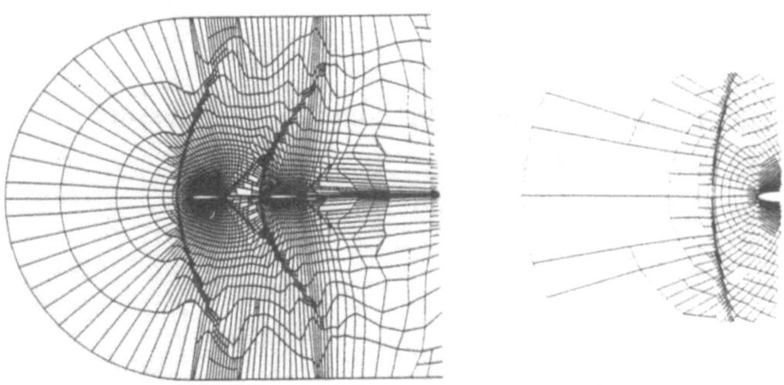

c) Node movement, point enrichment and derefinement to flowfield of Mach No. = 1.2, $\alpha = 0.0^{o}$

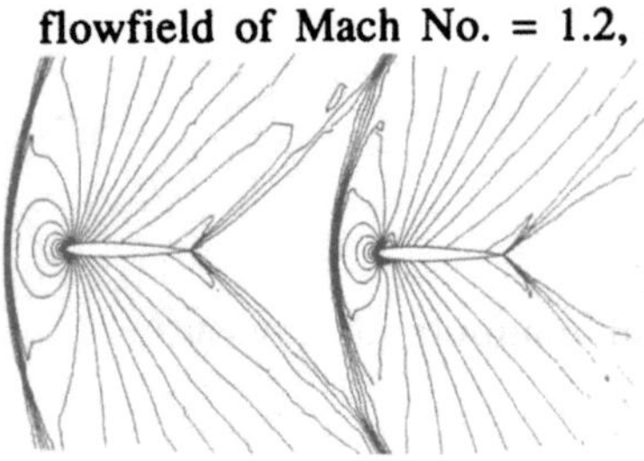

Flowfield contours of pressure

Figure 10. Multiblock structured grid for a tandem configuration.

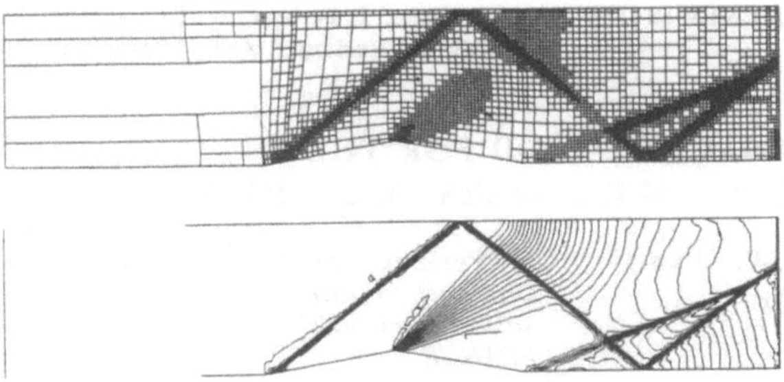

Contours of pressure

2 levels of ***node refinement*** and 3 levels of ***derefinement***

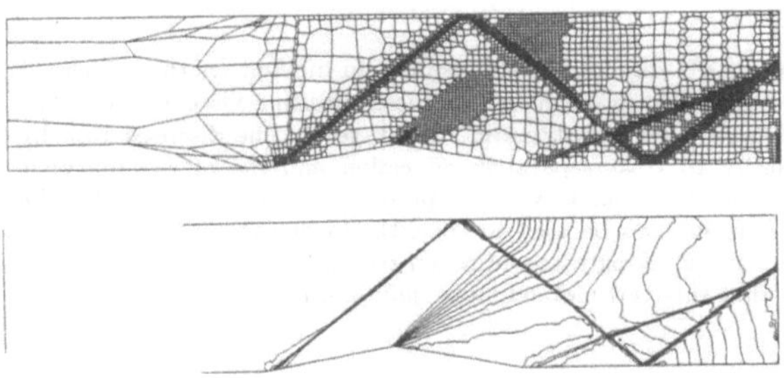

Contours of pressure

2 levels of ***node refinement***, 3 levels of ***derefinement*** and ***node movement***

Figure 11. Structured grid for Mach 2 flow over a double wedge in a channel.

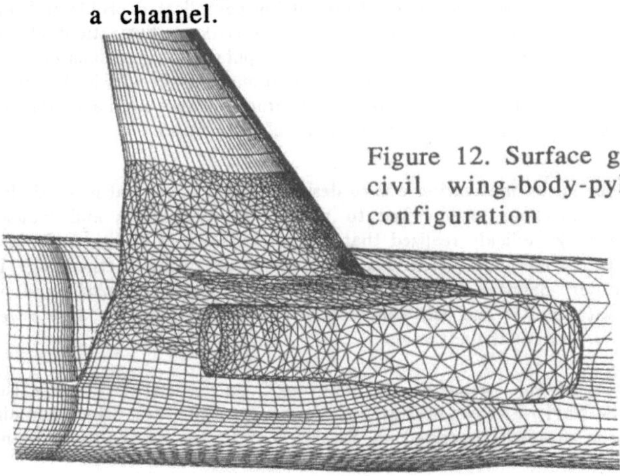

Figure 12. Surface grid on a civil wing-body-pylon-nacelle configuration

FINITE ELEMENTS IN THE 90's, E. Oñate, J. Periaux, A. Samuelsson (Eds.)
Springer-Verlag/CIMNE, Barcelona 1991

A FICTITIOUS DOMAIN METHOD FOR THE INCOMPRESSIBLE NAVIER-STOKES EQUATIONS

R. Glowinski and **T.W. Pan**
University of Houston,
Houston, Texas 77204,
USA

J. Periaux and **M. Ravachol**
Dassault Aviation,
Saint Cloud, 92214,
FRANCE

SUMMARY

In this article we briefly discuss a fictitious domain method for the numerical solution of the Navier-Stokes equations modelling incompressible viscous flow. The methodology described here takes a systematic advantage of time discretization by operator splitting in order to treat separately advection and incompressibility; it seems well suited to moving boundary flow problems. Indeed, due to the decoupling, fast elliptic solvers can be used to treat the incompressibility condition even if the original problem is taking place on a non regular geometry. Preliminary numerical results show that this new method looks quite promising.

INTRODUCTION

Beside his outstanding contributions to Structural Mechanics, Olek Zienkiewicz has shown during these last two decades a strong interest on *Computational Fluid Dynamics.* It makes then sense to us to dedicate him this article, which is concerned with *fictitious domain methods* for the numerical solution of the *incompressible Navier-Stokes equations. Fictitious domain methods* have shown recently a most interesting potential for solving complicated problems from Science and Engineering (see, for example, [1] for some striking illustrations of the above statement). One of the main reasons of this popularity of fictitious domain methods (they are sometimes called *domain imbedding methods*; cf. [2]) is that they allow the use of fairly *structured meshes* on a simple shape auxiliary domain containing the actual one, allowing therefore the use of *fast solvers.*

Indeed, our initial motivation was to design efficient numerical methods for the Navier-Stokes equations, well suited to *wavelet approximations* and *moving boundaries.* In fact, we quickly realized that the combination of *fictitious domain* and *operator splitting* methods which is at the foundation of our methodology is also of interest for classical approximations such as *finite elements* and is particularly advantageous if one uses the *method of characteristics*, or *high order upwinding schemes*, to treat the advection.

The content of this article is the following: In Section 1 we give the formulation of the Navier-Stokes problem. In Section 2 we give a fictitious domain formulation of the problem described in Section 1. In Section 3 we combine fictitious domain and operator splitting methods to decouple the numerical treatment of advection and incompressibility. In Section 4 we briefly discuss the solution of the Stokes problem by fictitious domain methods, and finally, in Section 5 we present the results of numerical experiments.

1. FORMULATION OF THE NAVIER-STOKES PROBLEM.

Using the notation of Figure 1.1, below, we consider the following Navier-Stokes equations

Figure 1.1

(1.1) $$\frac{\partial \mathbf{u}}{\partial t} - \nu \nabla^2 \mathbf{u} + (\mathbf{u}\cdot\nabla)\mathbf{u} + \nabla p = \mathbf{f} \textit{ in } \omega,$$

(1.2) $$\nabla \cdot \mathbf{u} = 0 \textit{ in } \omega \textit{ (incompressibility condition)},$$

(1.3) $$\mathbf{u} = \mathbf{g} \textit{ on } \gamma \textit{ (with } \int_\gamma \mathbf{g}\cdot\mathbf{n}\, d\gamma = 0),$$

(1.4) $$\mathbf{u}(x,0) = \mathbf{u}_0(x),\ x\in\omega, \textit{ with } \nabla\cdot\mathbf{u}_0 = 0.$$

In (1.1)−(1.4), $\omega\subset\mathbf{R}^d (d\geq 2)$, γ is the boundary of ω, $\mathbf{n}$ is the outward normal unit vector at γ, $\mathbf{u}=\{u_i\}_{i=1}^d$, $\nabla^2=\Delta=\sum_{i=1}^d \frac{\partial^2}{\partial x_i^2}$, and $(\mathbf{v}\cdot\nabla)\mathbf{w} = \left\{\sum_{j=1}^d v_j \frac{\partial w_i}{\partial x_j}\right\}_{i=1}^d$.

2. AN EQUIVALENCE RESULT INVOLVING A LAGRANGE MULTIPLIER.

We imbed ω in Ω as shown in Figure 1.1, and define

$$V_P = \left\{\mathbf{v} \mid \mathbf{v}\in(H^1(\Omega))^d,\ \mathbf{v} \textit{ periodic at } \Gamma\right\}$$

Next, we observe that if U_o is an extension of u_o which is divergence free in Ω, and if $\tilde{f}$ is an extension of f, we have equivalence between (1.1)–(1.4) and

$$(1.5)\quad \begin{cases} \int_\Omega \frac{\partial U}{\partial t}\cdot v dx + \nu\int_\Omega \nabla U\cdot\nabla v\, dx + \int_\Omega (U\cdot\nabla)U\cdot v dx - \int_\Omega P\nabla\cdot v dx = \\ \int_\Omega \tilde{f}\cdot v dx + \int_\gamma \lambda\cdot v d\gamma,\ \forall v\in V_P;\ U(t)\in V_P,\ t>0, \end{cases}$$

$$(1.6)\quad \nabla\cdot U = \textit{in } \Omega,$$

$$(1.7)\quad U(x,0) = U_o(x),\ x\in\Omega,$$

$$(1.8)\quad U = g \textit{ on } \gamma,$$

in the sense that

$$(1.9)\quad U|_\omega = u,\ P|_\omega = p.$$

Concerning the multiplier λ, its interpretation is very simple since it is equal to the jump of $\nu\frac{\partial U}{\partial n} - nP$ at γ. A closely related approach (limited to the steady Stokes problem) is discussed in [3]. We observe that the effect of the actual geometry is concentrated on $\int_\gamma \lambda\cdot v d\gamma$, in the right-hand side of (1.5), and on (1.8). From a practical point of view, the crucial issue will be the proper approximation of the multiplier space; we shall come back on this important issue in Sections 4 and 5.

3. ON A COMBINATION OF FICTITIOUS DOMAIN AND OPERATOR SPLITTING METHODS

In order to solve (1.5)–(1.8) we shall consider a time discretization by an *operator splitting method*, like the ones discussed in, e.g., [4]–[9].

For simplicity, we consider the time discretization of (1.5)–(1.8) by the *Peaceman-Rachford scheme* (cf. [6]). With $\Delta t>0$ a time discretization step, we obtain

$$(3.1)\quad U^o = U_o;$$

for $n\geq 0$, *knowing* U^n, *we compute* $\left\{U^{n+1/2}, P^{n+1/2}, \lambda^{n+1/2}\right\}$, *and then* U^{n+1} *by solving*

$$(3.2)_1\quad \begin{cases} \int_\Omega \frac{U^{n+1/2}-U^n}{\Delta t/2}\cdot v dx + \frac{\nu}{2}\int_\Omega \nabla U^{n+1/2}\cdot\nabla v\, dx - \int_\Omega P^{n+1/2}\nabla\cdot v dx - \int_\gamma \lambda^{n+1/2}\cdot v d\gamma \\ = \int_\Omega f^{n+1/2}\cdot v dx - \frac{\nu}{2}\int_\Omega \nabla U^n\cdot\nabla v\, dx - \int_\Omega (U^n\cdot\nabla)U^n\cdot v dx,\ \forall v\in V_P, \end{cases}$$

$(3.2)_2 \qquad \nabla\cdot U^{n+1/2} = 0 \ \textit{in}\ \Omega,$

$(3.2)_3 \qquad U^{n+1/2} = g^{n+1/2} \ \textit{on}\ \gamma,$

$(3.2)_4 \qquad U^{n+1/2}\in V_P,\ P^{n+1/2}\in L^2(\Omega),$

and then

$$(3.3)_1 \quad \begin{cases} \int_\Omega \frac{U^{n+1}-U^{n+1/2}}{\Delta t/2}\cdot v dx + \frac{\nu}{2}\int_\Omega \nabla U^{n+1}\cdot\nabla v dx + \int_\Omega (U^{n+1/2}\cdot\nabla)U^{n+1}\cdot v dx = \\ \int_\Omega f^{n+1}\cdot v dx + \int_\gamma \lambda^{n+1/2}\cdot v d\gamma + \int_\Omega P^{n+1/2}\nabla\cdot v dx - \frac{\nu}{2}\int_\Omega \nabla U^{n+1/2}\cdot\nabla v dx,\ \forall v\in V_P, \end{cases}$$

$(3.3)_2 \qquad U^{n+1}\in V_P.$

Due to page limitation we shall not give a detailed account on the solution of problems (3.2) and (3.3). However, we shall briefly discuss in Section 4 the solution of the *Stokes type problem* (3.2). Concerning problem (3.3) it is worth noticing that we have been taking advantage of the splitting to treat the advection without being concerned about the constraint $u=g$ at γ. Also due to the periodic boundary conditions, problem (3.3) is well suited to solution methods based on high order upwinding on regular meshes, or on the backward method of characteristics (see, e.g., [10]).

4. FICTITIOUS DOMAIN SOLUTION OF THE STOKES PROBLEM (3.2).

4.1 Generalities

Problem (3.2) is a particular case of

$$(4.1) \quad \begin{cases} \alpha\int_\Omega U\cdot v dx + \nu\int_\Omega \nabla U\cdot\nabla v dx - \int_\Omega P\nabla\cdot v dx - \int_\gamma \lambda\cdot v d\gamma \\ = \int_\Omega F\cdot v dx,\ \forall v\in V_P,\ U\in V_P, \end{cases}$$

$$(4.2) \quad \int_\Omega \nabla\cdot U q dx = 0,\ \forall q\in L^2(\Omega),\ P\in L^2(\Omega),$$

$$(4.3) \quad U = g \ \textit{on}\ \gamma,$$

where, in (4.1), $\alpha(>0)$ is the reciprocal of a timestep.

To solve the *saddle-point* system (4.1)–(4.3) we use in practice a *conjugate gradient* variant of the following algorithm:

$(4.4) \qquad \lambda^0\in\Lambda$ *is given;*

for $k\geq 0$, λ^k *being known in* Λ, *we solve the Stokes problem*

$$(4.5)\quad \begin{cases} \alpha\int_\Omega U^k\cdot v dx + \nu\int_\Omega \nabla U^k\cdot\nabla v dx - \int_\Omega P^k\nabla\cdot v dx = \\ \int_\Omega F\cdot v dx - \int_\gamma \lambda^k\cdot v d\gamma,\ \forall v\in V_P,\ U^k\in V_P, \end{cases}$$

$$(4.6)\quad \int_\Omega \nabla\cdot U^k q dx = 0,\ \forall q\in L^2(\Omega),\ P^k\in L^2(\Omega),$$

and next update λ^k *by*

$$(4.7)\quad \lambda^{k+1} = \lambda^k - \rho(U^k|_\gamma - g).$$

In (4.4)–(4.7), Λ is a well chosen multiplier space and ρ is a positive parameter which has to be taken sufficiently small. Concerning the solution of the Stokes problem (4.5)–(4.6) it will be discussed in Section 4.2.

4.2 Solution of the Stokes problem (4.5), (4.6).

The crucial step in our approach is the solution of the Stokes problem (4.5), (4.6). Problem (4.5), (4.6) is a particular case of

$$(4.8)\quad \alpha\int_\Omega U\cdot v dx + \nu\int_\Omega \nabla U\cdot\nabla v dx - \int_\Omega P\nabla\cdot v dx = L(v),\ \forall v\in V_P;\ U\in V_P,$$

$$(4.9)\quad \int_\Omega \nabla\cdot U q dx = 0,\ \forall q\in L^2(\Omega),\ P\in L^2(\Omega).$$

Suppose first that $L(v)=\int_\Omega f\cdot v dx$; the above problem is equivalent to

$$(4.10)\quad \alpha U - \nu\nabla^2 U + \nabla P = f \textit{ in } \Omega,$$

$$(4.11)\quad \nabla\cdot U = 0 \textit{ in } \Omega,$$

with U and P *periodic*. Solving problem (4.10), (4.11) is then quite easy since applying the divergence operator on both sides of (4.10) we obtain

$$(4.12)\quad \nabla^2 P = \nabla\cdot f \textit{ in } \Omega,$$

which has a *unique* periodic solution in $L^2(\Omega)/\mathbf{R}$. Next, we obtain U as the unique periodic solution of

$$(4.13)\quad \alpha U - \nu\nabla^2 U = f - \nabla P \textit{ in } \Omega.$$

Combining (4.12), (4.13) we obtain

$$(4.14)\quad \alpha(\nabla\cdot U) - \nu\nabla^2(\nabla\cdot U) = 0 \textit{ in } \Omega;$$

$\nabla\cdot U$ being itself periodic, (4.14) implies that $\nabla\cdot U = 0$.

ERRATUM

New Page 414a

If the linear functional $L(\cdot)$ is not associated to a function f like above, we can use the following iterative technique (introduced in [11]):

(4.15) $\quad \mathbf{P}^o$ *is given;*

then for $m \geq 0$, *assuming that* $\mathbf{P}^m$ *is known, we compute* $\mathbf{U}^m$ *and* $\mathbf{P}^{m+1}$ *as follows*

$$(4.16) \quad \int_\Omega (\alpha \mathbf{U}^m \cdot \mathbf{v} + \nu \nabla \mathbf{U}^m \cdot \nabla \mathbf{v}) dx = \int_\Omega \mathbf{P}^m \nabla \cdot \mathbf{v} dx + L(\mathbf{v}), \ \forall \mathbf{v} \in V_P; \ \mathbf{U}^m \in V_P,$$

$$(4.17) \quad \begin{cases} -\Delta \varphi^m = \nabla \cdot \mathbf{U}^m \ \text{in } \Omega, \\ \varphi^m \ \text{periodic}; \ \int_\Omega \varphi^m dx = 0, \end{cases}$$

$$(4.18) \quad \mathbf{P}^{m+1} = \mathbf{P}^m - (\alpha \varphi^m + \nu \nabla \cdot \mathbf{U}^m).$$

It has been shown in [11] (see also [12]) that $\forall P^o$, the above algorithm converges in *two* iterations in the sense that $\{P^1, \ U^1\} = \{P, U\}$. This remarkable property being a consequence of the periodicity.

Solving (4.16) and (4.17) is quite easy and can be done by many techniques including fast Poisson and Helmholtz solvers based on cyclic reduction, FFT, ...

5. COMPUTER IMPLEMENTATION AND NUMERICAL EXPERIMENTS.

We consider the test problem where $\omega = (.25, \ .75)^2$, $\Omega = (0,1)^2$, $u_o = 0$ and g defined as follows:

$$\begin{cases} \mathbf{g} = \{1, 0\} \ \text{on} \ \{\mathbf{x} | \mathbf{x} = \{x_1, x_2\}, x_2 = .75, .25 < x_1 < .75\}, \\ \mathbf{g} = 0 \ \text{elsewhere on} \ \partial\omega; \end{cases}$$

we are therefore considering a *wall driven cavity flow problem*. We use then a finite element triangulation $\mathcal{T}_h$ like the one in Figure 5.1 below to approximate velocity

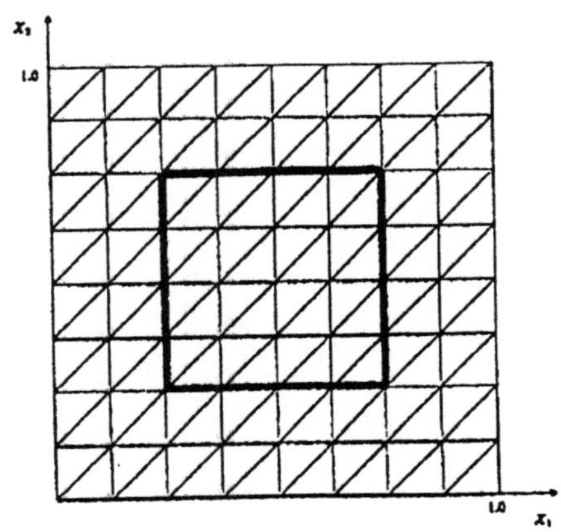

Figure 5.1 (h=1/8)

ERRATUM

Page 415

Eq (4.11) ⟶ (4.16)

Eq (4.12) ⟶ (4.17)

Page 411

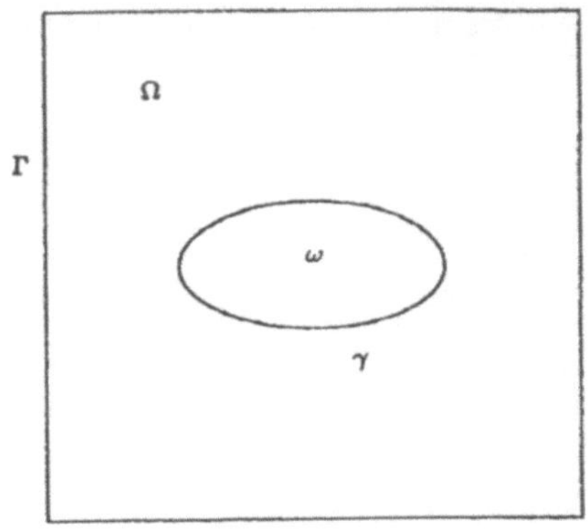

Figure 1.1

Page 415

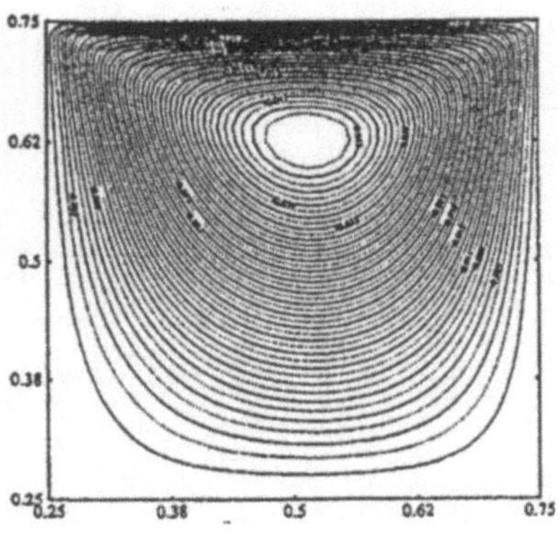

Figure 5.2
Streamlines ($\nu=10^{-2}$)

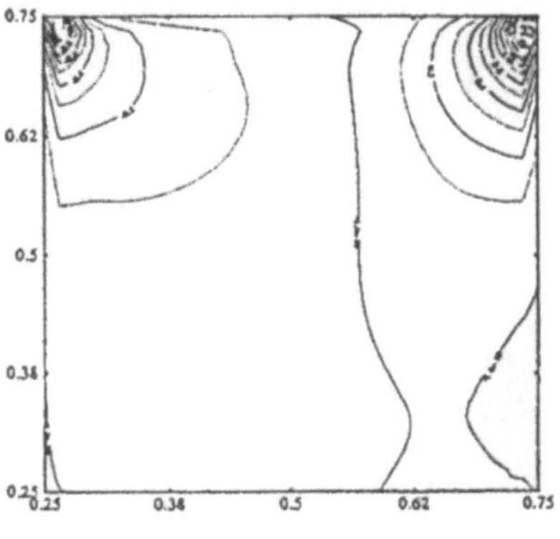

Figure 5.3
Contour lines of the pressure
($\nu=10^{-2}$)

THIS IS AND ADHESIVE SHEET. PLEASE CUT AND PASTE IN PROPER PLACE

We approximate then V_P by

$$V_{hP} = \left\{ v_h | v_h \in H^1_{hP} \times H^1_{hP} \right\}$$

where

$$H^1_{hP} = \left\{ \varphi_h | \varphi_h \in C^0(\bar{\Omega}),\ \varphi_h|_T \in P_1,\ \forall T \in \mathcal{T}_h,\ \varphi_h \textit{ periodic at } \Gamma \right\},$$

with P_1: space of the polynomials in x_1, x_2 of degree ≤ 1.

A traditional way of approximating the pressure is to take it in the space H^1_{2hP}, i.e., defined on a triangulation $\mathcal{T}_{2h}$ twice coarser than $\mathcal{T}_h$. Concerning the space Λ_h approximating Λ, we defined it by

$$\Lambda_h = \left\{ \mu_h | \mu_h \in (L^\infty(\partial\omega))^2,\ \mu_h \textit{ constant between the midpoints of the edges of } \mathcal{T}_h \textit{ supported by } \partial\omega \right\}.$$

Motivated by the fact that problem (4.11), (4.12) completed by periodic boundary conditions can be solved with any reasonable choice for approximating pressure and velocity (no compatibility condition then), we have been also testing the case where we use H^1_{hP} (instead of H^1_{2hP}) to approximate the pressure. The numerical experiments below have been obtained for $\nu = 10^{-2}$ and $h = 1/32$.

Figure 5.2 shows the streamlines of the steady state solution, while Figure 5.3 shows the contour lines of the pressure. The remarkable fact is that we can use either H^1_{hP} or H^1_{2hP} to approximate the pressure and obtain solutions which are free of spurious oscillations (and very close to each other).

Figure 5.2
Streamlines ($\nu = 10^{-2}$)

Figure 5.3
Contour lines of the pressure
($\nu = 10^{-2}$)

Further numerical experiments (and/or analysis) are needed to decide if the fact that we can use the same grid for pressure and velocity is due to the various symmetries in the problem or to a more general property, definitely resulting from the *periodic* embedding: Future will tell.

6. CONCLUSION

In this article we have briefly discussed fictitious domain methods for the incompressible Navier-Stokes equations. On the basis of preliminary numerical experiments the method looks promising particularly concerning the compatibility between pressure and velocity approximations.

ACKNOWLEDGMENTS

We would like to acknowledge the helpful comments and suggestions of the following individuals: C. Atamian, L. C. Cowsar, C. De la Foye, G. H. Golub, P. Joly, Y. Kuznetsov, A. Latto, W. Lawton, P. Le Tallec, J. L. Lions, P. L. Lions, G. Meurant, J. Pasciak, M. Ravachol, H. Resnikoff, H. Steve, J. Weiss, R. O. Wells, M. F. Wheeler, O. B. Widlund.

The support of the following corporations or institutions is also acknowledged: AWARE, Dassault Aviation, INRIA, University of Houston, Université Pierre et Marie Curie. We also benefited from the support of DARPA (Contracts AFOSR F49620-89-C-0125 and AFOSR-90-0334), DRET (Grant 89424) and NSF (Grants INT 8612680 and DMS 8822522). Finally, we would like to thank J. A. Wilson for the processing of this article.

References

[1] YOUNG, D. P., MELVIN, R. G., BIETERMAN, M. B., JOHNSON, F. T., SAMANTH, S.S., and BUSSOLETTI, J. E. - A locally refined finite rectangular grid finite element method. Application to Computational Physics, J. Comp. Physics, 92, 1-66 (1991).

[2] BUZBEE, B. L., DORR, F. W., GEORGE, J. A. and GOLUB, G. H. - The direct solution of the discrete Poisson equation on irregular regions, SIAM J. Num. Anal., 8, 722-736 (1971).

[3] BORGERS, C. - Domain imbedding methods for the Stokes equations, Num. Math., 57, 435-452 (1990)

[4] R. GLOWINSKI. - Numerical Methdos for Nonlinear Variational Problems, Springer-Verlag, New York, 1984.

[5] BRISTEAU, M. O., GLOWINSKI, R., MANTEL, B., PERIAUX, J. and PERRIER, P. - 'Numerical methods for incompressible and compressible Navier-Stokes problems,' Finite Element in Fluids, Vol. 6, Eds. Gallagher, R. H., Carey, G., Oden, J. T. and Zienkiewicz, O. C., Wiley, Chichester, 1985, p. 1.

[6] PEACEMAN, D. and RACHFORD, H. - The numerical solution of parabolic and elliptic differential equations, J. SIAM, 3, 28-41 (1955).

[7] GLOWINSKI, R. - 'Viscous flow simulation by finite element methods and related numerical techniques,' Progress and Supercomputing in Computational Fluid Dynamics, Eds. Murman, E. M. and Abarbanel, S. S., Birkhauser, Boston, 1985, p. 173.

[8] BRISTEAU, M. O., GLOWINSKI, R. and PERIAUX, J. - Numerical Methods for the Navier-Stokes equations, Comp. Phys. Rep., 6, 73-187 (1987).

[9] MARCHUK, G. I. - Methods of Numerical Mathematics, Springer, New York, 1975.

[10] PIRONNEAU, O. - Finite Element Methods for Fluids, J. Wiley, Chichester, 1989.

[11] CAHOUET, J. and CHABARD, J. P. - Some fast 3D solvers for the generalized Stokes problem, Int. J. Num. Meth. in Fluids, 8, 269-295 (1988).

[12] GLOWINSKI, R. - 'Finite element methods for the numerical simulation of incompressible viscous flow. Introduction to the control of the Navier-Stokes equations,' Proceedings of the AMS Conference on Vortex Dynamics and Vortex Methods, Seattle, July 1990, Eds. Anderson, C. and Greengard, C., AMS, Providence (to appear).

FINITE ELEMENTS IN THE 90's, E. Oñate, J. Periaux, A. Samuelsson (Eds.)
Springer-Verlag/CIMNE, Barcelona 1991

SIMPLE ELEMENTS AND LINELETS FOR INCOMPRESSIBLE FLOWS

R. Löhner
CMEE, SEAS, The George Washington University,
Washington, D.C. 20052,
USA

SUMMARY

The current trends for the simulation of large-scale incompressible flow fields using finite elements are discussed. The main items are: a) the use of simple elements through stabilization and analogy with LBB-satisfying elements, and b) the development of fast solvers for general grids.

1. INTRODUCTION

I still remember vividly the first time I met Olgierd C. Zienkiewicz. This was in the fall of 1982. Having just arrived from Germany with a fresh Diplom-title (MSc), I handed over to him my thesis which dealt with hybrid elements for the Stokes and Navier-Stokes equations [1]. In the thesis, some 15 element types with different combinations of shape-functions for the velocities and pressures were developed and tested. Remember these were the early 80's, when the LBB-condition [2] was just emerging from mathematical circles into the world at large. Testing new elements in this more experimental way was (and remains, see [3]) very common. Olgierd C. Zienkiewicz glanced through it, asked me how long it had taken me to write the thesis, and went on to discuss some of the recent developments that had taken place in Swansea. Among those were reduced integration [4], upwinding [5,6], and some tests of mixed elements for structural problems. The 80's were then dedicated to explore the use of finite elements for the simulation of high-speed compressible flows [7-10]. Here was a very well developed, traditional field, where even hundreds of engineers in industry were actively developing new codes. Industry at large would typically buy off-the-shelf codes to satisfy its needs for structural dynamics analysis. But due to the lack of adequate software or for some other historical reason, the aerospace industry was (and is) developing its own codes. The typical CFD code of that time was based on: structured grids, finite differences or volumes, central differences, artificial dampings, explicit timestepping and a high degree of vectorization [11]. Needless to say, it was a one-stencil (i.e. one-element) type code. In the majority of cases these codes were used to analyze transonic flow past airfoils or wings using the Euler equations. For these simple configurations 'near-optimal' structured grids could be constructed. During the 80's, we learned how to compete with finite elements in this arena. The development of adaptive refinement schemes [12,13] and automatic gridding techniques [14-16] in the second half of the decade then firmly established these schemes in the aerospace community.

For incompressible flow simulations, the early 90's find us in somewhat the same situation as the early 80's. As before, we face formidable competition from advanced structured grid, finite difference codes (e.g.[17]). We have, on the other hand, learned how to live with dominant first order derivatives and artificial dampings. What sets incompressible flows apart is the divergence constraint with its two consequences: a) the use of special elements b) the inifinite speed of sound, which requires implicit timestepping schemes for the pressure. For practical reasons, it may also be advantageous to use implicit schemes for the advective and diffusive terms. As an example, consider a typical boundary layer grid with very elongated elements aligned with the attached flow. As soon as separation occurs, an explicit Courant-Friedrichs-Levy (CFL) criterion for the velocities would impose time-step sizes that are orders of magnitude smaller than in the attached case (see Figure 1).

The rest of the paper is organized as follows. First, the question of element technology is addressed. Then, a new preconditioner for fast implicit solvers is developed. Finally, some practical examples are included that show the usefulness of the proposed methodology.

2. ELEMENT TECHNOLOGY

In order to define the notation used we start by recalling the incompressible Navier-Stokes equations

$$\mathbf{v}_{,t} + \mathbf{v}\nabla\mathbf{v} + \nabla p = \frac{1}{Re}\nabla^2\mathbf{v} \; , \tag{1a}$$

$$\nabla \cdot \mathbf{v} = 0 \; . \tag{1b}$$

Here p denotes the pressure, $\mathbf{v}$ the velocity vector and Re the Reynolds-number, and the pressure p has been normalized by the (constant) density ρ. The main difficulty for the design of practical elements stems from Eqn.(1b). Applying Eqn.(1b) to (1a), we obtain

$$\nabla \cdot \mathbf{v}\nabla\mathbf{v} + \nabla \cdot \nabla p = 0 \; . \tag{2}$$

On the other hand, a simple Galerkin discretization to Eqn.(1) would result in

$$\mathbf{C}^t\mathbf{M}^{-1}\left[\mathbf{A}\mathbf{v} + \mathbf{C}\mathbf{p} + \mathbf{K}\mathbf{v}\right] = 0 \; , \tag{3}$$

where $\mathbf{C}, \mathbf{K}, \mathbf{A}$ denote the matrices resulting from the discretization of the gradient/divergence, Laplace and advection operators respectively, and $\mathbf{M}$ the lumped mass-matrix. The potential problem now becomes apparent: the resulting discretization of the Poisson-operator for the pressures in Eqn.(2) via

$$\nabla \cdot \nabla p \approx \mathbf{C}^t\mathbf{M}^{-1}\mathbf{C}\mathbf{p} \tag{4}$$

may indeed be unstable. Table 1 lists the resulting operators in 1-D for different choices of approximations of velocity and pressure.

Table 1 Resulting Operators for Different Velocity/ Pressure Combinations

Velocity	Pressure	Operator
p1	q0	(0,-1, 2,-1, 0)
p1	p1	(-1, 0, 2, 0,-1)
iso-p1	p1	(-1,-1, 4,-1,-1)

One can clearly the see the unstable decoupling of nearest neighbors that occurs for the p1/p1 element. The p1/p0 element reproduces exactly the usual $(-1, 2, -1)$-Laplacian discretization, whereas the p1/iso-p1 element adds additional stabilizing terms to the unstable discretization resulting from the p1/p1 element. These stabilizing terms are equivalent to an artificial viscosity for the divergence equation. When going from 1-D to 2/3-D, the possibilities for choosing different combinations of shape-functions for pressures and velocities widen considerably. As an example, consider the p1/iso-p1 element in 1-D. Its 2-D extension may be accomplished either in the form of the p1/p1+bubble (mini) element (see [18]), or by using macro-element shape-functions for the pressures (p1/iso-p1). All of these possible combinations will simply lead to different artificial viscosities for the divergence-constraint. The sooner we realize this, the better off we are. This equivalence between a modified divergence equation with artificial viscosity of the form

$$\nabla \cdot \mathbf{v} = \beta\left(\nabla^2 p + \nabla \cdot \mathbf{v} \cdot \nabla\mathbf{v}\right) \; , \tag{5}$$

and the p1/p1+bubble element was shown analytically in [19]. Thus, given the divergence-constraint, two alternate routes may be taken:

a) spend more effort per element with the original (simpler) equations by using different shape-function spaces, or
b) spend more effort per equation with a simpler element.

Practical production codes place a requirement for simplicity. A complex element, however good, will always be less favoured by the user as well as by those maintaining and upgrading codes. Therefore, in the present case, the last option was chosen.

2.1 An Alternative Derivation

As with compressible flows, many of the difficulties encountered by first discretizing in space and then in time may be circumvented by proceeding in the reverse order. Therefore, we will re-derive the discretization of the divergence-equation in time using Taylor-Galerkin concepts. Re-stating the divergence constraint as

$$c^{-2}p_{,t} + \nabla \cdot \mathbf{v} = 0 \ . \tag{6}$$

where c is the speed of sound, the following Taylor-expansion is obtained:

$$c^{-2}\Delta p = c^{-2}\Delta t \ p_{,t}|^{n} + c^{-2}\frac{\Delta t^2}{2}p_{,tt}|^{n+\theta} \ . \tag{7}$$

Inserting Eqn.(6) repeatedly into (7), using (1a), ignoring any spatial derivatives of order higher than two, and taking the limit $c \to \infty$ yields

$$\Delta t \ \nabla \cdot \mathbf{v} = \frac{\Delta t^2}{2}\left[\nabla^2 p + \nabla \cdot \mathbf{v} \cdot \nabla \mathbf{v}\right]^{n+\theta} \ . \tag{8}$$

Observe that the same Laplacian 'pressure-diffusion' as encountered in Eqn.(5) now appears naturally on the right-hand side.

3. ELLIPTIC SOLVERS BASED ON LINELETS

As we require implicit timestepping, we will face the solution of large coupled systems of equations. For most incompressible flow solvers these will be elliptic Poisson-type problems. The fastest solvers for this class of problems are unstructured multigrid solvers [20,21]. They require good smoothers, as well as efficient intergrid transfer operators. The construction of good smoothers tends to be difficult for the very stretched and distorted grids encountered in high Reynolds-number applicatons. Furthermore, for a production environment, the construction and handling of the multiple grid systems must be made automatic. A one-grid method, due to its simplicity and 'black-box' character, appears more suitable for the non-expert user community. The main reason for the speed of multigrid solvers is the very fast transport of information through the grid. This allows the residuals to propagate quickly to the boundaries of the computational domain. This 'transport feature' of multigrid solvers must be kept if one is to produce a competitive scheme.
An alternate procedure for propagating information throughout the grid is the use of line-relaxation or approximate factorization [22,23]. If the freedom is given on how to choose 'lines' in the mesh, information can indeed be propagated in a fast way to the boundaries. Moreover, the direct solver being employed for the line-relaxation is working at its peak efficiency in terms of operations per (useless) fill-in.

Another possibility is the use of incomplete LU decompositions as preconditioners. The main problem we see in incomplete LU decompositions is that the effectiveness of the preconditioner depends on the numbering of the points [24]. This is to be expected, as there exists a very tight coupling for the points numbered sequentally, and a much looser (fill-in ignoring) coupling between all other nearest neighbors to a point. The use of standard renumbering techniques, such as the reverse Cuthill-McKee [25] algorithm, have been reported to improve this situation [24]. On the other hand, it is conceivable that a point renumbering obtained from such an algorithm may be an undesirable one in light of physics and mesh stiffness. Moreover, the flow of information will be inhibited at 'bifurcation points' in the renumbering, as the tight coupling will give way to loose coupling for at least one branch of the bifurcation.
The concept of lines translates to snakes [26] in the context of an unstructured grid. The corresponding relaxation scheme to solve

$$\mathbf{K}\mathbf{u} = \mathbf{r} \ , \tag{9}$$

where $\mathbf{K}$ is the matrix describing the elliptic problem, $\mathbf{u}$ the desired vector of unknowns, and $\mathbf{r}$ the right hand side vector, becomes

$$\mathbf{K}_1\mathbf{K}_2\Delta\mathbf{u} = \mathbf{r} - \mathbf{K}\mathbf{u} \ . \tag{10}$$

Here $\mathbf{K}_1, \mathbf{K}_2$ denote the entries of $\mathbf{K}$ for the active point-point combinations that define the snake. Observe that this formulation is not dimensionally consistent. Therefore, a better formulation is to use

$$\mathbf{K} = \mathbf{D}^{\frac{1}{2}}(\mathbf{I}+\mathbf{E})\mathbf{D}^{\frac{1}{2}} \approx \mathbf{D}^{\frac{1}{2}}(\mathbf{I}+\mathbf{E}_1)(\mathbf{I}+\mathbf{E}_2)\mathbf{D}^{\frac{1}{2}} , \quad (11)$$

where $\mathbf{D}, \mathbf{E}$ denote the diagonal and off-diagonal entries of $\mathbf{K}$. Thus, the relaxation scheme becomes

$$\mathbf{D}^{\frac{1}{2}}(\mathbf{I}+\mathbf{E}_1)(\mathbf{I}+\mathbf{E}_2)\mathbf{D}^{\frac{1}{2}}\Delta\mathbf{u} = \mathbf{r} - \mathbf{K}\mathbf{u} . \quad (12)$$

In order to achieve higher rates of convergence, we use Eqn.(12) within a preconditioned Conjugate gradient algorithm [27].

Because an unstructured grid usually does not possess an equal number of gridpoints along a certain direction, the resulting snakes may often exhibit folding (see Fig. 2b). This implies that the information flow from the domain to the boundary may be slowed down considerably. This is not important for smoothers, but crucial for the preconditioners required in one-grid solvers. In order to obtain a steady flow of information towards the boundaries in all directions, we reconnect the snake in the direction it intended to continue wherever it folds. This gives rise to a more complex structure, which we call linelet (see Fig. 2c). Whereas the storage requirements of snakes are fixed ($3N$, where N is the number of unknowns), the storage requirements of linelets depend on the structure of the mesh and the renumbering chosen.

3.1 Construction of Linelets

The construction of a proper set of linelets is crucial in order to obtain a good preconditioner. A linelet is defined as a set of active sides. It is these active sides, or point-point combinations, that are going to be used to form the entries in $\mathbf{K}_1, \mathbf{K}_2$. In order to build a linelet, we assume given:

- A preferred direction $\mathbf{d}_i$ at each point of the mesh;
- The sides of the mesh;
- The sides emanating from each point of the mesh;
- A list of 'forbidden' sides, i.e. sides that are not supposed to be used for the current linelet.

The aim is to touch every point of the mesh with the linelet in such a way that the inversion of the matrix $\mathbf{K}_i$ corresponding to this linelet requires minimum effort. In order to arrive at a set of active sides to define the present linelet, the following steps are taken:

Pass 1: in a first pass over the mesh, every point is connected to its best possible neighbor along allowed sides. The decision is based on the scalar product of the side with the preferred direction $\mathbf{d}$ at the given point. The sides obtained in this way are marked and stored.

Pass 2: in a second pass over the mesh, we add active neighbor connections to those points that are only connected to one neighboring point. Special care is required for boundary points. The sides obtained in this way are marked and stored.

Pass 3: in a third pass over the mesh, we identify points where 'sharp turns' in the current list of active sides occur. At these points, further sides sides are activated and added in order to obtain a uniform propagation of information in the desired direction.

Pass 4: finally, 3/4-combinations are removed where possible. This is done in order to reduce the bandwidth without penalizing the flow of information in the desired directions.

After having the list of sides defining the linelet, the points are renumbered so that the matrix inversion effort is minimized. We have used variations of the reverse Cuthill-McKee [25] algorithm. The entries into the matrix corresponding to the linelet are stored using the skyline system [28]. This reduces the required memory to a minimum, and facilitates the use of standard finite-element equation solvers.

4. NUMERICAL EXAMPLES

4.1 Circular Cylinder: This classic test-case was computed to verify the temporal accuracy of the method. The Reynolds-number was set to $Re = 100$. At this Reynolds-number, the maximum variation of Strouhal-number with Reynolds-number is observed [29]. The mesh, as well as the corresponding system of linelets and the solution obtained are shown in Figure 3.

The correlation of Strouhal-numbers computed with those observed experimentally is given in Table 2. As one can see, the agreement is exceptional.

Table 2 Comparison of Strouhal-Numbers

Re-nr.	Timestep	Steps/Cycle	Num. Str.	Exp. Str.
100	0.05	120	0.167	0.167
140	0.05	113	0.177	0.177
1000	0.05	96	0.208	0.208

4.2 Circular Arc Cascade: this is a 2-D, viscous, steady-state case at a Reynolds-number of $Re = 10^3$. The problem set-up and the boundary conditions were taken from Mansour and Hamed [30]. The thickness to chord ratio was set to 0.2 and the pitch to chord ratio was 2.0. The grid employed, consisting of 5,948 p1/p1 elements and shown in Figure 4a, was taylored to be similar to that used in [30]. The results obtained after 200 steps, corresponding to a residual decrease of 4 orders of magnitude, are shown in Figures 4b-d. Flow separation occurs at about 85% of the chord length. These results are almost graphically undistinguishable from those reported in [30].

4.3 Airfoil: This case was run using the Euler-option, i.e. no viscous terms were included. Local timestepping was used to accelerate the solution to steady state. The airfoil is a Wortman FX63-137, and the angle of attack was set to $\alpha = 5$ deg. The mesh, linelets, and pressure obtained after 200 steps (3 orders of magnitude decrease in the residual) are shown in Figure 5.

4.4 Laminar Flat Plate: The laminar boundary layer of a flat plate was computed for a Reynolds-number of $Re = 10^6$. The mesh employed, as well as the system of linelets and the solution obtained are displayed in Figure 6.

4.5 Turbulent Flat Plate: The turbulent boundary layer of a flat plate was computed for a Reynolds number of $Re = 0.75 \cdot 10^6$. Figure 7 shows the mesh employed, the system of linelets, as well the solution obtained. The development of the boundary layer is visible from Figure 7d, were a comparison with the boundary layer edge computed from the seventh power law is included. Convergence was obtained after 400 time steps.

The storage requirements for the linelet-preconditioner varied considerably for the cases shown above. Denoting by Np the number of points in the grid, the storage requirements ranged from $O(2Np)$ for the flat plates to $O(3Np)$ for the cylinder and to $O(10Np)$ for the airfoil. The high storage requirement for the airfoil is due to mesh topology. We are currently investigating new renumbering algorithms to reduce these storage requirements further.

5. CONCLUSIONS

A simple finite element for incompressible flows was developed. The element is based on a modified continuity equation, which adds stabilizing terms in the form of a pressure Laplacian. This element may be derived either from the p1/p1+bubble-element analytically [19], or through by discretizing the equations first in time and then in space using Taylor-Galerkin concepts.
The fast solution of the elliptic systems arising when timestepping the incompressible equations was made possible though the use of a new preconditioner based on linelets. Although the preconditioner was derived from line-relexation techniques, it is more general in that it allows grid-independent lines to be formed. Thus, physiscs-based 'lines' can be defined. This preconditioner leads to a significant increase in performance with a modest increase in memory requirements. However, we feel that more work is required in this area.
The areas most likely to see active development during the 90's will be:

- Turbulence Modelling: there is no point in having a fast solver if the physics are not well represented. We have explored both the Baldwin-Lomax algebraic model [32], as well as the Baldwin-Barth one-equation model. We feel that the trend will have to be towards libraries of turbulence models. The user then picks the model best suited for the application at hand.

- Automatic Gridding: as before, there is no point having a fast solver if it takes 6 months to grid a new problem. Truly automatic gridding in 3-D, i.e. a grid for a new problem of arbitrary complexity in a matter of minutes, will remain a pacing item through the 90's.
- Adaptive Grids for High-Re Flows: following the success of self-adaptive schemes for inviscid, compressible flows, one would expect huge payoffs for high-Re viscous cases. Pioneering work by Mavriplis [21] already has shown promising results. Automation and user-friendliness has, howerer (let's be honest) remained elusive, and much work is still required.
- Parallel Computing: there is no doubt that parallel machines will dominate the supercomputer market after the mid-90's. Their potential is just beggining to be realized. What most researchers don't seem to understand is the revolutionary nature of what awaits us. There is no point in just porting a flow solver to a parallel environment. The whole CFD process (modelling, gridding, solving, adapting, displaying) will have to be ported.

6. ACKNOWLEDGEMENTS

I would like to acknowledge the large number of valuable discussions with Olgierd C. Zienkiewicz that I had over the years. These led to many of the ideas reported in this paper. Moreover, it is a pleasure to acknowledge fruitful discussions with Profs. Ken Morgan and Jaime Peraire.

This work was supported by ONR under Grant N00014-90-J-1416, with Dr. Spiro Lekoudis as the technical monitor, and the Laboratory for Computational Physics and Fluid Dynamics of the Naval Research Laboratory.

7. REFERENCES

[1] R. Löhner - Theoretische und Numerische Untersuchungen zur Gemischt-Hybriden Elementformulierung der Navier-Stokes-Gleichungen für Stationäre und Instationäre Strömungen; MSc Thesis, TU Braunschweig (1982).

[2] M.D. Gunzburger - Mathematical Aspects of Finite Element Methods for Incompressible Viscous Flows. *Finite Elements: Theory and Application* (Dwoyer, Hussaini and Voigt eds.), Springer Verlag, 124-150, (1987).

[3] Fortin - Simple Continuous Pressule Elements for Two- and Three- Dimensional Incompressible Flows; *Comp. Meth. Appl. Mech. Eng.* **62**, 47-69 (1987).

[4] O.C. Zienkiewicz and R. Taylor - On Reduced Integration; Rep INME, Swansea, (1982).

[5] D.W. Kelly, S. Nakazawa, O.C. Zienkiewicz and J.C. Heinrich - A Note on Anisotropic Balancing Dissipation in Finite Element Approximation to Convection Diffusion Problems; *Int. J. Num. Meth. Eng.* **15**, 1705-1711 (1980).

[6] A.N. Brooks and T.J.R. Hughes - Streamline Upwind/Petrov Galerkin Formulations for Convection Dominated Flows with Particular Emphasis on the Incompressible Navier-Stokes Equations; *Comp. Meth. Appl. Mech. Eng.* **32**, 199-259 (1982).

[7] J. Donea - A Taylor Galerkin Method for Convective Transport Problems; *Int. J. Num. Meth. Eng.* **20**, 101-119 (1984).

[8] R. Löhner, K. Morgan and O.C. Zienkiewicz - The Solution of Nonlinear Systems of Hyperbolic Equations by the Finite Element Method; *Int. J. Num. Meth. Fluids* **4**, 1043-1063 (1984).

[9] R. Löhner, K. Morgan and O.C. Zienkiewicz - An Adaptive Finite Element Procedure for High Speed Flows; *Comp. Meth. Appl. Mech. Eng.* **51**, 441-465 (1985).

[10] R. Löhner, K. Morgan, J. Peraire and O.C. Zienkiewicz - Finite Element Methods for High Speed Flows; AIAA-85-1531-CP (1985).

[11] A. Jameson, W. Schmidt and E. Turkel - Numerical Solution of the Euler Equations by Finite Volume Methods using Runge-Kutta Time-Stepping Schemes; AIAA-Paper 81-1259 (1981).

[12] R. Löhner - An Adaptive Finite Element Scheme for Transient Problems in CFD; *Comp. Meth. Appl. Mech. Eng.* **61**, 323-338 (1987).

[13] J. Peraire, M. Vahdati, K. Morgan and O.C. Zienkiewicz - Adaptive Remeshing for Compressible Flow Computations; *J. Comp. Phys.* **72**, 449-466 (1987).

[14] J. Peraire, J. Peiro, L. Formaggia, K. Morgan and O.C. Zienkiewicz - Finite Element Euler Computations in Three Dimensions; AIAA-88-0032 (1988).

[15] R. Löhner and P. Parikh - Three-Dimensional Grid Generation by the Advancing Front Method. *Int. J. Num. Meth. Fluids* **8**, 1135-1149, (1988).

[16] T.J. Baker - Developments and Trends in Three-Dimensional Mesh Generation; *Appl. Num. Math.* **5**, 275-304 (1989).

[17] S. Yoon, D. Kwak and L. Chang - LU-SGS Implicit Algorithm for Three-Dimensional Incompressible Navier-Stokes Equations with Source Term; AIAA 89-1964-CP (1989).

[18] F. Thomasset - *Implementation of Finite Element Methods for Navier-Stokes Equations* p.87 ff, Springer-Verlag (1981).

[19] R. Löhner - A Fast Finite Element Solver for Incompressible Flows; AIAA-90-0398 (1990).

[20] R. Löhner and K. Morgan - An Unstructured Multigrid Method for Elliptic Problems; *Int. J. Num. Meth. Eng.* **24**, 101-115 (1987).

[21] D. Mavriplis - Euler and Navier-Stokes Computations for Two-Dimensional Geometries Using Unstructured Meshes; *ICASE Rep.* **90-3** (1990).

[22] W.R. Briley and H. McDonald - Solution of the Multi-Dimensional Compressible Navier-Stokes Equations by a Generalized Implicit Method. *J. Comp. Phys.* **21**, 372-397 (1977).

[23] R.M. Beam and R.F. Warming - An Implicit Finite Difference Algorithm for Hyperbolic Systess in Conservation-Law Form. *J. Comp. Phys.* **22**, 87-110 (1978).

[24] V. Venkatakrishnan and D. Mavriplis - Implicit Solvers for Unstructured Meshes; *NASA Ames Rep.* RNR-91-016 (1991).

[25] E. Cuthill and J. McKee - Reducing the Bandwidth of Sparse Symmetric Matrices; *Proc. ACM Nat. Conf.*, New York 1969, 157-172 (1969).

[26] O. Hassan, K. Morgan and J. Peraire - An Implicit Finite Element Method for High Speed Flows. AIAA-90-0402 (1990).

[27] M. Hestenes and E. Stiefel - Methods of Conjugate Gradients for Solving Linear Systems. *J. Res. Nat Bur. Standards* **49**, 409-436 (1952).

[28] O.C. Zienkiewicz - *The Finite Element Method* ; McGraw Hill (1982).

[29] H. Schlichting - *Boundary Layer Theory*. McGraw-Hill (1979).

[30] M.L. Mansour and A. Hamed -Implicit Solution of the Incompressible Navier-Stokes Equations on a Non-Staggered Grid; *J. Comp. Phys.* **86**, 147-167 (1990).

[31] B.S. Baldwin and H. Lomax - Thin Layer Approximation and Algebraic Model for Separated Turbulent Flows AIAA 78-257 (1978).

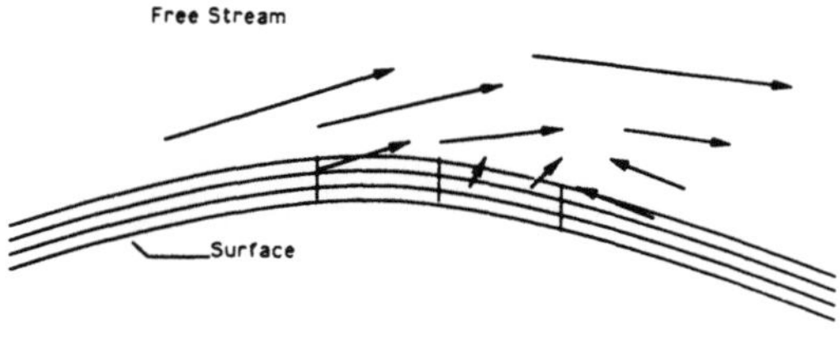

Figure 1: Separation point with typical boundary-layer grid Explicit schemes would imply very small timesteps in this situation

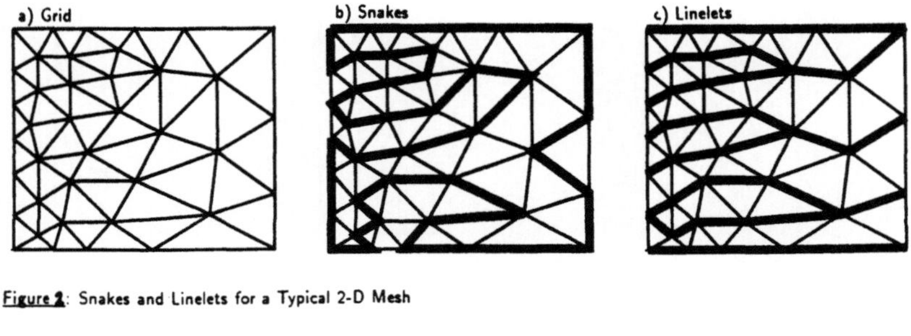

Figure 2: Snakes and Linelets for a Typical 2-D Mesh

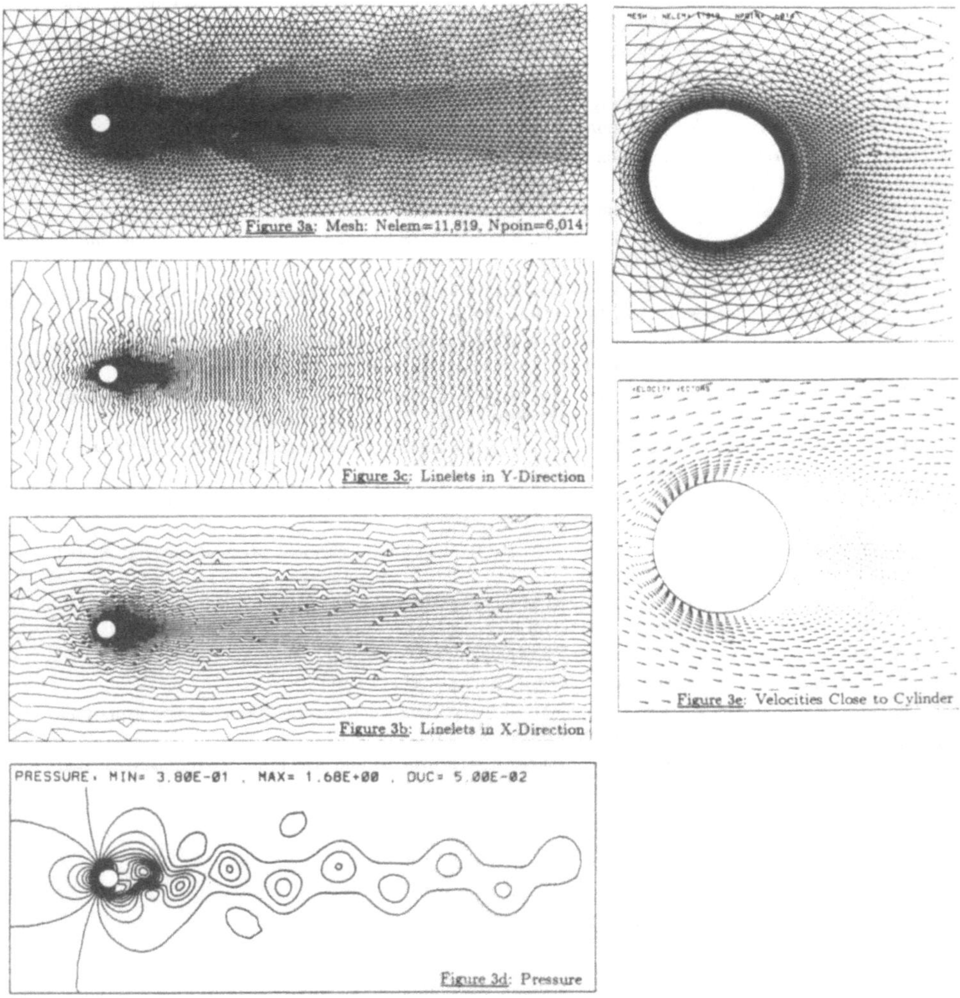

Figure 3a: Mesh: Nelem=11,819, Npoin=6,014

Figure 3c: Linelets in Y-Direction

Figure 3b: Linelets in X-Direction

Figure 3d: Pressure

Figure 3e: Velocities Close to Cylinder

Figure 3: von Karman Vortex Street (Re = 100)

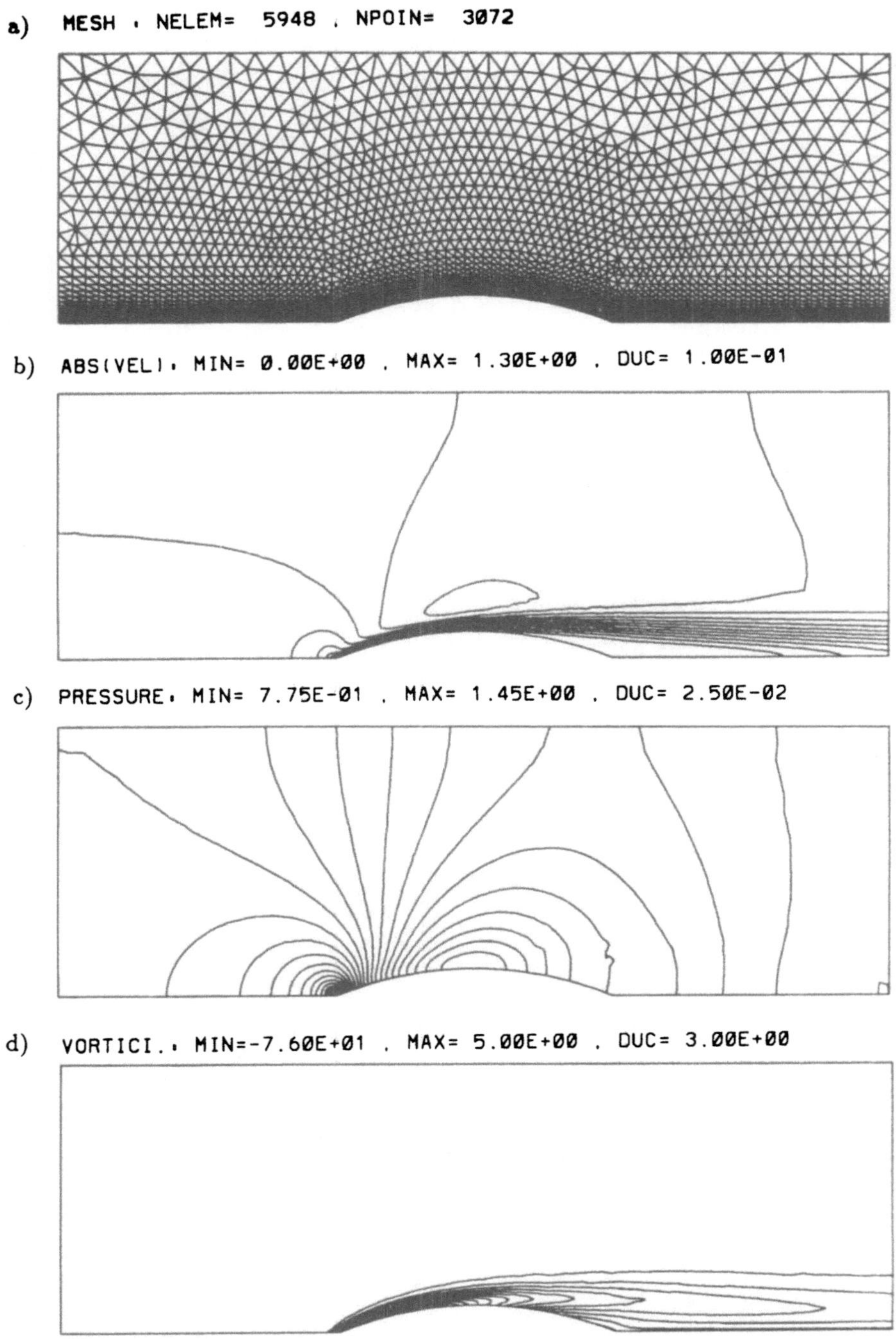

Figure 4: Circular Arc Cascade

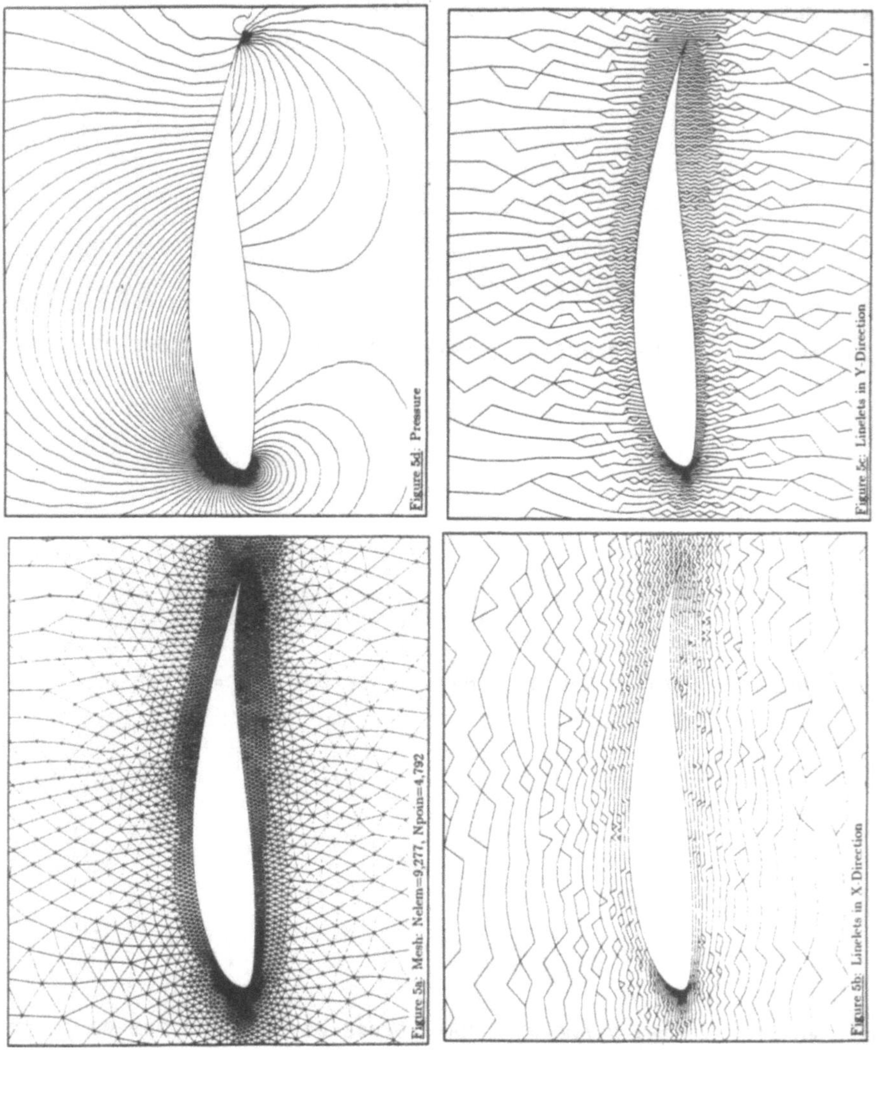

Figure 5: Wortman FX63-137 Airfoil

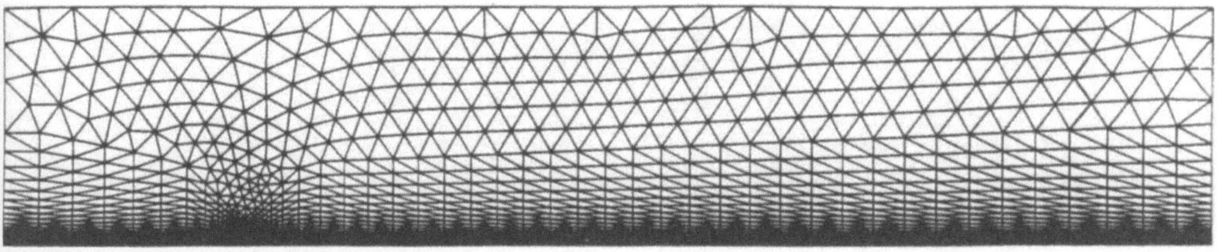

Figure 6a: Mesh: Nelem=4,801, Npoin=2,488

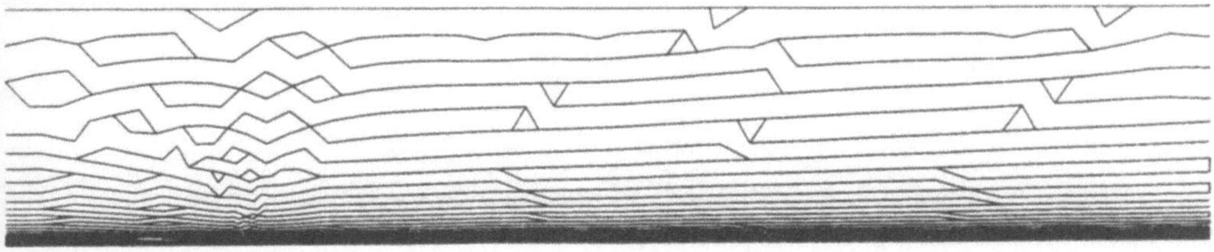

Figure 6b: Linelets in X-Direction

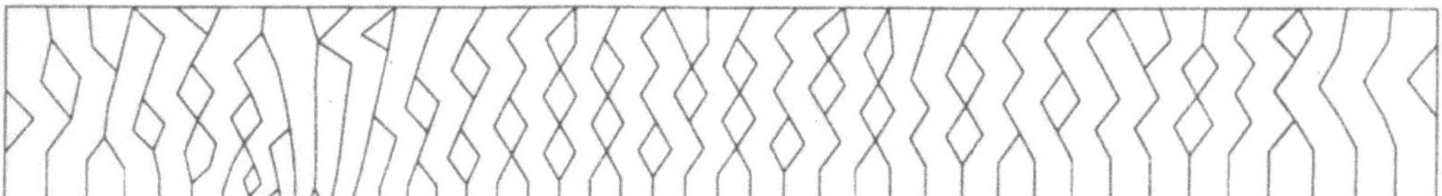

Figure 6c: Linelets in Y-Direction

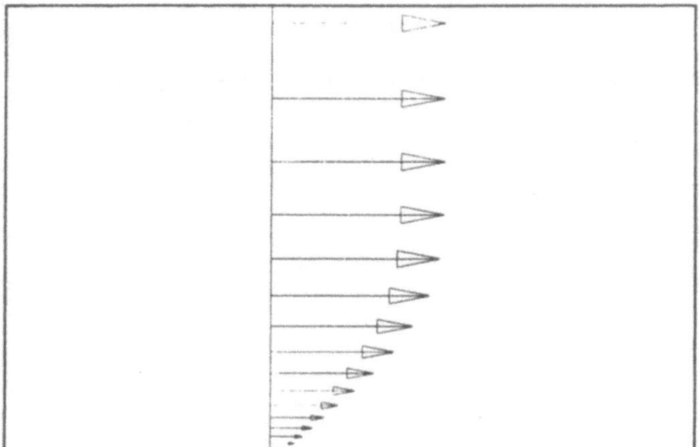

Figure 6d: Velocity Profile at Exit

Figure 6: Laminar Flat Plate ($Re = 10^6$)

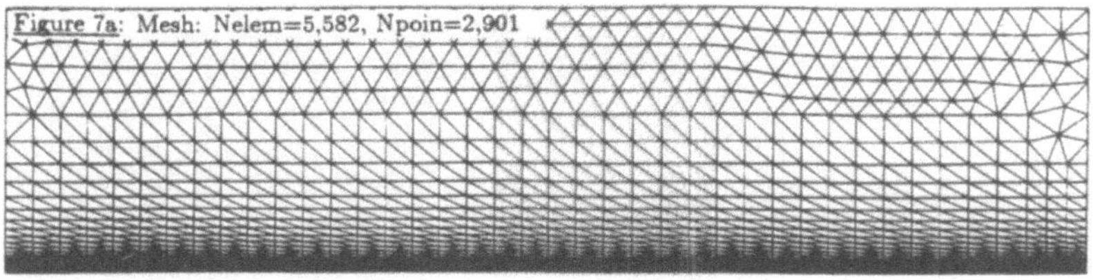

Figure 7a: Mesh: Nelem=5,582, Npoin=2,901

Figure 7b: Linelets in X-Direction

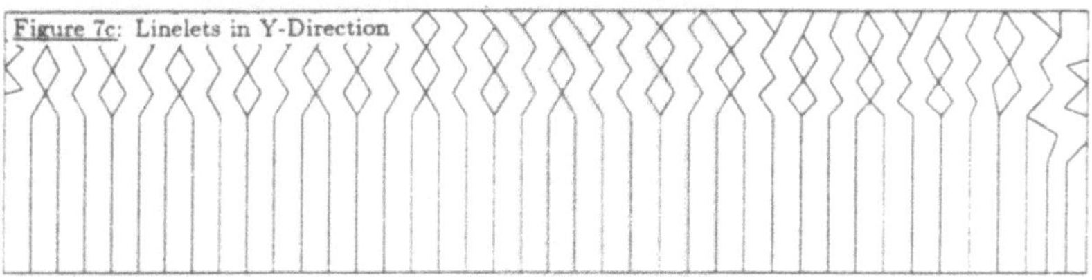

Figure 7c: Linelets in Y-Direction

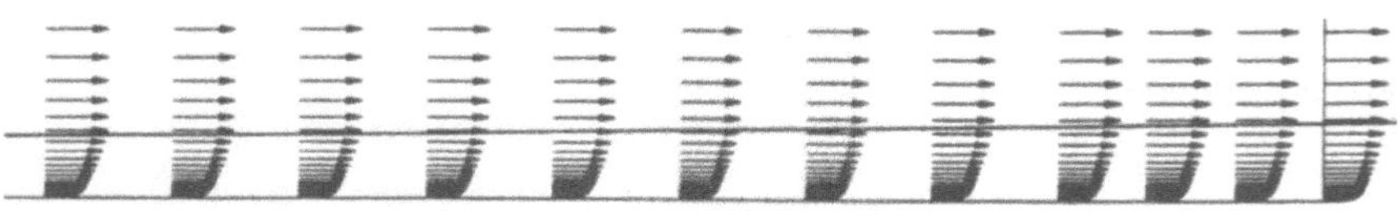

Figure 7d: Velocities, Showing 1/7-th Power Law Boundary Layer Edge

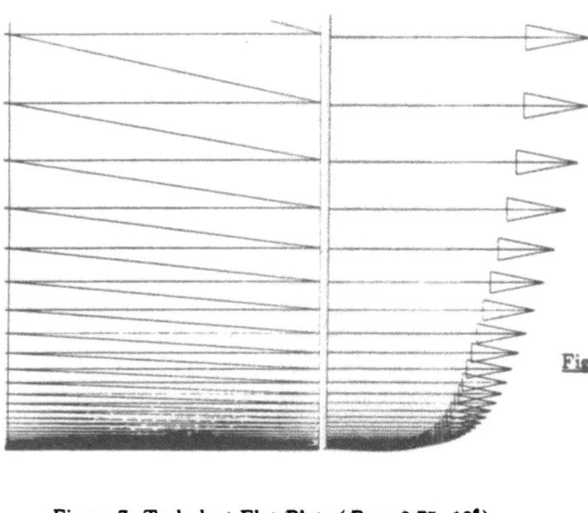

Figure 7e: Velocity Profile at Exit

Figure 7: Turbulent Flat Plate ($Re = 0.75 \cdot 10^6$)

FINITE ELEMENTS IN THE 90's, E. Oñate, J. Periaux, A. Samuelsson (Eds.)
Springer-Verlag/CIMNE, Barcelona 1991

FINITE ELEMENT ANALYSIS OF TSUNAMI WAVE PROPAGATION USING OPTIMAL MESH

K. Kashiyama and **M. Kawahara**
Department of Civil Engineering, Chuo University,
1-13-27 Kasuga, Bunkyo-ku, Tokyo 112,
JAPAN

SUMMARY

The tsunami accompanying the Kanto earthquake in 1923 is simulated by the finite element method in this paper. The automatic mesh generation method for shallow water flow is applied to prepare the mesh data. The key feature of this method is that the finite element mesh can be generated so that the element Courant number is to be nearly constant in the whole domain. It follows that the numerical accuracy and stability can be improved automatically. The selective lumping finite element method is applied for the discretization. The present method is shown to be a useful and powerful tool for the analysis of tsunami wave propagation.

1. INTRODUCTION

A number of numerical method have been presented to analyze the tsunami waves [1-7]. These numerical methods are usefully applied for the simulation of tsunami waves in order to estimate the damage caused by tsunami waves. The tsunami warning system based on the numerical method has been also developed in recent years [7]. However, in case that the configuration of the boundary and the water depth are complicated, the following problems are pointed out. First of all, in order to compute accurately, it is indicated that the appropriate partition of element must be realized in accordance with the variation of water depth. Secondly, it takes a lot of time to prepare the finite element mesh data with water depth, since there is no useful automatic mesh generation method for shallow water flow analysis which needs the data of water depth. In order to overcome these problems, Kashiyama and Okada [8] developed the automatic mesh generation method for shallow water flow based on the three node triangular element. The key feature of this method is that the finite element mesh can be generated so that the element Courant number is to be nearly constant in the whole domain. As this feature produces that the ratio between the element size and the wave length is to be constant, it is possible to generate the finite element mesh in accordance with the variation of wave length. It follows that not only the optimal mesh idealization can be realized automatically, but the numerical stability and accuracy can be improved.

In this paper, the tsunami wave propagation accompanying the Kanto earthquake in 1923 is simulated by the finite element method using the optimal mesh. The linearlized shallow water equation considering the bottom deformation is used for the basic equations and the treatment method of open boundary condition [5,6] is used. The selective lumping finite element method [9,10,11], which is developed by Kawahara et al., is usefully applied for the discretization of space and time. The computed results are compared with the observed results.

2. BASIC EQUATIONS

The tsunami wave propagation can be modeled by the shallow water equation which is obtained by the conservation equation of mass and momentum. The linearlized shallow water equation including the bottom deformation is used for the basic equations as (see Figure 1):

$$\frac{\partial U_i}{\partial t} + g\zeta_{,i} = 0 \qquad (1)$$

$$\frac{\partial(\zeta - b)}{\partial t} + hU_{i,i} = 0 \qquad (2)$$

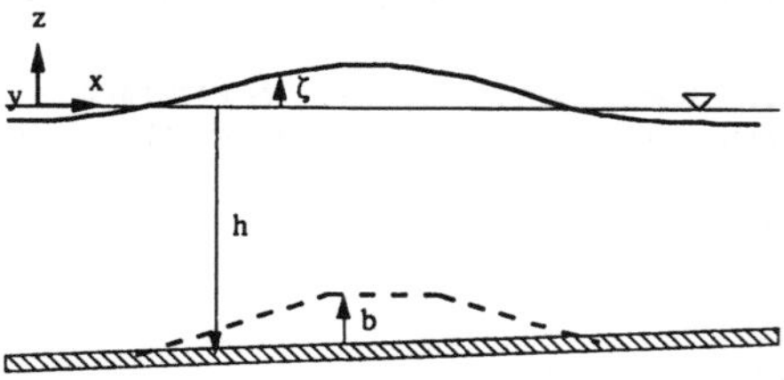

Figure 1 Co-ordinate system

where U_i is the mean current velocity, ζ is the water elevation from still water level, g and b denote the gravitational acceleration and the vertical bottom deformation respectively. The effects of Coriolis force, bottom friction and eddy viscosity are all neglected because these are insignificant for the tsunami propagation.

For the boundary conditions, the following conditions are applied. On the boundary of solid boundary Γ_s, the fully reflection condition is applied as follows (see Figure 2):

$$U_n = 0 \qquad \text{on } \Gamma_s \qquad (3)$$

where U_n denotes the normal velocity to the boundary Γ_s. On the open boundary Γ_o, the following continuity conditions are applied:

$$U_i = \bar{U}_i \qquad \text{on } \Gamma_o \qquad (4)$$

$$\zeta = \bar{\zeta} \qquad \text{on } \Gamma_o \qquad (5)$$

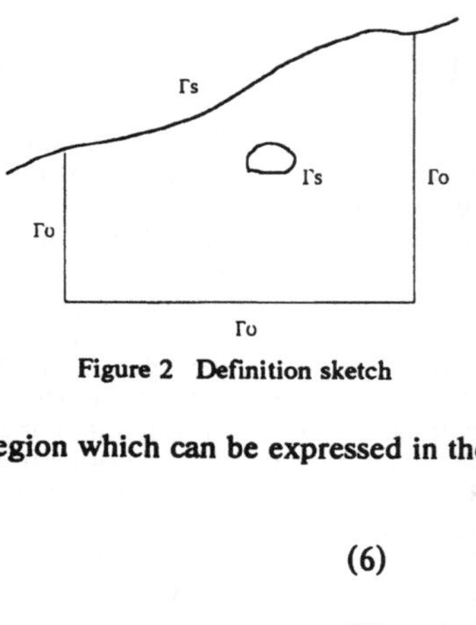

Figure 2 Definition sketch

where the overbar indicates the value of outer region which can be expressed in the following form [5]:

$$\bar{U}_i = U_i' \exp(-j\omega t)\exp(jk_i x_i) \qquad (6)$$

$$\bar{\zeta} = \zeta' \exp(-j\omega t)\exp(jk_i x_i) \qquad (7)$$

in which U_i' and ζ' represent the amplitude of velocity and water elevation, ω, k_i and j denote the angular frequency, the components of wavenumber and the imaginary unit respectively. The details of the treatment method of open boundary condition are given in the reference [5].

3. FINITE ELEMENT METHOD

For the discretization of the spatial unknown variables, the standard Galerkin finite element method has been successfully applied. A linear polynomial function based on a three node triangular element is used for the interpolation of current velocity and water elevation. The Galerkin procedure leads to the following finite element equations [6].

$$M_{\alpha\beta}\dot{U}_{\beta i} - H_{\alpha\beta i}\zeta_{\beta} - S_{\alpha i\mu}W_{\mu\lambda}^{-1}T_{\lambda\gamma}\dot{\omega}_{\gamma} = 0 \tag{8}$$

$$M_{\alpha\beta}(\dot{\zeta}_{\beta} - \dot{b}_{\beta}) - I_{\alpha i\beta}U_{\beta i} + R_{\alpha\mu}W_{\mu\lambda}^{-1}T_{\lambda\gamma}\omega_{\gamma} = 0 \tag{9}$$

Superposing equation (8) and (9) at all nodal points in the whole flow field, the final finite element equations can be derived. In order to solve the discretized finite element equations with appropriate initial and boundary conditions, the selective lumping two step explicit scheme, which has been presented by Kawahara et. al. [9,10,11], is employed in this paper. The details of this method are given in the reference [9].

4. AUTOMATIC MESH GENERATION

In order to compute tsunami wave propagation accurately and rapidly, it is necessary to develop the automatic mesh generation method for shallow water flow which needs the data of water depth. The data of water depth is really important for not only numerical accuracy but also numerical stability. Some automatic mesh generation method for shallow water flow have been developed in recent years [12,13,14]. However, no method has taken account of the numerical accuracy and stability. Kashiyama and Okada [8] developed a new automatic mesh generation method for shallow water flow analysis. The key feature of this method is that the finite element mesh can be generated so that the element Courant number is to be nearly constant in the whole domain as:

$$C=U\frac{\Delta t}{\Delta x}=\text{const.} \tag{10}$$

where C denotes the Courant number, U is the velocity, Δt is the time increment of numerical integration in time, Δx is the element size. The velocity U is assumed to be the sum of the phase velocity U_w and the current velocity U_c. As the relation $U_w >> U_c$ can be held, the current velocity can be neglected. The velocity U_w can be given as:

$$U_w=\sqrt{gh} \tag{11}$$

where g is the gravitational acceleration, h is the water depth. This is the phase velocity of long waves. Introducing equation (11) into (10), the following condition can be obtained as:

$$\frac{L}{\Delta x}=\text{const.} \tag{12}$$

where L denotes the wave length. Therefore, it is possible to generate the finite element mesh in accordance with the variation of wave length by using the condition (10).

The present mesh generation method is a sort of node distribution method [13,15], namely, the nodes on the boundary of the analytical region are first entered and the interior nodes are generated subsequently so that the element Courant number is to be nearly constant in the whole domain. The details of this method are given in the reference [8].

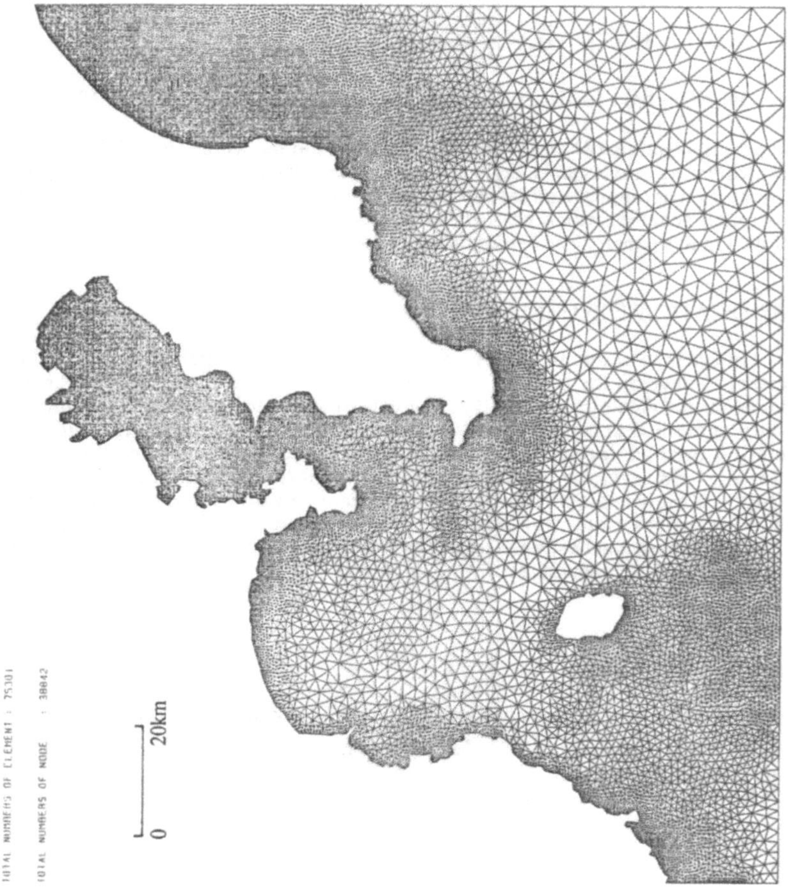

Figure 3 finite element idealization

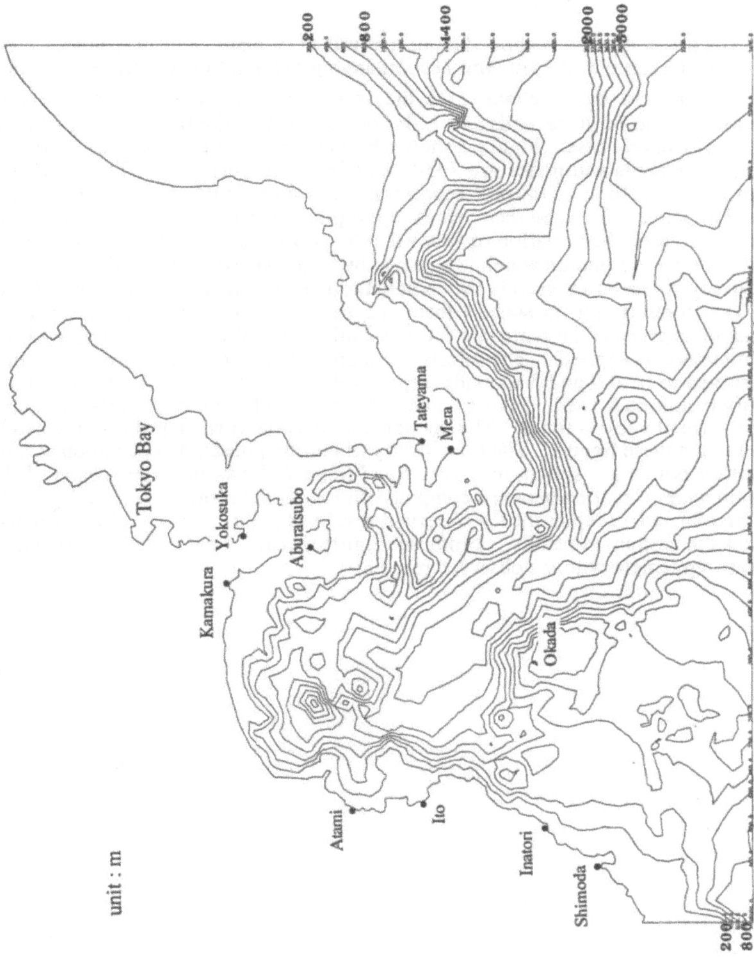

Figure 4 Water depth diagram

5. TSUNAMI WAVE ANALYSIS

The preset method is applied to the analysis of tsunami wave propagation. For the numerical example, the tsunami accompanying the Kanto earthquake in 1923 is chosen. In order to make the mesh data, the time increment Δt and the node interval δ [8] are assumed to be 20 seconds and 150m respectively. The total number of input nodes is 1641 in this case. Figure 3 shows the finite element idealization. The total number of elements and nodes are 75301 and 38842 respectively. The minimum element size is 150m. Figure 4 represents the water depth diagram and some observation points. It can be seen that the appropriate partition of elements is realized in accordance with the variation of water depth. Figure 5 shows the distribution of vertical bottom deformation which is obtained by the surveyed results. The value 1/24 of the surveyed results are used in the computation in according with the study of Aida [4]. The computation is started from the state of still water and the bottom deformation is given at the same rate for first 2 minutes. The selective lumping finite element method is applied. The selective lumping parameter and the time increment are assumed to be 0.9 and 5 seconds respectively. The treatment method of open boundary condition is used. Figure 6 and 7 show the bird's-eye view of the computed tsunami wave at various times. The phenomena of tsunami wave propagation can be clearly seen in these figures. The time history of the computed water elevation at some observation points are shown in Figure 8. Figure 9 shows the comparison of the maximum water elevation between computed and observed results. In this figure, the white circular marks represent the observed results [4] and the black circular marks represent the computed results. From this figure, it can be seen that the computed results are agreement with the observed results.

figure 5 Distribution of vertical bottom deformation

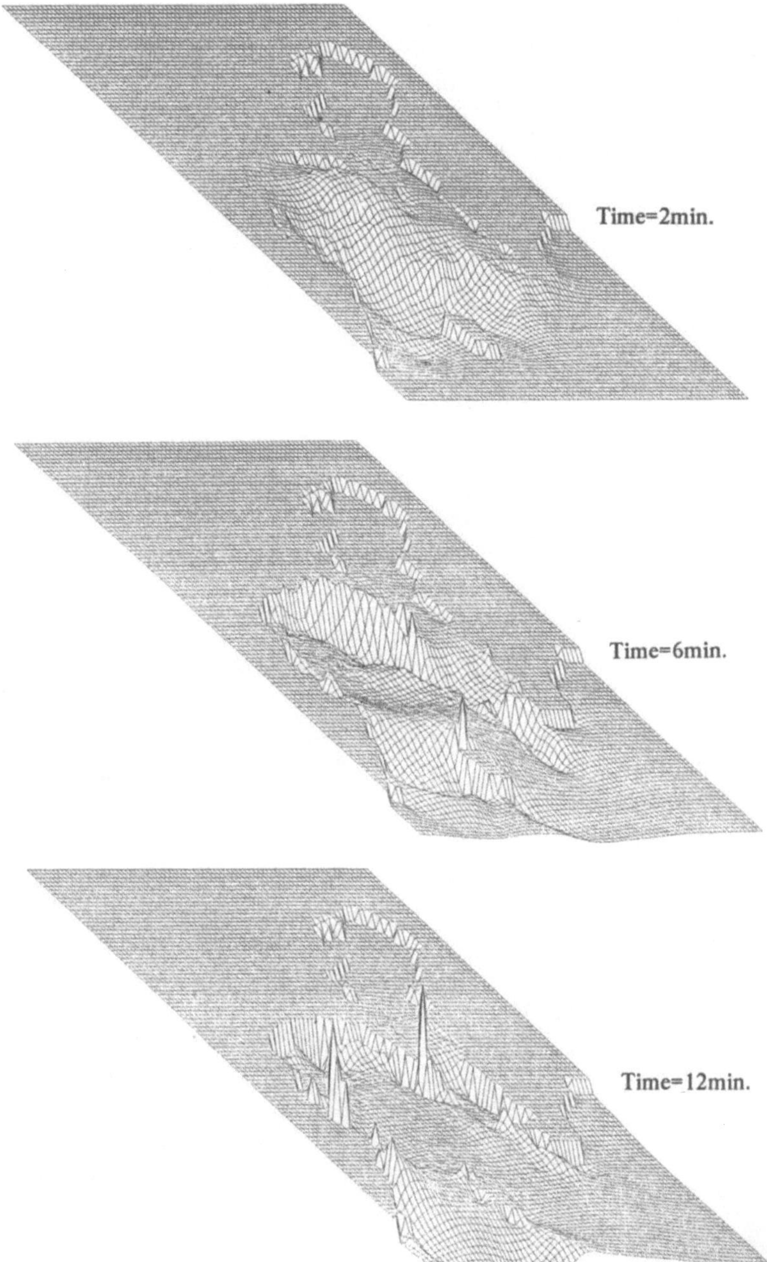

Figure 6 Computed tsunami waves (1)

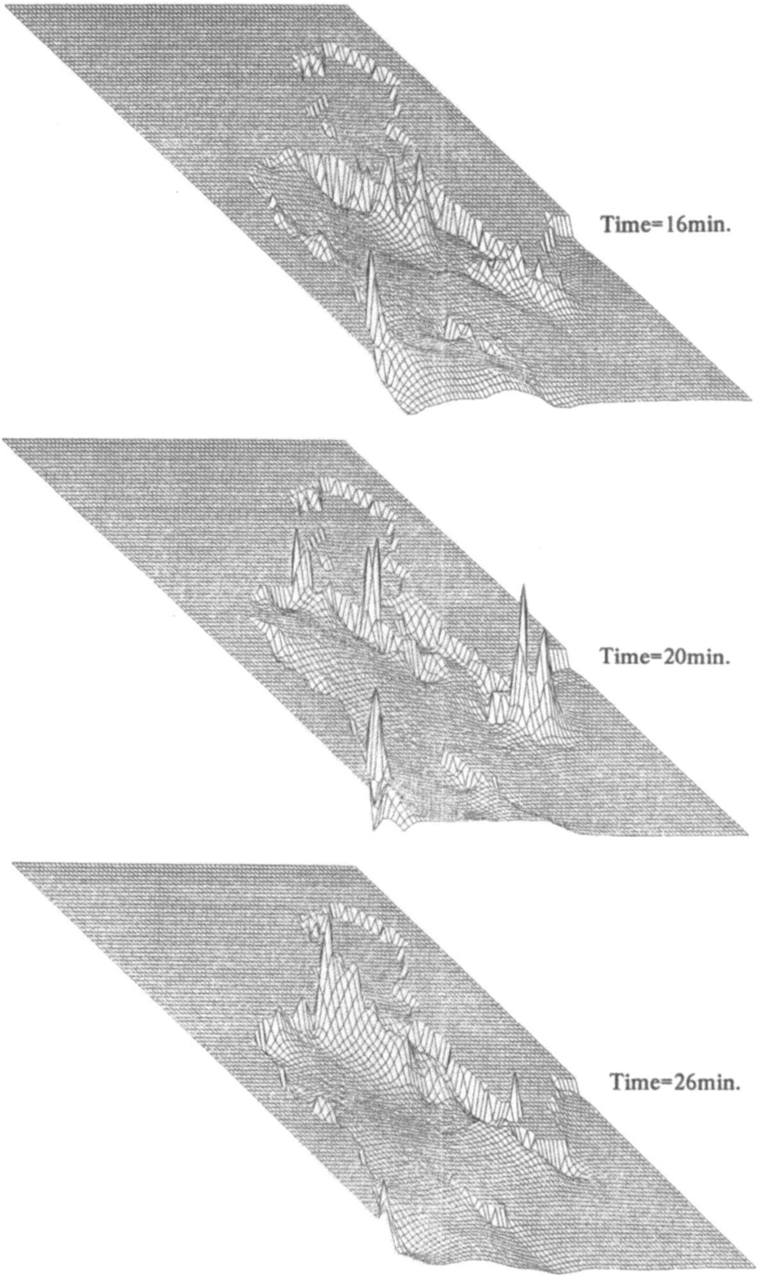

Figure 7 Computed tsunami waves (2)

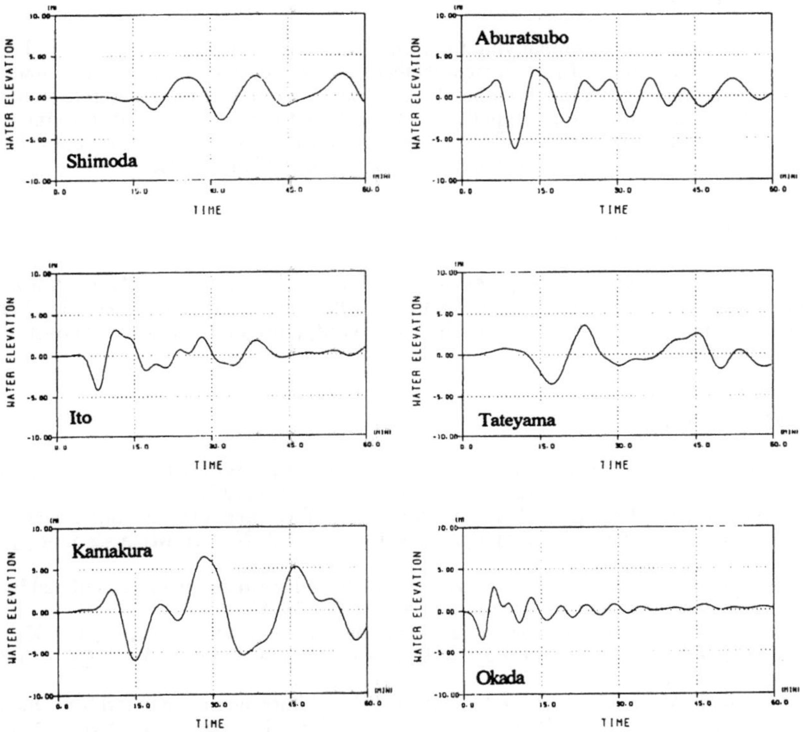

Figure 8 Time history of computed water elevation

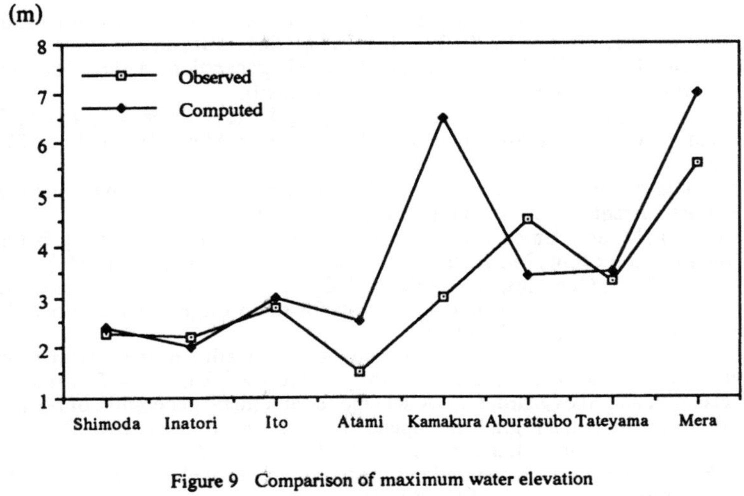

Figure 9 Comparison of maximum water elevation

6. CONCLUSION

The tsunami accompanying the Kanto earthquake in 1923 has been simulated by the selective lumping finite element method using the optimal mesh. The computed results have been compared with the observed data. From these comparative studies, it is concluded that the present method is a useful and powerful tool for studying the tsunami wave propagation. The computational time until 60 minutes (720 steps) later was about 40 minutes using the FACOM VP-30 of Chuo University.

ACKNOWLEDGEMENT

The authors wish to express their gratitude to Mr. Toshio Kodama, research associate of Chuo university, for his earnest discussions, and to Mr. Takeshi Imai and Mr. Hiroaki Yamada, student of Chuo university, for their help in performing computer runs.

REFERENCES

1. Kawahara, M., - 'Tsunami and storm surge', **Finite Element Handbook**, Ed. Kardestuncer, H., 3.89-3.92, McGraw-Hill, London, (1987).
2. Kawahara, M., Takeuchi, N. and Yoshida, T., - Two step explicit finite element method for tsunami wave propagation analysis, Int. J. Numer. Methods Eng., 12, 331-351 (1978).
3. Houston, J.R., - Interaction of tsunamis with the Hawaiian islands calculated by a finite element model, J. Phys. Occan., 8, 93-102 (1978).
4. Aida, I., - A numerical experiment for the Tsunami accompanying the Kanto earthquake of 1923, Bull. Earthq. Res. Inst., University of Tokyo, 48, 73-86 (1970). (in Japanese).
5. Kawahara, M., Kodama, T. and Kinoshita, M., - Finite element method for tsunami wave propagation analysis considering the open boundary condition, Comp. Math. Appl., 16, 139-152 (1988).
6. Kodama, T., Imai, T. and Kawahara, M., - A computational technique dealing with the open boundary condition and its application to tsunami wave propagation analysis, Proc. of 2nd Int. Conf. on Comp. Modelling of Ocean Eng., Barcelona (1991) (in press)
7. Shuto, N., Goto, C. and Imamura, I., - Use of numerical simulation as a means of tsunami waring, Proc. of JSCE, 393, 181-189 (1988) (in Japanese)
8. Kashiyama, K. and Okada, T., - Automatic mesh generation method for shallow water flow', Int. J. Num. Methods Fluids (submitted).
9. Kawahara, M., Hirano, H., Tsubota, K. and Inagaki, K., - Selective lumping finite element method for shallow water flow, Int. J. Numer. Methods. Fluids, 2, 89-112 (1982).
10. Kawahara, M. and Kashiyama, K., - Selective lumping finite element method for nearshore current, Int. J. Numer. Methods. Fluids, 4, 71-97 (1984).
11. Kawahara, M. and Umetsu, T., - Finite element method for moving boundary problems in river flow, Int. J. Numer. Methods Fluids, 6, 365-386 (1986)
12. Thacker, W.C., Gonzales, A. and Putland, G.E., - A method for automating the construction of irregular computational grids for storm surges forecast models', J. Comp. Physics, 37, 371-387 (1980).
13. Westwood, I.J. and Holtz, K.-P, - Automatic optimization of irregular meshes for natural flow computation, Proc. Int. Conf. on Hydrosoft 86, 423-425 (1986).
14. Nielsen, P. and Skovgaard, O., A scheme for automatic generation of boundary - fitted depth- and depth-gradient-dependent grids in arbitrary two dimensional regions, Int. J. Numer. Methods Fluids, 10, 741-752 (1990).
15. Lo, S.H., - A new mesh generation scheme for arbitrary planner domains, Int. J. Numer. Methods Eng., 21, 1403-1426 (1985).

FINITE ELEMENTS IN THE 90'S, E. Oñate, J. Periaux, A. Samuelsson (Eds.)
Springer-Verlag/CIMNE, Barcelona 1991

FINITE ELEMENT MODELLING OF MOULD FILLING

R.W. Lewis, A.S. Usmani and J.T. Cross
Institute for Numerical methods in Engineering,
University College of Swansea,
UNITED KINGDOM

Abstract

The flow associated with the pouring of molten metal into moulds has been modelled on the basis of certain simplifying assumptions. A finite element based Navier-Stokes equation solver has been used to analyse the flow. The velocities obtained from the Navier-Stokes equation solver are used to advect a pseudo-concentration function for modelling the fluid front. A simple problem with a known analytical solution is solved first to test the model. Subsequently, several more examples of varied geometrical configuration are modelled to demonstrate the applicability of the proposed model.

1 Introduction

Mould filling forms an integral part of the casting process and this provides in important motivation for developing an appropriate mathematical model. The most obvious advantage of preceding a thermal analysis of a casting with a mould filling analysis is that a realistic (non-uniform) initial temperature will be available. Furthermore, such an analysis can be used to obtain information about the flow patterns during the filling process and potential trouble spots with the height, dimensions and position of risers and the gating system may be identified. Until a suitable configuration is achieved, all this can be done on a computer screen, without undertaking any experimental work.

The method used for the modelling of mould filling in this work is based upon the volume of fluid method (VOF) [1], which is also called the pseudo-concentration method [2] or saturation function method [3, 4]. This method in turn is based upon the well known marker and cell (MAC) technique which extracts a high price in terms of computer storage and CPU time, especially for 3-D applications. The VOF method whilst retaining the conceptual simplicty of the MAC technique demands considerably less computing resources as the marker particles are replaced by a nodal variable. The pseudo-concentration or VOF method involves the use of a smooth function which is advected using the velocities from the solution of the Navier-Stokes equations. A particular value of this function can be chosen to represent the free fluid surface, which can then be tracked for each timestep. Both FEM and FD/FV methods have been used in modelling this problem. Thompson [2] has used pseudo-concentrations with the finite element method to follow creeping viscous and viscoplastic flows in the context of several practical applications including injection moulding, folding of rock strata in the earths crust etc. Dhatt *et. al.* [5] have tackled the mould filling problem in metal castings using the finite element method. Hwang and Stoehr [6] have used the FD based MAC technique to model the flow of metal in castings. Minaie *et. al.* [7] have used the VOF (pseudo-concentration) technique to model the flow in die castings with a finite difference discretisation of the governing equations.

The flow of molten metal into a casting is invariably turbulent, yet there is justification for modelling it as laminar. The time taken for a mould to fill is generally very short and the need for time averaging in turbulence modelling means that the flow must have a finite time to develop before the simulation of it is worthwhile[7]. It must also be borne in mind that the modelling of mould filling is, at the moment, not an end in itself but a means to a more accurate thermal analysis of a casting, and as such the formidable expense of modelling turbulent flow is difficult to warrant at this stage.

2 Governing differential equations and boundary conditions

The Navier-Stokes equations for incompressible flow in two-dimensions are used to model the flow. The equations are stated as,

$$\rho \frac{\mathbf{Dv}}{\mathbf{D}t} = \boldsymbol{\nabla} \cdot \boldsymbol{\tau} + \rho \mathbf{g} \tag{1}$$

Here, $\boldsymbol{\tau}$ is the sress tensor, ρ and $\mathbf{g}$ are density and gravitational acceleration, and $\frac{\mathbf{D}}{\mathbf{D}t}$ represents the total or substantial derivative. The mass conservation (continuity equation) is used to enforce incompressibility,

$$\boldsymbol{\nabla} \cdot \mathbf{v} = 0 \tag{2}$$

where, $\mathbf{v}$ represents velocity. These equations have been used extensively for modelling incompressible flow and further details can be found in abundance in the relevent literature, such as [8, 9].

Dirichlet boundary conditions for the Navier Stokes equations consist of specified velocities at the boundaries. Neumann boundary conditions can be normal and tangential traction forces, specified as below,

$$f_n = -P + 2\mu \frac{\partial u_n}{\partial n} \tag{3}$$

$$f_\tau = \mu \left(\frac{\partial u_n}{\partial \tau} + \frac{\partial u_\tau}{\partial n} \right) \tag{4}$$

n and τ are the unit normal and tangent vectors respectively, P is the pressure, μ is the dynamic viscosity.

3 Finite element formulation

The conventional Galerkin weighted residual technique with mixed velocity-pressure formulation has been used here to obtain the spatial discretisation. Detailed theoretical treatment of the procedure can be found in standard texts, such as [10, 11].

The field variables involved in the governing equations, *i.e.* velocities ($\mathbf{v}$) and pressure (P) are approximated over each element in terms of the nodal values by appropriate shape functions. The six noded triangular element has been used with quadratic velocity interpolation and linear pressure interpolation. This element is not optimal in terms of the enforcement of the incompressibility condition, or, in other words, is relatively under-constrained, which leads to poor velocity solution on coarse grids and for difficult problems [12]. Special elements with better constraint properties are discussed in [12]. Here this element has been chosen for the reason of optimal convergence rate and a proven record of performance in problems of incompressible flow.

Approximating equations (1) and (2) and the boundary conditions (3) and (4) using the appropriate shape functions, the final set of spatially discretized equations in a fully coupled form can be written as follows,

$$\mathbf{M}\dot{\boldsymbol{\theta}} + \mathbf{K}\boldsymbol{\theta} = \mathbf{F} \tag{5}$$

where, $\boldsymbol{\theta}$ represents all the variables. Equation (5) may be written in an expanded matrix form as

$$\begin{bmatrix} \rho\mathbf{M}_u & 0 & 0 \\ 0 & 0 & 0 \\ 0 & 0 & \rho\mathbf{M}_v \end{bmatrix} \begin{pmatrix} \dot{\mathbf{u}} \\ \dot{\mathbf{P}} \\ \dot{\mathbf{v}} \end{pmatrix} + \begin{bmatrix} \mathbf{K}_{uu} & \mathbf{C}_u & \mathbf{K}_{uv} \\ \mathbf{C}_u^T & 0 & \mathbf{C}_v^T \\ \mathbf{K}_{vu} & \mathbf{C}_v & \mathbf{K}_{vv} \end{bmatrix} \begin{pmatrix} \mathbf{u} \\ \mathbf{P} \\ \mathbf{v} \end{pmatrix} = \begin{pmatrix} \mathbf{F}_u \\ 0 \\ \mathbf{F}_v \end{pmatrix}$$

The exact expansions of the components of the above matrices may be found from literature such as [12, 10, 11].

The spatial discretization as above, creates a set of first order ordinary differential equations with respect to time. The first order system of equations represented by (5), are discretized in time using the implicit backward difference method. which is written as,

$$\left[\frac{\mathbf{M}_{n+1}}{\Delta t} + \mathbf{K}_{n+1}\right](\boldsymbol{\theta}_{n+1}) = \left[\frac{\mathbf{M}_{n+1}}{\Delta t}\right](\boldsymbol{\theta}_n) + (\mathbf{F}_{n+1}) \tag{6}$$

Different material properties are used for the elements in the fluid filled region and the empty region. Fictitious values of density and viscosity are used for the fluid to achieve a low *Re* to keep the flow in the laminar regime. The density and viscosity close to that of air are used for the elements in the empty region. For the elements containing the fluid front the properties are averaged according to the full and empty fractions of the element by interpolation at the integration points.

4 Fluid front tracking

As mentioned earler, a pseudo-concentration function is used to track the free fluid surface. We represent this function by $F(x,y,t)$ for 2-D flow, and write the first order pure advection equation which conserves the function $F(x,y,t)$ as,

$$\frac{DF}{Dt} = 0 \tag{7}$$

The equation indicates that the function F moves with the flow, therefore, if a particular value of this function, say F_c, is associated with the free fluid surface [2, 5], the free surface can be tracked over the timesteps by simply plotting the contour of F_c at each timestep. In most applications, F is assumed to be a step function since, in the real physical problem, F exhibits step-like behaviour separating the fluid region from the empty region. F is normally assigned a value of unity in the fluid region and zero in the empty region. Here, it has been preferred to use a continous function to avoid the numerical difficulties in advecting a step function. The value of $F(x,y,t) = 0.0 = F_c$ is used to mark the free surface, while $F(x,y,t) > 0.0$ indicates the fluid region and $F(x,y,t) < 0.0$ indicates the empty region. As this is a hyperbolic or pure advection equation, the boundary values of F are required only at the nodes where the fluid enters the cavity.

4.1 FEM implementation of the pseudo-concentration equation

Equation (7) may be discretised using the Galerkin finite element method. The continuous function F is approximated spatially by nodal values F_i using the same shape functions N_i over each element e as for the velocities. Applying the Galerkin form of the weighted residual procedure we obtain the discretised version of equation (7) in the form,

$$\mathbf{M}_F\dot{F} + \mathbf{K}_F F = 0 \tag{8}$$

where,

$$\begin{aligned}\mathbf{M}_F &= \int_\Omega N_i N_j \, d\Omega \\ \mathbf{K}_F &= \int_\Omega \left(N_i N_k u_k \frac{\partial N_j}{\partial x} + N_i N_k v_k \frac{\partial N_j}{\partial y} \right) d\Omega\end{aligned} \tag{9}$$

Equation (8) is discretised in time using the backward Euler method. After solving the Navier-Stokes equation according to the procedure mentioned earlier, equation (8) is solved at each timestep using the velocities obtained from the Navier-Stokes solution, which results in an updated pseudo-concentration field. The position of the free front of the fluid is coincident with the contour of F_c or $F(x,y,t) = 0.0$. The material properties of the elements are modified according to the new position of the fluid front.

4.2 Smoothing of the pseudo-concentration function

If the function F is maintained unmodified over several timesteps, it may begin to lose its smoothness and distorted regions of high gradient may develop, causing numerical difficulties. As it is necessary to preserve the values of F only at those positions which identify the free surface, it is possible to modify the pseudo-concentration field on either side of the F_c contour. This modification renders the F-field smooth once again, with first derivatives (slopes) of small numerical values, thus promoting well conditioned matrices. The smoothing method used here, proposed by Thompson [2], involves increasing the value of F on one side of F_c in direct proportion to the distance from it, similarly decreasing F on the other side of F_c. This can be written as,

$$F(x,y) = F_c + \xi A\|d\| \tag{10}$$

Where ξ is a gradient control parameter, $A = +1$ for fluid filled regions and $A = -1$ for empty region, while $\|d\|$ is the absolute minimum distance from the free surface.

4.3 Flow boundary conditions on the walls of the mould cavity

As mentioned earlier, the Von Karman's laws of the wall are used to obtain the flow boundary conditions at the walls of the cavity. These laws are normally applied to turbulent flow problems, here, as in [5], they are used to obtain tangential tractions at the walls of the mould cavity. The velocity normal to the walls is specified as zero. The tangential tractions are calculated from

$$f_\tau = \rho \frac{| u_w | u_w}{(u^+)^2} \tag{11}$$

where u_w is the tangential velocity to the wall, and

$$u^+ = 5.5 + 2.5 \ln y^+$$

An appropriate value of y^+ must be assumed. Gao *et. al.* [5] use a value of y^+ equal 100.0, values between 30 to 100 have been rather arbitrarily used here. The tractions obtained from the above formula are so low that the value of y^+ used does not affect the final analysis siginificantly, however, this aspect deserves greater attention. The inflow boundary conditions are assumed to be traction free ($f_\tau = 0.0$ and $f_n = 0.0$), if the flow is driven by gravity. A specified normal traction or a specified normal velocity at the inlet may also be used to drive the flow. The outflow boundary conditions are always specified as traction free.

5 Numerical examples

Before trying to model the filling of complicated mould shapes it would seem appropriate to model a benchmark type problem for which an analytical solution is available. The first of the three examples is just such a problem and has been used by Gao *et. al.* [5], who call it 'filling of a horizontal cavity'. It deals with tracking a segment of the profile of poisseuille flow between two parallel plates and is a good benchmark test for the front tracking algorithm, as the the movement of the fluid front can be determined analytically in both space and time. Figure 1 shows the problem definition for this example.

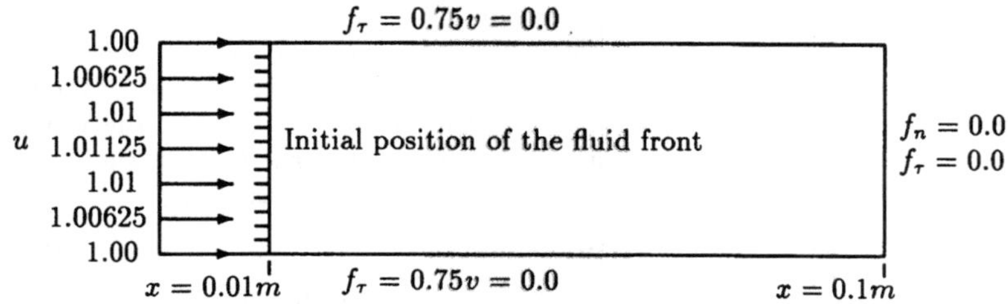

Figure 1: Problem definition for front tracking benchmark test.

The geometry, boundary conditions and material property data corresponds to that of reference [5]. The length of the domain is $0.10m$ and the height is $0.03m$.

The flow is driven by a constant pressure gradient ($\frac{dP}{dx}$) of 50.0 in the horizontal direction. The shear stress at the top and bottom can be found by the relation,

$$f_\tau = y\frac{dP}{dx}$$

where, y is measured from the centre. The velocity profile distribution is given by,

$$u(y) = \frac{1}{2\mu}\frac{dP}{dx}\left(y^2 - y_{max}^2\right)$$

The velocities obtained using the above relation are specified as the inlet boundary condition as shown in Figure 1. The vertical velocity v is specified as zero at the inlet nodes. The value of F is specified as 1.0 at the inlet boundary. The material properties for the fluid filled region and the empty region are as follows.

$$\rho_{full} = 1000.0\frac{kg}{m^3} \qquad \mu_{full} = 0.5\frac{kg}{m.s}$$
$$\rho_{empt} = 1.0\frac{kg}{m^3} \qquad \mu_{empt} = 0.005\frac{kg}{m.s}$$

The results are shown in Figure 2 in the form of the front position for a few timesteps. The position of the front is predicted to an accuracy of within one percent.

In the second, more realistic example, the value of F is specified as 1.0 at the inlet as before. The flow boundary conditions consist of specifying

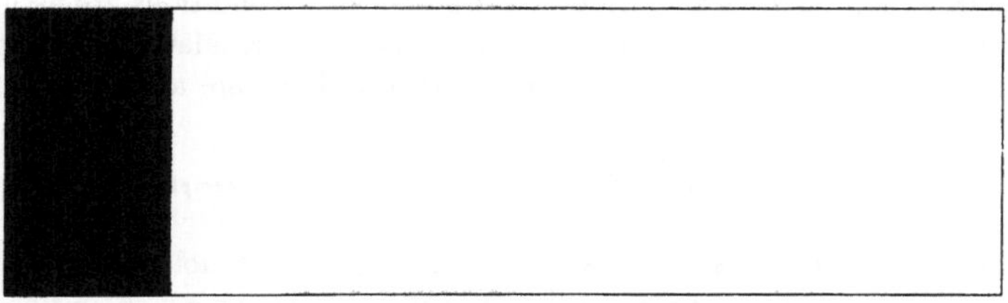

Time = 0.01

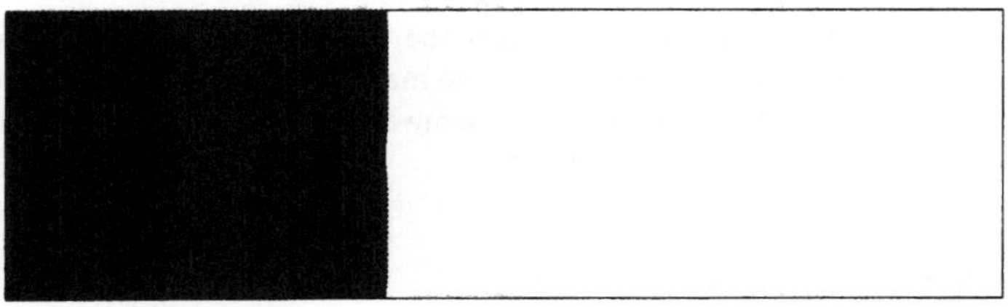

Time = 0.03

Time = 0.06

Time = 0.0844

Figure 2: Fluid front profiles for the benchmark problem.

all velocities normal to the walls of the system to be zero. For the the tangential direction to the walls, frictional slip boundary conditions are assumed. The traction force resulting from slipping friction is calculated according to equation (11). The material properties for this problem are as follows.

$$\rho_{full} = 150.0\frac{kg}{m^3} \quad \mu_{full} = 0.2\frac{kg}{m.s} \quad \rho_{empt} = 0.1\frac{kg}{m^3} \quad \mu_{empt} = 0.02\frac{kg}{m.s}$$

The flow properties are selected as usual, to keep the Reynold's number (Re) well in the laminar range for smooth flow and faster convergence, in line with the arguments presented earlier. Figure 3 shows the results of the filling analysis in the form of front positions at selected timesteps.

In the last of the three examples the melt encounters a core which forces the front to break into two parts. The material properties and boundary conditions for this problem are again the same as for the previous problem and the filling results are shown in Figure 4.

6 Conclusions

From the examples solved here, it can be concluded that a finite element flow program based on the procedure mentioned earlier, when used with a front tracking algorithm such as the VOF method or the pseudo-concentration method based on the pure advection equation, provides an effective means of solving problems such as that of mould filling. This method can be used in other applications as well, such as forming processes involving creeping viscous flows, multiphase flow etc. The main advantage of this method is that the moving front does not have to be explicitly determined, neither do the boundary conditions on the front have to be explicitly satisfied.

References

[1] C.W.Hirt and B.D.Nichols. Volume of fluid (vof) method for the dynamics of free boundaries. *Journal of Computational Physics*, 39:201–225, 1981.

[2] E.Thompson. Use of pseudo-concentrations to follow creeping viscous flows during transient analysis. *International Journal for Numerical Methods in Fluids*, 6:749–761, 1986.

[3] T.J.Smith and D.B.Welbourn. The integration of geometric modelling with finite element analysis for the computer-aided design of castings. *Applied Scientific Research*, 44:139–160, 1987.

[4] R.W.Lewis, K.Morgan, E.D.L.Pugh, and T.J.Smith. A note on discontinuous numerical solutions of the kinematic wave equation. *International Journal for Numerical Methods in Engineering*, 20:555–563, 1984.

[5] D.M.Gao, G.Dhatt, J.Belanger, and A.B.Cheikh. A finite element simulation of metal flow in moulds. In *Sixth International Conference for Numerical Methods in Thermal Problems*, Swansea, U.K., July 1989. Pineridge Press, Swansea.

[6] W.S.Hwang and R.A.Stoehr. Molten metal flow prediction for complete solidification analysis of near net shape castings. *Materials Science and Technology*, 4:240–250, 1988.

[7] B.Minaie, K.A.Stelson, and V.R.Voller. Fluid flow and solidification modelling of die castings. *ASME, Modelling of Materials Processing, MD*, 3:35–50, 1987.

[8] H.Schlichting. *Boundary-Layer Theory.* McGraw-Hill Book Company, 1968.

[9] D.K.Edwards, V.E.Denny, and A.F.Mills. *Transfer Processes - An Introduction to Diffusion, Convection and Radiation.* McGraw-Hill Book Company, 1973.

[10] O.C.Zienkiewicz. *The Finite Element Method.* McGraw-Hill Book Company (UK) Limited, London, 1977.

[11] T.J.R.Hughes. *The Finite Element Method - Linear Static and Dynamic Finite Element Analysis.* Prentice-Hall International, Inc., Englewood Cliffs, New Jersey 07632, 1987.

[12] P.M.Gresho, R.L.Lee, and R.L.Sani. On the time-dependent solution of the incompressible Navier-Stokes equations in two and three dimensions. In *Recent Advances in Numerical Methods in Fluids*, volume 1. Pineridge Press Limited, Swansea, 1980.

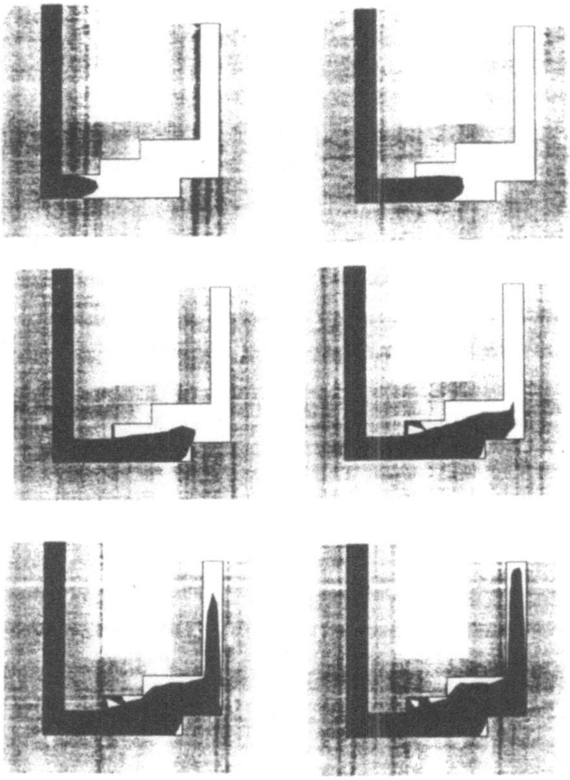

Figure 3: Front positions at selected timesteps for the pseudo-realistic problem.

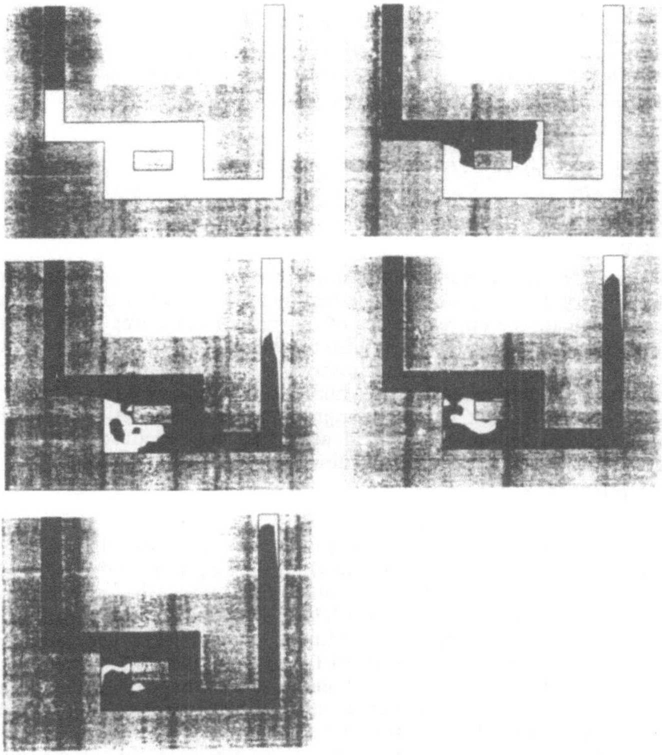

Figure 4: Front positions at selected timesteps for the multiple front problem.

FINITE ELEMENTS IN THE 90's, E. Oñate, J. Periaux, A. Samuelsson (Eds.)
Springer-Verlag/CIMNE, Barcelona 1991

THE COMPUTATION OF AERODYNAMIC FLOWS USING UNSTRUCTURED MESHES

J. Peraire and J. Peiró
Department of Aeronautics,
Imperial College, Swansea SW7 2BY,
UNITED KINGDOM

K. Morgan
Department of Civil Engineering,
University College, Swansea SA2 8PP,
UNITED KINGDOM

O. Hassan
CDR, Innovation Centre, University College,
Swansea SA2 8PP,
UNITED KINGDOM

SUMMARY

Major progress has been made in recent years on the development of unstructured mesh methods for the solution of compressible aerodynamic flows of practical interest. In this paper we indicate the current status of our work in this area by outlining a low storage, computationally efficient algorithm for the solution of the compressible Euler equations on unstructured tetrahedral meshes. The performance of the algorithm is demonstrated by including the results of a numerical analysis of a flow past an installed aero-engine nacelle. Recent work on improving the computational efficiency by incorporating an unstructured multigrid acceleration procedure is described and demonstrated for a complete twin engined aircraft configuration.

1. INTRODUCTION

We felt it appropriate on this occasion to describe the current status of our work on unstructured mesh methods for the solution of compressible inviscid flows of industrial interest. We started on this work at Swansea in the early eighties in collaboration with Professor O.C. Zienkiewicz. Our interest in compressible flows arose from research sponsorship made available by the Aerothermal Loads Branch of NASA Langley Research Center, while our finite element background naturally meant that unstructured meshes were considered from the outset. Our work has, almost exclusively, used meshes of triangles or tetrahedra and we have developed a variety of flow algorithms [1-4] suitable for implementation on such meshes. Major progress has been made in the area of mesh generation and adaptivity and we have pioneered the use of a fully automatic approach which is based upon an advancing front method [5,6].

This work has led to the development of the computer systems LISA and FLITE3D for the simulation of inviscid flows using unstructured meshes. With these systems it is possible to quickly produce solutions for flows involving geometries which can be as complex as a complete aircraft. The approach therefore becomes very attractive to the analyst involved in the design phase of aerodynamic components. In this paper, the construction of a modern unstructured mesh flow algorithm is described. The Euler flow solver which is developed is a Galerkin finite element scheme, with the explicit addition of artificial viscosity. The steady state flow conditions of interest are computed by advancing the transient form of the equations using a Runge-Kutta type time stepping process.The lack of structure in the mesh means that the solver will employ indirect addressing, thus making it subject to an overhead in terms of both computer time and memory when compared with similar solvers implemented on structured meshes. A side based data structure is employed for the three dimensional mesh, which has the effect of reducing the memory demands over those made by the more traditional element based data structure of the type familiar to finite element practitioners. This data structure appears to have certain features in common with that which has been proposed recently by other workers in this area [7,8]. The performance of the algorithm is demonstrated by including the results of a simulation [9] of a flow past an installed aero-engine nacelle for a civil transport aircraft. This problem has been traditionally handled within the aerospace industry by performing high speed wind tunnel model tests to optimise the power plant/airframe configuration. However, numerical techniques are now being increasingly employed to enhance the value of such tests, by reducing the number of configuration builds and by aiding the interpretation of the experimental measurements.

Although the side based data structure is also found to reduce the computer time requirements of the flow solver, the total CPU time necessary to produce a converged solution to a problem involving steady inviscid flow past an aerodynamic component can still be considerable. In the structured grid context, equivalent solution algorithms have been effectively accelerated by the use of multigrid methods [10], which employ a sequence of successively coarser nested meshes. When the grid which is to be employed for the analysis is unstructured, the multigrid concept has to be extended [11,12,8,13] as the grids which are employed are now not nested i.e. the nodes on one grid do not coincide with nodes on the next coarsest or the next finest grid. An unstructured multigrid procedure is added to the flow solver and the problem of an inviscid flow past a complete twin engined aircraft configuration is simulated. It is shown that the computational performance of the flow solver is dramatically improved, thus significantly increasing the attractiveness of the method for the aerodynamic analyst who is faced with the problem of rapidly assessing the effects of design changes.

2. THE GOVERNING EQUATIONS

The equations governing 3D compressible unsteady inviscid flow are considered in the conservation form

$$\frac{\partial \mathbf{U}}{\partial t} + \sum_{j=1}^{3} \frac{\partial \mathbf{F}^j}{\partial x_j} = 0 \qquad (1)$$

where $\mathbf{U}$, the unknown vector of conserved variables, and $\mathbf{F}^j$, the corresponding inviscid flux vectors, are defined by

$$\mathbf{U} = \begin{bmatrix} \rho \\ \rho u_1 \\ \rho u_2 \\ \rho u_3 \\ \rho \varepsilon \end{bmatrix} \qquad \mathbf{F}^j(\mathbf{U}) = \begin{bmatrix} \rho u_j \\ \rho u_1 u_j + p\delta_{1j} \\ \rho u_2 u_j + p\delta_{2j} \\ \rho u_3 u_j + p\delta_{3j} \\ (\rho\varepsilon + p) u_j \end{bmatrix} \qquad (2)$$

Here δ_{ij} is the Kronecker delta, ρ, p and ε denote the density, pressure and total specific energy of the fluid respectively and u_j; $j = 1, 2, 3$ are the components of the fluid velocity in the directions x_j; $j = 1, 2, 3$ of a cartesian coordinate system. The set of equations is completed by the addition of the perfect gas equation of state

$$p = (\gamma - 1)\, \rho \,(\varepsilon - 0.5 \sum_{j=1}^{3} u_j u_j) \qquad (3)$$

where γ (= 1.4) is the ratio of the specific heats.

Steady state solutions of these equations can be obtained by advancing the solution in time until stationary conditions are obtained. The numerical scheme which will be employed is developed by following a procedure which involves initially a Galerkin finite element discretisation in space followed by an independent discretisation of the time dimension.

3. AN UNSTRUCTURED MESH FLOW ALGORITHM

3.1 Mesh Generation

The spatial domain, Ω, of interest is discretised into an assembly of four noded linear tetrahedral elements. Here, this discretisation is accomplished by means of an advancing front based mesh generation system [5,6]. The first stage in this process consists in using a two dimensional advancing front method to produce a triangulation of the boundary, Γ, of the computational domain and the resulting set of triangular faces forms the initial front for the three dimensional generator. The interior of the domain is then filled with tetrahedral elements by using the three dimensional advancing front method. An essential feature of the adopted approach is the manner in which the size and shape of the generated elements is controlled by user specified distribution functions. The generation process is fully automatic and computational domains of general geometric shape can be rapidly meshed [14,15].

3.2 Variational Formulation

The identification of an appropriate variational formulation of the governing equation is the starting point in the construction of a finite element approximation within the spatial domain, Ω, for all $t > 0$. A suitable weak formulation for the problem defined in equation (1) would be : find $\mathbf{U}(\mathbf{x},t)$ such that

$$\int_\Omega \frac{\partial \mathbf{U}}{\partial t} W \, d\Omega = \sum_{j=1}^{3} \int_\Omega \mathbf{F}^j \frac{\partial W}{\partial x_j} \, d\Omega - \sum_{j=1}^{3} \int_\Gamma \mathbf{F}^j n^j W \, d\Gamma \tag{4}$$

for every suitable weighting function $W(\mathbf{x})$ and for all $t > 0$. In this equation, n^j denotes the j-th component of the unit outward normal vector, $\mathbf{n}$, to the boundary Γ of Ω. With the spatial domain Ω represented by an unstructured assembly of four noded tetrahedral elements, a piecewise linear approximate solution $\mathbf{U}^*$ is sought in the form

$$\mathbf{U}^*(\mathbf{x},t) = \sum_J \mathbf{U}_J(t) \, N_J(\mathbf{x}) \tag{5}$$

where the summation extends over each node J ($1 \le J \le n_p$) in the mesh, $\mathbf{U}_J(t)$ denotes the (unknown) value of the approximation $\mathbf{U}^*$ at node J at time t, and $N_J(\mathbf{x})$ is the standard linear finite element shape function associated with node J. Using the variational statement of equation (4), the Galerkin approximate solution is constructed as the function $\mathbf{U}^*$ which is such that

$$\int_\Omega \frac{\partial \mathbf{U}^*}{\partial t} N_I \, d\Omega = \sum_{j=1}^{3} \int_\Omega \mathbf{F}^j(\mathbf{U}^*) \frac{\partial N_I}{\partial x_j} \, d\Omega - \sum_{j=1}^{3} \int_\Gamma \mathbf{F}^j(\mathbf{U}^*) \, n^j N_I \, d\Gamma \tag{6}$$

for each N_I ($1 \le I \le n_p$) and for all $t > 0$. The integrals appearing here can be evaluated by summing individual element contributions and the compact support of the shape function N_I means that this equation can then be written as

$$\sum_{e \in I} \int_{\Omega_e} \frac{\partial \mathbf{U}^*}{\partial t} N_I \, d\Omega = \sum_{j=1}^{3} \sum_{e \in I} \int_{\Omega_e} \mathbf{F}^j(\mathbf{U}^*) \frac{\partial N_I}{\partial x_j} \, d\Omega - \sum_{j=1}^{3} \sum_{e \in I} \int_{\Gamma_e} \mathbf{F}^j(\mathbf{U}^*) \, n^j N_I \, d\Gamma \tag{7}$$

where the summation extends only over those elements Ω_e which contain node I. Inserting the assumed form for U^* into this equation, the left hand side integrals can be evaluated exactly to give

$$\sum_{e \in I} \int_{\Omega_e} \frac{\partial U^*}{\partial t} N_I \, d\Omega = \sum_{e \in I} \left[\int_{\Omega_e} N_I N_J \, d\Omega \right] \frac{dU_J}{dt} = M_{IJ} \frac{dU_J}{dt} \tag{8}$$

where **M** denotes the standard finite element consistent mass matrix. In the development of the algorithm, **M** is replaced by the lumped (diagonal) mass matrix, $\mathbf{M}_L$. The integrals appearing on the right hand side of equation (7) are evaluated approximately, by using a Lobatto type quadrature over each tetrahedral element e.g. for an element E, with nodes I, J, K and L, the domain integral is approximated as

$$\int_{\Omega_E} F^j \frac{\partial N_I}{\partial x_j} \, d\Omega \approx \frac{\Omega_E}{4} \left. \frac{\partial N_I}{\partial x_j} \right|_E \{ F^j_I + F^j_J + F^j_K + F^j_L \} \tag{9}$$

where Ω_E is the volume of element E. A standard finite element code would be organised in this fashion and would obtain the discrete form of equation (7), at each node I in the mesh, by looping over the individual elements and sending element contributions to the appropriate nodes. The associated data structure required in this case would consist of the list of the nodes I, J, K and L belonging to each element E in the mesh.

However, equation (7) can be reinterpreted and evaluated differently. Suppose that node I in the mesh is directly connected, by sides of the mesh, to the mI nodes J_1, J_2, ..., J_{mI}. Here, node I will be assumed to be an interior node and the procedure presented below will need to be suitably modified for nodes lying on the computational boundary. By using the results of equations (8) and (9) it can be readily shown that equation (7) can be written in the form

$$[M_L]_I \, \Delta U_I = \Delta t \sum_{j=1}^{3} \sum_{s=1}^{mI} C^j_{IJ_s} (F^j_I + F^j_{J_s}) \tag{10}$$

where ΔU_I is the increment in U_I over a time step Δt and the coefficients in this expression are such that

$$\sum_{s=1}^{mI} C^j_{IJ_s} = 0 \qquad \text{for all } j \tag{11a}$$

$$C^j_{IJ_s} = - C^j_{J_sI} \qquad \text{for all } j \text{ and } J_s \tag{11b}$$

If we define the vector C_{IJ_s} by the expression

$$C_{IJ_s} = (C^1_{IJ_s}, C^2_{IJ_s}, C^3_{IJ_s}) \tag{12}$$

and let $\mathcal{L}_{IJ_s}$ denote the modulus of C_{IJ_s} and $\mathcal{S}_{IJ_s}$ denote a unit vector in the direction of C_{IJ_s}, then equation (10) can be written as

$$[M_L]_I \, \Delta U_I = \Delta t \sum_{s=1}^{mI} \mathcal{L}_{IJ_s} (\mathcal{F}_{I_s} + \mathcal{F}_{J_s}) \tag{13}$$

where

$$\mathcal{F}_{I_s} = (F_I^1, F_I^2, F_I^3) . \mathcal{S}_{IJ_s} \qquad \mathcal{F}_{J_s} = (F_J^1, F_J^2, F_J^3) . \mathcal{S}_{IJ_s} \tag{14}$$

The alternative method of obtaining the discrete form of equation (7) is now apparent and takes the form of looping over each side in the mesh and sending side contributions to the appropriate nodes. The associated data structure required in this case would consist of the list of the nodes I and Js for each side in the mesh. For a general tetrahedral mesh, the ratio of elements to points is approximately equal to 6 and the ratio of sides to points is approximately equal to 7. Thus, the element based data structure will require 24 storage locations per node whereas the edge based data structure requires only 14. It is apparent that such a side based data structure is the most compact of all the various possible alternatives which can be used for unstructured meshes [7,8]. In addition, it will be demonstrated below that this data structure also proves to be extremely useful in the process of construction of artificial dissipation models.

3.3 Artificial Dissipation

The condition of equation (11a) means that a time stepping scheme of the form of equation (13) is linearly unstable. This situation can be remedied by the explicit addition of artificial viscosity to the algorithm. A stable first order upwind scheme can be constructed by using, for example, Roe's approximate Riemann solver [16] to replace equation (13) by

$$[M_L]_I \, \Delta U_I = \Delta t \sum_{s=1}^{mI} \mathcal{L}_{IJ_s} \left[\mathcal{F}_{I_s} + \mathcal{F}_{J_s} - |A_{IJ_s}| \; \{ U_{J_s} - U_I \} \right] \tag{15}$$

where $|A_{IJ_s}|$ denotes the standard Roe matrix in the direction $\mathcal{S}_{IJ_s}$. A possible simplification here is to replace this expression by

$$[M_L]_I \, \Delta U_I = \Delta t \sum_{s=1}^{mI} \mathcal{L}_{IJ_s} \left[\mathcal{F}_{I_s} + \mathcal{F}_{J_s} - |\lambda_{IJ_s}| \; \{ U_{J_s} - U_I \} \right] \tag{16}$$

where λ_{IJ_s} is the maximum eigenvalue of the matrix A_{IJ_s} and is computed as

$$\lambda_{IJ_s} = | \, u_{IJ_s} . \mathcal{S}_{IJ_s} \, | + a_{IJ_s} \tag{17}$$

where a denotes the speed of sound. A first order scheme of this type is of little use practically but it can be used to suggest the way in which a higher order dissipation should be constructed. If σ denotes a coordinate along the side between nodes I and Js of the mesh, then a first order approximation to $U_{J_s} - U_I$ can be obtained from

$$U_{J_s} - U_I \approx \frac{\partial U}{\partial \sigma} . \Delta\sigma \tag{18}$$

The right hand side of equation (18) is computed by using the nodal values of the gradient of U which are computed directly according to expressions such as

$$[M_L]_I \left. \frac{\partial U}{\partial x_j} \right|_I = \sum_{s=1}^{mI} C^j_{IJ_s} \, (F^j_I + F^j_{J_s}) \tag{19}$$

The time stepping scheme of equation (16) is now modified to read

$$[M_L]_I \, \Delta U_I = \Delta t \sum_{s=1}^{mI} \mathcal{L}_{IJ_s} \left[\mathcal{F}_{I_s} + \mathcal{F}_{J_s} - |\lambda_{IJ_s}| \left\{ U_{J_s} - U_I - \left[\frac{\partial U}{\partial \sigma} \cdot \sigma \right]_{IJ_s} \right\} \right] \quad (20)$$

in which a higher order form of the diffusion operator has now been constructed. In fact, in one space dimension, it is readily shown that the right hand side here represents a centered approximation to the flux derivative plus the addition of a standard fourth order artificial diffusion operator. This is known to lead to a stable higher order accuracy solution scheme [10,17].

An additional smoothing procedure is required if flows involving discontinuities are to be successfully modelled [10,17]. Here this is accomplished by adding a pressure switch [1,2] to the diffusion of equation (20) in the form

$$[M_L]_I \, \Delta U_I = \Delta t \sum_{s=1}^{mI} \mathcal{L}_{IJ_s} \left[\mathcal{F}_{I_s} + \mathcal{F}_{J_s} - |\lambda_{IJ_s}| \left\{ U_{J_s} - U_I - (1-P_{IJ_s}) \left[\frac{\partial U}{\partial \sigma} \cdot \sigma \right]_{IJ_s} \right\} \right] \quad (21)$$

where P_{IJ_s} denotes a side based pressure switch coefficient. The pressure switch is constructed such that $0 \le P_{IJ_s} \le 1$, with $P_{IJ_s} \approx 1$ in the vicinity of discontinuities and $P_{IJ_s} \approx 0$ in smooth regions of the flow.

3.4 Time Discretisation

Equation (21), which represents the discrete equation at a general node I of the mesh, can be expressed as

$$\Delta U_I = \Delta t \, [M_L]_I^{-1} \, \{ Q_I(U) + D_I(U) \} \; = \Delta t \, [M_L]_I^{-1} \, R_I(U) \quad (22)$$

where

$$Q_I(U) = \sum_{s=1}^{mI} \mathcal{L}_{IJ_s} \left[\mathcal{F}_{I_s} + \mathcal{F}_{J_s} \right]$$

$$D_I(U) = -\sum_{s=1}^{mI} \mathcal{L}_{IJ_s} \, |\lambda_{IJ_s}| \left\{ U_{J_s} - U_I - (1-P_{IJ_s}) \left[\frac{\partial U}{\partial \sigma} \cdot \sigma \right]_{IJ_s} \right. \quad (23)$$

In this form, the contributions to the right hand side made by the basic algorithm and the explicitly added diffusion can be clearly identified. Assuming that the nodal values U_I^n are known at time t_n, the solution is advanced over a time step Δt, to time t_{n+1}, by an explicit hybrid multi-stage time stepping scheme e.g. with ΔU_I defined to be equal to the change in the value of the solution **U** at node I over a time step, an m-stage version of the scheme can be written as

$$U_I^{(0)} = U_I^n$$

$$U_I^{(p)} = U_I^{(0)} - \alpha_p \, \Delta t \; [M_L]_I^{-1} \; R_I(U_I^{(p-1)}) \qquad p=1,2,...,m \quad (24)$$

$$U_I^{n+1} = U_I^{(m)}$$

with the parameters α_p assigned appropriate values. In the interests of computational efficiency, the diffusion contribution **D** to the right hand side vector **R** is not recalculated at every stage of this process. Since the correct modelling of the transient development of the flow is not of interest here, local time stepping is employed to accelerate the convergence behaviour. This is implemented by specifying a constant Courant number, CFL, throughout the mesh and computing the local time step Δt_I for node I in the mesh, according to an energy stability analysis [16], using the relation

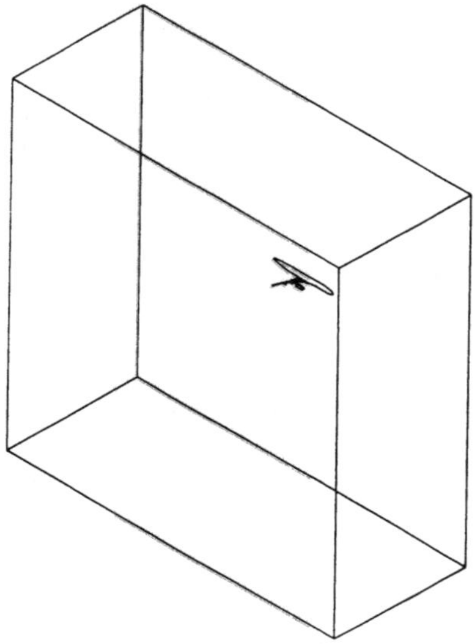

Figure 1
Wireframe of the computational domain employed in the analysis of an inviscid flow over a wing/body/pylon/nacelle configuration. Taken from [9].

$$\Delta t_I = 2 \ \mathrm{CFL} \ [\mathbf{M}_L]_I \left[\sum_{s=1}^{mI} \mathcal{L}_{IJ_s} \, |\lambda_{IJ_s}| \right]^{-1} \tag{25}$$

where a is the local speed of sound. Residual smoothing is also employed to increase the size of the allowable value of CFL [10,17].

Far field boundary conditions are applied by using a linearised characteristic analysis in the direction normal to the boundary to correct the computed values U_I^{n+1} obtained at far field boundary nodes I. The inviscid wall boundary condition of zero normal velocity is imposed simply by projection. An engine inlet can be modelled as a plane of constant static pressure, where the constant is adjusted, as the computation proceeds, so as to obtain a prescribed mass flow rate. At an engine outlet, the flow direction, the mass flow and total pressure of the fluid can be prescribed.

4. FLOW PAST AN INSTALLED AERO-ENGINE NACELLE

Results of a numerical simulation [9] of a transonic inviscid flow past a realistic fuselage/wing/pylon/nacelle configuration are included here to demonstrate the numerical performance of the unstructured mesh based solution procedure. The experiments were made with a turbine powered simulator to represent the flow into the engine and the jet emerging from the exhaust. The assumed symmetry of the problem means that only one half of the configuration need be considered. The definition of the computational domain is completed by the addition of the symmetry plane and planar far field boundaries, as shown in the wireframe of Figure 1. Following the triangulation of the computational boundaries, the aircraft surface is represented by an assembly of 38,330 triangles with 19,167 nodal points. Two different views of the surface triangulation of the model are given in Figure 2. Starting from this collection of triangular faces, the computational domain is filled with tetrahedra by the advancing front method, simultaneously creating interior nodal points and elements. The generated 3D mesh in this case consists of 592,380 tetrahedra and 112,198 nodes.

For the calculation [9], a free stream Mach number of 0.801 and an angle of attack of 2.738 degrees were assumed. For the simulation of the flow within the engine, a jet pressure ratio, an engine mass flow rate and a jet total temperature were computed by assuming that the separate core and fan streams in the experiment mix together completely within the turbine powered simulator, before exhausting through the nozzle. Starting from free stream conditions, the solution was advanced for 2,500 time steps, during which time the residual was reduced by 4 orders of magnitude. The computed distribution of pressure contours on the surface of the model is shown in Figure 3. In the experimental configuration, pressure gauges were distributed at fixed sections over the surface of the wing, and also on the nacelle. A comparison of the computed and the experimentally observed values of the pressure coefficient at a section along the wing and on the surface of the nacelle is given in Figure 4. The agreement between the numerically predicted values and the experimentally determined pressure data is very good, apart from the over prediction of pressures on the upper surface of the wing. The discrepancies between the experimentally observed pressure

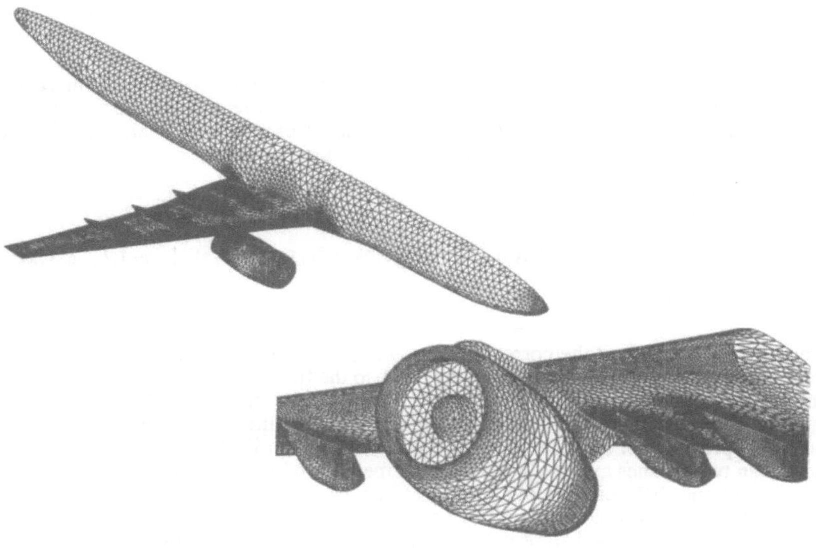

Figure 2
Two views of the surface triangulation employed in the analysis of an inviscid flow over a wing/body/pylon/nacelle configuration. Taken from [9].

distributions and the predictions of the Euler flow solver on the wing upper surface exist in regions where significant viscous effects are known to occur.

The important feature to be highlighted here is the routine way in which a complete analysis of this type can be accomplished within a time scale of just a few days.

5. MULTIGRID EXTENSION

It is well known that the use of an explicit multistage time stepping solution procedure for the Euler equations provides an attractive environment for the effective implementation of multigrid acceleration [10]. The basic idea behind the use of multigrid for the solution of a hyperbolic equation system is to use a sequence of successively coarser grids to compute corrections to the fine mesh solution. The coarse grid equations are also advanced with an explicit multistage scheme, making use of the

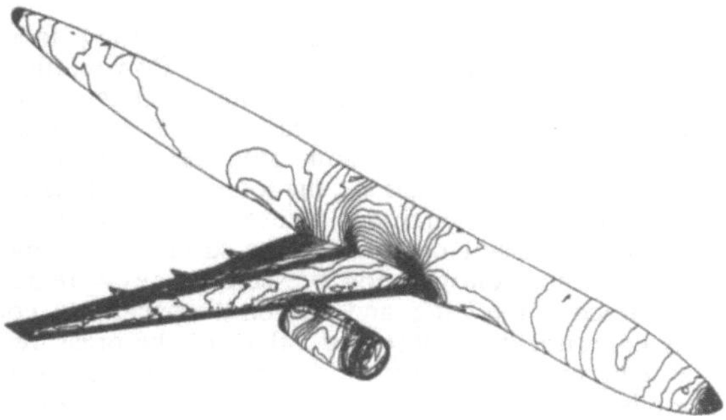

Figure 3
Computed pressure contours in the analysis of an inviscid flow over a wing/body/pylon/nacelle configuration. Free stream Mach number = 0.801. Angle of attack = 2.738 degrees. Taken from [9].

fact that the allowable time step is now larger than that which can be used on the finest mesh. For example, when two meshes are employed, the fine grid solution U_{If}^{n+1} and the corresponding residual $R_{If}(U^{n+1})$, computed as in equation (22), are transferred onto the coarse mesh as U_{Ic}^{n} and R_{Ic}^{n} and the time stepping scheme

$$U_{Ic}^{(0)} = U_{Ic}^{n}$$

$$U_{Ic}^{(p)} = U_{Ic}^{(0)} - \beta_p \Delta t_{Ic} [M_L]_{Ic}^{-1} \{ R(U_{Ic}^{(p-1)}) + R_{Ic}^{n} - R(U_{Ic}^{n}) \} \quad p=1,2,\ldots,q \qquad (26)$$

$$U_{Ic}^{n+1} = U_{Ic}^{(m)}$$

is employed to transport the corrections on this grid. When the coarse mesh steps are completed, the corrections are interpolated back to the fine mesh and used to update the fine mesh solution. It is important to observe that the coarse mesh algorithm of equation (26) ensures that the converged solution on the fine mesh is unaltered by the multigrid process i.e. the coarse mesh correction is zero if $R_{If}(U^{n+1})$ is zero. When more than two meshes are employed, various methods of extending this procedure can be adopted and different cycling strategies can be devised. A recursive algorithm, which allows for the use of an arbitrary number, *nmesh*, of meshes, in either a V or a W cycle, may be described algorithmically with the help of a stack *csw* as follows :

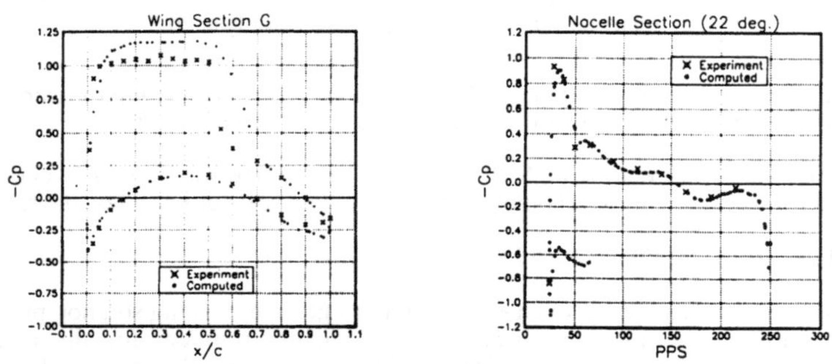

Figure 4
Comparison between the computed and experimentally measured pressure coefficient in the analysis of an inviscid flow over a wing/body/pylon/nacelle configuration. Free stream Mach number = 0.801. Angle of attack = 2.738 degrees. Taken from [9].

```
START CYCLE
          csw(1:nmesh) = 0
          ipmesh = 1
   1000   if(ipmesh.lt.nmesh) then
                 Advance on mesh ipmesh      (pre-smoothing)
                 csw(ipmesh) = csw(ipmesh)+1
                 Transfer - U, R - from mesh ipmesh to mesh ipmesh+1
                 ipmesh = ipmesh +1
                 go to 1000
          endif
          Advance on mesh ipmesh
   2000   if(ipmesh.gt.1) then
                 Transfer - ΔU - from mesh ipmesh to mesh ipmesh-1
                 ipmesh = ipmesh -1
                 Advance on mesh ipmesh      (post-smoothing)
          endif
          if (ipmesh.gt.1) then
                 if(csw(ipmesh).eq.icycle) then
                        csw(ipmesh) = 0
                        go to 2000
                 endif
                 go to 1000
          endif
   END CYCLE
```

This results in a V cycle if the parameter *icycle* is given the value 1 and a W cycle if *icycle* is equal to 2.

The transfer of information from a grid A to a grid B can be accomplished by either direct interpolation or by a conservative transfer. It is apparent that procedures of these types are readily devised when the meshes A and B are structured and nested. However, these operations are not easily accomplished when they are being performed between two general tetrahedral meshes. To illustrate the way in which such operations may be accomplished, consider the transfer of a function **f**. If we let IA and IB denote general nodes on mesh A and mesh B respectively, then a direct interpolation requires a knowledge of the element in mesh A within which node IB is located together with the values of the shape functions on mesh A at node IB. With this information, the direct interpolation can then be represented by the formula

$$\mathbf{f}_{IB} = \sum_{IA} \mathbf{f}_{IA} \, [N_{IA}]_{IB} \tag{27}$$

A conservative transfer requires the knowledge of the element in mesh B which contains node IA together with the values of the shape functions on mesh B at node IA. The transfer is accomplished as

$$\mathbf{f}_{IB} = [\mathbf{M}_L]_{IB}^{-1} \sum_{IA} [\mathbf{M}_L]_{IA} \, \mathbf{f}_{IA} \, [N_{IB}]_{IA} \tag{28}$$

The information required to accomplish these transfers is readily found using an alternate digital tree [19] similar to that used for the process of three dimensional unstructured mesh generation. In the present implementation, the fine to coarse mesh transfer of the unknowns is performed by direct interpolation, whereas conservative transfer is used for the residuals. The coarse to fine mesh transfer of the updates is accomplished by direct interpolation.

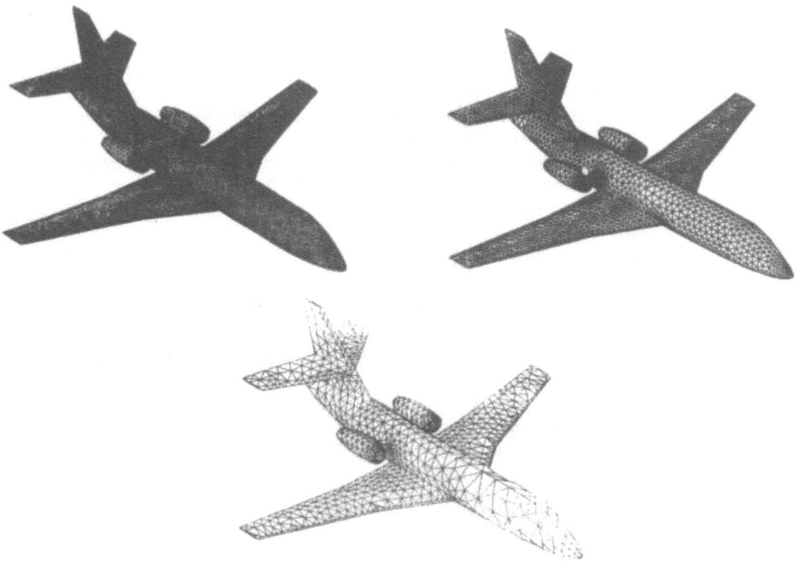

Figure 5
Surface triangulations for the sequence of meshes employed in the simulation of a transonic flow past a Dassault Falcon.

6. FLOW ABOUT A TWIN ENGINED AIRCRAFT

As an illustration of the performance of the multigrid acceleration procedure, a transonic flow about a twin engined Dassault Falcon has been computed. Three meshes were employed, consisting of approximately 850,000, 136,000 and 37,000 elements respectively. The surface triangulations for these meshes is shown in Figure 5. For the simulation, a free stream Mach number of 0.85 was assumed and the angle of attack was taken to be two degrees. Flow through conditions were imposed at the engine. A three stage time stepping scheme and a double W cycle, of the form showed in figure 6, was employed, with one pre- and one post-smoothing step used on each mesh. A plot showing the of the value of the logarithm of the density residual versus the number of fine grid residual evaluations is shown in Figure 7. The computed steady state distribution of pressure contours on the aircraft surface in shown in Figure 8. This solution was obtained in about 20 minutes of cpu time on a modern supercomputer using a single processor.

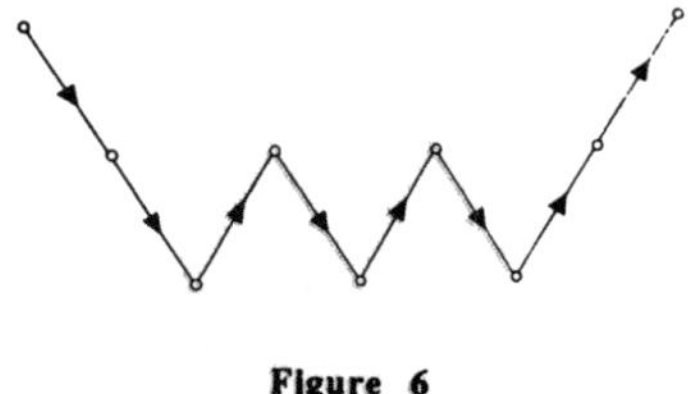

Figure 6
Schematic of the W cycle employed in the multigrid procedure.

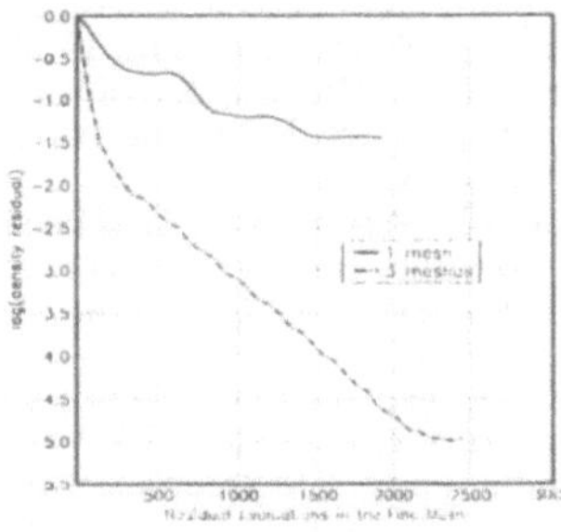

Figure 7
The magnitude of the logarithm of the density residual plotted against the number of fine mesh residual evaluations when both a single mesh and a sequence of meshes is employed in the simulation of a transonic flow past a Dassault Falcon. Free stream Mach number = 0.85. Angle of attack = 2 degrees.

7. CONCLUSIONS

It has been demonstrated that an unstructured mesh system provides a powerful method for the analysis of 3D inviscid transonic aerodynamic flows of current industrial interest. The utility of the method has been improved by the adoption of a side based data structure which leads to an unstructured mesh flow analysis code with minimum storage requirements. The practical usefulness of the method has been dramatically improved by the successful incorporation of a multigrid acceleration procedure for the analysis of steady state problems. It can be expected that, over the next few years, the unstructured approach will be increasingly used by the aerospace industry for the inviscid simulation of practical flows. However, more challenging applications, including the simulation of viscous flows and flows involving moving bodies, will have to be successfully addressed by researchers if the widespread acceptance of the approach is to be achieved.

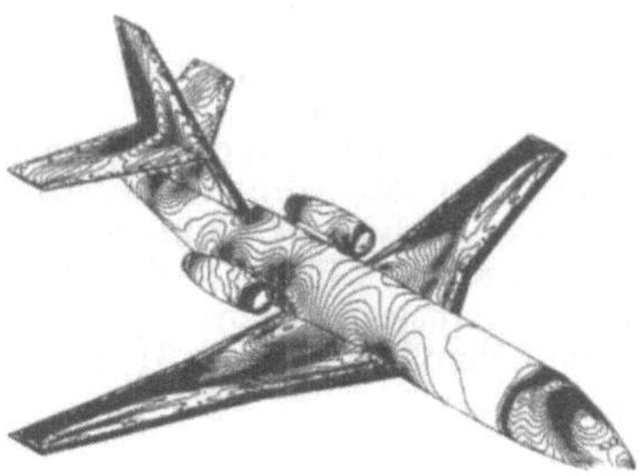

Figure 8
Computed steady state pressure contours from a simulation of a transonic flow past a Dassault Falcon. Free stream Mach number = 0.85. Angle of attack = 2 degrees.

REFERENCES

[1] J. Peraire, J. Peiró, L. Formaggia, K. Morgan and O.C. Zienkiewicz, 'Finite element Euler computations in three dimensions', Int.J.Num.Meth.Engng. 26, 2135-2159, 1988

[2] J. Peraire, K. Morgan, J. Peiró and O.C. Zienkiewicz, 'An adaptive finite element method for high speed flows', AIAA Paper 87-0558, 1987
[3] R.R. Thareja, J.R. Stewart, O. Hassan, K. Morgan and J. Peraire, 'A point implicit unstructured grid solver for the Euler and Navier-Stokes equations', Int. J. Num. Meth. Fluids 9, 405-425, 1989
[4] M. Vahdati, K. Morgan, J. Peraire and O. Hassan, 'A cell-vertex upwind unstructured grid solution procedure for high speed compressible viscous flow', in Proc. Royal Aeronautical Society Int. Conf. on Hypersonic Aerodynamics, Manchester, 12.1-12.22, 1989
[5] J. Peraire, M. Vahdati, K. Morgan and O.C. Zienkiewicz, 'Adaptive remeshing for compressible flow computations, J. Comp. Phys. 72, 547-574, 1987
[6] J. Peraire, K. Morgan and J. Peiró, 'Unstructured finite element mesh generation and adaptive procedures for CFD', in AGARD Conference Proceedings No. 464, 18.1-18.12, AGARD Paris, 1990
[7] T.J. Barth, 'Numerical aspects of computing viscous high Reynolds number flows on unstructured meshes', AIAA Paper 91-0721, 1991
[8] D.J. Mavriplis, 'Three dimensional unstructured multigrid for the Euler equations', AIAA Paper 91-1549, 1991
[9] J. Peiró, J. Peraire, K. Morgan, O. Hassan and N. Birch, 'The numerical simulation of flow about installed aero-engine nacelles using a finite element solver on unstructured meshes', submitted to Aero. J., 1991
[10] A. Jameson, 'Transonic flow calculations', Princeton University Report MAE 1751, 1984
[11] R. Löhner and K. Morgan, 'Unstructured multigrid methods for elliptic problems', Int.J.Num.Meth.Engng. 24, 101-115, 1987
[12] D.J. Mavriplis, 'Multigrid solution of the two-dimensional Euler equations on unstructured triangular meshes', AIAA J. 26, 824-831, 1988
[13] J. Peraire, J. Peiró and K. Morgan, 'A 3D finite element multigrid solver for the Euler equations', accepted for presentation at the AIAA Aerospace Sciences Meeting, Reno, 1992
[14] J. Peraire, K. Morgan and J. Peiró, 'Unstructured mesh methods for CFD', von Karman Institute for Fluid Dynamics Lecture Series, Brussels, 1990
[15] K. Morgan, J. Peraire, J. Peiró and O. Hassan, 'The computation of three dimensional flows using unstructured grids', Comp. Meth. Appl. Mech. Engng. 87, 335-352, 1991
[16] P. Roe, 'Approximate Riemann solvers, parameter vectors and difference schemes', J.Comp. Phys. 43, 357-372, 1981
[17] A. Jameson and W. Schmidt, 'Some recent developments in numerical methods for transonic flows', Comp.Meth.Appl.Mech.Engng. 51, 467-493, 1985
[18] M. Giles, 'Energy stability analysis of multi-step methods on unstructured meshes', M.I.T. CFD Laboratory Report CFDL-TR-87-1, 1987
[19] J. Bonet and J. Peraire, 'An alternate digital tree algorithm for geometric searching and intersection problems', Int.J.Num.Meth.Engng. 31, 1-17, 1991

FINITE ELEMENTS IN THE 90's, E. Oñate, J. Periaux, A. Samuelsson (Eds.)
Springer-Verlag/CIMNE, Barcelona 1991

ON THE PENALTY METHOD FOR INCOMPRESSIBLE FLUIDS

J.C. Heinrich and **B.R. Dyne**
Department of Aerospace and Mechanical Engineering,
The University of Arizona,
Tucson, Arizona, 85721,
USA

ABSTRACT

The penalty function finite element method for viscous incompressible flows is re-examined and re-formulated based on physically consistent approximations to the mass conservation equation. This leads to modified forms of the method that satisfy all fluid conditions under the assumption of a slightly compressible flow.

Both steady-state and time-dependent algorithms, the latter either implicit or explicit, are obtained and developed from these formulations. The implementations are based on the four-noded isoparametric bilinear element. Examples are given that show, in particular, their superior performance when the pressure field must be calculated.

1. INTRODUCTION

The use of penalization techniques in numerical solutions of differential equations was introduced by Courant [1]. Their use became widespread in areas such as optimization and linear programming, but remained dormant for many years in the context of the Rayleigh-Ritz method. A brief history of the penalty method in finite element analysis is given by Reddy [2].

The method was first applied in finite element analysis for the solution of elasticity problems with incompressible solids by Herrman [3], who considered the limit of compressible materials as Poisson's ratio approached 0.5. Zienkiewicz [4] re-introduced the penalty method as a means to modify variational principles in finite element analysis and proposed its extension to problems involving creeping flow of incompressible fluids, which are governed by the Stokes equations. Nickel et al. [5] presented the first approximation of the penalty method to the Stokes equations; Zienkiewicz and Godbole [6, 7] applied the method to plastic flow of metals, and it was quickly recognized that the penalty formulation offered a powerful tool for the numerical solution of the Navier-Stokes equations for viscous incompressible fluids. Some of the early publications by Hughes et al. [8, 9], Marshall et al. [10], Heinrich and Marshall [11], and Reddy and Satake [12] dealt with the application of the penalty finite element method applied to the Navier-Stokes equations and buoyancy driven flows when the Boussinesq approximation is valid.

Initially, theoretical difficulties were encountered that were related mainly to the consistency of the approximation spaces used to discretize the velocity and pressure fields, the occurrence of spurious pressure modes, and mesh locking. These have been successfully resolved by Oden [13], Song et al. [14], Carey and Krishnan [15], and Reddy [2]. To date, an extensive body of literature has been generated on the penalty method applied to viscous incompressible flows, and the method is now firmly established.

Computational difficulties also arise in standard implementations of the penalty function approach when the penalty parameters are increased to satisfy constraints. Iterative procedures were devised by Salonen [16] and Felippa [17] that allow for the use of relatively small penalty parameters. In the context of incompressible flows, Zienkewicz et al. [18] proposed an iterative method for steady-state problems that, as we will show later, is equivalent to introducting a false transient in the calculation.

The standard form of the penalty function formulation for viscous incompressible flows relies on the introduction of a pseudo constitutive law of the form

$$p = -\lambda \operatorname{div} \mathbf{V} = -\lambda \frac{\partial u_i}{\partial x_i} , \tag{1}$$

where p is the pressure, λ is the penalty parameter, and **V** is the velocity vector.

Although this has ample justification in the case of Stokes flow, it lacks an adequate physical interpretation in convective flows. Furthermore, Fukumori and Wake [19] pointed out that Eq. (1) fails to express the trivial condition that the pressure in the fluid must be hydrostatic for a zero velocity field, i.e.,

$$p = \rho g h , \tag{2}$$

where ρ is the fluid density, g is the gravitational acceleration, and h is the depth.

In the next section, modified forms of Eq. (1) are established that satisfy all fluid conditions based on physical considerations, and the calculation of the pressure field using penalty formulations is discussed in §3. Using numerical examples, the improved performance of these formulations is shown in §4.

2. PHYSICALLY CONSISTENT PENALTY FORMULATIONS

As already mentioned in the previous section, penalty finite element methods based on Eq. (1), although justified in the case of Stokes flow, are physically inconsistent for convective flows and have been used for highly convective flows on the basis that "they work" very well. However, they only allow us to calculate the velocity field, and the pressure has to be obtained a posteriori using a Poisson solver if it is to be calculated accurately [20]. There are at least two ways to improve on this situation. The first is based on the constitutive equations for an isotropic Newtonian fluid. It was first presented by Fukumori and Wake [19] and can be used to directly obtain steady-state solutions. We will refer to this as the steady-state formulation. The second way is based on an acoustic approximation to the mass conservation equation and leads to a time-dependent equation for the pressure. We will refer to this as the time-dependent formulation.

2.1 Steady-State Formulation

The constitutive equations for an isotropic Newtonian fluid are given by the Stokes viscosity law as

$$\sigma_{ij} = \left[-p_s + \mu' \frac{\partial u_k}{\partial x_k}\right]\delta_{ij} + 2\mu \, D_{ij} , \tag{3}$$

where σ_{ij} denotes the stress tensor; p_s is the hydrostatic or thermodynamic pressure; δ_{ij} is the Kronecker delta function; D_{ij} is the deviatoric component of stress,

$$D_{ij} = \left[\frac{\partial u_i}{\partial x_j} + \frac{\partial u_j}{\partial x_i}\right] ; \tag{4}$$

μ and μ' are the first and second coefficients of viscosity, respectively; and u_i is the velocity component in the x_i direction.

The mean pressure is defined by

$$p = - \frac{1}{3} \sigma_{ii} , \tag{5}$$

which is usually taken to be the same as the hydrostatic pressure. In general, however, the two pressures are different. In particular, for a slightly compressible fluid, replacing Eq. (3) into Eq. (5) we obtain

$$p = p_s - \left[\mu' + \frac{2}{3}\mu\right] \text{div } \mathbf{V} . \tag{6}$$

If we assume that $p = p_s$ in Eq. (6), we recover the familiar relation

$$\mu' = - \frac{2}{3} \mu . \tag{7}$$

We can define

$$\lambda = \mu' + \frac{2}{3} \mu \tag{8}$$

and write a modified penalty relation of the form

$$p - p_s = -\lambda \text{ div } \mathbf{V} , \tag{9}$$

where now the penalty parameter is the bulk viscosity of the fluid. Equation (9) satisfies all the important limiting relations for real fluids, in particular,

i. If $p = p_s$, then $\lambda = 0$ and Eq. (7) holds.

ii. If the fluid is incompressible, λ must be large and $\lim_{\lambda \to \infty} \lambda \text{ div } \mathbf{V} = \text{finite}$.

Physically, this approach replaces the fully incompressible constraint by a slight compressibility of the form $(p_s - p)/\lambda$. For most fluids, $p - p_s$ is a small quantity already, which means that λ need not assume an excessively large value for the constraint to be satisfied reasonably well.

The most dramatic effect of this formulation as opposed to that given by Eq. (1) occurs in the case of flows involving a free surface, where all three components of ∇p_s may be non-zero. In fact, Eq. (9) indicates that calculations of flows with free surfaces using a penalty function formulation based on Eq. (1) are incorrect.

2.2 Time-Dependent Formulation

The approach given in §2.1 is based on the use of the constitutive equations for the Stokes viscosity law in order to obtain a penalty function formulation that is physically consistent. In this section, we consider a different approach based on approximating the continuity equation using a barotropic equation of state of the form $\rho = \rho(p)$.

Consider the continuity equation

$$\frac{\partial \rho}{\partial t} + \mathbf{V} \cdot \text{grad } \rho + \rho \text{ div } \mathbf{V} . \tag{10}$$

If we assume a slightly compressible flow, i.e., $\rho = \rho_0 + \rho'$ and $\rho' << \rho_0$, where ρ_0 is the reference density, and use the barotropic equation of state in linear form,

$$p - p_0 = \left[\frac{\partial p}{\partial \rho}\right]_0 (\rho - \rho_0) , \tag{11}$$

we can re-write the continuity equation as

$$\frac{\partial p}{\partial t} + \mathbf{V} \cdot \text{grad } p + p_0 c_0^2 \text{ div } \mathbf{V} = 0 , \tag{12}$$

where $c_0^2 \equiv (\partial p/\partial \rho)_0$ is the speed of sound in the reference state. We now non-dimensionalize Eq. (12) using a reference length L, a reference velocity U, and a reference pressure $p^* = \rho_0 U^2$, and choosing $\tau = L/U$ as the time scale. The equation becomes

$$\frac{\partial p}{\partial t} + \mathbf{V} \cdot \text{grad } p + \frac{1}{M^2} \text{ div } \mathbf{V} = 0 , \tag{13}$$

where M is a pseudo Mach number, $M = U/c_0$.

From analytical solutions for acoustic waves of small amplitude, it can be further shown that the term $\mathbf{V} \cdot \text{grad } p$ will be negligible as long as U is small compared to c_0, that is for low Mach number flows. We can then write

$$\frac{\partial p}{\partial t} = - \frac{1}{M^2} \text{ div } \mathbf{V} . \tag{14}$$

This last approximation has also been validated numerically. If we now discretize in time using first-order time differencing, we can write

$$p^{n+1} = p^n - \frac{\Delta t}{M^2} \text{ div } \mathbf{V} , \tag{15}$$

where the divergence term can be treated implicitly or explicitly. Equation (15) can be re-written as

$$p^{n+1} - p^n = -\lambda \text{ div } \mathbf{V} , \tag{16}$$

with the penalty parameter λ now given by $\lambda = \Delta t/M^2$, and it relates pressure differences to the divergence of the velocity field in a manner similar in form to Eq. (9), but conceptually quite different.

Expression (16) also satisfies all limiting relations. One needs to keep in mind, however, that Eq. (12) must be used if $\text{div } \mathbf{V} \equiv 0$. Furthermore, when the flow reaches a steady state, Eq. (16) reduces to

$$\text{div } \mathbf{V} = 0 , \tag{17}$$

which is the statement of incompressibility.

Equations (9) and (16) are both physically consistent and have been obtained from widely different considerations. In fact, the penalty parameter λ has a completely different meaning in each of them. However, they share the fact that they relate the divergence of the velocity field to pressure differences rather than to the pressure itself, as does Eq. (1).

The second viscosity, μ', is a controversial quantity; it is actually known to be positive for most liquids, rather than $-2\mu/3$, and is often much larger than μ [21]. For slightly compressible fluids, μ' can be related numerically to the bulk modulus, which is of the order of 2.2×10^9 Pa for water, and would give an estimate for λ in Eq. (9). On the other hand, the speed of sound in water is approximately 1500 m/s, but the penalty parameter to be used in Eq. (16) also depends on the time step and the characteristic velocity of the flow under consideration.

3. NAVIER-STOKES EQUATIONS AND PRESSURE CALCULATIONS

The Navier-Stokes or momentum equation for a general linear, Newtonian viscous fluid is

$$\rho \frac{DV}{Dt} = -\nabla p + \frac{\partial}{\partial x_j}\left[\mu D_{ij} + \delta_{ij}\mu' \frac{\partial u_k}{\partial x_k}\right] + \rho g \,. \tag{18}$$

where ∇ is the del operator, $\nabla \equiv (\partial/\partial x, \partial/\partial y, \partial/\partial z)^T$, and D/Dt denotes the material derivative. The standard equation for compressible flow is recovered from Eq. (18) if we define μ' according to Eq. (7). If the fluid is incompressible and the viscosity constant, Eq. (18) takes the form

$$\rho \frac{DV}{Dt} = -\nabla p + \mu\nabla^2 V + \rho g \,. \tag{19}$$

Here, we will restrict ourselves to the finite element approximation of Eq. (16) and (19) in two spatial dimensions using bilinear isoparametric elements to discretize the velocity field in space and calculating piecewise constant pressures that are constant over an element.

The weak form of Eq. (19) over an open, connected domain Ω with boundary Γ can be written as

$$\int_\Omega \left[W\rho \frac{Du_i}{Dt} + \mu \frac{\partial W}{\partial x_j}\frac{\partial u_i}{\partial x_j} + \lambda \frac{\partial W}{\partial x_i}\frac{\partial u_j}{\partial x_j}\right] d\Omega = \int_\Omega \left[\frac{\partial W}{\partial x_i} p^n + W\rho g\right] d\Omega + \int_\Gamma W\sigma_{ij} n_j \, d\Gamma \,, \tag{20}$$

where n_j denotes the components of the unit outward normal to Γ and W is the weighting function. In Eq. (20), the viscous term takes a different form if we are fully consistent with a slightly compressible formulation and do not use the incompressibility condition there. However, numerical experiments indicate no significant differences among the different forms, and Eq. (20) is computationally more efficient. We are also assuming that Eq. (20) will be used to calculate the solution at time level t_{n+1}, hence, u_i^n and p^n are known quantities at the previous time level.

The divergence term in Eq. (20) must be reduced-integrated [13] in order to avoid locking. This is also related to the fact that the use of full integration in the divergence term of Eq. (20), when bilinear elements are used to interpolate the velocities, produces a set of equations that represents a statement of conservation of mass only after all elements surrounding a node have been assembled. However, for any one element in particular, the contributions to the divergence are not zero. On the other hand, reduced integration of the penalty term produces a statement of conservation both at the assembled and element levels, which is fully compatible with the treatment of the pressure equation, Eq. (16), as explained below. Finally, in the present algorithm, the convective terms in Eq. (20) are weighted using a Petrov-Galerkin formulation as described by Henrich and Yu [22] in order to prevent spatial instabilities in the solution.

The pressure equation, Eq. (16), is weighted using piecewise constant weighting functions and is integrated using a one-point reduced integration formula to advance from time level t_n to time level t_{n+1} after the velocities have been updated; i.e., in an element e we have

$$p^{n+1} = p^n - \frac{\lambda}{A^{(e)}} \int_{\Omega^{(e)}} \frac{\partial u_i^{n+1}}{\partial x_i} \, d\Omega \, , \tag{21}$$

where $A^{(e)}$ is the element's area and $\Omega^{(e)}$ is the restriction of Ω to element e. A fully implicit method for pressure is defined this way.

Explicit algorithms can also be constructed from Eqs. (16) and (20). However, our experience indicates that they are extremely sensitive to initial and boundary conditions and, therefore, we do not recommend them at this time. It is also interesting to note that the present scheme is equivalent to the iterative method of Zienkiewicz et al. [18] that was designed to obtain steady-state solutions of the Navier-Stokes equations using a small penalty parameter in order to obtain better conditioned matrices, except the time derivative terms in the momentum equations were not considered there. However, Eq. (21) is the same as Eq. (25a) in their work if the superscript n is interpreted as an iteration rather than as a time level index and if the penalty parameter is re-interpreted accordingly.

4. EXAMPLES

We now present some examples of calculations performed using the penalty formulation given by Eqs. (20)-(21). The first example is a simple channel flow. The walls of the channel are parallel, a parabolic velocity profile is prescribed at the inlet, and a zero normal stress condition at the outlet. The channel is 10 units long in the x-direction and 1 in the y-direction. The Reynolds number is 100. The pressure gradient in the x-direction is given by

$$\frac{\partial p}{\partial x} = - \frac{8U_{max}}{Re} = -0.08 \, ,$$

where U_{max} is set equal to 1. Table 1 shows results for several different meshes using the time-dependent formulation (TDF) given by Eqs. (20)-(21) and solutions obtained using the Pressure Poisson Equation (PPE) method of Sohn and Heinrich [22]. It can be observed that the present formulation yields very accurate results, even for coarse meshes, and that it takes a very refined grid for the PPE method to yield similar accuracy.

The second example involves flow over a backward-facing step at Reynolds number Re = 500 based on the entry length. The geometry and boundary conditions are shown in Fig. 1(a), together with the irregular mesh containing 1,999 elements and 2,129 nodes. Figure 1(b, c) shows the pressure field and stream

Table 1. Calculated pressure gradients for flow in a channel.

Case	Elements n_y	Elements n_x	Method[a]	$\partial P/\partial x$	% Error
1	10	50	TDF	-0.07836	2.05
	10	50	PPE	-0.07228	9.65
2	20	50	TDF	-0.07844	1.95
	20	50	PPE	-0.07612	4.85
3	20	100	PPE	-0.07612	4.85
4	40	50	TDF	-0.07858	1.78
	40	50	PPE	-0.07804	2.45

[a] TDF = Time-Dependent Formulation; PPE = Pressure Poisson Equation Method

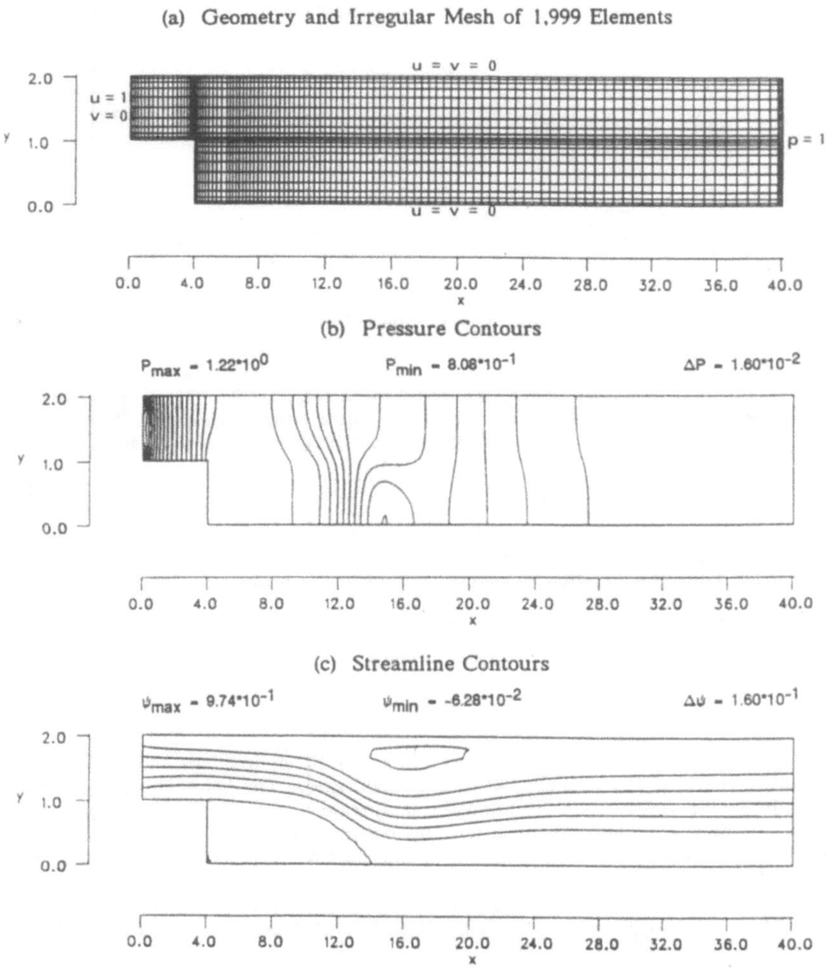

Fig. 1. Flow over a backward-facing step at Re = 500.

function at steady-state. It should be noted that imposition of the pressure boundary condition is very easily achieved in the TDF. Figure 2 shows the pressure along the top and bottom walls. It can be seen that all the expected features are captured and no oscillations in pressure occur.

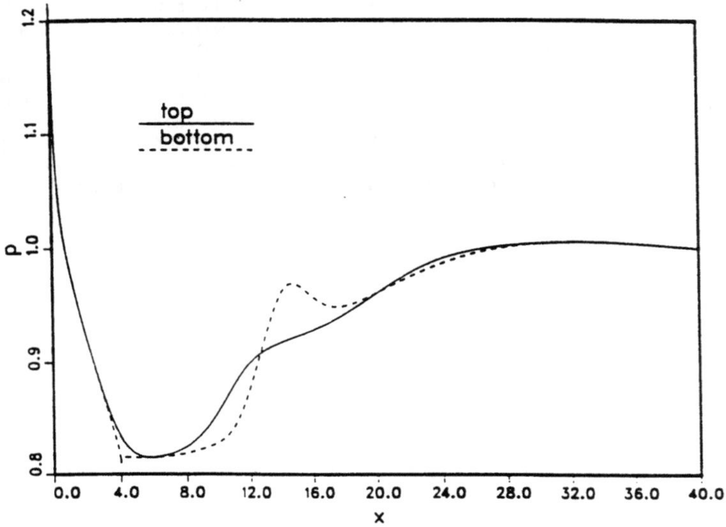

Fig. 2. Pressure along top and bottom walls for flow over a backward-facing step at Re = 500.

In the two examples above, the body force term in Eq. (20) was zero. The next example involves natural convection in a square enclosure and incorporates the body force term with the density defined from the linear equation of state,

$$\rho = \rho_0 + \beta(T - T_0) , \tag{22}$$

where T denotes temperature and β is the coefficient of thermal expansion; ρ_0 and T_0 are the reference density and temperature, respectively. The temperature is calculated from the equation of energy written in the form

$$\frac{DT}{Dt} = \alpha \frac{\partial^2 T}{\partial x_i \partial x_i} , \tag{23}$$

where α is the thermal diffusion coefficient.

Results of a calculation for $Ra = 10^4$ and $Pr = 0.72$ are shown in Fig. 3. Figure 3(a) shows the irregular 40 × 40 element grid. The boundary conditions are no slip on the walls ($u = v = 0$) and the temperature is set to $T = 0.5$ at $x = 0$ and $T = -0.5$ at $x = 1$, with adiabatic top and bottom boundaries. Figure 3(b, c) show the temperature and flow fields at steady state, obtained after starting with homogeneous initial conditions. Finally, Figure 3(d) shows the calculated pressure field. These results are in full agreement with previously calculated solutions for the flow and temperature fields [10, 12] and for the pressure fields calculated by Sohn and Heinrich [20].

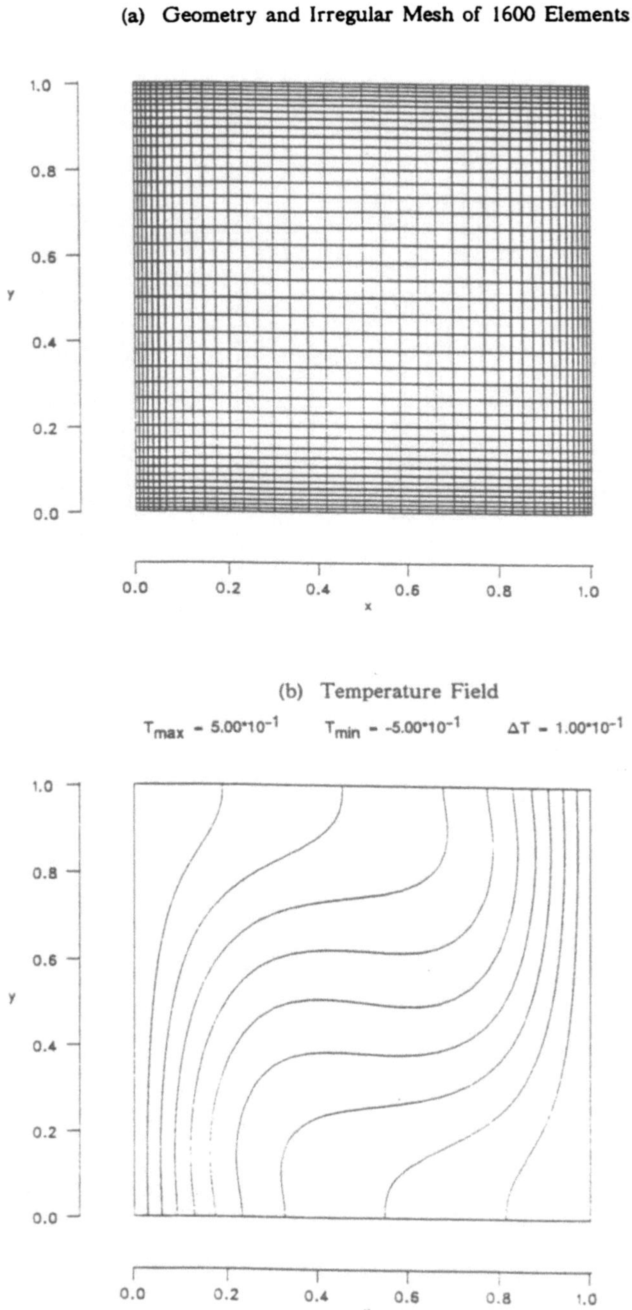

Fig. 3. Natural convection in a square cavity at Ra = 10^4 and Pr = 0.72.

Fig. 3.—Continued.

(c) Flow Field

maximum velocity magnitude = $2.33 \cdot 10^{-1}$

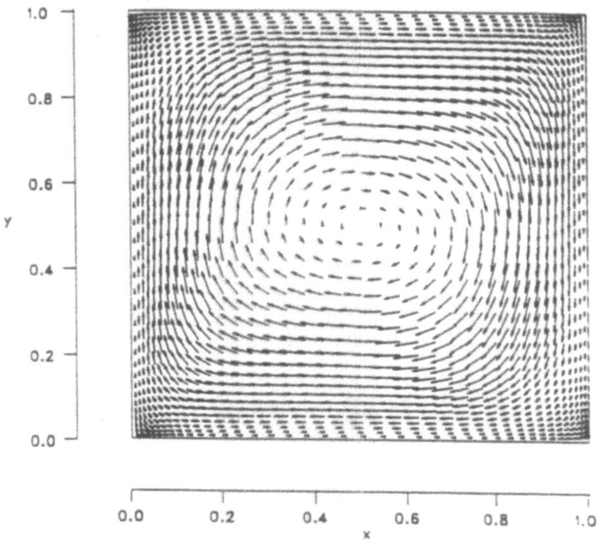

(d) Pressure Contours

$P_{max} = 1.14 \cdot 10^3$ $P_{min} = -1.25 \cdot 10^2$ $\Delta P = 7.50 \cdot 10^1$

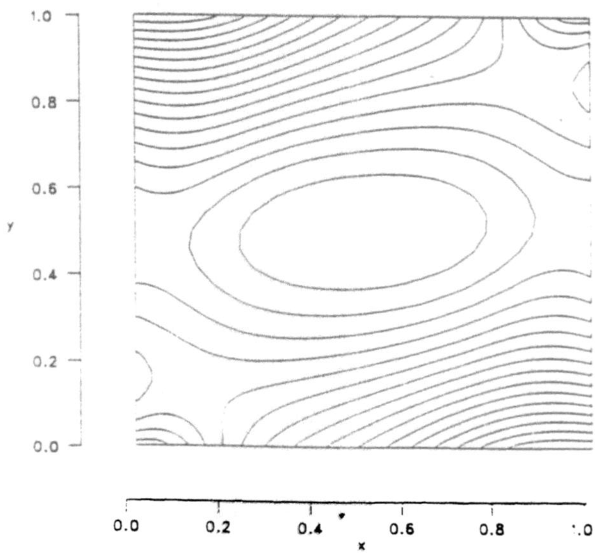

5. CONCLUSIONS

The penalty method has been reformulated based on physical considerations, and algorithms have been constructed that show improved performance when compared to the standard penalty formulation, which does not satisfy all fluid properties.

The algorithms produced still take full advantage of the elimination of the pressure as a variable in the momentum equations, and the pressure can be calculated explicitly at every time step. The improved accuracy in the pressure calculations using this formulation has been demonstrated through examples that also show no oscillations in the computed pressure fields.

REFERENCES

1. COURANT, R. – Variational methods for the solution of problems of equilibrium and vibrations, Bull. Amer. Math. Soc., 49, 1-23 (1943).

2. REDDY, J.N. – On penalty function methods in the finite element analysis of flow problems, Int. J. Num. Meth. Fluids, 2, 151-171 (1982).

3. HERRMAN, L.R. – Elasticity equations for incompressible or nearly incompressible materials by a variational theorem, AIAA J., 3, 1896-1900 (1965).

4. ZIENKIEWICZ, O.C. – 'Constrained variational principles and penalty function methods in finite element analysis,' Lecture Notes in Mathematics No. 363, Ed Watson, G.A., Springer Verlag, 1974, pp. 207-214.

5. NICKEL, R.E., TANNER, R.I. and CASWELL, B. – The solution of viscous, incompressible jet and free-surface flows using finite element methods, J. Fluid Mech., 65, 189-206 (1974).

6. ZIENKIEWICZ, O.C. and GODBOLE, P.N. – A penalty function approach to problems of plastic flow of metals with large surface deformations, J. Strain Anal., 10, 180-183 (1975).

7. ZIENKIEWICZ, O.C. and GODBOLE, P.N. – 'Viscous, incompressible flow with special reference to non-Newtonian (plastic) fluids,' Finite Elements in Fluids, Vol. 1, Ed. Gallagher, R.H., Oden, J.T., Taylor, C. and Zienkiewicz, O.C., Wiley, 1975, pp. 25-55.

8. HUGHES, T.J.R., TAYLOR, R.L. and LEVY, J.F. – 'High Reynolds number, steady, incompressible flows by a finite element method,' Finite Elements in Fluids, Vol. 3, Ed. Gallagher, R.H., Zienkiewicz, O.C., Oden, J.T., Morandi-Cechi, M. and Taylor, C., Wiley, 1978, pp. 55-72.

9. HUGHES, T.J.R., LIU, W.K. and BROOKS, A. – Finite element analysis of incompressible viscous flows by the penalty function formulation, J. Comp. Phys., 30, 1-60 (1979).

10. MARSHALL, R.S., HEINRICH, J.C. and ZIENKIEWICZ, O.C. – Natural convection in a square enclosure by a finite element, penalty function method using primitive fluid variables, Num. Heat Trans., 1, 315-330 (1978).

11. HEINRICH, J.C. and MARSHALL, R.S. – Viscous incompressible flow by a penalty function finite element method, Comp. and Fluids, 9, 73-83 (1980).

12. REDDY, J.N. and SATAKE, A. – A comparison of various finite-element models of natural convection in enclosures, J. Heat Trans., 102, 659-666 (1980).

13. ODEN, J.T. – 'The RIP method for Stokesian flows,' Finite Elements in Fluids, Vol. 4, Ed. Gallagher, R.H., Norrie, D.N., Oden, J.T. and Zienkiewicz, O.C., Wiley, 1982, pp. 305-318.

14. SONG, Y.J., ODEN, J.T. and KIKUCHI, N. – Discrete LBB conditions for RIP-finite element methods, TICOM Report 80-7, Texas Inst. for Computational Mechanics, Austin, Texas, 1980.

15 CAREY, G.F. and KRISHNAN, R. – Penalty approximation of Stokes flow, Comp. Meth. Appl. Mech. and Eng., 35, 169-206 (1982).

16. SALONEN, E.M. – An iterative penalty function method in structural analysis, Int. J. Num. Meth. Eng., 10, 413-421 (1976).

17. FELIPPA, C.A. – Iterative procedures for improving penalty function solutions of algebraic systems, Int. Journal Num. Meth. Eng., 12, 821-836 (1976).

18. ZIENKIEWICZ, O.C., VILOTTE, J.P., TOYOSHIMA, S. and NAKAZAWA, S. – Iterative method for constrained and mixed approximation; an inexpensive improvement on F.E.M. performance, Comp. Meth. Appl. Mech. Eng., 51, 3-29 (1985).

19. FUKUMORI, E. and WAKE, A. – A note on the penalty function method applied to the Navier-Stokes equations, Res. Rept., Dept. Civil Eng., Univ., Buffalo, New York, 1986.

20. SOHN, J.L. and HEINRICH, J.C. – A Poisson equation formulation for pressure calculations in penalty finite element models for viscous incompressible flows, Int. J. Num. Meth. Eng., 30, 349-361 (1990).

21. WHITE, F.M. – Viscous Fluid Flow, McGraw-Hill, New York, 1974.

22. HEINRICH, J.C. and YU, C.-C. – Finite element simulations of buoyancy-driven flows with emphasis on natural convection in a horizontal circular cylinder, Comp. Meth. Appl. Mech. Eng., 69, 1-27 (1988).

SECTION 4

GENERAL APPLICATIONS OF THE FINITE ELEMENT METHOD

FINITE ELEMENTS IN THE 90's, E. Oñate, J. Periaux, A. Samuelsson (Eds.)
Springer-Verlag/CIMNE, Barcelona 1991

THE FINITE ELEMENT METHOD IN THE 1990'S: A PERSONAL PERSPECTIVE

S. N. Atluri
Computational Modeling & Infrastucture Rehabilitation Center,
Georgia Institute of Technology,
Atlanta, GA 30332-0356,
U.S.A.

SUMMARY

The primary objective of this paper is to pay the most sincere tribute to a close friend, and a giant of the discipline of engineering analysis in this century, Professor Olek Zienkiewicz, who is the single most important individual in the history of the finite element method [13] that revolutionalized the engineering world and introduced new vocabulary ('Computational Mechanics') into the technical literature. A far less important objective of this paper is to give a highly personalized and subjective view of how the finite element method will evolve and be applied in the coming decade or so.

ELEMENT TECHNOLOGY

Even after nearly 35 years of the introduction of a 4-node quadrilateral membrane element, the quest for the 'perfect' 4-node quad still continues, and hopefully will end in the 1990's! Drilling degrees of freedom and the use of mixed variational principles with unsymmetric stresses and rotations as variables, and hybrid-mixed finite element methods in general, are likely to form the basis of these 'perfect' quads. (See [4], for example). The quest for perfect '3-D' and 'plate and shell' elements will continue for the foreseeable future. Here again, the mixed element technology will dominate.

While some researchers that work with integral equation methods may vehemently argue that the boundary element method is radically different from the finite element method, these two and other alternate domain descritization methods would become more unified in 1990's. In the 'finite element method' the trial and test function spaces are similar, and the algebraic equations are developed from a symmetric weak form, involving equal-order differentials of both the trial and test functions. In the integral equation approach, however, the test functions are a 'better' class of functions (more differentiable) than the trial functions. The test functions in this approach are the fundamental solutions (ie., for a Dirae delta forcing function) to the linear part of the differential operator of the problem, in infinite space, and hence do not obey *any* conditions at the boundaries of the given finite-shaped domain.

The algebraic equations are developed from the integral equation, which can be considered as an 'unsymmetric' weak form involving higher-order differentials of the test functions. [12, 11, 6 and 2].

While the integral equation methods were considered, in the development of the boundary element method, to be limited to linear problems, their extension to the fully nonlinear (material and geometric) problems is likely to be dominant in the 1990's [7]. Early evidence indicates that, while the usual 'dimensional' reduction is lost, the integral equation methods for nonlinear problems are quite competitive in terms of computational cost and accuracy. The spatial discretization method for nonlinear integral equations may be described as the 'Field-Boundary Element Method'. The development of the Field-Boundary-Element method for nonlinear problems of plates and shells, and for strain-localization problems, is likely to be pursued intensively in the 1990's.

ADAPTIVE MESH REFINEMENT AND A POSTERIORI ERROR ESTIMATIONS

Zienkiewicz's pioneering work in this area [14] has already started an explosion of research in this area, and has lead to the concept of 'smart' or 'autonomous' algorithms for finite element computations. In a recent report on future research directions [5] in computational mechanics, this topic heads the list of topics that are recommended for major funding by the U.S. research agencies. The development of software with 'smart algorithms' is likely to dominate the finite element field in the 1990's.

ENGINEERED & SMART MATERIALS

Recently, the U.S. Department of Defense, and the Aerospace Industries Association of America, have independently developed lists of 20 critical technologies for the next decade and beyond. It is heartening to note that 'engineered materials', 'computational fluid dynamics' and 'computational sciences' are three topics that are common to both the lists.

The history of humanity has always been labeled by the dominant materials of that era - the Stone Age, the Bronze Age, etc. In that light, the contemporary human history will most likely be characterized as the 'Age of the Engineered Materials'. Fiber-reinforced laminated composite materials for aerospace applications, metal-matrix composites; ceramic composites for high-temperature and high-strain-rate applications, are but some examples of engineered materials. The role of finite elements is not just in the thermo-mechanical stress analysis of the *engineered* materials, but more significantly in *computational micromechanics* aimed at *engineering* such materials. By 'engineering a material' we mean *tailoring* a material so as to optimize its strength, stiffness, fracture toughness, vibrational response, noise radiation and absorption and acoustic response.

The *meso* (atomic level) mechanics, and *micro*-mechanics of the composite materials becomes much more important than the *macro*-mechanical stress analysis. In this way, finite elements are likely to become indispensible tools in the science of developing materials and thus additional terminology - 'computational meso-mechanics' and 'computational micromechanics' is likely to come in vogue in the 1990's.

The thin-film technology of microelectronics and avionics, would also necessitate computational modeling at the atomic level, as is the discipline of wear and tribology.

An area of material mechanics - with intense use of computational meso, - micro, - and micromechanics, that will be pursued at Georgia Tech in the first half of the 1990's is the phenomenon of liquid metal embrittlement (LME) and solid metal embrittlement (SME). These are forms of environmental cracking that occurs when a metal, particularly in its liquid state, penetrates the host material to cause a drastic reduction in fracture energy. LME failures can be catastrophic - with crack speeds as high as 100 cm/s - quite different from other forms of corrosion and environmental embrittlement where the process is one of slower material degradation over months or years. The failures associated with LME and various: (i) joining or material fabrication processes with stainless steel or other nickel bearing alloys where a contamination metal (zinc) may be found in a liquid state due to the high temperatures; (ii) wear processes such as bearing locations where metals of different composition (copper and steel, say) are constantly in contact; (iii) nuclear systems, where transmutation products such as rubidium from the decay of Krypton 85 causes severe embrittlement of nickel-chrome/copper alloys, fission products from water reactor operations such as cadmium on zircalloy fuel cladding and liquid metal coolant such as lithium embrittlement of containment metals containing nickel and copper. Four thrusts to understand LME and SME are being made:

(i) lattice modeling of the liquid metal species interaction with the host lattice to distinguish between crack-tip tensile decohesion and crack dislocation transport of the embrittlement species into the solid. Band structures calculations of the host solid due the presence of embrittling species will be carried out.

(ii) Using the embrittled state material properties such as the embrittlement - modified flow stress along a slip plane, continuum scale computational modeling will be carried out, of the single-crystal embrittlement fracture with multiple active slip systems. Gallium and mercury are chosen as the embrittling species on zinc (HCP) and aluminum (FCC) single crystals. Lithium is chosen for FCC Ni-Cu of Cr-Mo simple crystals.

(iii) Using the single crystal slip-system constitutive equations, as modified for the presence of embrittling species, self-consistent bi-crystalline models will be developed.

(iv) Finite element simulations of a crack tip in a single grain, with the crack-tip being allowed to progress to a grain boundary which may contain precipitate species. The embrittlement crack will proceed along a path to minimize energy, depending on the relative energy states of the grain boundary (intergranular) to that verse the strained slip line in an adjoining crystal (intragranular). Inherent to this process will be the modeling of the diffusion of the embrittlement species into the solid along the crack plane for the polycrystalline aggregate.

Thus the LME and SME studies will involve computational modeling from the atomic to the continuum (single crystal and poly-crystalline) levels.

In a recent study, the U.S. National Science Foundation has set the following priorities, among others, for computational solid mechanics:

1. Methods for calculation of three dimensional plastic instabilities and failure phenomena. Enhanced and enriched element finite element methods seem promising in this regard. The Field Boundary Element Method will also prove very useful in these studies.

2. Adaptive methods for the analysis of strain localization phenomena. This can include adaptive mesh refinement strategies and, perhaps, adaptive enrichment strategies. These need to be applicable to localization and failure under the complex deformation histories encountered in metal-forming processes.

3. The development of computational methods for the transition from localization to failure. An element vanishing technique has been developed for porous plastic constitutive relations, but this method, or alternative methods, need to be developed for a range of failure modes.

4. The development of a firm mathematical basis for the accuracy and convergence of grid based numerical methods in localization problems.

5. The exploitation of new computer architectures, particularly massive parallel machines, to enable complex problems involving localization and failure with microstructurally based constitutive relations to be solved.

Thus, local material instability problems will provide significant challenges to the finite element research community in the coming few years.

Other areas of intense activity in the coming decade will be 'Smart Structures & Materials', and 'Bio-mimetics'. Smart or autonomous structures will have embedded sensors/actuators for the purpose of control of deformation, damage, and fracture. For instance, in space-structures such as the large space-antennae that are contemplated, the control of the shape of the antenna to retain its pointing accuracy is of importance. Thus, active controls are necessary to damp out quickly any deformations. Likewise, embedded sensors may be used to detect local failure modes such as delamination etc. in laminated composites, while thermo-mechanical induced-strain actuators can be used to control the spread of such delamination. Also, the use of shape-memory alloys to design intelligent structures, that may self-detect crack-growth and control such growth, is currently being explored. The problems for finite element computational mechanics in the analysis and design of smart-structures are many:

The analysis of the mechanical coupling between the sensor/actuators and the underlying structures.

The Design of nonlinear control algorithms to implement active control on very large-order dynamical systems.

'Smart materials' may loosely be defined as those that exhibit a memory of their deformation history and/or those that produce electrical effects or undergo phase transformations under deformation. Shape-memory alloys (Nitinol), piezoelectric ceramics, and zirconia toughened ceramics are examples of such materials. Zirconia undergoes dilatational phase transformation from tetragonal to monoclinic structure due to applied hydrostatic stress. When this is properly accounted for in the computational micromechanical model [9], the macroscopic constitutive relation of the ceramic exhibits strain-softening, and is responsible for a substantial increase in the fracture toughness of the ceramic. Thus, once again, computational micro-mechanics and computational macro-mechanics play significant roles in *engineering* the ceramic composite so as to tailor its fracture toughness [8]. 'Bio-mimetics' is an emerging discipline that seeks to use knowledge gained in understanding the natural biological materials, in engineering composite materials.

INFRASTRUCTURE REHABILITATION

The emergence of civil aviation as a means of mass transportation is primarily due to the large-scale construction of jet airplanes in the past 30 years or so. A large number of these jet planes is currently operating at or beyond their designed fatigue lives. Thus, the structural integrity of these aging airplanes [3] has become an issue of major concern globally. Considering the cost of a new commercial jet transport plane (approx. 200 million), and that there are about 5000 aging airplanes in service int he world today, the economic benefits of developing technologies to enhance the lives of these airplanes is immediately apparent. The rehabilitation of these aging airplanes involves a host of computational mechanics problems [3; 1], such as the analysis of multiple-site-damage near a row of fastener holes in a lap-jointed fuselage skin; repair of cracks in stiffened fuselage structures using composite patches [1]; and plasticity effects on crack-bulging and crack-flapping in fuselages.

Likewise, the advent of heavier commercial jet planes (with a gross weight of 1.3 million pounds!) brings with it, the need for strengthening and/or rehabilitating the runways and pavements at airports throughout the world. The life enhancement of the airport pavements brings again a host of challenging computational mechanics problems: the analysis of asphaltic concrete or PCC pavements resting on semi-infinite spaces of subgrade and soil, subjected to more or less concentrated moving loads. Pavement materials exhibit visco-elasticity, strain-softening and continuum damage of the micro-cracking type. To analyze layered media with highly peculiar inelastic material behavior, and subject to moving loads, is indeed a major task. To incorporate the results of stress and deformation in models for damage such as rutting, is indeed a challenging computational mechanics problem. Such problems of rehabilitation would become increasingly the focus of many researchers in the coming decade.

CLOSURE

In the spirit of the scientific work of Professor Olek Zienkiewicz who brought his technical ingenuity to bear on a variety of engineering problems over the past 30 years, I have tried to identify some problems of interest that may yet benefit from the attention of his creative mind. I look forward, with admiration, to my continued interactions with this great man of science. He is not only a great engineer, but a compassionate, caring individual. I record here my most sincere homage to Professor Zienkiewicz, whom I consider a close friend, on the occasion of his 70th birthday. May God bless him with a long life.

ACKNOWLEDGEMENT

This somewhat rambling narratory more or less reflects my research interests at present and in the near future. This research has been and/or will be supported by the Air Force Office of Scientific Research; Office of Naval Research; NASA Langley Research Center; U.S. Army Research Office; the Federal Aviation Administration; the National Science Foundation; and the Idaho National Laboratories. For this, I am grateful. I thank Drs. Spencer Wu, Yapa Rajapakse, Jerold Housner, Tony Chou and Sam Sampath for their past and continued encouragement. Finally, it is a pleasure to thank Ms. Brenda Bruce, an important contributor to all my current technical work.

REFERENCES

1. ATLURI, S.N. "Analytical Methods for Structural Integrity" Plenary Lecture, Int. Conf. on Aircraft Damage Assessment and Repair, Melbourne, Australia, 26-28 Aug. 1991.

2. ATLURI, S.N. and OKADA, H. "Field-Boundary Element Methods for Geometrical and Material Nonlinear Problems of Solid & Structural Mechanics" to appear in Advances, in Nonlinear Computational Mechanics in Aerospace Engineering (Ed: S.N. Atluri) Progress in Aero & Astro Series, AIAA, NY, 1991.

3. ATLURI, S.N. SAMPATH, S.G. and TONG, P. (Eds.) Structural Integrity of Aging Airplanes, Springer-Verlag, 495 pages, 1991.

4. IURA, M. and ATLURI, S.N. "Geometrically Nonlinear Membrane Elements with Drilling Degrees of Freedom" in Computational Mechanics '91 (Atluri, Beskos, Jones, and Yagawa, Eds.) pp. 393-399, ICES Publications, Atlanta, 1991.

5. ODEN, J.T."Theory and Implementations of High-Order Adaptive hp-Methods for the Analysis of Incompressible Viscous Flows" to appear in Advances in Nonlinear Computational Mechanics in Aerospace Engineering (Ed: S.N. Atluri) Progress in Aero & Astro Series, AIAA, NY, 1991.

6. O'DONOGHUE, P.E. and ATLURI, S.N. "Field-Boundary Element Approach to the Large Deflections of Thin Plates" Computers & Structures, 27, 427-435 (1987).

7. OKADA, H., RAJIYAH, H. and ATLURI, S.N. "A Full Tangent Stiffness Field-Boundary-Element Formulation for Geometric and Material Nonlinear Problems of Solid Mechanics" Int. Jnl. Num. Meth. in Engg (1990).

8. OKADA, H., TAMURA, T., RAMAKRISHNAN, N., ATLURI, S.N. and EPSTEIN, J.S. "Analysis of Toughening of Magnesia Partially Stabilized Zirconia, Due to Dilatational Phase Transformation" Acta Metallurgica et Materialia, submitted (1991).

9. RAMAKRISHNAN, N., OKADA, H. and ATLURI, S.N. "Computer Simulation of Transformation Induced Plasticity Using Finite Element Method" Acta Metallurgica et Materialia, 39, 1297-1308 (1991).

10. WU, J., ZHU, J.Z., SZMELTER, J. and ZIENKIEWICZ, O.C. "Error Estimation and Adaptivity in Navier-Stokes Incompressible Flows" Computational Mechanics, 6, 259 (1990).

11. ZHANG, J-D. and ATLURI S.N. "Post-Buckling Analysis of Shallow Shells by the Field-Boundary-Element Method" Int. Jnl. Num. Meth. in Engg., 26, 586-615 (1988).

12. ZHANG, J-D. and ATLURI, S.N. "A Boundary/Interior Element Method for Quasistatic and Transient Response Analysis of Shallow Shells" Computers & Structures, 24, 213-224 (1986).

13. ZIENKIEWICZ, O.C. - The Finite Element Methods in Engineering Science, McGraw-Hill, 513 pages, 1971.

FINITE ELEMENTS IN THE 90's, E. Oñate, J. Periaux, A. Samuelsson (Eds.)
Springer-Verlag/CIMNE, Barcelona 1991

CONCERNING THE FUTURE OF CIVIL ENGINEERING

E. R. de Arantes e Oliveira
Laboratório Nacional de Engenharia Civil,
Avenida do Brasil, 101 - 1799 Lisboa Codex,
PORTUGAL

O.C.Zienkiewicz is one of those who, in our time, more have influenced the evolution of civil engineering.

His outstanding work in Computational Mechanics definitely contributed to open the path for a future which the present paper concerns.

1 - Although there is a large risk involved in trying to fit into a formula the vast world of civil engineering, I think it is worth starting to define its aim: the planning, design, construction, maintenance and rehabilitation of major fixed equipments which are known as works.

As large, and therefore costly equipments are involved, the future of civil engineering will be determined by large-scale demand and this will, in itself, be influenced by the evolution of supply, that is by technological innovation.

Investment in large-scale fixed equipments will be affected, as it always has been in history, by the conceptions in force in Society.

Accordingly, in advanced societies, the evolution of civil engineering will increasingly be conditioned by respect for values such as the right to life and the happiness of citizens, and preservation and enhancement of the natural and cultural heritage.

The real needs of people, which to a great extent will dictate demand, will have less and less to do with mere survival and more and more with the quality of life.

For this reason, and because large equipments have a very significant influence on the environment, the planning and conception of works will increasingly be based on environmental impact studies .This means that environmental policies will be determinant in the future of civil engineering.

It may be noted that the quality of life inspires and is at the same time one of the outstanding aspects of a philosofy (especially adequate for post-industrial societies) based on the promotion of quality.

2 - The planning and design of works present themselves to the modern civil engineer as an optimization process and, therefore, as a choice of the optimum solution from a set of possible solutions.

Such choices will tend to be preceded by increasingly fine cost--efficacy analyses, involving considerations that are not exclusively economic or utilitarian, and this will mean taking into account cultural, social, moral or psychological factors that are hard to quantify.

Owing to the limitations of resources to which advanced societies are particularly sensitive, the reduction of costs will call for the rationalization aimed by the theory of safety, i.e, for conceiving works so as to minimize their overall costs, i.e., construction costs + operation costs + costs associated with lack of quality (including in the latter costs for prevention of dammages and for evaluating quality, as well as costs connected with the risks associated with failure to comply with the essential requirements).

Costs, however, even when generalized, must only be used for an overall orientation.

It must be recognized indeed that the progress of civil engineering requires not only the attribution of more realistic values to the parameters that quantify safety, which will call for better knowledge of the behaviour of works and the forces likely to act on them, but also a better quality management, which involves the implementation of procedures for controlling all the phases, from the conception to actual use of the works.

3 - Knowledge of the behaviour of works means systematic monitoring of works of the same type already constructed, and the preparation of models incorporating the result of that monitoring.

In the engineering of structures, such models nowadays are almost exclusively mathematical. In hydraulics, physical models are still preponderant but mathematical models are increasingly being used.

It is likely that in the future the mathematical models will be capable of simulating the different kinds of failure of works and enabling us to analyse by computer the behaviour of works subjected to actions likely to cause them.

4 - The quality of works is closely connected with the quality of the construction materials.

As regards the latter, it is to be expected that concretes will continue to hold first place; it is unlikely that they will be replaced, but simply improved.

In concretes, as in steels, it will be dangerous to increase their strength if this increase leads to greater brittleness. High-performance materials are of more interest than high-strength materials.

This leads to the problem of durability, which involves not only the evolution in time of the properties of the materials, but also the adaptability of the works to new requirements.

Works cannot be constructed for ever and the initial cost of a construction is in principle - failing a qualitative leap forward in what technology has to offer - as great as its expectancy of life is long.

Nowadays, however, we are fully aware that, broadly speaking, the overall cost of works reaches its minimum for values of the respective durability that are greater than those which until recently were considered acceptable.

It is the need to guarantee the use of works during the longest periods for which they will have to be constructed that causes preoccupation with their adaptability, and this involves aspects connected with their inspection, maintenance, repair, adaptation and demolition.

As regards materials, it is to be noted that reinforced concrete constructions, even more than steel constructions, give an impression of being perennial that does not correspond to reality. It is known that throughout the life of the work the aggregates gradualy react chemically with the cement and that, owing to these reactions, at long term the concrete tends to be permeable and, at still longer term, to lose cohesion.

Large structure in simple concrete, such as concrete dams, are thus threatened with unserviceability and even collapse.

Reactions to the environment, moreover, mean that the concrete, initially basic, tends to become neutral, and in reinforced concrete works this will lead to chemical attack on the reinforcements.

Special techniques based on the incorporation of molecules with dimensions in the order of a few "angstroms" may make concrete more compact and therefore less permeable. Painting of the surfaces may also prolong its life.

It is possible, however, that natural rocks and bricks, owing to their capacity for withstanding the aggressivity of the environment, in particular marine environment, may later be reinstated as construction materials.

As regards plastics, it can be said that steel components may come to be partly replaced by plastic components, which are less subject to corrosion and easier to execute and assemble. Since, however, they are less durable than steel components, they will have to be periodically substituted, and this suggests that they will mainly be used in light constructions.

The civil engineers of the 21st Century will be faced with major rehabilitation works, or even with the large-scale demolition of constructions which today seem to be of unquestionable quality.

Hence it seems that a possible diminution in investments in new works in the advanced countries will be largely offset not only by the construction of new works in the less developed countries, but also by the rehabilitation of infrastructures that now exist ,but are increasingly degraded, in the former.

Natural materials such as soils and rocks, whose durability is generally grater than that of manufactured materials, and whose evolution accompanies the rhytm of Nature, will also benefit from the fact of their behaviour becoming better known and the availability of suitable techniques for making use of them.

Modern technologies nowadays make possible underground works of large dimensions, for the most diverse purposes. Pressure from urban development, industry and of an environmental kind will mean opening more and bigger cavities in the subsoil. In works of this type, the material will be used in its natural deposit form.

Great progress is also to be expected as regards foundations at the bottom of the sea, even at great depths.

5 - In speaking of the seas and oceans, it must be mentioned that port structures (harbours, terminals, etc.) will continue to be infrastructures of vital importance on an international plane. It is thus possible to look forward to reasonable development in this sector.

Problems of the coastline will grow exponentially, mingling with those of estuaries and lagoons. These are preferential areas for human occupation and economical development, and have great ecological sensitivity and environmental interest.

The much discussed possiblity of a rise in the level of the seas, due to the greenhouse effect, will lead to growing attention to those areas, in particular to coastal protection works.

Inland water media, both in the component of rivers, brooks and other surface watercourses, and in that of groundwater, represent an important resource for domestic, public, industrial and agricultural supply, discharge of wastewater, energy production, river navigation, leisure and recreation. Their rational use will orientate the development of hydraulic civil engineering in the sector.

It is important to bear in mind that the future of hydraulics is closely bound up with the world water market, in which quality problems have been gradually gaining weight, although without the quantitative aspects being neglected.

Basic sanitation, which covers water supply and the discharge and control of wastewater from urban and industrial centres, is vital for public health and comfort, as well as for the protection of the environment. Development of the sector is unquestionable and challenges are to be expected as regards the desalination of seawater, increasingly sophisticated depollution techniques and the reuse of wastewater for irrigation.

6 - As regards the actions on works, it can be said that civil engineers will have to be more and more aware that rather than oppose the forces of Nature, they should channel them for man's benefit and this, of course, calls for a better knowledge of them.

The conception of the engineer fighting against the environment is out of date. Struggling against Nature represents, at the best, a waste of energy and this means that the better the works are integrated in the environment, the more economical they should be, in principle.

That is to say that progress in civil engineering calls for a better knowledge of Nature.

7 - Knowledge of Nature, however, is not enough. It is necessary for civil engineering to be equipped with results and methods originating both in the sciences on which it has traditionally been based and on other sciences, both hard and soft.

Mention may be made, inter alia, of architecture and town-planning, urban anthropology and psychology, the sciences of computation and information, medicine and public health, economics and management, the sciences of education, sociology, demography, the sciences of the environment, the juridical sciences and, as Ferry Borges reminds us (see "Ciência e Construção", LNEC Report, 1991), ethics.

8 - Reference to so ample a range of knowledge does not, of course, require that civil engineers shall master all the sciences, but it will oblige them to work in pluridisciplinary teams of a more and more complex kind.

It will also oblige them to resort systematically to instruments of the expert system type, and other intelligent systems whose application in civil engineering is still in the development stage.

Then there are the sciences, techniques and industries of information, whose evolution will have important economic and social consequences affecting the whole productive sector and, therefore, the construction industry.

The use of such instruments will not, however, preclude the need for an integrated training of a different type from that which has hitherto been given by engineering schools and, generally speaking, by the educational system.

The fact is that the civil engineer of the future must be a technician who is informed about the progress in his field and open to cooperation with professionals of other fields, but he will have to be more than a technician: a cultured and socially responsible man.

ACKNOWLEDGEMENT: The author wishes to express his thanks for the help received in preparing this article from various Heads of Department of LNEC and, in particular, from Dr. Julio Ferry Borges.

FINITE ELEMENTS IN THE 90's, E. Oñate, J. Periaux, A. Samuelsson (Eds.)
Springer-Verlag/CIMNE, Barcelona 1991

REFLECTIONS ON TECHNICAL JOURNAL EDITING

R. H. Gallagher
Clarkson University,
Potsdam, New York 13699-5500,
U.S.A.

Olgierd Zienkiewicz's career is remarkable in its own right, in terms of the wealth of contributions he has made to an astonishing variety of scientific fields. It is still more remarkable when one reflects on the continually expanding output and the level of sophistication of this output, even as he reaches his seventieth birthday. Thus, it is fitting that this volume contain a full measure of erudite contributions, written by friends and former collaborators. The reader, no doubt, avidly anticipates the opportunity to sample a high level of theory and advanced formulations.

Our contribution is offered with apologies for its departure from the regimen of such works. It's development began with the objective of setting out a broad discussion of modern technical journals -- their origins, their present status, and their future. The limitations of space preclude that treatment here, although we do briefly touch upon these matters. Mainly, we reflect on the value and limitations of technical journals, their editors, and their publishers.

Despite the mismatch of this contribution and others in this volume, this paper is not entirely out of place. One of our honoree's most significant and, perhaps in the long run, his most enduring contributions is the creation, almost 25 years ago of the *International Journal for Numerical Methods in Engineering*. His unstinting devotion to the high standards of technical periodicals, in this and other editorial capacities, should be an inspiration to every scientist and engineer.

SOME FACTS AND FIGURES ABOUT JOURNALS

A debate has raged for many years concerning the origins of finite elements. Some people advance the "romantic" notion that finite elements began with principles enunciated by Archimedes. Others assert that it began, for all practical purposes, with a series of papers in technical journals in the 1950's. Similarly, for scientific periodicals, one might assert that it began with cave dwellers etching their newest ideas onto rocks. The reality, however, is that the first scientific periodicals appeared in the mid-1600's.

The very first was the Journal des Scavans, published in 1665 by de Sallo, a counsellor of the French court of parliament. Within a few months, in that same year, the Philosophical Transactions appeared. Both have persisted to this day.

The number of different technical journals grew substantially in the subsequent centuries, but never so dramatically as in the past 20-30 years. "The Journal that Ate the Library" is not the name of a science-fiction novel but the eye-catching title of an article written a few years ago about the cost dilemnas surrounding technical journals.

Today, many individual scientists and engineers subscribe to five to ten journals. The library of Clarkson, a technological university, subscribes to 500 scientific journals. At the opposite extreme, Harvard University library lists (for 1989-90), over 103,000 journal titles. The total expenditures of this library in that year were on the order of $46 million. Its journal subscription cost to Harvard is most likely in the range of $10 million to $20 million! The aggregate cost of journal subscriptions, worldwide, is over a billion dollars.

Journal acquisition costs at US universities have been recognized to be in a critical state for at least five years. This is one of the dominant issues in the financing of academic institutions. Needless to say, the financial situation in academe is exceptionally tight, and technical journal subscription budgets are targets for reduction. As this is written, a news article appeared which stated that the University of Tennessee has cancelled 1,200 of a total of 8,000 journal subscriptions. Thus, there is an evident problem with respect to technical journals. University librarians assert that the key problem with respect to serial publications is the proliferation of new titles. The counter argument is to the effect that technology has advanced explosively, with the periodic creation of new fields. It is an issue about which one could speak volumes (no pun intended), but for that very reason we will not delve into it here. Rather, we remark on some virtues and faults of journals in themselves.

WHY JOURNALS?

The system of periodical technical literature, otherwise known as "serials" or "journals," whose development has been massive in the past three decades, has been the key medium for the exchange of information in the modern technological revolution.

Other media play a role, to be sure. Books, for example, generally have a different function than serials. They must be prepared with a care that generally excludes rapid appearance, and the sales penetration they must achieve means that they cater to an audience that has less interest in developments at the frontier of progress. At the other extreme we have laboratory reports. These quite definitely have an impact. Unfortunately, they reach a very small audience and have given way, to a large extent, to published journals.

Scientific journals play a central role in the modern technological revolution. By providing an outlet for published papers the scientific journal enables scholars to bring to public attention the results of their endeavors. Those who read the journals are able to make use of those accomplishments or even to move them ahead a few steps further, a process characterized by Isaac Newton as progressing by "climbing on the shoulders of giants."

The process through which journal articles make their contributions is not well understood. We do not, as I see it, have a cadre of specialists reading their own copies of the journals as they appear, with the subsequent transformation of those results into practice and further research. The key issue here is contemplation. That is, the researcher must be able to read journal contents carefully and consider them in a quiet environment.

To make contemplation workable, photocopying is essential. The vast majority of the subscriptions of most scientific journals are taken by libraries. The careful reading that I refer to takes place over weeks and months, once photocopying has been accomplished. The photocopying process, therefore, makes the existing scheme of journals attractive to the researcher. This is quite opposite to the prevailing view that photocopying is the bane of authors and publishers. It may be the cornerstone of the success that their efforts enjoy.

There are other functions fulfilled by journals, besides being stepping stones for further research. A plague of modern research is duplication of effort. The scientific periodical literature has the distinct function of making researchers aware of ground that has already been trodden. At the same time we must note that our ability to identify duplication still needs to be strengthened. Two developments in my acquaintance have appeared in literature no less than seven separate times in the past ten years.

Many journals owe their existence to the fact that a community of scholars and practitioners does not believe that its interests are being served adequately by existing publications. Once upon a time this resulted from there simply not being a journal in the field represented by the new publication. The policies of established publications have also shaped the changes that have taken place.

The *International Journal for Numerical Methods in Engineering* is a case in point. During the 1960's it was not uncommon for scientific journals to challenge the legitimacy of papers that dealt significantly with electronic computation. Computers and all of their evil works were suspect and were likely to represent a debasement of the archival literature. A host of new journals devoted to computer applications then emerged, and I daresay they are today in the forefront of technical literature.

WHERE JOURNALS FALL SHORT

So much for the clear-cut virtues of technical journals. I would now like to take up some criticisms leveled against them.

To be realistic we must include the phenomenon of pressure on the individual to publish. Up until 1950 the presence of a Ph.D. on a faculty was a rare circumstance. Now, it is a universal requirement for academic employment. U.S. engineering colleges alone recruit 1000 Ph.D.'s to their faculties each year; in science it is probably twice as many. Today, there are over 20,000 Ph.D.'s on engineering faculties of U.S. schools. If each of them produced, on the average, just one published paper each year, the output would fill five hundred technical periodicals each year.

This brings us to a widely criticized motivation for publication, the establishment of a record for academic advancement. The charge is often made that higher education has gotten off the track on account of an inordinate fixation on publication record as the principal criterion for promotion and tenure. Scholars may be motivated to rush as many articles into print as is possible, with no concern about quality or the audience to be reached. Undoubtedly, there are circumstances where this is the case.

Another motivation for publishing in periodical journals is to establish priority, with no real concern for the audience. In still others it represents a thinly disguised sales overture for a product of the author, a peril that has grown as computer software has become more marketable.

There are a number of other legitimate concerns at the operational level. I would like to comment about four: speed of publication, the quality of journal contents, editorial responsibility, and oversight.

We are all concerned about the rapidity with which the results of scientific research reaches the hands of the user. Although the fastest way to learn about ongoing research is to visit other laboratories and attend technical meetings, this approach has limited feasibility. We must rely on technical journals.

Technical journals, however, have a delay of 1-2 years between submittal and publication. The editor has control of the review cycle, which might take anywhere from three to nine months. The author spends up to three months in revisions. Finally, depending on the backlog of the journal in question, the paper is six to twelve months in press. Specialists see the paper in the library some months later.

It is clear, then, that there is a strong impetus to reduce the length of this cycle. One approach is through 'Communications' journals, which feature shorter papers. These, as I see it, can only reduce the review portion of the cycle, and they may eventually fall prey to the problem of backlog. Alternatively, the refereeing segment of journals that feature full length papers could be shortened. There are, however, considerable perils in the quality of the refereeing process when it is rushed.

Quality in the selection of papers for publication is a critical subject. In the best of all worlds, every submitted paper is scrutinized by two, or even three or four specialists. The editor then decides on acceptance or rejection, without favoritism or fear of recrimination on the part of the author.

In the real world, however, there are often actions which contravene the goals of the manuscript selection process. Leading figures in the field are "stroked" by accepting their manuscripts without review. Disappointed authors apply pressures to reverse rejection decisions, noting that they and their colleagues will take future business to any one of the many competing journals. We could add to this list.

These are the failings of journal editors. Publishers bear a responsibility as well. They need to do more to educate the editors of the journals they sponsor, perhaps to go as far as defining a "Code of Ethics." They need to identify bad practices and then progress to the construction of a set of principles to be adhered to by all reputable publications.

Journal editors often try to avoid conflict and this is a factor that influences the quality of the manuscripts that are published. In many ways this is unfortunate, because conflict is not merely a way of identifying error and setting out on the proper path, but it has the effect of clarifying issues for the readership. We cannot, of course, avoid conflict entirely, and here is where we again confront one of the major limitations of editorial practice.

Finally, I come to the matter of the oversight of the publications that now exist and those which will appear in the future. In other words, how is the publisher to know that the editor is doing his or her job? Is the editor fair in adjudicating disputes? Is the editor prompt in the acknowledgements that are necessary in all phases of the publication cycle? How was the editor chosen for this task in the first place?

Here, again, I believe that industry-wide standards need to be set out. Norms for the quality and the time period of the review process should be delineated. Responsibilities should be defined for advisory editors. Perhaps terms should be set for the editorship. Policies regarding reprints might be agreed upon.

What do we stand to gain from standards? We bear the obligation to search for the means to strengthen the quality of papers that appear in our scientific journals, and I believe that journal quality control has been neglected. If reputable publishers are known to adhere to a common set of editorial standards then librarians will be reluctant to subscribe to journals that do not meet those standards.

FUTURE DEVELOPMENTS

Futuristic developments are often touted as having the potential of altering dramatically the nature of scientific communication. These are expected to transform technical journals as we know them today. I wish to avoid commenting upon them here; the topic is complex and worthy of a much lengthier paper.

I would like to remark, however, on certain developments at the universities which I believe will have direct impact on scientific and technical publishing. These remarks are associated with the so-called personal computer revolution.

Now, in every case, the above action will have been preceded by universal faculty ownership of personal computers. As recently as five years ago, faculty were unreceptive to personal computer utilization -- perhaps 30 made use of them. Today, personal computers are more widespread and virtually everyone employs and is comfortable with them. Moreover, within the near future, if they have not done so already, institutions will be equipped for networking of these devices within the campus and for communication with the other campuses.

Recognizing that a high percentage of U.S. scientific and engineering research takes place at relatively few schools, say at 100 universities, I believe that new modes of communication of research results based on personal computer and work station networks will emerge. Who will be in the forefront of these developments? My guess is that it will emanate from faculty collaboratives, supported at the outset by federal research agency funding. Yet, the editors and publishers of scientific periodicals have a wealth of experience to contribute to this, and it is my earnest hope that they will not only contribute, but exercise leadership.

FINITE ELEMENTS IN THE 90's, E. Oñate, J. Periaux, A. Samuelsson (Eds.)
Springer-Verlag/CIMNE, Barcelona 1991

THE QUENCH PROPAGATION IN SUPERCONDUCTING MAGNETS. A FINITE ELEMENT SOLUTION

L. Bottura
The NET Team,
Boltzmannst. 2, D-8046 Garching,
GERMANY

SUMMARY

Superconducting magnets are in an unstable thermal equilibrium. Deviations from the equilibrium can cause a propagating thermal wave driven by non-linear Joule heat generation. The analysis of this process, called the quench of a magnet, is a vital design requirement. This paper deals with the numerical analysis of the quench propagation in superconducting coils with 3-D, arbitrary shape based on the finite element solution of a set of parallel 1-D problems represented by the compressible flow of the coolant, heat conduction and non-linear heat generation.

1. INTRODUCTION

Superconductivity occupies a growing area of importance in the generation of magnetic fields for both small and large scale applications. Its main appeal is obvious: the resistive losses associated with the flow of current in normal conductors disappear in a class of elements, compounds or alloys called superconductors. This allows the generation of large magnetic fields at very high current density, thus increasing compactness and cost effectiveness. However, there are strict limitations in the performance of the superconducting materials. The first -and more evident- is the operating temperature, which is required to be below a critical value T_c to have superconductivity. The typical range of T_c of *technically relevant* materials (i.e. of common use in industry) is between 10 and 20 K. Similar limits exist on the maximum magnetic field produced, or critical field B_c, and on the maximum current density which can be carried , or critical current density J_c. In addition, there is a relation among these limits: the increase of any couple of parameters, temperature field or current density, produces a decrease in the third.

This relation is usually represented in a very effective form in the $J_c(B,T)$ surface such as in Fig. 1. Whenever one of the limits is exceeded, i.e. when the operating point is above the $J_c(B,T)$ surface, the superconductor exhibits a phase change into the normal conducting state. In a current carrying superconductor an irreversible process begins: the onset of the Joule heating produces a temperature increase in the superconductor and prevents the recovery of the superconducting state. In these conditions the magnet must be switched-off to prevent damages due to over heating or over voltages as a consequence of the development of an

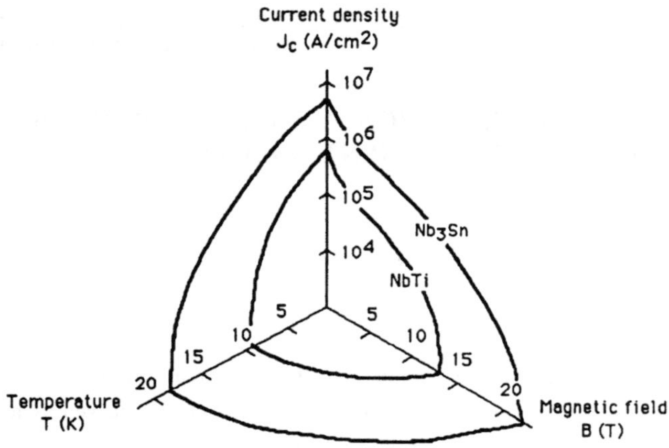

Fig. 1: Jc(B,T) surfaces for commonly used type II superconductors, the intermetallic compound Nb3Sn and the alloy NbTi

internal resistance. This unstable process is called *quench* of the superconducting magnets and usually ends with the dump of the magnetic energy stored in the magnetic field onto a discharge resistor. In the process are involved (at least in first approximation !) non-linear three dimensional heat conduction in the winding pack and heat convection through one dimensional compressible flow of the coolant, with non-linear Joule heat generation coupled to the evolution of the current in the coil and to the magnitude and distribution of the magnetic field. Complexity is not a missing ingredient.

The analysis of the quench propagation has always been a matter of fundamental importance for the design of the protection scheme of a superconducting magnet [1] . This is true for instance for the large superconducting magnets which are under design for fusion machines of the next generation [2] . These magnets will be a considerable investment compared to the cost of the whole plant, and will store an amount of energy in the order of several tens of GJ. Therefore there is an increasing interest in the issues of safety and protection of these systems

Because the magnet manufacturing techniques are various and the resulting geometries are difficult to generalize, the attention is restricted here to large size magnets manufactured using composite conductors cooled by the forced flow of gaseous, supercritical helium. A schematic view of the cross section of the conductor and of the typical winding geometry considerd is given in Fig. 2 and Fig. 3

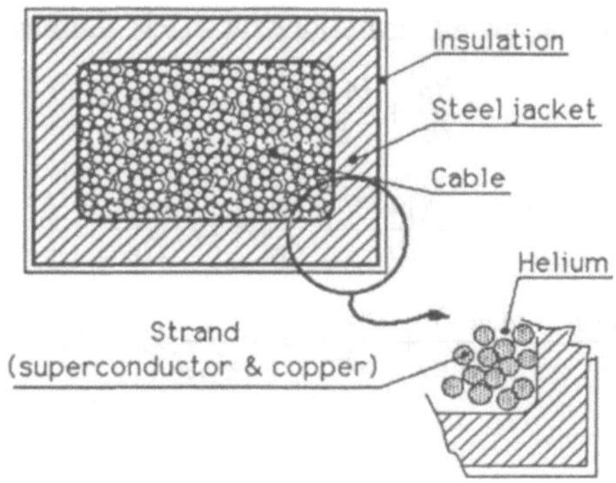

Fig. 2: Schematic view of the cross section of the conductor configuration assumed

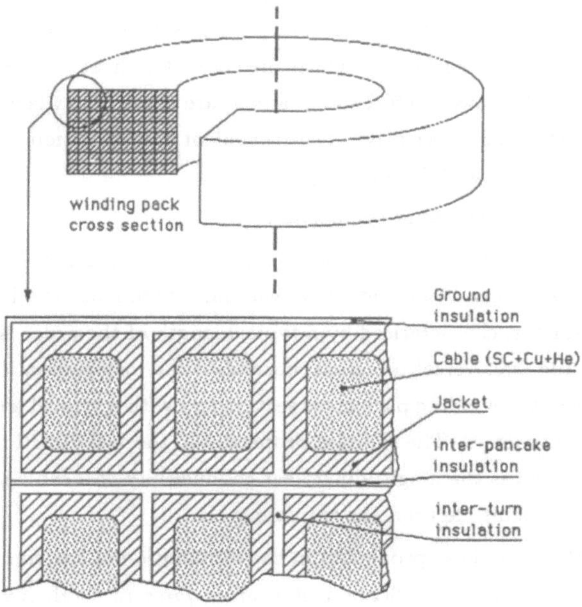

Fig. 3: Schematic view of the winding pack configuration assumed

2.EQUATIONS

For a force-flow cooled conductor the helium flow can be regarded as one dimensional along the cooling channels with a good degree of approximation. We assume throughout this work that the flow always remains in single-phase, a condition satisfied when supercritical helium is used as coolant. The analysis of the quench propagation requires the solution of the following equations:

(a) fluid flow and energy transport equations for a 1-D channel

$$\frac{\partial \rho}{\partial t} + \frac{\partial(\rho v)}{\partial x} = 0 \tag{1}$$

$$\frac{\partial(\rho v)}{\partial t} + \frac{\partial(\rho v^2)}{\partial x} + \frac{\partial p}{\partial x} = 2f\frac{\rho v|v|}{D_h} \tag{2}$$

$$\frac{\partial(\rho e)}{\partial t} + \frac{\partial(\rho e v)}{\partial x} + \frac{\partial(p v)}{\partial x} = h\frac{S}{A_{he}}(T_s - T_{he}) \tag{3}$$

$$e = i + \frac{v^2}{2} \tag{4}$$

The set of unknowns is the density ρ, the velocity v, the helium pressure p and temperature T_{he}, the total and internal energy e and i respectively. In above A_{he} is the cross sectional area of the helium conduit, D_h is its hydraulic diameter and S is the perimeter of the surface on which the convective heat transfer from the solid takes place. The two coefficients f and h are the friction factor and the heat transfer coefficients, descriptive of the turbulent flow conditions. A complex state equation relates p and T to i and ρ.

(b) Heat conduction in the solid composite :

The winding pack of a force-flow cooled superconducting coil presents a large anisotropy, where we can clearly identify two main directions: the longitudinal axis of the conductor, characterised by a typical length of the order of hundreds of meters, and the transversal direction, normal to the longitudinal axis, which lies in the plane of the winding pack cross section and with a typical length of the order of some centimeters. The thermal properties in the two directions are also largely non-isotropic as the longitudinal heat conductivity of a cable is larger by several orders of magnitude than the transverse one.

It is natural to use these properties in the formulation of the computational model. In fact, the mere application of a *brute force* method, involving full meshing the three dimensional domain, would be highly inefficient in terms of computational costs and could give serious numerical difficulties. The approach used here is to discretize the coil using a set of 1-D channels, and to couple these using an appropriate model of the transverse thermal resistance and heat capacity of the winding pack. A schematic view of this model is given in Fig. 4. For the present purpose it is convenient to deal only with the one dimensional form of the heat conduction equation, written along the longitudinal conductor axis x, valid in the hypothesis of an homogeneous component with cross section A_i, heat capacity C_i, thermal conductivity K_i and temperature T_i:

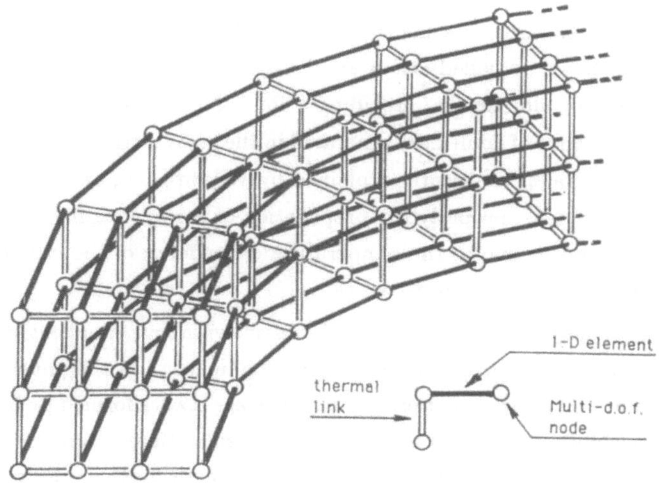

Fig. 4: Approximation of the original 3-D problem with a set of coupled 1-D channels thermally linked in the cross sections of the winding pack

$$A_i \rho C_i \frac{\partial T_i}{\partial t} - A_i \frac{\partial}{\partial x}(K_i \frac{\partial T_i}{\partial x}) =$$

$$h_1 S_{i-he}(T_{he} - T_i) + \sum_j h_{ij} S_{ij}(T_j - T_i) + A_i \dot{Q}_i \tag{5}$$

The source terms on the r.h.s. of eq. (5) are respectively the heat exchange with the helium at temperature T_{he} through an heat exchange coefficient h_1 and a contact surface S_{i-he}, heat exchange with another conductor j at temperature T_j through an heat exchange coefficient h_{ij} and a contact surface S_{ij}, and the density of heating power (external or Joule) per unit length.

(c) Joule heat generation: a special attention in the superconducting composite is given to the Joule heat generation term. The Joule heating is a function

$$\dot{Q}_J = \dot{Q}_J(T, B, J) \tag{6}$$

of the temperature T and the field B, in turn functions of position and time, and of the operating current density J in the conductor, dependent only on time through the solution of circuital equations. The Joule heating $\dot{Q}_J$ is related to the current carrying capability of the conductor specified by the critical surface $J_c(B,T)$, of which two typical examples are reported in Fig. 1. The calculation of the Joule heating requires the evaluation of the ***current sharing*** from the superconductor to the so called ***stabilizer***, a material with low electrical resistivity such as copper or alluminium, providing a shunt for the excess current and decreasing the amount of Joule heat generated in the cross section of the cable. The details of this process are clearly explained by Wilson [1] , and must be taken into account to model properly the propagation of the quench front. The whole process is non-linear due to the temperature dependence of the critical current density and of the material resistivities.

(d) Circuit equations and magnetic field: As shown above, the heat production term depends on the current flowing in the conductor, and this must be determined at any time by the solution of the electrical network formed by inductively coupled coils and the external electrical components. Here a non-linearity is intrinsical in the coil resistance. Finally, the magnetic field in the coil can be computed using standard techniques.

3. MODEL

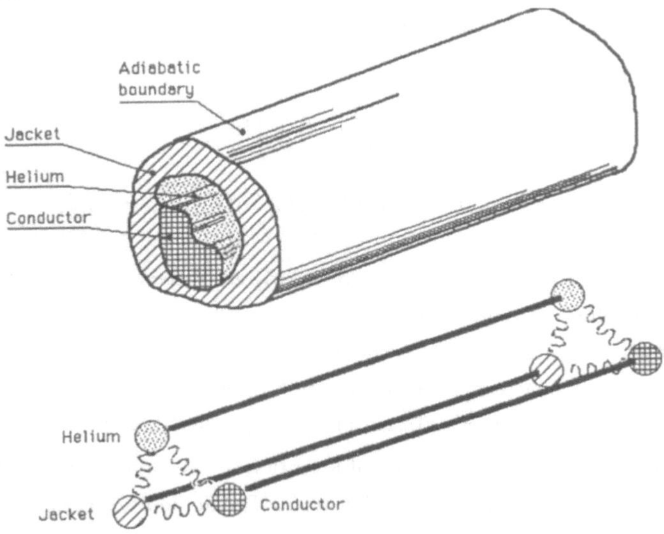

Fig. 5: Schematic view of the 1-D model used for the conductor in the longitudinal direction

The basis of the quench analysis model is a 1-D multi-dof element shown in Fig. 5. In the case considered here it consists of a 1-D representation of the cable in its longitudinal direction comprising strands, helium and jacket considered as three independent components. The reason for this break-down is that the stabilizer (copper) in the superconducting strand has a conductivity several orders of magnitude greater than that of the other components in the whole temperature range (from cryogenic temperature to room temperature). On the other hand the dominating heat capacities are those of the helium in the low temperature range (4 to 30 K) and those of steel and insulating materials (epoxy-resins) in the high temperature range (30 K to room temperature). Because of these differences in the material properties, it is not possible to lump all heat capacities and thermal conductivities into one node, as this would suppress most of the dynamic of the response of the cable cross section.

These 1-D elements must be coupled in transverse direction to simulate the 3-D heat conduction. The are several possibilities to derive equivalent lumped or distributed parameters for the thermal constants among nodes in transverse direction. Eigenvalue analysis [3] or model reduction techniques can be used to reduce a complex model, such as a 3-D mesh, to simpler equivalents. Engineering common sense can also give good and cheap guidelines. Looking at the winding pack, one sees that the cable spaces are thermally separated by a composite slab structure. A quick estimate gives for the composite a diffusion time constant of the order of 100 ms, while the quench propagation is expected to develop on the time scale of several seconds. Transient effects in the composite slab can therefore be neglected, and an equivalent non-linear thermal resistance can be substituted to simulate the coupling. This procedure was tested against the more sophisticated, exact (and expensive) eigenvalue analysis of a typical transient response, and was shown to give indeed very good answer [4] .

A detailed view of the resulting model for the coil is given in Fig. 6, where the structure already anticipated in Fig. 4 is expanded.

4. SOLUTION STRATEGY

Non-linearity was the keyword in the choice of the solution methodology for the coupled problem. To offset the complexity introduced by the material properties, geometry and coupling, explicit solvers were used as far as possible. Obviously this gave stability limitations in the time stepping and in the coupling procedures. Automatic time step selection, based on the strictest stability limitation, was used to optimise the run-time but still guarantee stable solutions. When computationally convenient the explicit treatment was substituted with a semi-implicit (linearised) treatment to relax the stability limits.

In particular, the 1-D compressible fluid flow was solved using the Swansea two-step algorithm [5,6] . The two-step form was preferred to the single step one

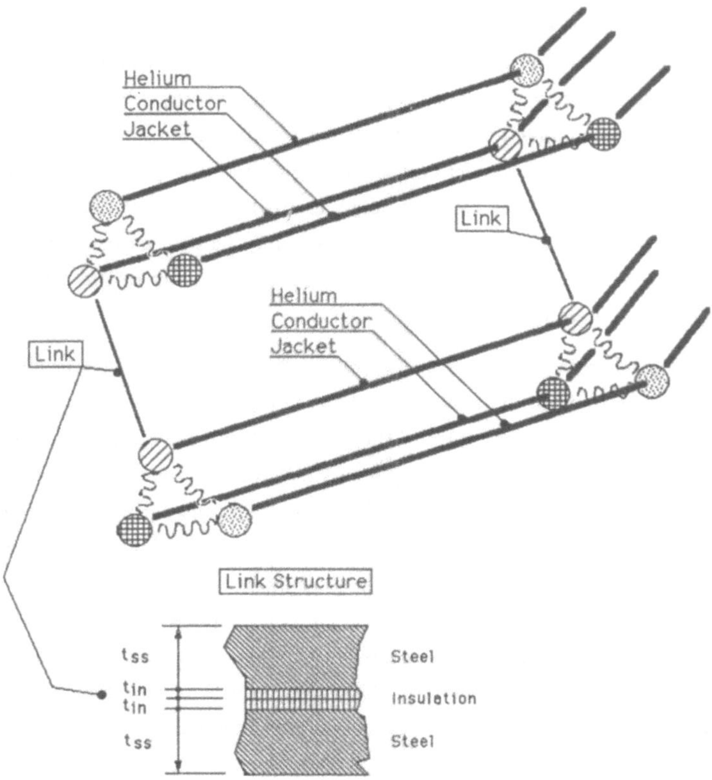

Fig. 6: Schematic view of the details of the coupled 1-D channels equivalent to the initial 3-D problem. The internal structure of the coupling links is shown

because it required the evaluation of the fluxes only, avoiding the calculation of derivatives of the material properties, a process which can be extremely expensive for helium at low temperature. Because of the presence of a non-linear source term a stability condition on the source number [4] had to be added to the usual limit on the Courant number [5] . Typically, the limits on the time step were of the order of 0.1 ms due to the momentum source, 100 ms due to the energy source and 1 ms due to the Courant limit.

For the heat conduction in 1-D (along the conductor length) the equations were linearised, taking the material properties at time n, and solved implicitly using a three level algorithm [7] . Matrix storage and inversion took into account the sparse structure, and proper node numbering was used to minimise the bandwidth. The coupling between the solid and the helium had also to be treated implicitly to avoid a drastic decrease of the stability limit on the time step due to the small thermal constant of the cable components compared to the large heat transfer coefficient to the helium. On the other hand the thermal coupling among the cables in the winding could be treated explicitly because of the longer time constants involved (typically 100 ms).

Finally, a circuit solver and a magnetic field calculation were coupled explicitly in the time stepping. Again the large time constants of the circuits analysed (of the order of seconds) allowed the explicit form without any penalty on the chosen stable time step.

5. AN EXAMPLE OF APPLICATION

One of the checks of the model was based on a quench propagation experiment performed on a large scale coil, the EURATOM-Large Coil Task (LCT) magnet, during the tests at Oak Ridge in the frame of the international Large Coil Program [8] . In this experiment a quench was induced in one turn of the coil, and the coil voltages were monitored for a time sufficiently long to see the propagation of the quench between two adjacent turns. The 3-D model was used to simulate the LCT geometry, initial conditions and the evolution of the quench.

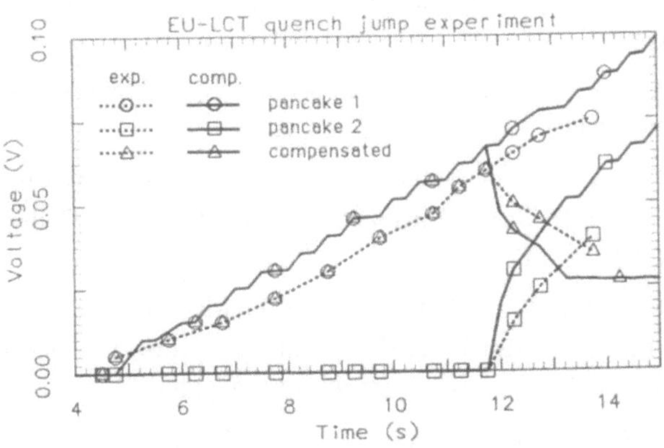

Fig. 7: Computed and experimentally measured voltages in the EU-LCT quench-jump experiment

The voltage traces computed by the code are compared in Fig. 7 to the values measured during the experiment. These last have been offset in Fig. 7 so that the quench in the adjacent turn starts at the same time. The voltage increase rate computed is very close to the measured one, as well as the quench jump time. The differences on the voltage might possibly due to uncertainties in the copper resistivity. Note also how the compensated voltage (difference of the voltages in the two portions of the coil) reaches a peak and then decreases as observed experimentally. The experiment can be said to agree very well with the simulation...

6. CONCLUSIONS

The model developed and the solution procedure are implemented in a computer code which is being used on a nearly daily basis for the prediction of quench evolutions. We consider a great success the ability of performing these highly non-linear simulations in a complex 3-D composite with a speed which allows the routine use of the code at moderate CPU costs. As it has been shown, excellent accuracy has been achieved by the use of finite element based solvers for the fluid flow and the heat conduction. All this was the fruit of an intense collaboration between the apprentice (with a medium engineering sense) and an excellent teacher (with a vaste experience in the numeric). I was the apprentice. There should be no need to mention who the teacher was: Olek Zienkiewicz.

REFERENCES

1. WILSON, M.N., Superconducting Magnets, Clarendon Press Oxford, 1983.
2. The Next European Torus (NET), Special Issue of Fusion Technology, **14**, 1, (1988).
3. GUYAN, R.J., Reduction of Stiffness and Mass Matrices, AIAA Journal, **3**, 380 , (1965).
4. BOTTURA, L., The Numerical Solution of Quench in Superconducting Magnets, Ph.D. Thesis, University of Wales, University College of Swansea, Wales, UK, (1991).
5. Peraire, J., PEIRO, J., FORMAGGIA, L., MORGAN, K., ZIENKIEWICZ, O.C., Finite Element Euler Computations in three Dimensions, Int. Jour. for Num. Meth. in Eng.ng, **26**, 2135-2159, (1988).
6. ZIENKIEWICZ, O.C., TAYLOR, R.L., The Finite Element Method, 4th Edition, Volumes 1 and 2, McGraw Hill, 1989-1991.
7. LEES, M., A Linear Three-Level Difference Scheme for Quasi Linear Parabolic Equations, Maths. Comp., **20**, 516-522, (1966).
8. BEARD, D.S., et al. editors, The IAEA Large Coil Task, Special issue of Fusion Engineering and Design, **7**, (1988).

FINITE ELEMENTS IN THE 90's, E. Oñate, J. Periaux, A. Samuelsson (Eds.)
Springer-Verlag/CIMNE, Barcelona 1991

FINITE ELEMENTS AND MAGNETIC SCALAR POTENTIALS

J. Simkin and **C.W. Trowbridge**
Vector Fields Ltd,
24 Bankside, Kidlington, Oxford OX5 1JE,
UNITED KINGDOM

SUMMARY

Scalar potentials are widely used in the solution of magnetostatic field problems. This paper discusses the underlying theory and the difficulties that must be overcome in order to exploit to the full the substantial gains in economy and efficiency that are expected.

1. INTRODUCTION

1.1. The Field Equations

The classical equations governing the physical behaviour of electromagnetic fields over the frequency range *DC to Light* are Maxwell's equations:

$$\left.\begin{aligned} &\nabla\cdot\mathbf{D}=\rho && \text{(Gauss's Law)}\\ &\nabla\cdot\mathbf{B}=0 && \\ &\nabla\times\mathbf{E}=-\frac{\partial\mathbf{B}}{\partial t} && \text{(Faraday's Law)}\\ &\nabla\times\mathbf{H}=\mathbf{J}+\frac{\partial\mathbf{D}}{\partial t} && \text{(Ampere's Law)}\end{aligned}\right\} \quad (1)$$

These equations relate the magnetic flux density (**B**), the electric field intensity (**E**), the magnetic field intensity (**H**), the electric field displacement (**D**) with the electric charge density (ρ) and electric current density (**J**). The field vectors are not independent since they are further related by the material constitutive properties: $\mathbf{B} = \mu\mathbf{H}$, $\mathbf{D} = \varepsilon\mathbf{E}$, and $\mathbf{J} = \sigma\mathbf{E}$ where μ, ε and σ are the material permeability, permittivity and conductivity respectively. In practice these quantities may often be field dependent, and furthermore, some materials will exhibit both anisotropic and hysteretic effects.

1.2. The Magnetostatics Case

Significant simplification of equations (1) arise if the fields are independent of time thus the electric and magnetic fields are decoupled and the field vectors satisfy:

$$\left.\begin{aligned} &\nabla\cdot\mathbf{B}=0\\ &\nabla\times\mathbf{H}=\mathbf{J}\end{aligned}\right\} \quad (2)$$

The field vectors must satisfy certain conditions of continuity at all material interfaces. These conditions are derived from the limiting forms of equations (1) and are given by:

$$\left.\begin{array}{l}(\mathbf{B}_2-\mathbf{B}_1)\cdot\mathbf{n}=0\\(\mathbf{H}_2-\mathbf{H}_1)\times\mathbf{n}=0\end{array}\right\}\tag{3}$$

where the vector **n** is the normal to the interface.

2. THE REDUCED SCALAR POTENTIAL APPROACH

In magnetostatics the magnetic field **H** can be partioned into two fields[1,2]; namely, the field generated by any prescribed current sources present $\mathbf{H}_S$, and the field arising from induced magnetism in ferromagnetic materials $\mathbf{H}_m$. Thus,

$$\mathbf{H}=\mathbf{H}_m+\mathbf{H}_s\tag{4}$$

In Eq.(4), $\mathbf{H}_S$ is given by the Biot Savart Law, i.e.

$$\mathbf{H}_s(\mathbf{r})=\frac{1}{4\pi}\int_\Omega \mathbf{J}(\mathbf{r}')\times\nabla\left(\frac{1}{R}\right)d\Omega\tag{5}$$

where $R=|\mathbf{r}'-\mathbf{r}|$ and **J** the prescribed current density. From Eq. (2) it can be seen that $\nabla\times\mathbf{H}_m=0$ and it follows that $\mathbf{H}_m=-\nabla\phi$ and that Eq. (4) becomes:

$$\mathbf{H}=-\nabla\phi+\mathbf{H}_s\tag{6}$$

the scalar quantity ϕ is known as the reduced potential [3]. In many cases Eq.(5) can be integrated to give an analytic expression for $\mathbf{H}_S$; for complicated current paths, the expression can be integrated by a combination of analytic and numerical quadrature. Any permanent magnet sources can be represented by a modified form of the constitutive relation , i.e.

$$\mathbf{B}=\mu(\mathbf{H})(\mathbf{H}-\mathbf{H}_c)\tag{7}$$

where μ is now a non-linear function of **H** (in general a tensor), and $\mathbf{H}_c$ is the coercive field for the material. In *soft* magnetic materials, the coercive field intensity is normally assumed to be zero. The governing equation for the reduced scalar potential, ϕ, is obtained by taking the divergence of Eq.(7), i.e.

$$\nabla\cdot\mu\nabla\phi=-\nabla\cdot\mu\mathbf{H}_c+\nabla\cdot\mu\mathbf{H}_s\tag{8}$$

There are pitfalls in seeking solutions of Eq. (8) directly since in magnetic materials the two parts of the field $\mathbf{H}_S$ and $\mathbf{H}_m$ tend to be of similar magnitude and opposite in direction so that cancellation occurs in computing the resultant field **H** giving a serious loss in accuracy. This fact is the motivation and justification of the considerable body of work reported in the literature since 1978.

3. TOTAL SCALAR POTENTIAL

For regions where there are no current sources the total field **H** can be represented by scalar potential directly since in this case, $\nabla \times \mathbf{H} = \mathbf{0}$, and it follows that,

$$\mathbf{H} = -\nabla \psi \tag{9}$$

where ψ is known as the total scalar potential. The governing equation for regions without currents is given by,

$$\nabla \cdot \mu \nabla \psi = -\nabla \cdot \mu \mathbf{H}_c \tag{10}$$

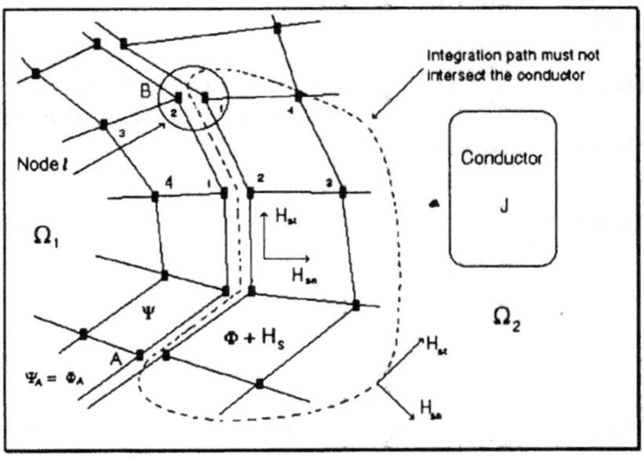

Figure 1. *Source Scalar Potentials and Integration Paths*

It is clear that the total scalar potential could be used to avoid the cancellation errors, unfortunately this potential is not defined in conductors where $\nabla \times \mathbf{H} \neq \mathbf{0}$ thus violating Eq.(9). In a numerical algorithm however, the two potentials can be combined to avoid the cancellation associated with reduced potentials and yet allow the inclusion of currents[3]. In regions that contain currents, see figure (1), the reduce potential should be used, and elsewhere the total scalar potential. The solutions can be coupled at the interfaces of the regions by applying the continuity conditions, equations (3), thus if material region 1 contains all the ferromagnetic and permanent magnet sources and region 2 all the conductor sources then,

$$\mu_1\left(-\frac{\partial \psi}{\partial n} - \mathbf{H}_{cn}\right)_1 = \mu_2\left(\mathbf{H}_{sn} - \frac{\partial \phi}{\partial n}\right)_2 \tag{11}$$

$$\left(-\frac{\partial \psi}{\partial t}\right)_1 = \left(-\frac{\partial \phi}{\partial t} + \mathbf{H}_{st}\right)_2 \tag{12}$$

where **n** and **t** are the outward normal and tangent directions respectively. Eq. (12) can be integrated over any path that does not intersect current regions or create a loop linking a current to give an integral relationship between potentials at two points A and B, i.e.

$$\psi_B = \phi_B - \int_A^B \mathbf{H}_s \cdot d\mathbf{s} \tag{13}$$

where $\mathbf{H_S}$ is obtained explicitly from Eq. (5) and at the point A, $\psi = \phi$, see Figure (1).

4. FINITE ELEMENT SOLUTIONS

It is convenient to specify a model problem as follows:

$$\nabla \cdot \mu \nabla \varphi = Q \qquad \mathbf{r} \in \Omega \tag{14}$$

here φ denotes either the reduced (ϕ) or total scalar (ψ) potential subject to boundary conditions:

$$\varphi = \overline{\varphi} \quad (\mathbf{r} \in \Gamma_1) \tag{15}$$

$$\mu \frac{\partial \varphi}{\partial n} = \overline{p} \quad (\mathbf{r} \in \Gamma_2) \tag{16}$$

where $\Gamma = \Gamma_1 + \Gamma_2$ on the surface of a domain Ω. The imposition of Equations (15) and (16) ensure that solutions to Eq. (14) are unique. Following the standard procedure with the finite element method[4] the solution is approximated by a set of basis functions:

$$\varphi = \tilde{u} = \sum N_i u_i \tag{17}$$

and the weighted residual at node i is given by

$$R_i = \int_\Omega W_i(\nabla \cdot \mu \nabla \tilde{u} - Q) d\Omega + \int_{\Gamma_2} \overline{W}_i \left(\mu \frac{\partial \tilde{u}}{\partial n} - \overline{p} \right) d\Gamma + \int_{\Gamma_1} \overline{\overline{W}}_i (\tilde{u} - \overline{\varphi}) d\Gamma = 0 \tag{18}$$

where the $W_i, \overline{W}$, and $\overline{\overline{W}}_i$ are an arbitrary set of weighting functions. Integrating by parts the first term in Eqn. (18), in the usual manner, to reduce the order of continuity on the functions gives:

$$\int_\Omega W_i(\nabla \cdot \mu \nabla \tilde{u} - Q)\, d\Omega = -\int_\Omega \nabla W_i \cdot \mu \nabla \tilde{u}\, d\Omega + \int_\Gamma W_i \mu \frac{\partial \tilde{u}}{\partial n} d\Gamma - \int_\Omega W_i\, d\Omega \tag{19}$$

and by choosing, $\overline{W}_i = -W_i$ to eliminate the normal gradient term along the boundary Γ_2, Eqn (18) becomes :

$$\begin{aligned} R_i = &-\int_\Omega \nabla W_i \cdot \mu \nabla \tilde{u}\, d\Omega - \int_\Omega W_i Q\, d\Omega + \int_{\Gamma_2} W_i \overline{p}\, d\Gamma \\ &+ \int_{\Gamma_1} W_i \mu \frac{\partial \tilde{u}}{\partial n} d\Gamma + \int_{\Gamma_1} \overline{\overline{W}}_i (\tilde{u} - \overline{\varphi}) d\Gamma \end{aligned} \tag{20}$$

Using the Galerkin method the weighting functions are identified with the basis functions, i.e. $W_i = N_i$ and as the basis functions are local to the elements containing the nodal parameters u Eqn (20) defines a set of algebraic equations based on the nodes. The boundary condition on Γ_1 is usually enforced and therefore the corresponding surface integrals are eliminated. The element *stiffness* matrix is given by:

$$K_{ij} u_i = C_i \tag{21}$$

with

$$K_{ij} = \int_\Omega \nabla N_i \cdot \mu \nabla N_j \, d\Omega \tag{22}$$

the matrix **K** is sparse and symmetric for this particular choice of weighting function. The right-hand side vector is given by:

$$C_i = \int_\Omega N_i Q \, d\Omega \tag{23}$$

4.1. The Two-Scalar Potential Formulation

In order to avoid cancellation errors two regions as in Figure (1) can be constructed. Region 1 consists of those parts of the problem which can be represented by the total scalar potential, i.e. regions where the current density is zero, and region 2 which contains the source conductors which can be represented by the reduced scalar potential. Furthermore, without too much loss in generality, region 2 can be assumed to have constant permeability (usually free space with $\mu = \mu_0$). This topology conforms to a large range of problems encountered in practise. Each region is divided into finite elements , except that now the problem will not be completely defined, since on the interface boundary between the two regions both ψ and ϕ are unknown. This indeterminacy can be resolved by application of the interface conditions, Equations (11) and (13). For simplicity the contribution from hard magnetic materials will be omitted (i.e. $\mathbf{H}_c = 0$ in Equations (8) and (10)) although it is perfectly straightforward to include these effects[5]. Applying the weighted residual method and integrating by parts to each region separately and setting the total residual to zero, and then by making use of the interface conditions, Equations (11), (13) the final residual equation is :

$$\int_{\Omega_1} \nabla W_i \cdot \mu \nabla \psi \, d\Omega + \int_{\Omega_2} \nabla W_i \cdot \mu \nabla \phi \, d\Omega = -\int_{\Gamma_l} W_i \left[-\mu \frac{\partial \psi}{\partial n} + \frac{\partial \phi}{\partial n} \right] d\Gamma = -\int_{\Gamma_l} W_i \mathbf{H}_s \cdot \mathbf{n} \, d\Gamma \tag{25}$$

In deriving Eq. (25) it should be noticed that the normal directions over the interface Γ_l , for the two regions, are of opposite sign, and that it has been assumed that there is continuity of the weight functions between the two regions. After applying the Galerkin method to Eq. (25) with a finite element discretization, the coefficient matrix is identical to Eq. (22) with appropriate permeability. At the interface Γ_l either ψ or ϕ can be eliminated by Eq. (13) . Eliminating ϕ results in a right-hand side term for a node on the interface given by:

$$C = Kg - h \tag{26}$$

where K is the element matrix, Eq.(22), and :

$$g=\int_A^B H_n\,dt \tag{27}$$

$$h=\int_{t_1}^{t_2} N_i H_{sn}\,dt \tag{28}$$

see Figure (1).

4.2. The Reduced Scalar Potential Formulation

If cancellation is not expected then Eq. (8) can be solved over the entire problem space but considerable care is needed. For example at interfaces between regions of differing permeability the weighted residual expression of Eq. (8) , after the continuity conditions between the regions, is:

$$R_i = -\int_{\Omega_1} \nabla W_i \cdot \mu_1 \nabla\phi\, d\Omega - \int_{\Omega_2} \nabla W_i \cdot \mu_2 \nabla\phi\, d\Omega + \int_{\Gamma_i} W_i H_{sn}(\mu_1 - \mu_2)\, d\Gamma \tag{29}$$

In addition to the extra term it should be noticed that the source fields have to be computed at all mesh points in the permeable regions. In practise the formulation is more stable if the volume term involving $\mathbf{H}_S$ is integrated by parts to give a natural surface condition.

4.3. The Difference Potential Formulation

This method introduces two scalar potentials which must be evaluated sequentially [6] . The first potential ϕ^{∞} is calculated with the permeability assumed infinite. In this case **H** is zero in ferromagnetic materials and is immaterial. With reference to Figure (1), if region 1 only contains ferromagnetic materials then the equations defining the first potential are as follows:

$$\left.\begin{array}{ll} \nabla^2\phi^{\infty}=0 & \mathbf{r}\in\Omega_2 \\ \mathbf{H}=\mathbf{H}_s-\nabla\phi^{\infty} & \end{array}\right\} \\ \mathbf{H}\times\mathbf{n}=\mathbf{0} \quad \mathbf{r}\in\Gamma_1 \tag{30}$$

The gradient of the second scalar potential ϕ^{-} is defined as the difference between the infinite permeability solution and the real solution since this field is irrational . The defining equations are:

$$\left.\begin{array}{ll} \mathbf{H}=\mathbf{H}_s-\nabla\phi^{\infty}+\nabla\phi^{-} & \mathbf{r}\in\Omega_2 \\ \mathbf{H}=\nabla\phi^{-} & \mathbf{r}\in\Omega_1 \end{array}\right\} \tag{31}$$

5. DISCUSSION

In this study the various alternatives for computing magnetostatic fields using scalar potentials have been reviewed. It is clear that the basic problem with using a single reduced scalar potential in all regions is purely numeric both with regard to accuracy and economy. The first point arises from the different spaces used to interpolate the scalar potential and source field respectively. The second from need to compute the source fields at each and every node .

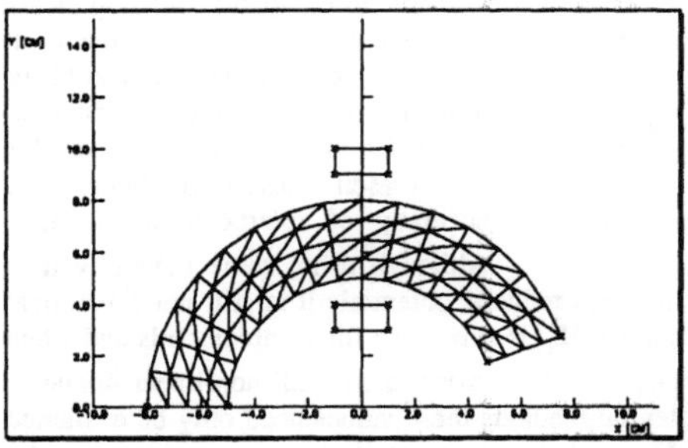

Figure 2. *C-Magnet Problem--iron elements*

Thus the effectiveness of a single reduced potential method depends critically on the sub-space of functions used to interpolate the coil fields $\mathbf{H}_S$ [7] , and in particular the relation between these functions and the finite element potential solution space. Implementations of the reduced potential method, Eq. (8), usually use exact evaluation of $\mathbf{H}_S$ at all points by analytic integration of the Biot-Savart expression for the field from a defined current distribution, Eq. (5).To some extent the whole problem of cancellation can be circumvented. To achieve this the space of functions used to interpolate $\mathbf{H}_S$ should be the same as that describing the gradients of the reduced potentials. If the two parts of the total field are not supported on the same space of functions then the cancellation between the two parts gives rise to large oscillations in the total field over each element. In order to demonstrate these effects Figures (3) and (4) show the fields in a simple model problem Figure (2). Figure (3) shows the expected field variation compared to the reduced potential results (which are plotted as targets) using the exact $\mathbf{H}_S$ from Biot-Savart and in addition the detail of the field variation is shown across 2 elements by lines joining the targets. This shows the gross fluctuations across elements when the two parts of the field are represented on very different spaces of functions. In Figure (4) the source field is approximated by centroid values, the wild variations have been removed, but the solution accuracy has been spoiled by the poor interpolation of $\mathbf{H}_S$. Slightly better results are obtained using the bi-linear interpolation of $\mathbf{H}_S$. The results are compared against an *exact* solution to the problem obtained by using the combined potential method [3].

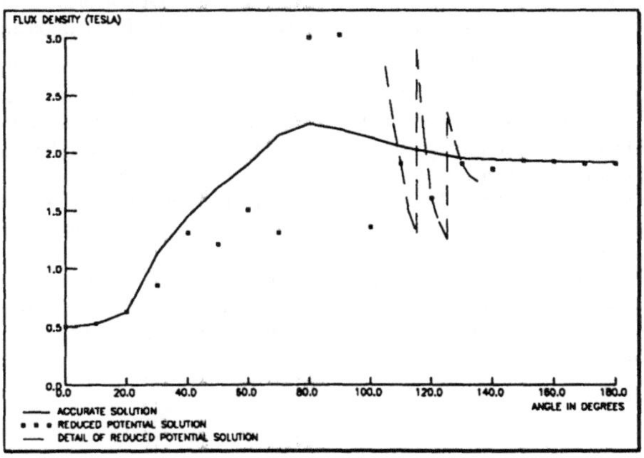

Figure 3. *Comparison of Reduced with Total/Reduced Potentials*

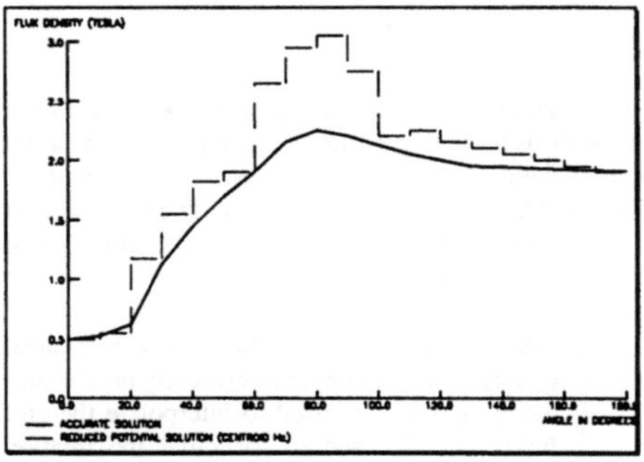

Figure 4. *Coil Fields approximated by value at centroid*

In all cases it is possible to reduce the cancellation errors to almost insignificant levels by representing $\mathbf{H}_S$ on the space of functions that support the gradients of the reduced potentials. In this case the cancellation only causes a problem when the differences become sensitive to computer rounding errors and the calculated fields are the same as those obtained in the combined total and reduced potential method, except for the variations expected for the reason described below ie. accuracy of interpolation of $\mathbf{H}_S$. Although using some ingenuity it is possible to solve the cancellation problems arising with reduced potentials it is still not a practical method for 3D magnetostatic fields. The integration for $\mathbf{H}_S$ is expensive for complex coils and when using a reduced potential method these integrations must be performed over all non-linear domains, whereas with the combined reduced and total potential methods these values need only be evaluated in the reduced potential regions. The number of $\mathbf{H}_S$ evaluations can easily be decreased by an order of magnitude using a reasonable combination of reduced and total potential spaces. Furthermore, using numerical basis function to support $\mathbf{H}_S$ implies that the elements must be capable of interpolating $\mathbf{H}_S$ with good accuracy, otherwise as shown in Figure (4) the solution suffers. The result of this is that the finite elements must, to some extent, be shaped to fit the coils. This can be almost intractable for a complex coil system, and is completely unnecessary with the combined total and reduced potential method. For most practical applications it can be said that,from a numerical perspective, it is better to compute the small difference quantity directly, i.e. in this case $-\nabla\phi + \mathbf{H}_s = -\nabla\psi$, instead of subtracting two large numbers which at best is inelegant and expensive but also susceptible to error. Nevertheless, on occasion, difficulties may be experienced when the current sources are not balanced (i.e. total net current is not zero). In this case the line integral of field along a closed path inside a connected 'iron' circuit is not zero and the above reasoning advocating the calculation of the difference quantity directly may not apply and it should be possible to obtain good results using the reduce scalar potential.

The difference method, see Equations (30,31) has been proposed to avoid the perceived algorithmic complexity of the total/reduced method, and it has been suggested that it avoids costly calculations of $\mathbf{H}_S$. In fact the total and reduced potential method only requires that $\mathbf{H}_S$ is computed at the interface between total and reduced potentials (in the same way that the *difference* method requires $\mathbf{H}_S$ over the iron surface) and the interface can be selected to minimise the cost (whereas it is fixed in the *difference* method). However the *difference* method does not avoid cancellation errors in free space. Consider a coil shielded by a ferromagnetic box, the field outside the box will be much smaller than the field from the coil alone hence there is a strong cancellation; in the extreme case of a closed infinite permeability box the cancellation will be catastrophic. The problem may be avoided if the box is closed, but it cannot be avoided with an open box. The total potential method allows almost complete freedom in the choice of potential type, the only restriction being that reduced potentials must be used in the coil volumes. This allows minimisation of the costs incurred in calculating with optimal accuracy (avoiding cancellation or calculating the main part of the field by exact integration with only a perturbation modelled by the finite elements). Intelligent use of the potentials is a great benefit of the method.

6. POSTSCRIPT

The work on scalar potentials in magnetostatics reviewed in this paper had its origins in the very fruitful and happy collaboration between the authors and O.C. Zienkiewicz in the 1970's. The authors always visited Swansea with great expectations that were always fulfilled.

7. REFERENCES

1. NEWMAN, M.J., TURNER, L.R. and TROWBRIDGE, C.W. - GFUN: An Interactive Program as an aid to Magnet Design, Ed. Winterbottom, Y., Proc. International Conference on Magnet Technology (MT4), Brookhaven National Laboratory, 1972, p. 617.

2. ZIENKIEWICZ, O.C, LYNESS, J. and OWEN, D.R.J. - Three Dimensional Magnetic Field Determination using a Scalar Potential--- a finite element solution. IEEE (Mag.Trans) **13**, 1649-56, (1977)

3. SIMKIN, J. and TROWBRIDGE, C.W. - On the use of the total scalar potential in the numerical solution of field problems in electromagnetics, Int. Journal for Num. Meth. in Eng., **14**, 423-440 (1979).

4. ZIENKIEWICZ, O.C - The Finite Element Method, McGraw-Hill, Maidenhead ,U.K. 1977

5.SIMKIN, J. and TROWBRIDGE, C.W. - Three Dimensional non-linear Electromagnetic Field computations using Scalar Potentials, Proc. IEE, **127**, No 6, (1980)

6.MAYERGOYZ, I.D, CHARI, M.V.K., and D'ANGELO, J.D.- A New Scalar Potential Formulation for Three-Dimensional Magnetostatic Problems, IEEE(Mag Trans) **23**, No 6, 3889-94, (1987)

7.SIMKIN, J. - Magnetostatic Fields in 3D, Vector Fields Ltd. News Letter, **4**, No 2, (1988)

FINITE ELEMENTS IN THE 90's, E. Oñate, J. Periaux, A. Samuelsson (Eds.)
Springer-Verlag/CIMNE, Barcelona 1991

UNITY AND HIERARCHY IN MODELLING. OLEK ZIENKIEWICZ'S CONTRIBUTIONS AND IMPACT

Chandra S. Desai
Department of Civil Engineering and Engineering Mechanics,
University of Arizona,
Tucson, Arizona,
USA

At the time of Olek's 70th birthday celebrations, I decided to express briefly, my subjective views, on what I have admired and learnt from him, particularly in the pursuit of unity and hierarchy, be it in numerical modelling or constitutive modelling for material behavior.

I studied for my master's and doctoral degrees at Rice University, Houston, and at the University of Texas at Austin. At Rice, I developed an interest in engineering numerical analysis through lectures from Professors W.J. Austin and A.S. Veletsos; this interest was kindled toward the finite element method while listening to a seminar at Rice by Professor Ray W. Clough. During 1966-1968 I was a doctoral student at the University of Texas at Austin, working under the direction of Professors Lymon C. Reese and Hudson Matlock; both have been pioneers in the application of numerical techniques for soil-structure interaction problems. There I found the opportunity to study and learn civil engineering, mechanics, applied mathematics and numerical analysis, although my research topics involved problems of foundations and stability of deep boreholes in rocks. This approach strengthened my inherent tendency to look beyond (geomechanics), with an appreciation of the generality of numerical techniques and material response. I have been guided by the feeling, howsoever incomplete its attributes may be, that hopefully the secrets that Nature holds are not too many, and maybe there exists unity or a collection of unities for diverse phenomena!

Some of Olek Zienkiewicz's publications, including his first text book on the subject of the finite element method (FEM), came around the time when I was a graduate student. From the very early stages, Olek's contributions represented a unifying viewpoint; they covered a wide range of problems in civil, mechanical, aerospace and electrical engineering. He saw the underlying unity in apparently different looking problems and applied to them the FEM with ingenuity and versatility. It is interesting to note that Olek's doctoral research involved the classical dam analysis problem in geomechanics.

As I watched the tremendous depth and breadth in the great range of topics that Olek's work covered, I received significant reinforcement in my own tendency to work in a number of interrelated areas in civil engineering, dynamics, flow through porous media, mass transport and some mechanical problems, and most recently some problems in electronics and space materials.

Olek is indeed a pioneer of finite element methods, and its most active worker over the last three decades or so. Undoubtedly, he has been the best exponent of the FEM. The appreciation of the capabilities of this method by the engineering and scientific communities, and its wider acceptance by them owe greatly to Olek's work and contributions.

When John Abel and I coauthored *Introduction to the Finite Element Method,* I was quite apprehensive about its reception, particularly by the original workers like Olek. We were pleasantly surprised when we received a kind letter from Olek with a note that stated, as I recall, *"now I have competition"*.

Olek appreciates the work of others, particularly of upcoming and younger persons. I have seen him, often in conferences, discussing the work of younger persons, with special kindness and appreciation. This human quality has helped, I believe, quite a few persons to pursue work in the FEM and a number of other areas.

As the FEM reached a level of maturity, a number of areas such as material behavior, numerical characteristics of associated algorithms, reliability of methods and robust time integration schemes for nonlinear (dynamics) problems assumed increasing significance and importance. Olek has been a pioneer and has made original contributions in these areas.

I would particularly like to identify constitutive modelling, which is vital for realistic solutions of problems, with computational techniques. Olek's contributions toward development and application of appropriate and realistic constitutive models in numerical techniques have been in the forefront. His perception of the unity and hierarchy is highly evident, as his contributions span many materials: metals, composites, polymers and geologic. My own work towards unified constitutive models with appropriate laboratory testing including development of some new test devices, and implementation in numerical techniques for nonlinear problems has endeavored to seek the unity and hierarchy. This endeavor by my students and myself has led to the development of unified ideas called the disturbed state concept (DSC) and hierarchical single surface (HSS) approach for the characterization of solids and discontinuities (interfaces and joints). These concepts have been applied to geologic materials, concrete, metals, ceramics, and interfaces in soil-structure interaction problems and semiconductor chip substrates, and joints in rocks and concrete. I would have probably written a short paper on this unified modelling approach. However, instead, I decided to write this subjective narrative to express my appreciation and respect for Olek's work and his human qualities.

I remain in awe when I think of Olek's monumental contributions and their impact on many engineering and scientific disciplines. I feel a deep sense of esteem and friendship to Olek, when I think of how he has inspired and supported me and many others. And here, appreciation of unity and hierarchy in modelling, even if their pursuit is limited greatly by the limits of human comprehension, stands out in the front of my thinking.

FINITE ELEMENTS IN THE 90's, E. Oñate, J. Periaux, A. Samuelsson (Eds.)
Springer-Verlag/CIMNE, Barcelona 1991

A FULLY CONSISTENT FORMULATION OF EARLY-TIME APPROXIMATIONS FOR ACOUSTIC MEDIA

T.L. Geers
Center for Acoustics, Mechanics and Materials,
Department of Mechanical Engineering,
University of Colorado,
Boulder, CO, 80309-0427,
USA

SUMMARY

A previous derivation of early-time approximations (ETA's) to treat the transient fluid-structure interaction for a body submerged in an infinite acoustic meduim is extended. The extension accounts for variation of the wet-surface fields in the vicinity of the field point. This produces a third-order ETA that does not collapse to the crude plane-wave approximation where the wet surface is locally flat. Specific-acoustic-impedance curves are presented for a spherical surface that demonstrate modest superiority of the third-order ETA over the second-order ETA for even a curvature-dominated geometry.

1. INTRODUCTION

In 1980, Felippa published an elegant formulation for constructing early-time approximations (ETA's) to treat the transient interaction between the wet surface of a body and an acoustic medium [1]. The formulation produced the plane-wave approximation (ETA_1) and the curved-wave approximation (ETA_2) constructed earlier for a circular cylindrical geometry by Mindlin and Bleich [2] and Haywood [3], respectively, and extended heuristically to general geometries by Bedrosian and Dimaggio [4]. Felippa's results for the first and second order ETA's were later confirmed by Nicolas-Vullierme [5] in the frequency domain and by Geers and Zhang [6] using ray acoustics.

Interestingly, no ETA produced by Felippa's formulation accounts for field variation over the wet surface, each step to a higher-order ETA involving the incorporation of additional surface-geometry parameters. Hence, as noted by Kriegsmann and Scandrett [7], all ETA's thus formulated for a flat surface collapse to the plane wave approximation, ETA_1. As seen below, this deficiency is removed by simple expansion of the surface fields in Taylor series about the field point of interest, which introduces surface-variation terms of the kind seen in paraxial approximations (see, e.g., [8]).

Here, attention is focused on ETA_3, which is the lowest-order ETA that incorporates a surface-variation term. This permits a clear exposition of the salient points, avoiding the proliferation of terms that characterizes the construction of higher order ETA's. The simplicity of ETA_3 facilitates the formulation of third-order *doubly asymptotic approximations*, [9-13, 5, 6]. DAA's are more robust than ETA's, fitting early-time and late-time approximations in order to span the entire time domain.

2. FORMULATION

Felippa's formulation [1] begins with Kirchoff's classical retarded potential formula for the smooth wet surface of a body in contact with an acoustic medium (see, *e.g.*, [14], [15])

$$2\pi\varphi_F = \int_S [R^{-1}\langle u\rangle - R^{-2}\frac{\partial R}{\partial n}(\langle\varphi\rangle + \frac{R}{c}\langle\dot{\varphi}\rangle)]dS \tag{1}$$

where φ and u are the velocity potential and normal fluid-particle velocity, respectively (the latter defined positive going *into* the fluid), R is the straight-line distance between an integration point P and the field point of interest F, n is the surface normal at the integration point (defined positive going into the fluid), the overdot denotes time differentiation, and the angular brackets indicate that the expression within, which is a function of space and time, is evaluated at the *retarded time* t - R/c. The geometry for (1) is shown in Figure 1, which is taken from [1]. Here we take x and y as lying in the planes of principal curvature for the surface at F.

Felippa provides in [1] expressions needed to evaluate (1); we write his expressions as

$$\begin{aligned} R^{-1}dS &= [1+\frac{1}{2}A_\theta^2R^2+O(R^3)]dRd\theta \\ R^{-2}\frac{\partial R}{\partial n}dS &= -[A+2BR+O(R^2)]dRd\theta \end{aligned} \tag{2}$$

where A, A_θ, and B are functions of the local polar angle θ in the plane tangent to the surface at F (Figure 1); it may be shown that, even though the two equations in (2) are expressed to different orders in R, they indeed yield a consistent formulation. In contrast to [1], we admit variation of the fields φ and u over the surface S, expanding them in local Taylor series in terms of the curvilinear coordinates ξ and ζ, which are defined by the intersections with the wet surface of the x-z plane and y-z plane, respectively. This yields for φ

$$\begin{aligned} \varphi = \varphi_F &+ (\frac{\partial\varphi}{\partial\xi})_F\,\xi + (\frac{\partial\varphi}{\partial\zeta})_F\,\zeta \\ &+ \frac{1}{2}(\frac{\partial^2\varphi}{\partial\xi^2})_F\,\xi^2 + (\frac{\partial^2\varphi}{\partial\xi\partial\zeta})_F\,\zeta\xi + \frac{1}{2}(\frac{\partial^2\varphi}{\partial\zeta^2})_F\,\zeta^2 + \dots \end{aligned} \tag{3}$$

with a similar expression holding for u. But it is clear that

$$\begin{aligned} &(\frac{\partial\varphi}{\partial\xi})_F = (\frac{\partial\varphi}{\partial x})_F\,, \quad (\frac{\partial^2\varphi}{\partial\xi^2})_F = (\frac{\partial^2\varphi}{\partial x^2})_F\,, \text{ etc.} \\ &\xi = R\cos\theta[1+O(R^2)], \quad \zeta = R\sin\theta[1+O(R^2)] \end{aligned} \tag{4}$$

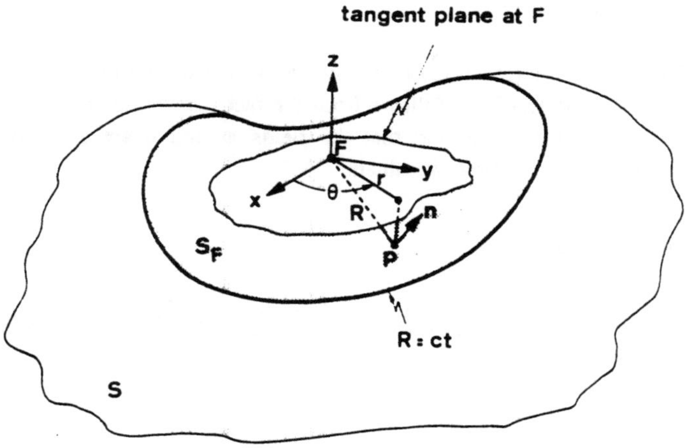

Figure 1. Local Geometry around a Field Point

so (3) yields

$$\varphi = \varphi_F + [(\frac{\partial\varphi}{\partial x})_F \cos\theta + (\frac{\partial\varphi}{\partial y})_F \sin\theta]R[1+O(R^2)]$$
$$+[\frac{1}{2}(\frac{\partial^2\varphi}{\partial x^2})_F \cos^2\theta + (\frac{\partial^2\varphi}{\partial x\partial y})_F \sin\theta\cos\theta + \frac{1}{2}(\frac{\partial^2\varphi}{\partial y^2})_F \cos^2\theta]R^2[1+O(R^2)] \tag{5}$$

utilizing this in conjuction with (2), we obtain the integrand expressions

$$R^{-1}\langle u\rangle dS = \{\langle u\rangle_F + (\langle\frac{\partial u}{\partial x}\rangle_F \cos\theta + \langle\frac{\partial u}{\partial y}\rangle_F \sin\theta)R[1+O(R^2)]$$
$$+(\frac{1}{2}\langle\frac{\partial^2 u}{\partial x^2}\rangle_F \cos^2\theta + \langle\frac{\partial^2 u}{\partial x\partial y}\rangle_F \sin\theta\cos\theta + \frac{1}{2}\langle\frac{\partial^2 u}{\partial y^2}\rangle_F \sin^2\theta)$$
$$\cdot R^2[1+O(R^2)] + O(R^3)\}[1+\frac{1}{2}A_\theta^2R^2+O(R^3)]\,dR\,d\theta$$

(6)

$$R^{-2}\frac{\partial R}{\partial n}\langle\varphi\rangle dS = -\{\langle\varphi\rangle_F + (\langle\frac{\partial\varphi}{\partial x}\rangle_F \cos\theta + \langle\frac{\partial\varphi}{\partial y}\rangle_F \sin\theta)R[1+O(R^2)] + O(R^2)\}$$
$$\cdot[A+2BR+O(R^2)]\,dR\,d\theta$$

3. EVALUATION

The surface integral in (1) may now be evaluated over the domain $0 \le \theta \le 2\pi$ and $0 \le R \le ct$. The last limit follows from the fact that the retarded time t-R/c is non-negative only for $R \le ct$ and that the fields φ and u are zero for $t < 0$. Evaluating the θ-integral first, we obtain

$$\int_0^{ct}\int_0^{2\pi} R^{-1}<u>dS =$$

$$2\pi\int_0^{ct} \{<u>_F + [\frac{1}{2}\bar{A}_\theta^2<u>_F + \frac{1}{4}(<\frac{\partial^2 u}{\partial x^2}>_F + <\frac{\partial^2 u}{\partial y^2}>_F)]R^2 + O(R^3)\}dR$$

$$\int_0^{ct}\int_0^{2\pi} R^{-2}\frac{\partial R}{\partial n}<\varphi>dS = -2\pi\int_0^{ct}[(\bar{A} + 2\bar{B}R)<\varphi>_F + O(R^2)]dR \tag{7}$$

where an overbar indicates an average value over the domain $0 \le \theta \le 2\pi$. Making use of (7) in (1), we find

$$\varphi_F = \int_0^{ct}[<u>_F + (\frac{1}{2}\bar{A}_\theta^2<u>_F + \frac{1}{4}<\nabla_S^2 u>_F)R^2 + O(R^3)]\,dR$$
$$+ \int_0^{ct}[(\bar{A} + 2\bar{B}R)(<\varphi>_F + \frac{R}{c}<\dot{\varphi}>_F) + O(R^2)]dR \tag{8}$$

where ∇_S^2 is the surface Laplacian.

In [1], Felippa evaluated the integrals appearing in (8), as well as the geometric parameters $\bar{A}$, $\bar{A}_\theta$, and $\bar{B}$. We therefore employ these evaluations in (8), which allows us to drop the F-subscripts. Furthermore, we multiply the resulting equation through by ρ, differentiate it, and utilize the fact that $p = \rho\dot{\varphi}$ to obtain

$$p - c\kappa p^{*} + ... = \rho c u + \frac{1}{2}\rho c^3(\kappa - \tau + \nabla_S^2)u^{**} + ... \tag{9}$$

where each asterisk denotes an integration in time, and κ and τ are the wet surface's mean curvature and total curvature, respectively [16]. We note in (9) the effect of field variation over the wet surface through the term involving ∇_S^2.

4. EARLY-TIME APPROXIMATIONS

We can obtain three early-time approximations from (9). First, retaining only the terms that involve no temporal integrations, we obtain the *first-order ETA*

$$p = \rho c u \tag{10}$$

Next, retaining only the terms through one temporal integration and then differentiating, we obtain the *second-order ETA*

$$\dot{p} - c\kappa p = \rho c \dot{u} \tag{11}$$

Finally, retaining only the terms through two temporal integrations and then double-differentiating, we obtain the *third-order ETA*

$$\ddot{p} - c\kappa\dot{p} = \rho c \ddot{u} + \frac{1}{2}\rho c^3(\kappa^2 - \tau + \nabla_S^2)u \tag{12}$$

The results (10) and (11) are in agreement with those found by previous investigators. The third ETA, however, disagrees with Felippa's result (his Eq. 24) in two ways. First, the surface-variation term is missing in his ETA_3, and second, his ETA_3 contains a third term in p that is not present in (12). The former difference derives from our surface Taylor series expansion of the field quantities. The latter difference derives from his including in (9) a term involving three temporal integrations in p; by our method of counting, such inclusion would produce ETA_4.

5. MODAL IMPEDANCE CURVES FOR A SPHERE

For a spherical wet surface, $\kappa = -a^{-1}$ and $\kappa^2 - \tau = 0$; furthermore, for axisymmetric surface motion given by $\varphi = \varphi_n P_n(\cos\phi)$ and $u = u_n P_n(\cos\phi)$, where $P_n(\ )$ is the Legendre polynimial of order n and ϕ is the meridional angle, $\nabla_S^2 u = -n(n+1)u$. Thus we have for each spherical harmonic

ETA_1^n:

$$p_n = \rho c u_n$$

ETA_2^n:

$$\dot{p}_n + (c/a)p_n = \rho c \dot{u}_n \tag{13}$$

ETA_3^n:

$$\ddot{p}_n + (c/a)\dot{p}_n = \rho c[\ddot{u}_n - \frac{1}{2}n(n+1)(c/a)^2 u_n]$$

If we assume sinusoidal motion $e^{i\omega t}$, we can define a modal specific acoustic impedance $z_n(\omega) = p_n(\omega)/u_n(\omega)$ for each spherical harmonic as

$$z_n(\omega) \doteq \rho c \rho_n(\omega) + i\omega\rho a \mu_n(\omega) \tag{14}$$

where ρ_n and μ_n are nondimensional, modal, specific acoustic resistance and inertia, respectively.

Figures 2 and 3 show ETA_2^n, ETA_3^n and exact specific acoustic resistance and inertia curves for n = 1 and n = 2, respectively; the exact curves are taken from [11]. No curves are shown for ETA_1 because $\rho_1(\omega) = 1$ and $\mu_1(\omega) = 0$ for this approximation; also no figure is provided for n = 0 because both ETA_2^n and ETA_3^n are exact in this case. The figures show that ETA_3 produces resistance curves that are more accurate than the corresponding curves produced by ETA_2 for $\omega a/nc > 1.5$, but produces inertia curves that are slightly less accurate than those produced by ETA_2. Because resistance effects dominate at high frequencies, however, the resistance curves are more important than the inertia curves; hence the two figures indicate that ETA_3 is slightly superior to ETA_2. This superiority should be much more pronounced for surface geometries with flat regions.

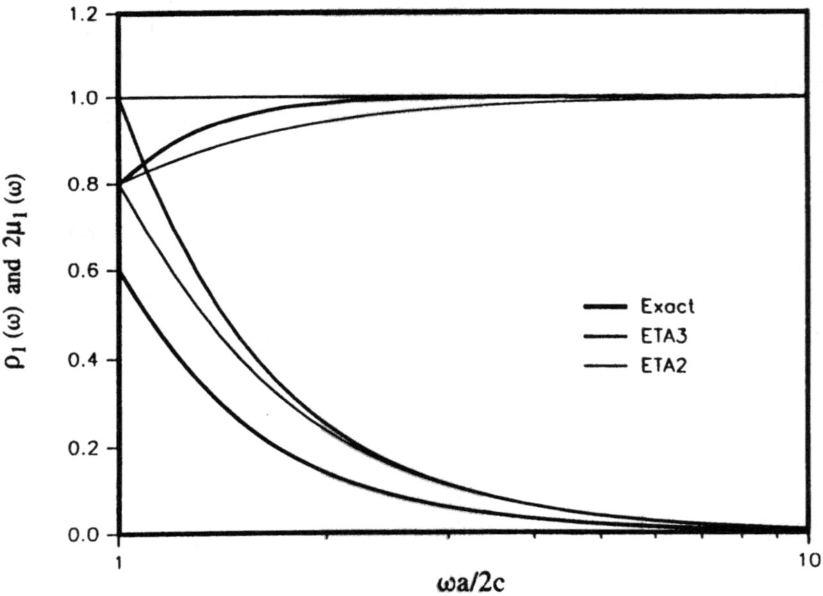

Figure 2. Specific Acoustic Resistance (Upper Curves) and Specific Acoustic Inertia (Lower Curves) for the Axisymmetric n = 1 Motion of a Spherical Surface

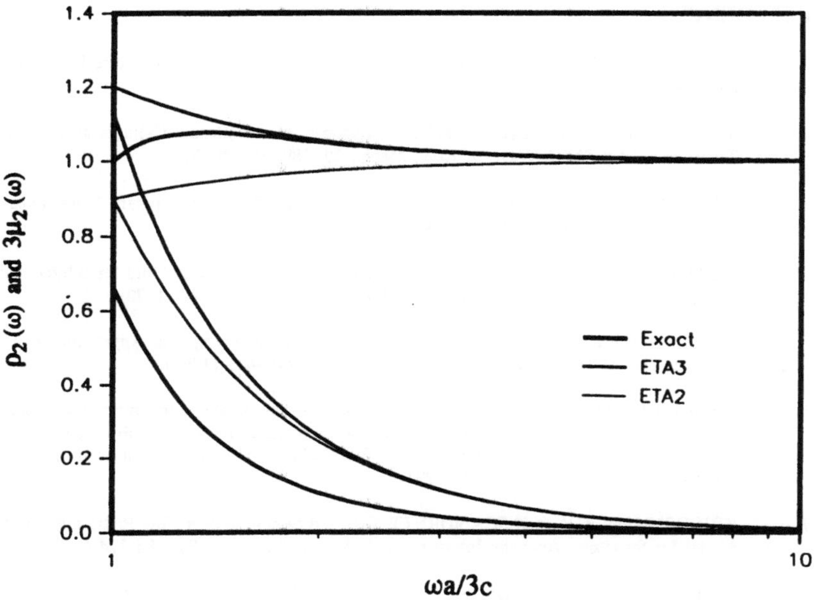

Figure 3. Specific Acoustic Resistance (Upper Curves) and Specific Acoustic Inertia (Lower Curves) for the Axisymmetric n = 2 Motion of a Spherical Surface

6. CONCLUSION

The expansion of surface fields in Taylor series about the field point has produced a surface-variation term in the third-order early-time approximation for the dynamic interaction between a submerged body and surrounding acoustic fluid. This prevents ETA_3 from collapsing to ETA_1 when the wet surface is locally flat. It will be interesting to observe the contribution of this term to the performance of a third-order doubly asymptotic approximation in problems where surface-curvature effects do not dominate field-variation effects.

ACKNOWLEDGEMENT

The author thanks Mohamad Hasheminejad for his help in preparing the manuscript of this paper.

REFERENCES

1. FELIPPA, C.A. - A Family of Early Time Approximations for Fluid Structure Interaction, J. Appl. Mech., 47, 703-708 (1980).

2. MINDLIN, R.D. and BLEICH, H.H. - Response of an Elastic Cylindrical Shell to a Transverse, Step Shock Wave, J. Appl. Mech., 20, 189-195 (1953).

3. HAYWOOD, J.H. - Response of an Elastic Cylindrical Shell to a Pressure Pulse, Quart. J. Mech. & Appl. Math., 11, Part 2, 129-141 (1958).

4. BEDROSIAN, B. and DIMAGGIO, F.L. - Acoustic Approximations in Fluid-Shell Interactions, J. Eng. Mech. Div., Proc. Am. Soc. Civil Eng., 98 (EM3), 731-742 (1972).

5. NICOLAS-VULLIERME, B. - A Contribution to Doubly Asymptotic Approximations: An Operator Top-down Derivation, J. Vib. Acoust., 113, 409-415 (1991).

6. GEERS, T.L. and ZHANG, P. - Doubly Asymptotic Approximations for Internal and External Acoustic Domains: Formulation, Dynamic Response of Structures to High-Energy Excitations, Ed. Geers, T.L. and Shin, Y.S., AMD-Vol. 127/PVP-Vol. 225, American Society of Mechanical Engineers, New York, 1991.

7. KRIEGSMANN, G.A. and SCANDRETT, C.L. - Assessment of a New Radiation Damping Model for Structural Acoustic Interactions, J. Acoust. Soc. Am., 86, 788-794 (1989).

8. COHEN, M. and JENNINGS, P.C. - "Silent Boundary Methods for Transient Analysis", Computational Methods for Transient Analysis, Ed. Belytschko, T. and Hughes, T.J.R., Elsevier Science Publishers, Amsterdam, 1983, pp. 301-360.

9. GEERS, T.L. - Residual Potential and Approximate Methods for Three-dimensional Fluid-structure Interaction Problems, J. Acoust. Soc. Am., 49, 1505-1510 (1971).

10. GEERS, T.L. - Shock Response Analysis of Submerged Structures, Shock and Vibration Bulletin, 44, Supp. 3, 17-32 (1974).

11. GEERS, T.L. - Doubly Asymptotic Approximation for Transient Motions of Submerged Structures, J. Acoust. Soc. Am., 64, 1500-1508 (1978).

12. FELIPPA, C.A. "Top-down derivation of Doubly Asymptotic Approximations for Structure-Fluid Interaction Analysis", Innovative Numerical Analysis for the Engineering Sciences, Ed. Shaw, R.P. et al., U.P. of Virginia, Charlottesville, 1980, pp 79-88.

13. GEERS, T.L. and FELIPPA, C.A. - Doubly Asymptotic Approximations for Vibration Analysis of Submerged Structures, J. Acoust. Soc. Am. 73, 1152-1159 (1983).

14. BAKER, B.B. and COPSON, E.T., - The Mathematical Theory of Huygen's Principle, Clarendon Press, Oxford, 1939.

15. SOBOLEV, S.L., - Partial Differential Equations of Mathematical Physics, Pergamon Press, London, 1964.

16. STRUIK, D.J. - Lectures on Classical Differential Geometry, Addison-Wesley, Reading, 1961.

FINITE ELEMENTS IN THE 90'S, E. Oñate, J. Periaux, A. Samuelsson (Eds.)
Springer-Verlag/CIMNE, Barcelona 1991

APPLICATION OF A POLYDESCRIPTIVE APPROACH: BOND GRAPHS AND FINITE ELEMENTS TO DESIGN ANALYSIS

M.A. Bossak
Institute of Mechanics and Design,
Warsaw University of Technology,
Narbutta 85, 02-524 Warszawa,
POLAND

SUMMARY

In design analyses of complex machinery, structural elements are most conveniently modelled using Finite Element Method, control systems are customarily described with Transfer Functions, power transmissions or suspension systems are frequently modelled with the aid of the Bond Graph technique.

For simulating the behaviour of the complex machinery as a whole it is natural to accept a multitude of formalisms as a fact of life, which leads immediately to the concept of polydescriptive approach.

A simulation system which uses such an approach is presented.

CONCURRENT DESIGN

For decades the design process has not changed much although the tools employed have changed significantly from the slide-rule to the workstation.

Today's market demands new products which are complex, have great variety and are of high quality. The design specifications for these products often exceed the today's engineering experience and require revolutionary rather than evolutionary concepts. Products also have shorter life cycles. The competitive nature of market exerts a great preasure on delivering a new product as quickly as possible, to avoid the risk of loosing the customers. To meet these challenges it is important to consider in the early design stage the following aspects of the product |4|:

- quality,
- reliability,
- maintainability,
- manufacturability,
- schedulability.

To achieve this, the concurrent design (also called simultaneous or parallel design) approach has been proposed. The basic idea of it is to consider, in parallel, all the requirements of the product's life cycle at the design phase rather than to consider them sequentially in the traditional process of designing.

Consequentially, engineering design and product development become complex processes involving an interaction (synergetic combination) between many skills and disciplines. As a result of this approach it is possible to move decissions on design changes early in the development cycle (Fig.1) |5|.

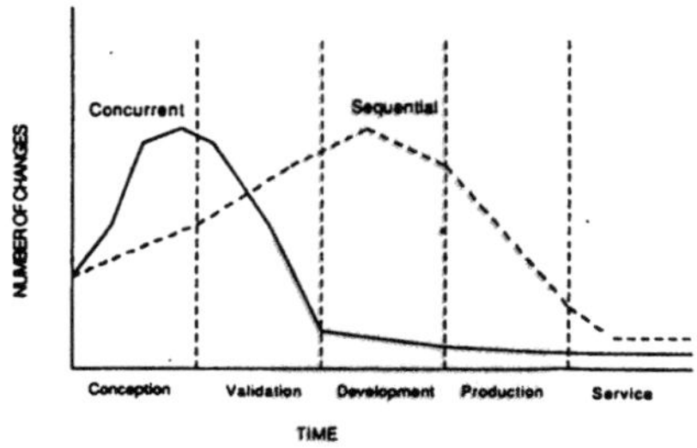

Fig.1 Number of changes vs time for concurrent and sequential engineering

The necessary condition for estimating the quantitive characteristics of the product, during the design phase, is elaboration of adequate design criteria. The design criteria are estabilished upon the data obtained from experience, analyses, test as well as competitors, and define what and how should be calculated (how - means method and accuracy).

As a consequence, at the heart of the design analysis is the mathematical model of the product, created according to the design criteria.

The model is used for simulating the performance of the designed product, to verify it and to improve it iteratively.

POLYDESCRIPTIVE APPROACH

The need for a polydescriptive approach to modelling and analysis of complex physical dynamic systems has been already mentioned in |2|.

In design analyses of complex machinery, structural elements are most conveniently modelled using Finite Element Method (FEM) |7|, control systems are customairly described with Transfer Functions (TF), power transmissions or suspension systems are frequently modelled with the aid of the Bond Graph (BG) technique |6|.

Although it is in principle possible, and sometimes even instructive, to use some specific modelling technique outside its customary field - in general, however, such modelling is troublesome and difficult and the resulting models cumbersome. Furthermore, there exists psychological barrier which prohibits, for instance, control engineers from readily using a technique which is not familiar to them, like FEM or BG.

Most of the present-day dynamic simulation languages try to ease these problems by providing language facilities of sufficient generality to make possible relatively simple modelling of even very complex systems. It must be realized, however, thet even the most modern and advanced simulation languages when applied to modelling of, say, a complex dynamic system incorporating a large structural part, bring considerable difficulties, a lot of work and additional computations when compared with the "natural" formalism for structural problems, which is FEM.

Instead to trying to build simulation software around the single-formalism or single-language concepts - it is more natural to "let all the flowers blossom" and accept a multitude of formalisms as a fact of life, which leads immediately to the concept of polydescriptive simulation systems.

COMPUTER AIDED MODELLING AND SIMULATION SYSTEM UNISYS/80

The system is designed as a general-purpose system for modelling and simulation of complex dynamic systems. It allows the user to develop and combine differently described submodels, and it generates from such a mixed-type parts a unifed description in terms of bond graph notation.

The unique feature of UNISYS/80 is thus, polydescri-

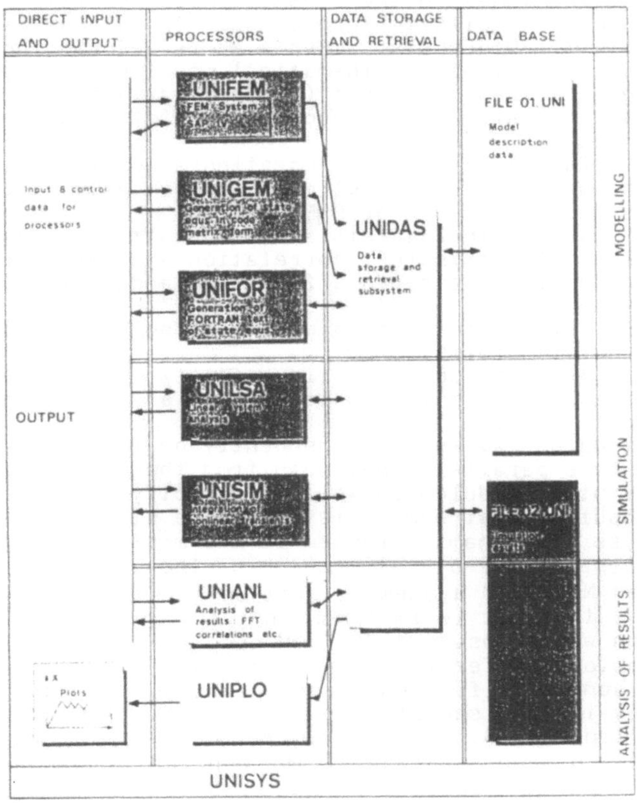

Fig.2 The overall scheme of the UNISYS/80 system

ptiveness, i.e. the ability of accepting simultaneously various subsystem description formalisms, such as:

- BOND GRAPH (BG) NOTATION, as a unfied description language for various types of lumped parameter systems (mechanical, electrical, hydraulic);
- FINITE ELEMENT (FE) DISCRETIZATION, widely accepted as the most adequate technique for continuous systems;
- TRANSFER FUNCTIONS (TF) as well as state space description, mostly used in control engineering;
- explicite sets of ordinary differential equations (ODE) or differential-algebraic equations (DAE).

The present version of the UNISYS/80 system consists of the following modules:

- UNIFEM is a Finite Element Method oriented processor, which translates a FE model generated via an external FEM system into the form, which is acceptable for UNIGEM processor,
- UNIGEM and UNIFOR modules generate a mathematical model of a complete system in the form of ODE/DAE set and the vector of I/O variables. While the UNIGEM generates the model in the form of a code matrix, UNIFOR produces an explicite form of ODE/DAE set as the FORTRAN text. UNIGEM also accepts macro description.
- UNILSA linearizes a mathematical model of the dynamic system, reduces its order (if necessary) and performs linear analysis. It may accept transfer functions as an input.
- UNISYM is a package of integration algorithms, which may be used to solve nonlinear, numerically stiff, large and sparce ODE or DAE systems.
- UNIANL allows the user to analys resulting transients by means of ststistics, correlation and intercorrelation functions, coherence functions and Fast Fourier Transform.
- UNIPLO allows the user to present results in plotted or tabulated forms.
- UNIDAS executes communication operations, whenever it is necessary to communicate between any module and the data base.

All modules may work independently or in a sequence as they have parallel input both from the data base generated by other modules as well as directly from user-defined input data stream. This structure allows the user to analyse his dynamic problem in the most efficient way. Fig.2 presents the overall scheme of the UNISYS/80 system.

The UNISYS/80 system is written in standard FORTRAN IV language, thus, it is implementable on any computer installation which posses a FORTRAN compiler. Since computer installations differ from one to another in the realization of some specific functions/operations UNISYS is supported by its own specific interface with FORTRAN compiler, called UNILIB, which is written for every computer installation type separately. The application concept of UNILIB is presented by Fig.3.

Numerous cases of dynamic system analysis have been performed with UNISYS, primarily at the Computer Aided Engineering Centre of the Industrial Institute of Construction Machinery, where the system was developed. These include mainly typical machine assemlies like hydraulically operated mechanical clutches, hydrostatic transmission systems and their elements, suspensions, industrial robot elements and control systems.

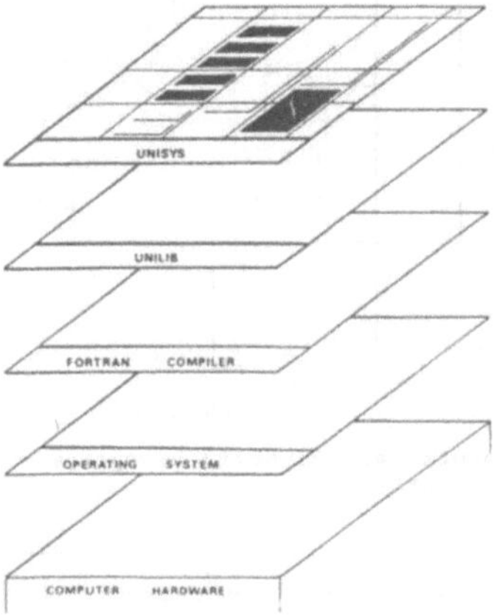

Fig.3 The concept of UNILIB

One of the examples is the design analysis of a power transmission and control system in the modern telescoping crane (Fig.4)

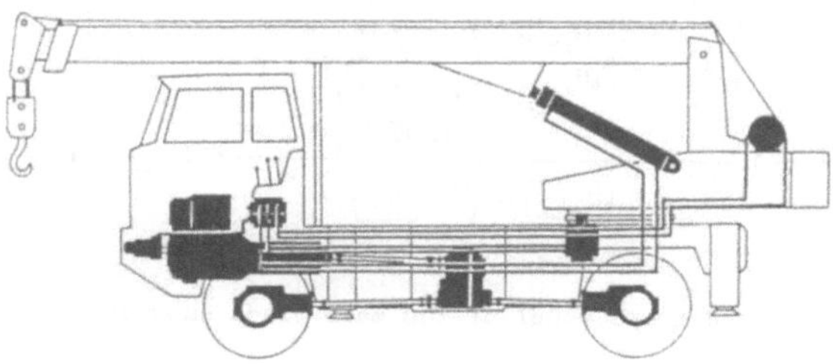

Fig.4 Power transmision and control system in the modern telescoping crane

The representative hydraulic subsystem - main winch subsystem presented in Fig.5 was modelled with bond graphs (Fig.6).

Results of simulation are given in Fig.7 prepared by means of UNIPLO module: where P1 and P2 are pressures in pipes before and behind hydraulic motor respectively. This example illustrates the influence of stiffness of those pipes on the pressure variation (P1,P2).

Above analysis allowed to eliminate harmfull dynamic effect through the use of a proper combination of stiff and elastic pipes.

Simulation analysis allows the designer to evaluate properties (in qualitative as well as quantitative way) of the control system at much lower cost and in shorter time than in a case of the totaly experimental work.

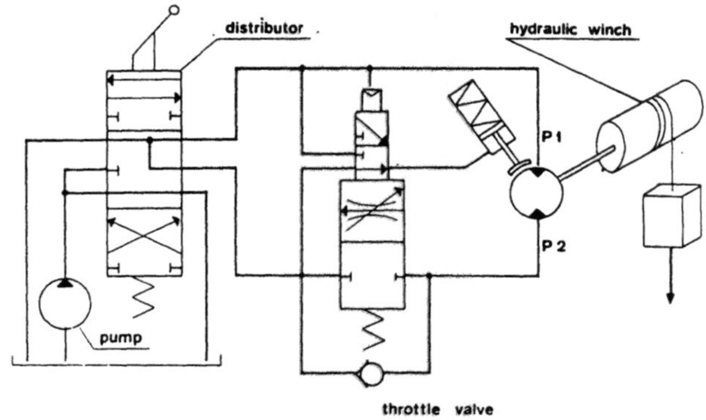

Fig.5 Representative hydraulic subsystem - main winch subsystem

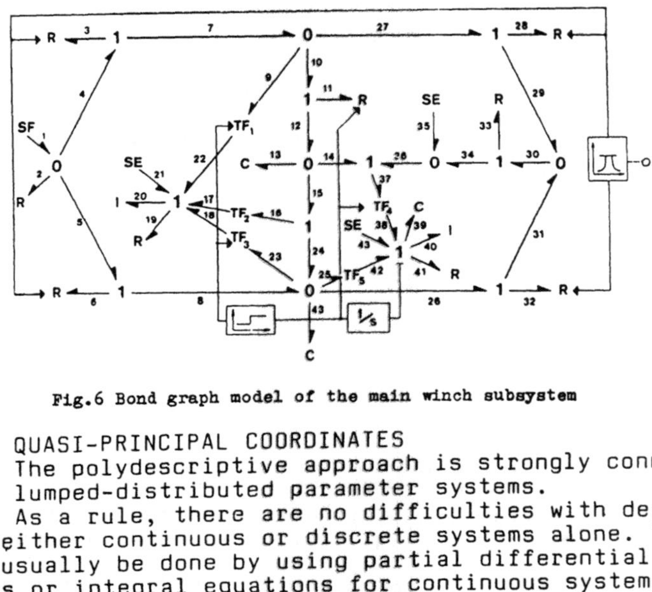

Fig.6 Bond graph model of the main winch subsystem

QUASI-PRINCIPAL COORDINATES

The polydescriptive approach is strongly connected with lumped-distributed parameter systems.

As a rule, there are no difficulties with describing either continuous or discrete systems alone. This can usually be done by using partial differential equations or integral equations for continuous systems and ordinary differential equations for discrete ones. But there is a lack of uniform and at the same time effective method for describing the continuous-discrete systems. Direct application of one of the above descriptions leads to the difficulties of the mathematical nature and as a rule is not effective.

These difficulties can be avoided if one uses so called quasi-principal coordinates |1|. They are a combination of the principal coordinates used for describing the continuous part of the system and physical (usually cartesian) coordinates applied to the discrete part.

Application of finite elements for modelling the continuous part (i.e. usage of discrete description) allows to describe the continuous-discrete system in a uniform way. The transformation of coordinates, build in such a way that it changes the coordinates describing the continuos part of the system to the principal coordinates and does not change the coordinates for the discrete part leads to the so called quasi-principal coordinates. They enable to describe (with sufficient accuracy) the behaviour of continuous-discrete systems using only a few degreese of freedom.

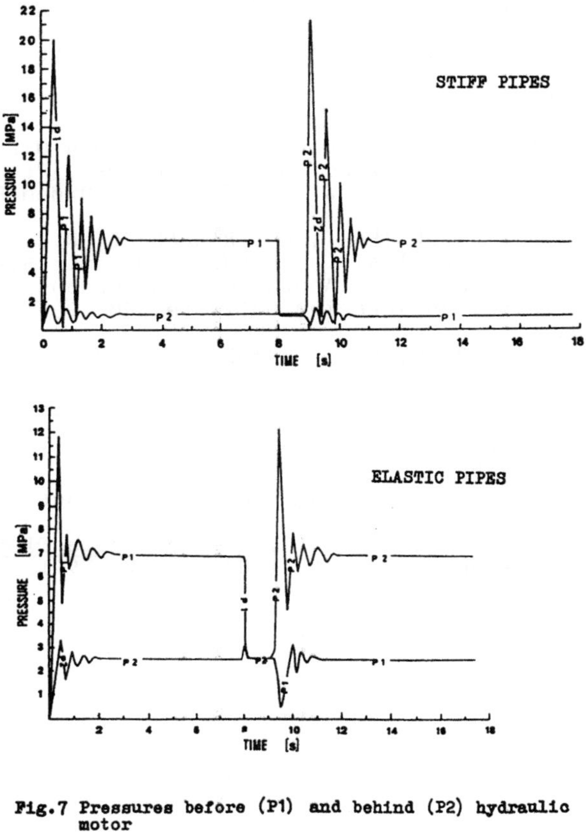

Fig.7 Pressures before (P1) and behind (P2) hydraulic motor

CONCLUSIONS

It has to be noted that the entire concept of linking FEM with other descriptions is not in any way limited to structural problems. FEM is presently the best elaborated and most widely used technique for many distributed parameter dynamic phenomena, mathematically described by partial differential equations (PDE). As the other formalisms used in UNISYS (BG, TF) are basically designated to attack lumped-parameter class of phenomena resulting in ordinary differential equations (ODE) UNISYS can be considered as an ODE/PDE automatic model generator and problem solver, thus allowing the user to analyse a very wide class of problems and physical systems. Questions arising in the dynamic behaviour of field - type phenomena (e.g. thermal, chemical, biological and ecological systems) controlled by some, even quite complex, control and powering mechanisms can be thus analysed in that way.

The approach may be used in other cases where mixed formalisms and/or lumped-distributed parameter models are useful for instance, recently, efforts aimed at using it for mechatronics |3| and biological systems (an artificial kidney - patient system dynamic model) have been made.

REFERENCES

1. BOSSAK, M.A., - 'Modelling of the Continuous-Discrete Systems Using Quasi-Principal Coordinates', Proceedings of XIII International FEM Congress, Baden-Baden, Nov.1984, 253-264.
2. BOSSAK, M.A., GRABOWIECKI, K., ROSENBERG, R.C., ZGORZELSKI, M., - 'Analysis of Physical Dynamic Systems Using a Polydescriptive Approach: Bond Graphs and Finite Elements', Engineering Software, CML Publications, 1981.
3. BRADLEY, D.A., BURD, N.C., DAWSON, D., LOADER, A.J., - Mechatronics, Chapman and Hall, 1991.
4. KUSIAK, A., - Intelligent Manufacturing Systems, Prentice-Hall, 1990.
5. SINGH, J., KEE, W.,T., - Designing it Right the First Time, Vision, Vol.7, No.2, 1991.
6. THOMA, J.,U., - Simulation by Bondgraphs, Springer-Verlag, 1990.
7. ZIENKIEWICZ, O.,C., TYLOR, R., - The Finite Element Method, McGraw-Hill, Vol.1, 1988; Vol.2, 1991.

FINITE ELEMENTS IN THE 90's, E. Oñate, J. Periaux, A. Samuelsson (Eds.)
Springer-Verlag/CIMNE, Barcelona 1991

A UNIFIED APPROACH TO THE DYNAMIC ANALYSIS OF SOILS AS MULTIPHASE MEDIA

W. Wunderlich
Technical University Munich, GERMANY

H. Cramer
Ruhr-University Bochum, GERMANY

Summary: The paper presents a unified approach to the formulation of soil dynamic problems including the multiphase nature of fully or partially saturated soils. The formulation is based on the general equations of the theory of mixtures. Different approximations and corresponding finite element models are discussed. The soil skeleton is assumed to have an elastic plastic behavior. For noncohesive soils an elastic plastic material model for cyclic loading is described. It includes the essential features of the solid skeleton that are of importance for the interaction of the different soil components. The paper presents the governing equations and their numerical implementation. Numerical applications are included.

1. Introduction

The behavior of saturated and unsaturated soils is a very important aspect in a large number of soil dynamic problems. The interaction between the different solid, liquid and gaseous soil components plays an important role with respect to the overall soil behavior. The evolution of the pressure in the pore fluids has considerable influence on the effective stresses in the solid skeleton. On the other hand the volumetric behavior of the solid skeleton is strongly coupled with a pore fluid pressure build up.

Therefore, a full modelling of the soil as a multiphase system by computational techniques is necessary, if nonlinear behavior can occur as in the case of liquefaction.

In the present paper we are dealing with

- a mathematical model for soil as a multiphase system for fluid-gas-skeleton interaction,
- a constitutive model for noncohesive soils to simulate the behavior under varying stresses (dilatancy, Bauschinger effect),
- the development of efficient numerical models by the finite element method.

The mathematical model of the soil system is based on the theory of mixtures [1,2]. As particular cases we obtain the equations governing the behavior of fully and partially saturated soils [3] as a generalization of Biot's [4] theory of porous media.

The constitutive equations of the solid skeleton are derived from an elastic plastic formulation with combined kinematic and isotropic frictional hardening. It is based on a multisurface model first proposed in [5,6] where the strain-dilatancy relation [7] is incorporated in order to simulate the volumetric behavior realistically.

The resulting nonlinear differential equations are the basis of the numerical models using finite elements. Several particular cases and approximations like fully saturated soils and soils under undrained conditions are obtained from the general multiphase formulation.

2. Mathematical model for soils as multiphase systems

2.1 Mass and momentum balance equations

The soil is considered as a three-phase medium composed of solid, liquid and gaseous constituents. Therefore, the fundamental mass and momentum balance equations apply not only for the mixture as a whole but also for each constituent. Under isothermal conditions the equations of motion and the equations for the balance of mass are obtained (not sum over α)

$$\bar{\rho}^{\alpha} \frac{D^{\alpha}}{dt} v_i^{\alpha} = \bar{\sigma}^{\alpha}_{ji,j} + \bar{p}_i^{\alpha} + \hat{p}_i^{\alpha} \, , \tag{2.1}$$

$$\frac{D^{\alpha}}{dt} \bar{\rho}^{\alpha} + \bar{\rho}^{\alpha} v^{\alpha}_{i,i} = 0 \, . \tag{2.2}$$

where $\bar{\rho}^{\alpha}$ and $\bar{\sigma}^{\alpha}_{ji}$ represent the mass and stresses of the α-constituent per unit of the mixture volume. v_i^{α} are the velocities of the α-phase, $\bar{p}_i^{\alpha}$ denote external bodyforces and $\hat{p}_i^{\alpha}$ are local interaction forces between the constituents being in equilibrium with each other

$$\sum_{\alpha} \hat{p}_i^{\alpha} = 0 \, . \tag{2.3}$$

Here the general superscript α should be replaced by s, f or g to represent the equations and quantities of the solid, fluid and gaseous or air phase respectively.

By the definition of the volume fraction n^{α} of the constituent α, the mean densities $\bar{\rho}^{\alpha}$ can be expressed by the true densities

$$\bar{\rho}^{\alpha} = n^{\alpha} \rho^{\alpha} \tag{2.4}$$

$$\text{with } n^{\alpha} = \frac{dV^{\alpha}}{dV}, \qquad \sum_{\alpha} n^{\alpha} = n^s + n^f + n^g = 1 \, .$$

To arrive at a notation that is more common in soil mechanics the volume fractions are expressed by the porosity n and the degree of water or fluid saturation s given by

$$n = n^f + n^g \, , \qquad (2.5) \qquad\qquad s^f = \frac{n^f}{n} = s \, , \qquad (2.6)$$

$$s^g = \frac{n^g}{n} = 1-s \, , \qquad (2.7) \qquad\qquad s^f + s^g = 1 \, . \qquad (2.8)$$

2.2 Kinematic variables

The velocity of the solid v_i^s is defined by the time derivative of the soil displacement u_i in the context of a Lagrangean description

$$v_i^s = \dot{u}_i \, . \qquad (2.9)$$

The velocities of the pore-water and the pore-air are splitted up into a sum of the soil velocity and the diffusion velocity relative to the grain matrix

$$v_i^f = \dot{u}_i + \tilde{v}_i^f \, , \qquad (2.10) \qquad\qquad v_i^g = \dot{u}_i + \tilde{v}_i^g \, . \qquad (2.11)$$

The relative velocities $\tilde{v}_i^\alpha$ on the other hand can be expressed by the filtration (Darcy) velocities q^α

$$q_i^f = n \, s \, \tilde{v}_i^f \, , \qquad (2.12) \qquad\qquad q_i^g = n(1-s) \, \tilde{v}_i^g \, . \qquad (2.13)$$

Thus the velocities of the liquid and gas phase are defined by

$$v_i^f = \dot{u}_i + \frac{1}{ns} q_i^f \, , \qquad (2.14) \qquad\qquad v_i^g = \dot{u}_i + \frac{1}{n(1-s)} q_i^g \, . \qquad (2.15)$$

2.3 Effective stress states

Assuming that the viscous part of the stresses in a fluid phase is generally negligible, the stresses in the liquid and gaseous phase are fully defined by the pore-water pressure p^f and the pore-air pressure p^g

$$\bar{\sigma}_{ij}^f = n^f \, p^f \, \delta_{ij} = n \, s \, p^f \, \delta_{ij} \, , \qquad (2.16)$$

$$\bar{\sigma}_{ij}^g = n^g \, p^g \, \delta_{ij} = n(1-s) \, p^g \, \delta_{ij} \, . \qquad (2.17)$$

Under the simplifying assumption that each solid grain is incompressible the intergranular stresses are given by a generalization of Therzaqi's effective stress concept [8] through

$$\begin{aligned} \bar{\sigma}_{ij}^s &= \sigma'_{ij} + n^s (s^f \, p^f + s^g \, p^g) \, \delta_{ij} \\ &= \sigma'_{ij} + (1-n)(s \, p^f + (1-s) p^g) \, \delta_{ij} \end{aligned} \qquad (2.18)$$

where σ'_{ij} are the effective stress components.

2.4 Constitutive relations

The equations given above have to be completed by a set of constitutive relations and state equations. The constitutive law for the solid skeleton is represented by a time independent elastic plastic stress strain relation

$$\dot{\sigma}'_{ij} = C^{ep}_{ijkl}\,\dot{\varepsilon}_{kl} \qquad (2.19) \qquad\qquad \text{with } \dot{\varepsilon}_{kl} = \frac{1}{2}(\dot{u}_{k,l} + \dot{u}_{l,k}) \,. \qquad (2.20)$$

The effective elastic plastic material tensor C^{ep} depends on the current stress state and the deformation path. A special formulation for noncohesive soils is given below.

The interaction forces between the solid skeleton and the pore-water and pore-air, respectively, are defined by

$$\hat{p}^f_i = -(ns)_{,i}\, p^f - \xi^f_{ij}\, \tilde{v}^f_j \,, \qquad (2.21)$$

$$\hat{p}^g_i = -(n(1-s))_{,i}\, p^g - \xi^g_{ij}\, \tilde{v}^g_j \,, \qquad (2.22)$$

$$\hat{p}^s_i = -(\hat{p}^f_i + \hat{p}^g_i) \,. \qquad (2.23)$$

The first part of equations (2.21) and (2.22) results from geometric variations of the solid skeleton and represents buoyancy forces. The second terms account for viscous effects due to fluid flow. They can be expressed by a generalization of Darcy's empirical flow law

$$\xi^f_{ij}\, \tilde{v}^f_j = \mu^f \frac{n\, s}{k(n)\; k^f_r(s)}\, q^f_i = \frac{ns}{\bar{k}^f}\, q^f_i , \qquad (2.24)$$

$$\xi^g_{ij}\, \tilde{v}^g_j = \mu^g \frac{n(1-s)}{k(n)\; k^g_r(s)}\, q^g_i = \frac{n(1-s)}{\bar{k}^g}\, q^g_i \,. \qquad (2.25)$$

The parameters μ^f and μ^g denote the dynamic viscosities of the pore-water and the pore-air respectively. For isotropic materials $k(n)$, $k^f_r(n)$ and $k^g_r(n)$ define the permeabilities depending on the porosity and the degree of saturation.

Under the simplifying assumption that the water phase and the air phase are coherently distributed in the unsaturated zone, the water pressure and the air pressure are in equilibrium with the capillary pressure p^c

$$p^c = p^g - p^f \,. \qquad (2.26)$$

The capillary pressure is defined as a function of the degree of saturation $p^c = p^c(S)$ or vice versa known from laboratory tests. Thus we obtain the change of saturation by

$$\dot{s} = \frac{\partial s}{\partial p^c}\, \dot{p}^c = S_p(\dot{p}^g - \dot{p}^f) \,. \qquad (2.27)$$

The single solid grains as well as the pore fluid are assumed to be incompressible whereas the pore-air is assumed to be an ideal gas.

$$\dot{\rho}^s = \dot{\rho}^f = 0 \,, \qquad (2.28) \qquad\qquad \frac{\dot{\rho}^g}{\rho^g} = \frac{\dot{p}^g}{p^g} = -\frac{1}{K^g}\, p^g \,. \qquad (2.29)$$

2.5 Resulting system of differential equations

Generally the equations given above can be further simplified under the assumptions that the inertia forces in the air phase can be omitted as well as all convective acceleration terms due to the small filtration velocities of the water- and air-phase.

By introducing the substitutions

$$\dot{w}_i = g_i^f , \quad (2.30) \qquad \dot{h} = p^g \quad (2.31)$$

we arrive at the following system of differential equations:

$$\bar{\rho}\ddot{u}_i + \rho^f \ddot{w}_i - \sigma'_{ji,j} - s\, p^f_{,i} - (1-s)\, p^g_{,i} = \bar{\rho}\, b_i \quad (2.32)$$

with $\bar{\rho} = (1-n)\,\rho^s + s\, n\, \rho^f$

$$\rho^f \ddot{u}_i + \frac{1}{ns}\rho^f \ddot{w}_i + \frac{1}{\bar{k}^f}\dot{w}_i - p^f_{,i} = \rho^f\, b_i \quad (2.33)$$

$$\bar{k}^g\, \dot{h}_{,i} - \bar{k}^g\, p_{,i} = 0 \quad (2.34) \qquad \dot{n} = \dot{u}_{i,i} \quad (2.35)$$

$$s\,\dot{u}_{i,i} + \dot{w}_{i,i} - n\, S_p\, \dot{p}^f + n\, S_p\, \dot{p}^g = 0 \quad (2.36)$$

$$(1-s)\,\dot{u}_{i,i} + (\bar{k}^g\, \dot{h}_{,i})_{,i} - \frac{\bar{k}^g}{K^g}\, p^g_{,i}\, \dot{h}_i + n\, S_p\, \dot{p}^f - \left(\frac{n(1-s)}{K^g} + n\, S_p\right)\dot{p}^g = 0 . \quad (2.37)$$

The equations of motion of the whole mixture resulting from the sum of the equations of motion of all constituents are given by equation (2.32), those for the pore-water and the pore-air are represented through (2.33) and (2.34). The continuity equations for the solid skeleton, the pore-water and the pore-air are given by (2.35), (2.36) and (2.37), respectively.

Another more simplified solution of the problem may be obtained for the case that the air is assumed to be at atmospheric pressure or equal to zero. Then the degree of saturation is a function of the pore-water pressure only and merely the equations of motion and the continuity equations for the solid skeleton and the pore-water are necessary.

3. A constitutive model for noncohesive soils

In the present paper the essential relations of a constitutive model for noncohesive soils under cyclic loading are described. A more detailed derivation will be found in [9]. The model is based on multisurface plasticity with combined isotropic kinematic hardening.

In order to define volumetric and shear hardening by different mechanisms we assume that the total strain increment can be written as the sum of the elastic and plastic parts

$$\dot{\varepsilon}_{ij} = \dot{\varepsilon}^{el}_{ij} + \dot{\varepsilon}^{pl,v}_{ij} + \dot{\varepsilon}^{pl,s}_{ij} , \qquad (3.1)$$

where $\dot{\varepsilon}^{pl,v}_{ij}$ and $\dot{\varepsilon}^{pl,s}_{ij}$ denote the plastic strain rates due to the volumetric and the shear hardening mechanism, respectively.

The volumetric part is formulated using a cap model with isotropic hardening and an associated flow rule.

$$F^v = \sigma_m + \kappa^v \leq 0 \qquad \text{with } \sigma_m = \frac{1}{3} I_1' = \frac{1}{3} \sigma'_{ij} \delta_{ij} , \qquad (3.2)$$

$$\dot{\varepsilon}^{pl,v}_{ij} = \frac{1}{3H^v} \delta_{ij} \dot{\kappa}^v \qquad \text{with } H^v = K_o + K_1(-\sigma_m)^p . \qquad (3.3)$$

Plastic deformations induced by shear loading are based on a bounding surface plasticity model [5,6]. Shear failure is defined by a modified Coulomb criterion with a smooth yield surface in terms of the invariants σ_m, S and θ (Lode-angle)

$$F^f = S/g(\theta) + \alpha \sigma_m \qquad (3.4)$$

$$\text{with } S = \surd J_2 , \quad J_2 = \frac{1}{2} s_{ij} s_{ij} , \quad s_{ij} = \sigma'_{ij} - \delta_{ij} \sigma_m .$$

The parameter α can be expressed by the angle of friction, and $g(\theta)$ defines the shape of the yield surface in the π-plane. A convenient function that allows reasonable approximations by convex surfaces for all materials of interest is given by

$$g(\theta) = \gamma \left(\frac{2}{(1-\sin 3\theta) + \gamma^4 (1+\sin 3\theta)}\right)^{.25} . \qquad (3.5)$$

The bounding or memory surface defined by F^m describes the maximum stress state reached during the load history

$$F^m = S/g(\theta) + \kappa \alpha \sigma_m . \qquad (3.6)$$

The hardening function κ defines the isotropic expansion of the memory surface, that separates the regions of primary loading and of unloading/reloading. Its evolution is assumed to be a hyperbolic function of the plastic deviatoric strains ξ, that are accumulated during the primary loading processes

$$\kappa = \frac{1}{R_f} \frac{\xi}{(A^q + \xi^q)^{1/q}} ; \qquad \kappa \leq 1 . \qquad (3.7)$$

This defines the hardening parameter usually denoted by H during primary loading as a function of κ ($H = H^m(\kappa)$). The yield surface F^s which moves kinematically within the memory surface is expressed by the function

$$F^s = \bar{S}/g(\bar{\theta}) + \kappa^s \alpha \sigma_m , \qquad (3.8)$$

$$\text{with } \bar{S} = \surd \bar{J}_2 , \quad \bar{J}_2 = \frac{1}{2} \bar{s}_{ij} \bar{s}_{ij} , \quad \bar{s}_{ij} = s_{ij} + \kappa_{ij} \alpha \sigma_m .$$

$\bar{\theta}$ denotes a 'Lode-angle' corresponding to the tensor $\bar{s}_{ij}$. The evolution of the isotropic and kinematic hardening functions κ^s and κ_{ij} is given as a function of κ and the normalized difference

d_{ij} [5,6,9] between the conjugate or image stress s'_{ij} on the memory surface $F^m = 0$ and the current stress deviator s_{ij}

$$d_{ij} = \frac{s'_{ij} - s_{ij}}{-\alpha\sigma_m} \tag{3.9}$$

The hardening parameter for reloading H^s is thus approximated by the function

$$H^s(\kappa,d) = H^m(\kappa - d) \left(\frac{d^{in}}{d^{in} - d}\right)^r , \tag{3.10}$$

$$\text{with } d = d_{ij}\, n_{ij} , \quad d^{in} = d^{in}_{ij}\, n_{ij} , \quad n_{ij} = \frac{\partial F^s}{\partial \sigma'_{ij}} .$$

The plastic strain increments are obtained by a nonassociated flow rule and a plastic potential based on a stress-dilatancy relation [7]

$$G^s = F^s - \bar{\alpha}\, \sigma_m , \quad \dot{\varepsilon}^{pl,s}_{ij} = \frac{\partial G^s}{\partial \sigma'_{ij}} \dot{\lambda} , \tag{3.11}$$

in which $\bar{\alpha}$ defines a surface in stress space (characteristic line [10]) that separates regions where plastic compaction or plastic dilatation occurs. Using these relations the incremental tangential material tensor C^{ep} can be determined by common procedures.

4. Finite element discretization

4.1 Discrete form of the differential equations

With the usual spatial finite element approach the unknown variables u_i, w_i, p^f, h and p^g are discretized by

$$\begin{aligned} u_i &= \mathbf{N}_u\, \hat{\mathbf{u}}_i , \quad w_i = \mathbf{N}_w\, \hat{\mathbf{w}}_i , \\ p^f &= \mathbf{N}_f\, \hat{\mathbf{p}}^f , \quad h = \mathbf{N}_h\, \hat{\mathbf{h}} , \quad p^g = \mathbf{N}_g\, \hat{\mathbf{p}}^g . \end{aligned} \tag{4.1}$$

The common weighted residual approach then results in the global form of the tangential equations

$$\mathbf{M}\, \ddot{\mathbf{z}} + \mathbf{C}\, \dot{\mathbf{z}} + \mathbf{K}\, \Delta\mathbf{z} = \bar{\mathbf{P}} - \bar{\mathbf{F}} \tag{4.2}$$

where $\mathbf{z} = \{\hat{\mathbf{u}}\ \ \hat{\mathbf{w}}\ \ \hat{\mathbf{p}}^f\ \ \hat{\mathbf{h}}\ \ \hat{\mathbf{p}}^g\}$ represents the vector of the unknown nodal degrees of freedom. This discrete system of second order differential equations can be treated by standard time stepping algorithms. The system matrices have the following structure:

$$\mathbf{M} = \begin{bmatrix} \mathbf{M}_{ss} & \mathbf{M}^T_{sf} & 0 & 0 & 0 \\ \mathbf{M}_{sf} & \mathbf{M}_{ff} & 0 & 0 & 0 \\ 0 & 0 & 0 & 0 & 0 \\ 0 & 0 & 0 & 0 & 0 \\ 0 & 0 & 0 & 0 & 0 \end{bmatrix} \qquad \mathbf{C} = \begin{bmatrix} 0 & 0 & 0 & 0 & 0 \\ 0 & \mathbf{C}_{ff} & 0 & 0 & 0 \\ 0 & 0 & 0 & 0 & 0 \\ 0 & 0 & 0 & \mathbf{C}_{gg} & 0 \\ 0 & 0 & 0 & 0 & 0 \end{bmatrix}$$

$$
\mathbf{K} = \begin{bmatrix} \mathbf{K}_{ss} & 0 & \mathbf{L}_{sf}^T & 0 & \mathbf{L}_{sg}^T \\ 0 & 0 & \mathbf{L}_{ff}^T & 0 & 0 \\ \mathbf{L}_{sf} & \mathbf{L}_{ff} & -\mathbf{F}_{ff} & 0 & \mathbf{F}_{fg} \\ 0 & 0 & 0 & 0 & -\mathbf{C}_{gg} \\ \mathbf{L}_{sg} & 0 & \mathbf{F}_{fg}^T & -\bar{\mathbf{C}}_{gg} & -\mathbf{F}_{gg} \end{bmatrix}
$$

The mass matrices $\mathbf{M}_{ss}$, $\mathbf{M}_{sf}$ and $\mathbf{M}_{ff}$ result from the inertia terms of the solid skeleton and the pore fluid, and the viscous damping matrices $\mathbf{C}_{ff}$ and $\mathbf{C}_{gg}$ from the discretization of Darcy's flow law. The flexibility matrices $\mathbf{F}_{ff}$, $\mathbf{F}_{fg}$ and $\mathbf{F}_{gg}$ arise from the compressibility of the pore-air and of the saturation-capillary pressure relation, while $\mathbf{L}_{sf}$, $\mathbf{L}_{ff}$ and $\mathbf{L}_{sg}$ represent the discrete form of the divergence terms in the continuity equations. Using quadrilateral 9-node elements, the variables u_i and w_i are approximated by biquadratic functions whereas a bilinear approximation is used for p^f, p^g and h in order to avoid spatial oscillations.

The equations given above represent the most general case of a three phase mixture. Several simplified problems and approximations are included. As already mentioned the air pressure may often be assumed to be at atmospheric pressure or equal to zero. This leads to a system where the third and fourth block of equations together with the model variables $\hat{\mathbf{h}}$ and $\hat{\mathbf{p}}^g$ are omitted. Similar equations are obtained for a fully saturated soil for which all terms in $\mathbf{F}_{ff}$ are equal to zero. A further possible approximation are undrained conditions for which fluid flow ($\hat{\mathbf{w}}_i = \mathbf{0}$) and the second block of equations can be neglected.

5. Numerical results

The interactions between the pore fluids and the solid skeleton are of great influence even if an elastic stress strain relation may be assumed. This has been demonstrated in [3] for fully and partially saturated soils. If nonlinear behavior and damage effects can occur they are still more important. This is particularly true if soil interaction with possible liquefaction occurs.

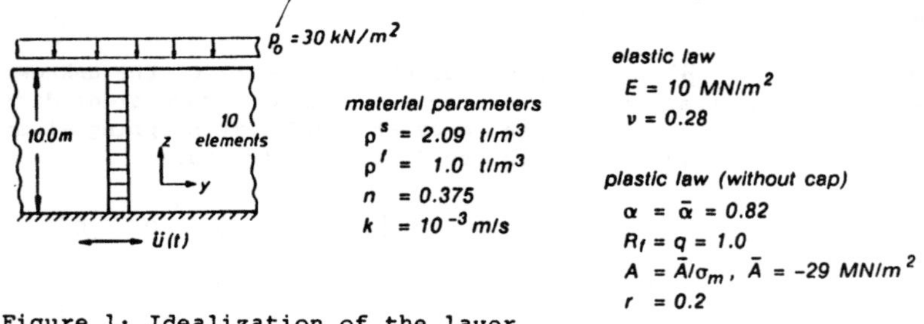

Figure 1: Idealization of the layer

To demonstrate this, the behavior of a layer of sand subjected to earthquake loading is investigated under different conditions. The time history of the ground acceleration is taken from the earthquake of Friaul. The material properties assumed to describe the soil behavior are representative for a very loose sand.

In the first analysis a dry soil is considered. The time histories of the displacements at the top of the layer are shown in figure 2b. The soil undergoes permanent vertical settlements due to a volumetric compaction caused by cyclic plastic shear deformations.

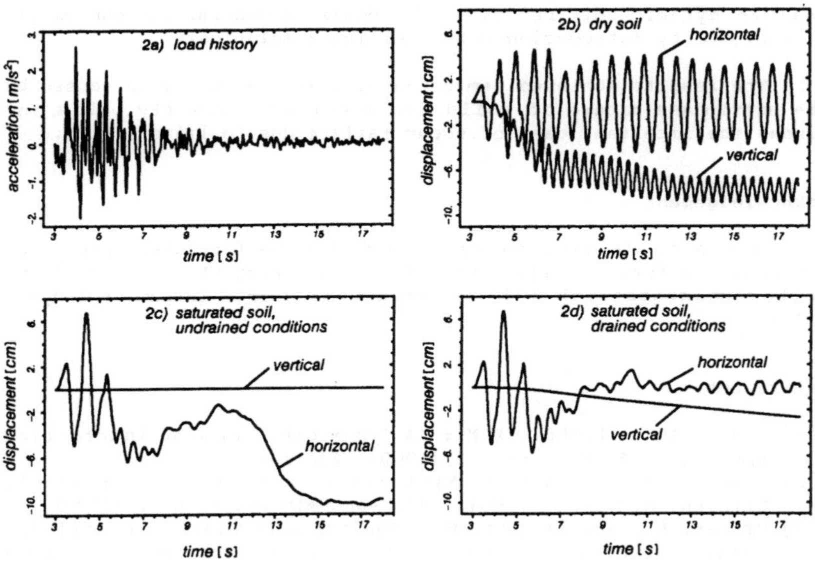

Figure 2: Base acceleration and surface displacements

In a saturated soil the contraction of the solid skeleton results in an increase of the pore-water pressure. This reduces the contact forces between the particles and hence may lead to failure due to liquefaction. This is demonstrated by the results of a second analysis, where a fully saturated soil under undrained conditions is considered. Fig. 2c shows that after a time period of about eight seconds the system exhibits pronounced horizontal deformations caused by shear failure.

On the other hand possible fluid flow will reduce the pore water pressure. This is taken into account in a third analysis. The time histories of the surface displacement (Fig. 2d) show that the results under drained and undrained conditions are nearly identical during the first two seconds. After this period a consolidation process due to fluid flow takes place that reduces the pore water pressure. Thus no failure of the layer occurs.

6. Conclusions

For the dynamic analysis of saturated and partially saturated soils a complete modelling by computational techniques is generally necessary. This is particularly true if nonlinear behavior can occur.

The mathematical model for soils as a three phase medium, the constitutive law for the behavior of noncohesive soils and the numerical schemes proposed in this paper provide an adequate tool to analyze dynamic soil behavior. The model is general and can easily be improved by introducing other nonlinear relations.

The results achieved demonstrate, that the phenomena caused by the multiphase nature of soils are modelled correctly. This includes the important case of shear failure due to liquefaction.

Acknowledgement

The authors greatfully acknowledge the financial support of the Deutsche Forschungsgemeinschaft through project C2 of the 'Sonderforschungsbereich 151' on Tragwerksdynamik (structural dynamics).

References

[1] Green, A.E., Naghdi, P.M.: 'A dynamical theory of interacting continua', J. Eng. Sci. 3 (1965), 231-241.
[2] Bowen, R.M.: 'Theory of Mixtures', in 'Continuum Physics', Eringen, A.C. (ed.), Vol. III, Academic Press, N.Y. (1979).
[3] Cramer, H., Wunderlich, W.: 'Multiphase models in soil dynamics', Proc. European Conf. on Structural Dynamics, Eurodyn '90, Bochum (1990), 165-172.
[4] Biot, R.M.: 'Theory of propagation of elastic waves in a fluid saturated porous solid', J. Acous. Soc. Am. 28 (1956), 168-178.
[5] Dafalias, Y.F., Popov, E.P.: 'A model of nonlinearly hardening materials for complex loading', Acta Mechanica 21 (1975), 173-192.
[6] Dafalias, Y.F.: 'Bounding surface plasticity. I: Mathematical foundation and hypoelasticity', J. Eng. Mech. Div., ASCE, Vol. 112 (1986), 966-987.
[7] Rowe, P.W.: 'The stress-dilatancy relation for static equilibrium of an assembly of particles in contact', Proc. R. Soc. A 269 (1962), 500-527.
[8] Therzagi, K.: 'Erdbaumechanik auf Bodenphysikalischer Grundlage', Franz Deuticke, Vienna (1925).
[9] Cramer, H., Wunderlich, W.: 'Elastic plastic anisotropic bounding surface models for soils with hyperbolic frictional hardening' (in preparation).
[10] Luong, M.P.: 'Mechanical aspects and thermal effects of cohesionsless soils under cyclic and transient loading', in 'Deformation and Failure of Granular Material', Vermeer, P.A., Luger, H.J. (eds.), Wiley (1982), 239-246.

SECTION 5

MATHEMATICAL ASPECTS OF THE FINITE ELEMENT METHOD

FINITE ELEMENTS IN THE 90's, E. Oñate, J. Periaux, A. Samuelsson (Eds.)
Springer-Verlag/CIMNE, Barcelona 1991

A NOTE ON THE DERIVATION OF WEAK FORMS

E-M. Salonen
Helsinki University of Technology
Institute of Mechanics
Otakaari 1, SF-02150 Espoo,
FINLAND

SUMMARY

The note is concerned with the didactics of the finite element method. The derivation of the standard weak form connected with the one-dimensional steady heat conduction problem is considered. Certain jump terms, usually left out in the process of integration by parts are taken into account. This means that some alterations must also be made in the conventional starting point equations. The manner of derivation described is recommended for use in teaching because it may help in understanding the properties of the finite element method.

1. INTRODUCTION

Let us consider the one-dimensional steady state heat conduction equation

$$\frac{\mathrm{d}}{\mathrm{d}x}\left(k\frac{\mathrm{d}\phi}{\mathrm{d}x}\right)+Q=0, \qquad 0<x<L \tag{1}$$

with the boundary conditions

$$\phi(0)-\bar{\phi}_0=0, \qquad \phi(L)-\bar{\phi}_L=0 \tag{2}$$

The notation used above and in the sequel follows closely that employed in the work [1] and thus needs no explanation here.

The standard way to derive the weak form corresponding to Eqs. (1) and (2) presented in the textbooks usually goes more or less in the following way. First the statement

$$\int_0^L v\left[\frac{\mathrm{d}}{\mathrm{d}x}\left(k\frac{\mathrm{d}\phi}{\mathrm{d}x}\right)+Q\right]\mathrm{d}x=0 \tag{3}$$

is written down. Thereafter, the integration by parts formula for two functions g and h

$$\int_0^L g\frac{\mathrm{d}h}{\mathrm{d}x}\mathrm{d}x=(gh)_L-(gh)_0-\int_0^L\frac{\mathrm{d}g}{\mathrm{d}x}h\mathrm{d}x \tag{4}$$

is applied with $g \triangleq v$, $h \triangleq k\mathrm{d}\phi/\mathrm{d}x$ giving

$$\int_0^L v\frac{\mathrm{d}}{\mathrm{d}x}\left(k\frac{\mathrm{d}\phi}{\mathrm{d}x}\right)\mathrm{d}x=\left(vk\frac{\mathrm{d}\phi}{\mathrm{d}x}\right)_L-\left(vk\frac{\mathrm{d}\phi}{\mathrm{d}x}\right)_0-\int_0^L\frac{\mathrm{d}v}{\mathrm{d}x}k\frac{\mathrm{d}\phi}{\mathrm{d}x}\mathrm{d}x \tag{5}$$

By setting further

$$v(0)=0,\quad v(L)=0 \tag{6}$$

the standard weak form

$$\int_0^L\left(k\frac{\mathrm{d}v}{\mathrm{d}x}\frac{\mathrm{d}\phi}{\mathrm{d}x}-vQ\right)\mathrm{d}x=0 \tag{7}$$

is arrived at. A typical discrete equation obtained by the Galerkin method is thus

$$\sum_i\left(\int_0^L k\frac{\mathrm{d}N_j}{\mathrm{d}x}\frac{\mathrm{d}N_i}{\mathrm{d}x}\mathrm{d}x\right)a_i-\int_0^L N_jQ\,\mathrm{d}x=0 \tag{8}$$

A conscientious student may feel that something is wrong in the development. He remembers from mathematics that a certain degree of smoothness is demanded for the integration by parts formula to be valid and suggests that to be on the safe side the manipulation should be performed separately in each element to take into account possible discontinuities of derivatives at the element interfaces.

2. AN ATTEMPT FOR A MORE PRECISE DERIVATION

Let us try to respond to the critique expressed in the preceding paragraph. To continue we change the notation somewhat so that the results are more easily extendible to two- or three-dimensional cases. The domain $(0,L)=\Omega$ is divided into subdomains or elements $\Omega^e, e=1,2,\ldots$. As the integrals in the finite element method are evaluated in practice in an element by element way, it is natural to write the left-hand side of Eq. 3 as a summation:

$$\sum_e\int_{\Omega^e}v\left[\frac{\mathrm{d}}{\mathrm{d}x}\left(k\frac{\mathrm{d}\phi}{\mathrm{d}x}\right)+Q\right]\mathrm{d}\Omega=0 \tag{9}$$

The integration by parts formula to be used is now

$$\sum_e\int_{\Omega^e}g\frac{\mathrm{d}h}{\mathrm{d}x}\mathrm{d}\Omega=(gh)_L-(gh)_0-\sum_I[[gh]]_I-\sum_e\int_{\Omega^e}\frac{\mathrm{d}g}{\mathrm{d}x}h\,\mathrm{d}\Omega \tag{10}$$

It is obtained by applying Eq. (4) separately in each element and by then summing the contributions. The meaning of the jump bracket notation is $[[f]]=f^+-f^-$ where the minus and plus values refer to the left- and right-handed limit values of function f at the element interface points $I=1,2,\ldots$.

The more careful version of Eq. (5) is thus

$$\sum_e \int_{\Omega^e} v \frac{\mathrm{d}}{\mathrm{d}x}\left(k\frac{\mathrm{d}\phi}{\mathrm{d}x}\right)\mathrm{d}\Omega = \left(vk\frac{\mathrm{d}\phi}{\mathrm{d}x}\right)_L - \left(vk\frac{\mathrm{d}\phi}{\mathrm{d}x}\right)_0$$

$$-\sum_I v_I \left[\left[k\frac{\mathrm{d}\phi}{\mathrm{d}x}\right]\right]_I - \sum_e \int_{\Omega^e} \frac{\mathrm{d}v}{\mathrm{d}x} k \frac{\mathrm{d}\phi}{\mathrm{d}x}\mathrm{d}\Omega \tag{11}$$

As is seen the weighting function v is assumed here to be continuous so that at each point I $v = v^+ = v^-$ and v can be taken outside the jump brackets. When formula (11) is applied in Eq. (9), and Eqs. (6) are taken into account we obtain – as we first might think – the more correct weak form

$$\sum_e \int_{\Omega^e} \left(k\frac{\mathrm{d}v}{\mathrm{d}x}\frac{\mathrm{d}\phi}{\mathrm{d}x} - vQ\right)\mathrm{d}\Omega + \sum_I v_I \left[\left[k\frac{\mathrm{d}\phi}{\mathrm{d}x}\right]\right]_I = 0 \tag{12}$$

Comparison with the standard form (7) shows the difference with respect to the jump terms. In spite of the honesty behind the attempt the present weak form is found, however, to be completely useless. For instance, taking the case k = constant, $Q = 0$ and by using linear two noded elements, discrete equations of the type $0 = 0$ are obtained.

3. THE REMEDY

It is fortunately possible to find a respectable solution to the dilemma posed by the worried student. Let us replace the heat conduction differential equation (1) with the formulation

$$\frac{\mathrm{d}}{\mathrm{d}x}\left(k\frac{\mathrm{d}\phi}{\mathrm{d}x}\right) + Q = 0 \qquad \text{in } \Omega^e, \quad e = 1,2,\ldots \tag{13}$$

$$\left[\left[k\frac{\mathrm{d}\phi}{\mathrm{d}x}\right]\right] = 0 \qquad \text{for } I = 1,2,\ldots \tag{14}$$

This means that we have anticipated that the solution may not be smooth, particularly at the element interfaces – say because k has a jump – and we thus replace the field equation at these points with the continuity condition of the heat flux which follows from the physics of energy conservation.

The logical starting point for the derivation of the weak statement based on Eqs. (13) and (14) is

$$\sum_e \int_{\Omega^e} v\left[\frac{\mathrm{d}}{\mathrm{d}x}\left(k\frac{\mathrm{d}\phi}{\mathrm{d}x}\right) + Q\right]\mathrm{d}\Omega + \sum_I w_I \left[\left[k\frac{\mathrm{d}\phi}{\mathrm{d}x}\right]\right]_I = 0 \tag{15}$$

Here the quantities w_I are still arbitrary weighting constants. However, a look at the results of the previous section shows that the specific selection

$$w_I = v(x_I) \equiv v_I \tag{16}$$

giving the starting statement

$$\sum_{e}\int_{\Omega^e} v\left[\frac{\mathrm{d}}{\mathrm{d}x}\left(k\frac{\mathrm{d}\phi}{\mathrm{d}x}\right)+Q\right]\mathrm{d}\Omega+\sum_{I} v_I\left[\left[k\frac{\mathrm{d}\phi}{\mathrm{d}x}\right]\right]_I=0 \tag{17}$$

now leads to the correct standard weak form (7) written here as

$$\sum_{e}\int_{\Omega^e}\left(k\frac{\mathrm{d}v}{\mathrm{d}x}\frac{\mathrm{d}\phi}{\mathrm{d}x}-vQ\right)\mathrm{d}\Omega=0 \tag{18}$$

4. CONCLUDING REMARKS

Generalizing rather boldly from the experience gained by studying the simple heat conduction problem we would like to advocate certain themes which could be of some didactic value in the teaching of the finite element method.

It seems that to understand well what is happening during the manipulations with the integration by parts formulae it is a good habit to write all the weak statements – and functionals – right from the start in an element by element way. After all, one fundamental defining feature of the finite element method is, Zienkiewicz [2]:

"(b) the n-equations governing the behaviour of the whole system, i.e.

$$F_j(a_i)=0 \qquad j=1-n \tag{1.1}$$

can be assembled by the simple process of addition of terms contributed from all subdomains (or elements) which divide the system into physically identifiable entities (without overlap or exclusion). Thus

$$F_j=\sum F_j^e \tag{1.2}$$

where F_j^e is the element contribution to the quantity under consideration."

The above fact should thus preferably be transparent in all the possible weak statement versions before and after the integrations by parts manipulations.

A second theme connected closely with the previous one is the importance of changing the original problem setting by enhancing the governing field equations (assumed now valid only inside the elements) with suitable jump conditions on the element interfaces. This emphasis seems also quite natural in connection with the finite element method. One advantage of the finite element method over classical ones is expressly its ability to cope with non-smooth solutions too, at least if the irregularities are confined to the element interfaces. It is thus logical also to try to introduce the governing jump conditions right from the beginning explicitly into the formulation of the problem.

Some of the relevant jump conditions are obtainable from the conservation laws of mass, momentum etc. of continuum mechanics. These laws deal with finite material bodies and both the governing differential equations and the jump conditions can be derived in well-known ways from them.

One additional point in favour of introducing the jumps is that they emerge in any case when the errors of the finite element solutions are studied.

The themes discussed above are developed further in articles [3] and [4] and in the report [5]. The textbook [6] also contains parts dealing with somewhat similar subjects.

ACKNOWLEDGEMENTS

The author wishes to thank Mr. J. Freund and Ms. S. Räsänen for useful comments on the original manuscript.

REFERENCES

1. ZIENKIEWICZ, O.C. and TAYLOR, R.L. – **The Finite Element Method**, 4th ed., Vol. 1, McGraw-Hill, London, 1989.

2. ZIENKIEWICZ, O.C. – 'Why Finite Elements?' Finite Elements in Fluids, Vol. 1, Ed. Gallagher, R.H., Oden, J.T., Taylor, C. and Zienkiewicz, O.C., Wiley, London, 1975, p. 2.

3. FREUND, J. and SALONEN, E-M. – On Weighting Functions for Weak Forms, Journal of Structural Mechanics, 23, 18-61 (1990) (in Finnish).

4. FREUND, J. and SALONEN, E-M. – 'A Logic for Simple Petrov - Galerkin Weighting Functions, Second World Congress on Computational Mechanics, August 27-31, 1990, Stuttgart, FRG, p. 902-905.

5. FREUND, J. and SALONEN, E-M. – Notes on Petrov-Galerkin Weighting Functions in Connection with the Scalar Convection-Diffusion Equation, Institute of Mechanics, Helsinki University of Technology, Report No. 28 (1990), 38 p.

6. HUGHES, T.J.R. – **The Finite Element Method: Linear Static and Dynamic Finite Element Analysis**, Prentice-Hall, New Jersey, 1987.

FINITE ELEMENTS IN THE 90's, E. Oñate, J. Periaux, A. Samuelsson (Eds.)
Springer-Verlag/CIMNE, Barcelona 1991

THE INDIVIDUAL ELEMENT TEST REVISITED

C. Militello and C. A. Felippa
Department of Aerospace Engineering Sciences,
and Center for Space Structures and Controls,
University of Colorado at Boulder,
Boulder, Colorado 80309-0429,
USA

SUMMARY

The subject of the patch test for finite elements retains several unsettled aspects. In particular, the issue of one-element versus multielement tests needs clarification. Following a brief historical review, we present the individual element test (IET) of Bergan and Hanssen in an expanded context that encompasses several important classes of new elements. The relationship of the IET to the multielement forms A, B and C of the patch test and to the single element test are clarified.

1. BACKGROUND

The patch test for convergence is a fascinating area in the development of nonconforming finite element methods. It grew up of the brilliant intuition of Bruce Irons. Initially developed in the mid-1960s at Rolls Royce and then at the Swansea group headed by Olek Zienkiewicz, by the early 1970s the test had became a powerful and practical tool for evaluating and checking nonconforming elements. And yet today it remains a controversial issue: accepted by most finite element developers while ignored by others, welcomed by element programmers, distrusted by mathematicians. For tracing down the origins of the test there is no better source than a 1973 survey article by Irons and Razzaque [12]. Added remarks to the quoted material are inserted in footnotes, and reference numbers have been altered to match those of the present paper.

Origins of the Patch Test

In 1965 even engineering intuition dared not predict the behavior of certain finite elements. Experience force those engineers who doubted it to admit that interelement continuity was important: the senior author[1] believed that it was necessary for convergence. It is not known which ideas inspired a numerical experiment by Tocher and Kapur [25], which demonstrated convergence within 0.3% in a biharmonic problem of plate bending, using equal rectangular elements with 1, x, y, x^2, xy, y^2, y^2, x^3, x^2y, xy^2, y^3, and x^3y and xy^3, as functional basis. The nodal variable of this Ari Adini rectangle [1] are w, $\partial w/\partial x$ and $\partial w/\partial y$ at the four corners, and this element guarantees only C^0 conformity.

[1] Bruce Irons

Some months later, research at Rolls-Royce on the Zienkiewicz nonconforming triangle [2], — a similar plate-bending element[2]— clarified the situation. Three elements with C^1 continuity were simultaneously available, and, because the shape function subroutine used for numerical integration had been exhaustively tested, the results were trustworthy. It was observed: (a) that every problem giving constant curvature over the whole domain was accurately solved by the conforming elements whatever the mesh pattern, as was expected, and (b) that the nonconforming element was also successful, but only for one particular mesh pattern.[3]

Thus the patch test was born. For if the external nodes of any sub-assembly of a successful assembly of elements are given prescribed values corresponding to an arbitrary state of constant curvature, then the internal nodes must obediently take their correct values. (An internal node is defined as one completely surrounded by elements.) Conversely, if two overlapping patches can reproduce any given state of constant curvature, they should combine into a larger successful patch, provided that every external node lost is internal to one of the original patches. For such nodes are in equilibrium at their correct values, and should behave correctly as internal nodes of the extended patch. In an unsuccessful patch test, the internal nodes take unsuitable values, which introduce interelement discontinuities. The errors in deflection may be slight, but the errors in curvature may be ±20%. We must recognize two distinct types of errors:

(i) The finite element equations would not be exactly satisfied by the correct values at the internal nodes — in structural terms, we have disequilibrium;

(ii) The answers are nonunique because the matrix of coefficients **K** is semidefinite.

Role of the Patch Test

Clearly the patch test provides a *necessary* condition for convergence with fine mesh. We are less confident that it provides a *sufficient* condition. The argument is that if the mesh is fine, the patches are also small. Over any patch the correct solution gives almost uniform conditions to which the patch is known to respond correctly — provided that the small perturbations from uniform conditions do not cause a disproportionate response in the patch: we hope to prevent this by insisting that **K** is positive definite.

The patch test is invaluable to the research worker. Already, it has made respectable

(i) Elements that do not conform,
(ii) Elements that contain singularities,
(iii) Elements that are approximately integrated,
(iv) Elements that have no clear physical basis.

In short, the patch test will help a research worker to exploit and justify his wildest ideas. It largely restores the freedom enjoyed by the early unsophisticated experimenters.

The late 1960s and early 1970s were a period of unquestionable success for the test. That optimism is evident in the article quoted above, and prompted Gilbert Strang to develop a mathematical version popularized in the Strang-Fix textbook [21].

Confidence was shaken in the late 1970s by several developments. Numerical experiments, for example, those of Sander and Beckers [20] suggested that the test is not necessary for convergence, thus disproving Irons' belief stated above.

---

[2] This element is that identified by 'BCIZ' in the present paper

[3] The bending element test referred to in this sentence appears in the Addendum to [2]. This Addendum was not part of the original paper presented at the First Wright-Patterson Conference held in September 1965; it was added to the Proceedings that appeared in 1966. The name "patch test" will not be found there; see the Appendix of [21] for further historical details.

Then a counterexample by Stummel [22] purported to show that the test is not even sufficient.[4] This motivated defensive responses by Irons [13] shortly before his untimely death, and by Taylor, Simo, Zienkiewicz and Chan [24]. These papers tried to set out the engineering version of the test on a more precise basis.

Despite these ruminations many questions persist, as noted in the lucid review article by Griffiths and Mitchell [11]. Some of them are listed below.

Q1. What is a patch? Is it the ensemble of all possible meshes? Are some meshes excluded? Can these meshes contain different types of elements?

Q2. The test was originally developed for harmonic and biharmonic compressible elasticity problems, for which the concept of "constant strains" or "constant curvatures" is unambiguous. But what is the equivalent concept for shells? Even Reissner-Mindlin plates (which lead to the so-called C^0 elements) pose difficulties.

Q3. What are the modifications required for incompressible media? Is the test applicable to dynamic or nonlinear problems?

Q4. Are single-element versions of the test equivalent to the conventional, multielement versions?

Q5. Is the test restricted to nonconforming assumed-displacement elements? Can it be extended to encompass assumed-stress or assumed-strain mixed and hybrid elements? (For initial attempts in this direction, see [10])

The following treatment is aimed primarily at answering the last two questions. No position as to the mathematical relevance of the test is taken.

2. THE INDIVIDUAL ELEMENT TEST

Because of practical difficulties incurred in testing all possible patches there have been efforts directed toward translating the original test into statements involving a single element. These will be collectively called ***one-element tests***.

The first step along this path was taken by Strang [21], who using integration by parts recast the original test in terms of "jump" contour integrals over element interfaces. An updated account is given by Griffiths and Mitchell [11], who remark that Strang's test can be passed in three different ways:

JCS: Jump integrals cancel over common sides of adjacent elements (*e.g.* DeVeubeke's 3-midside-node triangle, Morley's plate elements).

JOS: Jump integrals cancel over opposite element sides (*e.g.* Wilson's incompatible plane rectangle [26]).

---

[4] Stummel has constructed [23] a ***generalized patch test*** that is mathematically impeccable in that it provides necessary and sufficient conditions for convergence. Unfortunately such test lacks important side benefits of Irons' patch test, such as element checkout by computer, because it is administered as a mathematically limiting process in function spaces. Furthermore, it does not apply to a mixture of different element types, or to situations such as a side shared by more than two elements.

JEC: Jump integrals cancel over the element contour (see examples in [11]).

Another important development, not so well publicized as Strang's, was undertaken by Bergan and coworkers at Trondheim over the period 1975–1984. The so called ***individual element test***, or IET, was proposed by Bergan and Hanssen [4] in 1975. The underlying goal was to establish a test that could be directly carried out on the stiffness equations of a ***single*** element — an obvious improvement over the multielement form. In addition the test was to be constructive, *i.e.*, used as an ***a priori*** guide during element formulation, rather than as a post-facto check.

The IET has a simple physical motivation: to demand pairwise cancellation of tractions among adjacent elements that are subjected to a common uniform stress state. This is precisely the 'JCS' case of the Strang test noted above. Because of this inclusion, the IET is said to be a ***strong*** version of the patch test in the following sense: any element passing the IET also verifies the conventional multielement form of the patch test, but the converse is not necessarily true.

The IET has formed the basis of the free formulation (FF) later developed by Bergan and Nygård [6]. It has also played an important part in the development of high performance finite elements undertaken by the authors [7–9,15–18].

In an important paper written in response to Stummel's counterexample, Taylor, Simo, Zienkiewicz and Chan [24] defined multielement patch tests in more precise terms, introducing the so-called A, B and C versions. They also discussed a one-element test called the "single element test," herein abbreviated to SET. They used the BCIZ plate bending element [2] to show that an element may pass the SET but fail multielement versions, and consequently that tests involving single elements are to be viewed with caution. In what follow we try to clarify this apparent contradiction and to establish precisely what the individual element test entails. In particular ***it is shown that the IET contains a crucial condition that the SET lacks***, and that the two tests are not therefore equivalent.

Furthermore, we extend the IET to conditions beyond those considered by Bergan and Nygård by including elements with unknown internal displacement fields. The most important sources of such elements are: stress-assumed hybrids, and elements constructed through the assumed natural strain (ANS) and assumed natural deviatoric strain (ANDES) formulations.

3. ASSUMPTIONS FOR ELEMENT CONSTRUCTION

Suppose that we want to test an individual element of volume V and boundary S with exterior normal $\mathbf{n}$. The element satisfies the following assumptions.

A1. The element shares displacement degrees of freedom collected in $\mathbf{v}$ (the so called ***visible*** degrees of freedom) with adjacent elements. The boundary displacement field $\mathbf{d}$ is uniquely determined by $\mathbf{v}$ as

$$\mathbf{d} = \mathbf{N}_d \mathbf{v}, \tag{1}$$

where $\mathbf{N}_d$ are boundary shape functions.

The term "boundary displacement field" is meant to include normal derivatives (side rotations or slopes) in bending problems. More generally, in a problem governed by a variational principle of index $m > 1$, $\mathbf{d}$ includes normal derivatives up to the $(m-1)^{th}$ order.

This assumption says nothing about the internal displacement field $\mathbf{u}$. In free-formulation elements $\mathbf{u}$ is known and agrees with $\mathbf{d}$ only at the nodes. In the ANS [3,14,19,] and ANDES [8,9,17,18] formulations, $\mathbf{u}$ is unknown because the deviatoric strain field ϵ_h introduced in A2 below is not generally integrable.

A2. The strain field ϵ within the element is expressible as

$$\epsilon = \mathbf{B}\mathbf{v}, \tag{2}$$

which admits the following decomposition into mean and deviatoric parts:

$$\epsilon = \bar{\epsilon} + \epsilon_h = \overline{\mathbf{B}}\,\mathbf{v} + \mathbf{B}_h\mathbf{v} = (\overline{\mathbf{B}} + \mathbf{B}_h)\mathbf{v}, \tag{3}$$

where

$$\bar{\epsilon} = \frac{1}{V}\int_V \epsilon\, dV, \qquad \epsilon_h = \epsilon - \bar{\epsilon}. \tag{4}$$

We note that

$$\int_V \mathbf{B}_h\, dV = \mathbf{0}, \qquad \int_V \mathbf{A}\mathbf{B}_h\, dV = \mathbf{0}. \tag{5}$$

in which $\mathbf{A}$ is an arbitrary matrix constant over the element.

Subscript h stands for "high order." The strain field ϵ_h is not generally integrable, that is, associable with an internal displacement field $\mathbf{u}$ such that $\epsilon_h = \mathbf{D}\mathbf{u}$, where $\mathbf{D} = \frac{1}{2}(\nabla + \nabla^T)$ is the symmetric vector gradient operator. On the other hand, the mean strain field $\bar{\epsilon}$, being constant, is integrable, as discussed under assumption A4.

A3. Suppose the element is under a constant stress state $\boldsymbol{\sigma}_0$. Then a nodal force system $\mathbf{p}_0$ conjugate to $\mathbf{v}$ in the sense of virtual work develops. These forces are connected to $\mathbf{v}$ through the relation

$$\mathbf{p}_0 = \mathbf{L}\boldsymbol{\sigma}_0, \qquad \mathbf{L} = \int_S \mathbf{N}_{dn}\, dS, \tag{6}$$

where $\mathbf{L}$ is called the *force lumping matrix* and $\mathbf{N}_{dn}$ denotes the projection of the shape functions $\mathbf{N}_d$ over the normal to the element side.

Matrix $\mathbf{L}$ was introduced by Bergan and coworkers in their studies leading to the free formulation [5,6], and plays a crucial role in the individual element test.

A4. The constant stress field $\boldsymbol{\sigma}_0$ is associated with a given displacement field called $\mathbf{u}_{rc}$, such that the associated strain and stress fields are

$$\bar{\epsilon}^u = \mathbf{D}(\mathbf{u}_{rc}), \qquad \sigma^0 = \mathbf{E}\bar{\epsilon}^u, \tag{7}$$

where $\mathbf{E}$ is the symmetric matrix of elastic moduli, assumed constant over the element. This constitutive assumption excludes incompressibility, which must receive special treatment.

Field $\mathbf{u}_{rc}$ cannot be immediately linked to $\mathbf{v}$ because it spans a subspace of the possible boundary motions. We must start by expressing $\mathbf{u}_{rc}$ in the modal or generalized-coordinate form

$$\mathbf{u}_{rc} = \mathbf{N}^q_{rc}\mathbf{q}_{rc}, \tag{8}$$

where $\mathbf{N}^q_{rc}$ are modal functions and $\mathbf{q}_{rc}$ their amplitudes. The projection $\mathbf{v}_{rc}$ of $\mathbf{u}_{rc}$ over the space of boundary motions spanned by $\mathbf{v}$ can be most easily obtained by collocation, that is, evaluating $\mathbf{u}_{rc}$ at the nodal points. This process yields

$$\mathbf{v}_{rc} = \mathbf{G}_{rc}\mathbf{q}_{rc}, \tag{9}$$

in which $\mathbf{G}_{rc}$ will generally be a rectangular matrix with more rows than columns.

Subscripts r and c mean that $\mathbf{u}_{rc}$ is supposed to include rigid-body and constant-strain modes. In mathematical terms, $\mathbf{u}_{rc}$ is a polynomial of degree $m-1$ when the variational index is m.

4. THE STIFFNESS MATRIX

Under the previous assumptions, the stiffness matrix is given by

$$\mathbf{K} = \int_V \mathbf{B}^T\mathbf{E}\mathbf{B}\, dV. \tag{10}$$

Using the strain decomposition (3), $\mathbf{K}$ splits as follows:

$$\mathbf{K} = \mathbf{K}_b + \mathbf{K}_h = \int_V \overline{\mathbf{B}}^T\mathbf{E}\overline{\mathbf{B}}\, dV + \int_V \mathbf{B}_h^T\mathbf{E}\mathbf{B}_h\, dV = V\overline{\mathbf{B}}^T\mathbf{E}\overline{\mathbf{B}} + \int_V \mathbf{B}_h^T\mathbf{E}\mathbf{B}_h\, dV \tag{11}$$

because of the *energy orthogonality condition*

$$\int_V \overline{\mathbf{B}}^T\mathbf{E}\mathbf{B}_h\, dV = \mathbf{0}, \tag{12}$$

which results on taking $\mathbf{A} = \overline{\mathbf{B}}^T\mathbf{E}$ on the second of (5). Matrices $\mathbf{K}_b$ and $\mathbf{K}_h$ receive the name of *basic stiffness* and *higher order stiffness*, respectively.

5. FIRST CONDITION: CONSTANT STRAIN STATES

Bergan and Nygård [6] state two constraints for FF elements, which taken together represent the satisfaction of the IET. The first one is

$$\mathbf{K}\,\mathbf{v}_{rc} = \mathbf{L}\,\boldsymbol{\sigma}_0, \tag{13}$$

which is essentially an equilibrium statement at the element level. Premultiplying (13) by $\mathbf{v}^T_{rc}$ we get $\mathbf{v}^T_{rc}\mathbf{K}\mathbf{v}_{rc} = \mathbf{v}^T_{rc}\mathbf{L}\boldsymbol{\sigma}_0$, which on introducing (9) and (11) becomes

$$\mathbf{q}^T_{rc}\mathbf{G}^T_{rc}\overline{\mathbf{B}}^T\mathbf{E}\overline{\mathbf{B}}\mathbf{G}_{rc}\mathbf{q}_{rc}V + \mathbf{q}^T_{rc}\mathbf{G}^T_{rc}\mathbf{K}_h\mathbf{G}_{rc}\mathbf{q}_{rc} = \mathbf{q}^T_{rc}\mathbf{G}^T_{rc}\mathbf{L}\boldsymbol{\sigma}_0. \tag{14}$$

If we request that $\mathbf{K}_h$ cannot contribute to the internal energy under a constant strain state we must have $\mathbf{G}_{rc}^T\mathbf{K}_h\mathbf{G}_{rc} = \mathbf{0}$, or, since $\mathbf{K}_h = \int_V \mathbf{B}_h\mathbf{E}\mathbf{B}_h\,dV$,

$$\mathbf{B}_h\mathbf{G}_{rc} = \mathbf{0}, \tag{15}$$

This may be called the *higher order strain cancellation condition.* Taking into account that $\boldsymbol{\sigma}_0 = \mathbf{E}\mathbf{B}_{rc}^q\mathbf{q}_{rc}$, where $\mathbf{B}_{rc}^q = \mathbf{D}(\mathbf{N}_{rc}^q)$, the above equation becomes $\overline{\mathbf{B}}^T\mathbf{E}\overline{\mathbf{B}}\mathbf{G}_{rc} = V^{-1}\,\mathbf{L}\mathbf{E}\mathbf{B}_{rc}^q$. This can be split into $\overline{\mathbf{B}}^T = V^{-1}\,\mathbf{L}$ and $\overline{\mathbf{B}}\mathbf{G}_{rc} = \mathbf{B}_{rc}^q$. Replacing the former in the latter we obtain

$$V^{-1}\,\mathbf{L}^T\mathbf{G}_{rc} = \mathbf{B}_{rc}^q. \tag{16}$$

These conditions were introduced by Bergan and Nygård [6] in the context of the free formulation. They state that equation (16) should be used to check that the matrix $\mathbf{L}$ is correct. Then (16) is the first consistency constraint on $\mathbf{L}$. Equations (15) and (16) are necessary *in order that a single element, which is in equilibrium, be capable of copying a constant strain state.* To prove that they are also sufficient conditions is straightforward.

An important consequence of (16) can be investigated as follows. Rewrite it as $V^{-1}\,\mathbf{L}^T\mathbf{G}_{rc}\,\mathbf{q}_{rc} = \mathbf{D}\mathbf{u}_{rc}$. Multiplying both sides by $\boldsymbol{\sigma}_0^T$ and integrating over the element volume we obtain

$$\boldsymbol{\sigma}_0^T\mathbf{L}^T\mathbf{G}_{rc}\,\mathbf{q}_{rc} = \int_V \boldsymbol{\sigma}_0^T\mathbf{D}\mathbf{u}_{rc}\,dV. \tag{17}$$

Integration by parts of the right hand side yields

$$\boldsymbol{\sigma}_0^T\mathbf{L}^T\mathbf{G}_{rc}\,\mathbf{q}_{rc} = \boldsymbol{\sigma}_0^T\int_S \mathbf{N}_{un}\,dS\,\mathbf{q}_{rc}, \tag{18}$$

where $\mathbf{N}_{un}$ are the the projections of the modal functions over the normal to the element side. From the definition of $\mathbf{L}$ in (6) we conclude that

$$\int_S \mathbf{N}_{dn}^T\,dS\,\mathbf{G}_{rc}\mathbf{q}_{rc} = \int_S \mathbf{N}_{un}\,dS\,\mathbf{q}_{rc}. \tag{19}$$

This result may be stated as follows: the force lumping produced by the boundary displacement field should be energy consistent (in the sense of virtual work) with that produced by the displacement field $\mathbf{u}_{rc}$ over the element side. Although $\mathbf{B}_{rc}^q$ is unique for a given problem, since $\mathbf{G}_{rc}$ is generally a rectangular matrix, equation (16) clearly shows that $\mathbf{L}$ is not necessarily unique. Examples that illustrate this property may be found in [17]. $\mathbf{L}$ is unique for simplex elements where we have the same number of nodal connectors $\mathbf{v}$ and rc-modal amplitudes $\mathbf{q}_{rc}$, because in this case $\mathbf{G}_{rc}$ is square and non-singular. For these elements the total and basic stiffness matrices coalesce. An obvious example is provided by the constant strain triangle (CST).

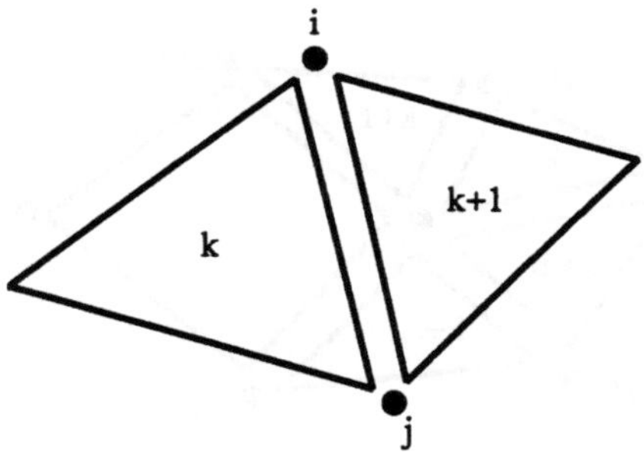

Figure 1: A common side $i-j$ shared by two elements

6. SECOND CONDITION: PAIRWISE FORCE CANCELLATION

Quoting from [4]: "The basis for the individual element test is that the element, when interacting with its neighbors, should be capable of identically reproducing an arbitrary rigid-body/constant strain field ... The interelement forces transferred at nodes should cancel out in a pairwise manner for adjacent elements during such state." This establishes a second key constraint on $\mathbf{L}$.

Suppose we have a side $i-j$ joining elements k and $k+1$, as illustrated in Figure 1. The second condition requires that

$$\int_i^j \mathbf{N}_{dn}^k \, dS = \int_i^j \mathbf{N}_{dn}^{k+1} \, dS. \tag{20}$$

The easiest way of enforcing this condition is by choosing a boundary displacement $\mathbf{d}$ that is uniquely defined over $i-j$ by degrees of freedom on that interface. This rule can be extended to cases in which more than two elements share a side, as is the case in many practical structures. Note that (20) does not involve the internal displacement field in any way. Consequently it establishes the *mixability* of elements of different types (for example, FF with ANDES elements). The SET discussed in [24] omits this important condition.

7. MULTIELEMENT PATCH TESTS AND THE IET

Bergan and coworkers called conditions (13) and (20) the IET. We now prove that if the element under consideration satisfies these conditions, it will also pass the so-called forms A and B of the multielement patch test [24]. Furthermore, if the element is rank sufficient it will also pass form C.

Let us consider the assemblage of elements shown in Figure 2 as a patch. The global displacement field consistent with a constant strain field is $\mathbf{v}_{rc}^g = \mathbf{G}_{rc}^g \mathbf{q}_{rc}$ The stiffness matrix of the kth element satisfies equation (13), or its equivalent global form

$$(\mathbf{P}^k)^T \mathbf{K}^k \mathbf{P}^k \mathbf{v}_{rc}^g = (\mathbf{P}^k)^T \mathbf{L}^k \boldsymbol{\sigma}_0, \tag{21}$$

where $\mathbf{P}^k$ are Boolean localization matrices. Upon assembly we obtain

$$\mathbf{K}^g \mathbf{v}_{rc}^g = \mathbf{L}^g \boldsymbol{\sigma}_0 = \mathbf{p}, \tag{22}$$

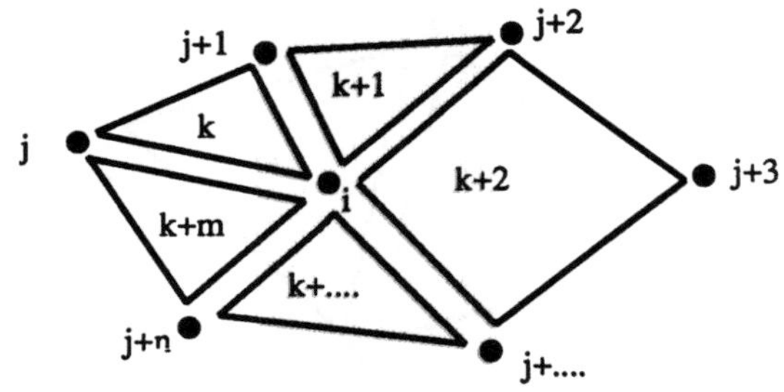

Figure 2: An assemblage of elements

but because of the satisfaction of condition (20) the force vector $\mathbf{p}$ has only components in nodes $j, \ldots j+n$. Then, for the ith internal degree of freedom we have

$$K^g_{ij} v^g_{rcj} = p_i = 0, \tag{23}$$

which is the statement of the form A of the patch test. If an element satisfies A, form B is also satisfied because from (22) we can obtain the displacement of the internal node i as:

$$v_{rc_i} = (K^g_{ii})^{-1}(p_i - K_{ij} v^g_{rc_j}), \qquad j \neq i. \tag{24}$$

Because the element satisfies (13), $\mathbf{v}_{rc}$ can be obtained if upon removing the rigid body motions $\mathbf{K}^k$ is nonsingular and can be inverted. Consequently $\mathbf{K}^k$ should be rank sufficient in order to satisfy form C.

8. CONCLUSIONS

It has been shown that the IET constraints plus rank sufficiency provide sufficient conditions to pass any form of the multielement patch test. The main practical advantages of the IET are:

1. By applying rules (13) and (20) elements can be constructed that will pass any multielement patch test *a priori*, provided that they are rank sufficient, while being capable of copying constant strain states. No such possibility exists in the conventional patch test, which must be necessarily applied *a posteriori*.
2. Element mixability is immediately established.
3. A "surgical operation" can be established to "sanitize" elements that fail the IET, as discussed in the Appendix of [17]. The operation essentially amounts to the replacement of the basic stiffness.

The price paid for these advantages is that the test is occasionally stronger than strictly necessary. For example, the BCIZ nonconforming triangle [2] fails the IET but passes the multielement test for certain mesh configurations.

A potential difficulty in the application of the IET to *existing* elements is the need for extracting the force-lumping matrix **L**. This matrix may not be readily available and, as mentioned in Section 5, is not necessarily unique.

Finally, as remarked in several places, the present statement of the IET is not restricted to the free formulation, and has actually been used in this expanded form for constructing high-performance elements based on the ANDES formulation [8, 9,17,18].

ACKNOWLEDGEMENT

The work reported here has been supported by NASA Langley Research Center under Grant NAG1-756.

REFERENCES

1. ADINI, A. - *Analysis of Shell Structures by the Finite Element Method*, Ph. D. Dissertation, Dept. of Civil Engineering, University of California, Berkeley, CA, 1961.

2. BAZELEY, G.P., CHEUNG, Y.K., IRONS, B.M. and ZIENKIEWICZ, O.C. - Triangular Elements in Plate Bending - Conforming and Nonconforming Solutions, *Proceedings 1st Conference on Matrix Methods in Structural Mechanics*, AFFDL-TR-66-80, Air Force Institute of Technology, Dayton, Ohio, 1966, p. 547.

3. BATHE, K.J. and DVORKIN, E.N. - A Four-node Plate Bending Element Based on Mindlin/Reissner Plate Theory and a Mixed Interpolation, Int. J. Numer. Meth. Engrg., **21**, 367–383 (1985)

4. BERGAN, P.G. and HANSSEN, L. - A New Approach for Deriving 'Good' Finite Elements, MAFELAP II Conference, Brunel University, 1975, in *The Mathematics of Finite Elements and Applications - Volume II*, ed. by J. R. Whiteman, Academic Press, London, 1976.

5. BERGAN, P.G. - Finite Elements Based on Energy Orthogonal Functions, Int. J. Num. Meth. Engrg., **15**, 1141–1555 (1980)

6. BERGAN, P.G. and NYGÅRD, M.K. - Finite Elements with Increased Freedom in Choosing Shape Functions, Int. J. Num. Meth. Engrg., **20**, 643–664 (1984)

7. FELIPPA, C.A. - Parametrized Multifield Variational Principles in Elasticity: II. Hybrid Functionals and the Free Formulation, Communications in Applied Numerical Methods, **5**, 79–88 (1989)

8. FELIPPA, C.A. and MILITELLO, C. - Variational Formulation of High Performance Finite Elements: Parametrized Variational Principles, Computers & Structures, **36**, 1–11 (1990)

9. FELIPPA, C.A. and MILITELLO, C. - 'Developments in Variational Methods for High Performance Plate and Shell Elements', in *Analytical and Computational Models for Shells*, CED Vol. 3, Eds. A. K. Noor, T. Belytschko and J. C. Simo, The American Society of Mechanical Engineers, ASME, New York, 1989, p. 191.

10. FRAEIJS DE VEUBEKE, B.M. - Variational Principles and the Patch Test, Int. J. Numer. Meth. Engrg, **8**, 783–801 (1974)

11. GRIFFITHS, D.F. and MITCHELL, A.R. - 'Nonconforming Elements', in *The Mathematical Basis of Finite Element Methods*, Ed. Griffiths, D. F., Clarendon Press, Oxford, 1984, p. 41.

12. IRONS, B.M. and RAZZAQUE, A. – 'Experiences with the Patch Test for Convergence of Finite Elements', in *The Mathematical Foundations of the Finite Element Method with Applications to Partial Differential Equations*, Ed. Aziz, A. K., Academic Press, New York, 1972, p. 557.

13. IRONS, B.M. and LOIKKANEN, M. – An Engineer's Defense of the Patch Test, Int. J. Numer. Meth. Engrg., 19, 1391–1401 (1983)

14. MacNEAL, R.H. – Derivation of Element Stiffness Matrices by Assumed Strain Distribution, Nuclear Engrg. Design, 70, 3–12 (1978)

15. MILITELLO, C. and FELIPPA, C.A. – A Variational Justification of the Assumed Natural Strain Formulation of Finite Elements: I. Variational Principles, Computers & Structures, 34, 431–438 (1990)

16. MILITELLO, C. and FELIPPA, C.A. – A Variational Justification of the Assumed Natural Strain Formulation of Finite Elements: II. The Four Node C^0 Plate Element, Computers & Structures, 34, 439–444 (1990)

17. MILITELLO, C. and FELIPPA, C.A. – The First ANDES Elements: 9-DOF Plate Bending Triangles, to appear in Comp. Meth. Appl. Mech. Engrg. (1991)

18. MILITELLO, C. – *Application of Parametrized Variational Principles to the Finite Element Method*, Ph. D. Dissertation, Department of Aerospace Engineering Sciences, University of Colorado, Boulder, CO, 1991.

19. PARK, K.C. and STANLEY, G.M. – A Curved C^0 Shell Element Based on Assumed Natural-Coordinate Strains, Journal of Applied Mechanics, 53, 278–290 (1986)

20. SANDER, G. and BECKERS, P. – 'The Influence of the Choice of Connectors in the Finite Element Method', in *The Mathematical Aspects of the Finite Element Method*, Lecture Notes in Mathematics, Vol. 606, Springer-Verlag, Berlin, 1977, p. 316.

21. STRANG, G. and FIX, G. – *An Analysis of the Finite Element Method*, Prentice-Hall, Englewood Cliffs, N.J., 1973.

22. STUMMEL, F. – The Limitations of the Patch Test, Int. J. Numer. Meth. Engrg., 15, 177–188 (1980)

23. STUMMEL, F. – The Generalized Patch Test, SIAM J. Numer. Anal., 16, 449–471 (1979)

24. TAYLOR, R.L., SIMO, J.C., ZIENKIEWICZ, O.C. and CHAN, A.C. – The Patch Test: A Condition for Assessing FEM Convergence, Int. J. Numer. Meth. Engrg, 22, 39–62 (1986)

25. TOCHER, J.L. and KAPUR, K.K. – Discussion of 'Basis for derivation of matrices for the direct stiffness method' by R. J. Melosh, J. AIAA, 3, 1215–1216 (1965)

26. WILSON, E.L., TAYLOR, R.L., DOHERTY, W.P. and GHABOUSSI, J. – 'Incompatible Displacement Models', in *Numerical and Computer Models in Structural Mechanics*, Eds. S. J. Fenves *et.al.*, Academic Press, New York, 1973, p. 43.

FINITE ELEMENTS IN THE 90's, E. Oñate, J. Periaux, A. Samuelsson (Eds.)
Springer-Verlag/CIMNE, Barcelona 1991

DUALITIES IN MINIMUM SURFACES

B. Tabarrok and **L. Tong**
Department of Mechanical Engineering,
University of Victoria,
Victoria, B.C.,
CANADA V8W 3P6

Summary Through a Legendre transformation the complementary form of the classical variational functional for minimum surfaces is derived. The two functionals are then used for development of finite element models for solution of the minimum surface problem. Some examples bring to light the existence of geometrically different surfaces with identical surface areas.These are dual surfaces. The extremum nature of the two functionals gives rise to upper and lower bound solutions, bracketing the exact solution.

1 Introduction

The celebrated minimum surface problem may be stated as follows: find the surface of the minimum area bounded by a given closed curve in space. This problem was first cast into a variational statement by Lagrange and later Euler showed that all such surfaces must be saddle-shaped, with their mean curvature equal to zero. Minimum surfaces find applications in design of tension structures.

Since the governing equations for minimum surfaces are nonlinear partial differential equations, they do not admit general closed form solutions. However Plateau found "physical" solutions for a large variety of boundary contours by dipping wire frames of various shapes into soap film solutions. From numerical point of view the minimum surface problem is an excellent candidate for the finite element method. This versatile method has been developed for solutions of numerous physical and mathematical problems by many researchers and most notably by Zienkiewicz [1]. The minimum surface problem however has also been analyzed by other methods[2-8].

2 The Classical Formulation

Let a surface S, with its bounding curve ∂S, be described by a position vector $\mathbf{r}$ as

$$\mathbf{r} = x\ \mathbf{e}_1 + y\ \mathbf{e}_2 + z\ \mathbf{e}_3 \tag{1}$$

where $\mathbf{e}_i (i = 1, 2, 3)$ are the unit vectors of the 3D Euclidean space. Since the tip of $\mathbf{r}$ must stay on the surface, x, y and z in Eq (1) are not independent; they are related through the equation of the surface which we describe parametrically as

$$x = f_1(\alpha, \beta) \qquad y = f_2(\alpha, \beta) \qquad z = f_3(\alpha, \beta) \tag{2}$$

The functions $f_i (i = 1, 2, 3)$ are single valued and continuous functions of parameters α and β. Thus

$$\mathbf{r} = f_1\,\mathbf{e}_1 + f_2\,\mathbf{e}_2 + f_3\,\mathbf{e}_3 \tag{3}$$

and

$$d\mathbf{r} = \frac{\partial \mathbf{r}}{\partial \alpha} d\alpha + \frac{\partial \mathbf{r}}{\partial \beta} d\beta \tag{4}$$

further

$$d\mathbf{r} \cdot d\mathbf{r} = E d\alpha^2 + 2F d\alpha d\beta + G d\beta^2 \tag{5}$$

where E, F and G are the first fundamental surface coefficients given by

$$E = \frac{\partial \mathbf{r}}{\partial \alpha} \cdot \frac{\partial \mathbf{r}}{\partial \alpha} \qquad F = \frac{\partial \mathbf{r}}{\partial \alpha} \cdot \frac{\partial \mathbf{r}}{\partial \beta} \qquad G = \frac{\partial \mathbf{r}}{\partial \beta} \cdot \frac{\partial \mathbf{r}}{\partial \beta} \tag{6}$$

Consider now a small region bounded by curves $(\alpha\ , \alpha + d\alpha)$, and $(\beta\ , \beta + d\beta)$. The area of this parallelogram is given by

$$a = |\frac{\partial \mathbf{r}}{\partial \alpha} \times \frac{\partial \mathbf{r}}{\partial \beta}| d\alpha d\beta = \sqrt{EG - F^2}\ d\alpha d\beta \tag{7}$$

For sake of simplicity we employ orthogonal curvilinear coordinates and thereby we set F equal to zero. In that case, the total area bounded by a close curve ∂S may be written as

$$A = \int_\alpha \int_\beta\ a\ \ d\alpha d\beta = \int_\alpha \int_\beta \sqrt{(\mathbf{r}_{,\alpha} \cdot \mathbf{r}_{,\alpha})(\mathbf{r}_{,\beta} \cdot \mathbf{r}_{,\beta})} d\alpha d\beta \tag{8}$$

Now minimising A, with respect to $\mathbf{r}$, we have

$$\delta A = -\int_\alpha \int_\beta \{\frac{\partial}{\partial \alpha}[\sqrt{\frac{(\mathbf{r}_{,\beta} \cdot \mathbf{r}_{,\beta})}{(\mathbf{r}_{,\alpha} \cdot \mathbf{r}_{,\alpha})}}\ \mathbf{r}_{,\alpha}] + \frac{\partial}{\partial \beta}[\sqrt{\frac{(\mathbf{r}_{,\alpha} \cdot \mathbf{r}_{,\alpha})}{(\mathbf{r}_{,\beta} \cdot \mathbf{r}_{,\beta})}}\ \mathbf{r}_{,\beta}]\} \cdot \delta \mathbf{r} d\alpha d\beta$$
$$+ \oint_{\partial S} (n_\alpha \sqrt{\frac{(\mathbf{r}_{,\beta} \cdot \mathbf{r}_{,\beta})}{(\mathbf{r}_{,\alpha} \cdot \mathbf{r}_{,\alpha})}}\ \mathbf{r}_{,\alpha} + n_\beta \sqrt{\frac{(\mathbf{r}_{,\alpha} \cdot \mathbf{r}_{,\alpha})}{(\mathbf{r}_{,\beta} \cdot \mathbf{r}_{,\beta})}}\ \mathbf{r}_{,\beta}) \cdot \delta \mathbf{r} ds \quad = 0 \tag{9}$$

Since $\mathbf{r}$ is prescribed as $\mathbf{r}_p$ along the boundary, then $\delta \mathbf{r} = 0$ along ∂S and equation (8) yields the following Euler-Lagrange equation in the domain

$$\frac{\partial}{\partial \alpha}[\sqrt{\frac{(\mathbf{r}_{,\beta} \cdot \mathbf{r}_{,\beta})}{(\mathbf{r}_{,\alpha} \cdot \mathbf{r}_{,\alpha})}}\ \mathbf{r}_{,\alpha}] + \frac{\partial}{\partial \beta}[\sqrt{\frac{(\mathbf{r}_{,\alpha} \cdot \mathbf{r}_{,\alpha})}{(\mathbf{r}_{,\beta} \cdot \mathbf{r}_{,\beta})}}\ \mathbf{r}_{,\beta}] = 0 \tag{10}$$

3 Complementary Formulation

To develop the complementary formulation, let us define the following pseudo stress vectors

$$\boldsymbol{\sigma} = \sqrt{\frac{(\mathbf{r}_{,\beta} \cdot \mathbf{r}_{,\beta})}{(\mathbf{r}_{,\alpha} \cdot \mathbf{r}_{,\alpha})}}\, \mathbf{r}_{,\alpha} \qquad \boldsymbol{\tau} = \sqrt{\frac{(\mathbf{r}_{,\alpha} \cdot \mathbf{r}_{,\alpha})}{(\mathbf{r}_{,\beta} \cdot \mathbf{r}_{,\beta})}}\, \mathbf{r}_{,\beta} \tag{11}$$

or on inversion

$$\mathbf{r}_{,\alpha} = \sqrt{\frac{(\boldsymbol{\tau} \cdot \boldsymbol{\tau})}{(\boldsymbol{\sigma} \cdot \boldsymbol{\sigma})}}\, \boldsymbol{\sigma} \quad \mathbf{r}_{,\beta} = \sqrt{\frac{(\boldsymbol{\sigma} \cdot \boldsymbol{\sigma})}{(\boldsymbol{\tau} \cdot \boldsymbol{\tau})}}\, \boldsymbol{\tau} \tag{12}$$

These equations have the appearance of some constitutive relations connecting the pseudo stress vectors to the gradients of the position vector. A further analogy with the theory of elasticity is brought to light if the condition for minimum area, as given in Eq (10), is expressed in terms of the stress vectors as follows;

$$\frac{\partial \boldsymbol{\sigma}}{\partial \alpha} + \frac{\partial \boldsymbol{\tau}}{\partial \beta} = 0 \tag{13}$$

Equation (13) has the appearance of an equilibrium equation.

Now let us develop a complementary expression for the infinitesimal area, through a Legendre transformation, as follows;

$$a^* = \boldsymbol{\sigma} \cdot \mathbf{r}_{,\alpha} + \boldsymbol{\tau} \cdot \mathbf{r}_{,\beta} - a \tag{14}$$

On integrating this expression by parts, we obtain

$$\int_\alpha \int_\beta a \; d\alpha d\beta + \int_\alpha \int_\beta a^* \; d\alpha d\beta + \int_\alpha \int_\beta (\boldsymbol{\sigma}_{,\alpha} + \boldsymbol{\tau}_{,\beta}) \cdot \mathbf{r} \; d\alpha d\beta - \oint_{\partial S} (n_\alpha \boldsymbol{\sigma} + n_\beta \boldsymbol{\tau}) \cdot \mathbf{r} ds = 0 \tag{15}$$

The first term is of course the area A. The third term can be eliminated if $\boldsymbol{\sigma}$ and $\boldsymbol{\tau}$ are constrained such that equilibrium Eq (13) is identically satisfied. Then we may express Eq (15) as

$$A + A^* = 0 \tag{16}$$

where

$$A^* = \int_\alpha \int_\beta a^* \; d\alpha d\beta - \oint_{\partial S} (\mathbf{r}_p \cdot \boldsymbol{\sigma}_n) ds \tag{17}$$

where we have expressed $\boldsymbol{\sigma}_n$ along ∂S as

$$\boldsymbol{\sigma}_n = n_\alpha \boldsymbol{\sigma} + n_\beta \boldsymbol{\tau} \tag{18}$$

Equation (16) also reveals that when A is minimised, A^* is maximised. Evidently A^* will be function of $\boldsymbol{\sigma}$ and $\boldsymbol{\tau}$. However these variables are not independent and are required to satisfy the equilibrium equation (13) identically. To this end we introduce a vector stress function $\boldsymbol{\phi}(\alpha,\beta)$ as follows;

$$\boldsymbol{\phi}(\alpha,\beta) = \boldsymbol{\phi}_x(\alpha,\beta)\mathbf{e}_1 + \boldsymbol{\phi}_y(\alpha,\beta)\mathbf{e}_2 + \boldsymbol{\phi}_z(\alpha,\beta)\mathbf{e}_3 \tag{19}$$

such that

$$\boldsymbol{\sigma} = -\frac{\partial\boldsymbol{\phi}}{\partial\beta} \qquad \boldsymbol{\tau} = \frac{\partial\boldsymbol{\phi}}{\partial\alpha} \tag{20}$$

From this definition it is evident that only the gradients of $\boldsymbol{\phi}$ control the surface A^* and we may arbitrarily set $\boldsymbol{\phi}$ equal to a constant, say zero, at one point on the surface. Now eliminating the variable $\mathbf{r}$ from a^* in Eq (14), via (12) and expressing $\boldsymbol{\sigma}$ and $\boldsymbol{\tau}$ in terms of $\boldsymbol{\phi}$, we may write A^* as;

$$A^* = \int_\alpha\int_\beta \sqrt{(\boldsymbol{\phi}_{,\alpha}\cdot\boldsymbol{\phi}_{,\alpha})(\boldsymbol{\phi}_{,\beta}\cdot\boldsymbol{\phi}_{,\beta})}\ d\alpha d\beta - \oint_{\partial S}(\mathbf{r}_p\cdot\boldsymbol{\phi}_{,s})\ ds \tag{21}$$

Since the area integral in Eq (21) is identical in form to that in Eq (8) one can see that the Euler-Lagrange equation of the above functional will be identical, in form, to that given in Eq (10). Thus the $\boldsymbol{\phi}$ and $\mathbf{r}$ surfaces are governed by the same differential equation. However the boundary conditions for the two surfaces are different. For the $\mathbf{r}$ surface the boundary is prescribed as $\mathbf{r}_p$ along ∂S, while for the $\boldsymbol{\phi}$ surface the boundary condition is of the natural type, i.e., $\boldsymbol{\phi}$ is free along ∂S. This boundary condition found from extremum condition of the functional in Eq (21) is given by

$$n_\alpha\sqrt{\frac{(\boldsymbol{\phi}_{,\beta}\cdot\boldsymbol{\phi}_{,\beta})}{(\boldsymbol{\phi}_{,\alpha}\cdot\boldsymbol{\phi}_{,\alpha})}}\ \boldsymbol{\phi}_{,\alpha} + n_\beta\sqrt{\frac{(\boldsymbol{\phi}_{,\alpha}\cdot\boldsymbol{\phi}_{,\alpha})}{(\boldsymbol{\phi}_{,\beta}\cdot\boldsymbol{\phi}_{,\beta})}}\ \boldsymbol{\phi}_{,\beta} + \mathbf{r}_{p,s} = 0 \tag{22}$$

Through eqs.(12), we may readily interpret the Euler-Lagrange equation and the boundary condition for the $\boldsymbol{\phi}$ surface. The boundary condition, in Eq (12), implies that

$$n_\alpha\mathbf{r}_{,\beta} - n_\beta\mathbf{r}_{,\alpha} + \mathbf{r}_{p,s} = 0 \tag{23}$$

The governing equation (10), when expressed in terms of $\boldsymbol{\phi}$, implies that

$$\frac{\partial}{\partial\alpha}(\mathbf{r}_{,\beta}) + \frac{\partial}{\partial\beta}(-\mathbf{r}_{,\alpha}) = 0 \tag{24}$$

Now we can see that the extremum conditions for A^* express some compatibility requirements along the boundary and in the domain.

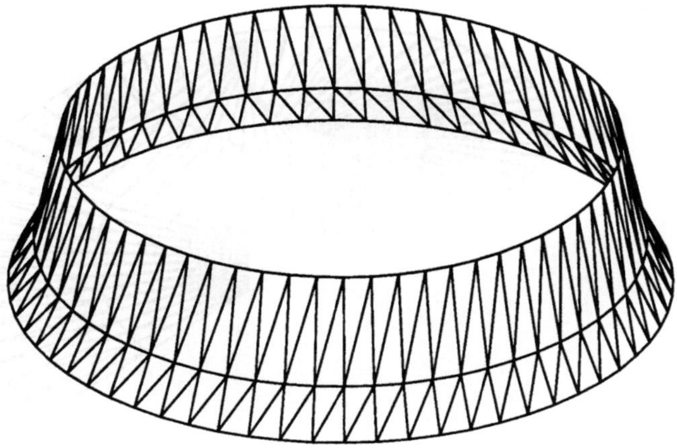

Figure 1: Surface of half catenoid

4 Illustrative Examples

Using the functionals in Eqs (8) and (21), triangular elements with linear shape functions were developed. Utilizing the Newton-Raphson method for solution we obtain the system (incremental) equations as follows

$$[K_i]\{\Delta V_i\} = \{R_i\} \tag{25}$$

where $[K_i]$ is the tangent stiffness matrix in the ith step, i.e., $[K_i]$ is formed by using values of $\{V_i\}$ in the previous step ; $\{\Delta V_i\}$ is the increment of $\{V\}$ at the ith step and, $\{V\}$ stands for $\mathbf{r}$ for the classical formulation A and $\boldsymbol{\phi}$ for the complementary formulation A^*. For formulation A, R_i represents the unbalanced force of the previous step; For the complementary fomulation A^*, R_i denotes the unbalanced terms plus the term from the line integral if the element is along the boundary. Following the conventional finite element procedures, the system stiffness matrix and load vector may be assembled and the algebraic equations may be solved using a linear equation solver. For formulation A, once the boundary conditions are imposed, convergent solutions are obtained relatively easily, while for the complementary formulation A^*, the tangent matrix $[K]$ is singular due to the presence of three rigid body modes. These may be removed by setting (x, y, z) components of $\boldsymbol{\phi}$, at one node, equal to zero.

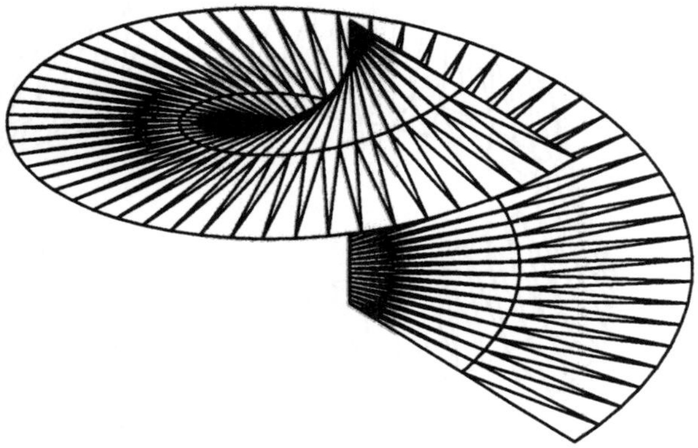

Figure 2: Complemetary surface corresponding to half catenoid

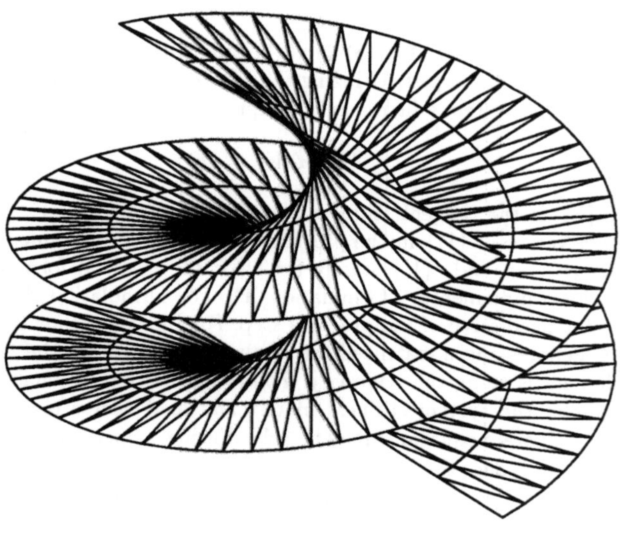

Figure 3: Complemetary surface corresponding to complete catenoid

Using this procedure two simple examples are studied. The first example is that of a catenoid, as shown in Fig.1 and analysed in Ref.[8]. Using the same data as in Ref.[8], numerical rèsults are presented in Table 2. It is worth pointing out that using 256 elements we obtain almost the same results for A and A^* as those in Ref.[8] using 864 elements. For this problem, our present ϕ surface is similar to that in Ref.[8]. But our staircase like surface for ϕ has zero inner radius as shown in Fig.2. The complementary surface corresponding to complete catenoid is shown in Fig.3.

Table 2. Surface Area for the Catenoid

No. of Elements	256	384	Exact solution
A	3.0089	3.0031	2.9923
A^*	-2.9694	-2.9803	

The second example is a saddle surface as shown in Fig. 4. The side lengths along x and y directions are 2.0, the heights of the four corners are 0.0 and 2.0, alternatively. Numerical results are presented in Table 3. The complementary surface for ϕ is shown in Fig. 5.

Table 3. Area for the Saddle-shaped Surface

No. of Elements	8	32	72	128	162
A	5.8284	5.4010	5.3213	5.2932	5.2855
A^*	-4.7582	-5.1481	-5.2378	-5.2669	-5.2721

From these two illustrative examples, one can see that convergence occurs from above and below for the classical A and the complementary A^* formulations. In both examples it was found that the complementary solution is quite sensitive to the starting values of the iteration process. This is a consequence of the fact that in the examples analysed the boundary conditions were entirely prescribed in terms of the position vector $\mathbf{r}$, and ϕ was free along ∂S.

5 Cocluding Remarks

From the classical variational statement for minimal surfaces, a complementary formulation has been derived and expressed in curvilinear coordinates. Illustrative examples show that when both formulations (A and A^*) are used, upper and lower bounds to the exact value of the minimum area may be determined.

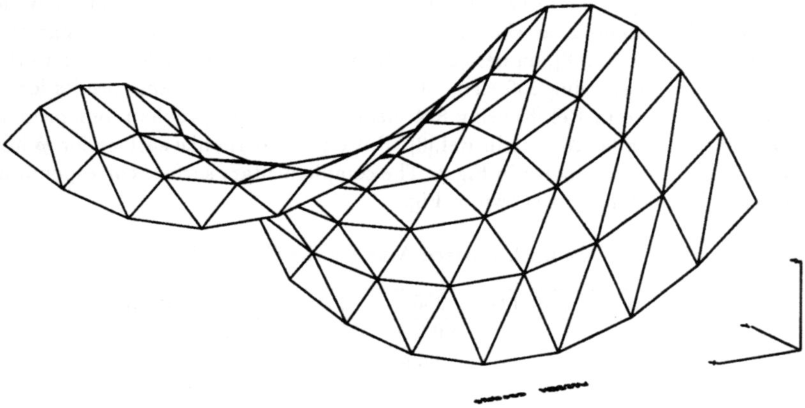

Figure 4: Saddle surface

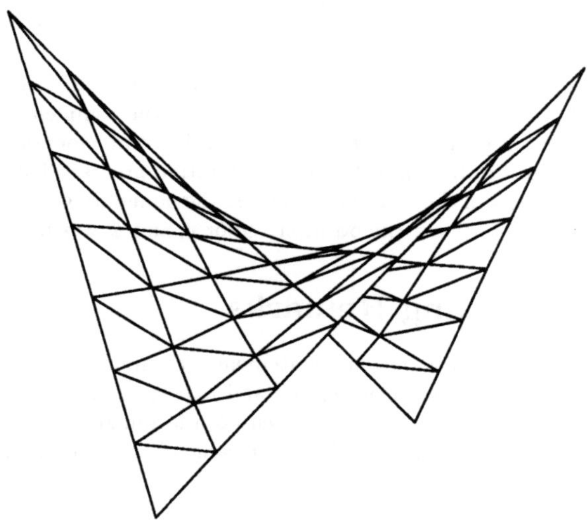

Figure 5: Complemetary surface corresponding to saddle surface

References

[1] O. C. Zienkiewicz, **The Finite Element Method**, McGraw-Hill, London, 1977

[2] D. Greenspan,'On approximating extremals of functions. I: The method and examples for boundary value problems',ICC Bull.,**4**,99-120(1965)

[3] D. Greenspan,'On approximating extremals of functions. II: Theory and generaliszation related to boundary value problems for nonlinear differential equations', Int. J. Eng. Sci., **5**,571-588(1967)

[4] P. Conus,'Numerical solutions of the minimal surfaces', Math. Comp., **21**,340-350 (1967)

[5] M. Hinata, M. Shimasaki and T. Kiyono,'Numerical solution of Plateau's problem', Math. Comp., **19**,35-58 (1977)

[6] H.J. Wagner,'A contribution to the numerical approximation of minimal surfaces', Computing, **19**,35-58 (1977)

[7] D.E. Hewgill,'Computing surfaces of constant mean curvature with singularities', Computing, **32**,81-92 (1984)

[8] B. Tabarrok and Y. Xiong,'Some variational formulations for minimal surfaces', to appear in ACTA MECHANICA

FINITE ELEMENTS IN THE 90's, E. Oñate, J. Periaux, A. Samuelsson (Eds.)
Springer-Verlag/CIMNE, Barcelona 1991

MULTIDISCIPLINARY FINITE ELEMENT MODELING AND SIMULATION OF AEROSPACE VEHICLES

K. K. Gupta
NASA Ames Research Center,
Dryden Flight Research Facility,
Edwards, California,
U.S.A.

SUMMARY

The paper presents details of a numerical formulation and the associated general-purpose finite element software developed for the integrated solution of complex engineering systems, encompassing such disciplines as structures, unsteady aerodynamics, and feedback controls. This is achieved by transforming the equations of motion from the frequency to the Laplace domain and formulating a state-space matrix relationship, the related matrices being further augmented by the analog elements such as actuators and filters as well as digital or analog controllers.

Numerical results pertaining to the X-29A aircraft, obtained by utilizing the associated computer program STARS, are compared to relevant flight test results and they are presented here in some detail. Such results, involving damping and frequency responses, relate to the stability characteristics of the aircraft and testify to the efficacy of the present solution procedure.

INTRODUCTION

Analysis of many complex practical structures, such as a highly maneuverable modern aircraft, is multidisciplinary in nature due to the expected interaction of relevant systems during its various phases of operation; the finite element method serves as an ideal technique for unifying[1] such diverse disciplines as structures, aerodynamics, and controls engineering, among others. In such cases, it is vital to evolve solution procedures and associated analysis tools to accurately predict the stability characteristics of such structures within the entire flight envelope. Routine utilization of such a tool enables, for example, an effective control law design that meets the specified requirements in such areas as stability and handling qualities and in general proves to be vital in ensuring a safe and optimum design of the overall flight structural system.

An interdisciplinary solution procedure[2] is first presented in which structural discretization is achieved by the usual finite element method. Resulting stiffness and inertia matrices are next utilized for the free vibration analysis enabling them to be expressed in the generalized coordinate system in the frequency domain. The unsteady aerodynamic forces, both in the subsonic and supersonic regimes, are computed in the frequency domain by the panel method such as the doublet lattice[3] and the constant pressure[4] procedures, respectively. Related aerodynamic forces, computed for a number of reduced frequency values, are next curve fitted using Padé polynomials along with a least squares technique and the entire matrix equation of motion is cast in the Laplace domain (s-plane). These matrices are then utilized for the state-space formulation yielding a set of matrices that completely define the aero-structural characteristics of the plant.

The plant state-space matrices are next augmented by analog control elements such as notch filters, actuators, and controllers. In the case of a digital controller, a hybrid technique is used to augment the same. Frequency responses of the open- or closed-loop system in the shape of gain and phase margins, as well as damping and frequency values may then be routinely computed that are indicative of the aeroelastic and aeroservoelastic stability characteristics. Such information is vital for the safe design of many important practical structures like aircraft and spacecraft.

This paper first presents a summary of a numerical formulation of the integrated linear aero-structural-control analysis procedure incorporated in the STARS[2] (*ST*ructural *A*nalysis *R*outine*S*) program. Some relevant aeroelastic and aeroservoelastic analysis results pertaining to the X-29A aircraft are next correlated with the flight test results, testifying to the efficacy and usefulness of the current solution procedure and the associated analysis tool.

NUMERICAL PROCEDURES

Associated matrix equation of motion of a structure, such as an aircraft, discretized by the finite element method, has the following form

$$\mathbf{M}\ddot{\mathbf{q}} + \mathbf{C}\dot{\mathbf{q}} + \mathbf{K}\mathbf{q} + \bar{q}\mathbf{A}_e(\mathbf{k})\mathbf{q} = \mathbf{P}(t) \qquad (1)$$

in which relevant terms are defined as follows:

- $\mathbf{M}$ inertia matrix,
- $\mathbf{C}$ damping matrix,
- $\mathbf{K}$ elastic stiffness matrix,
- $\bar{q}$ dynamic pressure $1/2\rho V^2$, ρ and V being the air density and true airspeed, respectively,
- k reduced frequency $\omega b/V$, ω and b being the natural frequency and wing semichord length, respectively,
- $A_e(k)$ aerodynamic influence coefficient matrix for a given Mach number M_∞ and set of k_i values,
- $\mathbf{q}$ displacement vector,
- $\mathbf{P}(t)$ external forcing function, and
- s Laplace variable (= $i^*\omega$, i^* being $\sqrt{-1}$).

A solution[2] of the related free vibration problem

$$\mathbf{M}\ddot{\mathbf{q}} + \mathbf{k}\mathbf{q} = \mathbf{0} \qquad (2)$$

yields the desired roots ω and vectors Φ. Next, applying a transformation

$$\mathbf{q} = \Phi\eta \qquad (3)$$

to equation (1) and premultiplying both sides by Φ^T, the generalized equation of motion is derived as

$$\hat{\mathbf{M}}\ddot{\eta} + \hat{\mathbf{C}}\dot{\eta} + \hat{\mathbf{K}}\eta + \bar{q}\mathbf{Q}(\mathbf{k})\eta = \hat{\mathbf{P}}(t) \qquad (4)$$

in which $\hat{\mathbf{M}} = \Phi^T\mathbf{M}\Phi$, etc., the modal matrix $\Phi = [\Phi_r\ \Phi_e\ \Phi_\delta]$ and the generalized coordinate $\eta = [\eta_r\ \eta_e\ \eta_\delta]$, incorporating rigid body, elastic, and control surface motions, respectively.

Expressing the generalized aerodynamic force matrix **Q**(k) as Padé polynomials[5] in i*k (=i*ωb/V = sb/V) results in

$$\mathbf{Q}(k) = \mathbf{A}_0 + i{*}k\mathbf{A}_1 + (i{*}k)^2\mathbf{A}_2 + \frac{i{*}k}{i{*}k+\beta_1}\mathbf{A}_3 + \frac{i{*}k}{i{*}k+\beta_2}\mathbf{A}_4 + \ldots \tag{5}$$

where β_j are the aerodynamic lag terms, and the coefficients $\mathbf{A}_0, \mathbf{A}_1, \ldots$, are determined by the least squares method using the aerodynamic data computed for a number of k_i values. Substituting equation (5) into equation (4) and performing some matrix manipulations[6] yields the state space matrix equation in time domain

$$\dot{\hat{\mathbf{X}}} = \hat{\mathbf{A}}\hat{\mathbf{X}} + \hat{\mathbf{B}}\mathbf{u} \tag{6}$$

in which

$$\hat{\mathbf{X}} = \left(\eta_r\ \eta_e\ \dot{\eta}_r\ \dot{\eta}_e\ \mathbf{X}_1\ \mathbf{X}_2\right),\quad \mathbf{u} = \left(\eta_\delta\ \dot{\eta}_\delta\right),\quad \mathbf{X}_j = \frac{s\eta}{\left[s+\left(\frac{V}{b}\right)\beta_j\right]}$$

the relevant matrices and vectors[6] being defined as

- $\hat{\mathbf{A}}$ plant dynamics matrix
- $\hat{\mathbf{B}}$ control surface influence matrix
- $\hat{\mathbf{X}}$ generalized coordinates in inertial frame
- **u** control surface motion input into plant

and in which the terms $\hat{\mathbf{A}}\hat{\mathbf{X}}$ and $\hat{\mathbf{B}}\mathbf{u}$ represent, for an aircraft as for example, the airplane dynamics and forcing function on airplane due to control surface motion, respectively.

Coordinate Transformation

To incorporate control laws and feedback, it is necessary to transform equation (6) from the earth-fixed (inertial) to the body-fixed coordinate system. Since no transformations are applied to elastic and aerodynamic lag state vectors, a transformation of the form

$$\begin{aligned}\dot{\mathbf{X}} &= \tilde{\mathbf{T}}_2^{-1}\left(\hat{\mathbf{A}}\tilde{\mathbf{T}}_1 - \tilde{\mathbf{T}}_3\right)\mathbf{X} + \tilde{\mathbf{T}}_2^{-1}\hat{\mathbf{B}}\mathbf{u} \\ &= \mathbf{A}\mathbf{X} + \mathbf{B}\mathbf{u}\end{aligned} \tag{7}$$

in which

$$\tilde{\mathbf{T}}_1 = \begin{bmatrix} \mathbf{T}_1 & \mathbf{0} \\ \mathbf{0} & \mathbf{I} \end{bmatrix} \quad \text{etc.,}$$

$\mathbf{T}_1$ being the 12 x 12 coordinate transformation matrix, yields the required state-space equation in body coordinate system. A detailed description of the transformation procedure is given in reference 6.

Aeroelastic Instability

Such a stability analysis involves solution of the relevant eigenvalue problem

$$|\mathbf{A} - \lambda \mathbf{I}| = 0 \tag{8}$$

for various dynamic pressures. An instability of the system (flutter and divergence) is indicated by a change in sign of the real part of λ from negative to positive. In the particular case in which the corresponding frequency, the imaginary part of the root λ, also approaches the zero value, it is indicative of divergence.

Determination of Sensor Outputs

The structural nodal displacements are related to the generalized coordinates by equation (3), and the related sensor motion can be expressed as

$$\begin{aligned} q_s &= \mathbf{T}_s \Phi \eta \\ &= \mathbf{C}_0 \mathbf{X} \end{aligned}$$

where $\mathbf{C}_0 = \begin{bmatrix} \mathbf{T}_s\Phi & 0 & 0 & 0 \end{bmatrix}$, and in which $\mathbf{T}_s$ is an interpolation matrix. Similar relations may be derived for sensor velocities and acceleration which are appropriately substituted in equation (7) to yield

$$\mathbf{y} = \mathbf{C}\mathbf{X} + \mathbf{D}\mathbf{u} \tag{9}$$

which is the required sensor output relationship, the matrices $\mathbf{C}$ and $\mathbf{D}$ signifying output at sensor due to body and control surface motions, respectively.

Augmentation of Analog Elements and Controller

The complete state-space formulation for an aircraft incorporating structural and aeroelastic effects is represented by equations (7) and (9), derived earlier. To conduct an aeroservoelastic analysis, it is essential to augment such a formulation with associated analog elements such as actuators, sensors, notch and pre-filters along with the controller. Thus denoting the state-space equations of one such typical element as below

$$\dot{\mathbf{X}}^{(i)} = \mathbf{A}^{(i)}\mathbf{X}^{(i)} + \mathbf{B}^{(i)}\mathbf{u}^{(i)} \tag{10}$$

$$\mathbf{y}^{(i)} = \mathbf{C}^{(i)}\mathbf{X}^{(i)} + \mathbf{D}^{(i)}\mathbf{u}^{(i)} \tag{11}$$

these can be augmented to such original equations (7) and (9) as

$$\begin{bmatrix} \dot{X} \\ \dot{X}^{(i)} \end{bmatrix} = \begin{bmatrix} A & 0 \\ B^{(i)}C & A^{(i)} \end{bmatrix} \begin{bmatrix} X \\ X^{(i)} \end{bmatrix} + \begin{bmatrix} B \\ B^{(i)}D \end{bmatrix} [u] \tag{12}$$

or

$$\dot{X}_{(i)} = A_{(i)}X_{(i)} + B_{(i)}u \tag{13}$$

noting that $u^{(1)} = y$. Also, similarly,

$$y_{(i)} = C_{(i)}X_{(i)} + D_{(i)}u \tag{14}$$

becomes the new sensor output expression.

Any analog element, including a controller, can be augmented as above. Figure 1 shows a typical feedback control system. For such a

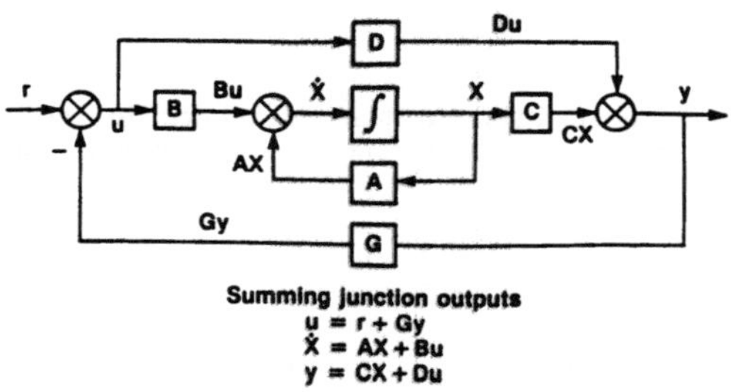

Fig. 1 Feedback control system

system, the three sets of relevant matrix equations are as follows

$$\dot{X} = AX + Bu \tag{15}$$

$$y = CX + Du \tag{15a}$$

$$u = r - Gy \tag{15b}$$

where equation (15b) is the feedback equation. To achieve frequency response characteristics of an aircraft with a feedback control system, a Laplace transformation is applied to equations (15), yielding

$$\begin{aligned} y(s) &= \left[C(sI - A)^{-1}B + D\right]u(s) \\ &= H(s)u(s) \end{aligned} \tag{16}$$

H(s) being the equivalent open-loop (loop-gain) transfer function with the analog controller or the open-loop transfer function without the controller. To obtain the closed-loop frequency response relationship, equations (16) and (15), are suitably manipulated to yield

$$\begin{aligned} y(s) &= (H(s)[I + G(s)H(s)]^{-1})r(s) \\ &= \hat{H}(s)r(s) \end{aligned} \tag{17}$$

in which $\hat{H}(s)$ is the desired closed-loop transfer function. The frequency responses plots can be simply obtained from the transfer matrices $\mathbf{H}(s)$ or $\hat{\mathbf{H}}(s)$, as the case may be. Associated damping and frequency values for the system, for the loop-gain or open-loop case, may also be calculated by solving the eigenvalue problem of the relevant $\mathbf{A}$ matrix for various k_i values and observing the changes in sign of the real part of an eigenvalue.

In the presence of a digital controller, a hybrid approach[7] is adopted for the frequency response solution. Thus, if $\mathbf{A}'$, $\mathbf{B}'$, $\mathbf{C}'$, and $\mathbf{D}'$ are the state-space matrices associated with the controller, the related transfer function is simply given by

$$\mathbf{G}(z) = \mathbf{C}'[z\mathbf{I} - \mathbf{A}']^{-1}\mathbf{B}' + \mathbf{D}' \tag{18}$$

and the frequency response relationship for the hybrid analog/digital system can be written as

$$\mathbf{y}(s) = \mathbf{G}(z)_{[\text{at } z = e^{sT}]} \left\{\frac{\mathbf{H}(s)[ZOH]}{T}\right\} \mathbf{u}(s) \tag{19}$$

$$= \mathbf{H}^*(s)\mathbf{u}(s) \tag{19a}$$

in which

$\mathbf{H}(s)$ is the open-loop transfer function for the plant and other analog elements

[ZOH] is the zero order hold complex expression $\left(= e^{-s\tau} \left(\frac{1 - e^{-sT}}{s}\right)\right)$

and where $\mathbf{H}^*(s)$ is now the equivalent open-loop (loop-gain) transfer function of the hybrid system. The closed-loop frequency response relationship may be obtained as before by using equations (19a) and (15)

$$\begin{aligned} \mathbf{y}(s) &= \left\{\mathbf{H}(s)[\mathbf{I} + \mathbf{G}(s)\mathbf{H}(s)]^{-1}\right\}\mathbf{r}(s) \\ &= \hat{\mathbf{H}}^*(s)\mathbf{r}(s) \end{aligned} \tag{20}$$

To compute the damping and frequencies, the analog plant dynamics matrix $\mathbf{A}$ is first transformed into the z-plane by standard discretization procedure which is next augmented to $\mathbf{A}'$. Appropriate eigenproblem solution of the final matrix yields the required results, as before.

The STARS program has been extended to include capabilities representative of formulations presented in this section.

NUMERICAL EXAMPLES

The results of the extensive integrated aero-structural-control analysis[6] of the X-29A forward swept wing research aircraft[8,9] (fig. 2) by the STARS program are presented herein in some detail. Such an

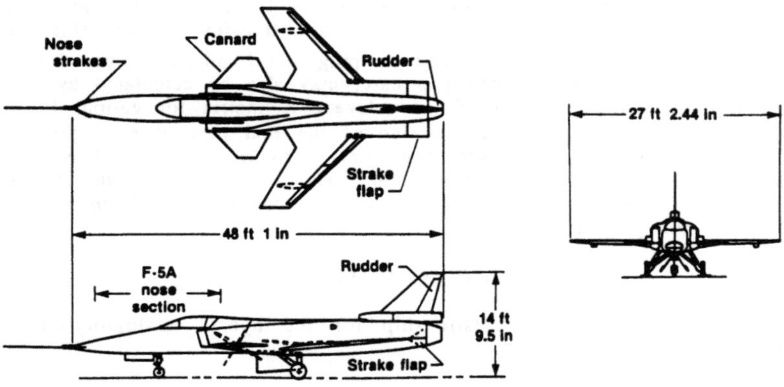

Fig. 2 X-29A forward swept wing aircraft

aircraft is characterized by aeroelastically-tailored thin wings with composite covers designed to eliminate structural divergence within the desired flight envelope without any weight penalty. Since the aircraft is up to 35 percent statically unstable, it requires appropriate feedback flight controls for augmented static stability. Significant improvement in structural efficiency and aerodynamic performance is obtained by suitably combining these diverse technologies. However, this may also cause adverse dynamic interaction of the flight controls with flexible structure if such an integration is not achieved appropriately. Thus, an ASE analysis assumes a rather important role in such an integration process. A comparison of various phases of ASE analysis and flight test results of the X-29A aircraft is presented next.

Structural Dynamic and Aeroelasticity Results

The reduced finite element dynamic model of the symmetric half of the aircraft is shown in Figure 3. A block Lanczos[2] procedure was adopted to solve the free vibration problem, and a comparison of such results with that obtained from ground vibration survey (GVS) is depicted in Table 1. Figure 4 depicts the unsteady aerodynamics grid employed for extensive aeroelastic analysis of the aircraft. Such flutter and divergence analyses were performed by three different techniques, the k, p-k, and the ASE methods, for the purpose of detailed correlation, utilizing analytical as well as experimental GVS modes. Table 2 summarizes some of these results, pertaining to the k and ASE methods.

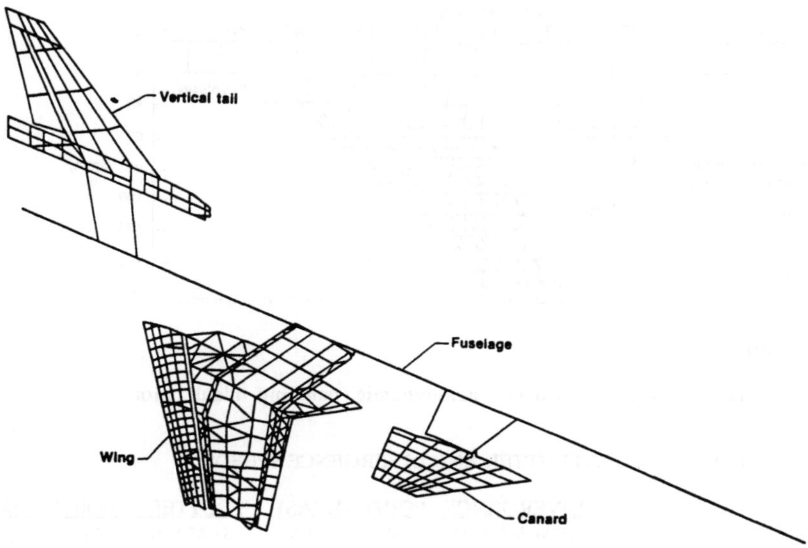

Fig. 3 Finite element dynamics model of a symmetric half of X-29A aircraft.

TABLE 1. X-29A FREE VIBRATION ANALYSIS RESULTS, SYMMETRIC (S) AND ANTISYMMETRIC (A/S) MODES*

Mode No.	Primary Motion	Natural Frequency (Hz) STARS	GVS
1, 2, 3	Rigid body	0.0	
4	Wing First Bending (S)	8.96	8.61
5	Wing First Bending (A/S)	10.08	11.3
6	Fuselage First Bending (A/S)	12.35	12.5
7	Fuselage First Vertical Bending (S)	12.87	11.65
8	Vertical Fin First Bending (A/S)	17.18	15.2

* Lightweight fuel condition

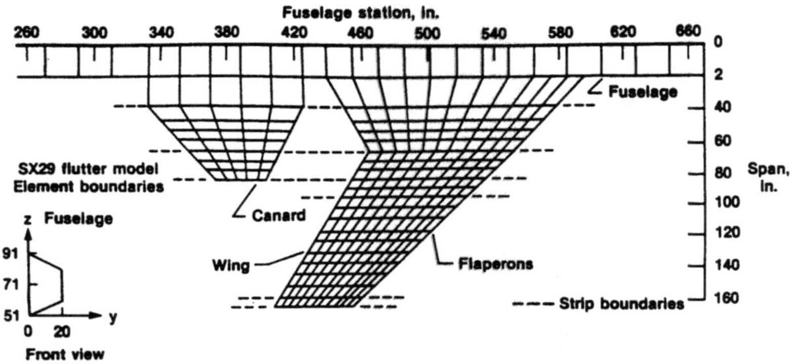

Fig. 4 X-29A unsteady aerodynamic paneling idealization.

TABLE 2. X-29A FLUTTER AND DIVERGENCE SPEEDS

	DIVERGENCE POINT (KEAS)			FLUTTER POINT (KEAS)		
	STARS (k)	STARS (ASE)	GVS	STARS (k)	STARS (ASE)	GVS
Wing First Bending (A/S)	833	819	865			
Wing First Bending (S)	838	833	808			
Fuselage First Bending (S)				848	797	924
Canard Pitch (A/S)	912	910	1017			

Aeroservoelastic Results

To evaluate the pattern of interaction between the X-29A aeroelastic dynamics and a control system, the plant matrices that include effects of rigid body, elastic, and control modes along with two unsteady aerodynamic lag states are augmented with elements of the control system. Equivalent open-loop (loop-gain) and closed-loop frequency responses were computed and stability margins determined for various flight conditions with normal digital mode control systems[10] augmented to the aircraft states. These results were found to compare well with wind tunnel and flight test results. Figure 5, for example, depicts a typical comparison of X-29A longitudinal loop-gain results between STARS analysis and flight test data for 0.9 mach at sea level.

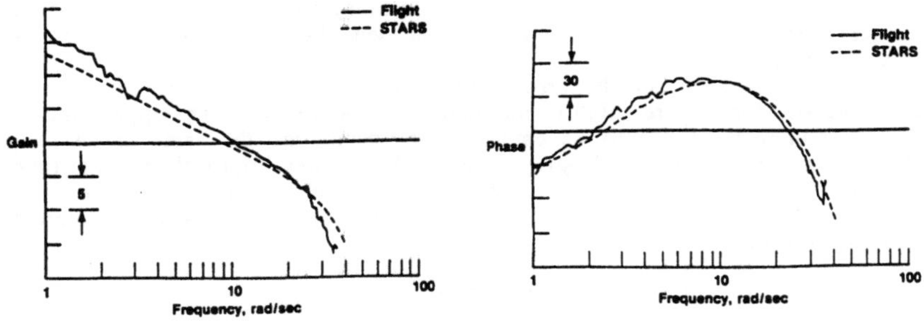

Fig. 5 Longitudinal loop gains from flight data and STARS at 1.2 Mach/ 30,000 altitude.

Figure 6 portrays typical comparisons of such flight measured modal results with STARS calculated data for the typical antisymmetric fin bending modes; results for the symmetric case having been published elsewhere[11]. STARS results are obtained by calculating the eigenvalues of the appropriate state-space matrix of the flexible model augmented to the digital control system. Such comparisons for the fully augmented aircraft dynamics indicate the accurate stability prediction capability of the present procedure and the resulting STARS program.

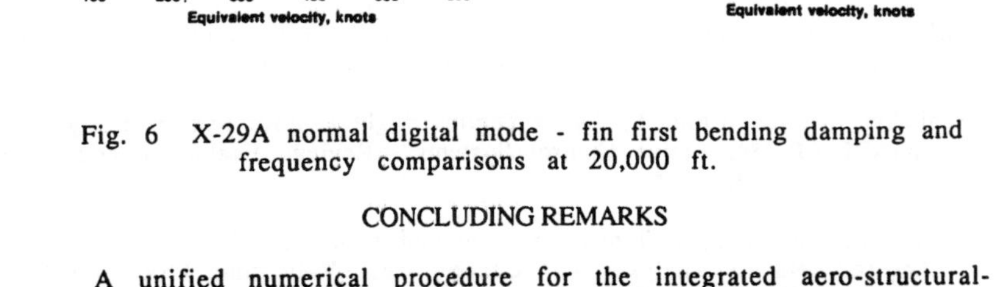

Fig. 6 X-29A normal digital mode - fin first bending damping and frequency comparisons at 20,000 ft.

CONCLUDING REMARKS

A unified numerical procedure for the integrated aero-structural-control analysis of practical engineering systems has been presented. Implementation of these capabilities into the general-purpose finite element computer program STARS is also described in some detail. Effectiveness of such a program in predicting stability characteristics of a highly unstable modern aircraft has also been demonstrated both in subsonic and supersonic flow regimes. When both flexible and rigid-body stability are satisfactorily achieved through successive analyses using such tools as the STARS program, the control system structure may then be further modified to meet performance and sensitivity conditions for body-axis control.

For transonic and hypersonic flow regimes, the unsteady aerodynamic forces are highly nonlinear in nature, and an appropriate CFD formulation[12] has to be employed for computing the same in an accurate manner. A step-by-step integrated solution approach in the time domain may then be utilized to predict stability characteristics of the vehicle. The STARS program is currently being extended in this important area of analysis.

REFERENCES

1. Zienkiewicz, O. C., *The Finite Element Method in Engineering Science*, McGraw-Hill, U.K., 1989.

2. Gupta, K. K., "STARS - An Integrated General-Purpose Finite Element Structural, Aeroelastic, and Aeroservoelastic Analysis Computer Program," NASA TM-101709, January 1991.

3. Taylor, R. F., Miller, K. L., and Brockman, R. A., "A Procedure for Flutter Analysis of FASTOP-3 Compatible Mathematical Models, Volume I - Theory and Application," AFWAL-TR81-3063, June 1981.

4. Appa, K., "Constant Pressure Panel Method for Supersonic Unsteady Airload Analysis," *J. Aircraft*, **24**, 1987, pp. 696-702.

5. Abel, I., "An Analytical Technique for Predicting the Characteristics of a Flexible Wing Equipped with an Active Flutter-Suppression System and Comparison with Wind-Tunnel Data," NASA TP-1367, 1979.

6. Gupta, K. K., Brenner, M. J., and Voelker, L. S., "Development of an Integrated Aeroservoelastic Analysis Program and Correlation with Test Data," NASA TP-3120, May 1991.

7. Whitbeck, R. F., Didaleusky D. G. J., and Hofman, L. G., "Frequency Response of Digitally Controlled Systems," *J. Guidance Control*, **4**, 1981, pp. 423-427.

8. Sefic, Walter J., and Maxwell, Cleo M., "X-29A Technology Demonstrator Flight Test Program Overview," NASA TP-86809, 1986.

9. Whitaker, A., and Chin, J., "X-29 Digital Flight Control System Design," AGARD-CP-384, Active Control Systems - Review, Evaluation and Projections, Oct. 1984.

10. Zislin, A., Laurie, E., Wilkinson, K., and Goldstein, R., "X-29 Aeroservoelastic Analysis and Ground Test Validation Procedures," AIAA-85-3091, Oct. 1985.

11. Gupta, K. K., Brenner, M. J., and Voelker, L. S., "Integrated Aeroservoelastic Analysis Capability with X-29A Comparisons," *J. Aircraft*, **26**, 1989, pp. 84-90.

12. Peraire, J., Formaggia, L., Morgan, K., Zienkiewicz, O. C., "Finite Element Euler Computations in Three Dimensions," *Inter. J. Numer. Methods Eng.*, **26**, 1988, pp. 2135-2159.

FINITE ELEMENTS IN THE 90's, E. Oñate, J. Periaux, A. Samuelsson (Eds.)
Springer-Verlag/CIMNE, Barcelona 1991

A NEW APPROACH TO STABILITY ANALYSIS OF PARTITIONED SYSTEMS

T. Belytschko and Y.Y. Lu
Department of Civil Engineering,
Robert R. McCormick School of Engineering and Applied Science,
Northwestern University, Evanston, Illinois, 60208,
USA

SUMMARY

A methodology is presented for analyzing the stability of elemental explicit-implicit partitions of multi-step time integrators, and is applied to develop stability criteria for elemental partitions of the generalized trapezoidal rule for first order systems and Newmark β-method for second order systems. The methodology developed here is readily applicable to other explicit-implicit methods, since in contrast to energy methods, it does not involve any serendipity. Through the use of the theorems and methodologies presented here, any elemental explicit-implicit method can be analyzed in a straightforward manner.

1. INTRODUCTION

The stability of explicit-implicit methods has usually been proven by energy methods (see, e.g. [1-4]). However, the construction of energy proofs is difficult, and energy stability proofs have not been found explicit-implicit partitions of the Hilber-Hughes-Taylor (HHT) α-method[5-6], for example.

In this paper, a technique used earlier in [7,10] will be used to obtain stability proofs on elemental partitions of explicit-implicit integration. The proof is based on an extension of Iron's element eigenvalue inequality to quasi-symmetric matrices, which enables the amplification factor of the explicit-implicit integrator to be bounded by eigenvalues related to element-level matrices. As a result, spectral methods can be used to obtain sufficient conditions for stability of explicit-implicit integrators.

In Section 2, we first illustrate the method by a stability analysis for explicit-implicit integration of first-order systems by the generalized trapezoidal rule. This context provides a simple setting to clearly illustrate the essential features of the proof. In Section 3, the method is used to prove stability for explicit-implicit partitions of the equations of motion integrated by a predictor-corrector [3], form of the Newmark β-method.

2. FIRST ORDER SYSTEMS

The stability proofs will be based on an extension of the element eigenvalue inequalities [7,8,9]. This theorem states the following:

Consider a system consisting of a set of elements $\mathbf{S}$. Each element has eigenvalues μ_i^e which satisfy

$$\mathbf{A}^e \mathbf{x}_i^e = \mu_i^e \mathbf{B}^e \mathbf{x}_i^e, \quad i=1,\cdots,N_e, \qquad e \in \mathbf{S}, \tag{2.1}$$

where N_e is the order of $\mathbf{A}^e$ and $\mathbf{B}^e$. The system matrices $\mathbf{A}$ and $\mathbf{B}$ are obtained by the finite element assembly process and are given by

$$\mathbf{A} = \mathop{\mathcal{A}}_{e} \mathbf{A}^e \qquad e \in \mathbf{S}, \tag{2.2}$$

$$\mathbf{B} = \mathop{\mathcal{A}}_{e} \mathbf{B}^e \qquad e \in \mathbf{S}, \tag{2.3}$$

where $\mathcal{A}$ is the finite element assembly operator. The system eigenvalues μ_i satisfy

$$\mathbf{A}\, \mathbf{x}_i = \mu_i\, \mathbf{B}\, \mathbf{x}_i\ , \tag{2.4}$$

where the matrices $\mathbf{A}$, $\mathbf{A}^e$, $\mathbf{B}$ and $\mathbf{B}^e$ are symmetric.

Theorem 1. The maximum absolute system eigenvalue $|\mu|_{max}$ is bounded by the maximum absolute element eigenvalue $|\mu^E|_{max}$, i.e.

$$|\mu|_{max} \leq |\mu^E|_{max} \tag{2.5}$$

where

$$|\mu|_{max} = \max_i |\mu_i| \tag{2.6}$$

$$|\mu^E|_{max} = \max_{e,i} |\mu_i^e| \tag{2.7}$$

We now consider a semidiscretization which leads to a system of first-order ordinary differential equations; these arise out of parabolic partial differential equations such as heat-conduction. The linear forms of these semidiscretized equations can be written as

$$\mathbf{M}\dot{\mathbf{u}} + \mathbf{K}\mathbf{u} = \mathbf{f}^{ext} \tag{2.8}$$

In the Hughes-Liu elemental explicit-implicit partition, the domain is subdivided into an explicit domain E and an implicit domain I. The solution of these equations by the generalized trapezoidal rule consists of the following

$$\mathbf{M}\, \dot{\mathbf{u}}_{n+1} + \mathbf{K}^E\, \tilde{\mathbf{u}}_{n+1} + \mathbf{K}^I\, \mathbf{u}_{n+1} = \mathbf{f}_{n+1}^{ext} \tag{2.9}$$

$$\tilde{\mathbf{u}}_{n+1} = \mathbf{u}_n + (1-\alpha)\Delta t \dot{\mathbf{u}}_n \tag{2.10}$$

$$\mathbf{u}_{n+1} = \tilde{\mathbf{u}}_{n+1} + \alpha \Delta t \dot{\mathbf{u}}_{n+1}. \tag{2.11}$$

Here Δt is the time step and α an integration parameter, $0 \leq \alpha \leq 1$; the mesh has been subdivided into explicit and implicit elements and

$$\mathbf{K}^E = \mathop{\mathcal{A}}_{\substack{e \\ e\in E}} \mathbf{K}_e \qquad \mathbf{K}^I = \mathop{\mathcal{A}}_{\substack{e \\ e\in I}} \mathbf{K}_e \tag{2.12}$$

To obtain a proof of stability, we will write the homogeneous ($\mathbf{f}^{ext} = 0$) equations (2.9) in amplification matrix form. We note that from (2.11) that

$$\tilde{\mathbf{u}}_{n+1} = \mathbf{u}_{n+1} - \alpha\Delta t\dot{\mathbf{u}}_{n+1} \tag{2.13}$$

so using (2.9) and (2.13) gives

$$\tilde{\mathbf{M}}\,\dot{\mathbf{u}}_{n+1} + \mathbf{K}\,\mathbf{u}_{n+1} = 0 \tag{2.14}$$

where

$$\tilde{\mathbf{M}} = \mathbf{M} - \alpha\Delta t\mathbf{K}^{E} \tag{2.15}$$

The amplification matrix form of the integrator is obtained by eliminating $\tilde{\mathbf{u}}_{n+1}$ from (2.10) and (2.11):

$$\mathbf{u}_{n+1} = \mathbf{u}_n + (1-\alpha)\Delta t\,\dot{\mathbf{u}}_n + \alpha\Delta t\,\dot{\mathbf{u}}_{n+1} \tag{2.16}$$

and substituting in (2.14), some rearrangement yields

$$\mathbf{B}\mathbf{u}_{n+1} = \mathbf{A}\mathbf{u}_n \tag{2.17}$$

where

$$\mathbf{A} = (\tilde{\mathbf{M}} - (1-\alpha)\Delta t\,\mathbf{K})$$

$$\mathbf{B} = (\tilde{\mathbf{M}} + \alpha\Delta t\mathbf{K})$$

The above integration will be stable if the eigenvalues μ_i of the eigenproblem

$$\mathbf{A}\,\mathbf{x}_i = \mu_i\,\mathbf{B}\,\mathbf{x}_i \tag{2.18}$$

all satisfy

$$|\mu_i| \le 1 \tag{2.19}$$

Using theorem 1, it then follows that it is sufficient for stability that the eigenvalue of

$$\mathbf{A}_e\,\mathbf{x}_i^e = \mu_i^e\,\mathbf{B}_e\,\mathbf{x}_i^e \tag{2.20a}$$

satisfy

$$|\mu_i^e| \le 1 \quad \forall\, i \text{ and } \forall\, e \tag{2.20b}$$

where

$$\mathbf{A}_e = (\tilde{\mathbf{M}}_e - (1-\alpha)\Delta t\,\mathbf{K}_e) \tag{2.21a}$$

$$\mathbf{B}_e = (\tilde{\mathbf{M}}_e + \alpha\Delta t\mathbf{K}_e) \tag{2.21b}$$

So from (2.15) and (2.21)

$$\mathbf{A}_e = \begin{cases} \mathbf{M}_e - \Delta t\,\mathbf{K}_e & \text{if } e \in E \\ \mathbf{M}_e - (1-\alpha)\Delta t\,\mathbf{K}_e & \text{if } e \in I \end{cases} \tag{2.22}$$

$$\mathbf{B}_e = \begin{cases} \mathbf{M}_e & \text{if } e \in E \\ \mathbf{M}_e + \alpha\Delta t\,\mathbf{K}_e & \text{if } e \in I \end{cases} \tag{2.23}$$

It can be seen from (2.22-2.23) that the element-level amplification matrices for the explicit and implicit elements are in fact identical to those of the forward Euler and generalized trapezoidal rule, so nothing further remains to be done. However, for completeness, in addition to stating the sufficient stability conditions, we shall briefly outline their proof.

Stability Condition for Element Explicit-Implicit Partition on (2.14): The integration procedure given by (2.9)-(2.11) is stable if

$$\Delta t \leq \frac{2}{\lambda^e_{max}} \qquad \forall\ e \in E \tag{2.24}$$

$$\left| \frac{1-(1-\alpha)\Delta t \lambda^e_{max}}{1+\alpha\Delta t\, \lambda^e_{max}} \right| \leq 1 \qquad \forall\ e \in I \tag{2.25}$$

where λ^e_{max} is the maximum eigenvalue of

$$\mathbf{K}_e\, \mathbf{y}^e_i = \lambda^e\, \mathbf{M}_e\, \mathbf{y}^e_i \qquad i=1, \text{ to } N_e \tag{2.26}$$

To show this result for $e \in E$, we write (2.18), which from (2.22-2.23), can be written as

$$(\mathbf{M}_e - \Delta t\, \mathbf{K}_e)\, \mathbf{x}^e_i = \mu^e_i\, \mathbf{M}_e\, \mathbf{x}^e_i \tag{2.27}$$

The above matrices can be diagonalized by expanding $\mathbf{x}^e_i$ in terms of $\mathbf{y}^e_i$ and premultiplying by $\mathbf{y}^e_j$. The result gives the diagonal form of the above,

$$1 - \Delta t\, \lambda^e_i = \mu^e_i \qquad i=1, \text{ to } N_e \tag{2.28}$$

and enforcing (2.20) on $\lambda^e_{max} \equiv \max\limits_{e,i} \lambda^e_i$ gives

$$\Delta t \leq \frac{2}{\lambda^e_{max}} \qquad e \in E \tag{2.29}$$

The stability condition on the implicit elements can be developed similarly and given as follows:

$$\Delta t \leq \frac{2}{(1-2\alpha)\lambda^e_{max}}, \qquad e \in I \tag{2.30}$$

3. SECOND ORDER SYSTEMS

We now consider the second-order ordinary differential equations which are found in structural dynamics and transient continuum mechanics. The semidiscrete momentum equations are

$$\mathbf{Ma} + \mathbf{Cv} + \mathbf{Ku} = \mathbf{f} \tag{3.1}$$

where $\mathbf{M}$ is the mass matrix, $\mathbf{C}$ the damping matrix and $\mathbf{K}$ the stiffness; $\mathbf{u}$ are the nodal displacements and

$$\mathbf{v} = \dot{\mathbf{u}} \tag{3.2}$$

$$\mathbf{a} = \dot{\mathbf{v}} \tag{3.3}$$

The Hughes-Liu predictor-corrector form of the Newmark β-integrator can be written as

$$\tilde{\mathbf{u}}_{n+1} = \mathbf{u}_n + \Delta t\, \mathbf{v}_n + \Delta t^2(\tfrac{1}{2} - \beta)\, \mathbf{a}_n \tag{3.4}$$

$$\tilde{\mathbf{v}}_{n+1} = \mathbf{v}_n + \Delta t\,(1 - \gamma)\, \mathbf{a}_n \tag{3.5}$$

$$\mathbf{u}_{n+1} = \tilde{\mathbf{u}}_{n+1} + \beta\, \Delta t^2\, \mathbf{a}_{n+1} \tag{3.6}$$

$$\mathbf{v}_{n+1} = \tilde{\mathbf{v}}_{n+1} + \gamma\, \Delta t\, \mathbf{a}_{n+1} \tag{3.7}$$

The momentum equations are for the explicit-implicit partition are

$$\mathbf{M}\, \mathbf{a}_{n+1} + \mathbf{C}^E\, \tilde{\mathbf{v}}_{n+1} + \mathbf{C}^I\, \mathbf{v}_{n+1} + \mathbf{K}^E\, \tilde{\mathbf{u}}_{n+1} + \mathbf{K}^I\, \mathbf{u}_{n+1} = \mathbf{f}_{n+1} \tag{3.8}$$

Eliminating the "predictor" variables from (3.8) with the use of (3.6-3.7) gives

$$\tilde{\mathbf{M}}\mathbf{a}_{n+1} + \mathbf{C}\mathbf{v}_{n+1} + \mathbf{K}\mathbf{u}_{n+1} = \mathbf{f}_{n+1} \tag{3.9}$$

where

$$\tilde{\mathbf{M}} = \mathbf{M} - \gamma\, \Delta t\, \mathbf{C}^E - \beta\, \Delta t^2 \mathbf{K}^E \tag{3.10}$$

We now place the integrator in amplification matrix form by eliminating the predictor quantities from (3.7) and using (3.9), which gives

$$\begin{bmatrix} \tilde{\mathbf{M}} + \beta\Delta t^2\mathbf{K} & \beta\Delta t^2\mathbf{C} \\ \gamma\Delta t\mathbf{K} & \tilde{\mathbf{M}} - \gamma\Delta t\mathbf{C} \end{bmatrix} \begin{Bmatrix} \mathbf{u}_{n+1} \\ \mathbf{v}_{n+1} \end{Bmatrix} = \begin{bmatrix} \tilde{\mathbf{M}} - \Delta t^2(\tfrac{1}{2} - \beta)\mathbf{K} & \Delta t\tilde{\mathbf{M}} - \Delta t^2(\tfrac{1}{2} - \beta)\mathbf{C} \\ (\gamma - 1)\Delta t\mathbf{K} & \tilde{\mathbf{M}} - \gamma\Delta t\mathbf{C} \end{bmatrix} \begin{Bmatrix} \mathbf{u}_n \\ \mathbf{v}_n \end{Bmatrix} \tag{3.11}$$

Note that the effect of explicit elements is now embodied completely in the matrix $\tilde{\mathbf{M}}$.

The integration procedure is stable provided that

$$|\mu_{max}| \le 1 \tag{3.12}$$

where μ_{max} is the maximum eigenvalue of

$$\mathbf{A}\, \mathbf{x}_i = \mu_i\, \mathbf{B}\, \mathbf{x}_i \tag{3.13}$$

and $|\ |$ is the modulus of the complex number, **A** and **B** are defined in (3.11).

We now use a generalization of Theorem 1 which states the following:

Theorem 2. If $\mathbf{A}_{\alpha\beta}$ and $\mathbf{B}_{\alpha\beta}$ are assembled from $\mathbf{A}^e_{\alpha\beta}$ and $\mathbf{B}^e_{\alpha\beta}$, $(\mathbf{A}^e_{\alpha\beta}) = (\mathbf{A}^e_{\alpha\beta})^T$, $(\mathbf{B}^e_{\alpha\beta}) = (\mathbf{B}^e_{\alpha\beta})^T$, and the submatrices have the same eigenvectors; Let $\mu(\mathbf{A}_{\alpha\beta},\mathbf{B}_{\alpha\beta})$ and $\mu^e(\mathbf{A}^e_{\alpha\beta},\mathbf{B}^e_{\alpha\beta})$ be the characteristic roots of the following eigenproblems respectively,

$$\begin{bmatrix} \mathbf{B}_{11} & \mathbf{B}_{12} \\ \mathbf{B}_{21} & \mathbf{B}_{22} \end{bmatrix} \mathbf{x}_i = \mu_i \begin{bmatrix} \mathbf{A}_{11} & \mathbf{A}_{12} \\ \mathbf{A}_{21} & \mathbf{A}_{22} \end{bmatrix} \mathbf{x}_i \tag{3.14}$$

and

$$\begin{bmatrix} \mathbf{B}^e_{11} & \mathbf{B}^e_{12} \\ \mathbf{B}^e_{21} & \mathbf{B}^e_{22} \end{bmatrix} \mathbf{x}^e_i = \mu^e_i \begin{bmatrix} \mathbf{A}^e_{11} & \mathbf{A}^e_{12} \\ \mathbf{A}^e_{21} & \mathbf{A}^e_{22} \end{bmatrix} \mathbf{x}^e_i \tag{3.15}$$

if $\mu(\mathbf{A}_{\alpha\beta},\mathbf{B}_{\alpha\beta})$ and $\mu^e(\mathbf{A}^e_{\alpha\beta},\mathbf{B}^e_{\alpha\beta})$ are the quasi-convex functions, then

$$\max_i |\mu_i| \leq \max_{\forall e,i} |\mu^e_i| \equiv \mu^E_{max} \tag{3.16}$$

Remark: We call elements consisting of symmetric submatrices such as the above quasi-symmetric.

The proof of the theorem 2 is based on quasi-convexity conditions and is given in the Appendix. We now outline the proof of the stability for the case of no damping. Since Eq.(3.16) shows the eigenvalue of the amplification matrix is bounded by the amplification matrix of the element, as in the previous Section, we write the element-level amplification equations. If element e is explicit (in E) then we have

$$\begin{bmatrix} \mathbf{M}_e-\frac{1}{2}\Delta t^2\mathbf{K}_e & \Delta t(\mathbf{M}_e-\beta\Delta t^2\mathbf{K}_e) \\ (\gamma-1)\Delta t\mathbf{K}_e & \mathbf{M}_e-\beta\Delta t^2\mathbf{K}_e \end{bmatrix} \mathbf{x}_e = \mu^e \begin{bmatrix} \mathbf{M}_e & \mathbf{0} \\ \gamma\Delta t\mathbf{K}_e & \mathbf{M}_e-\beta\Delta t^2\mathbf{K}_e \end{bmatrix} \mathbf{x}_e \tag{3.17}$$

For implicit elements, we have

$$\begin{bmatrix} \mathbf{M}_e-\Delta t^2(\frac{1}{2}-\beta)\mathbf{K}_e & \Delta t\,\mathbf{M}_e \\ (\gamma-1)\Delta t\mathbf{K}_e & \mathbf{M}_e \end{bmatrix} \mathbf{x}_e = \mu^e \begin{bmatrix} \mathbf{M}_e+\beta\Delta t^2\mathbf{K}_e & \mathbf{0} \\ \gamma\Delta t\mathbf{K}_e & \mathbf{M}_e \end{bmatrix} \mathbf{x}_e \tag{3.18}$$

By comparing (3.18) with the amplification matrix for the standard Newmark β-method, it can be seen that the two are identical, so the stability conditions are those of the Newmark β-method expressed on the implicit element:

$$\gamma \geq \frac{1}{2} \tag{3.19a}$$

$$\text{if } 2\beta - \gamma \geq 0, \text{ unconditional stability} \tag{3.19b}$$

$$\text{if } 2\beta - \gamma \leq 0, \ \Delta t^2 \lambda^e_{max} \leq \frac{2}{\gamma-2\beta} \quad \forall e \in I \tag{3.19c}$$

where λ^e_{max} is the maximum eigenvalue of (2.26).

For the explicit elements, the criterion is independent of β and is simply

$$\Delta t^2 \lambda^e_{max} \leq \frac{2}{\gamma} \tag{3.20}$$

REFERENCES

1. Belytschko T. and Mullen R. - Mesh Partitions of Explicit-Implicit Time Integration, Proceedings, U.S.-Germany Symposium on Formulations and Computational Algorithms in Finite Element Analysis, Massachussetts Institute of Technology, Cambridge, Mass. 1976.

2. Belytschko T. and Mullen R. - Stability of Explicit-Implicit Mesh Partitions in Time Integration," International Journal for Numerical Methods in Engineering, 12, 1575-1586 (1978).

3. Hughes T.J.R. and Liu W.K. - Implicit-Explicit Finite Elements in Transient Analysis: Stability Theory, Journal of Applied Mechanics, 45, 371-374 (1978).

4. Park K.C. - Partitional Transient Analysis Procedures for Coupled-Field Problems: Stability Analysis, Journal of Applied Mechanics, 47, 370-376 (1980).

5. Hilber H.M., Hughes T.J.R., and Taylor R.L. - Improved Numerical Dissipation for Time Integration Algorithms in Structural Dynamics, Earthquake Engineering and Structural Dynamics, 5, 283-292 (1977).

6. Miranda I., Ferencz R.M. and Hughes T.J.R. - An improved implicit-explicit time integration method for structural dynamics, Earthquake Engrg. Struct. Dyn., 18, 643-653 (1989).

7. Belytschko T., Smolinski P. and Liu W.K. - Stability of Multi-time Step Partitioned Integrators for First-order Finite Element Systems, Computer Methods in Applied Mechanics and Engineering, 49, 281-297 (1985).

8. Flanagan D.P. and Belytschko T. - Simultaneous relaxation in structural dynamics, J. Engrg. Mech. Div. ASCE, 107, 1039-1055 (1981).

9. Irons B.M. - Applications of a theorem on eigenvalues to finite element problems, CR/132/70, University of Wales, Dept. of Civil Engineering, Swansea, 1970.

10. Belytschko T. and Lu Y.Y. - Stability analyses of elemental explicit-implicit partitions by Fourier methods, to appear in Comp. Methods Appl. Mech. Eng., 1990.

APPENDIX

This Appendix gives the proof of the theorem 2 for the particular case where **A** and **B** are given by (3.17) and (3.18). To facilitate the proof, we only consider the amplification matrix without damping:

$$\mu \begin{bmatrix} \tilde{\mathbf{M}}+\beta\Delta t^2\mathbf{K} & \mathbf{0} \\ \gamma\Delta t\mathbf{K} & \tilde{\mathbf{M}} \end{bmatrix} \mathbf{x} = \begin{bmatrix} \tilde{\mathbf{M}}-\Delta t^2(\frac{1}{2}-\beta)\mathbf{K} & \Delta t\tilde{\mathbf{M}} \\ (\gamma-1)\Delta t\mathbf{K} & \tilde{\mathbf{M}} \end{bmatrix} \mathbf{x} \tag{A-1}$$

We also have the elemental amplification matrix form:

$$\mu^e \begin{bmatrix} \tilde{\mathbf{M}}^e+\beta\Delta t^2\mathbf{K}^e & \mathbf{0} \\ \gamma\Delta t\mathbf{K}^e & \tilde{\mathbf{M}}^e \end{bmatrix} \mathbf{x}^e = \begin{bmatrix} \tilde{\mathbf{M}}^e-\Delta t^2(\frac{1}{2}-\beta)\mathbf{K}^e & \Delta t\tilde{\mathbf{M}}^e \\ (\gamma-1)\Delta t\mathbf{K}^e & \tilde{\mathbf{M}}^e \end{bmatrix} \mathbf{x}^e \tag{A-2}$$

Let λ and λ^e be the eigenvalues which satisfy the following equations

$$\lambda_i \tilde{\mathbf{M}}\, \mathbf{y}_i = \mathbf{K}\, \mathbf{y}_i \qquad \lambda_i^e \tilde{\mathbf{M}}^e\, \mathbf{y}_i^e = \mathbf{K}^e\, \mathbf{y}_i^e \tag{A-3}$$

respectively. By using the theorem 1 given in [10], the characteristic equations associated with (A-1) and (A-2) are respectively,

$$(1+\beta h^2\lambda_i)\,\mu^2 + [-2+(\tfrac{1}{2}-2\beta+\gamma)h^2\lambda_i]\,\mu + 1 + (\tfrac{1}{2}+\beta-\gamma)h^2\lambda_i = 0 \tag{A-4}$$

$$(1+\beta h^2\lambda_i^e)\,\mu^{e2} + [-2+(\tfrac{1}{2}-2\beta+\gamma)h^2\lambda_i^e]\,\mu^e + 1 + (\tfrac{1}{2}+\beta-\gamma)h^2\lambda_i^e = 0 \tag{A-5}$$

in which $h = \Delta t$. From the above, we can see that μ and μ^e satisfy the same equation such that

$$(1+\beta\,\bar{\lambda})\,\bar{\mu}^2 + [-2+(\tfrac{1}{2}-2\beta+\gamma)\bar{\lambda}]\,\bar{\mu} + 1 + (\tfrac{1}{2}+\beta-\gamma)\bar{\lambda} = 0 \tag{A-6}$$

where $h = \Delta t$ and $\bar{\lambda} = h^2\lambda$; $\bar{\mu} = \mu$, when $\bar{\lambda} = \lambda_i$, and $\bar{\mu} = \mu^e$, when $\bar{\lambda} = h^2\lambda_i^e$.

By solving (A-6), we have

$$\bar{\mu} = [2-(\tfrac{1}{2}-2\beta+\gamma)\bar{\lambda} \pm \{(\tfrac{1}{4}-4\beta+\gamma+\gamma^2)\bar{\lambda}^2 - 4\bar{\lambda}\}^{1/2}]/\{2(1+\beta\bar{\lambda})\} \tag{A-7}$$

Considering the function $F(\bar{\lambda}) = \max[\,|\bar{\mu}_+|, |\bar{\mu}_-|\,]$ which is defined on the closed interval $\bar{\lambda} \in \Omega = [\,\bar{\lambda}^e_{min}, \bar{\lambda}^e_{max}]$, we have

$$|\mu|_{max} = \max F(\bar{\lambda}) \qquad \bar{\lambda} \in \bar{\lambda}_i, \quad \forall\, i \tag{A-8}$$

$$|\mu^e|_{max} = \max F(\bar{\lambda}) \qquad \bar{\lambda} \in \bar{\lambda}_i^e, \quad \forall\, i \tag{A-9}$$

In order to prove the eigenvalue inequality (3.16), we only need prove that the function $F(\bar{\lambda})$ must attain the maximum value at the end point of the closed interval Ω since we have the element eigenvalue inequality $\bar{\lambda}^e_{min} \le h^2\lambda_i \le \bar{\lambda}^e_{max}$ from [8,9]. We now consider the function $F(\bar{\lambda})$ in three cases as follows:

1) If $\bar{\mu}$ is complex on the closed interval $\bar{\lambda} \in \Omega$, we have

$$F^2 = |\,(1+(\tfrac{1}{2}+\beta-\gamma)\bar{\lambda}\,)/(1+\beta\bar{\lambda}\,)\,| \tag{A-10}$$

Let

$$f(\bar{\lambda}) = (1+(\tfrac{1}{2}+\beta-\gamma)\bar{\lambda}\,)/(1+\beta\bar{\lambda}\,) \tag{A-11}$$

Since

$$\frac{df}{d\bar{\lambda}} = \frac{1-2\gamma}{2(1+\beta\bar{\lambda})} \tag{A-12}$$

$f(\bar{\lambda})$ is a monotone decreasing function if $\gamma \geq \frac{1}{2}$ or a monotone increasing function if $\gamma \leq \frac{1}{2}$, and hence $|f(\bar{\lambda})|$ is maximum at the end point of the closed interval Ω, i.e.,

$$|\bar{\mu}|^2_{max} = \max\{\, F^2(\bar{\lambda})|_{\bar{\lambda}=\bar{\lambda}^e_{min}},\; F^2(\bar{\lambda})|_{\bar{\lambda}=\bar{\lambda}^e_{max}} \} \tag{A-13}$$

so the eigenvalue inequality (3.16) holds if $\bar{\mu}$ is complex.

2) If $\bar{\mu}$ is real on the closed interval $\bar{\lambda} \in \Omega$, we have

$$F(\bar{\lambda})=[2-(\tfrac{1}{2}-2\beta+\gamma)\bar{\lambda}+\{(\tfrac{1}{4}-4\beta+\gamma+\gamma^2)\bar{\lambda}^2-4\bar{\lambda}\}^{1/2}]/\{2|1+\beta\bar{\lambda}|\} \quad \text{if } 2-(\tfrac{1}{2}-2\beta+\gamma)\bar{\lambda} \geq 0 \tag{A-14}$$

$$F(\bar{\lambda})=[(\tfrac{1}{2}-2\beta+\gamma)\bar{\lambda}-2+\{(\tfrac{1}{4}-4\beta+\gamma+\gamma^2)\bar{\lambda}^2-4\bar{\lambda}\}^{1/2}]/\{2|1+\beta\bar{\lambda}|\} \quad \text{if } 2-(\tfrac{1}{2}-2\beta+\gamma)\bar{\lambda} < 0 \tag{A-15}$$

For Eq.(A-14), we have

$$\frac{dF}{d\bar{\lambda}} = \frac{-(\frac{1}{2}+\gamma)\{(\frac{1}{4}-4\beta+\gamma+\gamma^2)\bar{\lambda}^2-4\bar{\lambda}\}^{1/2}+(\frac{1}{2}+\gamma)^2\bar{\lambda}-2(1+\beta\bar{\lambda})}{2(1+\beta\bar{\lambda})^2\{(\frac{1}{4}-4\beta+\gamma+\gamma^2)\bar{\lambda}^2-4\bar{\lambda}\}^{1/2}} \tag{A-16}$$

Since $\bar{\mu}$ is real, i.e.,

$$(\tfrac{1}{4}-4\beta+\gamma+\gamma^2)\bar{\lambda}^2-4\bar{\lambda} = (\tfrac{1}{2}+\gamma)^2\bar{\lambda}^2-4(1+\beta\bar{\lambda})\bar{\lambda} \geq 0 \tag{A-17}$$

it is easy to show that

$$(\tfrac{1}{2}+\gamma)^2\bar{\lambda}-2(1+\beta\bar{\lambda}) \geq (\tfrac{1}{2}+\gamma)\{(\tfrac{1}{4}-4\beta+\gamma+\gamma^2)\bar{\lambda}^2-4\bar{\lambda}\}^{1/2} \tag{A-18}$$

If $(\frac{1}{2}+\gamma) \leq 0$, it is true because the right hand side of the above inequality is negative, but its left hand side is positive. If $(\frac{1}{2}+\gamma) > 0$, we have

$$[(\tfrac{1}{2}+\gamma)^2\bar{\lambda}-2(1+\beta\bar{\lambda})]^2-(\tfrac{1}{2}+\gamma)^2\{(\tfrac{1}{4}-4\beta+\gamma+\gamma^2)\bar{\lambda}^2-4\bar{\lambda}\} = 4\,(1+\beta\bar{\lambda})^2 \geq 0 \tag{A-19}$$

so Eq.(A-18) is also true, and hence we have

$$\frac{dF}{d\bar{\lambda}} \geq 0 \qquad \bar{\lambda} \in \Omega \tag{A-20}$$

Similarly, we can prove that the function $F(\bar{\lambda})$ defined in (A-15) also satisfies (A-20), which implies that function $F(\bar{\lambda})$ is a monotone increasing function. Since $F(\bar{\lambda})$ is always a positive for each $\bar{\lambda} \in \Omega$, it get a maximum value only at the end point $\bar{\lambda} = \bar{\lambda}^{e}_{max}$, i.e.,

$$|\bar{\mu}|_{max} = F(\bar{\lambda})|_{\bar{\lambda}=\bar{\lambda}^{e}_{max}} \tag{A-21}$$

so the eigenvalue inequality (3.16) also holds if $\bar{\mu}$ is real.

3) If $\bar{\mu}$ is real or complex on the closed interval $\bar{\lambda} \in \Omega$, we divide $\Omega=[\bar{\lambda}^{e}_{min}, \bar{\lambda}^{e}_{max}]$ into two subdomains $\Omega_1 = [\bar{\lambda}^{e}_{min}, \bar{\lambda}_m]$ and $\Omega_2 = [\bar{\lambda}_m, \bar{\lambda}^{e}_{max}]$. When $\bar{\mu}$ is real, we have from (A-17),

$$(\tfrac{1}{4} - 4\beta + \gamma + \gamma^2)\bar{\lambda} \geq 4 \tag{A-22}$$

it follows that

$$\bar{\lambda}_m = \bar{\lambda} \geq \frac{4}{\frac{1}{4} - 4\beta + \gamma + \gamma^2} \qquad \text{with } \tfrac{1}{4} - 4\beta + \gamma + \gamma^2 > 0 \tag{A-23}$$

The above equation implies that $\bar{\mu}$ is real only on the subdomain Ω_2. From the results discussed in case 2, we know that $|\bar{\mu}|$ may get a maximum value only at the right end point of the closed interval Ω_2 if $\bar{\mu}$ is real, i.e.,

$$|\bar{\mu}|_r = F(\bar{\lambda})|_{\bar{\lambda}=\bar{\lambda}^{e}_{max}} \geq F(\bar{\lambda}) \geq F(\bar{\lambda})|_{\bar{\lambda}=\bar{\lambda}_m} \qquad \bar{\lambda} \in \Omega_2 \tag{A-24}$$

On the other hand, from the results discussed in case 1, we also know that $|\bar{\mu}|$ may get a maximum value only at the end point of the closed interval Ω_1 if $\bar{\mu}$ is complex, i.e.,

$$|\bar{\mu}|_c = \max\{F(\bar{\lambda})|_{\bar{\lambda}=\bar{\lambda}^{e}_{min}}, F(\bar{\lambda})|_{\bar{\lambda}=\bar{\lambda}_m}\} \geq F(\bar{\lambda}) \qquad \bar{\lambda} \in \Omega_1 \tag{A-25}$$

Combining (A-24) and (A-25) yields

$$|\bar{\mu}|_{max} = \max\{|\bar{\mu}|_r, |\bar{\mu}|_c\} = \max\{F(\bar{\lambda})|_{\bar{\lambda}=\bar{\lambda}^{e}_{min}}, F(\bar{\lambda})|_{\bar{\lambda}=\bar{\lambda}^{e}_{max}}\} \tag{A-26}$$

and hence the eigenvalue inequality (3.16) also holds if $\bar{\mu}$ is real or complex on the closed interval Ω. The proof is completed.

FINITE ELEMENTS IN THE 90's, E. Oñate, J. Periaux, A. Samuelsson (Eds.)
Springer-Verlag/CIMNE, Barcelona 1991

FURTHER TESTS ON THE DERIVATIVE RECOVERY TECHNIQUE AND A POSTERIORI ERROR ESTIMATOR

J.Z. Zhu
Institute of Numerical Method in Engineering,
University College of Swansea,
Singleton Park, Swansea SA2 8PP,
UNITED KINGDOM

This paper is dedicated to Professor Olek Zienkiewicz on the occasion of his 70th birthday

SUMMARY

Further numerical tests are presented for the finite element derivative recovery technique recently developed by Zienkiewicz and Zhu [1]. The performance of the *a posteriori* error estimator using the recovered derivatives is also demonstrated.

1. INTRODUCTION

Very recently, a general finite element derivative recovery technique which includs two similar, however different, recovery procedures has been developed by Zienkiewicz and Zhu [1]. The general idea and the implementation of the derivative recovery technique is simple and the technique can be applied to any finite element approximations. The procedure 1 of the recovery technique is termed as *discrete superconvergent recovery procedure.* The procedure 2 is called *local L_2 projection recovery procedure.* It has been demonstrated, by numerical experiments, that the recovered nodal values of the derivatives by the discrete superconvergent recovery procedure are superconvergent. One order higher accuracy is achieved by the procedure for the derivatives of linear and cubic elements. Two order higher accuracy is achieved for the derivatives with quadratic elements. In particular, a $O(h^4)$ convergence of the nodal values of the derivatives for quadratic triangular element is reported for the first time. As a direct application, the discrete superconvergent recovery procedure is used in the computation of the Zienkiewicz-Zhu (also known as Z^2) *a posteriori* error estimator [2]. It was shown that, with retrieved superconvergent solutions, reliable and asymptotically exact energy norm error estimation and pointwise error estimation have been achieved. It was also stated in Reference 1 that the local L_2 projection recovery procedure only provides superconvergent nodal value for linear and cubic elements but not for quadratic elements although the recovered solution is substantially more accurate.

The main objective of this paper is to give numerical evidence on the convergence of the recovered derivatives by the local L_2 projection recovery procedure and the performance of the *a posteriori* error estimator using such recovery procedure. Some general issues regarding the recovery procedure are also addressed.

2. THE LOCAL L_2 PROJECTION RECOVERY PROCEDURE

A comprehensive discussion on the derivative recovery technique of both discrete superconvergent recovery procedure and local L_2 projection recovery procedure is given in [1]. Here only the local L_2 projection recovery procedure will be considered. For more details of the basic ideas and the implementation aspects of the recovery technique the readers are referred to Reference 1.

Assume that $\boldsymbol{\sigma}_h$ is the gradient (stress or flux) of the finite element solution, $\boldsymbol{\sigma}^*$ is the recovered solution defined by the basis function $\mathbf{N}$ and nodal values $\bar{\boldsymbol{\sigma}}^*$

$$\boldsymbol{\sigma}^* = \mathbf{N}\bar{\boldsymbol{\sigma}}^* \tag{1}$$

where $\mathbf{N}$ are the same basis functions as the ones used in the finite element approximation. The nodal values $\bar{\boldsymbol{\sigma}}^*$ are calculated by the following procedure.

Considering a vertex node i of a particular mesh (interelement node in one dimension) and the element patch formed by the m elements surrounding node i (see Figure 1 for element patches of 6 node and 8 node elements). A polynomial $\boldsymbol{\sigma}_p^*$ of the same order with the basis function $\mathbf{N}$ is defined on the element patch and is termed as smoothing function. It has the form of

$$\boldsymbol{\sigma}_p^* = \mathbf{Pa} \tag{2}$$

where $\mathbf{a}$ is a set of unknown parameters and $\mathbf{P}$ contains the appropriate polynomial terms, e.g. for a quadratic quadrilateral element $\mathbf{P}$ may contain eight terms of polynomial and has the form of

$$\mathbf{P} = [1, x, y, xy, x^2, y^2, x^2y, xy^2] \tag{3}$$

Remark How many terms of polynomial should be used in the smoothing function is a question of choosing the best mathematical model for the smoothing function. Here the order of the polynomial for the smoothing function is chosen to be the same as the basis functions used in the finite element (displacement) approximation. The reason is that we expect the recovered derivatives (stresses) will locally possess the same order of accuracy as that of the primary variables (displacements).

The unknown values $\mathbf{a}$ in equation (2) may be determined by requiring the functional

$$\begin{aligned} F(\mathbf{a}) &= \int_{\Omega_*} (\boldsymbol{\sigma}_h - \boldsymbol{\sigma}_p^*)^2 d\Omega \\ &= \int_{\Omega_*} (\boldsymbol{\sigma}_h - \mathbf{Pa})^2 d\Omega \end{aligned} \tag{4}$$

to be minimized, where $\Omega_s = \bigcup_{j=1}^{m} \Omega_m$, is the domain of the element patch surrounding vertex node i.

The minimization condition of $F(\mathbf{a})$ results in

$$\int_{\Omega_s} \mathbf{P}^T \mathbf{P} d\Omega \mathbf{a} = \int_{\Omega_s} \mathbf{P}^T \sigma_h d\Omega \tag{5}$$

in matrix form, this can be written as

$$\mathbf{A}\mathbf{a} = \mathbf{b} \tag{6}$$

with

$$\mathbf{A} = \int_{\Omega_s} \mathbf{P}^T \mathbf{P} d\Omega \quad \text{and} \quad \mathbf{b} = \int_{\Omega_s} \mathbf{P}^T \sigma_h d\Omega \tag{7}$$

The order of matrix **A** is obviously the same as the number of the terms of the polynomial used in **P**.

Solving equation (6), we obtain a and therefore σ_p^*. The recovered nodal values $\bar{\sigma}^*$ at vertex node i and other interelement nodes inside of the element patch can be readily obtained by directly substitute the node coordinates into equation (2). Indeed the recovered values of σ_p^* at any point of the element patch can be calculated in the same manner, we here however are only interested in the calculation of the nodal values.

The above recovery procedure continues over every vertex node. The recovered nodal values at each vertex node are only calculated once, the nodal values for the interelement side nodes and element interior nodes will be calculated several times corresponding to different element type and different (interior or boundary) vertex nodes. An averaging of these values will be taken as their final values.

Remark The averaging process for the nodal values of face nodes and element interior nodes is not intended to improve the accuracy. The systematic manner of the calculation however simplifies the implementation of the procedure.

Remark There are two methods to calculate the recovered nodal values at boundary nodes. The first method is that the recovery procedure is only performed over the interior vertex nodes. The boundary nodal values are calculate from the element patch which contains the corresponding boundary nodes. In the second method, the recovery procedure will be performed over all the vertex nodes except the ones surrounded only by one element. The latter will have to be used for the elements which have all their vertex nodes being boundary nodes.

Remark In solving equation (6), if the mesh is irregular or if the vertex node is a boundary node, matrix **A** may not be positive definite. This implies that

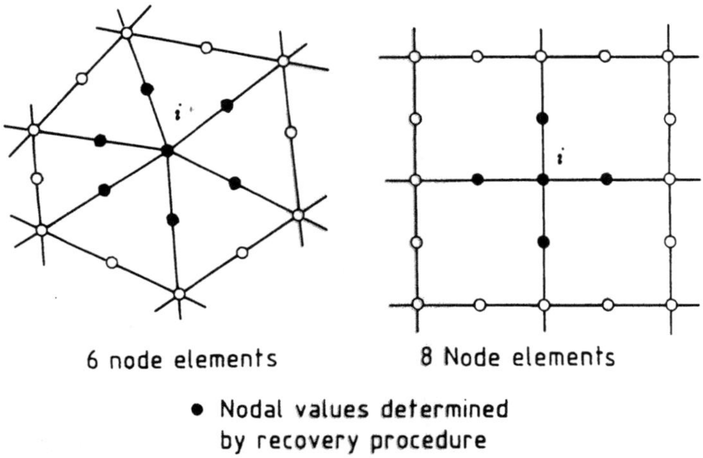

Figure 1. Element patches used in the derivative recovery procedure.

too many polynomial terms have been used in the smoothing function. Because the polynomial terms used in **P** have a hierarchical structure, i.e. the higher order $\boldsymbol{\sigma}_p^*$ includes the lower order ones, the program can be easily designed in a way such that the number of polynomial terms can be automatically reduced and only the highest order of positive definite submatrix of **A** is solved.

As equation (6) is only solved once for each vertex node, the procedure is obviously inexpensive.

Using the retrieved nodal values $\bar{\boldsymbol{\sigma}}^*$, the recovered solution $\boldsymbol{\sigma}^*$ is readily constructed by Equation (1).

3. A POSTERIORI ERROR ESTIMATOR

The recovered solution $\boldsymbol{\sigma}^*$ of equation (1) can be used in the Zienkiewicz-Zhu error estimator to assess the accuracy of the finite element approximation. It has been shown that if $\boldsymbol{\sigma}^*$ is superconvergent, the error estimator will be asymptotically exact [2]. It has also been demonstrated in the case that $\boldsymbol{\sigma}^*$ is not superconvergent but consistently much more accurate than $\boldsymbol{\sigma}_h$, the error estimator will aslo provide reliable estimation of the exact error of the approximation. Only energy norm error estimation is considered in this paper. The exact error in energy norm is denoted by $\| \mathbf{e} \|$ and the energy norm error estimator is expressed by $\| \bar{\mathrm{e}} \|$. For more details on the basic ideas and the theory of the Zienkiewicz-Zhu error estimator we refer to [2,3]. For the estimation of pointwise error the readers are also referred to [2].

4. NUMERICAL STUDIES

The accuracy and the convergence of recovered derivatives achieved by local L_2 projection procedure and the reliability of the error estimator using the recovered solution will be examined in this section. The convergence of $\boldsymbol{\sigma}_h$ and $\boldsymbol{\sigma}^*$ is examined at nodal points. The maximum value of the error of $\boldsymbol{\sigma}_h$ at node point where $\boldsymbol{\sigma}_h$ is discontinuous is taken in the convergence study. The effectivity index which is the ratio of the estimated error and the exact error

$$\theta = \frac{\| \bar{\mathbf{e}} \|}{\| \mathbf{e} \|} \tag{8}$$

is used to examine the reliability of the error estimator. For a reliable error estimator θ should be always close to unity.

To compare the performance of the recovery procedures, the results of the global L_2 projection of [4] and the bilinear extrapolation of [5] are also presented. These results are represented by superscript L and HC respectively, e.g. $\boldsymbol{\sigma}^L$ and $\boldsymbol{\sigma}^{HC}$.

Example 1.

Consider a one dimensional two-point boundary value problem

$$-\frac{d}{dx}\left(\frac{du}{dx}\right) + u = f \qquad \text{on } I = (0,1) \tag{9}$$

with boundary conditions

$$u(0) = 0 \text{ and } u(1) = 0 \tag{10}$$

The function f is chosen so that the exact solution is of the form

$$u = x^2 - \frac{sh4x}{sh4} \tag{11}$$

and we define

$$\sigma = \frac{du}{dx} \tag{12}$$

A sequence of uniform refinements is employed in the finite element analysis using linear, quadratic and cubic elements.

Pointwise error of σ_h, σ^* and σ^L are examined at an interelement point $x = 0.5$ where the exact solution is $\sigma = 0.4485588705$. Numerical results of using linear, quadratic and cubic elements are presented in Table 1(a), 1(b) and 1(c) respectively. We observe, from Table 1(a) and 1(c), that both σ^* and σ^L are superconvergent for linear and cubic elements with $O(h^{p+1})$ convergence. It is also found, from Table 1(b), that neither σ^* nor σ^L is superconvergent for quadratic element, however that σ^* is significantly more accurate than σ_h and σ^L.

h	$\lvert\sigma - \sigma_h\rvert$	$\lvert\sigma - \sigma^*\rvert$	$\lvert\sigma - \sigma^L\rvert$
1/2	0.7753(00)	0.4486(00)	0.4485(00)
1/4	0.1291(00)	0.8501(−1)	0.1035(00)
1/8	0.3010(−1)	0.1822(−1)	0.8451(−2)
1/16	0.8882(−2)	0.4437(−2)	0.1361(−2)
1/32	0.3139(−2)	0.1102(−2)	0.3370(−3)
1/64	0.1270(−2)	0.2750(−3)	0.8416(−4)
1/128	0.5635(−3)	0.6873(−4)	0.2103(−4)
Rate of convergence	1	2	2

Table 1(a). Convergence of σ_h, σ^* and σ^L at $x = 0.5$. Linear element. Example 1.

h	$\lvert\sigma - \sigma_h\rvert$	$\lvert\sigma - \sigma^*\rvert$	$\lvert\sigma - \sigma^L\rvert$
1/2	0.5314(00)	0.2541(−1)	0.3098(00)
1/4	0.7668(−1)	0.2903(−2)	0.6276(−1)
1/8	0.1474(−1)	0.1261(−2)	0.1272(−1)
1/16	0.3247(−2)	0.3481(−3)	0.2946(−2)
1/32	0.7630(−3)	0.8907(−4)	0.7226(−3)
1/64	0.1850(−3)	0.2240(−4)	0.1798(−3)
1/128	0.4556(−4)	0.5607(−5)	0.4489(−4)
Rate of convergence	2	2	2

Table 1(b). Convergence of σ_h, σ^* and σ^L at $x = 0.5$. Quadratic element. Example 1.

h	$\lvert\sigma - \sigma_h\rvert$	$\lvert\sigma - \sigma^*\rvert$	$\lvert\sigma - \sigma^L\rvert$
1/2	0.1031(00)	0.3805(−1)	0.4540(−1)
1/4	0.7486(−2)	0.2098(−2)	0.8810(−3)
1/8	0.7186(−3)	0.1270(−3)	0.8428(−4)
1/16	0.7882(−4)	0.7874(−5)	0.5369(−5)
1/32	0.9232(−5)	0.4912(−6)	0.3363(−6)
1/64	0.1117(−5)	0.3068(−7)	0.2103(−7)
Rate of convergence	3	4	4

Table 1(c). Convergence of σ_h, σ^* and σ^L at $x = 0.5$. Cubic element. Example 1.

	θ^*		
h	Linear element	Quadratic element	Cubic element
1/2	0.315477	0.882174	1.280225
1/4	0.738324	0.835094	0.987410
1/8	0.862150	0.936571	0.963938
1/16	0.928239	1.006588	0.985223
1/32	0.963116	1.041881	1.001245
1/64	0.981245	1.057887	1.008216
1/128	0.990534	1.065119	

Table 2. Global effectivity index of Z^2 error estimator using the local L_2 projection recovery procedure. Example 1.

The performance of the error estimator using the recovered solution by the local L_2 projection recovery procedure for linear, quadratic and cubic elements are reported in Table 2. It can be seen that the effectivity index θ^* is very close to one for all three elements while the mesh is refined. It is therefore demonstrated that a reliable error estimation has been achieved.

Further numerical tests over a wide range of elements have confirmed that the above observations are also true for elements of higher dimensions. In the following we shall report the numerical tests of the recovery procedure for 8 node Serendipity element.

Example 2.

A two dimensional problem is considered with governing equation

$$-\Delta u = f \qquad \text{in } \Omega \tag{13}$$

and boundary condition

$$u = 0 \qquad \text{on } \partial\Omega \tag{14}$$

where Ω is a unit square domain $\Omega = (0,1) \times (0,1)$ and f is constructed to correspond to the exact solution

$$u(x,y) = x(1-x)y(1-y)(1+2x+7y) \tag{15}$$

This problem has been used by Zlamal [6] to demonstrate the superconvergence of the derivatives at Gauss points for Serendipity elements.

Here we define

$$\sigma = \nabla u \tag{16}$$

with

$$\sigma_x = \frac{\partial u}{\partial x} \quad \text{and} \quad \sigma_y = \frac{\partial u}{\partial y} \tag{17}$$

h	$\lvert\sigma - \sigma_{x_h}\rvert$	$\lvert\sigma - \sigma_x^*\rvert$	$\lvert\sigma - \sigma_x^{HC}\rvert$
1/4	0.603870(−2)	0.118705(−2)	0.149724(−1)
1/8	0.271645(−2)	0.128587(−3)	0.361427(−2)
1/12	0.118086(−2)	0.479221(−4)	0.159930(−2)
1/16	0.687360(−3)	0.252046(−4)	0.897703(−3)
1/20	0.444705(−3)	0.155949(−4)	0.573958(−3)
Rate of convergence	2	2	2

Table 3(a). Convergence of σ_{x_h}, σ_x^* and σ_x^{HC} at (0.25, 0.25). 8 node element. Example 2.

h	$\lvert\sigma - \sigma_{y_h}\rvert$	$\lvert\sigma - \sigma_y^*\rvert$	$\lvert\sigma - \sigma_y^{HC}\rvert$
1/4	0.410063(−1)	0.840456(−2)	0.693626(−1)
1/8	0.101845(−1)	0.159968(−2)	0.167264(−1)
1/12	0.448771(−2)	0.652312(−3)	0.740548(−2)
1/16	0.253078(−2)	0.353542(−3)	0.415923(−2)
1/20	0.162183(−2)	0.221373(−3)	0.265990(−2)
Rate of convergence	2	2	2

Table 3(b). Convergence of σ_{y_h}, σ_y^* and σ_y^{HC} at (0.25, 0.25). 8 node element. Example 2.

The local error of the recovered derivatives $\boldsymbol{\sigma}^*$ and $\boldsymbol{\sigma}^{HC}$ is examined at a vertex nodal point with coordinates of (0.25,0.25) where the exact solution has the values of $\sigma_x = 0.375$ and $\sigma_y = 0.55078125$.

The numerical results of the convergence of σ_{x_h}, σ_x^*, σ_x^{HC} and σ_{y_h}, σ_y^*, σ_y^{HC} for 8 node element are presented in Table 3(a) and 3(b) respectively. An eight term polynomial is used in the smoothing function. The performance of the error estimator using the recovered solutions is demonstrated in Table 4. It is again observed, from Table 3(a) and 3(b), that $\boldsymbol{\sigma}^*$ is much more accurate than $\boldsymbol{\sigma}_h$ and $\boldsymbol{\sigma}^{HC}$ but is not superconvergent with $O(h^p)$ convergence. As for one dimensional problem, the effectivity index of the error estimator using the recovered solution by local L_2 projection recovery procedure is consistently close to unity, the error estimation is therefore accurate and the error estimator is reliable.

h	θ^*	θ^{HC}
1/4	1.118697	1.332874
1/8	1.065637	1.468899
1/12	1.062058	1.490541
1/16	1.062927	1.498526
1/20	1.063950	1.501506

Table 4. Global effectivity index of Z^2 error estimator for 8 node element. Example 2.

5. CONCLUDING REMARKS

The results presented in this paper provide numerical evidence for the statements made in Reference 1 on the finite element derivative *local L_2 projection recovery procedure.* It demonstrates that the procedure produces superconvergent nodal values of derivatives for linear and cubic elements but does not result in superconvergence of derivative nodal values for quadratic elements though considerable improvement of the accuracy has been achieved. However, when the recovery procedure is used in the computation of the Zienkiewicz-Zhu *a posteriori* error estimator, reliable error estimation can always be achieved including for quadratic elements.

6. ACKNOWLEDGEMENT

The author would like to take this opportunity to express his deep appreciation of his teacher and a friend Professor Olek Zienkiewicz, whose knowledge, enthusiasm and stimulative discussions have contributed much to this research and indeed have influenced the author's research in the finite element method.

REFERENCES

1. Zienkiewicz,O.C. and Zhu,J.Z. – Superconvergent derivative recovery techniques and a posteriori error estimation in the finite element method, Part 1 : *A* general superconvergent recovery technique, *Report of INME, University College of Swansea, CR*/671/91, *also Int. J. Numer. Methods Eng.*, to appear.

2. Zienkiewicz,O.C. and Zhu,J.Z. – Superconvergent derivative recovery techniques and a posteriori error estimation in the finite element method, Part 2: The Zienkiewicz-Zhu error estimator, *Report of INME, University College of Swansea, CR*/672/91, *also Int. J. Numer. Methods Eng.*, to appear.

3. Zienkiewicz,O.C. and Zhu,J.Z. – A simple error estimator and adaptive procedure for practical engineering analysis, *Int. J. Numer. Methods Eng.,* **24**, 337-357, (1987).

4. Oden,J.T. and Brauchli,H.J. – On the calculation of consistent stress distributions in finite element applications, *Int. J. Numer. Methods Eng.,* **3**, 317-325, (1971).

5. Hinton,E. and Campbell,J.S. – Local and global smoothing of discontinues finite element functions using a least square method, *Int. J. Numer. Methods Eng.,* **8**, 461-480, (1974).

6. Zlamal,M. – Superconvergence and reduced integration in the finite element method, *Math. Comp.* **32**, 663-685, (1978).

FINITE ELEMENTS IN THE 90's, E. Oñate, J. Periaux, A. Samuelsson (Eds.)
Springer-Verlag/CIMNE, Barcelona 1991

OLEK AND A MATHEMATICIAN

I. Babuska
Institute for Physical Science and Technology,
University of Maryland,
College Park, Maryland, 20742,
USA

Summary: Paper elaborates on the relation of Olek with a mathematician.

1. Olek is an engineer, I am a mathematician with an engineering background. We were in touch for the first time when I was heading the computational analysis team for the construction of ORLIK [1] the largest dam in Czechoslovakia. This was many years before the finite element and computer era. Since then we have had numerous encounters especially after 1968, when Olek was already the leader of finite elements in engineering and I was involved in the mathematics of finite elements. Since then, tremendous progress in the engineering and mathematics of the finite elements has been made and these two fields have influenced each other.

An interesting question arises: How does the work in FEM of an engineer and of a mathematician differ? Roughly speaking an engineer, and especially such an outstanding one as Olek, relies on his perfect intuition, physical reasoning and numerical experimentation applied to problems reflecting the practice. He does not usually place too much emphasis on mathematical rigor, exactly formulated definitions and assumptions. What is essential is whether *it* works on the practical examples and in engineering practice. A mathematician, on the other hand, emphasizes mathematical rigor, exactly formulated definition and assumptions. What is essential for him is whether *it* is completely "clean", whether *it* works in *all* cases characterized by the assumptions made, etc. In contrast to the engineer for whom an *example* is essential, a mathematician finds that a *counterexample* plays an important role. Such counterexample could indicate various fine points of the strictly logical concept and could give insight in the needed assumptions, etc. These counterexamples could and often are "ridiculous" from an engineering point of view.

Once Olek and I had a discussion. I have shown him such a counterexample illustrating my point. Olek said: Ivo, who could be interested in such a "perverse" case. A discussion continued when I claimed that my countraexample fully satisfies the definition and the assumption. Olek retorted that no *reasonable* engineer will interpret the notion in such a formal way as I did. Well, the mathematicians, by their nature, often differ from "reasonably" thinking engineers.

I would like now to address some of these aspects in a concrete setting of the a-posteriori error estimation in the finite elements.

2. Since the advent of the a-posteriori error estimation and adaptive approaches in finite elements Olek and myself have been very interested in this subject. In fact we jointly organized an international meeting [2] devoted to this subject and wrote joint papers [3]. Of course, because of our nature - of an engineer and of a mathematician - some of our emphases were different.

I would like to illustrate this difference in the approaches on this topic.

A couple of years ago, Olek and Zhu [4] proposed an estimator which became popular in the engineering community. Hence I will address this estimator in a most simple setting.

Let us consider the one dimensional problem

(1) $$-u''(x) = f(x), \qquad x \in (0,1) = I,$$

(2) $$u(0) = u(1) = 0$$

and assume that f is such that $u \in H_0^1(I)$, i.e., $\int_0^1 u'^2 dx < \infty$ and (2) holds. More particularly let

(3) $$u(x) = x^p - 1, \qquad p > \frac{1}{2}.$$

Let us consider the usual finite element method with continuous piecewise linear solution $u_{FE}(x)$ on the mesh

(4) $$0 = x_0 < x_1 < \dots < x_{n-1} < x_n = 1$$

with

(5) $$h_i = x_i - x_{i-1}, \quad I_i = (x_{i-1}, x_i), \quad i = 1, \dots, n, \quad h = \max h_i.$$

It is well known that u_{FE} is the linear interpolant of $u(x)$, i.e.,

(6) $$u(x_i) = u_{FE}(x_i), \quad i = 0, \dots, n$$

We will be interested in the error $e = u - u_{FE}$ measured in the energy norm

(7) $$\|e\|_{E(I)} = \left[\int_0^1 (e')^2\, dx\right]^{1/2}.$$

Let us define the elemental *error indicator* η_i such that

(8) $$\eta_i^2 \approx \int_{x_{i-1}}^{x_i} (e')^2\, dx$$

and the *error estimator* $\mathcal{E}$

(9) $$\mathcal{E} = \left(\sum_{i=1}^{\eta} \eta_i^2\right)^{1/2} \approx \|e\|_{E(I)}.$$

The quality of the estimator is measured by its effectivity index ξ (resp. the elemental effectivity index ξ_i) where

(10) $$\xi = \frac{\mathcal{E}}{\|e\|_E}, \qquad \xi_i = \frac{\eta_i}{\|e\|_{E(I_i)}}.$$

The estimator is called *asymptotically exact* relatively to the family $\mathcal{P}$ of the solutions if $\xi \longrightarrow 1$ as $h \longrightarrow 0$ for *any* $u \in \mathcal{P}$ and *any* mesh. If $|\xi - 1| \le K$ for any $h \le h_0$, the estimator is called K *correct*. If in addition for a $u \in \mathcal{P}$ and a mesh with $h \le h_0$, $|\xi - 1| = K$ then the estimator is *strictly* K correct.

Let us now describe two very similar estimators, the Zienkiewicz - Zhu estimator [4], (briefly ZZ estimator) and Babuška - Miller estimator [5], [6], [7], (briefly BM estimator) and the principles for deriving them and their comparison. This will illustrate the difference between the thinking of an engineer and a mathematician.

Let us first explain Olek's indicator and the major intuitive reasoning for its derivation. First we observe that u'_{FE} is discontinuous and is constant on I_i, $i = 1,\dots,n$. In contrast u' is continuous on I. Hence the main idea is to construct by a *smoothing technique* a continuous piecewise linear function $\varphi(x)$ such that

$$\int_0^1 (u'-\varphi)^2\, dx \ll \int_0^1 (u'-u'_{FE})\, dx$$

which yields

$$\int_0^1 (u'_{FE}-\varphi)^2\, dx \approx \int_0^1 (u'-u'_{FE})^2\, dx$$

The ZZ constructions of $\varphi(x)$ is as follows. Consider $I_{i-1} \cup I_i$ and find $\psi_i(x)$ which is continuous on $\bar{I}_{i-1} \cup \bar{I}_i$, linear on I_{i-1} and I_i and which minimizes

(11) $$\int_{x_{i-2}}^{x_i} (u'_{FE}-\psi_i)^2\, dx.$$

Then we let $\varphi(x_{i-1}) = \psi_i(x_{i-1})$.

On I_i then we have

$$\varphi(x) = u'_{FE} + \beta_{i-1}^{ZZ} [\![u'_{FE}]\!]_{i-1}\, \Phi_{i-1}(x) + \alpha_i^{ZZ} [\![u'_{FE}]\!]_i\, \Phi_i(x)$$

where

$$[u'_{FE}]_i = u'_{FE} \mid I_{i+1} - u'_{FE} \mid I_i, \quad i = 1,\dots,n-1,$$

$$[u'_{FE}]_1 = [u'_{FE}]_n = 0,$$

Φ_{i-1}, Φ_i are the linear shape functions on I_i,

$$\beta^{ZZ}_{i-1} = -\frac{h_{i-1}}{h_{i-1}+h_i}, \quad \alpha^{ZZ}_i = \frac{h_{i+1}}{h_i+h_{i+1}}. \tag{13}$$

We can also write

$$\varphi(x_i) = u'_{FE}(x_i-0) + \rho^{ZZ}_i(u'_{FE}(x_i+0) - u'_{FE}(x_i-0)$$

with

$$\rho^{ZZ}_i = \frac{1}{1+\Delta_i}, \qquad \Delta_i = \frac{h_i}{h_{i+1}} \tag{14}$$

The idea of the smoothing and the heuristic expectation of the success is obvious. The numerical experiments made by Olek confirmed this expectation.

On the other hand the BM estimator has the same form but

$$\beta^{BM}_{i-1} = -\frac{h_i}{h_{i-1}+h_i}, \qquad \alpha^{BM}_i = \frac{h_i}{h_i+h_{i+1}} \tag{15}$$

and

$$\rho^{BM} = \frac{1}{1+\Delta_i^{-1}}. \tag{16}$$

The main idea of the derivation of BM estimator was quite different. The aim was to get such an estimator which is *asymptotically exact* relatively to the family $\mathcal{P}$ of smooth functions.

Obviously for $\Delta_i = 1$, the ZZ and BM estimator are identical, but not for $\Delta_i \neq 1$.

It is easy to prove.

<u>Theorem 1</u>. [8]. The BM estimator is asymptotically exact relatively to the functions having 3 bounded derivatives, while ZZ estimator is not. □

It is also possible to construct as counterexample, a sequence of non-uniform meshes and a smooth u such that $\limsup_{h\to 0} |\xi-1| > K$ for K arbitrarily large. Hence the Z-Z estimator can be very inaccurate.

Does it mean that ZZ estimator is inferior? Not at all! (For more see [8].) To show it let us consider the problem (1) (2) with the solution given by (3) and the mesh

$$x_i = \left(\frac{i}{n}\right)^{\beta}, \qquad i = 0,\dots,n.$$

In Table 1 we report the true error measured in the energy norm on I_j, and the elemental effectivity index ξ_j for some p and β.

	p = 2.25, β = 2			p = 1.25, β = 2		
int	error	ZZ	BM	error	ZZ	BM
1	2.11 (-4)	4.024	1.341	6.45 (-3)	1.172	0.391
2	1.67 (-3)	1.807	1.050	7.93 (-3)	0.937	1.441
3	4.57 (-3)	1.352	1.019	8.02 (-3)	0.968	1.198
4	8.93 (-3)	1.194	1.010	8.04 (-3)	0.983	1.103
5	1.47 (-2)	1.122	1.006	8.05 (-3)	0.990	1.063
6	2.20 (-2)	1.083	1.004	8.06 (-3)	0.993	1.042
7	3.07 (-2)	1.060	1.003	8.06 (-3)	0.995	1.030
8	4.09 (-2)	1.045	1.002	8.06 (-3)	0.996	1.023
9	5.25 (-2)	1.036	1.002	8.06 (-3)	0.997	1.018
10	6.56 (-2)	0.872	0.974	8.07 (-3)	0.970	1.084

Table 1. The elemental effectivity indices.

For the first element I_1 and the last one I_{10} the effectivity index is influenced by the choice $[\![u'_{FE}]\!]_1 = 0$ and $[\![u'_{FE}]\!]_n = 0$.

We see interesting feature. For the *smooth solution* and the mesh with $\beta = 2$ BM indicator performs better. (This follows from Theorem 1.) *In contrast*, for the solution which has singular behavior (of the corner type in 2 dimensions) and for the *adaptively* constructed meshes which equilibrate the elemental errors, ZZ estimator is superior over BM estimator. Olek's intuition and numerical experiments were related to this situation.

We see that the mathematical goal chosen for the derivation of BM estimator was well formulated and achieved but was not completely relevant for practical use. Hence based on Olek's results it is possible to reformulate the mathematical goal: To construct the estimator which is optimal for equilibrated meshes. For detail see [8]. We have

Theorem 2. [8]. Let $u''(x) > 0$ (or $u''(x) < 0$) (and other technical assumptions given in [8] hold). Then if the mesh is equilibrated with respect to the energy norm the (only) choice

$$\rho_i = \frac{1}{1 + \Delta_i^{1/2}}$$

yield an asymptotically exact estimator. □

In addition for the more unsmooth solution (for more see [8])

$$\rho_i = \frac{1}{1+\Delta_i^{\gamma}} \qquad \gamma > 1/2$$

is preferable. (Note that $\rho^{ZZ} = 1$, $\rho^{BM} = -1$.) In Table 2 we show the performance of the indicator for $\gamma = 1, 1/2$ and equilibrated mesh.

	p = 2.25, β = 6/7			p = 1.25, β = 2		
int	error	γ = 1	γ = 1/2	error	γ = 1	γ = 1/2
1	2.11 (-2)	0.95182	1.00696	6.45 (-3)	1.17157	0.99033
2	2.04 (-2)	1.04344	1.02352	7.93 (-3)	0.93675	0.90257
3	2.04 (-2)	1.00638	1.00288	8.02 (-3)	0.96772	0.97650
4	2.04 (-2)	1.00292	1.00128	8.04 (-3)	0.98300	0.98921
5	2.04 (-2)	1.00169	1.00074	8.05 (-3)	0.98962	0.99374
6	2.04 (-2)	1.00111	1.00048	8.06 (-3)	0.99302	0.99590
7	2.04 (-2)	1.00079	1.00034	8.06 (-3)	0.99499	0.99710
8	2.04 (-2)	1.00059	1.00025	8.06 (-3)	0.99623	0.99784
9	2.04 (-2)	1.00046	1.00020	8.06 (-3)	0.99707	0.99833
10	2.04 (-2)	1.00414	1.00018	8.07 (-3)	0.96997	0.99848

Table 2. The elemental effectivity indices for equilibrated meshes.

Note that the indicators on I_i, $i = 1,10$ are influenced by the same effect as in Table 1.

We see that in fact $\gamma = 1/2$ is better here than $\gamma = 1$ (the ZZ estimator).

I have illustrated one concrete example of what I said earlier. Olek's great intuition applied to the relevant engineering applications gives very good results. On the other hand a mathematician who exactly formulates the exact conditions, the assumptions and the goals is able by mathematical understanding to construct an optimal estimator which performs under *these conditions* better than the one intuitively constructed. Unfortunately I am unable at present to generalize the results in the one dimensional setting to two and three dimensions.

I would like to end this paper by congratulating Olek on his 70th birthday on his great intuition and his outstanding contribution to finite elements.

I look forward to many more of Olek's creative ideas and to our further collaboration.

REFERENCES

1. BABUŠKA, I. - The Development of Computational Mathematics in Czechoclovakia and the U.S.S.R. in **A History of Scientific Computing**, Ed. Nash, S.G., Addison-Wesley ACM Press, New York, 1990, pp 302-313.

2. BABUŠKA, I, ZIENKIEWICZ, O.C., GAGO, J.P., and OLIVEIRA, E.R. de A, eds. - **Accuracy Estimates and Adaptive Refinements in Finite Element Computations**, J. Wiley & Sons, New York, 1986.

3. HILLY, D.W., GAGO, J.P., ZIENKIEWICZ, O.C., and BABUŠKA, I. - A-posteriori error analysis and adaptive processes in the finite element method; Part I, Error Analysis; Part II, Adaptive method refinement, Int. Journal Num. Meth. in Eng., 19 (1983), pp. 1593-1619, 1621-1656.

4. ZIENKIEWICZ, O.C. and ZHU, J.Z. - A simple error estimator and adaptive procedure for practical engineering analysis, Int. Journal for Num. Meth. in Eng., **24**, 337-57 (1987).

5. BABUŠKA, I. and RHEINBOLDT, W.C. - A-posteriori error estimators in the finite element method, Int. Journal for Num. Meth. in Eng., 12, 1597-615 (1978).

6. BABUŠKA, I. and MILLER, A. - A-posteriori error estimates and adaptive techniques for the finite element method, Tech Note BN 968 (1981), Institute for Physical Science and Technology, University of Maryland, College Park, Maryland, 20742, U.S.A.

7. BABUŠKA, I. and MILLER, A. - A feedback finite element method with a-posteriori error estimation: Part I, The finite element method and some basic properties of the a-posteriori error estimator, Comp. Meth. Appl. Mech. Eng., **61**, 1-40 (1987).

8. BABUŠKA, I. and RODRIGUEZ, R. - The problem of the selection of an a-posteriori error indicator based on smoothening techniques, Tech Note BN-1126 (1991), Institute for Physical Science and Technology, University of Maryland, College Park, Maryland, 20742, U.S.A.

FINITE ELEMENTS IN THE 90's, E. Oñate, J. Periaux, A. Samuelsson (Eds.)
Springer-Verlag/CIMNE, Barcelona 1991

REMARKS ON APPROXIMATE CONTROLLABILITY FOR PARABOLIC SYSTEMS

J.L. Lions
Collège de France,
3, rue d'Ulm, 75005 Paris,
FRANCE

Dedicated to O.C. ZIENKIEWICZ

1. Introduction

Let Ω be an open set of $\mathbb{R}^n$ and let $\mathcal{O}$ be an open subset of Ω.

Let $T > 0$ be given once for all. It is the time horizon.

Let ***the state equation*** of the system we want to control be given by

$$\frac{\partial y}{\partial t} - \Delta y = v\chi_{\mathcal{O}} \quad \text{in } \Omega \times (0,T) \tag{1.1}$$

where

$$v \in L^2(\mathcal{O} \times (0,T)) \quad (v = \text{control}) \tag{1.2}$$

is to be found, and where in (1.1) $\chi_{\mathcal{O}}$ denotes the characteristic function of $\mathcal{O}$.

We assume the ***boundary condition***

$$y = 0 \quad \text{on} \quad \Sigma = \partial\Omega \times (0,T). \quad \square \tag{1.3}$$

A controllability problem in the above situation could be stated as follows : given two arbitrary functions y^0 and y^1 in $L^2(\Omega)$, is it possible to find v satisfying (1.2) such that ***the solution of*** (1.1) (1.3) ***and***

$$y|_{t=0} = y^0 \tag{1.4}$$

satisfies

(1.5) $\quad y|_{t=T} = y^1$.

But this is general ***impossible*** , since, by well known regularity properties of solutions of the Heat equation, $y|_{t=T}$ will be (no matter how v is chosen) C^∞ outside $\mathcal{O}$, hence an impossibility for (1.5) if y^1 is not (at least !) C^∞ outside $\mathcal{O}$ (if $\mathcal{O} = \Omega$ the problem is without interest). □

Therefore a natural idea is to relax (1.4) or (1.5) or the two conditions together. In order to fix ideas, we shall replace (1.4) (1.5) by

(1.4)* $\quad y|_{t=0} \in y^0 + \beta_0 B$

(1.5)* $\quad y|_{t=T} \in y^1 + \beta_1 B$

where β_0 , β_1 are given $\geqslant 0$, and where B denotes the unit ball of $L^2(\Omega)$.

The problem we want to study can now be stated as follows : ***find*** v ***such that***

(1.6) $\quad \frac{1}{2} \iint_{\mathcal{O}\times(0,T)} v^2 \, dx \, dt = \text{minimum}$

and such that (1.1) (1.3) (1.4)* (1.5)* ***are satisfied.*** □

The following remark is essential :

Remark 1.1.
Let $y(x,t\,;v)$ denote the solution of (1.1) (1.3) (1.4).

Then, when v spans $L^2(\mathcal{O}\times(0,T))$, $y(\,.\,,T\,;v)$ spans a ***dense*** subset of $L^2(\Omega)$.

The proof (well known) is by duality. There is no restriction in assuming that $y^0 = 0$. Let f be in $L^2(\Omega)$ such that

(1.7) $\quad \int_\Omega y(x,T\,;v)\, f(x)\, dx = 0 \quad \forall\ v \in L^2(\mathcal{O}\times(0,T))$.

We define ψ by

$$(1.8)\qquad -\frac{\partial\psi}{\partial t} - \Delta\psi = 0 \quad \text{in } \Omega \times (0,T),\ \psi(.\,,T) = f \quad , \quad \psi = 0 \text{ on } \Sigma\,.$$

Then, multiplying (1.8) by $y(x,t\,;v)$ and integrating by parts, we obtain

$$(1.9)\qquad \int_\Omega y(x,T\,;v)\, f(x)\,dx = \iint_{\mathcal{O}\times(0,T)} \psi\, v\, dx\, dt$$

so that (1.7) is equivalent to

$$(1.10)\qquad \psi = 0 \qquad \text{in} \qquad \mathcal{O} \times (0,T).$$

But by a classical result (which is valid for large classes of parabolic operators with variable coefficients, as proven by S. MIZOHATA [1]) (1.10) implies $\psi \equiv 0$, hence $f = 0$, and the density result is proven. □

It follows from Remark 1.1 that ***problem*** (1.6) ***makes sense*** (the inf. is finite) ***when*** $\beta_1 > 0$. We shall set

$$(1.11)\qquad \inf.\ \frac{1}{2} \iint_{\mathcal{O}\times(0,T)} v^2\, dx\, dt = j\,(\beta_0\,, y^0\,; \beta_1\,, y^1\,)$$

when (1.1) (1.3) (1.4)* (1.5)* are satisfied.

What Remark 1.1 proves is that

(1.12) $j(\beta_0\,, y^0\,; \beta_1\,, y^1)$ is finite if $\beta_1 > 0$, arbitrarily small and if $\beta_0 \geqslant 0$.

Remark 1.2.
If $\beta_0 = 0$, the fact that we can take β_1 as small as we want is referred to as ***approximate controllability.*** □

It is obvious that

$$(1.13)\qquad j(\beta_0\,, y^0\,; \beta_1\,, y^1\,) \leqslant j\,(0, y^0\,; \beta_1\,, y^1\,)\,.$$

In what follows we want to give a constuctive formula for

$j(\beta_0\,, y^0\,, \beta_1\,, y^1)$. (cf. (2.4) below). □

Remark 1.3.
The corresponding formula to (2.4) for $\beta_0 = 0$ has been given in J.L. LIONS [2]. □

Remark 1.4.
We have taken in (1.1) a ***distributed*** control. We could also consider similar questions for ***boundary controls,*** namely

$$(1.14) \qquad \frac{\partial y}{\partial t} - \Delta y = 0 \quad \text{in } \Omega \times (0,T),$$

$$(1.15) \qquad y = \begin{cases} v & \text{on } \Sigma_0 = \Gamma_0 \subset \Gamma = \partial\Omega \ , \\ 0 & \text{on } \Sigma \backslash \Sigma_0 . \end{cases}$$

There is then a slight technical difficulty. If $y|_{t=0} = y^0$ is given in $L^2(\Omega)$ and if $v \in L^2(\Sigma_0)$ then $y|_{t=T}$ ***is not*** in general in $L^2(\Omega)$ and one has to introduce more general Sobolev spaces. It is to avoid this (purely technical) difficulty that we confine ourselves here to distributed controls. □

Remark 1.5.
We emphasize the fact that (1.12) is true for ***arbitrary "small"*** $\mathcal{O}$ ***and arbitrary small*** T . □

Remark 1.6.
The situation would be ***entirely different*** for hyperbolic operators. A general method for this case has been introduced in J.L. LIONS [3] [4]. □

Remark 1.7.
Numerical applications of the techniques presented here have been made for the hyperbolic cases in R. GLOWINSKI , C.H. LI and J.L. LIONS [5] and R. GLOWINSKI and C.H. LI [6].

Numerical applications for ***parabolic cases*** are under progress, cf. R. GLOWINSKI [7]. □

Remark 1.8.
Problems similar to (1.11) for hyperbolic problems have been considered, for a situation that would correspond here to the case $\beta_0 = 0$, in J.L. LIONS [8]. □

Remark 1.9.

The method introduced below is restricked to ***linear problems.*** Exact or Approximate Controllability problems for ***non linear*** distributed systems are difficult. For some non linear ***hyperbolic*** problems, cf. E. ZUAZUA [9]. For some non linear ***parabolic*** problems cf. I. DIAZ [10]. A general conjecture is that for diffusion equations leading to turbulence (or chaos) there is ***approximate controllability.*** cf. J.L. LIONS [2] [11]. □

2. A duality formula.

Let ρ^0 be given in $L^2(\Omega)$. Let ρ be the solution of

$$(2.1) \qquad -\frac{\partial \rho}{\partial t} - \Delta \rho = 0 \qquad \text{in } \Omega \times (0,T),$$

$$(2.2) \qquad \rho(x,T) = \rho^0(x) \qquad \text{in } \Omega ,$$

$$(2.3) \qquad \rho = 0 \qquad \text{on} \qquad \Sigma = \Gamma \times (0,T) \quad , \quad \Gamma = \partial\Omega .$$

We have then, with the notations of the Introduction

$$(2.4) \qquad j(\beta_0 , y^0 ; \beta_1 , y^1) = - \inf. \Big[\frac{1}{2} \iint_{\mathcal{O}\times(0,T)} \rho^2 \, dx \, dt + \beta_0 \| \rho(0) \| + \beta_1 \|\rho^0 \| -(y^1 , \rho^0) + (y^0 , \rho(0))\Big].$$

In (2.4) $\rho = \rho(x, t ; \rho^0)$ denotes the solution of (2.1) (2;2) (2.3), $\| f \|$ = norm of f in $L^2(\Omega)$, $(f, g) = \int_\Omega fg \, dx$. □

Remark 2.1.

We can let β_0 ***tend to*** 0 in (2.4) . Except for y^1 in very particular classes, we ***cannot*** let β_1 tend to 0 . □

We now proceed with the proof of (2.4). Let us begin with some notations. Let us denote by $y(k, v)$ the solution of (1.1) (1.3) and

$$(2.5) \qquad y(. , t = 0) = k \quad , \quad k \in L^2(\Omega) .$$

Then

(2.6) $y(k,v) = \eta(k) + z(v)$

where

(2.7) $\frac{\partial \eta}{\partial t} - \Delta \eta = 0 \quad , \quad \eta(0) = k \quad , \quad \eta = 0 \quad \text{on } \Sigma$

(2.8) $\frac{\partial z}{\partial t} - \Delta z = 0\chi_\sigma \quad , \quad z(0) = 0 \quad , \quad z = 0 \quad \text{on } \Sigma .$

The constraints (1.4)* (1.5)* can now be written as

(2.9) $k \in y^0 + \beta_0 B$

(2.10) $z(T\,;v) + \eta(T\,;k) \in y^1 + \beta_1 B$.

We then introduce ***for*** k ***fixed***

(2.11) $m(k\,;\beta_1\,,y^1) = \inf_{v} \frac{1}{2} \iint_{\sigma\times(0,T)} v^2\,dx\,dt$,

for

(2.12) $z(T\,;v) \in y^1 - \eta(T\,;k) + \beta_1 B$.

We then have

(2.13) $j(\beta_0\,,y^0\,;\beta_1\,,y^1) = \inf_{k\in y^0+\beta_0 B} m(k\,;\beta_1\,,y^1)$. □

We now compute $m(k\,;\beta_1\,,y^1)$ by a duality argument. Let us introduce

(2.14)
$$F(v) = \frac{1}{2}\iint_{\sigma\times(0,T)} v^2\,dx\,dt ,$$

$$G(f) = \begin{cases} 0 & \text{if } f \in y^1 - \eta(T\,;k) + \beta_1 B , \\ +\infty & \text{otherwise} , \quad f \in L^2(\Omega) , \end{cases}$$

$$Lv = z(T\,;v), \quad L \in \mathcal{L}(L^2(\sigma\times(0,T))\,;L^2(\Omega)).$$

With these notations, one has

$$(2.15) \qquad m(k\,;\beta_1\,,y^1) = \inf_{v}\; F(v) + G(Lv).$$

We now use duality theory for convex proper functions. If $K(f)$ is such a function, its conjugate is given by

$$K^*(f) = \sup_{\hat f}\; [(f,\hat f) - K(\hat f)\,].$$

With this notation, one has

$$(2.16) \qquad \begin{aligned} &F^*(v) = F(v),\\ &G^*(f) = \sup\; (f, y^1 - \eta(T\,;k) + \beta_1 B) = (f, y^1 - \eta(T\,;k)) + \beta_1 \|f\|. \end{aligned}$$

According to a result of T.R. ROCKAFELLAR [12], one has

$$(2.17) \qquad m(k\,;\beta_1\,,y^1) = -\inf_{\rho^0 \in L^2(\Omega)} \; F^*(L^*\rho^0) + G^*(-\rho^0)\,,$$

where L^* = adjoint of L.

If one multiplies (2.1) by $z(v)$, one obtains after integrations by parts :

$$(2.18) \qquad (z(T\,;v)\,,\rho(T)) = \iint_{\mathcal{O}\times(0,T)} \rho\, v\, dx\, dt$$

i.e.

$$(2.19) \qquad (Lv\,,\rho^0) = \iint_{\mathcal{O}\times(0,T)} \rho\, v\, dx\, dt$$

hence

$$(2.20) \qquad L^*\rho^0 = \rho\,\chi_{\mathcal{O}}\,.$$

Using (2.16) and (2.20) in (2.17) gives :

$$(2.21) \qquad \begin{aligned} m(k\,;\beta_1\,,y^1) = -\inf_{\rho^0}\; [\frac{1}{2} \iint_{\mathcal{O}\times(0,T)} \rho^2\, dx\, dt + \beta_1 \|\rho^0\| -\\ - (\rho^0\,, y^1 - \eta(T\,;k))]\,. \end{aligned}$$

If we multiply (2.1) by $\eta(k)$ one obtains, after integrations by parts

$$(\rho^0 , \eta(T ; k)) = (\rho(0), k)$$

so that (2.21) gives

$$m(k ; \beta_1 , y^1) = - \inf_{\rho^0} \Big[\frac{1}{2} \iint_{\mathcal{O}\times(0,T)} \rho^2 \, dx\, dt + \beta_1 \| \rho^0 \| - (\rho^0 , y^1) + (\rho(0), k)\Big] . \tag{2.22}$$

We now use (2.13). We observe that

$$\sup_{k\in y^0 + \beta_0 B} (\rho(0),k) = \beta_0 \| \rho(0) \| + (\rho(0), y^0) . \tag{2.23}$$

Using (2.23) into (2.22) gives the desired result (2.4). □

Remark 2.2.
One can consider, by similar methods, other problems of the same type for operators with variable (smooth) coefficients, for higher order operators in space variables, for systems of equations, for other boundary conditions. The structure of the formulae will be unchanged. □

Bibliography.

1. MIZOHATA, S. - Unicité du prolongement des solutions pour quelques opérateurs différentiels paraboliques. Mem. Coll. Sc. Univ. Kyoto. 31 (3), 1958, p. 219-239.

2. LIONS, J.L. - Exact controllability for distributed systems. Some trends and some problems, Venice, Applied and Industrial Mathematics, p. 59-84 (1991) Kluwer Academic Publishers, Netherlands.

3. LIONS, J.L. - Exact controllability, Stabilization and perturbation for distributed systems, SIAM Rev. 30 (1988), p. 1-68.

4. LIONS, J.L. - **Contrôlabilité exacte. Perturbation et stabilisation de systèmes distribués.** Masson, Paris, R.M.A., Volume 8 and 9, 1988.

5. GLOWINSKI, R., LI, C.H. and LIONS, J.L., A numerical approach to the exact controllability of the wave equation, Japan J. of Applied Math. 7 (1990), p. 1-76.

6. GLOWINSKI, R. and LI, C.H., On the numerical implementation of the Hilbert Uniqueness Method for the exact boundary controllability of the wave equation. C.R.A.S. 311 (1990), p. 135-142.

7. GLOWINSKI, R., Boundary Controllability Problems for the wave and heat equations. To appear

8. LIONS, J.L., Remarks on approximate controllability, Israel J. of Math.

9. ZUAZUA, E., Exact controllability for the semi linear wave equation, J.M.P.A. 69 (1990), p. 1-31.

10. DIAZ, I., To appear.

11. LIONS, J.L., Are there connections between turbulence and controllability ?, Antibes, June 1990.

12. ROCKAFELLAR, T.R., Duality and stability in extremum problems involving convex functionals. Pac. J. Math. 21 (1967), p. 167-187.

FINITE ELEMENTS IN THE 90's, E. Oñate, J. Periaux, A. Samuelsson (Eds.)
Springer-Verlag/CIMNE, Barcelona 1991

ON A NEW FEASIBLE DIRECTIONS SOLUTION APPROACH IN CONSTRAINED OPTIMIZATION

J. Orkisz and M. Pazdanowski
Technological University of Cracow,
31-155 Kraków, ul. Warszawska 24,
POLAND

SUMMARY

The latest version of a new Feasible Directions Method for solution of nonlinear constrained optimization problems is presented. The method is designed to effectively analyze large convex problems. It is a search method that may start from an arbitrary point. It provides a search algorithm, at first for a feasible point, and later on, for the optimum point inside the feasible region. The search strategy is based on geometry, and is entirely different from the one used in the conventional FDM. This strategy is applied in two key procedures called "standard" and "auxiliary".

Besides the solution algorithm, the relevant computer program is also discussed, and results of numerical benchmarks are presented as well. A large scale application to a nonlinear shake-down problem is also given. Efficiency of both the method and computer program has been successfully tested versus the conventional FDM implemented in the professional ADS optimization software. All the results obtained show that the approach proposed may provide an effective tool for large scale convex analysis.

1. INTRODUCTION

Consider a nonlinear constrained optimization problem posed in the following form:
find:

$$\min_{\underline{x}} F(\underline{x}) \quad , \quad \underline{x} \in \mathbb{R}^n \tag{1.1}$$

subject to inequality constraints:

$$g_i(\underline{x}) \le 0 \quad , \quad i = 1,2, \ldots ,m \tag{1.2}$$

We assume here that both the objective function $F(\underline{x})$ and constraints $g_i(\underline{x})$ are convex, and may be nonlinear. Moreover, either $F(\underline{x})$ or the feasible region (Fig. 1):

$$\Omega_f = \Omega_1 \cap \Omega_2 \cap \ldots \cap \Omega_m \quad ,$$
$$\Omega_i = \{\underline{x} : g_i(\underline{x}) \le 0 \; , \; \underline{x} \in \mathbb{R}^n\} \quad , \; i = 1,2, \ldots , m. \tag{1.3}$$

are strictly convex in order to provide uniqueness of the optimal solution, if it exists.

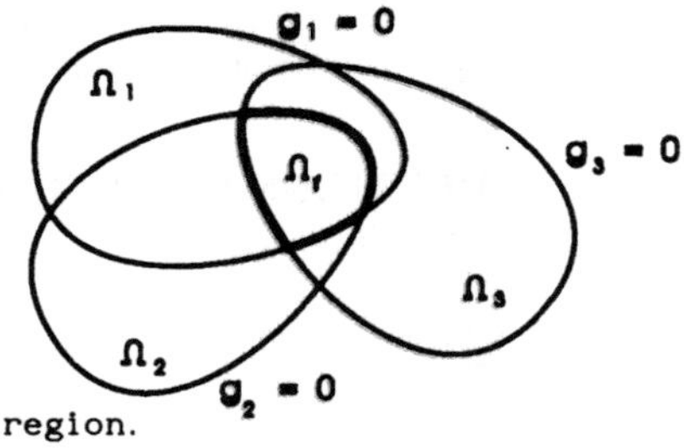

Fig. 1. Feasible region.

Nonlinear optimization problems of the type (1.1), (1.2) are usually solved using one of three methods:

1) classical Feasible Directions Method (FDM) [7,10];
2) successive linearization of the original optimization problem and sequential application of linear programming [7,9];
3) transformation of the original problem to the unconstrained one using a penalty concept and application of appropriate solution techniques [7,9].

The Feasible Directions Method procedure, in its conventional version starts from a feasible point $\underline{x}_f \in \Omega_f$, i.e. a point that has to satisfy all constraints. However, it may be difficult to find such a point in case of numerous optimization variables and constraints. Furthermore, in order to determine a direction of search, the method requires, in each step, laborious solution of a linear programming problem, and may also show poor convergence.

The effectiveness of the sequential linear programming decreases sharply with the dimension n of space $\mathbb{R}^n$ and number of nonlinear constraints.

The penalty function methods may lead to results highly dependent on the arbitrary choice of the proper penalty function coefficients. This effect is difficult to control in large optimization problems.

Due to the disadvantages of the above mentioned methods quite a new version of the Feasible Directions Method had been recently proposed [1-5]. It had been designed to eliminate at least some of the deficiencies described above. The main purpose of this paper is a presentation of the latest developments in that new method. A relevant computer program based on such an approach is also discussed here. This program had been specially designed to facilitate effective solution of large constrained optimization problems resulting from the shake-down analysis of residual stresses arising in elastic - perfectly plastic bodies under cyclic loadings. Results of several numerical tests done are also reported here.

2. ALGORITHM OF THE NEW FDM

As with the conventional FDM, the new one is also a search method. Search may start, however, from an arbitrary point $\underline{x}_0$. If this is not a feasible point (FP), i.e. $\underline{x}_0 \notin \Omega_f$ the method provides, at first, a search algorithm for a FP, and later on, for the optimum point $\underline{x}_*$ inside the feasible region (FR) until the Kuhn - Tucker optimality conditions [9] are satisfied.

The proposed method is based on a proper search through the domain. Thus the method is fully described when, for each iterative step, rules are given on how to compute the search direction $\underline{d}_q$, and the step size k_q , as well as the search termination criteria are set.

This search is done using a new, purely geometrically based strategy, entirely different from the conventional one. This strategy is presented and applied in two procedures called "standard" and "auxiliary", briefly described below.

The following denotations have been assumed in the subsequent part of this paper:

$$g_i^{I} \Leftrightarrow g_i \le 0 , \qquad i \in M^{I} \quad \text{- satisfied constr.,}$$
$$g_i^{II} \Leftrightarrow g_i > 0 , \qquad i \in M^{II} \quad \text{- violated constr.,}$$
$$g_i^{III} \Leftrightarrow -\varepsilon < g_i \le 0 , \ \varepsilon > 0 , \quad i \in M^{III} \quad \text{- active constr.}$$

It was also assumed that the dashed quantities are normalized, thus the denotation $\nabla\bar{g}$ is equal to the condition $|\nabla g_i| = 1$.

Search for a feasible point $\underline{x}_f$.

The objective is to find any feasible point $\underline{x}_f \in \Omega$, or to prove that such a point does not exist. When using "auxiliary" procedure during the area search, as opposed to the "standard" one, it is assumed that the subsequent point $\underline{x}_q$ cannot leave the area defined by the constraints $g_i^{I}(\underline{x}) \le 0$ already satisfied by the point $\underline{x}_{q-1}$.

"Standard" procedure.

The "standard" procedure is based on the following search strategy (Fig. 2):

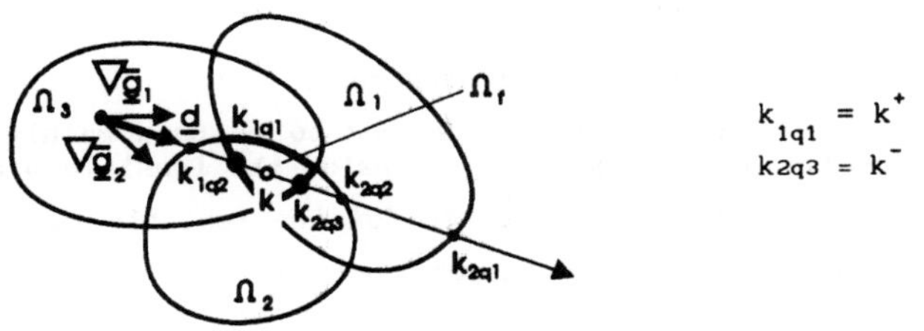

Fig. 2. Strategy of the "standard" procedure.

- an arbitrary starting point $\underline{x}_0$ is assumed,
- the search direction vector is defined as:

$$\underline{d}_q = -\nabla\bar{\underline{I}} \ , \quad \nabla\underline{I} = \Sigma\nabla\bar{g}_i^{II} \ , \quad i \in M^{II} , \tag{2.1}$$

$\nabla\bar{\underline{I}}$ represents here gradient of a new, auxiliary objective function, which temporarily replaces the original one;
- step size k_q to the subsequent point $\underline{x}_{q+1}$ is computed as:

$$k_q = \frac{1}{2}\cdot(k_q^{-} + k_q^{+}) \ , \tag{2.2}$$

where:

$$k_q^{+} = \sup_i k_{1q(i)} \ , \quad \text{at } k_{1q(i)} \ge 0 \ , \tag{2.3}$$

denotes the longest of the steps $k_{1q(i)}$ from the point $\underline{x}_q$ in the direction $\underline{\bar{d}}_q$ to the entry point of a line $\underline{x} = \underline{x}_q + k\cdot\underline{\bar{d}}_q$ into the region defined by the constraints $g_i(\underline{x}) \le 0$, $i \in M$, and:

$$k_q^- = \inf_i k_{2q(i)} \quad , \quad \text{at } k_{2q(i)} \ge 0 \quad , \tag{2.4}$$

denotes shortest of the steps $k_{2q(i)}$ from $\underline{x}_q$ to the exit point, so:

$$g_i(\underline{x}_q + k_{q(i)}\cdot\underline{\bar{d}}_q) = 0 \Rightarrow k_{1q(i)},\ k_{2q(i)} \ge k_{1q(i)}; \tag{2.5}$$

- the new point $\underline{x}_{q+1}$ is found such as:

$$\underline{x}_{q+1} = \underline{x}_q + k_q^-\cdot\underline{\bar{d}}_q \tag{2.6}$$

As long as $k_q^- \ne k_q^+$, the point $\underline{x}_{q+1}$ does not belong to any of the surfaces $g_i(\underline{x}) = 0$, although it may leave the area limited by the constraints $g_i^I(\underline{x}) < 0$ already satisfied by the point $\underline{x}_q$.

In the particular case, when constraints $g_i(\underline{x}) \le 0$ are of square form, the step sizes $k_{1q(i)}$, $k_{2q(i)}$ may be found through explicit formulas [1].

The search termination criteria in the "standard" procedure.

The "standard" procedure may give one of the three possible answers:

I - the feasible point $\underline{x}_f$ is found,

II - the feasible area does not exist,

III - there is no definite answer after the allowed number of iterations.

The appropriate search termination criteria are described in a more detailed way in reports [1].

"Auxiliary" procedure.

The "auxiliary" procedure is destined to be used mainly during the search for the optimum point $\underline{x}_*$ inside the feasible area, but it may also be used during the search for the feasible point $\underline{x}_f$, when the "standard" procedure happens not to be effective enough (case III) (Fig. 3).

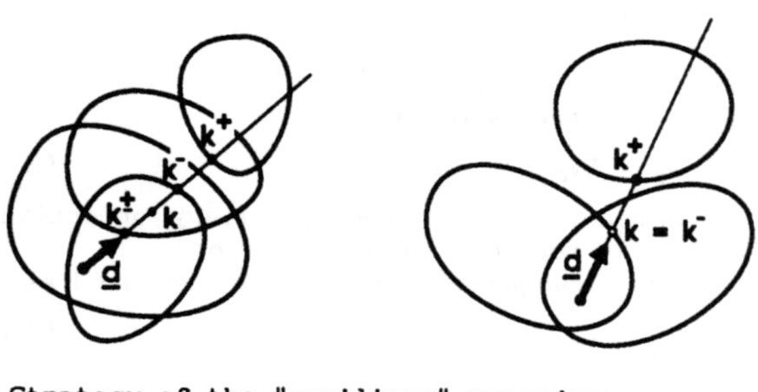

Fig. 3. Strategy of the "auxiliary" procedure.

a) $k^{\pm}$ exists b) $k^{\pm}$ does not exist

The following search strategy is applied then:

- starting point $\underline{x}_0$ is arbitrary, for instance it may be supplied by the "standard" procedure,
- it is assumed, that currently satisfied constraints $g_i^I(\underline{x}) \le 0$ define the current feasible area Ω^I, and search for the subsequent point $\underline{x}_{q+1}$ is conducted only inside this domain Ω^I or on its boundary. As a result, the subsequent points $\underline{x}_q$ may be found inside domains Ω_i, or on one or more than one of their common boundaries. In the first case the search direction vector $\underline{d}_q$ is found according to the formula (2.1). In all other cases the active constraints $g_i^{III}(\underline{x})=0$ may have influence on the search direction vector $\underline{d}_q$, according to the mutual orientation of the $\nabla\bar{\underline{I}}$ and $\nabla\bar{g}_i^{III}$ vectors. In this situation the step size is computed as

$$k_q = \begin{cases} \frac{1}{2}\cdot(k_q^- + k_q^{\pm}) & \text{, if } k_q^{\pm} \text{ exists (Fig. 3a),} \\ k_q^- & \text{, if } k_q^{\pm} \text{ does not exist (Fig. 3b),} \end{cases} \tag{2.7}$$

where

$$k_q^{\pm} = \sup_i k_{iq(i)}, \text{ at } k_{iq(i)} \ge 0 \text{ and } k_{iq(i)} < k^- , \tag{2.8}$$

denotes the farthest of the intersection points $k_{iq(i)}$ still belonging to the current feasible domain Ω^I.

- the new point $\underline{x}_{q+1}$ is again computed according to the formula (2.7).

Determination of the search direction $\underline{d}_q$ *with one active constraint* $g_i^{III}(\underline{x}_q) = 0$.

Depending on the mutual orientation of the normalized $\nabla\bar{\underline{I}}$ and $\nabla\bar{g}_i$ vectors, defined by the scalar product

$$\eta_i = \nabla\bar{\underline{I}}^T\cdot\nabla\bar{g}_i , \tag{2.9}$$

the following four situations may occur:

- $0 < \eta_i \le 1 \Rightarrow \underline{d}_{iq} = -\nabla\bar{\underline{I}}$, (2.10)
- $\eta_i = 0 \Rightarrow$ limit case, $-\nabla\bar{\underline{I}}$ is usable only when the $g_i(\underline{x})$ function is convex and not strictly convex,
- $-1 < \eta_i < 0 \Rightarrow \underline{d}_{iq} = -\dfrac{1 - \alpha\cdot\eta_i}{\alpha - \eta_i}\cdot\nabla\bar{\underline{I}} - \nabla\bar{g}_i$, (2.11)
- $\eta_i = -1 \Rightarrow$ the feasible and usable direction does not exist, i.e. the feasible area does not exist (Kuhn - Tucker conditions are satisfied).

Determination of the search direction $\underline{d}_q$ *with many active constraints* $g_i^{III}(\underline{x}_q) = 0$.

The point $\underline{x}_q$ belonging to more than one boundary $g_i(\underline{x}_q) = 0$ is currently considered. Now the three following situations may take place:

- $0 \leq \eta_i \leq 1$, $i \in M^{III}$ $\Rightarrow$ $\underline{d}_{Nq} = -\nabla\bar{\underline{I}}$,
- $-1 < \eta_i \leq 0$ for one and only one $i \in M^{III}$, then there is only one really active constraint and $\underline{d}_{Nq} = \underline{d}_{1q}$ is given by the formula (2.11) as before,
- all other cases. Two situations may happen then:
  = a direction that is both feasible and usable exists, or
  = such a direction does not exist; this means that the Kuhn - Tucker conditions are satisfied.

Solution in the directions space.

The last case may be reduced to the following problem: ascertain whether exists, and if it does exist, find both a usable and feasible direction vector $\underline{d}_{Nq}$, satisfying the constraints:

$$h_i = \underline{d}^T_{Nq} \cdot \nabla \bar{\underline{g}}^{III}_i (\underline{x}_q) < 0 \quad ,$$
$$h_T = \underline{d}^T_{Nq} \cdot \nabla \bar{\underline{I}} < 0 \quad . \tag{2.12}$$

This problem is formulated in the directions space.

The feasible region for this problem - if it does exist - is convex, but not strictly convex, because it is bounded by hiperplanes $h_i = 0$, $i \in M^{III}$ and $h_T = 0$. Once again we deal with a search for a point ($\underline{d}_{Nq}$) in the feasible domain (but now in the space $\underline{d}_N$). This problem may be solved by means of:

- classic linear programming approach [10];
- modified gradient projection method [9];
- search in the directions space, using procedure analogical to the one used for search in the decision variables space, as briefly described above, taking into account the linearity of constraints in this case [1] (Fig. 4).

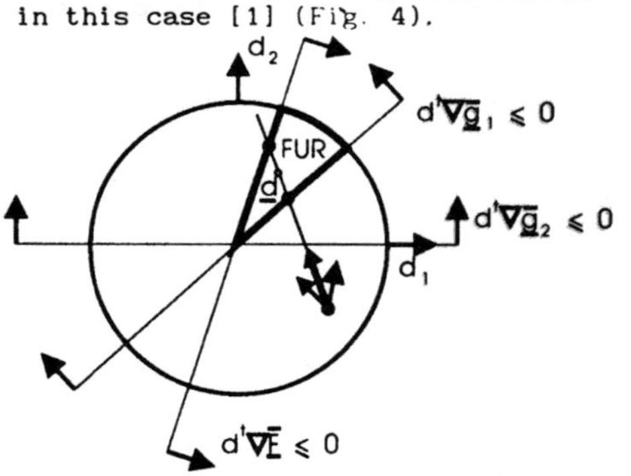

FUR - feasible and usable region

Fig. 4. Standard procedure in the space of directions vectors $\underline{d}_N$

Of those three above mentioned methods, the first one (linear programming) is logically the simplest, though the most time consuming in numerical implementation. The third one (search in the directions space) is the most complicated, but fastest in search for a feasible and usable direction. The method using modified gradient projection algorithm may be placed in between, having certain advantages as compared to both linear programming and search in subspace:

- is much faster than linear programming in computing the direction vector,

- provides the "best" direction vector resulting in better convergence in the decision variables space [1],
- is always used to determine, whether the Kuhn - Tucker optimality conditions are satisfied.

Search termination criteria in the "auxiliary" procedure.

The same three kinds of answer as in the "standard" procedure may be obtained here, namely:

I - the feasible point $\underline{x}_f$ is found,

II - the feasible area does not exist,

III - there is no definite answer after the allowed number of iterations.

The appropriate search termination criteria are described in detail in reports [1].

Search for the optimum point $\underline{x}_*$.

Let us consider situation, when a feasible point $\underline{x}_f$ was found. The search for the optimum point $\underline{x}_*$ may now be started inside the feasible domain Ω_f . The "auxiliary" procedure may now be used once again after the following necessary modifications:

- return to the original objective function F and its gradient defined as $\nabla \underline{F} = \underline{H}\cdot\underline{x}$, instead of the formula $\nabla \underline{I} = \Sigma\nabla \underline{g}_i^{IF}$, $i \in M^{II}$.
- modification of the routine used to evaluate the step size k_q , because the objective function F may reach optimum inside the feasible region now. Two strategies may be applied here:
  = strategy 1

$$k_q = \inf(\bar{k}_q, k_{Fq}) \ , \ \text{when } k_{Fq} > 0 \ , \tag{2.13}$$

  = strategy 2

$$k_q = \begin{cases} k_{Fq} \ , & \text{when } k_{Fq} > 0 \ , \ k_{Fq} < \bar{k}_q \ , \\ \bar{k}_q \ , & \text{when } k_{Fq} > \bar{k}_q \ , \text{ and the direction} \\ & \text{vector has been computed according} \\ & \text{to the formula (2.2),} \\ \frac{1}{2}\bar{k}_q \ , & \text{in all other cases.} \end{cases} \tag{2.14}$$

In all above presented formulas k_{Fq} is computed using the minimization condition:

$$\frac{d}{dk} F(\underline{x} + k\cdot\underline{d})\big|_q = 0 \Rightarrow k_{Fq} \tag{2.15}$$

In case of the objective function having square form, k_{Fq} may be obtained using explicit formulas [1,9].

The search termination criteria remain in principle unchanged, although

- case I denotes now only necessary though not satisfactory condition,
- case II denotes both necessary and satisfactory condition for the optimum point.

3. PROGRAM

The FDM optimization procedures package constitutes a numerical implementation of the above described algorithm and may be used to solve the nonlinear optimization problems described as:

find:

$$\min_{\underline{x}} \quad F(\underline{x}) = \frac{1}{2}\cdot\underline{x}^T\cdot\underline{H}\cdot\underline{x} + \underline{x}^T\cdot\underline{J} \quad , \quad \underline{H} = \underline{H}^T \quad , \qquad (3.1)$$

subject to inequality constraints:

$$g_i(\underline{x}) = \frac{1}{2}\cdot\underline{x}^T\cdot\underline{A}_i\cdot\underline{x} + \underline{x}^T\cdot\underline{B}_i + C_i \le 0 \quad , \quad \underline{A}_i = \underline{A}_i^T \quad , \qquad (3.2)$$

$$i = 1,2, \ldots ,m \ .$$

The following denotations hold for the above formulas:

$\underline{H}$, $\underline{A}_i$ - square, symmetrical matrices, of size n,

$\underline{J}$, $\underline{B}_i$ - vectors of size n,

C_i - scalars.

Executable version of this program under DOS (Disk Operating System) on PC occupies from about 150 kB (20 decision variables and 20 constraints) to about 510 kB (140 decision variables and 140 constraints) of memory, depending also on the size of the user supplied main program.

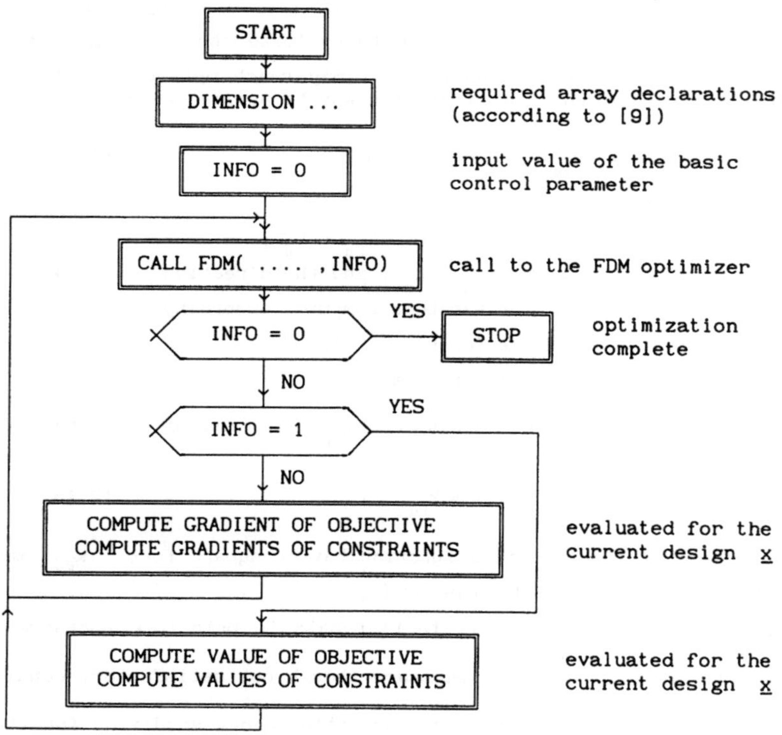

Fig. 5. Information flow logic in user supplied calling program.

The FDM optimizer is called by the user supplied main program and does not call any user supplied procedures by itself. When it is necessary to evaluate information related to gradients or values of the objective function or constraints, control is returned to the main program. The required information is evaluated and then the FDM optimizer is called again. Due to this organization of information flow logic, similar to the one used by the ADS package [8], great flexibility of user supplied code is obtained. Besides the number of possible errors in the package - user interface is limited.

It is possible to increase the optimizer effectiveness (i.e. convergence rate) using many control and steering parameters imbedded into the program. Those steering and control parameters have influence, among others, on computational accuracy, termination criterions, convergence criterions, amount and detail of printouts. In order to facilitate work for a user not familiar with the package, certain default values of those parameters were assumed. However, since those values were set to work properly for wide array of optimization problems, they may happen to be not perfectly suited and therefore not very effective for any particular optimization problem. Thus users acquainted with the program are strongly encouraged to experiment with those steering parameters values.

Detailed description of meaning, function and sets of allowed values for all the steering parameters, as well as the calling sequence may be found in the manual [6].

4. TESTS AND COMPARISONS

For the following tests, results obtained using the FDM were compared with results computed by the ADS professional optimization package [8], as well as known analytical solutions. Two particular test problems were considered:

Test no. 1.

In the $\mathbb{R}^n$ space the following problem is defined:

find

$$\min_{\underline{x}} \; F(\underline{x}) = x_1^2 + x_2^2 + \ldots + x_i^2 + \ldots + x_n^2 , \tag{4.1}$$

at

$$g_i(\underline{x}) = x_1^2 + x_2^2 + \ldots + x_i^2 + \ldots + x_n^2 - 2 \cdot x_i \cdot c + c^2 - 1 \le 0$$
$$i = 1, 2, \ldots, n . \tag{4.2}$$

The aim of this test is to find, whether for a given value of the offset parameter c, a feasible region defined in the n dimensional space as the common part of n spheres with radii equal to one (equations (4.2)), does exist, and if it does, to find a point inside the feasible region, that is closest to the coordinate system origin.

For any given space dimension n the maximum offset parameter c, for which a feasible region still exists, may be computed using the formula:

$$c \le c_{max} = \sqrt{\frac{n}{n-1}} , \tag{4.3}$$

For given values of n and c, coordinates of the optimum point may be found by the analytical formula:

$$x_{opt} = \frac{c - \sqrt{c^2 - n\cdot(c^2 - 1)}}{n} , \qquad (4.4)$$

Results of some tests made are presented in Table 1.

Sp. Dm.	Offset c / Theoretical optimum	Opt. Mth.	Evaluated Optimum $\underline{x}_{opt}$	No Itr.	Relt. Error [%]	Time [s]
2	1.000000	FDM	$0.45217\cdot10^{-12}$	4	-	7.2
	0.0000000000	ADS	$0.34694\cdot10^{-16}$	11	-	6.3
2	1.234500	FDM	0.2722827008	15	0.000	13.3
	0.2722827008	ADS	0.2722827008	34	0.000	43.7
2	1.414213	FDM	0.7064759001	8	$1\cdot10^{-7}$	9.0
	0.7064758989	ADS	0.7064758994	25	$7\cdot10^{-8}$	11.3
8	1.000000	FDM	$0.12401\cdot10^{-12}$	4	-	9.7
	0.0000000000	ADS	$0.44877\cdot10^{-16}$	13	-	13.4
8	1.023450	FDM	0.0257785853	36	0.000	44.8
	0.0257785853	ADS	0.0257785853	117	0.000	198.3
8	1.069044	FDM	0.1331548059	28	$3\cdot10^{-6}$	44.2
	0.1331548023	ADS	0.1331548041	93	$1\cdot10^{-6}$	101.5
32	1.000000	FDM	$0.45217\cdot10^{-12}$	4	-	46.6
	0.0000000000	ADS	$0.34694\cdot10^{-16}$	18	-	167.8
32	1.008642	FDM	0.0102819962	42	0.000	716.0
	0.0102819962	ADS	0.0102819962	152	0.000	1847.1
32	1.016001	FDM	0.0317186592	44	$7\cdot10^{-6}$	690.8
	0.0317186570	ADS	0.0317186575	118	$2\cdot10^{-6}$	1437.3

Table 1. Test results for various offset parameters in multidimensional space

In the tests for each space dimension n, the offset parameter value c (given in the second column of Table 1 together with appropriate optimum coordinate) was chosen in such a manner that:

- in the first test the feasible domain is of the maximum size and include the origin of the coordinate system (optimum point x_{opt} = 0.000000),
- in the second test the feasible domain is of an intermediate size between those in the tests one and three,
- in the third test the feasible domain is of the minimum size (here parameter c has seven significant digits of value c_{max} in formula (4.3)).

In the second column of Table 1, only the coordinate farthest from the exact solution vector $\underline{x}$ is given.

It is worth noting that relatively high number of iterations required to arrive at the optimum solution, results from the very high solution precision required (constraint thickness equals $1 \cdot 10^{-10}$ here). When the constraint thickness parameter was set to 0.0001, the number of iterations required to arrive at the optimum with the FDM optimizer never exceeded 20.

Test no 2.

Considered here is a thick walled cylinder under radially symmetrical loads (varying internal pressure p, axial force N and torsional moment M) exceeding the material yield limit. In order to find residual stresses resulting from these loads the new mathematical model and the Finite Difference discretization are used. More detailed description of this problem and formulas for the objective function, constraints and their gradients may be found, among others, in [4].

This problem has been also solved using both the FDM and ADS optimizers. The following results have been obtained:

Loading Parameters	Optimizer	Averaged Relative Error	Time [s]
M = 00.0000 N = 00.0000 p = 0.7207	FDM ADS	0.232 0.231	181.9 359.9
M = 64.3899 N = 00.0000 p = 0.0	FDM ADS	0.014 0.013	164.7 362.9
M = 45.2099 N = 27.0689 p = 0.0	FDM ADS	0.185 0.183	514.0 839.4

Table 2. Test results for various loading cases

Relative error was computed in reference to the known analytical solution [11], and then averaged over the number of nodes used in discretization (in tests shown here 9). Relative error in any of the nodes for any optimization method did not exceed 1.71%. Discrepancy between optimization methods for any given node did not exceed 0.02%, so this error is clearly attributable to the discretization method used (Finite Difference) and not to the internal optimizer inaccuracy (both optimizers seem to converge to the same point in the decision variables space).

5. CONCLUDING REMARKS

An updated version of the New Feasible Directions Method algorithm is briefly described in this paper, together with its numerical implementation. Numerical tests done have confirmed correctness of the algorithm proposed and numerical efficiency of the developed computer code. In particular:

- using the FDM optimizer on an IBM PC class computer one may obtain results of good quality (on IBM PC when computations are done in double precision it is possible to obtain precision in the order of $1 \cdot 10^{-10}$),
- computation time is quite competitive with existing optimizers; comparison with the professional ADS optimizer in an modified ver-

sion - faster than the original one, shows clear gain on computation time, 33 to 50% depending on the test done, correct results were also obtained for very small feasible regions, where the ADS code failed to find the solution,
- it is possible to solve large optimization problems in reasonable time (problems of residual stress analysis in a railroad rail with as many as 320 decision variables and 40 constraints have been successfully solved on the SUN 4/110 workstation).

Currently the algorithm and code are being worked on to add two new features:
- capability to include linear inequality constraints as well as quadratic ones in the problem formulation,
- incorporation of the QDR orthogonal decomposition process in the variables space optimization in order to obtain further improvement of program efficiency.

In the authors opinion the current version of the program constitutes an effective tool for convex analysis.

5. BIBLIOGRAPHY

[1] - Orkisz J. - Evaluation of Residual Stresses in Railroad Rails, DOT TSC, Cambridge Mass., 1984 and 1985 Progress Rpt.
[2] - Orkisz J. - A New Feasible Directions Algorithm for Effective Solution of Constrained Optimization Problem, Proc. 8th International Conference on Structural Mechanics in Reactor Technology, Brussels, Aug 1985.
[3] - Orkisz J. - A Method of Solution of Nonlinear Constrained Optimization Problems, Proc. 1st World Congress on Computational Mechanics, Austin, Sept. 1986.
[4] - Orkisz J., Orringer O., Holowinski M., Pazdanowski M., Cecot W. - Discrete Analysis of Actual Residual Stresses Resulting from Cyclic Loadings, Proc. Conf. "Frontiers in Computational Mechanics", Cambridge Mass., Apr. 1989.
[5] - Orkisz J., Pazdanowski M. - A New Feasible Directions Method in the Nonlinear Optimization, Proc. IX Conf. MKMK, Cracow, May 1989. (in Polish) and Proc. Conf. Residual Stress in Rails: Effects on Rail Integrity and Railroad Economics, Cracow, May 1990.
[6] - Pazdanowski M. - FDM Optimization Program, ver. 3.00 - User Manual, Technological University of Cracow, unpublished.
[7] - Vanderplaats G. N. - Numerical Optimization Techniques for Engineering Design with Applications, Mc Graw - Hill Publ., 1984.
[8] - Vanderplaats G. N. - ADS - a FORTRAN Program for Automated Design Synthesis, Naval Postgraduate School, Monterey Ca., May 1984.
[9] - Wismer D. A., Chattergy R. - Introduction to Nonlinear Optimization, North - Holland Publ., 1978.
[10] - Zoutendijk G. - Methods of Feasible Directions, Elsevier, Amsterdam, 1960.
[11] - Życzkowski M. - Combined Loadings in the Theory of Plasticity, PWN (Polish Scientific Publishers), Warsaw, 1981.

FINITE ELEMENTS IN THE 90'S, E. Oñate, J. Periaux, A. Samuelsson (Eds.)
Springer-Verlag/CIMNE, Barcelona 1991

HIERARCHICAL OR DOMAIN DECOMPOSITION PRECONDITIONING FOR THE *P*-VERSION FINITE ELEMENT METHOD

A. Craig
University of Durham,
CANADA

1 Introduction

The p−version of the finite element method (where we use relatively coarse meshes and high degree basis functions), has increased in popularity over recent years. This is due to several reasons, the main one being a greater return in accuracy in terms of the size of the problem. A notable secondary reason is is the ease with which we can design efficient p−adaptive codes. Much of the fundamental work in this area has been carried out by Zienkiewics [1, 8, 9, 10, 11, 12, 13, 14, 15, 16]. However, there are computational difficulties associated with the method. The one which we address in this paper is the deterioration of the conditioning of the discrete operator.

In the paper [8] the idea of hierarchical preconditioning was introduced. The basic idea behind this was to pose the finite element problem using a hierarchical basis, and then precondition the problem by taking the 'linear' part of the matrix together with the higher order diagonal terms as a simple preconditioning matrix. This idea worked extraordinarily well with a simple class of problems, but as can be easily seen degraded for more complicated ones. However, in recent papers[2, 3, 4, 5, 6] domain decomposition methods have been produced which result in highly parallel preconditioning techniques for the h− and p−versions. These methods are *precisely* hierarchical preconditioning for particular choices of basis functions. As will be seen, the choice of basis functions will be guided by the problem to be solved. A welcome side effect of this technique, and one which was not envisaged with the original idea, is that both the construction and the inversion of the preconditioning matrix can be carried out in parallel.

The theoretical and numerical results in these papers demonstrate that it is possible to reduce the growth in the condition number of the finite element stiffness matrix from polynomial to logarithmic in the number of degrees of freedom.

2 Preliminary results

The finite element method is based on a variational formulation of the problem: find $u \in H$:

$$a(u,v) = f(v), \ \forall v \in H, \tag{1}$$

where H is a Hilbert space, $a(\cdot,\cdot)$ is a symmetric bilinear form defined over $H \times H$, and f is a bounded linear functional defined over H. We assume that the standard conditions hold for such a problem to have a unique solution[7].

The finite element method consists of choosing a finite dimensional subspace $S \subset H$ and posing problem (1) on $S \times S$. Choosing a basis for S transforms the problem into a matrix problem

$$Ax = y \tag{2}$$

Our basic approach to solving (2) is the preconditioned conjugate gradient method. We construct a preconditioning form $c(\cdot,\cdot)$ such that: $\exists m_1, m_2 > 0$:

$$m_1 a(u,u) \le c(u,u) \le m_2 a(u,u) \ \forall u \in H. \tag{3}$$

The form $c(\cdot,\cdot)$ is also chosen so that a problem $c(u,v) = g(v)$ similar to (1) is easier to solve than the original. The number $\kappa = m_1/m_2$ is called *the relative condition number*. It is one of the aims of preconditioning that κ should be as small as possible[2]. The other desirable property for $c(\cdot,\cdot)$ is that, in our chosen basis, the resulting matrix should be easily invertible.

Let $\Omega \subset \Re^2$ be the region over which problem (1) is posed, and let K be a decomposition of Ω into finite elements satisfying the usual conditions ([2]). We can decompose the bilinear forms a and $c(\cdot,\cdot)$ in the obvious way into a sum of bilinear forms over elements (in matrix terms this is nothing more than the finite element assembly procedure), so

$$a(u,v) = \sum_{K \in \mathcal{K}} a_K(u^K, v^K),$$

and similarly for $c(\cdot,\cdot)$, where u^K is the restriction of u to K. Now it is obvious that, if $\exists m_1, m_2 > 0$:

$$m_1 a_K(u^K,u^K) \le c_K(u^K,u^K) \le m_2 a_K(u^K,u^K) \ \forall u^K \in H_K. \tag{4}$$

then (3) holds. This result is what gives us the key to calculating the preconditioner in parallel. We can carry out the calculations independently on each element K and assemble the preconditioner in the same way as the standard finite element stiffness matrix. We shall see later that because of the particular form we shall choose for the preconditioner then we can also carry out a major part of the inversion of the preconditioning matrix in parallel.

From now on we can therefore drop the index K and work over individual elements. We shall also assume from now on that K is the unit square. In [2] we give conditions, which are not restrictive, on K for the results which follow to be true.

3 Preconditioning forms

Now take V to be the space of complete polynomials of degree p defined over the unit square, denoted K. We shall describe some decompositions of functions $u \in V$ which will be used to construct the preconditioner. Firstly, we take a general decomposition

$$u = \sum_{i=1}^{m} u_i,$$

for some fixed m. Then our preconditioning form $c(\cdot, \cdot)$ will be defined as

$$c(u, v) = \sum_{i=1}^{m} a(u_i, v_i),$$

(i.e. the "diagonal" part of $c(\cdot, \cdot)$).

Now we can immediately bound $a(\cdot, \cdot)$ in terms of $c(\cdot, \cdot)$ by the Cauchy Schwarz inequality:

$$a(u, u) \leq mc(u, u).$$

Now if we assume that the decompositions are bounded, that is for each i, $i = 1, \ldots, m$ then $\exists b_i$, not necessarily independent of p, such that

$$a(u_i, u_i) \leq b_i a(u, u),$$

then

$$c(u, u) \leq (\sum_{i=1}^{m} b_i) a(u, u),$$

so combining the above results we have

$$\frac{1}{m} a(u, u) \leq c(u, u) \leq (\sum_{i=1}^{m} b_i) a(u, u). \tag{5}$$

It is seen from inequalities (5) that the relative condition number, for fixed m, depends only on how the decomposition is bounded. We can therefore take our lead from these inequalities and attempt to produce a decomposition so that the relevant b_is are small.

In order to precondition the problem we shall decompose our function u into components which are referred to as *vertex*, *side*, and *interior* functions. On the reference triangle, with nodes numbered $1, \ldots, 4$, we shall write u as

$$u = \sum_{i=1}^{4} u_{\mathcal{V},i} + \sum_{i=1}^{4} u_{\mathcal{S},i} + u_{\mathcal{I}},$$

(where $\mathcal{V}, \mathcal{S}$, and $\mathcal{I}$ refer respectively to vertex, side and interior). These functions satisfy the following conditions:

- $u_{\mathcal{V},i} = u$ at vertex i, is zero on the the opposite sides, and $u_{\mathcal{V},i}$ is bilinear.
- $u_{\mathcal{S},i} = u - \sum_{j=1}^{4} u_{\mathcal{V},i}$ on side i, is zero on the other three sides and the quantity $a(u_{\mathcal{S},i}, u_{\mathcal{S},i})$ is minimised.
- $u_{\mathcal{I}} = u - \sum_{i=1}^{4} u_{\mathcal{V},i} - \sum_{i=1}^{4} u_{\mathcal{S},i}$.

We then take our preconditioner to be:

$$c(u,v) = a(\sum u_{\mathcal{V},i}, \sum v_{\mathcal{V},i}) + \sum a(u_{\mathcal{S},i}, v_{\mathcal{S},i},) + a(u_{\mathcal{I}}, v_{\mathcal{I}}) \tag{6}$$

In this case it is immediate that $m = 6$, and we also show in [2] that for the preconditioner (6)

$$b_i \leq C(1 + \ln^2 p)$$

and so $\kappa = O(\ln^2 p)$

4 Implementation

The method has the following steps:

- Construction of the standard sets of basis functions, i.e. $\mathcal{V}$, $\mathcal{S}$, $\mathcal{I}$. The hierarchical basis functions fall naturally into these categories.
- Transformation of the basis functions into those which satisfy the above conditions. This is done in parallel on each element, and is in fact equivalent to static condensation of the interior functions. The resulting matrix representation of $c(\cdot,\cdot)$ is block diagonal, corresponding to a block for the vertex functions, four blocks for the side functions and (if not condensed) a block for the interior functions. It is easily seen that when the global preconditioning matrix is assembled that it retains this form, and the solution can be carried out in parallel for the blocks corresponding to interior and side functions, with a (small) global problem corresponding to the vertex functions.
- Implementation of conjugate gradients, either through the assembled matrix or through the local stiffness matrices.

Full numerical results are shown in [2], but in general, even with the associated overheads we observe speedup of the order of 85%.

5 Extensions

It is easy to show that this method will *not* work in three dimensions, the energy of the trilinear interpolant cannot be bounded by a logarithmic function of p times the energy of the original function. However, it can be shown that if instead of defining the vertex functions as above we define them as

- $u_{V,i} = u$ at vertex i, is zero on the the opposite sides, and $a(u_{V,i}, u_{V,i})$ is minimised.

then with a little extra work we can produce an algorithm which works well. For full details we refer to [2].

References

[1] Ainsworth, M., Zhu, J.Z., Craig, A.W. and Zienkiewicz, O.C., *Analysis of the Zienkiewicz-Zhu a posteriori error estimator in the finite element method.* Int. J. Num. Meth. Engng., 28, pp 2161-2174 (1989).

[2] Babuška, I., Craig, A., Mandel, J., Pitkäranta, J., *Efficient preconditioning for the p version finite element method in two dimensions.* SIAM J Num Anal, 28, 3 (1991).

[3] Bramble, J.H., Pasciak, J.E., Schatz, A.E., *The construction of preconditioners for elliptic problems by substructuring I.* Math. Comp. 47, 103 – 133 (1986).

[4] Bramble, J.H., Pasciak, J.E., Schatz, A.E., *The construction of preconditioners for elliptic problems by substructuring II.* Math. Comp. 49, 1 – 16 (1987).

[5] Bramble, J.H., Pasciak, J.E., Schatz, A.E., *The construction of preconditioners for elliptic problems by substructuring III.* Math. Comp. 51, 415 – 430 (1988).

[6] Bramble, J.H., Pasciak, J.E., Schatz, A.E., *The construction of preconditioners for elliptic problems by substructuring IV.* Math. Comp. 53, 1 – 24 (1989).

[7] Ciarlet, P.G., *The Finite Element Method for Elliptic Problems.* North-Holland (1978).

[8] Craig, A.W. and Zienkiewicz, O.C., *A multigrid algorithm using a hierarchical finite element basis.* In Multigrid Methods for Integral and Differential Equations, ed. D.J.Paddon and H.Holstein, Oxford University Press (1985).

[9] Craig, A.W., Zhu, J.Z. and Zienkiewicz, O.C., *Adaptive mesh refinement, a posteriori error estimates and multigrid algorithms using hierarchical finite element bases.* In MAFELAP 1984, ed. J.R.Whiteman, Academic Press (1985).

[10] Craig, A.W., Zhu, J.Z., Ainsworth, M. and Zienkiewicz, O.C., *h and h-p version error estimation and adaptive procedures: from theory to practice.* Invited paper for a special edition of Engineering with Computers, 5, pp 221-234 (1989).

[11] Zhu, J.Z., Zienkiewicz, O.C. and Craig, A.W., *Some new a posteriori error estimators for linear elastic problems.* In NUMETA 87, ed. J.Middleton and G.Pande, Pineridge Press (1987).

[12] Zienkiewicz, O.C. and Craig, A.W., *A posteriori error estimation and adaptive mesh refinement in the finite element method.* In The Mathematical Basis of Finite Element Methods, ed. D.Griffiths, Oxford University Press (1984).

[13] Zienkiewicz, O.C. and Craig, A.W., *Adaptive mesh refinement and a posteriori error estimation for the p-version of the finite element method.* In Adaptive Computational Methods for Partial Differential Equations, ed. I.Babuska, J.Chandra and J.Flaherty.

[14] Zienkiewicz, O.C. and Craig, A.W., *Adaptive refinement, error estimates, multigrid solution and hierarchic finite element concepts.* In Accuracy Estimates and Adaptive Refinements in Finite Element Computations, ed. I.Babuska, O.C.Zienkiewicz,

[15] Zienkiewicz, O.C., Craig, A.W., Zhu, J.Z. and Gallagher, R.H., *Adaptive analysis, refinement and shape optimisation: Some new possibilities.* In The Optimum Shape - Automated Structural Design, ed. J.A.Bennett and M.E.Botkin, Plenum Press (1986).

[16] Zienkiewicz, O.C., Zhu, J.Z., Craig, A.W. and Ainsworth, M., *Simple and practical error estimation and adaptivity: h and h-p version procedures.* In Adaptive Methods for Partial Differential Equations, ed. J.Flaherty and M.Shepherd, SIAM, (1990).

MIX
Papier aus verantwortungsvollen Quellen
Paper from responsible sources
FSC® C105338

If you have any concerns about our products,
you can contact us on
ProductSafety@springernature.com

In case Publisher is established outside the EU,
the EU authorized representative is:
Springer Nature Customer Service Center GmbH
Europaplatz 3, 69115 Heidelberg, Germany

Printed by Libri Plureos GmbH
in Hamburg, Germany